Lecture Notes in Bioinformatics   10208

Subseries of Lecture Notes in Computer Science

More information about this series at http://www.springer.com/series/5381

Ignacio Rojas · Francisco Ortuño (Eds.)

Bioinformatics and Biomedical Engineering

5th International Work-Conference, IWBBIO 2017
Granada, Spain, April 26–28, 2017
Proceedings, Part I

 Springer

Editors
Ignacio Rojas
Universidad de Granada
Granada
Spain

Francisco Ortuño
Universidad de Granada
Granada
Spain

ISSN 0302-9743   ISSN 1611-3349   (electronic)
Lecture Notes in Bioinformatics
ISBN 978-3-319-56147-9   ISBN 978-3-319-56148-6   (eBook)
DOI 10.1007/978-3-319-56148-6

Library of Congress Control Number: 2017935843

LNCS Sublibrary: SL8 – Bioinformatics

Printed on acid-free paper

This Springer imprint is published by Springer Nature
The registered company is Springer International Publishing AG
The registered company address is: Gewerbestrasse 11, 6330 Cham, Switzerland

Preface

We are proud to present the set of final accepted full papers for the third edition of the IWBBIO conference "International Work-Conference on Bioinformatics and Biomedical Engineering" held in Granada (Spain) during April 26–28, 2017.

The IWBBIO 2017 (International Work-Conference on Bioinformatics and Biomedical Engineering) seeks to provide a discussion forum for scientists, engineers, educators, and students about the latest ideas and realizations in the foundations, theory, models, and applications for interdisciplinary and multidisciplinary research encompassing the disciplines of computer science, mathematics, statistics, biology, bioinformatics, and biomedicine.

The aim of IWBBIO is to create a friendly environment that could lead to the establishment or strengthening of scientific collaborations and exchanges among attendees, and, therefore, IWBBIO 2017 solicited high-quality original research papers (including significant work in progress) on any aspect of bioinformatics, biomedicine, and biomedical engineering.

The following topics were especially encouraged: new computational techniques and methods in machine learning; data mining; text analysis; pattern recognition; data integration; genomics and evolution; next-generation sequencing data; protein and RNA structure; protein function and proteomics; medical informatics and translational bioinformatics; computational systems biology; modelling and simulation and their application in the life science domain, biomedicine, and biomedical engineering. The list of topics in the successive call for papers also evolved, resulting in the following list for the present edition:

1. **Computational proteomics.** Analysis of protein–protein interactions. Protein structure modelling. Analysis of protein functionality. Quantitative proteomics and PTMs. Clinical proteomics. Protein annotation. Data mining in proteomics.
2. **Next-generation sequencing and sequence analysis.** De novo sequencing, re-sequencing, and assembly. Expression estimation. Alternative splicing discovery. Pathway analysis. Chip-seq and RNA-Seq analysis. Metagenomics. SNPs prediction.
3. **High performance in bioinformatics.** Parallelization for biomedical analysis. Biomedical and biological databases. Data mining and biological text processing. Large-scale biomedical data integration. Biological and medical ontologies. Novel architecture and technologies (GPU, P2P, Grid etc.) for bioinformatics.
4. **Biomedicine.** Biomedical Computing. Personalized medicine. Nanomedicine. Medical education. Collaborative medicine. Biomedical signal analysis. Biomedicine in industry and society. Electrotherapy and radiotherapy.
5. **Biomedical engineering.** Computer-assisted surgery. Therapeutic engineering. Interactive 3D modelling. Clinical engineering. Telemedicine. Biosensors and data acquisition. Intelligent instrumentation. Patient monitoring. Biomedical robotics. Bio-nanotechnology. Genetic engineering.

6. **Computational systems for modelling biological processes**. Inference of bio-logical networks. Machine learning in bioinformatics. Classification for biomedical data. Microarray data analysis. Simulation and visualization of biological systems. Molecular evolution and phylogenetic modelling.

7. **Health care and diseases**. Computational support for clinical decisions. Image visualization and signal analysis. Disease control and diagnosis. Genome-phenome analysis. Biomarker identification. Drug design. Computational immunology.

8. **E-health**. E-health technology and devices. E-health information processing. Telemedicine/e-health application and services. Medical image processing. Video techniques for medical images. Integration of classical medicine and e-health.

After a careful peer-review and evaluation process (each submission was reviewed by at least two, and on average 3.2, Program Committee members or additional reviewers), 120 papers were accepted for oral, poster, or virtual presentation, according to the recommendations of the reviewers and the authors' preferences, and to be included in the LNBI proceedings.

During IWBBIO 2017, several Special Sessions were carried out. Special Sessions are a very useful tool to complement the regular program with new and emerging topics of particular interest for the participating community. Special Sessions that emphasize multi-disciplinary and transversal aspects as well as cutting-edge topics were especially encouraged and welcomed, and in IWBBIO 2017 they were the following:

– **SS1: Advances in Computational Intelligence for Critical Care**

Decision-making in health care in clinical environments is often made on the basis of multiple parameters and in the context of patient presentation, which includes the setting and the specific conditions related to the reason for admission and the procedures involved. The data used in clinical decision-making may originate from manifold sources and at multiple scales: devices in and around the patient, labo-ratory, blood tests, omics analyses, medical images, and ancillary information available both prior to and during the hospitalization.

Arguably, one of the most data-dependent clinical environments is the intensive care unit (ICU). The ICU environment cares for acutely ill patients. Many patients within ICU environments, and particularly surgical intensive care units (SICU), are technologically dependent on the life-sustaining devices that surround them. Some of these patients are indeed dependent for their very survival on technologies such as infusion pumps, mechanical ventilators, catheters and so on. Beyond treatment, assessment of prognosis in critical care and patient stratification combining different data sources are extremely important in a patient-centric environment.

With the advent and quick uptake of omics technologies in critical care, the use of data-based approaches for assistance in diagnosis and prognosis becomes para-mount. New approaches to data analysis are thus required, and some of the most interesting ones currently stem from the fields of computational intelligence (CI) and machine learning (ML). This session is particularly interested in the proposal of novel CI and ML approaches and in the discussion of the challenges for the application of the existing ones to problems in critical care.

Topics that are of interest to this session include (but are not necessarily limited to):

- Novel applications of existing CI and ML and advanced statistical methods in critical care
- Novel CI and ML techniques for critical care
- CI and ML-based methods to improve model interpretability in a critical care context, including data/model visualization techniques
- Novel CI and ML techniques for dealing with non-structured and heterogeneous data formats in critical care

Organizers: Dr. Alfredo Vellido, PhD, Department of Computer Science, Universitat Politécnica de Catalunya, BarcelonaTECH (UPC), Barcelona (Spain). Dr. Vicent Ribas, eHealth Department, EURECAT Technology Centre of Catalonia, Barcelona, Barcelona (Spain).

- **SS2: Time-Lapse Experiments and Multivariate Biostatistics**
  Biological samples are evolving in time, phases, periods, behavior. To be able to understand the dynamics, we need to perform time lapse experiments. Today's technique and measurement devices allow us to monitor numerous parameters in semi-controled environments during the experiment. The increase of measured data is enormous. The interpretation requires both qualitative and quantitative analysis. There are useful methods of biostatistics, multivariate data analysis, and artificial intelligence, namely, neural networks, genetic algorithms, and agent-based modeling, respectively.
  In this special section we will provide a discussion on broad examples from time-lapse experimental design through information and data acquisition, using methods of bioinformatics, biophysics, biostatistics, and artificial inteligence. The aim of this section is to present the possible increase in data interpretation and related methods.

Organizer: Dr. Jan Urban, Laboratory of Signal and Image Processing, Institute of Complex Systems, Faculty of Fisheries and Protection of Waters, University of South Bohemia.

- **SS3: Half-Day GATB Tutorial. The Genome Analysis Toolbox with de Bruijn Graph**
  The GATB programming day is an educational event organized by the GATB team. This free event is open to everyone who is familiar in C++ programming and wants to learn how to create NGS data analysis software.
  The tutorial has a focus on the high-performance GATB-Core library and is taught by the developers of this library.
  During this half-day tutorial, some of the following topics are explored:

- A theoretical introduction to GATB: the basic concepts.
- GATB-Core practical coding session 1: I/O operations on reads files.
- The GATB de Bruijn graph API.
- GATB-Core practical coding session 2: k-mer and graph APIs in action.
- GATB-Core practical coding session 3: writing a short read corrector tool.
- Q and A session: obtain answers from GATB experts.

Organizers: Dr. Dominique Lavenier, GenScale Team Leader, Inria/IRISA, Campus de Beaulieu, Rennes, France.
Dr. Patrick Durand, Inria, Genscale Team, Campus de Beaulieu, Rennes, France.

- **SS4: Medical Planning: Operating Theater Design and Its Impact on Cost, Area, and Workflow**
  The design of operating rooms is one of the most complicated tasks of hospital design because of its characteristics and requirements. Patients, staff, and tools should have determined passes through the operating suite. Many hospitals assume that the operating suite is the most important unit in the hospital because of its high–revenue. Arch design of these suites is a very critical point in solving an optimized problem in spaces, workflow of clean, dirty, and patient in/out in addition to staff together with their relations with adjacent departments. In this session, we illustrate the most common designs of operating suites and select the most suitable one to satisfy the effectiveness of the operating suite, maximizing throughput, minimizing the costs, and decreasing the required spaces related to available resources/possibilities. The design should comply with country guidelines, infection control rules, occupational safety and health, and satisfy the maximum benefits for patients and staff. A comparative study was performed on 15 hospitals and it recorded that the single input–output technique is the best design.
  Motivation and objectives for the session: Operating suite design is a very critical task owing to its impact. Biomedical engineers should participate in the design and review the workflow and available functions.

  Organizer: Dr. Khaled El-Sayed, Department of Electrical and Medical Engineering at Benha University, Egypt

- **SS5: Challenges Representing Large-Scale Biological Data**
  Visualization models have been shown to be remarkably important in the interpretation of datasets across many fields of study. In the context of bioinformatics and computational biology various tools have been proposed to visualize molecular data and help understand how biological systems work. Despite that, several challenges still persist when faced with complex and dynamic data and major advances are required to correctly manage the multiple dimensions of the data.
  The aim of this special session is to bring together researchers to present recent and ongoing research activities related to advances in visualization techniques and tools, focused on any major molecular biology problem, with the aim of allowing for the exchange and sharing of proposed ideas and experiences with novel visualization metaphors.
  Topics of interest include:

  - Genome and sequence data
  - Omics data (transcriptomics, proteomics, metabolomics)
  - Biological networks and pathways
  - Time-series data
  - Biomedical ontologies
  - Macromolecular complexes
  - Phylogenetic data

- Biomarker discovery
- Integration of image and omics data for systems biology
- Modeling and simulation of dynamical processes

Organizer: Prof. Joel P. Arrais FCTUC – University of Coimbra, Portugal.

- **SS6: Omics of Space Travelled Microbes – Bioinformatics and Biomedical Aspects**
  The National Research Council (NRC) Committee for the Decadal Survey on Biological and Physical Sciences in Space reported that "microbial species that are uncommon, or that have significantly increased or decreased in number, can be studied in a "microbial observatory" on the International Space Station (ISS)." As part of the microbial observatory effort the NRC decadal survey committee suggested that NASA should: "(a) capitalize on the technological maturity, low cost, and speed of genomic analyses and the rapid generation time of microbes to monitor the evolution of microbial genomic changes in response to the selective pressures present in the spaceflight environment; (b) study changes in microbial populations from the skin and feces of the astronauts, plant and plant growth media, and environmental samples taken from surfaces and the atmosphere of the ISS; and (c) establish an experimental program targeted at understanding the influence of the spaceflight environment on defined microbial populations."
  The proposed session discusses state-of-the-art molecular techniques, bioinformatics tools, and their benefit in answering the astronauts and others who live in closed systems.

Organizer: Dr. Kasthuri Venkateswaran, Senior Research Scientist, California Institute of Technology, Jet Propulsion Laboratory, Biotechnology and Planetary Protection Group, Pasadena, CA

- **SS7: Data-Driven Biology – New Tools, Techniques, and Resources**
  Advances in sequencing techniques have accelerated data generation at diverse regulatory levels in an unprecedented way. The challenge now is to integrate these data to understand regulation at a systems level. As the sequencing technologies evolve, new tools and resources follow, revealing new aspects of complex biological systems.
  This special session brings together experts from computational biology and machine learning to present recent advances in the development of new tools and resources using next-generation sequencing data including novel emerging fields such as single-cell transcriptomics. The session features an invited speaker and three/four short talks. To promote emerging leaders of the field, we select invited speakers who have gained their independence in recent years.

Organizer: Dr. Joshi Anagha, Division of Developmental Biology at the Roslin Institute, University of Edinburgh

- **SS8: Smart Sensor and Sensor-Network Architectures**
  There is a significant demand for tools and services supporting rehabilitation, well-being and healthy life styles while reducing the level of intrusiveness as well as increasing real-time available and reliable results. For example, self-monitoring

applications need to be improved to move beyond tracking exercise habits and capture a more comprehensive digital footprint of human behavior. This session focuses on primary parameter capturing devices and networks demonstrating advances in sensor development including a customized algorithmic shell research to support diagnostic decisions. Target domains are, for example, continuous differentiating between mental and physical stress, blood pressure monitoring, sleep quality monitoring, HRV etc.

Organizers: Prof. Dr. Natividad Martinez, Internet of Things Laboratory, Reutlingen University, Germany.
Prof. Dr. Juan Antonio Ortega, University of Seville, Spain.
Prof. Dr. Ralf Seepold, Ubiquitous Computing Lab, HTWG Konstanz, Germany.

- **SS9: High-Throughput Bioinformatic Tools for Genomics**
Genomics is concerned with the sequencing and analysis of an organism's genome. It is involved in the understanding of how every single gene can affect the entire genome. This goal is mainly afforded using the current, cost-effective, high-throughput sequencing technologies. These technologies produce a huge amount of data that usually require high-performance computing solutions and opens new ways for the study of genomics, but also transcriptomics, gene expression, and systems biology, among others. The continuous improvements and broader applications on sequencing technologies is producing an ongoing demand for improved high-throughput bioinformatics tools.

Therefore, we invite authors to submit original research, new tools or pipelines, or their update, and review articles on topics helping in the study of genomics in the wider sense, such as (but not limited to):

- Tools for data pre-processing (quality control and filtering)
- Tools for sequence mapping
- Tools for de novo assembly
- Tools for quality check of sequence assembling
- Tools for the comparison of two read libraries without an external reference
- Tools for genomic variants (such as variant calling or variant annotation)
- Tools for functional annotation: identification of domains, orthologues, genetic markers, controlled vocabulary (GO, KEGG, InterPro)
- Tools for biological enrichment in non-model organisms
- Tools for gene expression studies
- Tools for Chip-Seq data
- Tools for "big-data" analyses
- Tools for handling and editing complex workflows and pipelines
- Databases for bioinformatics

Organizers: Prof. M. Gonzalo Claros, Department of Molecular Biology and Biochemistry, University of Málaga, Spain.
Dr. Javier Pérez Florido, Bioinformatics Research Area, Fundación Progreso y Salud, Seville, Spain.

- **SS10: Systems Biology Approaches to Decipher Long Noncoding RNA–Protein Associations**

Long noncoding RNAs (lncRNAs) make up large amounts of the RNA and total genomic repertoire. Studies on the functional characterization of lncRNAs have resulted in data on interactions with their RNA peers, DNA, or proteins. Although there has been an increase in evidence on the link between lncRNAs and diverse human diseases, there is a dearth of lncRNA–protein association studies. Additionally, existing methods do not provide theories about the possible molecular causes of such associations linking to diseases. How such regulatory interactions between classes of lncRNAs and proteins would have a significant influence on the organism and disease remains a challenge. A good number of bioinformatics approaches have arisen in the recent past exploring these challenges. The idea behind this session is to bring together the wide gamut of researchers who have worked on these methods across different organisms.

The following are the sub-topics of the proposed session, which we would like to call for papers.

- LncRNAs in genomes: annotation and curation
- LncRNA–protein interactions leading to important diseases: systems Biology approaches
- Identifying lncRNAs with respect to their mechanism and dysregulation in diseases
- LncRNA databases and webservers
- Machine learning approaches and prediction servers

Organizer: Prashanth Suravajhala, PhD, Department of Biotechnology and Bioinformatics, Birla Institute of Scientific Research, India

- **SS11: Gamified Rehabilitation for Disabled People**

Gamification is a hot topic in many areas as it aims at motivating people to do things driven by different innate needs like the wish to accomplish tasks, to compete against others, or to gain something. These and other motivators are efficiently applied in computer games and could be extraordinarily useful in ensuring that patients perform their daily exercises regularly and have fun.

The idea of exergames (exercise games) is not new; the literature reveals that much effort has already been made and with great success. Nevertheless, most applications have been developed for special problems or diseases (i.e., stroke, parkinson, cerebral palsy etc.) and are not generally applicable. In general, people suffering from severe disabilities and chronic diseases are rarely addressed as a target group. Also, the focus is generally set on the medical achievements, which is correct, but the next step would be to enhance the fun factor because no tool is of much use if the patient is not using it because of boredom or demotivation.

The objective of this special session is to gather new ideas about the combination of need and fun, i.e., find ways to create exercise platforms that fit everybody's needs, provide access to the therapist for monitoring and configuration, while the patients benefit physically and mentally when having a good time.

Target groups would be people of:

- All ages, while focusing on younger people, who can be involved more easily but are less addressed in the literature
- All diseases, while focusing on chronic illness and severe disabilities (e.g., muscle dystrophies and atrophies)

The contributions should show advances in at least one of the following areas:

- Adaptability to users with all kinds of problems (e.g., possibility to configure the limbs used to play or playing with facial movements, wheelchair and standing modes, coping with muscle weaknesses etc.)
- Implementation of gaming techniques and special motivators
- Physical or mental exercises, aimed at rehabilitation or daily practice
- Understanding the users, awareness of their level of motivation, fatigue or progress and react accordingly

Organizer: Dr. Martina Eckert, Associate Professor at University of Madrid, Spain.

- **SS12: Modelling of Glucose Dynamics for Diabetes**
Diabetes is the eighth most common cause of death, while its treatment relies on technology to process continuously measured glucose levels.

Organizer: Dr. Tomas Koutny, Faculty of Applied Sciences, University of West Bohemia

- **SS13: Biological Network Analysis in Multi-omics Data Integration**
In many biological applications, multiple data types may be produced to determine the genetics, epigenetics, and microbiome affecting gene regulation and metabolism. Although producing multiple data types should provide a more complete description of the processes under study due to multiple factors such as study design (synchronization of data production, number of samples, varying conditions), the analysis may leave more unfulfilled promises than synergy expected from the wealth of data.
In this session, some of the following challenges are addressed:

- How to conduct meaningful meta-analysis on historical data.
- How to use biological knowledge (represented in reproducible and interoperable manner) in the analysis of large and sparse data sets more effectively.
- How to fill the gap between hypothesis-driven mechanistic studies, e.g., applying modelling to very well studied biochemical processes and data-driven hypothesis-free approaches. How omics data can help.
- Beyond meta-transcriptomics and metagenomics: integration and interpretation of microbiome and host data.

We would like to bring together communities concerned with these topics to present state-of-the-art and current cutting-edge developments, preferably work under construction or published within the past year.

- Objective 1: presentation and discussion of newest methods
- Objective 2: round-table discussions on the topics highlighted above and other related topics suggested by session participants

An additional topic that does not fit the proposed session but that I would love to see addressed is: How to improve open access to the data that is not next-generation sequencing (e.g., metabolomics, proteomics, plant phenotyping). For this an active participation of journal editors would be necessary to discuss opportunities to change journal publication policies.

Organizer: Dr. Wiktor Jurkowski, Jurkowski Group, Earlham Institute, Norwich Research Park, UK

- **SS14: Oncological Big Data and New Mathematical Tools**
Current scientific methods produce various omics data sets covering many cellular functions. However, these data sets are commonly processed separately owing to limited ways in how to connect different omics data together for a meaningful analysis. Moreover, it is currently a problem to integrate such data into mathematical models. We are entering the new era of biological research where the main problem is not to obtain the data but to process and analyze them. In this regard, a strong mathematical approach can be very effective (see J. Gunawardena's essay "Models in biology: accurate descriptions of our pathetic thinking," BMC Biology 2014, 12:29).
In this Special Session we focus on big data (omics and biological pathways) related to oncological research. General biological processes that are relevant to cancer can also be studied. Mathematical tools basically mean statistical learning (data mining, inference, prediction), modeling, and simulation. We want to place a special emphasis on causality. Closely tied to mathematical tools, efficient computational tools can be considered.

Organizers: Dr. Gregorio Rubio, Instituto de Matematica Multidisciplinar, Universitat Politecnica de Valencia, Valencia, Spain. Dr. Rafael Villanueva, Instituto de Matematica Multidisciplinar, Universitat Politecnica de Valencia, Valencia, Spain.

In this edition of IWBBIO, we were honored to have the following invited speakers:

1. Prof. Roderic Guigo, Coordinator of Bioinformatics and Genomics at Centre de Regulacio Genomica (CRG). Head of the Computational Biology of RNA Processing Group. Universitat Pompeu Fabra, Barcelona, Spain
2. Prof. Joaquin Dopazo, Director of the Computational Genomics Department, Centro de Investigación Príncipe Felipe- CIPF, Valencia, Spain
3. Prof. Jose Antonio Lorente, Director of Centre for Genomics and Oncological Research (GENYO). Professor of Legal and Forensic Medicine, University of Granada, Spain

It is important to note, that for the sake of consistency and readability of the book, the presented papers are classified under 16 chapters. The organization of the papers is in two volumes arranged following the topics list included in the call for papers. The first volume (LNBI 10208), entitled "Advances in Computational Intelligence: Part I" is divided into seven main parts and includes the contributions on:

1. Advances in computational intelligence for critical care
2. Bioinformatics for health care and diseases
3. Biomedical engineering
4. Biomedical image analysis
5. Biomedical signal analysis
6. Biomedicine
7. Challenges representing large-scale biological data

The second volume (LNBI 10209), entitled "Advances in Computational Intelligence: Part II" is divided into nine main parts and includes the contributions on:

1. Computational genomics
2. Computational proteomics
3. Computational systems for modelling biological processes
4. Data-driven biology: new tools, techniques, and resources
5. E-health
6. High-throughput bioinformatic tools for genomics
7. Oncological big data and new mathematical tools
8. Smart Sensor and sensor-network architectures
9. Time-lapse experiments and multivariate biostatistics

This fifth edition of IWBBIO was organized by the Universidad de Granada together with the Spanish Chapter of the IEEE Computational Intelligence Society. We wish to thank to our main sponsor and the Faculty of Science, Department of Computer Architecture and Computer Technology and CITIC-UGR, from the University of Granada, for their support and grants. We also wish to thank the Editors-in-Chief of different international journals for their interest in having special issues with the best papers of IWBBIO.

We would also like to express our gratitude to the members of the different committees for their support, collaboration, and good work. We especially thank the local Organizing Committee, Program Committee, the reviewers, and special session organizers. Finally, we want to thank Springer, and especially Alfred Hofmann and Anna Kramer, for their continuous support and cooperation.

April 2017   Ignacio Rojas
                                                            Francisco Ortuño

Organization

Steering Committee

Miguel A. Andrade	University of Mainz, Germany
Hesham H. Ali	University of Nebraska, USA
Oresti Baños	University of Twente, the Netherlands
Alfredo Benso	Politecnico di Torino, Italy
Giorgio Buttazzo	Superior School Sant'Anna, Italy
Mario Cannataro	University Magna Graecia of Catanzaro, Italy
Jose María Carazo	Spanish National Center for Biotechnology (CNB), Spain
Jose M. Cecilia	Universidad Católica San Antonio de Murcia, Spain
M. Gonzalo Claros	University of Malaga, Spain
Joaquin Dopazo	Research Center Principe Felipe, Spain
Werner Dubitzky	University of Ulster, UK
Afshin Fassihi	Universidad Católica San Antonio de Murcia, Spain
Jean-Fred Fontaine	University of Mainz, Germany
Humberto Gonzalez	University of the Basque Country, Spain
Concettina Guerra	College of Computing, Georgia Tech, USA
Andy Jenkinson	Karolinska Institute, Sweden
Craig E. Kapfer	Reutlingen University, Germany
Narsis Aftab Kiani	European Bioinformatics Institute, UK
Natividad Martinez	Reutlingen University, Germany
Marco Masseroli	Politechnical University of Milan, Italy
Federico Moran	Complutense University of Madrid, Spain
Cristian R. Munteanu	University of A Coruña, Spain
Jorge A. Naranjo	New York University, Abu Dhabi
Michael Ng	Hong Kong Baptist University, SAR China
Jose L. Oliver	University of Granada, Spain
Juan Antonio Ortega	University of Seville, Spain
Julio Ortega	University of Granada, Spain
Alejandro Pazos	University of A Coruña, Spain
Javier Perez Florido	Genomics and Bioinformatics Platform of Andalusia, Spain
Violeta I. Pérez Nueno	Inria Nancy Grand Est, LORIA, France
Horacio Pérez-Sánchez	Universidad Católica San Antonio de Murcia, Spain
Alberto Policriti	University of Udine, Italy
Omer F. Rana	Cardiff University, UK
M. Francesca Romano	Superior School Sant'Anna, Italy
Yvan Saeys	VIB - Ghent University, Belgium
Vicky Schneider	The Genome Analysis Centre, UK
Ralf Seepold	HTWG Konstanz, Germany

Mohammad Soruri	University of Birjand, Iran
Yoshiyuki Suzuki	Tokyo Metropolitan Institute of Medical Science, Japan
Oswaldo Trelles	University of Malaga, Spain
Renato Umeton	CytoSolve Inc., USA
Jan Urban	University of South Bohemia, Czech Republic
Alfredo Vellido	Polytechnic University of Catalonia, Spain

Program Committee

Jesus S. Aguilar
Carlos Alberola
Hisham Al-Mubaid
Rui Carlos Alves
Yuan An
Georgios Anagnostopoulos
Eduardo Andrés León
Antonia Aránega
Saúl Ares
Masanori Arita
Ruben Armañanzas
Joel P. Arrais
Patrizio Arrigo
O. Bamidele Awojoyogbe
Jaume Bacardit
Hazem Bahig
Pedro Ballester
Graham Balls
Ugo Bastolla
Sidahmed Benabderrahmane
Steffanny A. Bennett
Daniel Berrar
Mahua Bhattcharya
Concha Bielza
Armando Blanco
Ignacio Blanquer
Paola Bonizzoni
Christina Boucher
Hacene Boukari
David Breen
Fiona Browne
Dongbo Bu
Jeremy Buhler
Keith C.C.
Gabriel Caffarena
Anna Cai

Carlos Cano
Rita Casadio
Daniel Castillo
Ting-Fung Chan
Nagasuma Chandra
Kun-Mao Chao
Bernard Chen
Bolin Chen
Brian Chen
Chuming Chen
Jie Chen
Yuehui Chen
Jianlin Cheng
Shuai Cheng
I-Jen Chiang
Jung-Hsien Chiang
Young-Rae Cho
Justin Choi
Darrell Conklin
Clare Coveney
Aedin Culhane
Miguel Damas
Antoine Danchin
Bhaskar DasGupta
Ricardo De Matos
Guillermo de la Calle
Javier De Las Rivas
Fei Deng
Marie-Dominique Devignes
Ramón Diaz-Uriarte
Julie Dickerson
Ye Duan
Beatrice Duval
Saso Dzeroski
Khaled El-Sayed
Mamdoh Elsheshengy

Christian Exposito
Weixing Feng
Jose Jesús Fernandez
Gionata Fragomeni
Xiaoyong Fu
Juan Manuel Galvez
Alexandre G.de Brevern
Pugalenthi Ganesan
Jean Gao
Rodolfo Garcia
Mark Gerstein
Razvan Ghinea
Daniel Gonzalez Peña
Dianjing Guo
Jun-tao Guo
Maozu Guo
Christophe Guyeux
Michael Hackenberg
Michiaki Hamada
Xiyi Hang
Jin-Kao Hao
Nurit Haspel
Morihiro Hayashida
Jieyue He
Pheng-Ann Heng
Luis Javier Herrera
Pietro Hiram
Lynette Hirschman
Ralf Hofestadt
Vasant Honavar
Wen-Lian Hsu
Jun Hu
Xiaohua Hu
Jun Huan
Chun-Hsi Huang
Heng Huang
Jimmy Huang
Jingshan Huang
Seiya Imoto
Anthony J. Kusalik
Yanqing Ji
Xingpeng Jiang
Mingon Kang
Dong-Chul Kim
Dongsup Kim
Hyunsoo Kim

Sun Kim
Kengo Kinoshita
Ekaterina Kldiashvili
Jun Kong
Tomas Koutny
Natalio Krasnogor
Abhay Krishan
Marija Krstic-Demonacos
Sajeesh Kumar
Lukasz Kurgan
Stephen Kwok-Wing
Istvan Ladunga
T.W. Lam
Pedro Larrañaga
Dominique Lavenier
Jose Luis Lavin
Doheon Lee
Xiujuan Lei
André Leier
Kwong-Sak Leung
Chen Li
Dingcheng Li
Jing Li
Jinyan Li
Min Li
Xiaoli Li
Yanpeng Li
Li Liao
Hongfei Lin
Hongfang Liu
Jinze Liu
Xiaowen Liu
Xiong Liu
Zhenqiu Liu
Zhi-Ping Liu
Rémi Longuespée
Miguel Angel Lopez Gordo
Ernesto Lowy
Jose Luis
Suryani Lukman
Feng luo
Bernard M.E.
Bin Ma
Qin Ma
Malika Mahoui
Alberto Maria

Jason Wang
Jian Wang
Jianxin Wang
Jiayin Wang
Junbai Wang
Junwen Wang
Lipo Wang
Li-San Wang
Lusheng Wang
Yadong Wang
Yong Wang
Ka-Chun Wong
Ling-Yun Wu
Xintao Wu
Zhonghang Xia
Fang Xiang
Lei Xu
Zhong Xue
Patrick Xuechun
Hui Yang
Zhihao Yang

Jingkai Yu
Hong Yue
Erliang Zeng
Xue-Qiang Zeng
Aidong Zhang
Chi Zhang
Jin Zhang
Jingfen Zhang
Kaizhong Zhang
Shao-Wu Zhang
Xingan Zhang
Zhongming Zhao
Huiru Zheng
Bin Zhou
Shuigeng Zhou
Xuezhong Zhou
Daming Zhu
Dongxiao Zhu
Shanfeng Zhu
Xiaoqin Zou
Xiufen Zou

Additional Reviewers

Alquezar, Rene
Amaya Vazquez, Ivan
Belanche, Lluís
Browne, Fiona
Calabrese, Barbara
Cho, Ken
Coelho, Edgar
Datko, Patrick
Gaiduk, Maksym
Gonzalez-Abril, Luis
Ji, Guomin
Kang, Hong
Langa, Jorge
Liang, Zhewei
Martin-Guerrero, Jose D.

Moulin, Serge
Olier, Iván
Ortega-Martorell, Sandra
Ribas, Vicent
Scherz, Wilhelm Daniel
Sebastiani, Laura
Seong, Giun
Smith, Peter
Treepong, Panisa
Tu, Shikui
Valenzuela, Olga
Vizza, Patrizia
Wu, Yonghui
Yan, Shankai
Zucco, Chiara

Contents – Part I

Biomedical Engineering

Biomedical Image Analysis

Biomedical Signal Analysis

Biomedicine

Challenges Representing Large-Scale Biological Data

Contents – Part II

Computational Proteomics

Computational Systems for Modelling Biological Processes

Data Driven Biology - New Tools, Techniques and Resources

High-Throughput Bioinformatic Tools for Genomics

Oncological Big Data and New Mathematical Tools

Time Lapse Experiments and Multivariate Biostatistics

Advances in Computational Intelligence for Critical Care

Health Monitoring System Based on Parallel-APPROX SVM

Fahmi Ben Rejab$^{(\boxtimes)}$, Walid Ksiaâ, and Kaouther Nouira

BESTMOD, Institut Supérieur de Gestion de Tunis,
Université de Tunis, 41 Avenue de la Liberté, 2000 Le Bardo, Tunisie
fahmi.benrejab@gmail.com, ksiaawalid@gmail.com,
kaouther.nouira@planet.tn

Abstract. In this paper, we propose a new approach to deal with the high rate of false alarms generated by the health monitoring system (HMS) in intensive care units (ICU). We propose a new HMS based on a new classification method consisting of the parallel-Approx support vector machines (PASVM). The main aim of the new HMS denoted by PASVM-MS is to considerably reduce the rate of false alarms and to make a fast prediction in each new state of the patient. Besides, it overcomes the main issue of the existing HMS by proposing a classification model that considers the variation of the patient states over time. Also, the number of measured parameters have to be changed when patients are getting better by removing one or more variable each time. However, thresholds are stable and do not translate the states of patients over time, since all existing systems in ICU do not take into account of the patients' states evolution. Our proposal is able to generate an initial model classifying states of patients to normal and abnormal (critical) using the PASVM. Then, it updates its model by considering the evolution in the states of patients using PASVM especially when we deleting one variable. As a result, the new system gives what the medical staff wants as information and alarms relative to monitored patient.

Keywords: Health monitoring system · Parallel SVM · Approx SVM · False alarms · Intensive care unit

1 Introduction

Intensive care unit (ICU) is a special department in hospital devoted to patients whose conditions are life threating. It provides intensive care to patients and try to control critical states. ICU is equipped by sophisticated monitoring devices such as health monitoring system (HMS). The principal aim of HMS is to measure and alert medical staff when patient has a critical state. The current HMS function is based on threshold set by care-givers. An alarm is trigged when its limits are violated.

In many research studies [7,17] authors prove that there were an excessive number of alarms trigged by the HMS. This can affect the working conditions

© Springer International Publishing AG 2017
I. Rojas and F. Ortuño (Eds.): IWBBIO 2017, Part I, LNBI 10208, pp. 3–14, 2017.
DOI: 10.1007/978-3-319-56148-6_1

and make the patient state worst. The most trigged alarms are considered as false alarms. Studies have demonstrated that the presence of false or clinically insignificant alarms ranges from 80% to 99% [15]. The myriad of the alarm number creates a bad environment in ICU and reduces trust of care-givers to this system. Besides, it causes a continuous stress to both medical staff and patients. To avoid the noise of false alarms, the medical staff may silence, disable or even ignore the alarms [1]. By this behavior, true alarms are missed and patient will be in real danger. Furthermore, in [10] authors have reported that in Massachusetts General Hospital in 2010 the result of turning off alarms have caused the death of a patient.

Improving the HMS in ICU, reducing the number of false alarms, and increasing the positive ones have attracted the intention of many researchers over the past ten years. Several researchers have focused on this problem in ICU. We can mention the use of the digital signal processing in [6] where there is also a clinical validation study for two recently developed on-line signal filters, the use of trend extraction methodology based on the time evolution of signals in [8], and the use of the intelligent monitoring [14] detailed through the time series technology and multi-agent sub-systems. Moreover, there were several other studies [3,16] that have reported and detailed this issue and how to overcome it.

To this end, we propose in this paper a new HMS based on a new machine learning technique which is the parallel and Approx support vector machines (PASVM). The PASVM is a modified version of the standard SVM which can improve the model of classification in the test phase when we decrease the number of attributes. Our proposal reduces the number of ignored or ineffective alarms, makes a faster prediction in the patient's state, and takes into account the evolution of his state as well as the number of measured parameters over time.

The rest of the paper is structured as follows: Sect. 2 describes the health monitoring system in ICU. Section 3 provides an overview of the support vector machines and the PASVM. Section 4 illustrates our proposal which is the monitoring system based on PASVM. Section 5 details and analyzes the experimental results. Section 6 concludes the paper.

2   Health Monitoring System in ICU

Monitoring patient in critical care environments such as intensive care units (ICUs) and operating rooms involves estimating the status of the patient, reacting to events that may be life-threatening, and taking actions to bring the patient to a desired state. There is the use of medical devices when monitoring patients. Besides, the current HMS generates important data and information relative to the monitored patient. Each measured variable has a practical limits or threshold, when a parameter exceeds its limits an alarm is trigged.

Table 1. Some measured parameters

Medical parameters	Minimum value	Maximum value
Heart rate	50	120
Respiratory rate	5	25
Pulse rate	65	115
Saturated percentage of Oxygen in the blood	90	130

Table 1 shows an example of some measured parameters with their thresholds. Alarms are generated by crossing a given limit. Unfortunately, it is not the best method to indicate the patient states. There is not a consideration of the simultaneous evolution of different parameters. The information that the medical staff generally wants is the detection of critical changes in a patient conditions.

Studies have demonstrated that the majority of alarms created by patient monitoring systems have no clinical relevance. Borowski et al. [6] recorded monitoring data of 68 patients in a medical ICU. As a result, only 15% of all alarms were clinically relevant. Furthermore, there were two other studies that have illustrated this problem. The first one is the survey of German ICUs detailed in [1] where authors conclude that more than 50% of all alarms were irrelevant. The second one, is the national on-line survey on the effectiveness of clinical alarms which have reported the same results in [13]. Moreover, in [10], only 10% of all alarms were taken into account by care-givers and 50% of all relevant alarms were not correctly identified.

With the large number of false alarms, care-givers do not trust the used monitoring system anymore and are becoming desensitized. As a results, care-givers ignore the majority of alarms and consider this system as a measurement tool not as a monitoring one. Due to the high rate of false alarms, the sensitivity of the current HMS is not close to 100%. There were 75 life-threatening situations, where no alarm occurred, reported in the Federal Institute for Drugs and Medical Devices (BfArM) in Germany. Besides, missing true alarms caused, between 2002 and 2004, 237 deaths related to device alarms [11]. New solutions are needed to manage and process the continuous flow of information and to provide efficient and reliable decision support tools. The reduction of false alarms is the main purpose of many researches. We can mention the monitoring based on machine learning [4,5].

As a result, the future patients monitoring system has to allow the medical staff to be more confident in the HMS. It has not to be a simple measurement tool but a monitoring one by providing important medical information. In order to avoid the issues indicated above, the monitoring system should be improved. Hence, we propose a new HMS based on parallel and Approx support vector machines. This latter makes it possible to detect different patients' states as desired and needed by medical staff. The PASVM is trained from expert decisions and can simulate the expert task for new observations especially when we decrease the number of measured parameters.

3   Parallel and Approx Support Vector Machines

3.1   The Standard SVM

Support vector machines (SVM) has been introduced by Vapnik in 1995 [9]. The basic idea was to find an hyperplane which separates data into its two classes with a maximization of the margin.

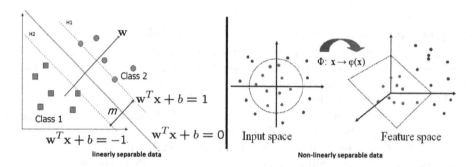

Fig. 1. The optimal hyperplane

1. **Case of linearly separable data:** Given a training set of observations $(x_i, y_i), i = \{1....m\}$ where $x_i \in R^n$ and $y \in \{1, -1\}$ (as illustrated in Fig. 1, left). Training SVM means solving the following optimization problem.

$$\begin{cases} \min \frac{1}{2}||w||^2 + C, w \in R^d, b \in R, \\ subject\ to \\ y_i(x_iw + b) \geq 1 for\ i = 1, ..., m. \end{cases} \tag{1}$$

After finding the w and b, we use the decision rule to classify the new observations:

$$f_{w,b}(x) = sign(w^T x_i + b). \tag{2}$$

2. **Case of non-linearly separable data:** SVM can also be used to separate classes that cannot be separated with a linear classifier (as shown in Fig. 1, right).

   To detect the hyperplane, we have to solve the optimization problem defined as follows using the slack variable ξ_i:

$$\min \frac{1}{2}||w||^2 + C \sum_{i=1}^{m} \xi_i, w \in R^d, b \in R, \tag{3}$$

subject to

$$y_i(x_iw + b) \geq 1 - \xi_i, \xi_i \geq 0 for\ i = \{1, ..., m\}. \tag{4}$$

In a previous work, we have proposed a monitoring system based on the SVM [2] and interesting results have been obtained. However, the system was tested in batch mode and it needs a high execution time to be trained with a large dataset. To overcome the SVM limitations, we aim to improve the proposed HMS by using the PASVM. These following subsections describe the PSVM and Approx SVM used in our proposal.

3.2   The Parallel SVM: The PSVM

The PSVM is an SVM extension, and it has the ability to train simultaneously many SVMs on a set of data chunks.

Theoretically, PSVM is represented by a distributed optimization. Thus, it includes the Quadratic Programming (QP), used but in a distributed approach. Indeed, the output will be two sets of support vectors, which means that two optimization problems will be involved. The new problem is the combination of the previous optimization problems, from the merged inputs [18].

W_i and G_i are respectively, the objective function, and the gradient of the objective function in the ith SVM [18]. Graf et al. gave the following formal definitions:

$$ W_i = -\frac{1}{2} \ \vec{\alpha}_i^T * Q_i \ \vec{\alpha}_i + \vec{e}_i^T \ \vec{\alpha}_i \qquad\qquad \vec{G}_i = -\vec{\alpha}_i^T * Q_i + \vec{e}_i^T \qquad (5) $$

Graf Peter et al., affirmed that initializing the merged results is initiating a combination of previous set of optimization, through a matrix:

$$ W_3 = -\frac{1}{2} \begin{bmatrix} \vec{\alpha}_1 \\ \vec{\alpha}_2 \end{bmatrix}^T \begin{bmatrix} Q_1 & Q_{12} \\ Q_{21} & Q_2 \end{bmatrix} \begin{bmatrix} \vec{\alpha}_1 \\ \vec{\alpha}_2 \end{bmatrix} + \begin{bmatrix} \vec{e}_1 \\ \vec{e}_2 \end{bmatrix}^T \begin{bmatrix} \vec{\alpha}_1 \\ \vec{\alpha}_2 \end{bmatrix} \qquad (6) $$

$$ \vec{G}_3 = - \begin{bmatrix} \vec{\alpha}_1 \\ \vec{\alpha}_2 \end{bmatrix}^T \begin{bmatrix} Q_1 & Q_{12} \\ Q_{21} & Q_2 \end{bmatrix} + \begin{bmatrix} \vec{e}_1 \\ \vec{e}_2 \end{bmatrix} \qquad (7) $$

3.3   The Approx SVM: The ASVM

The Approx SVM, proposed by Claesen et al. [19], is based on the use of Radial Basis Function (RBF) kernel. The RBF kernel has a simpler definition, when introducing a parameter γ: $\gamma = \dfrac{1}{2\sigma^2}$.

Thus, the RBF it is equal to:

$$ K(x, y) = \exp(-\gamma \parallel x - y \parallel^2) \qquad (8) $$

The hypothesis based on the RBF kernel, will become as follows:

$$ h(y) = \sum_{i=1}^{m} \alpha_i^* u_i . \exp(-\gamma \parallel x - y \parallel^2 + \omega_0^*) \qquad (9) $$

The Approx SVM keeps relying on the use of RBF kernel for lowering the run-time complexity, by introducing **the second-order Maclaurin series** [19].

The origin of the Maclaurin series is the **Taylor series** [20].

The first configuration, will be the inclusion of the Maclaurin series based on RBF kernel [19].

$$\exp(x = \sum_{k=0}^{\infty} \frac{1}{k!} x^k) \tag{10}$$

A new parameter, the **Hessian matrix**, was added, with applying the **gradient** upon it at the second order, which incites **the derivation of the Maclaurin series based on RBF kernels at the second order,** by following the approximation process below [19].

The Approximation Process. According to Claesen et al., given the expanded hypothesis:

$$h(y) = \sum_{i=1}^{m} \alpha_i^* u_i . \exp(-\gamma \parallel x \parallel^2) . \exp(-\gamma \parallel y \parallel^2) . \exp(2\gamma x_i^T y + \omega_0^*):$$

1. Reorder $h(y)$: $h(y) = \exp(-\gamma \parallel x \parallel^2) . (\sum_{i=1}^{m} \alpha_i^* u_i . \exp(-\gamma \parallel y \parallel^2) . \exp(2\gamma x_i^T y) + \omega_0^*)$
2. Suppose that: $g(y) = \sum_{i=1}^{m} \alpha_i^* u_i . \exp(-\gamma \parallel y \parallel^2) . \exp(2\gamma x_i^T y)$
   The second-order Maclaurin series is defined as follows [19]: $\exp(x \approx 1 + x + \frac{1}{2} x^2)$
   Its exponent is replaced by the one from $\exp(2\gamma x_i^T y)$ in $g(y)$ [19]:

$$\exp(2\gamma x_i^T y) \approx 1 + 2\gamma x_i^T z + 2\gamma^2 (x_i^T y)^2 \tag{11}$$

Thus we obtain a new $g(y)$ noted $\hat{g}(y)$:

$$\hat{g}(y) = \sum_{i=1}^{m} \alpha_i^* u_i . \exp(-\gamma \parallel y \parallel^2) . (1 + 2\gamma x_i^T y + 2\gamma^2 (x_i^T y)^2) \tag{12}$$

3. Equation (13), is a simpler formal definition of $\hat{g}(y)$ after being developed:

$$\hat{g}(y) = c + v^T . y + y.^T M . y \tag{13}$$

Each constant represents an inner products developed with the second inner product exponent, formal notations are given by Claesen et al. in [19]. The second-order hessian matrix is the third constant M in Eq. (13) [19].
4. As a consequence, we obtain the new hypothesis $\hat{h}(y)$, using the approximated function, based on the second order Maclaurin series, $\hat{g}(y)$:

$$\hat{h}(y) = \exp(-\gamma \parallel x \parallel^2) . \hat{g}(y) + \omega_0^* = \exp(-\gamma \parallel x \parallel^2) . (c + v^T . y + y.^T M . y) + \omega_0^* \tag{14}$$

A new constraint will be added in our optimization problem, to be able to generate the approximated hypothesis: $|2\gamma x_i^T y| < \frac{1}{2} \quad \forall i$ [19].

The absolute inner product is expressed otherwise, using the Cauchy-Schwarz inequality [19]. Combining the two previous equations in equation is a way to evaluate the validity of the approximation in terms of the support vector $\| x_M \|$ with maximal norm ($\forall i : \| x_M \| \geq \| x_i \|$):

$$\| x_M \|^2 \| y \|^2 < \frac{1}{16\gamma^2} \tag{15}$$

4 Monitoring System Based on Parallel and Approx SVM

This section presents our proposal consisting of the monitoring system using Parallel and Approx SVM (PASVM-MS). We explain the PASVM algorithm then, we move to presenting our new system.

The PASVM assumes as an input a PSVM model. While ASVM is not purposely proposed for big data classification, the experimental results proved its relevance, qualifying it as one of the right solutions for PSVM in classifying large data sets. The new objective function, which include the new constraint, yields to the generation of a gradient, which outputs either another gradient (intermediate layer) or a hypothesis (last layer).

Including the ASVM within yields: $\hat{h}(y) = \exp(-\gamma \| x \|^2).(c + v^T.y + y.^T M.y) + \omega_0^*$.

Our new Parallel and Approx SVM (PASVM) algorithm is presented as follows.

Data: Recorded medical parameter from the patient's health
Result: Prediction of the patient's health state
$NS=0$ //Number of splits;
//While NS is not a "power of two" number
while *!IsPowerOfTwo(NS)* **do**
| Set the number of splits NS
end
Split large medical data set into NS splits;
while $NS \neq 1$ **do**
  Train an SVM on each subset
  Merge each two neighbor SVs
  $NS=NS/2$
  if $NS=1$ **then**
  | PSVM-Model=Train SVM on the Last SVs
  end
end
New-Data=DeleteFeature(Data,Feature);
// Estimate the best γ, *maxnorm* and approximate PSVM model
Approx-analyse(New-Data);
PASVM=ASVM(PSVM-Model);
UpdateParameters(PASVM, γ^*, *maxnorm**);
Accuracy=Approx-predict(NewData,PASVM);

Based on the pseudo code of PASVM algorithm, we notice that the PASVM takes as input the original model of classification obtained in training phase using PSVM. When we have a new vector (set of measured variables), PSVM assigns to this vector a label equal to -1 if this vector represents an alarm or $+1$ otherwise. During the classification of new medical data, the expert can remove a measured parameters.

After modifying the number of medical parameters relative to the current monitored patient, we update the original model parameters using the Approx SVM. It improves the original model over time to adapt it to the new patient states.

In our case, data describe the measured medical parameters of patients that dynamically change and their states could be more or less critical over time. Our new monitoring system PASVM-MS is characterized by taking into account of all new states of patients over time especially the evolution of the number of attributes. It also updates the model built by the PSVM each time interval. These updates avoid the missing of important information and guarantee an easier and more effective monitoring.

5   Experiments

5.1   The Framework

In ICU, we can find different types of HMS such as apnea monitor, intracranial pressure monitor, and multi-parameter monitoring system. This latter is used, in this work, to measure several medical parameters in the same time and to indicate the state of patients.

We tested our proposal (PASVM-MS), using real-world datasets from MIMICII database, in Physiobank [12]. This database includes data from hemodynamically unstable patients hospitalized in 1996 in ICU of the cardiology division in the Teaching Hospital of Harvard Medical School.

We divide dataset of patients into training set and test set. We use the stratified sampling which assigns 70% of the data points to the training set and 30% of data points to the test set. We use the grid search technique based on the cross-validation, to set the appropriate parameters.

We used 7 databases relative to several patients containing different physiological parameters, such as Heart Rate (HR), Oxygen Saturation (SpO_2), Non-Invasive Blood Pressure (NBP), Respiratory rate (Resp), and Artery Blood Pressure (ABP). Table 2 details real-world databases taken from MIMICII where #Attributes and #Instances denote respectively the total number of measured parameters and the total number of instances for a specific database.

Table 2. Description of the used datasets

Databases	# Attributes	# Instances
patient 01	5	4606
patient 02	12	38252
patient 03	7	65535
patient 04	7	42188
patient 05	9	42188
patient 06	7	42188
patient 07	7	42185

5.2   Evaluation Criteria

To test our new monitoring system i.e. PASVM-MS, we use two evaluation criteria:

1. The false alarm reduction rate (FARR) [6] defined by:

$$FARR = \frac{Suppressed\,false\,alarms}{Total\,number\,of\,false\,alarms}. \tag{16}$$

2. The execution time (ET): The time needed to build the model. It is needed to show the importance of incremental learning when dealing with large datasets.

5.3   Results and Discussion

In this section, we report the results of our proposal using the evaluation criteria and then, we detail all obtained results, generated by following these steps:

1. Divide the patient dataset into a training set and a test set.
2. Build the initial model through the PSVM by using the training set.
3. In test phase, classify the observations using the obtained model, to alert medical staff when a patient has a critical state.
4. Delete a parameter (attribute). PASVM output the new model using ASVM.
5. Classify the remaining observations in the test phase.

The main problem of the current monitoring system is the high number of false positive alarms. In fact, it does not generates a high rate of false negative alarms since it has a high rate of sensitivity. The results obtained in Table 3 show the rate of suppressed alarms by the different proposed systems.

Table 3. Suppressed false alarms for different patients' datasets

Databases	FARR SVM-MS	FARR PASVM-MS
patient 01	25.5	25.45
patient 02	100	100
patient 03	99.99	100
patient 04	100	99.81
patient 05	100	99.98
patient 06	100	100
patient 07	100	100

The new system PASVM-MS output rates describing the full suppression of false alarms like the SVM.

Obviously, the current alarm system has a FARR equals to 0%. From Table 3, we can remark that these systems using PASVM and SVM have considerably improved the results. This capacity of reducing the number of false alarms is due the large dataset used to learn the SVM and PASVM. For example, for patient 06, the FARR achieved 100%.

The execution time to build a classification model are compared in Table 4.

Table 4. Execution time in seconds

Databases	ET training		ET test	
	SVM-MS	PASVM-MS	SVM-MS	PASVM-MS
patient 01	913	6771	347	7
patient 02	14258	12879	2397	93
patient 03	104306	84715	10713	111
patient 04	89005	119224	9879	82
patient 05	146734	310018	16560	84
patient 06	22692	28884	4027	85
patient 07	23216	19233	3963	83

From Table 4, the PASVM-MS provides prediction time about 7 patients' state, lower than the original SVM, even if 4 of the input data required a higher training time to output a PSVM model than the one from the SVM.

These results are explained by the ability of the PASVM to update the model of classification without retraining from the beginning when there are new information. However, the SVM needs to train the whole database from the beginning.

6   Conclusion

In this paper, we have avoided the main problems of monitoring system in intensive care unit (ICU). We have presented a new technique based on the SVM consisting of parallel and Approx SVM (PASVM). This new technique is used to propose a new monitoring system more efficient than the current one. The new system significantly improves the working conditions in ICU by detecting true positive alarms and reducing the frequency of false alarms. In addition, our proposal has the capacity to identify patients' critical states over time. It updates the initial classification model through new observations in order to follow patient states evolution. As a result, the new system based on the PASVM offers to the medical staff better working conditions and makes triggered alarms more significant. Furthermore, it helps care-givers to take the best decisions.

However, the monitoring system using PASVM needs an expert (i.e. a doctor) to update the measured variables of monitored patients. Thus, as a future work, we aim to propose a real-time monitoring system based on PASVM to be used in real-world situations.

References

1. Adamski, P.: About the Healthcare Technology Safety Institute, HTSI Webinar Serieson Alarm Systems Management. https://www.aami.org/meetings/webinars/2013/092513_HTSI_TJC_NPSG.html. Accessed 11 Jan 2014
2. Ben Rejab, F., Nouira, K., Trabelsi, A.: Support vector machines versus multi-layer perceptrons for reducing false alarms in intensive care units. Int. J. Comput. Appl. Found. Comput. Sci. **49**, 41–47 (2012)
3. Ben Rejab, F., Nouira, K., Trabelsi, A.: Incremental support vector machines for monitoring systems in intensive care unit. In: Science and Information, SAI 2013, pp. 496–501 (2013)
4. Ben Rejab, F., Nouira, K., Trabelsi, A.: On the use of the incremental support vector machines for monitoring systems in intensive care unit. In: Proceedings of the 2013 International Conference on Technological Advances in Electrical, Electronics and Computer Engineering, TAEECE 2013, pp. 266–270 (2013)
5. Ben Rejab, F., Nouira, K., Trabelsi, A.: Health monitoring systems using machine learning techniques. In: Chen, L., Kapoor, S., Bhatia, R. (eds.) Intelligent Systems for Science and Information. SCI, vol. 542, pp. 423–440. Springer, Heidelberg (2014). doi:10.1007/978-3-319-04702-7_24
6. Borowski, M., Siebig, S., Wrede, C., Imhoff, M.: Reducing false alarms of intensive care online monitoring systems: an evaluation of two signal extraction algorithms. Comput. Math. Methods Med. **2011**, 143480:1–143480:11 (2011)
7. Chambrin, M., Ravaux, P., Calvelo-Aros, D., Jaborska, A., Chopin, C., Boniface, B.: Multicentric study of monitoring alarms in the adult intensive care unit (ICU): a descriptive analysis. Intensive Care Med. **25**, 1360–1366 (1999)
8. Charbonnie, S., Gentil, S.: A trend-based alarm system to improve patient monitoring in intensive care units. Control Eng. Pract. **15**, 1039–1050 (2007)
9. Cortes, C., Vapnik, V.: Support vector networks. Mach. Learn. **20**, 273–297 (1995)
10. Cvach, M.: Monitor alarm fatigue: an integrative review. Biomed. Instrum. Technol. **46**, 268–277 (2012)

11. Food and Drug Administration: Alarming Monitor Problems: Preventing Med-icalErrors. FDA Patient Safety News, January 2011. www.accessdata.fda.gov/scripts/cdrh/cfdocs/psn/transcript.cfm?show=106#7. Accessed 27 July 2011
12. Goldberger, A., Amaral, L., Glass, L., Hausdorff, J., Ivanov, P., Mark, R., Mietus, J., Moody, G., Peng, C., Stanley, H.: PhysioBank, PhysioToolkit, and PhysioNet: components of a new research resource for complex physiologic signals. Circulation **101**, e215–e220 (2000)
13. Korniewicz, D., Clark, T., David, Y.: A national online survey on the effectiveness of clinical alarms. Am. J. Crit. Care **17**, 36–41 (2008)
14. Nouira, K., Trabelsi, A.: Intelligent monitoring system for intensive care units. J. Med. Syst. **36**, 2309–2318 (2011)
15. Schmid, F., Goepfert, M.S., Kuhnt, D., Eichhorn, V., Diedrichs, S., Reichen-spurner, H., Goetz, A.E., Reuter, D.A.: The wolf is crying in the operating room: patient monitor and anesthesia workstation alarming patterns during car-diac surgery. Anesth Analg. **112**, 78–83 (2011)
16. Siebig, S., Kuhls, S., Imhoff, M., Langgartner, J., Reng, M., Scholmerich, J., Gather, U., Wrede, C.E.: Collection of annotated data in a clinical validation study for alarm algorithms in intensive care-a methodologic framework. J. Crit. Care **25**, 128–135 (2010)
17. Tsien, C.: Reducing false alarms in the intensive care unit: a systematic comparison of four algorithms. In: Proceedings of the AMIA Annual Fall Symposium, American Medical Informatics Association (1997)
18. Hans, P.G., Cosatto, E., Bottou, L., Dourdanovic, I., Vapnik, V.: Parallel support vector machines: the cascade SVM. In: Advances in Neural Information Processing Systems (2004)
19. Claesen, M., De Smet, F., Suykens, J., De Moor, B.: Fast prediction with SVM models containing RBF kernels, CoRR. 1403.0736 (2014)
20. Milton, A., Stegun, I.A.: Handbook of Mathematical Functions: With Formulas, Graphs, and Mathematical Tables, vol. 55. Courier Corporation, North Chelmsford (1964)

Machine Learning for Critical Care: An Overview and a Sepsis Case Study

Alfredo Vellido[1]([✉]), Vicent Ribas[2], Carles Morales[1], Adolfo Ruiz Sanmartín[3], and Juan Carlos Ruiz-Rodríguez[3]

[1] Departament de Ciències de la Computació,
Universitat Politècnica de Catalunya, 08034 Barcelona, Spain
avellido@cs.upc.edu
[2] Data Analytics in Medicine, EureCat, Barcelona, Spain
vicent.ribas@eurecat.org
[3] Critical Care Deparment, Vall d'Hebron University Hospital,
Shock, Organ Dysfunction and Resuscitation (SODIR) Research Group,
Vall d' Hebron Research Institute (VHIR), Universitat Autònoma de Barcelona,
08035 Barcelona, Spain

Abstract. Biology in general and medicine and healthcare in particular are facing the critical challenge of exponentially increasing data availability. The core of this challenge is putting these data to work through computer-based knowledge extraction methods. In the medical context this could take the form of medical decision support systems for diagnosis, prognosis or general management. Arguably, one of the most data dependent clinical environments is the critical care unit and by extension the whole area of critical care. Fresh approaches to data analysis in critical care are required, and Computational Intelligence and Machine Learning methods have already shown their usefulness in tackling problems in the area. This brief paper aims to be an introduction to the use of such methods in critical care.

Keywords: Machine Learning · Critical care · Intensive care unit

1 Introduction

Biology is experiencing a tectonic shift from being a wet-laboratory-focus science towards becoming a data-centred one, in what could be considered as a pertinent example of the pervasive Big Data paradigm [1]. Medicine as a biological science does not escape from this transformation, which can be understood as the convergence of two factors: the fast evolution of information technology networked systems and the equally fast development of novel and increasingly sophisticated non-invasive data acquisition methods.

A further convergence to consider in this context is that of the increasing dependence of medicine on the omics sciences to fulfill the promises of truly personalized medicine. The omics sciences require large scale concerted efforts

© Springer International Publishing AG 2017
I. Rojas and F. Ortuño (Eds.): IWBBIO 2017, Part I, LNBI 10208, pp. 15–30, 2017.
DOI: 10.1007/978-3-319-56148-6_2

to guarantee the medical community reliable access to large, heterogeneous and ever-growing databases. A central concept to the tasks of managing such complex databases is that of data biocuration [2].

Data is also increasingly a core concern in healthcare and its availability to medical experts is the main building block for the the development of medical decision support systems (MDSS) [3–6]. Decision making in healthcare at clinical environments is often made on the basis of multiple parameters and in the context of patient presentation, which includes the setting and the specific conditions related to the reason for admission and the procedures involved. The data used in clinical decision-making may originate from manifold sources and at multiple scales: devices in and around the patient, laboratory, blood tests, omics analyses, medical images, and ancillary information available both prior to and during the hospitalization.

Arguably, one of the most data dependent clinical environments is the critical care department (CCD) in any of its forms: intensive care unit (ICU), pediatric intensive care unit (PICU), neonatal intensive care unit (NICU) or surgical intensive care units (SICU), and this involves very practical implications for MDSS at the point of care [7]. The ICU environments care for acutely ill patients. Many of their patients, and particularly SICU patients, are technologically dependent on the life-sustaining devices that surround them. Some of these patients are indeed dependent for their very survival on technologies such as infusion pumps, mechanical ventilators, catheters and so on. Beyond treatment, assessment of prognosis in critical care and patient stratification combining different data sources is extremely important in a patient-centric environment.

Of course, the assessment of clinical needs changes depending on the acuity of the patient and conditions present at the point of care. Changes in patient status drive the quantity of data captured within the bedside documentation, either through flow sheets or paper and electronic records. The team supporting the patient, though, ultimately must define what is required and, in order to support clinical decision making, it is also necessary to include other data from the electronic health record and monitoring devices. These include fluid intake and patient output, demographic information, laboratory blood draw assessments, medical images, and so on.

In any case, medical device connectivity in the ICU is essential for providing a complete clinical decision support framework. While electronic medical records in and of themselves offer enormous work flow benefits, the documentation and charting systems are only as good as the data they convey. Due diligence by care providers can be augmented by automated and validated data collection, achieved through a seamless form of medical device connectivity and interoperability that is supported both inside and outside the hospital premises, and that follows the patient throughout the assisting process.

Fresh approaches to data analysis tailored to the needs of the ICU environments are thus required, and some of the most interesting ones are currently stemming from the fields of Computational Intelligence (CI) and Machine Learning (ML), which have already shown its relevance as the basis for MDSS [8] and as tools to improve hospital inpatient care [9].

This short overview paper provides a non-exhaustive state-of-the-art on the use of CI, ML and statistical methods for data analysis in critical care environments. We emphasize the main advantages, limitations and potential challenges of these methods and illustrate all of these by focusing in a single major pathology that is commonplace at the ICU, namely sepsis.

2   Machine Learning and Computational Intelligence in Critical Care

ML, CI and, more in general, other advanced strategies for data analysis under the umbrella concept of Artificial Intelligence (AI) have of late demonstrated not just their promise, but their actual value in different fields of biology and health. They include, amongst others and not exhaustively, bioinformatics [10,11], genetics and genomics [12,13], clinical applications [14], medical decision support and clinical diagnosis [15,16], oncology [17,18], psychiatry and neurological disorders [19,20], or cytopathology [21].

Critical care might seem too narrow a field as to provide a particular perspective on the use of ML and CI. The situation is actually the opposite: this type of methods is being applied to critical care problems with a variety of approaches of astonishing depth and breadth.

In fact, a full review of such applications is well beyond the scope of this brief overview paper. A very recent discussion and review paper of this type can be found in [22]. It provides a very interesting point of view according to which researchers in the field should consider the need to focus as much in data-related challenges as in the development and application of appropriate data modelling techniques. That is, from a Data Mining perspective, we are advised to shift part of our focus from the data modelling stage to the data understanding and pre-processing stages. Authors in [22] consider three main challenges, namely *compartmentalization, corruption* and *complexity*. Compartmentalization would include problems related to data privacy and anonymization, data integration from potentially heterogeneous databases, and data harmonization in terms of consistent definition of concepts throughout databases. Corruption would involve different types of data errors, issues of data missingness and data imprecision (usually due to a lack of matching goals in the data acquisition and the data modelling processes). Finally, complexity, including issues of prediction, state estimation and data multi-modality. This latter challenge bridges the stages of data pre-processing and modelling.

Reviews in this field are not necessarily this recent and can be traced back to the early century in work by Hanson and Marshall [23], who already stated that the ICU environment is particularly suited to the deployment of AI-based analytical strategies due to the wealth of available data and the promise they hold of increased efficiency in inpatient care due to their specific characteristics.

Specific sub-fields of critical care, such as *alarm algorithms* in critical care monitoring have also been reviewed in some detail [24]. More recent work in this sub-field [25] proposed the use of Artificial Neural Networks (ANN) and Decision

Trees (DT) for the design of patient-specific alarm algorithms in real time. DTs [26] and Random Forests (RF) [27], an extension of DTs, have recently been proposed for the reduction of false cardiac arrhythmia alarms and also for the assessment of prognosis in Sepsis [28].

All these reviews reflect the broad palette of methods available to practitioners in critical care. They include, not exhaustively, ANN [29–31] and Support Vector Machines (SVM) [32] for mortality prediction, or Deep Reinforcement Learning [33] for medicine dosing. Other applications of Deep Learning (the current *reincarnation* of ANNs) include that for the unsupervised learning of phenotypical features in longitudinal sequences of serum uric acid measurements, in [34].

Other less standard methods include Bayesian Networks (BN), used in [35] for medicine dosing and in [36] for event detection in patient monitoring. Gaussian Processes (GP) have also been used in patient vital signals monitoring after surgery [37], amongst other applications. Unsupervised hierarchical clustering was used in [38] for the identification of physiologic patient states at the ICU.

One of the medical problems in critical care to which more attention has been paid from the point of view of ML and related techniques is that of the management of the sepsis pathology, mostly from the point of view of diagnosis and prognosis, which will be discussed in more detail in the following sections.

Fuzzy systems and rule extraction, mostly as strategies for increasing the interpretability and usability of the results have been proposed: A Fuzzy DSS for the management of post-surgical cardiac intensive care unit (CICU) patients was described in [39]. The problem of rule generation was addressed in [40,41], the latter together with an ANN.

Beyond [41], other studies have deployed ANNs for the study of Sepsis. Amongst them, [42] presented a clinical study examining Systemic Inflammatory Response Syndrome (SIRS) and Multiple organ dysfunction syndrome (MODS) in the ICU after cardiac and thoracic surgery. The initiatives related to the application of ANNs to the study of Sepsis have also resulted in expert systems such as the one called SES, described in [43], which was designed for the diagnosis of pathogens and prescription of antibiotics. Ross and co-workers [44] derived a system of ordinary differential equations together with an ANN model of inflammation and Septic Shock.

SVM models have also been used for the prediction of Sepsis. Kim and co-workers [45] applied them to study Sepsis in post-operative patients. Wang *et al.* [46] built a DSS for the diagnosis of Sepsis. Tang and colleagues [47] presented a SVM-based system for Sepsis and SIRS prediction from non-invasive cardiovascular spectrum analysis.

ML methods have also been used with varying success for the more specific problem of the prediction of mortality caused by Sepsis. A diagnostic system for Septic Shock based on ANNs (Radial Basis Functions -RBF- and supervised Growing Neural Gas) was presented in [48]. Also in this area, Brause and colleagues [49] applied an evolutionary algorithm to an RBF network (the MEDAN Project) to obtain, over a retrospective dataset, a set of predictive attributes for assessing mortality for Abdominal Sepsis. BN models were used in [50,51], kernel methods were used in [52] and Relevance Vector Machines (RVM), SVM variants with embedded feature selection, were used in [53].

The following section focuses exclusively and in some detail in some of the authors' work on the application of ML and related methods to the analysis of different problems in the management of sepsis.

3   Machine Learning for the Analysis of Sepsis as a Paradigmatic Critical Care Pathology

3.1   Sepsis: Some Basic Background

The official consensus definition of the sepsis pathology has evolved over the decades. The last consensus meeting held in 2016 provided new definitions for Sepsis and its complications [54]. With the objective of increasing the specificity in diagnosing Sepsis in clinical practice, the new definitions include organ dysfunction for the diagnosis. As a consequence, the term Severe Sepsis is no longer used and the use of the SIRS [55] for diagnosing Sepsis is also not recommended. Instead, the role of the Sequential Organ Failure (SOFA) [56] score becomes even more prominent in the diagnosis and management of Sepsis.

However, the calculation of the SOFA score can be time consuming since it requires to perform blood tests to assess coagulation, liver and renal function. Therefore, to make executive decisions and admit patients into the ICU, the quick SOFA (qSOFA) score has been defined, which assesses mental status, systolic blood pressure (SBP) and respiratory rate [54].

In this context, *Sepsis is defined as life-threatening organ dysfunction caused by a dysregulated host response to infection* and *Organ dysfunction can be identified as an acute change in total SOFA score ≥2 points consequent to the infection.* Septic Shock is defined as a subset of sepsis in which underlying circulatory and cellular/metabolic abnormalities are profound enough to substantially increase mortality [54].

Even though the current definitions are yet to be widely adopted in critical care processes, they also set the basis to validate the common ground to study the pathophysiology of Sepsis and its prognosis at multiple levels of analysis. They include the study of the inflammatory response during sepsis and its association with organ dysfunction both from already available clinical data and also at a transcriptomics, proteomics and metabolomics level for better understanding the underlying mechanisms of the process.

3.2   Review of the Definitions of Sepsis Through Causal Probabilistic Networks

This section describes a study for the identification of key prognostic factors in patients that suffered a septic shock during their stay in the ICU. The analyses presented here are based on the publicly available MEDAN database [57], which recorded data from 71 voluntary cooperating German ICUs between 1998 and 2002.

The data from the MEDAN database was processed with causal probabilistic networks (CPN). CPNs are recognized in bioinformatics and computational biology as relevant representations capable of modeling causal relationships more precisely than standard clustering or regression models. CPNs also have sound statistical foundations for inferential modeling [58] and for handling noise and missing data.

The implementation of the CPNs was based on the *Causal Explorer* public library [59] with the three-phase dependency analysis algorithm (TPDA). This algorithm consists of three phases: drafting, thickening and thinning. In the drafting phase, TPDA produces an initial set of edges based on a simpler test (basically just having sufficient pairwise mutual information). This first draft is a graph without loops. In the second phase, TPDA adds edges to the current graph when the pairs of nodes cannot be separated using a set of conditional independence tests. The graph produced by this phase will contain all the edges of the underlying dependency model. In the thinning phase each edge is examined and it will be removed if the two nodes of the edge are found to be conditionally independent. The threshold value for our TDPA implementation is 0.05.

For ICU admission, the graph presented in Fig. 1 was obtained. In this graph, the dependence relations show the link between between SIRS and the ICU outcome variable and point toward a strong relationship between the SIRS diagnosis and patient outcome in the ICU.

It is also important to analyse how would SIRS relate to organ dysfunction measured through the SOFA score (Fig. 2). The resulting TPDA graph also shows a strong relation between SIRS and SOFA and the statistical dependence of SIRS with the SOFA score is therefore clear. In this graph, it is also important to note the strong relation between the SOFA score and severity measured by the APACHE II [60] and SAPS II [61].

The results of applying Causal Explorer to the analysis of the MEDAN data set provide evidence to support the very recent official modifications in the clinical definition of sepsis [54], in the sense that SIRS relates to organ dysfunction and it is therefore convenient to give more prominence to the latter for its diagnosis. However, it is also important to note the role of the inflammatory response in patient outcomes and also in the physiopathology of Sepsis.

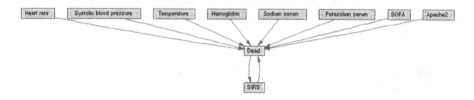

Fig. 1. Conditional independence map for data at ICU admission.

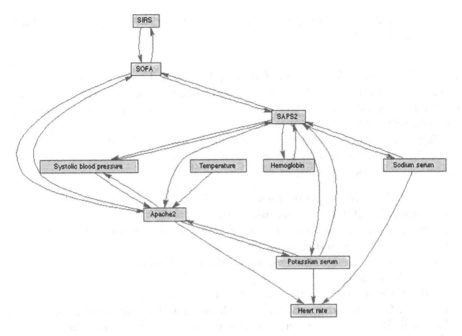

Fig. 2. Conditional independence map for organ dysfunction at ICU admission.

3.3   Finding Prognostic Factors Through Machine Learning

Here, we provide an overview of the use of a latent model-based feature extraction approach, namely Factor Analysis (FA), to obtain new sets of descriptors, or prognostic factors, for the prediction of mortality due to Sepsis. Within this framework, the reported experimental results are readily interpretable.

In the reported experiments [62], the obtained prognostic factors were used to predict mortality through standard Logistic Regression (LR), a method commonly used in medical applications [63,64] and widely trusted by clinicians. The prediction accuracy results herein reported improve on those obtained with current standard data descriptors and therefore provide support for the use of these new factors as risk-of-death predictors in ICU environments.

In this work, we resorted to a prospective observational cohort study of adult patients with severe sepsis. The study was conducted at the Critical Care Department of the Vall d'Hebron University Hospital (Barcelona, Spain), and it was approved by the Research Ethics Committee of the Hospital. The database consists of data from patients with severe sepsis and septic shock, collected at the ICU by the Research Group in Shock, Organic Dysfunction and Resuscitation (SODIR), between June, 2007 and December, 2010. During this period, 354 patients with severe sepsis (medical and surgical patients) were admitted in the ICU. The collected data show the worst values for all variables during the first 24 h of evolution for Severe Sepsis. Organ dysfunction was evaluated through the SOFA score system. Severity was evaluated through the APACHE II score.

A set of 34 features was used for the mortality prediction analyses. The full list can be found in [62].

Factor Interpretation from a Clinical Viewpoint. The application of FA resulted in a consistent 14-factor model of the original data set. The cumulative proportion of total (standardized) sample variance explained by this model was found to be 83.27%.

Taking into consideration the highest factor loadings (in absolute value) for every given variable, these factors were mapped into different easily interpretable clinical descriptors, explained as follows:

- Factor 1: Related to cardiovascular function and, more specifically, to the cardiovascular SOFA score and use of vasoactive drugs.
- Factor 2: Corresponds to haematologic function (as presented in the haematologic SOFA score and the total platelet count).
- Factor 3: Corresponds to respiratory function, Respiratory SOFA score and PaO_2/FiO_2 ratio.
- Factor 4: Corresponds to the use of mechanical ventilation and PPlateau.
- Factor 5: Corresponds to the 24 h SSC bundles and glycaemic indices.
- Factor 6: Related to the micro-organism producing the Sepsis and whether this sepsis polymicrobial or not.
- Factor 7: Corresponds to renal function measured by the SOFA score and total SOFA score.
- Factor 8: Corresponds to the administration of antibiotics and haemocultures taken during the first 6 h of ICU stay.
- Factor 9: Relates to the number of organs in dysfunction for a moderate SOFA and the total number of organs in dysfunction.
- Factor 10: Related to the hepatic function measured by the SOFA score.
- Factor 11: Corresponds to the CNS function measured by the SOFA score and the number of organs in dysfunction.
- Factor 12: Related to the loci of Sepsis and whether the infection is polymicrobial or not.
- Factor 13: Corresponds to the APACHE II score and worst lactate levels.
- Factor 14: Relates the total number of organs in dysfunction.

The factors obtained with this method are coherent with the SOFA score as a description and measure of organ failure and dysfunction [56], combined with the management guidelines defined by the Surviving Sepsis Campaign [65]. Therefore, it can be safely concluded that they are related to SOFA and the actions taken to mitigate this organ deterioration. This is a result of particular interest. One of the main challenges in mortality prediction is that of producing flexible models that can robustly fit the observed data without the need for unnecessary contextual assumptions, and in the presence of subtle interactions between covariates.

Mortality Prediction Using Logistic Regression over 14 Factors. Mortality prediction was then performed using the obtained 14-factor FA solution as starting point. The performance of the model was evaluated by 10-fold cross validation. Table 1 shows the coefficient estimates β, Z-Scores and maximum and minimum values resulting from fitting a LR model to the 14 factors (inputs) and the outcome in the ICU (output) and removing those factors yielding Z-Scores smaller than 1.96. The Z-Scores measure the effect of removing one factor from the model [66,67]. A Z-score greater than 1.96 in absolute value is significant at the 5% level and provides a measure of the relevance for the prediction of a given factor.

As shown in Table 1, factor 3, related to *Mechanical Ventilation* and *Pplateau*, shows the strongest effect together with factor 13, which is related to the APACHE II score. Factor 8 (Hepatic Function measured with the SOFA Score) and factor 10 (related to the number of Dysfunctional Organs) are also found to be relevant. It is worth noting at this stage that, with LR, the factors related to the *Surviving Sepsis Campaign* show no strong effect on mortality prediction. This result may be due to the low compliance with the *Surviving Sepsis Campaign Bundles* for the first 6 and 24 h of evolution (26.18% and 44.06% respectively for the ICU under study). However, it is interesting to note that factor 9 (antibiotic administration and haemocultures) presents a higher impact than that of factor 6 (24 h bundles with glycaemic indexes). For the ICU under analysis, 80.22% of patients received antibiotics during the first 6 h of evolution and 77.14% had haemocultures during the same period of time. In fact, timely administration of antibiotics and performance of haemocultures are considered critical to improving the prognosis of septic patients.

Regression on the 14 factors together with 10-Fold cross validation resulted in an Area Under the ROC Curve (AUC) of 0.78. A decision threshold of $\gamma = 0.68$ was automatically selected (for the maximization of the discrimination probability) to decide whether the patient survives. This 10-fold cross-validation experiment yielded an AUC of 0.78, an error rate of 0.24, a sensitivity of 0.65 and a specificity of 0.80. The results of LR over latent factors is presented in Table 1. This table also shows that the two most representative factors are F10 and F13, which correspond to organ dysfunction measured through the SOFA score and illness severity measured through the APACHE II score combined with the worst lactate levels.

Table 1. Results for LR over latent factors with 10-fold cross validation

	β coeff	MAX	MIN	Z-score
Intercept	1.22	1.53	.87	7.11
F4	−0.54	−0.23	−0.86	−3.38
F10	−0.69	−0.38	−1.05	−4.26
F9	−0.51	−0.21	−0.81	−3.36
F13	−0.49	−0.24	−0.74	−3.80

Comparison with Logistic Regression over a Selection of the Original Variables. Further experiments aimed to compare the predictive ability of the FA 14-factor solution with that of the original data variables were carried out. For that, the most significant clinical attributes were selected in a backward feature selection process (in our case, the backward feature selection removes those variables resulting in non-significative Z-scores). The selected attributes were: the total number of dysfunctional organs; the APACHE II score; and the worst lactate levels. The corresponding coefficients, maximum and minimum values and Z-scores for these three variables are presented in Table 2.

Table 2. Results for LR with 10-fold cross validation

	β coeff	MAX	MIN	Z-score
Intercept	4.20	3.11	5.29	7.56
APACHE II	−0.08	−0.13	−0.04	−3.77
Worst lact.	−0.25	−0.38	−0.11	−3.63

Regression on the most significant attributes together with 10-fold cross validation yielded an AUC of 0.75, a lower result than the one obtained with the FA solution. Following the procedure outlined in the previous subsection, a decision threshold of $\gamma = 0.68$ was automatically selected. This resulted in a prediction error over the test data of 0.3 (higher than the FA solution), a specificity of 0.72, and a sensitivity of 0.64.

Comparison with the APACHE II Mortality Score. The Risk-of-Death (ROD) formula based on the APACHE II score can be expressed as [60]:

$$\ln\left(\frac{ROD}{1 - ROD}\right) = -3.517 + 0.146 \cdot A + \epsilon \tag{1}$$

Where A is the APACHE II score and ϵ is a correction factor depending on clinical traits at admission in the ICU. For instance, if the patient has undergone post-emergency surgery, ϵ is set to 0.613. The application of this formula with a threshold of $\gamma = -0.25$ to the population under study yielded an error rate of 0.28 (higher than the FA solution), a sensitivity of 0.82 and a specificity of 0.55. The AUC was 0.70.

3.4   Sepsis Mortality Prediction from Observed Data

The previously reviewed studies analyzed dependence relations between the different variables and clinical traits and exploited their marginalization to study Sepsis and its prognosis through CIM, FA, LR and the APACHE II score. Here, we review the use of kernel methods to analyse the available data.

Table 3. Summary of prognosis indicators and their corresponding accuracies

Method	AUC	Error rate	Sens.	Spec.
LR-FA	0.78	0.24	0.65	0.80
LR	0.75	0.30	0.64	0.72
APACHE II	0.70	0.28	0.82	0.55
RVM	0.86	0.18	0.67	0.87
SVM-Quotient	0.89	0.18	0.70	0.86
SVM-Fisher	0.76	0.18	0.68	0.86
SVM-EXP	0.75	0.21	0.70	0.82
SVM-INV	0.62	0.22	0.70	0.82
SVM-CENT	0.75	0.21	0.70	0.82
SVM-GAUSS	0.83	0.24	0.65	0.81
SVM-LIN	0.62	0.26	0.62	0.78
SVM-POLY	0.69	0.28	0.71	0.76

In our approach [52], we first embedded the data in a suitable feature space, and then used algorithms based on linear algebra, geometry and statistics for inference. With this informal definition, it becomes apparent that all the methods used so far could be kernelized as long as we used the appropriate mappings, spaces, measures and topologies. Given the simplicity of the models used here (we only have multinomial and multivariate Gaussian distributions, which can be efficiently modelled algebraically by means of the Regular Exponential Family), we proposed to use a generative approach and exploit the inner data structure in order to build a set of efficient closed-form kernels best suited for these two distributions. In particular, we assessed the performance of the Quotient Basis Kernel (QBK) [52], the simplified Fisher kernel against other state-of-the art methods such as, support vector machines with a Gaussian, Polynomial and linear kernels, generative kernels based on the Jensen-Shannon metric (Centred, Inverse and Exponential kernels) [68] and RVM [53] as sepsis mortality predictors.

All model performances were evaluated over a random test population consisting of 15% of the available data. The models were trained with 10-fold cross-validation and with 70% of data reserved for training and 15% of the data reserved for validation. Table 3 shows the results for all the experiments.

4 Conclusions

In this short overview paper, we have described the increasingly complex challenge posed by the current data availability surplus in medicine in general and critical care in particular. ML and related advanced data analysis methods have provided evidence of their value in extracting usable knowledge from these data. The many different approaches to the use of these methods in the CCD have been surveyed and the main challenges they face have been outlined.

We have exemplified the potential of ML by focusing on the Sepsis pathology. Attention has been paid to the new definitions of Sepsis and the relevance of using SIRS in its diagnosis. To address this problem, we have used conditional independence models, which have shown a dependence of SIRS in both organ dysfunction measured through the SOFA score and ICU outcome. In the light of these results, it is our opinion that the SIRS should still be considered in the study of the pathophysiology of Sepsis.

One of the main limitations of the quantitative methods for the assessment of ROD currently in use at the ICU is their lack of specificity (i.e. the high number of false positive cases they incur), which not only puts an extra risk on an already severely affected patient population, but also results in an unnecessary burden for National Health Systems. In this regard, ML and related techniques can play an important role as they improve the overall performance by combining the indicators already in place with other clinical variables, which are routinely measured.

Attending to the nature of clinical data available in the ICU, it is possible to better assess prognosis through a proper embedding of the data. The techniques proposed in the case study of this overview are related to regular exponential families in general and to the multinomial and Gaussian exponential families that resulted in the generative kernels outlined in previous sections. The accuracy of routinely used methods for assessing prognosis of septic patients in the ICU (LR and the APACHE II) yield acceptable accuracy but their performance in terms of specificity is limited. The RVM and LR over latent factors have been shown to yielded an acceptable performance. However, the kernel methods presented the best balance in all these performance parameters. It is also important to note that the QBK and the simplified Fisher kernel yielded the best results, as shown in Table 3. It is also important to note that both QBK and simplified Fisher kernel can be represented through graphical models, increasing their interpretability.

Acknowledgments. This research was partially funded by Spanish TIN2016-79576-R research project and carried out under the Shockomics programme funded under the 7th Framework Program of the European Union (EU Grant 602706).

References

1. Marx, V.: Biology: the big challenges of big data. Nature **498**(7453), 255–260 (2013)
2. Howe, D., Costanzo, M., Fey, P., Gojobori, T., Hannick, L., Hide, W., Hill, D.P., Kania, R., Schaeffer, M., St Pierre, S., Twigger, S.: Big data: the future of biocuration. Nature **455**(7209), 47–50 (2008)
3. Fieschi, M., Dufour, J.C., Staccini, P., Gouvernet, J., Bouhaddou, O.: Medical decision support systems: old dilemmas and new paradigms. Methods Inf. Med. **42**(3), 190–198 (2003)
4. Bates, D.W., Kuperman, G.J., Wang, S., Gandhi, T., Kittler, A., Volk, L., Spurr, C., Khorasani, R., Tanasijevic, M., Middleton, B.T.: Commandments for effective clinical decision support: making the practice of evidence-based medicine a reality. J. Am. Med. Inform. Assoc. **10**(6), 523–530 (2003)

5. Sittig, D.F., Wright, A., Osheroff, J.A., Middleton, B., Teich, J.M., Ash, J.S., Campbell, E., Bates, D.W.: Grand challenges in clinical decision support. J. Biomed. Inform. **41**(2), 387–392 (2008)
6. Belle, A., Kon, M.A., Najarian, K.: Biomedical informatics for computer-aided decision support systems: a survey. Sci. World J. (2013). Article ID 769639
7. Wilk, S., Michalowski, W., O'Sullivan, D., Farion, K., Sayyad-Shirabad, J., Kuziemsky, C., Kukawka, B.: A task-based support architecture for developing point-of-care clinical decision support systems for the emergency department. Methods Inf. Med. **52**(1), 18–32 (2013)
8. Lisboa, P.J., Taktak, A.F.G.: The use of artificial neural networks in decision support in cancer: a systematic review. Neural Netw. **19**(4), 408–415 (2006)
9. Neill, D.B.: Using artificial intelligence to improve hospital inpatient care. IEEE Intell. Syst. **28**(2), 92–95 (2013)
10. Baldi, P., Brunak, S.: Bioinformatics: The Machine Learning Approach. MIT Press, Cambridge (2001)
11. Min, S., Lee, B., Yoon, S.: Deep Learning in Bioinformatics. Brief. Bioinform., bbw068 (2016)
12. Libbrecht, M.W., Noble, W.S.: Machine learning applications in genetics and genomics. Nat. Rev. Genet. **16**(6), 321–332 (2015)
13. Leung, M.K., Delong, A., Alipanahi, B., Frey, B.J.: Machine learning in genomic medicine: a review of computational problems and data sets. Proc. IEEE **104**(1), 176–197 (2016)
14. Lisboa, P.J.: A review of evidence of health benefit from artificial neural networks in medical intervention. Neural Netw. **15**(1), 11–39 (2002)
15. Wagholikar, K.B., Sundararajan, V., Deshpande, A.W.: Modeling paradigms for medical diagnostic decision support: a survey and future directions. J. Med. Syst. **36**(5), 3029–3049 (2012)
16. Vasilakos, A.V., Tang, Y., Yao, Y.: Neural networks for computer-aided diagnosis in medicine: a review. Neurocomputing **216**, 700–708 (2016)
17. Vellido, A., Lisboa, P.J.G.: Neural networks and other machine learning methods in cancer research. In: Sandoval, F., Prieto, A., Cabestany, J., Graña, M. (eds.) IWANN 2007. LNCS, vol. 4507, pp. 964–971. Springer, Heidelberg (2007). doi:10. 1007/978-3-540-73007-1_116
18. Lisboa, P.J.G., Vellido, A., Tagliaferri, R., Napolitano, F., Ceccarelli, M., Martín-Guerrero, J.D., Biganzoli, E.: Data mining in cancer research. IEEE Comput. Intell. Mag. **5**(1), 14–18 (2010)
19. Maroco, J., Silva, D., Rodrigues, A., Guerreiro, M., Santana, I., de Mendonça, A.: Data mining methods in the prediction of Dementia: a real-data comparison of the accuracy, sensitivity and specificity of linear discriminant analysis, logistic regression, neural networks, support vector machines, classification trees and random forests. BMC Res. Notes **4**(1), 299 (2011)
20. Iniesta, R., Stahl, D., McGuffin, P.: Machine learning, statistical learning and the future of biological research in psychiatry. Psychol. Med. **46**(12), 2455–2465 (2016)
21. Pouliakis, A., Karakitsou, E., Margari, N., Bountris, P., Haritou, M., Panayiotides, J., Koutsouris, D., Karakitsos, P.: Artificial neural networks as decision support tools in cytopathology: past, present, and future. Biomed. Eng. Comput. Biol. **7**, 1 (2016)
22. Johnson, A.E., Ghassemi, M.M., Nemati, S., Niehaus, K.E., Clifton, D.A., Clifford, G.D.: Machine learning and decision support in critical care. Proc. IEEE **104**(2), 444–466 (2016)

23. Hanson, C.W., Marshall, B.E.: Artificial intelligence applications in the intensive care unit. Crit. Care Med. **29**(2), 427–435 (2001)
24. Imhoff, M., Kuhls, S.: Alarm algorithms in critical care monitoring. Anesth. Analg. **102**(5), 1525–1537 (2006)
25. Zhang, Y., Szolovits, P.: Patient-specific learning in real time for adaptive monitoring in critical care. J. Biomed. Inform. **41**, 452–460 (2008)
26. Caballero, M., Mirsky, G.M.: Reduction of false cardiac arrhythmia alarms through the use of machine learning techniques. In: Proceedings of the Computing in Cardiology Conference (CinC), vol. 42, pp. 1169–1172 (2015)
27. Eerikäinen, L.M., Vanschoren, J., Rooijakkers, M.J., Vullings, R., Aarts, R.M.: Reduction of false arrhythmia alarms using signal selection and machine learning. Physiol. Meas. **37**(8), 1204 (2016)
28. Ribas, R.V., Caballero-López, J., Ruiz-Sanmartín, A., Ruiz-Rodríguez, J.C., Rello, J., Vellido, A.: On the use of decision trees for ICU outcome prediction in sepsis patients treated with statins. In: IEEE Symposium on Computational Intelligence and Data Mining (CIDM), pp. 37–43 (2011)
29. Dybowski, R., Weller, P., Chang, R., Gant, V.: Prediction of outcome in critically ill patients using artificial neural network synthesised by genetic algorithm. Lancet **347**(9009), 1146–1150 (1996)
30. Clermont, G., Angus, D., DiRusso, S., Griffin, M., Linde-Zwirble, W.: Predicting hospital mortality for patients in the intensive care unit: a comparison of artificial neural networks with logistic regression models. Crit. Care Med. **29**(2), 291–296 (2001)
31. Wong, L.S., Young, J.D.: A comparison of ICU mortality prediction using the APACHE II scoring system and artificial neural networks. Anaesthesia **54**(11), 1048–1054 (1999)
32. Citi, L., Barbieri, R.: Physionet 2012 challenge: predicting mortality of ICU patients using a cascaded SVM-GLM paradigm. Comput. Cardiol. **39**, 257–260 (2012)
33. Nemati, S., Adams, R.: Identifying outcome-discriminative dynamics in multivariate physiological cohort time series. In: Advanced State Space Methods for Neural and Clinical Data, p. 283. Cambridge University Press, Cambridge (2015)
34. Lasko, T.A., Denny, J.C., Levy, M.A.: Computational phenotype discovery using unsupervised feature learning over noisy, sparse, irregular clinical data. PLoS One **8**(6), e66341 (2013)
35. Nachimuthu, S.K., Wong, A., Haug, P.J.: Modeling glucose homeostasis and insulin dosing in an intensive care unit using dynamic Bayesian networks. In: Proceedings of the AMIA Annual Symposium, p. 532 (2010)
36. Laursen, P.: Event detection on patient monitoring data using causal probabilistic networks. Methods Inf. Med. **33**, 111–115 (1994)
37. Pimentel, M.A., Clifton, D.A., Tarassenko, L.: Gaussian process clustering for the functional characterisation of vital-sign trajectories. In: Proceedings of the 2013 IEEE International Workshop on Machine Learning for Signal Processing (MLSP), p. 1–6 (2013)
38. Cohen, M.J., Grossman, A.D., Morabito, D., Knudson, M.M., Butte, A.J., Manley, G.T.: Identification of complex metabolic states in critically injured patients using bioinformatic cluster analysis. Crit. Care **14**(1), R10 (2010)
39. Denai, M., Mahfouf, M., Ross, J.: A fuzzy decision support system for therapy administration in cardiovascular intensive care patients. In: Proceedings of the FUZZ-IEEE 2007, pp. 1–6 (2007)

40. Paetz, J.: Intersection based generalization rules for the analysis of symbolic septic shock patient data. In: Proceedings of the ICDM 2002, pp. 673–676 (2002)

41. Paetz, H.: Metric rule generation with septic shock patient data. In: Proceedings of the ICDM 2001, pp. 637–638 (2001)

42. Schuh, C.: Sepsis and septic shock analysis using neural networks. In: Proceedings of the Annual Meeting of the NAFIPS 2007, pp. 650–654 (2007)

43. Duhamel, A., Beuscart, R., Demongeot, J., Mouton, Y.: SES (septicemia expert system): knowledge validation from data analysis. In: Proceedings of the Annual International Conference of the IEEE Engineering in Medicine and Biology Society (EMBS), vol. 3, pp. 1400–1401 (1988)

44. Ross, J.J., Mason, D.G., Paterson, I.G., Linkens, D.A., Edwards, N.D.: Development of a knowledge-based simulator for haemodynamic support of septic shock. In: IEEE Colloquium on Simulation in Medicine (Ref. No. 1998/256), 3/1-3/4 (1998)

45. Kim, J., Blum, J., Scott, C.: Temporal features and Kernel methods for predicting sepsis in postoperative patients. Technical report, University of Michigan, USA (2010)

46. Wang, S.-L., Wu, F., Wang, B.-H.: Prediction of severe sepsis using SVM model. In: Arabnia, H. (ed.) Advances in Computational Biology. Advances in Experimental Medicine and Biology Series, vol. 680, pp. 75–81. Springer, New York (2010). doi:10.1007/978-1-4419-5913-3_9

47. Tang, C.H.H., Middleton, P.M., Savkin, A.V., Chan, G.S.H., Bishop, S., Lovell, N.H.: Non-invasive classification of severe sepsis and systemic inflammatory response syndrome using a nonlinear support vector machine: a preliminary study. Physiol. Meas. **31**, 775–793 (2010)

48. Brause, R., Hamker, F., Paetz, J.: Septic shock diagnosis by neural networks and rule based systems. In: Schmitt, M., Teodorescu, H.-N., Jain, A., Jain, A., Jain, S., Jain, L.C. (eds.) Computational Intelligence Processing in Medical Diagnosis. Studies in Fuzziness and Soft Computing, vol. 96, pp. 323–356. Springer, Heidelberg (2002)

49. Brause, R., Hanisch, E., Paetz, J., Arlt, B.: Neural networks for sepsis prediction - the MEDAN project. Journal fur Anasthesie und Intensivbehandlung **11**, 40–43 (2004)

50. Ribas, V.J., López, J.C., Sáez de Tejada, A., Ruiz-Rodríguez, J.C., Ruiz-Sanmartín, A., Rello, J., Vellido, A.: On the use of graphical models to study ICU outcome prediction in septic patients treated with statins. In: Biganzoli, E., Vellido, A., Ambrogi, F., Tagliaferri, R. (eds.) CIBB 2011. LNCS, vol. 7548, pp. 98–111. Springer, Heidelberg (2012). doi:10.1007/978-3-642-35686-5_9

51. Morales, C., Vellido, A., Ribas, V.: Applying conditional independence maps to improve sepsis prognosis. In: Data Mining in Biomedical Informatics and Healthcare (DMBIH) Workshop. IEEE International Conference on Data Mining (ICDM) (2016)

52. Ribas, V., Vellido, A., Romero, E., Ruiz-Rodríguez, J.C.: Sepsis mortality prediction with quotient basis Kernels. Artif. Intell. Med. **61**(1), 45–52 (2014)

53. Ribas, V., Ruiz-Rodríguez, J.C., Wojdel, A., Caballero-López, J., Ruiz-Sanmartín, A., Rello, J., Vellido, A.: Severe sepsis mortality prediction with relevance vector machines. In: Proceedings of the 33rd Annual International Conference of the IEEE Engineering in Medicine and Biology Society (EMBC 2011), pp. 100–103 (2011)

54. Singer, M., Deutschman, C.S., Seymour, C., et al.: The third international consensus definitions for sepsis and septic shock (sepsis-3). JAMA **315**(8), 801–810 (2016)

55. Levy, M.M., Fink, M.P., Marshall, J.C., et al.: 2001 SCCM/ESICM/ACCP/ ATS/SIS international sepsis definitions conference. Intensive Care Med. **29**, 530–538 (2003)
56. Vincent, J.L., Moreno, R., Takala, J., et al.: The SOFA (Sepsis-related Organ Failure Assessment) score to describe organ dysfunction/failure. Crit. Care. Med. **22**, 707–710 (1996)
57. Hanisch, E., Brause, R., Arlt, B., Paetz, J., Holzer, K.: The MEDAN database. Comput. Methods Programs Biomed. **75**(1), 23–30 (2003)
58. Neapolitan, R.E.: Probabilistic Methods for Bioinformatics: With an Introduction to Bayesian Networks. Morgan Kaufman, Burlington (2009)
59. Aliferis, C.F., Tsamardinos, I., Statnikov, A.R., Brown, L.E.: Causal explorer: a causal probabilistic network learning toolkit for biomedical discovery. METBMS **3**, 371–376 (2003)
60. Knaus, W.A., Draper, E.A., Wagner, D.P., Zimmerman, J.E.: APACHE II: a severity of disease classification system. Crit. Care Med. **13**, 818–829 (1985)
61. Le Gall, J.R., Neuman, F.H., Bleriot, J.P., Fulgencio, J.P., Garrigues, B., Gouzes, C., Lepage, E., Moine, P., Villers, D.: Mortality prediction using SAPS II: an update for French intensive care units. Crit. Care **9**(6), R645–R652 (2005)
62. Ribas, V.J., Vellido, A., Ruiz-Rodríguez, J.C., Rello, J.: Severe sepsis mortality prediction with logistic regression over latent factors. Expert Syst. Appl. **39**(2), 1937–1943 (2012)
63. Paliwal, M., Kumar, U.A.: Neural networks and statistical techniques: a review of applications. Expert Syst. Appl. **36**(1), 2–17 (2009)
64. Kurt, I., Ture, M., Kurum, A.T.: Comparing performances of logistic regression, classification and regression tree, and neural networks for predicting coronary artery disease. Expert Syst. Appl. **34**(1), 366–374 (2008)
65. Dellinger, R.P., Carlet, J.M., Masur, H., Gerlach, H., Calandra, T., Cohen, J., Gea-Banacloche, J., Keh, D., Marshall, J.C., Parker, M.R., Ramsay, G., Zimmerman, J.L., Vicent, J.L., Levy, M.M.: Surviving sepsis campaign guidelines for management of severe sepsis and septic shock. Intensive Care Med. **30**, 536–555 (2004)
66. Johnson, R.A., Wichern, D.W.: Applied Multivariate Statistical Analysis, 6th edn. Prentice Hall, Upper Saddle River (2007)
67. Friedman, J., Hastie, T., Tibshirani, R.: The Elements of Statistical Learning. Springer, New York (2008)
68. Ribas Ripoll, V.: On the intelligent management of sepsis. Ph.D. dissertation, Universitat Politcnica de Catalunya (UPC BarcelonaTech) (2013)
69. Hammersley, J.M., Clifford, P.: Markov fields on finite graphs and lattices. Unpublished Manuscript (1971). http://www.statslab.cam.ac.uk/grg/books/hammfest/ hamm-cliff.pdf

Bioinformatics for Healthcare and Diseases

A Meta-Review of Feature Selection Techniques in the Context of Microarray Data

Zahra Mungloo-Dilmohamud[1]([⊠]), Yasmina Jaufeerally-Fakim[1], and Carlos Peña-Reyes[2]

[1] University of Mauritius, Reduit, Mauritius
{z.mungloo,yasmina}@uom.ac.mu
[2] University of Applied Sciences Western Switzerland (HES-SO),
School of Business and Engineering Vaud (HEIG-VD),
Swiss Institute of Bioinformatics (SIB), CI4CB, Computational Intelligence
for Computational Biology Group, Yverdon, Switzerland
carlos.pena@heig-vd.ch

Abstract. Microarray technologies produce very large amounts of data that need to be classified for interpretation. Large data coupled with small sample sizes make it challenging for researchers to get useful information and therefore a lot of effort goes into the design and testing of feature selection tools; literature abounds with description of numerous methods. In this paper we select five representative review papers in the field of feature selection for microarray data in order to understand their underlying classification of methods. Finally, on this base, we propose an extended taxonomy for categorizing feature selection techniques and use it to classify the main methods presented in the selected reviews.

Keywords: Feature selection · Microarray data · Machine learning · Statistical methods

1 Introduction

With the advances in experimental techniques, biological and biomedical researchers increasingly rely on processing, analysing, and interpreting big amounts of diverse data from a variety of sources: ad-hoc experiments, gene expression profiles, raw sequences, single nucleotide polymorphisms, proteomics, metabolic pathways, clinical and personal-history and many others [1]. Investigators depend, for such analyses, on advanced tools and methods from the fields of bioinformatics and computational biology that must deal with heterogeneous data, such as long sequences, 3D structures, noisy images, and/or thousands of gene expressions values from microarrays.

Bioinformatics and computational biology are widely applied, for instance, in the search for new drugs, which relies on a thorough understanding of target diseases. Information on underlying disease mechanisms is obtained through different experimental methods, among which DNA microarrays occupy a preeminent place. DNA microarrays produce huge amounts of data usually containing a very large number of features, although associated with small sample sizes [2]. Hence, as explained below,

© Springer International Publishing AG 2017
I. Rojas and F. Ortuño (Eds.): IWBBIO 2017, Part I, LNBI 10208, pp. 33–49, 2017.
DOI: 10.1007/978-3-319-56148-6_3

feature selection tools are needed to select gene expression levels that are relevant for a given biological question. As the domain of feature selection is both well-established and dynamically evolving, several reviews have been published, even recently. Herein we intend to provide a global view of the feature selection (FS) domain by analyzing and summarizing five representative reviews. On the base of this work, we finally propose a synthetic taxonomy for better understanding and classifying feature selection methods.

1.1 Microarray Data

A transcriptome is the set of all RNA molecules, including mRNA, rRNA, tRNA, and non-coding RNA produced in one or a population of cells at any one time (also known as gene expression). Transcriptomics, the study of transcriptomes, focuses mainly on how transcript patterns are affected by development, disease, or environmental factors such as hormones, drugs, etc. Although several experimental techniques exist to measure gene expression, e.g., RT-qPCR, RNA-seq, and nanoSTRING®, DNA microarrays still constitute the largest source of gene expression data [3]. Microarray data is usually expressed as values assumed to be directly proportional to the amount of mRNA that has been transcribed from each gene. Usually, there are several thousands of raw gene expression values (up to 60'000) but only a few hundreds of samples available for analysis and classification. Such data distribution leads to hard statistical and analytical challenges [4] motivating some initial filtering to be performed to bring this number down to a few thousands [5].

1.2 The Need for Feature Selection in Microarray Data Analysis

Although feature selection is relevant in several domains, we concentrate here in its application to transcriptomics data with emphasis on selecting genes from microarray data. Gene expression data is used, among other areas, to build predictors of diseases (e.g., for diagnostic or prognostic purposes). Gene selection aims at identifying relevant subsets of genes that can act as predictive biomarkers for a given problem [6]. Given a number of features, the simplest algorithm would be to test all the possible combinations of features and keep the most predictive. Unfortunately, such an approach is, for any practical purpose, unfeasible. As the number of features increase, the computational cost of the task increases exponentially. For example, selecting 4 out of 1000 features would imply testing c(1000,4) = 4.1E10 different classifiers.

Vast amounts of irrelevant features affect learning algorithms at various levels. First, learning algorithms do not scale well—most of the time the number of training iterations required to attain a given performance grows exponentially with the number of irrelevant features [7]. Secondly, the classification accuracy is considerably degraded given the training set size [8, 9], even for advanced learning algorithms, such as Support Vector Machines (SVM), that generally scale well with the dimension of the feature space [10]. The third aspect relates to the execution time of the learning algorithm, which may become too high for some real-world applications. Another aspect to be considered is the problem of determining how many relevant features to select. Very often the user has to manually choose the number of features. Finally, financial cost should be taken into account. For example, in many diagnostic tests, it is important to make few gene expression measurements as each one implies a cost.

1.3   Why a Meta-Review

With the area of FS continuously evolving with the never-ending search for improved methods, a number of papers reviewing the state of the art on feature selection techniques have already been published. The present paper aims at presenting a critical analysis of some selected reviews, not from the point of view of the feature selection methods they contain, but more in terms of how they classify such methods according to specific criteria (for example, looking at how the methods perform the selection of the features, they might be classified as filters, wrappers, and embedded). The set of all classification criteria used by a review constitute, thus, its taxonomy.

2   Selected Reviews and Taxonomies

Five selected papers constitute the base of this meta-review: The first paper [11] is a very complete review that presents a seminal taxonomy of feature selection methods. It allows comparing if, and how, these methods have evolved over time. The second paper [12] is an in-depth study of filter techniques, very comprehensive in terms of filter techniques but that also contains a few wrapper and embedded methods. The third paper [13] is a review by Bolon-Canedo and team. They have published a number of papers in the field of FS from microarray data. The fourth paper [14] is a survey paper presenting different FS methods, some classifiers and the results obtained when using the FS techniques with the classifiers discussed using 7 different datasets. The fifth paper is a review paper is by [15] where the authors present a large number of methods and have sketched a classification for them.

2.1   Understanding Feature-Selection Taxonomies

According to [16], feature selection methods that search through the space of feature subsets must address three main concerns: (i) Where to start the search? It may start either with an empty set and adding features to it, i.e., the forward search, or using the full set of features and removing unwanted data, the backward search [17]. (ii) How to organise the search? The cost of exhaustive search may be prohibitive; it is thus necessary to define algorithms to search for feature subsets. Such algorithms might range from simple greedy search to advanced machine learning algorithms [18]. (iii) How to evaluate feature subsets? Adequate metrics are crucial to determine the relevance of the features in a given context. These latter may be categorised as strongly relevant, weakly relevant, and irrelevant [19]. The feature selection methods presented in the different reviews are classified according to how they deal with these three issues. The goal of this work is to analyse, extract, and formalize the systematic organization of these criteria—i.e., the taxonomy—that is contained, mainly implicitly, in the selected reviews. The resulting taxonomies are presented graphically so as to provide a clear and readable hierarchical view of the criteria. In addition, we propose a unified view of these taxonomies, intended at harmonizing the classification and facilitating the understanding, of feature selection techniques.

2.2   Saeys et al., Bioinformatics 2007 [11]

This is a very complete review on feature selection in bioinformatics that also includes a specific section on microarray data. Their taxonomy, presented in Fig. 1, is relatively simple, with emphasis on "selection management" as main criterion and its three categories: filter, wrapper, and embedded. Further classification criteria include feature interaction for filter methods, as well as search strategy and search space management for wrapper methods.

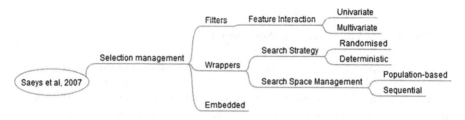

Fig. 1. Classification by Saeys et al. [11]

2.3   Lazar et al., IEEE 2012 [12]

This second review is an in-depth study of filter techniques that is, as stated by the authors, "an extended taxonomy inspired from [11]." As seen in Fig. 2, their proposed top-level criterion is the type of evaluation used when selecting features, giving place to

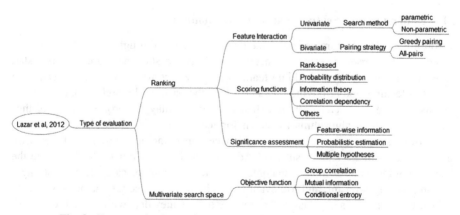

Fig. 2. Feature selection techniques classification from Lazar et al. [12]

two main classes: ranking and multivariate search space. The criteria used at the second level are feature interaction, scoring functions, and significance assessment of features for the ranking-based methods, as well as objective function for multivariate methods.

2.4   Bolon-Canedo et al., IS 2014 [13]

Even though this paper does not explicitly present a taxonomy, a detailed study suggests the implicit taxonomy shown in Fig. 3. At the top level methods can be classified based on three criteria: selection management, where hybrid methods are considered in addition to filter, wrappers, and embedded; type of evaluation, with ranking and subset classes; and class dimensionality, with three categories: binary, multiclass and multiple binary.

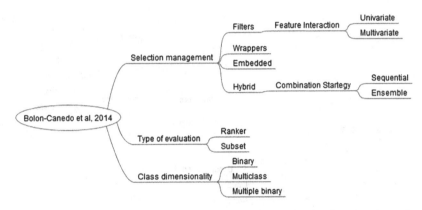

Fig. 3. Graphical representation of taxonomy in review paper, Bolon-Canedo et al.

2.5   Chandrashekar and Sahin, Computers and Electrical Engineering 2014 [14]

In this review the authors have focused mainly on the much documented filter, wrapper and embedded methods for supervised learning. But they have also briefly surveyed

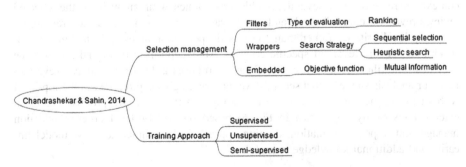

Fig. 4. Graphical representation of taxonomy in review paper, Chandrashekar et al. [14]

unsupervised and semi-supervised learning methods [20]. Out of all the papers reviewed, the term semi-supervised learning is only used in this review. Figure 4 shows the taxonomy that has been used in the paper.

2.6   Hira and Gillies, Advances in Bioinformatics 2015 [15]

This review presents a very detailed taxonomy, as shown in Fig. 5. At the top level, the methods are classified based according to four criteria: selection management of features, as in [11], training approach as discussed in [14], model linearity, with linear and non-linear as categories, and knowledge use, that refers to whether or not some prior knowledge—e.g., Gene Ontology, Protein-protein interactions, and pathways—are used to improve both feature selection and classification.

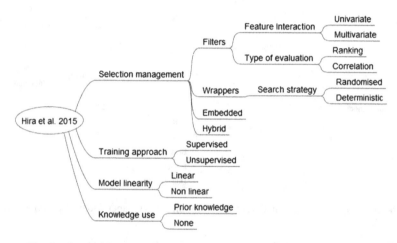

Fig. 5. Graphical representation of taxonomy in review paper, Hira et al.

3   Proposed Extended Taxonomy

As stated by Lazar et al. [12], building a taxonomy is not a trivial task and, for a given context, there may exist several possible taxonomies. Our analysis of the selected reviews reveals a chronological tendency towards longer taxonomies (i.e., relying on more top-level criteria) rather than towards deeper taxonomies (i.e., making use of several subsequent levels of specific criteria). An exception to that trend seems to be present in the filter-oriented review [12] as it proposes a deeper hierarchy. Nevertheless, our analysis suggests that second-level criteria might be promoted to the top-level without losing the correctness of the taxonomy. On this base, we propose herein an extended taxonomy (as shown in Fig. 6) based on six top-level criteria: selection management, type of evaluation, training approach, class dimensionality, model linearity and additional knowledge required.

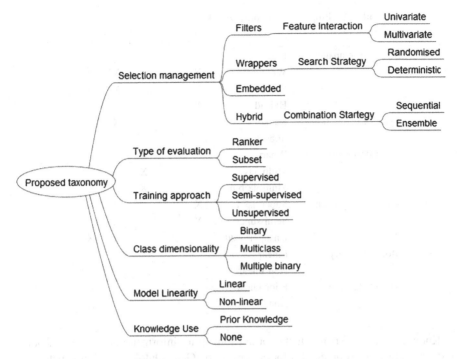

Fig. 6. Our proposed extended taxonomy

Selection management refers to the manner in which the different methods interact with the full set of features in order to perform selection. The four categories are: (1) Filters that select features based on intrinsic properties of the data rather than on classifiers. (2) Wrappers that use specific classifiers, as black-box predictors, to assess the relevance of the features to select them. (3) Embedded methods where the strategy for feature selection is built together with a given specific classifier, and (4) hybrid methods that combine, carefully, two or more approaches.

Type of evaluation refers to the way in which features are marked for maintain or elimination. Features may be ranked according to a given criterion or may be evaluated on the base of or more feature subsets to which they can belong.

The training approach refers to whether the selection is driven by presence or absence of class information. Supervised techniques are techniques where some prior labels are already available, unsupervised techniques do not make use of any labelling while semi-supervised methods use both labelled and unlabelled data for learning.

Class dimensionality refers to the number of classes being dealt with. In binary approaches the features can be classified into 2 classes, in multiclass approaches, they can be classified in more than 2 classes and in multiple binary methods binary problems are generated from multiple class datasets.

A model is linear when each term is either a constant or the product of a parameter and a predictor variable and it is nonlinear if it does not meet this definition.

Table 1. Proposed taxonomy vs taxonomy of review papers.

		[11]	[12]	[13]	[14]	[15]
Selection management	Filter	X	X	X	X	X
	Wrapper	X		X	X	X
	Embedded	X		X	X	X
	Hybrid	X		X		X
Type of evaluation	Ranker		X	X	X	
	Subset		X	X		
Class dimensionality	Binary			X		
	Multiclass			X		
	Multiple binary			X		
Training approach	Supervised	X			X	X
	Semi-supervised	X			X	
	Unsupervised				X	X
Model linearity	Linear					X
	Non-linear					X
Knowledge use	Prior knowledge					X
	None					X

Knowledge use refers to whether or not some prior information is available that can guide the selection. Sources of information can be Gene Ontology, PPI and pathways, among others.

To better understand how the proposed taxonomy summarizes the different views observed in the selected reviews, Table 1, below, illustrates whether or not these papers contain, explicitly or implicitly, the six proposed top-level criteria.

Finally, to promote understanding, and to facilitate choosing, feature selection methods, we present below, in Tables 2, 3, 4, 5, 6 and 7, how the methods found in each reviewed paper belong to the different categories proposed in the new extended taxonomy. Only the methods that have already been presented in the selected reviews have been classified given that they represent the state of the art and most representative approaches.

Table 2. Feature selection methods classified based on type of evaluation (ranker/subset).

Methods		
Ranker	BAHSIC [21] Md [22] M_FS [23] MASSIVE [24] PLS [25]	RFS [25, 26] RRFS [28] PAC-Bayes [29] Information gain [30]
Subset	Discretizer & filter [30, 31] EFA [33] MWMR [34] GA-KDE-Bayes [35] SPS [36]	FRFS [37] IFP [38] KP-SVM [39] Random forest [40]

Table 3. Feature selection methods classified based on class dimensionality.

Methods		
Binary	IFP [38]	EFA [33]
	PAC-Bayes [29]	MWMR [34]
	GA-KDE-Bayes [35]	
Multiclass	BAHSIC [21]	KP-SVM [39]
	Md [22]	Random forest [40]
	M_FS [23]	PLS [25]
	MASSIVE [24]	RFS [26, 27]
	SPS [36]	RRFS [28]
	FRFS [37]	
Multi binary	Student and Fujarewicz [25]	Discretizer & filter [30, 31]

Table 4. Feature selection methods classified based on the training approach.

Methods		
Supervised	SAM [41]	Nudge [51]
	GALGO [42]	Qvalue [52]
	GA-KNN [43]	Twilight [53]
	Rankgene [44]	Comparative marker selection [54]
	EDGE [45]	Bayesian networks [48, 49, 55, 56]
	GEPAS-Prophet [46]	Genetic algorithms [50, 51]
	DEDS [47]	SVM - RFE [58]
	RankProd [48]	Random forests [45, 46, 59, 60]
	Limma [49]	LASSO [61]
	Multtest [50]	
Semi-superv.	Law et al. [62]	Zhu [64]
	Xu et al. [63]	Zhao and Liu [65]
Unsuperv.	Clustering techniques [65–71]	Information gain [70]
	Cliff [69]	

Table 5. Feature selection methods classified based on model linearity.

Methods		
Linear	t-test [71]	SVM - RFE [58]
	CFS [41, 42]	Random forests [45, 46]
	Information gain [70]	LASSO [61]
Non linear	Bayesian networks [48, 49]	Genetic algorithms [50, 51]

Table 6. Feature selection methods classified based on selection management.

Methods			
Filter	Univariate	Regression [74] Gamma [75] t-test [71] F-test [76] ANOVA [71] Bayesian [77, 78] TNoM [79]	Wilcoxon rank sum [80] Random permutations [52, 81–83, 88–90] BSS/WSS [84] Rank products [32, 48] ReliefF [85] Information gain [30] Fold-change [86] Golub [87]
	Multivariate	Bivariate [88] CFS [72, 73] MRMR [89] USC [90]	Markov blanket filter [[91–94], 98–101] FCBF method [19]
Wrapper	Randomised	Simulated annealing SFS [95] SBE [95] Plus q take-away r [96]	Beam search [97] BIRS [98] GASVM [99]
	Deterministic	Randomized hill climbing [100] Genetic algorithms [101] Estimation of distribution [102]	GLGS [103–106, 109–112] LOOCSFS [107]
Embedded		Decision trees weighted naive Bayes [108] Weight vector of SVM [57, 115] KP-SVM [39] PAC-Bayes [29]	Random forest [40] FRFS [34] IFP [38] LASSO [61] SVM - RFE [58]
Hybrid	Ensemble	MFMW [110] CF-RFE [111]	FS-TGA [112] EF [113]
	Sequential	SVM-RFE and MRMR [114] R-m-GA [115] GADP [116]	

Table 7. Feature selection methods classified based on knowledge use.

Methods			
Prior knowledge	Gene ontology	Segal et al. [117] Kustra and Zagdanski [118]	Cheng et al. [119]
	PPI	Chuang et al. [120]	SAMBA [121]
	Pathways	Li and Li [122] Rapaport et al. [123]	Bandyopadhyay et al. [124]
No	All the methods mentioned in Tables 2, 3, 4, 5 and 6		

4 Conclusion

In this paper, we have analyzed five representative review papers on feature selection with the goal to extract, understand, and compare their, implicit or explicit, method classification criteria. The taxonomies obtained from these papers were used as base to propose an extended taxonomy that captures, as close as possible, the current state of the art in feature selection. To the best of our knowledge, this proposal is the first meta-review performed in the domain of feature selection.

References

1. Lacroix, Z., Critchlow, T.: Bioinformatics Managing Scientific Data. Academic Press, Cambridge (2003). 441 p.
2. Somorjai, R.L., Dolenko, B., Baumgartner, R.: Class prediction and discovery using gene microarray and proteomics mass spectroscopy data: curses, caveats, cautions. Bioinformatics **19**, 1484–1491 (2003). doi:10.1093/bioinformatics/btg182
3. Milward, E.A., Shahandeh, A., Heidari, M., et al.: Transcriptomics. Encycl. Cell Biol. 160–165 (2015). doi:10.1016/B978-0-12-394447-4.40029-5
4. Jirapech-Umpai, T., Aitken, S.: Feature selection and classification for microarray data analysis: evolutionary methods for identifying predictive genes. BMC Bioinform. **6**, 148 (2005). doi:10.1186/1471-2105-6-148
5. Guyon, I., Elisseeff, A.: An introduction to variable and feature selection. J. Mach. Learn. Res. **3**, 1157–1182 (2003). doi:10.1162/153244303322753616
6. Lai, C., Reinders, M.J.T., van't Veer, L.J., Wessels, L.F.: A comparison of univariate and multivariate gene selection techniques for classification of cancer datasets. BMC Bioinform. **7**, 235 (2006). doi:10.1186/1471-2105-7-235
7. Langley, P.A.T., Iba, W.: Average-case analysis of a nearest neighbor algorithm, pp. 889–894 (1993)
8. Almuallim, H., Dietterich, T.: Learning boolean concepts in the presence of many irrelevant features. AI **69**, 279–305 (1991)
9. Kira, K., Rendell, L.: The feature selection problem: traditional methods and a new algorithm. In: AAAI, pp. 129–134 (1992). doi:10.1016/S0031-3203(01)00046-2
10. Weston, J., Pavlidis, P., Cai, J., Grundy, W.N.: Gene functional classification from heterogeneous data. In: Proceedings of the Fifth Annual International Conference on Computational Molecular Biology, pp. 1–11 (2001)
11. Saeys, Y., Inza, I., Larranaga, P.: A review of feature selection techniques in bioinformatics. Bioinformatics **23**, 2507–2517 (2007). doi:10.1093/bioinformatics/btm344
12. Lazar, C., Taminau, J., Meganck, S., et al.: A survey on filter techniques for feature selection in gene expression microarray analysis. IEEE/ACM Trans. Comput. Biol. Bioinform. **9**, 1106–1119 (2012). doi:10.1109/TCBB.2012.33
13. Bolón-Canedo, V., Sánchez-Maroño, N., Alonso-Betanzos, A., et al.: A review of microarray datasets and applied feature selection methods. Inf. Sci. (Ny) **282**, 111–135 (2014). doi:10.1016/j.ins.2014.05.042
14. Chandrashekar, G., Sahin, F.: A survey on feature selection methods. Comput. Electr. Eng. **40**, 16–28 (2014). doi:10.1016/j.compeleceng.2013.11.024

15. Hira, Z.M., Gillies, D.F.: A review of feature selection and feature extraction methods applied on microarray data. Adv. Bioinform. (2015). doi:http://dx.doi.org/10.1155/2015/198363
16. Langley, P., Sage, S.: Induction of selective bayesian classifiers. In: Proceedings of the UAI-1994 (1994)
17. Liu, H., Motoda, H., Yu, L.: A selective sampling approach to active feature selection. Artif. Intell. **159**, 49–74 (2004). doi:10.1016/j.artint.2004.05.009
18. Cormen, T.H., Leiserson, C.E., Rivest, R.L.: Greedy algorithms (Chapter 17). In: Introduction to Algorithms (1990)
19. Yu, L., Liu, H.: Feature selection for high-dimensional data: a fast correlation-based filter solution. In: International Conference on Machine Learning, pp. 1–8 (2003). doi:10.1.1.68.2975
20. Bair, E., Tibshirani, R.: Semi-supervised methods to predict patient survival from gene expression data (2004). doi:10.1371/journal.pbio.0020108
21. Song, L., Smola, A., Gretton, A., et al.: Feature selection via dependence maximization. J. Mach. Learn. Res. **13**, 1393–1434 (2012). doi:10.1145/1273496.1273600
22. Bolon-Canedo, V., Seth, S., Sanchez-Marono, N., et al.: Statistical dependence measure for feature selection in microarray datasets. In: ESANN, pp. 27–29 (2011)
23. Lan, L., Vucetic, S.: Improving accuracy of microarray classification by a simple multi-task feature selection filter. Int. J. Data Mining Bioinform. **5**, 189–208 (2011)
24. Meyer, P.E., Schretter, C., Bontempi, G.: Information-theoretic feature selection in microarray data using variable complementarity. IEEE J. Sel. Top Sig. Process. **2**, 261–274 (2008). doi:10.1109/JSTSP.2008.923858
25. Student, S., Fujarewicz, K.: Stable feature selection and classification algorithms for multiclass microarray data. Biol. Direct. **7**, 33 (2012). doi:10.1186/1745-6150-7-33
26. Ferreira, A.J., Figueiredo, M.A.T.: Efficient feature selection filters for high-dimensional data. Pattern Recognit. Lett. **33**, 1794–1804 (2012). doi:10.1016/j.patrec.2012.05.019
27. Nie, F., Huang, H., Cai, X., Ding, C.H.: Efficient and robust feature selection via joint ℓ_2, 1-norms minimization. Adv. Neural Inf. Process. Syst. **23**, 1813–1821 (2010)
28. Ferreira, A.J., Figueiredo, M.A.T.: An unsupervised approach to feature discretization and selection. Pattern Recognit. **45**, 3048–3060 (2012). doi:10.1016/j.patcog.2011.12.008
29. Shah, M., Marchand, M., Corbeil, J.: Feature selection with conjunctions of decision stumps and learning from microarray data. IEEE Trans. Pattern Anal. Mach. Intell. **34**, 174–186 (2011). doi:10.1109/TPAMI.2011.82
30. Hall, M.A., Smith, L.A.: Practical feature subset selection for machine learning. Comput. Sci. **98**, 181–191 (1998)
31. Bolon-Canedo, V., Sanchez-Marono, N., Alonso-Betanzos, A.: On the effectiveness of discretization on gene selection of microarray data. In: 2010 International Joint Conference on Neural Networks, pp. 1–8. IEEE (2010)
32. Sanchez-Marono, N., Alonso-Betanzos, A., Garcia-Gonzalez, P., Bolon-Canedo, V.: Multiclass classifiers vs multiple binary classifiers using filters for feature selection. In: 2010 International Joint Conference on Neural Networks, pp. 1–8. IEEE (2010)
33. González Navarro, F.F., Muñoz, L.A.B.: Gene subset selection in microarray data using entropic filtering for cancer classification. Expert Syst. **26**, 113–124 (2009)
34. Wang, J., Wu, L., Kong, J., et al.: Maximum weight and minimum redundancy: a novel framework for feature subset selection. Pattern Recognit. **46**(6), 1616–1627 (2013)
35. Wanderley, M.F., Gardeux, V.: GA-KDE-Bayes: an evolutionary wrapper method based on non-parametric density estimation applied to bioinformatics problems. In: 21st European Symposium on Artificial Neural Networks-ESANN, pp. 24–26 (2013)

36. Sharma, A., Imoto, S., Miyano, S.: A top-r feature selection algorithm for microarray gene expression data. IEEE/ACM Trans. Comput. Biol. Bioinform. **9**, 754–764 (2012)
37. Wang, G., Song, Q., Xu, B., Zhou, Y.: Selecting feature subset for high dimensional data via the propositional FOIL rules. Pattern Recognit. **46**, 199–214 (2013). doi:10.1016/j. patcog.2012.07.028
38. Canul-Reich, J., Hall, L., Goldgof, D., Eschrich, S.: Iterative feature perturbation method as a gene selector for microarray data, pp. 1–25 (2012)
39. Maldonado, S., Weber, R., Basak, J.: Simultaneous feature selection and classification using kernel-penalized support vector machines. Inf. Sci. (Ny) **181**, 115–128 (2011). doi:10.1016/j.ins.2010.08.047
40. Anaissi, A., Kennedy, P.J., Goyal, M.: Feature selection of imbalanced gene expression microarray data. In: 2011 12th ACIS International Conference on Software Engineering, Artificial Intelligence, Networking, and Parallel/Distributed Computing (SNPD), pp. 73–78 (2011). doi:10.1109/SNPD.2011.12
41. Tusher, V.G., Tibshirani, R., Chu, G.: Significance analysis of microarrays applied to the ionizing radiation response. Proc. Natl. Acad. Sci. USA **98**, 5116–5121 (2001). doi:10. 1073/pnas.091062498
42. Trevino, V., Falciani, F.: GALGO: an R package for multivariate variable selection using genetic algorithms. Bioinformatics **22**, 1154–1156 (2006). doi:10.1093/bioinformatics/ btl074
43. Li, L., Weinberg, C.R., Darden, T.A., Pedersen, L.G.: Gene selection for sample classification based on gene expression data: study of sensitivity to choice of parameters of the GA/KNN method. Bioinformatics **17**, 1131–1142 (2001). doi:10.1093/bioinformatics/17.12.1131
44. Su, Y., Murali, T.M., Pavlovic, V., et al.: RankGene: identification of diagnostic genes based on expression data. Bioinformatics **19**, 1578–1579 (2003). doi:10.1093/bioinformatics/btg179
45. Leek, J.T., Monsen, E., Dabney, A.R., Storey, J.D.: EDGE: extraction and analysis of differential gene expression. Bioinformatics **22**, 507–508 (2006). doi:10.1093/bioinformatics/ btk005
46. Medina, I., Montaner, D., Tárraga, J., Dopazo, J.: Prophet, a web-based tool for class prediction using microarray data. Bioinformatics **23**, 390–391 (2007). doi:10.1093/bioinformatics/btl602
47. Yang, Y.H., Xiao, Y., Segal, M.R.: Identifying differentially expressed genes from microarray experiments via statistic synthesis. Bioinformatics **21**, 1084–1093 (2005). doi:10.1093/ bioinformatics/bti108
48. Breitling, R., Armengaud, P., Amtmann, A., Herzyk, P.: Rank products: a simple, yet powerful, new method to detect differentially regulated genes in replicated microarray experiments. FEBS Lett. **573**, 83–92 (2004). doi:10.1016/j.febslet.2004.07.055
49. Smyth, G.K.: Linear models and empirical Bayes methods for assessing differential expression in microarray experiments. Stat. Appl. Genet. Mol. Biol. **3**, 1–25 (2004). doi:10. 2202/1544-6115.1027
50. Dudoit, S.: Multiple hypothesis testing in microarray experiments multiple hypothesis testing in microarray experiments. Stat. Sci. **18**, 7–103 (2003)
51. Dean, N., Raftery, A.E.: Normal uniform mixture differential gene expression detection for cDNA microarrays. BMC Bioinform. **6**, 173 (2005). doi:10.1186/1471-2105-6-173
52. Storey, J.: A direct approach to false discovery rates on JSTOR. Wiley Online Libr. **64**, 479–498 (2002). doi:10.1111/1467-9868.00346
53. Scheid, S., Spang, R.: Twilight; a bioconductor package for estimating the local false discovery rate. Bioinformatics **21**, 2921–2922 (2005). doi:10.1093/bioinformatics/bti436
54. Gould, J., Getz, G., Monti, S., et al.: Comparative gene marker selection suite. Bioinformatics **22**, 1924–1925 (2006). doi:10.1093/bioinformatics/btl196

55. Hruschka, E.R., Hruschka, E.R., Ebecken, N.F.F.: Feature selection by Bayesian networks. In: Tawfik, A.Y., Goodwin, S.D. (eds.) AI 2004. LNCS (LNAI), vol. 3060, pp. 370–379. Springer, Heidelberg (2004). doi:10.1007/978-3-540-24840-8_26
56. Rau, A., Jaffrézic, F., Foulley, J.-L., Doerge, R.W.: An empirical Bayesian method for estimating biological networks from temporal microarray data. Stat. Appl. Genet. Mol. Biol. 9, Article 9 (2010). doi:10.2202/1544-6115.1513
57. Ooi, C.H., Tan, P.: Prediction for the analysis of gene expression data. Bioinformatics 19, 37–44 (2003)
58. Guyon, I., Weston, J., Barnhill, S., Vapnik, V.: Gene selection for cancer classification using support vector machines. Mach. Learn. (2002). doi:10.1023/A:1012487302797
59. Díaz-Uriarte, R., Alvarez de Andrés, S.: Gene selection and classification of microarray data using random forest. BMC Bioinform. 7, 3 (2006). doi:10.1186/1471-2105-7-3
60. Li, L., Jiang, W., Li, X., et al.: A robust hybrid between genetic algorithm and support vector machine for extracting an optimal feature gene subset. Genomics (2005). doi:10.1016/j.ygeno.2004.09.007
61. Ma, S., Song, X., Huang, J.: Supervised group Lasso with applications to microarray data analysis. BMC Bioinform. 8, 60 (2007). doi:10.1186/1471-2105-8-60
62. Law, M.H.C., Figueiredo, M.A.T., Jain, A.K.: Simultaneous feature selection and clustering using mixture models. IEEE Trans. Pattern Anal. Mach. Intell. 26, 1154–1166 (2004). doi:10.1109/TPAMI.2004.71
63. Xu, Z., King, I., Lyu, M.R.T., Jin, R.: Discriminative semi-supervised feature selection via manifold regularization. IEEE Trans. Neural Netw. 21, 1033–1047 (2010). doi:10.1109/TNN.2010.2047114
64. Zhu, X.: Semi-supervised learning literature survey contents. Sci. York 10, 10 (2008). doi:10.1.1.103.1693
65. Zhao, Z., Liu, H.: Semi-supervised feature selection via spectral analysis. In: Proceedings of the 7th SIAM International Conference on Data Mining, pp. 641–646 (2007)
66. Pudil, P., Novovičová, J., Choakjarernwanit, N., Kittler, J.: Feature selection based on the approximation of class densities by finite mixtures of special type. Pattern Recognit. 28, 1389–1398 (1995). doi:10.1016/0031-3203(94)00009-B
67. Mitra, P., Murthy, C.A., Pal, S.K.: Unsupervised feature selection using feature similarity. IEEE Trans. Pattern Anal. Mach. Intell. 24, 301–312 (2002). doi:10.1109/34.990133
68. Pal, S.K., De, R.K., Basak, J.: Unsupervised feature evaluation: a neuro-fuzzy approach. IEEE Trans. Neural Netw. 11, 366–376 (2000). doi:10.1109/72.839007
69. Xing, E.P., Karp, R.M.: CLIFF: clustering of high-dimensional microarray data via iterative feature filtering using normalized cuts. Bioinformatics 17(Suppl. 1), S306–S315 (2001). doi:10.1093/bioinformatics/17.suppl_1.S306
70. Yang, P., Zhou, B.B., Zhang, Z., Zomaya, A.Y.: A multi-filter enhanced genetic ensemble system for gene selection and sample classification of microarray data. BMC Bioinform. 11 (Suppl. 1), S5 (2010). doi:10.1186/1471-2105-11-S1-S5
71. Jafari, P., Azuaje, F.: An assessment of recently published gene expression data analyses: reporting experimental design and statistical factors. BMC Med. Inform. Decis. Mak. 6, 27 (2006). doi:10.1186/1472-6947-6-27
72. Wang, Y., Tetko, I.V., Hall, M.A., et al.: Gene selection from microarray data for cancer classification - a machine learning approach. Comput. Biol. Chem. 29, 37–46 (2005). doi:10.1016/j.compbiolchem.2004.11.001
73. Yeoh, E.-J., Ross, M.E., Shurtleff, S.A., et al.: Classification, subtype discovery, and prediction of outcome in pediatric acute lymphoblastic leukemia by gene expression profiling. Cancer Cell 1, 133–143 (2002)

74. Thomas, J.G., Olson, J.M., Tapscott, S.J., Zhao, L.P.: An efficient and robust statistical modeling approach to discover differentially expressed genes using genomic expression profiles. Genome Res. 1227–1236 (2001)
75. Newton, M.A., Kendziorski, C.M., Richmond, C.S., et al.: On differential variability of expression ratios: improving statistical inference about gene expression changes from microarray data. J. Comput. Biol. **8**, 37–52 (2001). doi:10.1089/106652701300099074
76. Bhanot, G., Alexe, G., Venkataraghavan, B., Levine, A.J.: A robust meta-classification strategy for cancer detection from MS data. Proteomics **6**, 592–604 (2006). doi:10.1002/pmic.200500192
77. Baldi, P., Long, A.D.: A Bayesian framework for the analysis of microarray expression data: regularized t-test and statistical inferences of gene changes. Bioinformatics **17**, 509–519 (2001). doi:10.1093/bioinformatics/17.6.509
78. Fox, R.J., Dimmic, M.W.: A two-sample Bayesian t-test for microarray data. BMC Bioinform. **7**, 126 (2006). doi:10.1186/1471-2105-7-126
79. Ben-Dor, A., Bruhn, L., Friedman, N., et al.: Tissue classification with gene expression profiles. J. Comput. Biol. **7**, 559–583 (2000). doi:10.1089/106652700750050943
80. Hart, T.C., Corby, P.M., Hauskrecht, M., et al.: Identification of microbial and proteomic biomarkers in early childhood caries. Int. J. Dent. **2011**, 196721 (2011). doi:10.1155/2011/196721
81. Efron, B., Tibshirani, R., Storey, J.D., Tusher, V.: Empirical Bayes analysis of a microarray experiment. J. Am. Stat. Assoc. **96**, 1151–1160 (2001). doi:10.1198/016214501753382129
82. Pan, W.: On the use of permutation in and the performance of a class of nonparametric methods to detect differential gene expression. Bioinformatics **19**, 1333–1340 (2003)
83. Park, P.J., Pagano, M., Bonetti, M.: A nonparametric scoring algorithm for identifying informative genes from microarray data. In: Pacific Symposium on Biocomputing, pp. 52–63 (2001)
84. Dudoit, S., Fridlyand, J., Speed, T.P.: Comparison of discrimination methods for the classification of tumors using gene expression data. J. Am. Stat. Assoc. **97**, 77–87 (2002). doi:10.1198/016214502753479248
85. Kononenko, I.: Estimating attributes: analysis and extensions of RELIEF. In: Bergadano, F., Raedt, L. (eds.) ECML 1994. LNCS, vol. 784, pp. 171–182. Springer, Heidelberg (1994). doi:10.1007/3-540-57868-4_57
86. DeRisi, J.L., Iyer, V.R., Brown, P.O.: Exploring the metabolic and genetic control of gene expression on a genomic scale. Science **278**, 680–686 (1997). doi:10.1126/science.278.5338.680
87. Golub, T.R., Slonim, D.K., Tamayo, P., et al.: Molecular classification of cancer: class discovery and class prediction by gene expression monitoring. Science **286**, 531–537 (1999). doi:10.1126/science.286.5439.531
88. Bo, T., Jonassen, I.: New feature subset selection procedures for classification of expression profiles. Genome Biol. **3**, RESEARCH0017 (2002)
89. Ding, C., Peng, H.: Minimum redundancy feature selection from microarray gene expression data. In: Proceedings of the IEEE Conference Computational Systems Bioinformatics, pp. 523–528 (2003)
90. Yeung, K.Y., Bumgarner, R.E.: Multiclass classification of microarray data with repeated measurements: application to cancer. Genome Biol. **4**, R83 (2003). doi:10.1186/gb-2003-4-12-r83
91. Koller, D., Sahami, M.: Toward optimal feature selection, pp. 284–292 (1996)
92. Gevaert, O., De Smet, F., Timmerman, D., et al.: Predicting the prognosis of breast cancer by integrating clinical and microarray data with Bayesian networks. Bioinformatics **22**, 184–190 (2006). doi:10.1093/bioinformatics/btl230

93. Mamitsuka, H.: Selecting features in microarray classification using ROC curves. Pattern Recogn. **39**, 2393–2404 (2006). doi:10.1016/j.patcog.2006.07.010

94. Xing, E.P., Jordan, M.I., Karp, R.M.: Feature selection for high-dimensional genomic microarray data. In: Proceedings of the Eighteenth International Conference on Machine Learning, pp 601–608. Morgan Kaufmann Publishers Inc., San Francisco (2001)

95. Kittler, J.: Pattern recognition and signal processing. In: Pattern Recognition Signal Processing, pp. 41–60. Sijthoff and Noordhoff, Alphen aan den Rijn, Netherlands (1978)

96. Ferri, F., et al.: Comparative study of techniques for large-scale feature selection. In: Pattern Recognition in Practice IV, Multiple Paradigms, Comparative Studies and Hybrid Systems, pp. 403–413. Elsevier, Amsterdam (1994)

97. Siedelecky, W., Sklansky, J.: On automatic feature selection. Int. J. Pattern Recognit. **2**, 197–220 (1998)

98. Ruiz, R., Riquelme, J.C., Aguilar-Ruiz, J.S.: Incremental wrapper-based gene selection from microarray data for cancer classification. Pattern Recognit. **39**, 2383–2392 (2006). doi:10.1016/j.patcog.2005.11.001

99. Perez, M., Marwala, T.: Microarray data feature selection using hybrid genetic algorithm simulated annealing. In: 2012 IEEE 27th Convention of Electrical & Electronics Engineers in Israel (IEEEI), pp. 1–5 (2012)

100. Skalak, D.B.: Prototype and feature selection by sampling and random mutation hill climbing algorithms (1994)

101. Holland, J.: Adaptation in Natural and Artificial Systems. University of Michigan Press, Ann Arbor (1975)

102. Inza, I., Larrañaga, P., Etxeberria, R., Sierra, B.: Feature subset selection by Bayesian networks based optimization. Artif. Intell. **123**, 157–184 (2000). doi:10.1016/S0004-3702 (00)00052-7

103. Chapelle, O., Vapnik, V., Bousquet, O., Mukherjee, S.: Choosing multiple parameters for support vector machines. Mach. Learn. (2002). doi:10.1023/A:1012450327387

104. Liu, Q., Sung, A.H., Chen, Z., et al.: Feature selection and classification of MAQC-II breast cancer and multiple myeloma microarray gene expression data (2009). doi:10.1371/journal. pone.0008250

105. Tang, E.K., Suganthan, P.N., Yao, X.: Gene selection algorithms for microarray data based on least squares support vector machine. BMC Bioinform. **7**, 1–16 (2006). doi:10.1186/ 1471-2105-7-95

106. Xia, X., Xing, H., Liu, X.: Analyzing kernel matrices for the identification of differentially expressed genes (2013). doi:10.1371/journal.pone.0081683

107. Ambroise, C., McLachlan, G.J.: Selection bias in gene extraction on the basis of microarray gene-expression data. Proc. Natl. Acad. Sci. USA **99**, 6562–6566 (2002). doi:10.1073/pnas. 102102699

108. Duda, R.O., Hart, P.E., Stork, D.G.: Pattern Classification. Wiley, New York (2001)

109. Weston, J., Elisseeff, A., Scholkopf, B., Tipping, M.: Use of the zero-norm with linear models and kernel methods. J. Mach. Learn. Res. **3**, 1439–1461 (2003). doi:10.1162/ 153244303322753751

110. Leung, Y., Hung, Y.: A multiple-filter-multiple-wrapper approach to gene selection and microarray data classification. IEEE/ACM Trans. Comput. Biol. Bioinform. **7**, 108–117 (2008). doi:10.1109/TCBB.2008.46

111. Yang, F., Mao, K.Z.: Robust feature selection for microarray data based on multicriterion fusion. IEEE/ACM Trans. Comput. Biol. Bioinform. **8**, 1080–1092 (2010). doi:10.1109/ TCBB.2010.103

112. Chuang, L., Yang, C., Wu, K., Yang, C.: A hybrid feature selection method for DNA microarray data. Comput. Biol. Med. **41**, 228–237 (2011). doi:10.1016/j.compbiomed.2011. 02.004

113. Bolón-Canedo, V., Sánchez-Maroño, N., Alonso-Betanzos, A.: An ensemble of filters and classifiers for microarray data classification. Pattern Recognit. **45**, 531–539 (2012). doi:10. 1016/j.patcog.2011.06.006

114. Mundra, P.A., Rajapakse, J.C.: SVM-RFE with MRMR filter for gene selection. IEEE Trans. Nanobiosci. **9**, 31–37 (2010). doi:10.1109/TNB.2009.2035284

115. Shreem, S.S., Abdullah, S., Nazri, M.Z.A., Alzaqebah, M.: Hybridizing ReliefF, MRMR filters and GA wrapper approaches for gene selection. J. Theor. Appl. Inf. Technol. **46**, 1034–1039 (2012)

116. Lee, C.-P., Leu, Y.: A novel hybrid feature selection method for microarray data analysis. Appl. Soft Comput. **11**, 208–213 (2011). doi:10.1016/j.asoc.2009.11.010

117. Segal, E., Pe'er, D., Regev, A., et al.: Learning module networks. J. Mach. Learn. Res. **6**, 557–588 (2005). doi:10.1016/j.febslet.2004.11.019

118. Kustra, R., Zagdanski, A.: Data-fusion in clustering microarray data: balancing discovery and interpretability. IEEE/ACM Trans. Comput. Biol. Bioinform. **7**, 50–63 (2010). doi:10. 1109/TCBB.2007.70267

119. Cheng, J., Cline, M., Martin, J., et al.: A knowledge-based clustering algorithm driven by gene ontology. J. Biopharm. Stat. **14**, 687–700 (2004)

120. Chuang, H.-Y., Lee, E., Liu, Y.-T., et al.: Network-based classification of breast cancer metastasis. Mol. Syst. Biol. **3**, 140 (2007). doi:10.1038/msb4100180

121. Tanay, A., Sharan, R., Shamir, R.: Discovering statistically significant biclusters in gene expression data. Bioinformatics **18**(Suppl. 1), S136–S144 (2002). doi:10.1093/bioinformatics/ 18.suppl_1.S136

122. Li, C., Li, H.: Network-constrained regularization and variable selection for analysis of genomic data. Bioinformatics **24**, 1175–1182 (2008). doi:10.1093/bioinformatics/btn081

123. Rapaport, F., Zinovyev, A., Dutreix, M., et al.: Classification of microarray data using gene networks. BMC Bioinform. **15**, 1–15 (2007). doi:10.1186/1471-2105-8-35

124. Bandyopadhyay, N., Kahveci, T., Goodison, S., et al.: Pathway-based feature selection algorithm for cancer microarray data (2009). doi:10.1155/2009/532989

Immune Network Technology on the Basis of Random Forest Algorithm for Computer-Aided Drug Design

Galina Samigulina[1] and Samigulina Zarina[2,3(✉)]

[1] Laboratory "Intellectual Control Systems and Forecasting",
Institute of Information and Computing Technologies, Str. Pushkeen 125,
050010 Almaty, Kazakhstan
galinasamigulina@mail.ru
[2] Department "Electrotechnics and Computer Science",
Kazakh British Technical University, Str. Tole bi 59,
050000 Almaty, Kazakhstan
zarinasamigulina@mail.ru
[3] Department "Automation and Control", Kazakh National Research Technical
University After K.I. Satpaev, Str. Satpaev 22, 050013 Almaty, Kazakhstan

Abstract. The article is devoted to immune network technology of new drugs sulfonamides properties prediction based on chemical structural information processing using the descriptor approach. Nowadays, the establishment of effective methods of QSAR (Quantitative Structure Activity Relationships) to predict the properties of new chemical compounds and directional computational molecular design of drug compounds are the most important and urgent tasks of bioinformatics. The article proposes a technology based on immune network modeling to determine the substance of new drug compounds. There was presented the algorithm for the modeling of the dependences "structure-property" on the example of sulfonamides with different duration of action based on the biological approach of artificial immune systems. Sulfonamides have been classified according to prognostic groups with different durations of action. As a method of selection of informative descriptors there was used Random Forest algorithm. The simulation results are presented in the software WEKA (Waikato Environment for Knowledge Analysis) and R Studio.

Keywords: Artificial intelligence · Artificial immune systems · Quantitative structure activity relationships · Descriptor approach · Sulfonamides · Random forest

1 Introduction

The development of non-traditional intellectual approaches for predicting the properties of new drugs and computer molecular design of chemical compounds with a specified set of properties are the most important and urgent problems of modern pharmacology.

Nowadays there are rapidly developing the researches on the creating of intelligent methods of QSAR (Quantitative Structure-Activity Relationships) and QSPR (Quantitative Structure-Property Relationships) based on artificial neural networks (NN) [1, 2],

© Springer International Publishing AG 2017
I. Rojas and F. Ortuño (Eds.): IWBBIO 2017, Part I, LNBI 10208, pp. 50–61, 2017.
DOI: 10.1007/978-3-319-56148-6_4

genetic algorithms (GA) [3, 4], artificial immune systems (AIS) and other approaches of bioinspired artificial intelligence (AI) [5]. These methods are based on determining the quantitative relationship between the structure of substances and their activity. These researches are aimed at reducing the time and cost of development of new drugs and are connected with the solution of some problems in bioinformatics, pharmacology, molecular biology, computer modeling, and others [6, 7]. Solving the problems of QSAR there must be processed massive amounts of data. The amount of virtual computer generated molecules for the last time was nearly a billion [8, 9]. Therefore, obviously experimental testing of all chemical compounds is practically impossible.

Application of modern AI techniques for predicting structure-property relation allows making directed molecular design of compounds with desired characteristics [10]. The NN were well approved. The article [11] shows the researches in the field of quantitative relationships between molecular structure and activity using two effective methods of non-linear regression and radial basis functions of the NN for the assessment of the quantitative relationship "structure-property" of the compounds. There were conducted the researches [12] in the field of computer modeling of QSAR in the sphere of design of drugs based on estrogen for the endocrine diseases treatment. In the article [13] there was presented the Bayesian regularized artificial neural network for modeling and forecasting of the activity necroptosis of the thiadiazole inhibitors and thiophene derivatives. The article [14] is devoted to the researches in pharmacology for the development of a dynamic neural network model consisting of several NN with different structure and for the obtaining the optimum output characteristics of drugs. The researches [15] propose the development of new drugs based on neuro-fuzzy approach. The article [3] shows the GA algorithm for the preparation of drugs using QSAR model based on Bayesian-regularized genetic neural networks and GA algorithm based on the Support Vector Method. There was presented the research [16] with the comparison of QSAR models based on a GA combination of the stepwise multiple regression and on the methods on the basis of artificial NN for the prediction of some derivatives of aromatic sulfonamides. The researches [17] describe the use of GA to solve the problem of creating new drugs for reducing the dimensionality, for model optimization, for conformational search and docking. Also there are conducted the articles [18] in the sphere of combined use of GA and SVM method in the research of VEGFR-2 inhibitors for the QSAR models. Along with the GA and NN for QSAR research in the article [19] there was used swarm intelligence. Due to the fact that the selection of the most informative variables in QSAR/QSPR modeling is an important task the article [20] presents a modified method of swarm optimization based on a multiple linear regression for the selection of a small subset of descriptors.

For the development of new drug compounds there is used a promising method of artificial intelligence - AIS approach [21]. A great contribution to the development of AIS was made by such scholars as: Dasgupta et al. [22], Timmis [23–25] and Castro and Zuben [26], Tarakanov [27–29], Nicosia [30], and others. There are several main areas of AIS: based on the clonal selection, on the theory of adverse selection and on the immune network. Firstly the AIS approach for computer molecular design of drugs based on the clonal selection was applied by Ivanciuc [5, 21]. The use of AIS based on immune network modeling of the structure-property relation was firstly proposed by Samigulina [31].

There is proposed following article structure: the second section presents experimental results and discussion, which includes a descriptor approach to describe the chemical structure information, basics of immune network technology and deals with the allocation of informative descriptors based on the Random Forest algorithm. The next section describes the results of simulation in software WEKA and RStudio, the third section presents a conclusion.

2   Descriptor Approach Describing the Chemical Structure Information of Drug Compounds

Descriptor approach is widely used for computer molecular design of drugs. The structure of chemical substance is described by the molecular descriptors of different levels, which include the features that characterize the physical, chemical and biological properties. There exist several directions of molecular descriptors including topology (based on graph theory), geometric characteristics (length, bond angle, etc.), quantum-chemical, thermodynamic properties, etc.

For the calculation of molecular descriptors there was developed a large number of modern software: ADRIANA.code (calculation of descriptors includes the physico-chemical properties of the descriptors, shape, size, auto correlation of interatomic 2D distance of weight distribution and auto correlation of the distance between the points of the surface); DRAGON; Molcon-Z; PaDEL-Descriptor. Software for modeling of chemical compounds: KNIME, RapidMiner, WEKA, Orange, TANAGRA, SciFinder (CAS), PubChem, ChEMBL, ChemSpider, Reaxys (Beilstein+Gmelin), NCI (National Cancer Institute Databases), Kegg, DrugBank, VCCLAB.ORG, VCCLAB.ORG, ISIDA QSPR, Molecular Operating Environment (MOE), RCDK, etc.

To select descriptors it is actual to use one of the largest databases of chemical compounds Mol-Instincts, which contains more than 2.85 million of chemicals, 2100 sets of data on the components and a total of more than 10 billion sets of chemical information database. Mol-Instincts data base presents information on sections: constants describing the organic compound (dipole moment, entropy, normal boiling point, etc.), the temperature dependence (fluid density, vapor pressure, etc.), molecular descriptors (topological descriptors, quantum-chemical, electrostatic, etc.), quantum information (the number of electrons, molecular orbitals, etc.), medical information (lipophilicity, hydrophobicity, etc.), spectral information (nuclear magnetic resonance, infra-red spectrum etc.), data analysis (3-D optimized geometry, bond length, angle, etc.).

In this article chemical compound descriptor database was developed on the basis of the resource Mol-Instincts.

2.1   Development of the Sulfonamides Descriptors Database

Let us consider the class of antimicrobial sulfonamide drug compounds as the studied chemical compounds. Sulfonamide drugs are the drugs containing a sulfamide group, mostly derivatives of benzosulfamide. These substances belong to the means of broad antibacterial spectrum. Sulfonamides are one of the oldest classes of antibacterial drugs. Currently, there were discovered and studied more than 6,000 derivatives of this group, but the practical application have about 20 drugs. Currently, sulfonamides are often used for QSAR studies to find new drug compounds [32]. All sulfa drugs, depending on the chemical structure are classified into:

1. The aliphatic derivatives: sulfonamide (streptocid); sodium sulfacetamide (sulfatsil sodium); urosulfan;
2. Heterocyclic derivatives: sulfadimetoksin; sulfalen;
3. The aromatic and heterocyclic derivatives: ftalazilsulfatiazol (ftalazol); salazodin (salazopiridazin).

Using Mol-Instincts resource there was designed a database of sulfonamide descriptors, consisting of a total of 30,075 copies. A fragment of the chemical compounds sulfonamide database and their 2D structures and specifications in SMILES format (Simplified Molecular Input Line Entry Specification) is presented in Table 1.

Table 1. A fragment of sulfonamides database

Substance	Chemical formula and 2D structure	SMILE format
Sulfathiazole	$C_9H_9N_3O_2S_2$	O=S(=O)(Nc1nccs1)c2ccc(N)cc2
Sulfamethazine	$C_{12}H_{14}N_4O_2S$	O=S(=O)(Nc1nc(cc(n1)C)C)c2ccc(N)cc2
Sulfamethoxy-pyridazin	$C_{11}H_{12}N_4O_3S$	COC1=NN=C(C=C1)NS(=O)(=O)C2=CC=C(C=C2)N
...	...	...
Sulfaquinoxaline	$C_{14}H_{12}N_4O_2S$	NC1=CC=C(C=C1)S(=O)(=O)NC1=NC2=CC=CC=C2N=C1

The values of descriptors and their name are represented in the database fragment in Table 2.

Table 2. A fragment of sulfonamides descriptor database

Substance	Mass	pI	logP	LogD	Mp	Ps	PlI	RI	...	BI
1	2	3	4	5	6	7	8	9	...	100
$C_9H_9N_3O_2S_2$	255.317	4.48	0.88	−2.44	24.19	85.08	48	7.58	...	1.63
$C_{12}H_{14}N_4O_2S$	278.33	4.51	0.43	−3.08	28.10	97.97	58	8.87	...	1.70
$C_{11}H_{12}N_4O_3S$	280.302	4.43	0.47	−2.94	27.15	107.20	56	9.01	...	1.62
$C_{12}H_{14}N_4O_2S$	278.33	5.56	0.91	−3.09	28.10	97.97	58	8.87	...	1.70
$C_{11}H_{12}N_4O_2S$	264.304	4.50	0.52	−2.94	26.35	97.97	54	8.47	...	1.76
$C_{10}H_{10}N_4O_2S$	250.277	4.50	0.39	−2.62	24.59	97.97	50	8.08	...	1.84
$C_{14}H_{12}N_4O_2S$	300.336	4.46	1.55	−3.60	32.05	97.97	66	10.0	...	1.33
...	...	...	...	...	...	...	...	...	...	...
$C_{12}H_{14}N_4O_4S$	310.329	4.43	1.26	−3.05	29.70	116.43	62	9.94	...	1.83

Table 2 shows the following types of descriptors: Mass, Pl (isoelectric point); logP, logD (lipophilicity and hydrophilicity); M_p (Molecular polarizability); P_s (polar surface area, the descriptor for the assessment of the transport properties of drugs); Pl_I (Platt index); R_I (Randic index); B_I (Balaban index), etc.

Considered sulfonamides can be classified as follows:

1 class. Sulfonamides of short action (less than 10 h);
2 class. Sulfonamides of average duration of action (10–24 h);
3 class. Long-acting sulfonamides (24–48 h).

Table 3 describes sulfonamides of different duration of action, the information provided by the WHO Collaborating Centre for Drug Statistics Methodology.

This classification can be used in solving the problem of computer molecular design of sulfonamide drugs using the research and predicting of relationships "structure-property" of chemical compounds based on AIS.

Table 3. Classification of sulfanomides on duration of action

Duration of action	Substance
Less than 10 h	Sulfamethizole, Sulfadimidine, Sulfapyridine, Sulfafurazole, Sulfanilamide, Sulfathiazole, Sulfathiourea
10–24 h	Sulfamethoxazole, Sulfadiazine, Sulfamoxole
24–48 h	Sulfadimethoxine, Sulfametomidine, Sulfametoxydiazine, Sulfaperin, Sulfamerazine, Sulfaphenazole, Sulfamazon

2.2   Immune Network Technology of Chemical Structural Sulfonamides Information Processing to Predict Dependence "Structure-Property"

Let us consider immune network technology of chemical information processing based on the approach of artificial immune systems.

Remark 1. The term artificial immune system refers to information technology, using the theoretical immunology concepts for various applications [27–30].

Remark 2. In the proposed AIS approach as a mathematical model [27–30] there are examined the time series, which correspond to the formal peptides (FP) and solves the problem of image recognition on the basis of mechanisms of molecular recognition between the two peptides (antigens and antibodies).

Algorithm 1 shows the main stages of immune network technology of chemical information processing.

Algorithm 1. Immune network technology of chemical information processing.

1. Selecting of chemical compounds for research.
2. Sulfonamides database development for the description of the structure of the chemical substances by the numerical parameters.
3. Description of the compounds structure based on descriptors.
4. The classification of chemical substances with the set properties with the use of descriptors.
5. Pre-processing of the descriptors (normalization, centering, restoration and etc.)
6. Construction of optimal immune network model based on multi-algorithmic approach (Random Forest, Factor Analysis, Genetic algorithms and etc.) [31, 33].
7. Image recognition based on AIS.
8. Forecasting the pharmacological properties of sulfonamides [34].
9. Selection of sulfa group compounds as drug candidates for further research.

The problem of image recognition using AIS is formulated as follows: after AIS training according to the standards $\Re$ for each class $\mathcal{K}$ it is necessary to determine to which class the image $\Im$ belongs that is based on the evaluation of proximity measure of the image to the standard (by the minimum binding energy obtaining). Let consider the image recognition algorithm using AIS [27–30].

Algorithm 2. Image recognition algorithm using AIS.

1. Determination of the number of $\mathcal{K}$ classes by experts;
2. Formation of standards matrix by the experts (which are obtained from the clotting of the corresponding time series of the descriptors) as follows: $\Re = \{\Re_1, \Re_2, \ldots, \Re_{\mathcal{K}}\}$, where $\mathcal{K}$- number of classes, AIS training with the teacher;
3. Formation of image matrix: $\Im = \{\Im_1, \Im_2, \ldots, \Im_n\}$, where n – number of images.
4. Determination of the binding energy ϖ_n between FP (antibodies and antigens) based on singular decomposition of matrices.
5. The minimum value of the binding energy is determined by the class to which belongs the image: $k : \varpi_k = \min\{\varpi_1, \varpi_2, \varpi_3, \ldots, \varpi_n\}$.
6. Estimation of the energy errors. This assessment is based on the properties of homologous proteins [35]. There is implemented a calculation of a Z - factor. The value of a Z-factor is determined by the average number of standard deviations between the energy of the native structure and the energy of a random chain placement.

7. Definition of the risk coefficients. There is carried out the recognition of protein native structure according to the homologies and determination of the reliability of the forecast on the basis of AIS, depending on the value of Z - factor. There are calculated risk factors: $K_R = |1 - Z_i|$, $i = \overline{1, m}$, where m- the number of homogenious peptides.
8. Estimation of the knowledge, decision-making.

The purpose of immune network training is in a choice of structure and parameters of the network, which will provide a minimum generalization error.

An important step of the immune network technology is the choice of informative descriptor, which is necessary to improve the accuracy of prediction and to reduce the chemical compounds processing time. To solve such problems there are well proven algorithms based on an ensemble of decision trees [36].

Therefore there is formulated following problem: it is necessary to solve the problem of selecting informative descriptors based on Random Forest algorithm describing the chemical compounds (sulfonamides) to build an optimal immune network model.

Remark 3. The term optimal structure of the immune network refers to a network built on the basis of the weighting factors of the assigned informative features that best characterize the state of the system. The criterion is the maximum preservation of information with a minimum amount of features.

2.3   Selecting Informative Descriptors Based on Random Forest Algorithm

Let us consider the Random Forest algorithm [36, 37], to solve the problem of selection of informative descriptors describing sulfonamides. The method is based on the construction of an ensemble of decision trees. Set the following definitions [38]: $T_i(x)$, $i= 1...B$, where T_i - an ensemble of decision trees, x - dimension descriptor vector N, B - amount of trees in the ensemble; $D = \{(x_1, y_1), (x_2 y_2), ..., (x_n y_n)\}$, where D - training data selection, y_n - class.

Further, from the original data sample there are selected n random objects with repetitions (bootstrap sample), accepting value D_i. Then there is built a decision tree for D_i (complete construction without cutting off branches) with algorithm repetition for B times.

The information content measure of the descriptors based on the Random Forest algorithm is made by the following Algorithm 3.

Algorithm 3. Feature selection algorithm on the basis of Random Forest.

1. Let x_i- a certain descriptor. It is necessary to build the random forest and get an evaluation of the probability of misclassification E_i by the Out-Of-Bag method, proposed by Breyman on the basis of observations outside the bootstrap sample. These samples are called Out-Of-Bag samples.
2. In Out-Of-Bag samples implement a random values permutation of the descriptor x_i for each tree of the constructed random forest.

3. According to the modified Out-Of-Bag there is determined an evaluation of the probability of misclassification $\hat{E}_i$.
4. Determination of information content of the descriptor is realized as the difference value: $I(x_i) = \max\left(0, \hat{E}_i - E_i\right)$.

The pseudocode of the Random Forest algorithm is presented below:

Input: dataset tree $T_i(x)$, number of trees B number of random features f
Output: D_i, a set of grown trees
Initialize D_i
For i = 1 to B do
$T_i \leftarrow$ bootstrap (T_i)
$T_i \leftarrow$ train DT(T_i, f)
add T_i to D_i
end for

3 Modeling Results

Let us consider the example of informative descriptors selection based on the Random Forest algorithm. Data processing is carried out with the help of software WEKA (Waikato Environment for Knowledge Analysis) and RStudio. As an example, for convenience of calculations, form the fragment of sulfonamides dimension database $(n \times m)$; $n = 1, \ldots, 15$; $m = 1, \ldots, 100$, where n - chemical substance, m - descriptor (Table 2). The selected fragment contains 1500 data copies. Visualizing of information sulfonamides descriptors after processing with the Random Forest algorithm is presented in Figs. 1 and 2.

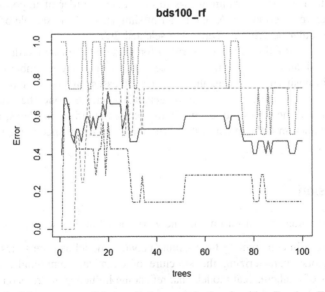

Fig. 1. Measurement of the prediction error in the algorithm Random Forest

bds100_rf

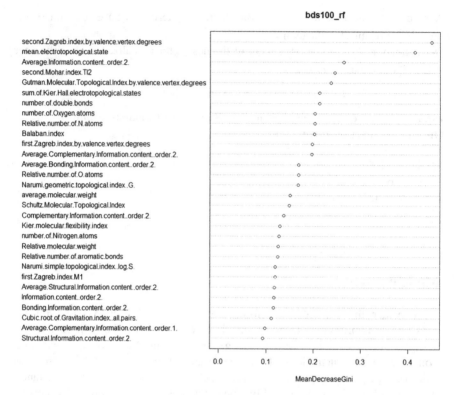

Fig. 2. Graph of the distribution degree of the importance of attributes (MeanDecreaseGini)

The Random Forest algorithm allows to rank data in order of importance and to select informative descriptions. As the result of simulation of the sample beyond 1500 copies of molecular descriptors there were selected 200.

Application of Random Forest algorithm for selecting informative descriptors has the following advantages: the ability to process data with a large number of attributes and copies, insensitivity to scale, the presence of assessing tools of the degree of importance of the variables, the ability to generalize, etc. Beyond the disadvantages there can be noted the possibility of retraining for some kind of problems, and a large size of the obtained models, in connection with which there are special requirements to computing resources.

4 Conclusion

Intelligent technology based on immune network modeling has several advantages:

- Ability to use in constructing the immune network model as time series consisting of descriptors characterizing the structure of chemical compounds, and as the parameters of mathematical models that reflect the influence of drugs on the human body;

- Reduction of time to study the immune network by constructing an optimal immune network model and reduction of the descriptors carrying significant errors;
- The system ability to analyze the hidden (latent) relationships between the descriptors and the underlying factors that affect them;
- The use of multi-algorithmic approach (Random Forest algorithm, the principal component method, neural networks, genetic algorithms, etc.) in the construction of the optimal immune network model;
- Recognition of peptides located on the border of the nonlinear classes (having similar structures) based on the properties of homologous proteins [13].
- Researchers are conducted under the grant KS MES RK №GR 0115RK00549 on the theme: Computer molecular design of drugs based on immune network modeling (2015–2017).

References

1. Montañez-Godínez, N., Martínez-Olguín, A.C., Deeb, O., Garduño-Juárez, R., Ramírez-Galicia, G.: QSAR/QSPR as an application of artificial neural networks. Artif. Neural Netw. **1260**, 319–333 (2014)
2. Baskin, I.I., Palyulin, V.A., Zefirov, N.S.: Neural networks in building QSAR models. Artif. Neural Netw. **458**, 133–154 (2009)
3. Fernandez, M., Caballero, J., Fernandez, L., Sarai, A.: Genetic algorithm optimization in drug design QSAR: Bayesian-regularized genetic neural networks (BRGNN) and genetic algorithm-optimized support vectors machines (GA-SVM). Mol. Divers. **1**, 269–289 (2011)
4. Sukumar, N., Prabhu, G., Saha, P.: Applications of genetic algorithms in QSAR/QSPR Modeling. In: Valadi, J., Siarry, P. (eds.) Applications of Metaheuristics in Process Engineering, pp. 315–324. Springer, Heidelberg (2014)
5. Ivanciuc, O.: Drug design with artificial intelligence methods. In: Meyers, R.A. (ed.) Encyclopedia of Complexity and Systems Science, pp. 2113–2139. Springer, New York (2009)
6. Andersson, D., Hillgren, M., Lindgren, C., Qian, W., Berg, L., Ekstrom, F., Linusson, A.: Benefits of statistical molecular design, covariance analysis, and reference models in QSAR: a case study on acetylcholinesterase. J. Comput. Aided Mol. Des. **3**, 199–215 (2014)
7. Macalino, S.J., Gosu, V., Hong, V., Choi, S.: Role of computer-aided drug design in modern drug discovery. Arch. Pharm. Res. **9**, 1686–1701 (2015)
8. Blum, L., Reymond, J.: 970 million drug like small molecules for virtual screening in the chemical universe database GDB -13. J. Am. Chem. Soc. **25**, 8732–8733 (2009)
9. Nonell-Canals, A., Mestres, J.: In silico target profiling of one billion molecules. Mol. Inform. **5**, 405–409 (2011)
10. Priest, A.C., Williamson, A.J., Cartwright, H.M.: The applications of artificial neural networks in the identification of quantitative structure-activity relationships for chemotherapeutic drug carcinogenicity. In: Cohen, P.R., Adams, N.M., Berthold, M.R. (eds.) IDA 2010. LNCS, vol. 6065, pp. 137–146. Springer, Heidelberg (2010). doi:10.1007/978-3-642-13062-5_14
11. Shahlaei, M., Madadkar-Sobhani, A., Fassihi, A., Saghaie, L., Arkan, E.: QSAR study of some CCR5 antagonists as anti-HIV agents using radial basis function neural network and general regression neural network on the basis of principal components. Med. Chem. Res. **10**, 3246–3262 (2012)

12. Ji, L., Wang, X.D., Luo, S., Qin, L., Yang, X., Liu, S., Wang, L.: QSAR study on estrogenic activity of structurally diverse compounds using generalized regression neural network. Sci. China, Ser. B: Chem. **7**, 677–683 (2008)

13. Chamjangali, A., Ashrafi, M.: QSAR study of necroptosis inhibitory activities (EC50) of thiadiazole and thiophene derivatives using Bayesian regularized. Artif. Neural Netw. Calc. Descr. **1**, 392–400 (2013)

14. Abraham, A., Grosan, C., Tigan, S.: Pharmaceutical drug design using dynamic connectionist ensemble networks. Stud. Comput. Intell. (SCI). **123**, 221–231 (2008)

15. Grasan, C., Abraham, A., Tigan, S.: Engineering drag desain using a multi-input multi-output neuro-fuzzy system. In: 8-th International Symposium on Symbol and Numeric Algorithms for Scientific Computing (SYNAC-2006), pp. 365–371. IEEE CS Press, Timisoara, Romania (2006)

16. Maleki, A., Daraei, H., Alae, L., Faraji, A.: Comparison of QSAR models based on combinations of genetic algorithm, stepwise multiple linear regression, and artificial neural network methods to predict of some derivatives of aromatic sulfonamides as carbonic anhydrase II inhibitors. Russ. J. Bioor. Chemi. **40**, 61–75 (2014)

17. Sukumar, N., Prabhu, G., Saha, P.: Applications of genetic algorithms in QSAR/QSPR modeling. In: Siarry, P., Valadi, J. (eds.) Applications of Metaheuristics in Process Engineering, pp. 55–68. Springer, Heidelberg (2014)

18. Nekoei, M., Mohammadhosseini, M., Pourbasheer, E.: QSAR study of VEGFR-2 inhibitors by using genetic algorithm-multiple linear regressions (GA-MLR) and genetic algorithm-support vector machine (GA-SVM): a comparative approach. Med. Chem. Res. **7**, 3037–3046 (2015)

19. Prakasvudhisarn, C., Lawtrakul, L.: Feature set selection in QSAR of 1-[(2-Hydroxyethoxy) methyl]-6-(phenylthio)thymine (HEPT) analogues by using swarm intelligence. Monatshefte für Chemie - Chemical Monthly **3**, 197–211 (2008)

20. Khajek, A., Modarres, H., Zeinoddini-Meymand, H.: Modified particle swarm optimization method for variable selection in QSAR/QSPR studies. Struct. Chem. **5**, 1401–1409 (2013)

21. Ivanciuc, O.: Artificial immune system classification of drug-induced torsade de pointes with AIRS (Artificial Immune Recognition System). J. Mol. Des. **5**, 488–502 (2006)

22. Dasgupta, D., Yu, S., Nino, F.: Recent advances in artificial immune systems: models and applications. Appl. Soft Comput. **2**, 1574–1587 (2011)

23. Timmis, J.: Artificial immune systems: today and tomorrow. Nat. Comput. **6**(1), 1–18 (2007)

24. Timmis, J., Hone, A., Stibor, T., Clark, E.: Theoretical advances in artificial immune systems. Theoret. Comput. Sci. **403**, 11–32 (2008)

25. Cutello, V., Nicosia, G., Pavone, M., Timmis, J.: An immune algorithm for protein structure prediction on lattice models. IEEE Trans. Evol. Comput. **1**, 101–117 (2007)

26. Castro, P.A.D., Zuben, F.J.: MOBAIS: a Bayesian artificial immune system for multi-objective optimization. In: Bentley, P.J., Lee, D., Jung, S. (eds.) ICARIS 2008. LNCS, vol. 5132, pp. 48–59. Springer, Heidelberg (2008). doi:10.1007/978-3-540-85072-4_5

27. Tarakanov, A.O., Tarakanov, Y.A.: A comparison of immune and neural computing for two real-life tasks of pattern recognition. In: Nicosia, G., Cutello, V., Bentley, Peter J., Timmis, J. (eds.) ICARIS 2004. LNCS, vol. 3239, pp. 236–249. Springer, Heidelberg (2004). doi:10.1007/978-3-540-30220-9_20

28. Tarakanov, A.O., Tarakanov, Y.A.: A comparison of immune and genetic algorithms for two real-life tasks of pattern recognition. Int. J. Unconv. Comput. **4**, 357–374 (2005)

29. Tarakanov, A.O., Borisova, A.V.: Formal immune networks: self-organization and real-world applications. In: Prokopenko, M. (ed.) Advances in Applied Self-organizing Systems, pp. 321–341. Springer, London (2013)

30. Tarakanov, A., Nicosia, G.: Foundations of immunocomputing. In: Proceedings of the 1-st IEEE Symposium on Foundations of Computational Intelligence (FOCI 2007), pp. 503–508. Honolulu, Hawaii (2007)
31. Samigulina G.A., Samigulina Z.I.: Development of the immune network modeling technology of computer molecular design of drugs. Certificate of state registration of intellectual property object in the Committee on Intellectual Property MJ RK. 473 (2011)
32. Sharma, B.K., Pilana, P., Sarbhai, K., Singh, P., Prabhakar, S.Y.: Chemometric descriptors in modeling the carbonic anhydrase inhibition activity of sulfonamide and sulfamate derivatives. Mol. Divers. 14(2), 371–384 (2010)
33. Samigulina, G.A., Samigullina, Z.I.: Construction of optimal immune network model for predicting the properties of the unknown drug compounds based on multi algorithmic approach. Probl. Inform. 2, 21–29 (2013)
34. Samigulina, G.A., Samigulina, Z.I., Wuizik, W., Krak, Y.: Prediction of «structure – property» dependence of new organic compounds on the basis of artificial immune systems. J. Autom. Inf. Sci. 4, 28–35 (2014)
35. Samigulina, G.A., Samigulina, Z.I.: Intellectual Systems of Forecasting and Control of Complex Objects Based on Artificial Immune Systems, p. 189. Science Book Publishing House, Yelm (2014)
36. Teixeira, A., Leal, J., Falcao, A.: Random forests for feature selection in QSPR models - an application for predicting standard enthalpy of formation of hydrocarbons. J. Cheminform. 5 (9), 1–15 (2013)
37. Chistyakov, S.P.: Random forests: review. Res. Karel. Res. Centre RAS 1, 117–136 (2013)
38. Breiman, L.: Random forest. Mach. Learn. 45(1), 5–32 (2001)

H-RACER: Hybrid RACER to Correct Substitution, Insertion, and Deletion Errors

Salma Gomaa, Nahla A. Belal[✉], and Yasser El-Sonbaty

College of Computing and Information Technology, AASTMT, Alexandria, Egypt
salma.gomaa89@gmail.com, {nahlabelal,yasser}@aast.edu

Abstract. The Next-Generation sequencing technologies produce large sets of short reads that may contain errors of different types. These errors represent a great obstacle to utilize data in sequencing projects; such as assemblers. Consequently, error correction is a vital process that aims to reduce the error rate. So, the correction of all errors types becomes very challenging. H-RACER is an error correcting tool for all types of errors (substitutions, insertions, and deletions) in a mixed set of reads. It mainly depends on RACER algorithm in detecting the error and correcting it. The major advantage presented by H-RACER is the correction of substitution errors as well as the insertions and deletions with the highest accuracy and the least time compared to other existing algorithms that specialize in correcting all types of errors.

Keywords: DNA sequencing · k-mer · Error correction · Substitution · InDels

1 Introduction

The Next-Generation sequencing (NGS) high-throughput technologies [11] were originally proposed in order to help make the vast analysis of genomes less expensive and more spread, this was doable by enhancing the sequencing time, where too many reads can be generated in a suitable time. But unfortunately, this type of sequencing introduced two painful issues; the first issue is that the read length becomes much shorter than the conventional sequencing, while the second issue is the decrement of the accuracy, where each erroneous nucleotide can be introduced to the read sequence via one of the three erroneous actions; which are substitution, insertion and deletion. The substitution takes place when the nucleotide is replaced with another erroneous one, while the insertion is when an erroneous nucleotide is newly inserted to the read sequence, and finally the deletion results due to the deletion of a nucleotide from the sequence.

The reads accuracy is a vital factor in all processes that can be applied to the output reads. As an example, the assembly of NGS reads can not be accomplished successfully until the reads errors are corrected or eliminated. So detecting and correcting (or eliminating) the reads errors is an essential step that should precede the assembly process. This step can be accomplished either by a

© Springer International Publishing AG 2017
I. Rojas and F. Ortuño (Eds.): IWBBIO 2017, Part I, LNBI 10208, pp. 62–73, 2017.
DOI: 10.1007/978-3-319-56148-6_5

standalone solution or implicitly within the assembly mechanism. The frequency and quality value of the nucleotide are the two main factors used in evaluating it to be erroneous or not [13].

This newly proposed error correction methodology aims to correct all types of errors (substitutions, insertions, and deletions) taking into consideration both accuracy and time. This methodology builds its correction decisions on RACER [4] (an already existing algorithm that handles substitution errors only) with some tuning to handle the insertions and deletions errors. This paper is organized as follows; existing methodologies for error correction are demonstrated in Sect. 2, then the proposed algorithm is introduced in Sect. 3. In Sect. 4, a comparison is set between the new methodology versus existing ones through experimental data. The conclusion is shown in Sect. 5.

2   Related Work

The error correction methodologies can be either an implicit process within the assembly methodology or a standalone solution that reproduces reads after correction. The assemblers that have embedded error correction are Euler [2], Velvet [15], AllPaths [1] and SOAP [6]. While the standalone methodologies are Coral [10], Quake [5], Reptile [14], HSHREC [9], HiTEC [3], RACER [4], Pollux [7], Parallel Error Correction with CUDA [12], and Error Corrector (EC) [8].

2.1   Euler

Euler [2] assembly method runs a filtration step called spectrum alignment that aims to classify the k-mers into two categories according to their frequencies all over the reads. Strong k-mers are the ones with high frequencies, while weak k-mers are the ones with lower frequencies. The correction takes place by executing a greedy exploration for base call substitutions aiming to reduce the weak k-mers count.

2.2   Velvet

Velvet's [15] tour bus algorithm uses breadth-first search (BFS), starting at nodes with multiple out-going edges, where candidate paths are traversed in step, moving ahead one node on all paths per iteration, until the path lengths exceed a threshold. Velvet removes the path representing fewer reads then re-aligns reads from the removed path to the remaining path.

2.3   AllPaths

AllPaths [1] uses a read-correcting preprocessor related to the spectral alignment in Euler, where the reads filtration is based on quality values, which is further used in correcting some substitutional errors.

2.4 SOAP

SOAP [6] filters and corrects reads using pre-set thresholds for k-mer frequencies. It removes bubbles with an algorithm like Velvet's tour bus, with higher read coverage determining the surviving path (read).

2.5 Coral

Coral [10] is able to correct all types of errors (substitution, insertion and deletion). The methodology is built on scoring the alignments between short reads, where each alignment runs on a base read with all the reads that have at least one common k-mer with this base read. Then the correction takes place for the misaligned positions depending on the number of times each letter occurs in addition to quality scores of the nucleotide.

2.6 Quake

Quake [5] determines a cut-off value which separates trusted k-mers from untrusted k-mers, using the distribution of k-mers based on their quality scores. The intersection of the untrusted k-mers is used to localize the search for an error in a read. Quake tries to evaluate the conditional probability of assigning the actual nucleotides of the sequenced fragment, given the observed nucleotides.

2.7 Reptile

Reptile [14] uses the spectral alignment approach used in Euler, with the quality score information if available. Trying to create approximate multiple alignments by considering all reads with pairwise hamming distance less than a pre-set threshold.

2.8 HSHREC

HSHREC [9] is able to correct all types of errors (substitution, insertion and deletion). The methodology depends on the alignment of a read with others using a suffix trie, where the edges are labelled with DNA letters and a node weight is the number of leaves in the sub-trie rooted at that node. On the down levels, a node with more than one child is considered to have a substitution error, while extra branching in the generalized suffix trie is caused by indels.

2.9 HiTEC

HiTEC [3] uses a suffix array that is built using a string of reads and their reverse complements. The correction of an erroneous nucleotide takes place with the letter that appears most at that position.

2.10 RACER

RACER [4] is able to correct data sets that have varying read lengths. Using a hash table that stores the total times each nucleotide appears before and after each k-mer, where the error is corrected via the counts.

2.11 Pollux

Pollux [7] calculates the k-mer frequencies in the entire set of reads. Identifying the discontinuities by comparing the frequencies of adjacent k-mers within reads, assuming that individual k-mers are not erroneous. The discontinuities within reads are used to find error locations and evaluate correctness. The correction is chosen to be the one that removes or minimizes the k-mer count discontinuity.

2.12 Parallel Error Correction with CUDA

Parallel Error Correction with CUDA [12] uses the spectrum alignment besides a voting algorithm for the single-mutation using each letter, hence errors can be fixed based on high values in the voting matrix.

2.13 Error Corrector

Error Corrector (EC) [8] an error correction algorithm for correcting short reads with substitution errors only. Using k-mers hashing tables to find the neighbours of each of the reads, where each read is corrected using its neighbours.

3 The Proposed Algorithm

H-RACER is the newly proposed approach for correcting all types of errors (substitution, insertion, and deletion). Although RACER is the fastest DNA error correction algorithm existent nowadays with a high accuracy, but it can not correct all types of errors, it can only correct substitutions. So, H-RACER is proposed in order to correct all types of errors. H-RACER follows the same algorithm of RACER in detecting errors and deciding their corrections, the newly added part in H-RACER is the detection of the error type in order to apply the correction properly for data sets with varying error types.

H-RACER detects the error type for an erroneous nucleotide by studying its correction value (obtained by RACER) against its neighbours, then decides the corrective action (substitute, insert, delete) according to the detected error type. Once the error detection and correction stages are done, H-RACER starts applying correction using its own methodology which mainly depends on detecting the error type in order to apply the detected correction with the proper action (substitution, insertion or deletion).

H-RACER starts the error type detection by looping on every nucleotide in the whole reads set, to check if this nucleotide is an erroneous one. For every erroneous nucleotide H-RACER checks the position of this nucleotide in the read not to be the last one in the read (so that there is at least one nucleotide following it) where the follower nucleotide is erroneous too. Then H-RACER starts to examine the erroneous and corrective values for both nucleotides (the current and its follower). H-RACER checks if the correction value of the current erroneous nucleotide is equal to the erroneous value of its follower, so it will be concluded that this current erroneous nucleotide is a result of an insertion erroneous action. Hence, H-RACER applies the correction as a deletion action for this current erroneous nucleotide. But, if it is found that the erroneous value of the current nucleotide is equal to the correction value of its erroneous follower, so it will be concluded that this current erroneous nucleotide is a result of a deletion erroneous action. Hence, H-RACER applies the correction as an insertion action at a position directly before the current erroneous nucleotide with the correction value of it (the current erroneous nucleotide). Otherwise, if there is not any criss-cross equality relation between the erroneous and correction values of the current nucleotide and its erroneous follower (i.e. the correction/erroneous value of the current erroneous nucleotide is not equal to the erroneous/correction value of its follower), then it will be concluded that this current erroneous is a result of a substitution erroneous action. Hence, H-RACER applies the corrective action as a substitution action for the erroneous value of the current nucleotide with its correction value.

On another side, if it is found that the current nucleotide is the last one in the read or its follower nucleotide is not erroneous, then H-RACER will check the position of this nucleotide in the read not to be the first in the read (so that there is at least one nucleotide that precedes it) where the precedent nucleotide is erroneous too. Then H-RACER starts to examine the erroneous and corrective values for both nucleotides (the current and its precedent). H-RACER checks if the correction value of the current erroneous nucleotide is equal to the erroneous value of its precedent, so it will be concluded that this current erroneous nucleotide is a result of an insertion erroneous action. Hence, H-RACER applies the correction as a deletion action for this current erroneous nucleotide. But, if it is found that the erroneous value of the current nucleotide is equal to the correction value of its erroneous precedent, so it will be concluded that this current erroneous nucleotide is a result of a deletion erroneous action. Hence, H-RACER applies the correction as an insertion action at a position directly after the current erroneous nucleotide with the correction value of it (the current erroneous nucleotide). Otherwise, if there is no criss-cross equality relation between the erroneous and correction values of the current nucleotide and its erroneous precedent (i.e. the correction/erroneous value of the current erroneous nucleotide is not equal to the erroneous/correction value of its precedent), then it will be concluded that this current erroneous is a result of a substitution erroneous action. Hence, H-RACER applies the corrective action as a substitution action for the erroneous value of the current nucleotide with its correction value. Finally, if H-RACER finds that the current nucleotide is either the last or the first in the read, or neither its follower nor precedent nucleotides are erroneous, then it will be concluded that this current erroneous is a result of a substitution erroneous action. Hence, H-RACER applies the corrective action as a substitution action for the erroneous value of the current nucleotide with its correction value.

H-RACER error detection algorithm has a complexity $O(r)$, where r is the number of reads. For more illustration check the examples shown below in Figs. 1 and 2, and the pseudo-code shown in Fig. 3.

1. Inserted Nucleotide - Deletion Correction
   (a) Read Sequence: AC<u>GT</u>···
   (b) Correction: AC**T**···
   (c) Tracing:
       i. <u>G</u> is an erroneous nucleotide
       ii. <u>G</u> is followed by an erroneous nucleotide <u>T</u>
       iii. <u>G</u>'s correction value is **T**
   (d) Conclusion: <u>G</u> is an erroneously inserted nucleotide
2. Deleted Nucleotide - Insertion Correction
   (a) Read Sequence: AC<u>GT</u>···
   (b) Correction: AC**AG**···
   (c) Tracing:
       i. <u>G</u> is an erroneous nucleotide
       ii. <u>G</u> is followed by an erroneous nucleotide <u>T</u>
       iii. <u>G</u>'s correction value is **A**
       iv. <u>T</u>'s correction value is **G**
   (d) Conclusion: **A** is an erroneously deleted nucleotide
3. Substituted Nucleotide - Substitution Correction
   (a) Read Sequence: AC<u>GT</u>···
   (b) Correction: AC**AC**···
   (c) Tracing:
       i. G is an erroneous nucleotide
       ii. <u>G</u> is followed by an erroneous nucleotide <u>T</u>
       iii. <u>G</u>'s correction value is **A**
       iv. <u>T</u>'s correction value is **C**
   (d) Conclusion: **A** is an erroneously substituted nucleotide with <u>G</u>

Note: The erroneous nucleotides are the underlined ones, while the nucleotides corrections are the ones in bold.

Fig. 1. H-RACER error type detection examples

Assume a read represented as: $r = s_1 s_2 \cdots s_n$
if s_i is erroneous
Then,

1. if s_{i+1} has error Then,
   (a) Delete s_i, if the correction of s_i equals to s_{i+1}
   (b) Insert the correction of s_i before s_i, if s_i equals to the correction of s_{i+1}
   (c) Substitute s_i with its correction, otherwise
2. if s_{i-1} has error Then,
   (a) Delete s_i, if the correction of s_i equals to s_{i-1}
   (b) Insert the correction of s_i after s_i, if s_i equals to the correction of s_{i-1}
   (c) Substitute s_i with its correction, otherwise

Fig. 2. H-RACER error type detection abstraction

```
for every read in reads do
    for every nuc in read do
        if has_error(nuc)
            if not_last(nuc) AND has_error(next(nuc))
                if correction(nuc) equals to next(nuc)
                    Delete nuc from read
                else if nuc equals to correction(next(nuc))
                    Insert correction(nuc) before nuc in read
                else
                    substitute nuc with correction(nuc) in read
                end if
            else if not_first(nuc) AND has_error(previous(nuc))
                if correction(nuc) equals to previous(nuc)
                    Delete nuc from read
                else if nuc equals to correction(previous(nuc))
                    Insert correction(nuc) after nuc in read
                else
                    substitute nuc with correction(nuc) in read
                end if
            else
                substitute nuc with correction(nuc) in read
            end if
        end if
    end for
end for
```

Fig. 3. H-RACER pseudo-code - $O(r)$

4   Evaluation

4.1   Datasets and Platform

The testing was performed on a wide variety of real data sets, shown below in Table 1, with different read length, genome size and coverage. It was preferred to use real data sets and to avoid any simulated ones as they do not offer a good indication of real life performance. All data sets were brought from the National Center for Biotechnology Information (NCBI).

All algorithms were executed on the same amazon elastic cloud (AWS EC2) instance with 32 vCPU and 244GiB RAM, with Linux (Ubuntu) operating system.

Table 1. Data sets used in evaluation

Name	Accession number	Genome	Genome length	Read length	Number of reads	Coverage
Lactococcus lactis	SRR088759	NC_013656.1	2,598,144	36	4,370,050	60.55
Treponema Pallidum	SRR361468	CP002376.1	1,139,417	35	7,133,663	219.13
E.coli 75a	SRR396536	NC_000913.2	4,639,675	75	3,454,048	55.83
E.coli 75b	SRR396532	NC_000913.2	4,639,675	75	4,341,061	70.17

4.2   Results

The comparisons, shown below in Tables 2, 3, 4, and 5, were established between H-RACER and algorithms specialized in correcting all types of errors (substitutions, insertions, and deletions). While the obtained measurements were brought via the verification code implemented by RACER, that has the advantage of avoiding the interference of mapping or assembling programs.

As shown below in the comparisons tables, H-RACER has the best results in accuracy and time. Actually, H-RACER aims to increase the genome reads accuracy, consequently, it aims to eliminate the errors existing in the genome reads. So, H-RACER avoided introducing errors to the reads by lowering the false positive rate and consequently increasing the specificity rate. On the other side, the false negative rate was negatively affected and the sensitivity rate was decreased as well. But, this approach did not negatively affect the accuracy, on contradictory, it resulted in getting the best accuracy.

In other words, the high accuracy of H-RACER mainly resulted from both, the remarkable lowering of false positive rate and the high raising of true negative rate. On the other side, both, the true positive and false negative rates were negatively affected. Hence, H-RACER does not have neither the highest true positive nor the lowest false positive rates. And this is what H-RACER follows in RACER's footsteps, where it is preferred to lower the algorithm sensitivity represented in raising the false negative rate and lowering the true positive rate rather than raising the false positive rate and lowering the true negative rate. And this makes sense, as enhancing the reads overall accuracy is the main vital target. So, corrective algorithms should not introduce errors (represented in false positive rate), but it should target a higher gain to get higher accuracy, and so does RACER followed by H-RACER.

Table 2. Evaluation comparison table for Lactococcus lactis

	Coral	Pollux	HSHSREC	H-RACER
True positive	15,396,336	25,325,532	25,537,644	21,237,660
False positive	2,039,148	7,720,920	6,053,580	19,656
False negative	11,413,764	1,484,568	1,272,456	5,572,440
True negative	128,472,552	122,790,780	124,458,120	130,492,044
Sensitivity	57.43%	94.46%	95.25%	79.22%
Specificity	98.44%	94.08%	95.36%	99.98%
Gain	49.82%	65.66%	72.67%	79.14%
Accuracy	91.45%	94.15%	95.34%	96.45%
Time in minutes	5	3	15	1

Table 3. Evaluation comparison table for Treponema Pallidum

	Coral	Pollux	HSHSREC	H-RACER
True positive	25,553,185	63,845,425	64,381,905	56,277,270
False positive	3,462,165	8,832,320	8,133,895	223,405
False negative	41,547,065	3,254,825	2,718,345	10,822,980
True negative	179,115,790	173,745,635	174,444,060	182,354,550
Sensitivity	38.08%	95.15%	95.95%	83.87%
Specificity	98.10%	95.16%	95.55%	99.88%
Gain	32.92%	81.99%	83.83%	83.54%
Accuracy	81.97%	95.16%	95.65%	95.58%
Time in minutes	12	3	22	2

The comparisons with the other algorithms, shown below in Tables 2, 3, 4, and 5, prove that although the other algorithms have higher sensitivity than H-RACER, but all of them do not explicitly beat H-RACER accuracy, except for "Treponema Pallidum", where HSHREC beats H-RACER's accuracy by 0.07% and this is due to the high coverage rate of "Treponema Pallidum" that will be explained below.

For the short genome "Lactococcus Lactis", illustrated in Table 2, H-RACER shows the best results in specificity, gain, accuracy and time compared to CORAL, Pollux and HSHREC, although the sensitivity of H-RACER is not the best. But H-RACER gains the highest accuracy by lowering the false positive rate (as explained above).

For "Treponema Pallidum", the short genome with a very high coverage (the average number of reads representing a given nucleotide in the reconstructed sequence) as illustrated in Tables 1 and 3, H-RACER shows a lowering in the true positive rate with a raising in the false negative one rather than the expected rates. This is due to the very high coverage of "Treponema Pallidum" that increases the ambiguity for H-RACER in detecting the proper correction for some nucleotides, leading to a lowering in the true positive rate and consequently in the accuracy. But, by comparing such an accuracy with others, as illustrated in Table 3, it is obvious that H-RACER shows the best results in specificity, gain, accuracy and time compared to CORAL and Pollux. While HSHREC is the only algorithm that beats H-RACER's gain and accuracy (for "Treponema Pallidum" only) with a very little difference rates (0.29% and 0.07% respectively). But, H-RACER accomplished such a correction in the best time compared to all others including HSHREC's (with a very remarkable ratio) and the best specificity as well.

For long genomes "E.coli 75a" and "E.coli 75b", illustrated in Tables 4 and 5, H-RACER obviously shows the best results in accuracy with a very perfect time, while CORAL and Pollux show lower accuracy with too much longer time, but HSHREC's running throws exception for such genomes, as SHREC (and consequently HSHREC) requires a very large space [4,9], so it is unable to run successfully for the larger genomes on the specified machine.

Table 4. Evaluation comparison table for E.coli 75a

	Coral	Pollux	HSHSREC	H-RACER
True positive	26,434,125	79,984,425	N/A	76,325,475
False positive	5,549,925	31,675,650	N/A	33,000
False negative	73,707,075	20,164,125	N/A	23,823,075
True negative	153,362,475	127,229,400	N/A	158,872,050
Sensitivity	26.40%	79.87%	N/A	76.21%
Specificity	96.51%	80.07%	N/A	99.98%
Gain	20.85%	48.24%	N/A	76.18%
Accuracy	69.40%	79.99%	N/A	90.79%
Time in minutes	9	16	N/A	1

Table 5. Evaluation comparison table for E.coli 75b

	Coral	Pollux	HSHSREC	H-RACER
True positive	13,312,725	99,375,600	N/A	81,059,700
False positive	3,681,450	37,779,750	N/A	35,925
False negative	108,494,025	22,439,925	N/A	40,755,825
True negative	200,091,375	165,984,300	N/A	203,728,125
Sensitivity	10.93%	81.58%	N/A	66.54%
Specificity	98.19%	81.46%	N/A	99.98%
Gain	7.91%	50.56%	N/A	66.51%
Accuracy	65.55%	81.50%	N/A	87.47%
Time in minutes	13	21	N/A	2

Finally, the remarkable great difference in time between H-RACER and the rest of the algorithms is due to the bitwise orientation in implementation (inherited from RACER), and also H-RACER keeps RACER's complexity, consequently, H-RACER gains the advantage of having the best time.

5   Conclusion

H-RACER comes up with the advantage of correcting different error types which were missed in RACER. H-RACER followed in the footsteps of RACER's implementation in order to acquire the major advantages of RACER in both aspects performance and time, then added its elegant algorithm in detecting the errors types and properly applying their corrections. Consequently, H-RACER shows great results in both performance and time compared to existing algorithms specialized in correcting all types of errors for large and small genome lengths. And by comparing H-RACER with existing algorithms specialized in correcting all types of errors, it is proved that H-RACER is the fastest with the highest accuracy algorithm.

Finally, H-RACER algorithm has been implemented in C/C++ as an open source program.

Available at: drive.google.com/open?id=0B7Otgzz7lZlldkE2ZFVwcU5qN1U

References

1. Butler, J., MacCallum, I., Kleber, M., et al.: ALLPATHS: de novo assembly of whole-genome shotgun microreads. Genome Res. **18**, 810–820 (2008)
2. Chaisson, M., Pevzner, P., Tang, M.: Fragment assembly with short reads. Bioinformatics **20**, 2067–2074 (2004)
3. Ilie, L., Fazayeli, F., Ilie, S.: HiTEC: accurate error correction in high-throughput sequencing data. Bioinformatics **27**, 295–302 (2011)
4. Ilie, L., Molnar, M.: RACER: rapid and accurate correction of errors in reads. Bioinformatics **19**, 2490–2493 (2013)
5. Kelley, D., Schatz, M., Salzberg, S.: Quake: quality-aware detection and correction of sequencing errors. Genome Biol. **11**, R116 (2010)
6. Li, R., Zhu, H., Ruan, J., et al.: De novo assembly of human genomes with massively parallel short read sequencing. Genome Res. **20**, 265–272 (2010)
7. Marinier, E., Brown, D., McConkey, B.: Pollux: platform independent error correction of single and mixed genomes. BMC Bioinform. **15**, 435 (2015)
8. Saha, S., Rajasekaran, S.: EC: an efficient error correction algorithm for short reads. BMC Bioinform. **16**(Suppl 17), S2 (2015)
9. Salmela, L.: Correction of sequencing errors in a mixed set of reads. Bioinformatics **26**, 1284–1290 (2010)
10. Salmela, L., Schrder, J.: Correcting errors in short reads by multiple alignments. Bioinformatics **27**, 1455–1461 (2011)
11. Shendure, J., Ji, H.: Next-generation dna sequencing. Nat. Biotechnol. **26**, 1135–1145 (2008)

12. Shi, H., Schmidt, B., Liu, W., et al.: A parallel algorithm for error correction in high throughput short-read data on CUDA-enabled graphics hardware. J. Comput. Biol **17**, 603–615 (2010)
13. Yang, X., Chockalingam, S., Aluru, S.: A survey of error-correction methods for next-generation sequencing. Brief. Bioinform. **14**, 56–66 (2013)
14. Yang, X., Dorman, K., Aluru, S.: Reptile: representative tiling for short read error correction. Bioinformatics **26**, 2526–2533 (2010)
15. Zerbino, D., Birney, E.: Velvet: algorithms for de novo short read assembly using De Bruijn graphs. Genome Res. **18**, 821–829 (2008)

A Multi-sensor Approach for Biomimetic Control of a Robotic Prosthetic Hand

Jeetinder Ghataurah[1], Diego Ferigo[1], Lukas-Karim Merhi[1],
Brittany Pousett[2], and Carlo Menon[1(✉)]

[1] MENRVA Research Group, Simon Fraser University,
School of Engineering Science, Burnaby, BC, Canada
cmenon@sfu.ca
[2] Barber Prosthetics Clinic, Vancouver, BC, Canada

Abstract. Robotic prosthetic hands with five digits have become commercially available however their use is limited to a few grip patterns due to the unnatural and unreliable human-machine interface (HMI). The research community has addressed this problem extensively by investigating Pattern Recognition (PR) based surface-electromyography (sEMG) control. This control strategy has been recently commercialized however has yet to show clinical adoption. One of the reasons identified in the literature is due to the sEMG signals that are affected by sweating, electrode shift, ambient noise, fatigue, cross-talk between adjacent muscles, signal drifting, and force level variation. Hence recently the scientific community has started proposing multi-modal sensing techniques as a solution.

This study aims to investigate the use of multi-modal sensor approach to control a robotic prosthetic hand by investigating the sparsely studied sensing mechanism called Force Myography (FMG) as a synergist to the conventional technique of sEMG. FMG uses pressure sensors on the surface of a limb to detect the volumetric changes in the underlying musculotendinous complex. This paper presents a custom prosthetic prototype instrumented with sEMG and FMG sensors and tested by a participant with a transradial amputation. Results demonstrate that this multi-sensor approach has the potential to be a valid HMI for prosthesis control.

Keywords: Force sensitive resistors (FSR) · Robotic prosthetic · Residual limb imaging · LDA · Electromyography (EMG) · Amputee

1 Introduction

Individuals with upper extremity amputations can be equipped with bionic hands, which are conventionally controlled using sEMG signals acquired from the users' residual limbs (Ravindra and Castellini 2014; Atzori and Müller 2015). Although available bionic hands have multiple Degrees of Freedom (DOFs) and offer high levels of dexterity (Atzori and Müller 2015), EMG-driven hands, have as high as a 75% rejection rate by the end-users (Biddiss and Chau 2007) because one of the reasons is that the control is based on non-intuitive series of muscle contractions and requires an extensive training period with experts (Castellini et al. 2014). In order to reduce this rejection rate, individuals with amputations need to be equipped with a control interface

© Springer International Publishing AG 2017
I. Rojas and F. Ortuño (Eds.): IWBBIO 2017, Part I, LNBI 10208, pp. 74–84, 2017.
DOI: 10.1007/978-3-319-56148-6_6

that is intuitive and reliable, and allows the users to dexterously and naturally control their bionic hands.

This study aims to investigate the feasibility of increasing robustness of natural control of upper extremity prosthesis by using FMG as a synergist to the conventional sEMG. This was accomplished by embedding multiple FSRs and two EMG electrodes in a custom prosthetic socket. The intended upper-extremity residual limb motions were captured using the FMG and sEMG sensors to distinguish between different grip patterns using pattern recognition, and then to control a commercially available robotic prosthetic hand, Bebionic3 hand by the Steeper Group UK. These advanced control strategies would potentially increase the adoption of the robotic prosthetic hand and thus provide a better quality of life to amputee users.

2 Background

When a limb is replaced with prosthesis, the user and the prosthesis start an intimate relationship in which human-machine interface is a fundamental component. Unfortunately, the link between the user and their prosthesis is the weakest link in the chain. The conventional non-invasive HMI is based on surface-Electromyography (Atzori and Müller 2015a).

2.1 Surface Electromyography

sEMG sensors detect the electrical potentials created by the motor units from the human skin surface of skeletal muscles. These sEMG signals have a wide variety of uses such as identifying neuromuscular or motor control diseases to various kinesiology studies.

EMG sensors have been used to control robotic prostheses since the 1950s (Wirta et al. 1963). Most commercially available prostheses use control strategies based upon open-close state detected by two sEMG sensors. This control strategy is robust for commercial myoelectric two-configuration grippers. Unfortunately for advanced robotic prostheses this control strategy not only limits the DOFs but also requires non-intuitive Morse code like sequences to obtain multiple DOFs, which requires an extra mental interpretation between the intended gesture to the open-close control command (Yang et al. 2014).

The research community has addressed this problem extensively by investigating Pattern Recognition (PR) based surface-electromyography (sEMG) control (Castellini et al. 2014). This control strategy has been recently commercialized however has yet to show clinical adoption (Atzori and Müller 2015). One of the reasons identified in the literature is due to the sEMG signals that are affected by sweating, electrode shift, ambient noise, fatigue, cross-talk between adjacent muscles, signal drifting, and force level variation (Merletti et al. 2010; Atzori et al. 2014; Al-Timemy et al. 2015). Therefore several alternative sensing techniques including pressure sensors have recently been explored (Fang et al. 2015). Ravindra and et al. compared three non-invasive HMIs (sEMG, ultrasound imaging, pressure sensing) and concluded that pressure sensing represented a valid alternative/augmentation to sEMG because of its potential to provide the highest PR prediction accuracy, signal stability over time, wearability, simplicity in socket embedding, and affordability of cost.

2.2 Pressure Sensing Techniques

The use of pressure sensors as a sensing technique has been termed in various ways in the literature such as: residual kinetic imaging (Phillips and Craelius 2005), pressure distribution map (Radmand et al. 2014a), topographical force map (Ravindra and Castellini 2014), force myography, and muscle pressure mapping (Wininger et al. 2008; Li et al. 2012; Cho et al. 2016).

It is a technique involving the use of different types of pressure sensors such as Force Sensitive Resistors (FSRs) on the surface of the limb to detect the volumetric changes in the underlying musculo-tendonous complex. Each term shares the same fundamental principle and are described in the following section.

3 Materials and Methods

3.1 Subject

A participant with a transradial amputation, was recruited via Barber Prosthetics Clinic (BPC) to help assessing the feasibility of the control technology developed in this study. The participant acquired his transradial amputation in 1980 and has used a body-powered prosthesis since. Although significant time has passed since the amputation, the test pilot is extremely active and has ample sensation and control in his residual limb. All the methods within this study were in compliance with the declaration of Helsinki and were approved by the Simon Fraser University (SFU) Office of Research Ethics (#2015s0143).

3.2 Prosthetic Prototype

A conventional socket design with a locking liner was fabricated by Barber Prosthetics Clinic (BPC) and connected to the BeBionic3 robotic hand to be donned by the participant with a transradial amputation. The custom socket is comprised of a transparent thermoplastic inner and outer socket. The outer socket houses an opening for easy accessibility of the internal electronics, cut-outs for the sEMG sensors, slits for the FSR bands and is covered using a 3D printed enclosure. The setup is shown in Fig. 1.

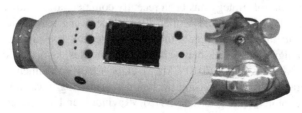

Fig. 1. Prosthesis with 3D printed housing - distal end (Left), proximal end (Right)

3.3 FSR Bands

Five FSR Bands were used in total. FSR bands were custom-printed and each contains 16 FSR sensors (individual sensors have same properties as FSR 402 from Interlink Electronics). Four bands were placed linearly from the proximal to the distal of the inner socket and the fifth band was placed circumstantially. This provided a total of 48 active FSR sensors whereby active is defined as in contact with the residual limb, the non-active sensors were not sampled.

3.4 EMG Sensors

OttoBock 13E68 sEMG electrodes were used in the study which provide an amplified, bandpass-filtered and Root-Mean-Square (RMS) rectified version of the raw sEMG signal. The electrodes amplification gain was set at about 14,000. Two extrusions were created in the socket such that the sEMG sensors would receive the highest muscle activity, and highest signal extraction, from the extensor and flexor muscle bellies on the residual limb. Thus, optimizing EMG positioning over FMG positioning.

3.5 Data Acquisition Platform

The 3D printed enclosure contained a BeagleBone Black (BBB) Rev C development board with a custom made cape/shield. The sEMG sensors were directly connected to the cape while the FSR bands were interfaced with vertical zero-insertion force (ZIF) connectors. Since the frequency of human hand motion is typically less than 4.5 Hz (Xiong and Quek 2006; Amft et al. 2006), a 10 Hz sampling frequency was deemed sufficient. Each FSR band was supplied through a BBB general purpose input \output (GPIO) with 3.3 volts. Bands' outputs were connected to voltage dividers using 5 $k\Omega$ and 6 $k\Omega$ resistors and individual sensors on a band were sampled individually (in series). The network used is similar to that of Rasouli et al.

4 Experimental Protocol

4.1 Protocol

It has been determined that most functional activities of daily living can be performed through four primary grips: power (also known as force), tripod, finger point, and the key grips (Peerdeman et al. 2011; Yang et al. 2014). In this study three different grip sets are examined. First, the six grips containing the four primary grips above and also relax hand open, and open palm grips. The second set contains only the opposed-thumb primary grips: relax hand open, open palm, power, and tripod. Finally, the third set contained the non-opposed thumb primary grips: relax hand open, open palm, finger point, and key grip. The opposed and non-opposed grip sets were examined because they match the grip patterns made available by the Bebioinc3 opposed and non-opposed hand configurations.

The experiment included three trials, each having three repetitions. Each repetition included 11 grip patterns performed in a stationary position with the pilot's elbow off

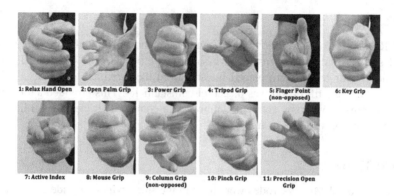

Fig. 2. 11 grip patterns. Grip patterns 3 to 6 are considered the primary grip patterns (Peerdeman et al. 2011; Yang et al. 2014). Grip patterns 1 to 4 constitute the opposed-thumb primary grips.

the table, and a 5-min rest period between repetitions. The 11 grip patterns used in this study are shown in Fig. 2.

The FSR and sEMG sensors were sampled at 10 Hz and 100 samples, respectively. For each grip pattern the test pilot first created the bilateral mirrored contraction for the desired grip pattern and once he felt it was stable he initiated data acquisition by pressing a button on the user interface (UI) with his sound limb. Each recording lasted 10 s.

4.1.1 Trial 1: Without Robotic Prosthetic Hand

In Trial 1, three repetitions were performed without connecting the prosthesis to the BeBionic3 robotic hand. This trial served two purposes: to investigate the effects the weight of the hand has on classification accuracy and to increase the pilot's residual limb volume due to muscle fatigue.

4.1.2 Trial 2: Complete System

In Trial 2, three repetitions were performed with the prosthesis connected to the BeBionic3 robotic hand. These trials closely resemble a real-life scenario.

4.1.3 Trial 3: EMG Threshold Force Levels

In Trial 3, three repetitions were performed with the complete system. Various force levels were used in different repetitions to establish the low, medium, and high force levels from the sEMG sensor values to investigate using EMG force level to trigger FMG pattern recognition analysis. In Trial 3 Repetition 1, for the low force (41.6% of maximal voluntary contraction (MVC)), the pilot provided the minimal amount of force he felt required to create the desired grip pattern with his residual limb. In Trial 3 Repetition 2, the medium force (64.5% of MVC), the pilot created each grip pattern with his residual limb using the normal amount of force he would use with his sound limb. In Trial 3 Repetition 3, the high force, the pilot used his maximum strength when creating each grip pattern with his residual limb. Since each grip pattern in Trial 3 Repetition 3 was performed with maximal force, it was used as the MVC level.

A collected sample has the structure (FSR1, ..., FSR48, EMG1, EMG2). The data from the last trial was used to create 3 different force thresholds for data filtering. For each repetition, the mean EMG value was obtained from both EMG sensors during each of the 11 grip patterns. Once the 3 thresholds were created, it could be used to filter the raw data of Trials 1 and 2 to remove the FSR bias values associated with an EMG value below the set threshold. The filtered data could then be classified to investigate the effects of EMG threshold filtering on classification accuracy.

4.2   Machine Learning Algorithm

To classify the grip patterns, a Linear Discriminant Analysis (LDA) model was used offline. LDA is a robust machine learning algorithm that finds a linear combination of features to separate two or more classes of objects. Additionally, LDA has been identified as to have real-time applicability and provides better or equal classification results when compared to more complex algorithms (E. Scheme and Englehart 2011; Scheme et al. 2013; Zhang et al. 2013; Amsuss et al. 2014). For Trials 1 and 2, Leave-One-Out Cross-Validation (LOOCV) method was used to obtain the classification accuracy. To implement LOOCV for each trial, the first repetition was used as the test data set against the second and third repetitions which were pooled and used as the training data set. The data sets were fed into the LDA model and the classification accuracy was recorded. This process was repeated to test the accuracies of repetition 2 then 3. The average of the three classification accuracies were then averaged as the classification accuracy of the trial.

5   Experimental Results

5.1   Single FSR Bands

Data from a single FSR band was used to classify 4 and 11 grip patterns, the resulting classification accuracies are shown in Table 1. The average single band classification

Table 1. Single FSR band classification accuracy

	4 grips [opposed]		11 grips	
	Without hand [%]	With hand [%]	Without hand [%]	With hand [%]
Band 1 (7 active sensors)	53.0	35.3	30.3	15.3
Band 2 (9 active sensors)	59.2	44.5	31.6	19.8
Band 3 (6 active sensors)	60.5	42.4	25.2	33.2
Band 4 (10 active sensors)	78.3	40.9	33.0	28.7
Band 5 (16 active sensors)	68.0	51.1	39.9	31.7
Single band average	63.8 ± 9.7	42.8 ± 5.7	32.0 ± 5.3	25.7 ± 7.8

accuracy for 4 (opposed) grip patterns was 63.8% and 42.8% without and with the robotic hand respectively. The average single band classification accuracy for 11 grip patterns was 32.0% and 25.7% without and with the robotic hand respectively.

5.2 Multiple FSR Bands

In trial 2, data from a different number of FSR bands was used to classify 4 and 11 grip patterns. The average accuracy resulting from using all permutations of different number of bands can be seen in Table 2.

Table 2. Multiple FSR band classification accuracy. The variance was calculated considering all the possible permutations of the number of active bands (e.g. [write the simplest example])

	4 grips [non-opposed][%]	11 grips [%]
1 Band average	42.8 ± 5.7	25.7 ± 7.8
2 Bands average	51.3 ± 5.7	33.6 ± 6.0
3 Bands average	60.0 ± 4.3	37.6 ± 6.5
4 Bands average	68.4 ± 4.4	41.5 ± 6.2
5 Bands	76.2	50.4

5.3 EMG Features

For these results the values from the two sEMG sensors were treated as additional features and they were included with all 5 FSR bands in Trial 2 data for classification. The results for the accuracy with and without EMGs and with and without EMG filtering for the different grip pattern groups can be seen in Table 3. In addition, confusion matrices were created for the 4 grips (non-opposed and opposed) to understand the increase in classification accuracy. These can be seen in Fig. 3 where the x axis is the actual class and the y axis is the machine learning algorithm prediction.

Table 3. Classification Accuracy of 5 FSR Bands without EMG, with EMG as Features, and with EMG as Features and Medium Threshold Filtration

	4 grips [%] (opposed)	4 grips [%] (non-opposed)	6 grips [%]	11 grips [%]
5 FSR bands	76.2	60.3	62.4	50.4
5 FSR bands + EMG	81.9	60.0	63.3	50.6
5 FSR bands + EMG + MedT	93.5	93.3	63.4	47.7

5.4 EMG Filtration

Trial 3 created 3 sets of grip pattern specific threshold levels to filter the data of Trial 2 in order to increase classification accuracies. During all three trials the average EMG values were recorded for both EMG sensors per grip pattern. For low, medium and high filtration, several adaptations of the thresholds were applied such as using: only EMG 1

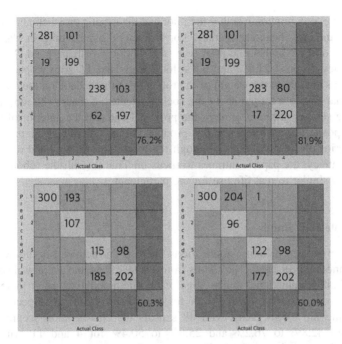

Fig. 3. 5 FSR band 4 grip opposed and non-opposed with and without EMG as features confusion matrix; opposed w/o EMG (top left), opposed w/EMG (top right), non-opposed w/o EMG (bottom left), and non-opposed w/EMG (bottom right); (1 - relax hand open, 2 - open palm grip, 3 – power Grip, 4 – tripod grip, 5 – finger point, and 6 – column grip

threshold, only EMG 2 threshold, EMG 1 or EMG 2 threshold, and EMG 1 and EMG 2 thresholds. It was found that a medium level threshold, assessed as in Sect. 4.1.3, for both EMG sensors was optimal. The accuracy results from classification using all 5 FSR bands, both EMG sensors as features and as a filter can be seen in Table 3.

6 Discussion

6.1 Single FSR Band

The FSR single analysis shows the validity of the positioning for the 5 FSR bands. Looking at Table 1, we can see that the classification accuracy does not differ significantly with the standard deviation ranging between 5.3% to 9.7% depending on the configuration. The classification accuracy differences between the bands can be attributed to the placement and the number of active FSR sensors per band. In addition, the comparatively lower accuracy results for FSR Band # 1 can also be attributed to the specific style of socket chosen. Band #1 is positioned between the flexor and extensor on the exterior surface and thus when the elbow was bent into the stationary position a space was formed between the liner and the socket near the proximal end of the socket.

The testing configuration whether the pilot had the robotic prosthetic hand attached or not alters the classification accuracy. For any single band a drop in classification accuracy was observed when the pilot wore the robotic prosthetic hand. The classification accuracy dropped 21.0% and 6.2% for 4 grips and 11 grips respectively. The accuracy drops are due to the weight of the robotic prosthetic hand on the distal end of the socket that causes a torque about the socket's center of mass. The torque applied pre-loads the sensors at the exterior of the proximal end and the anterior of the distal end. The opposite occurs where a gap is produced at the anterior of the proximal end and the exterior of the distal end. Although the robotic prosthetic hand reduces classification accuracy, the paper continues to focus on Trial 2 (with the robotic hand) as they are most similar to real life use for the participant.

6.2   Multiple FSR Bands

For 4 or 11 grips grip pattern classifications, there is an increased accuracy with the addition of another FSR band. The results from Table 2 show the average classification accuracy of all permutations. On average, each additional FSR band increased the classification accuracy by 8.4% and 6.2% for 4 and 11 grip patterns respectively. As a result of using all 5 FSR bands with 48 sensors the classification accuracy nearly doubled from 42.8% to 76.2% and 25.8% to 50.4% for 4 and 11 grip patterns respectively. The remainder of this discussion will mainly focus on the four primary grips as they are also the first four grips of the BeBionic3 robotic hand.

6.3   EMG Features

The two EMG sensors' data were added to the 48 FSR sensors' data and used as additional features to increase classification accuracy. From Table 3 we see that the classification accuracy for 4 grip (opposed) increased from 76.2% to 81.9% due to the addition of the EMG sensors as features. Investigating the confusion matrices in Fig. 3, we see a significant reduction in misclassification between grips 3 and 4, force and tripod respectively. From Fig. 3, it can be found that the grip pattern classification accuracy increased from 79.3% to 94.3% and 65.7% to 73.3% for force and tripod grips respectively. By adding the EMG sensors as additional features the increased classification accuracy of 81.9% was achieved for the four chosen classes.

6.4   EMG Filtration

It was found that using both EMG 1 and EMG 2 values for medium thresholds increased the 4 grips (non-opposed) classification accuracy using 5 FSR bands and both EMGs as input features from 81.9% to 93.5%. In comparison, the 11 grips with 5 FSR bands and two EMGs as input features decreased slightly: from 50.6% to 47.7%. In both cases, the data set size was reduced by an average of 79.0% with the medium threshold level. Even with the smaller data set, the 11 grips classification accuracy dropped only by 2.9% yet the 4 grips accuracy rose 11.6%.

7 Conclusions

For this study, a prosthetic socket was produced for our pilot with a transradial amputation. Five customized FSR bands were embedded in the socket providing 48 sensors that were used to create a pressure map in addition to two sEMG sensors. 11 different grip patterns were examined and classified offline using LDA machine-learning algorithm. The average accuracies using single FSR bands were 42.8% and 25.8% for 4 (opposed) and 11 grips respectively. By using all 5 FSR bands the accuracy rose from 42.8% to 76.2% and 25.8% to 50.4% for 4 (opposed) and 11 grips respectively. Using both sEMG sensors as additional features to the 5 FSR bands, accuracy changed from 76.2% to 81.9%, 60.3% to 60.0%, 62.4% to 63.3%, and 50.4% to 50.6% for 4 (opposed), 4 (non-opposed), 6, and 11 grips respectively. Finally, by applying data filtration of the previous configuration using EMG 1 and EMG 2 medium level thresholds, the accuracy changed from 81.9% to 93.5%, 60.0% to 93.3%, 63.3% to 63.4%, and 50.6% to 47.7% for 4 (opposed), 4 (non-opposed), 6, and 11 grips respectively. The results indicate the potential benefit of a multi-sensor approach for controlling robotic prostheses.

References

Al-Timemy, A., Khushaba, R., Bugmann, G., Escudero, J.: Improving the performance against force variation of EMG controlled multifunctional upper-limb prostheses for transradial amputees. IEEE Trans. Neural Syst. Rehabil. Eng. **24**, 650–661 (2015). doi:10.1109/TNSRE. 2015.2445634

Amft, O., Junker, H., Lukowicz, P., Tröster, G., Schuster, C.: Sensing muscle activities with body-worn sensors. In: Proceedings - BSN 2006: International Workshop on Wearable and Implantable Body Sensor Networks, vol. 2006, pp. 138–141. IEEE, Cambridge (2006). http://doi.org/10.1109/BSN.2006.48

Amsuss, S., Goebel, P.M., Jiang, N., Graimann, B., Paredes, L., Farina, D.: Self-correcting pattern recognition system of surface EMG signals for upper limb prosthesis control. IEEE Trans. Biomed. Eng. **61**(4), 1167–1176 (2014). doi:10.1109/TBME.2013.2296274

Atzori, M., Gijsberts, A., Castellini, C., Caputo, B., Hager, A.-G.M., Elsig, S., Müller, H.: Electromyography data for non-invasive naturally-controlled robotic hand prostheses. Sci. Data **1**, 140053 (2014). doi:10.1038/sdata.2014.53

Atzori, M., Müller, H.: Control capabilities of myoelectric robotic prostheses by hand amputees: a scientific research and market overview. Front. Syst. Neurosci. **9**, 162 (2015). http://doi.org/10.3389/fnsys.2015.00162

Biddiss, E.A., Chau, T.T.: Upper limb prosthesis use and abandonment: a survey of the last 25 years. Prosthet. Orthot. Int. **31**(3), 236–257 (2007). doi:10.1080/03093640600994581

Castellini, C., Artemiadis, P., Wininger, M., Ajoudani, A., Alimusaj, M., Bicchi, A., Scheme, E.: Proceedings of the first workshop on peripheral machine interfaces: going beyond traditional surface electromyography. Front. Neurorobot. **8**(AUG), Article no. 22(2014). http://doi.org/10.3389/fnbot.2014.00022

Cho, E., Chen, R., Merhi, L., Xiao, Z., Pousett, B., Menon, C.: Force myography to control robotic upper extremity prostheses: a feasibility study. Front. Bioeng. Biotechnol. **4**(March), 1–12 (2016). doi:10.3389/fbioe.2016.00018

Fang, Y., Hettiarachchi, N., Zhou, D., Liu, H.: Multi-modal sensing techniques for interfacing hand prostheses: a review. IEEE Sens. J. **15**(11), 6065–6076 (2015). doi:10.1109/JSEN.2015. 2450211

Li, N., Yang, D., Jiang, L., Liu, H., Cai, H.: Combined use of FSR sensor array and SVM classifier for finger motion recognition based on pressure distribution map. J. Bionic Eng. **9** (1), 39–47 (2012). doi:10.1016/S1672-6529(11)60095-4

Merletti, R., Aventaggiato, M., Botter, A., Holobar, A., Marateb, H., Vieira, T.M.M.: Advances in surface EMG: recent progress in detection and processing techniques. Crit. Rev. Biomed. Eng. **38**(4), 305–345 (2010). doi:10.1615/CritRevBiomedEng.v38.i4.10

Peerdeman, B., Boere, D., Witteveen, H., Huis in't Veld, R., Hermens, H., Stramigioli, S., Misra, S.: Myoelectric forearm prostheses: state of the art from a user-centered perspective. J. Rehabil. Res. Dev. **48**(6), 719 (2011). doi:10.1682/JRRD.2010.08.0161

Phillips, S.L., Craelius, W.: Residual kinetic imaging: a versatile interface for prosthetic control. Robotica **23**(3), 277–282 (2005). doi:10.1017/S0263574704001298

Radmand, A., Scheme, E., Englehard, K.: High resolution muscle pressure mapping for upper limb prosthetic control. In: Proceeding of MEC - Myoelectric Control Symposium, 19–22 August, pp. 189–193 (2014a)

Rasouli, M., Ghosh, R., Lee, W.W., Thakor, N.V, Kukreja, S.: Stable force-myographic control of a prosthetic hand using incremental learning. In: 37th Annual International Conference of the IEEE Engineering in Medicine and Biology Society (EMBC), pp. 4828–4831. IEEE (2015). http://doi.org/10.1109/EMBC.2015.7319474

Ravindra, V., Castellini, C.: A comparative analysis of three non-invasive human-machine interfaces for the disabled. Front. Neurorobot. **8**(October), 1–10 (2014). doi:10.3389/fnbot. 2014.00024

Scheme, E., Englehart, K.: Electromyogram pattern recognition for control of powered upper-limb prostheses: state of the art and challenges for clinical use. J. Rehabil. Res. Dev. **48**(6), 643 (2011). doi:10.1682/JRRD.2010.09.0177

Scheme, E.J., Hudgins, B.S., Englehart, K.B.: Confidence-based rejection for improved pattern recognition myoelectric control. IEEE Trans. Biomed. Eng. **60**(6), 1563–1570 (2013). doi:10. 1109/TBME.2013.2238939

Wininger, M., Kim, N.-H., Craelius, W.: Pressure signature of forearm as predictor of grip force. J. Rehabil. Res. Dev. **45**(6), 883–892 (2008). doi:10.1682/JRRD.2007.11.0187

Wirta, R.W., Taylor, D.R., Wirta, R.W., Wirta, R.W., Finley, F.R.: Pattern-recognition arm prothesis: a historical perspective—a final report. Nonr **4292**, 1–28 (1963)

Xiong, Y., Quek, F.: Hand motion gesture frequency properties and multimodal discourse analysis. Int. J. Comput. Vis. **69**(3), 353–371 (2006)

Yang, D., Jiang, L., Huang, Q., Liu, R., Liu, H.: Experimental study of an EMG-controlled 5-DOF anthropomorphic prosthetic hand for motion restoration. J. Intell. Robot. Syst. **76**(3–4), 427–441 (2014). doi:10.1007/s10846-014-0037-6

Zhang, H., Zhao, Y., Yao, F., Xu, L., Shang, P., Li, G.: An adaptation strategy of using LDA classifier for EMG pattern recognition. In: Conference Proceedings: Annual International Conference of the IEEE Engineering in Medicine and Biology Society. IEEE Engineering in Medicine and Biology Society. Annual Conference, vol. 2013, pp. 4267–4270. IEEE (2013). http://doi.org/10.1109/EMBC.2013.6610488

Feature Selection in Multiple Linear Regression Problems with Fewer Samples Than Features

Paul Schmude[(⊠)]

Sonovum AG, Perlickstraße 5, 04103 Leipzig, Germany
paul.schmude@sonovum.com

Abstract. Feature selection is of utmost importance when it comes to problems with large p (number of features) and small n (number of samples). Using too many features for a final model will most probably result in overfitting. There are many possibilities to select a subset of features to represent the data, this paper illustrates correlation filters, forward selection and genetic algorithm for feature selection and PCA and PLS as transformation methods. The methods are tested on three artificial data sets and one data set from an ultrasound study. Results show that no method excels for all problems and every method gives different insights into the data. The greedy style forward selection usually overfits and shows the largest difference between training and testing data, the PLS and PCA perform worse on the artificial data, but better for the ultrasound data.

Keywords: Overfitting · Feature selection · Filter method · Correlation · PCA · PLS · Forward selection · Genetic algorithm

1 Introduction

In life sciences the money and time spent on experiments usually scales with the sample size. Since methods like MRI or blood analysis are expensive, one might want to gather as many data as possible from every single subject, which results in a data set with a high number of features and a comparably small number of samples. Such a large p, small n problem arose during the development of a new medical device using acoustocerebrography [3]. Studies on humans are usually expensive and time consuming, so only 114 samples were collected. The corresponding data set will be introduced in Sect. 2.2, accompanied by three artificial data sets with a known best solution.

The main difficulty facing data sets like the above is overfitting, which has two different sides. On one hand, if too many features are included it is very likely to get a good model, even if the added features only contain noise. The feature set has to be limited to only contain absolutely necessary items, also known as Occam's Razor. As stated in the Cambridge Dictionary of Statistics [1] the number of features has to be limited to a reasonable amount regarding the data. A good rule of thumb is to have ten times more samples than variables in the final model.

© Springer International Publishing AG 2017
I. Rojas and F. Ortuño (Eds.): IWBBIO 2017, Part I, LNBI 10208, pp. 85–95, 2017.
DOI: 10.1007/978-3-319-56148-6_7

The second manifestation of overfitting is the loss of predictive power due to a misleading selection of features based on the training set. It is possible to build a model based on only a few features that fits all training instances perfectly, but fails to reflect the real properties of the studied phenomenon.

Since the data sets are large, the methods are automated and not influenced by any prior domain knowledge. This method was criticized by Henderson and Velleman "The data analyst knows more than the computer." [2], which might holds for many data sets, but in some cases the underlying physical, chemical or biological properties are unknown and since evaluating all possible combinations of features is often computationally impossible, automated searching methods are necessary. The problem setup has been widely researched, for example in the context of CART [4], Bayesion regression [5] or penalized discriminant analysis [6].

2   Data

2.1   Friedman Based

For more objectivity, artificial data sets with known best solutions will used to compare the feature selection procedures. The following regression problems were originally proposed by Friedman [7] and Breiman [8] and are modified to fit the setup of having many variables with few samples. The modifications involve the addition of noised versions of the important features, squares of the important features and uniformly and normally distributed noise. The resulting data sets contain 1000 continuous features and 100 samples. The data were simulated using the mlbench [21] package from R and the *runif* and *rnorm* functions from R.

Friedman 1. The original regression problem uses ten independent, uniformly distributed variables on the interval $[0,1]$, with 5 variables being useful. Let $x_1, ..., x_5$ be the 5 important variables, then the dependent variable is formed by

$$y = 10\sin(\pi x_1 x_2) + 20(x_3 - 0.5)^2 + 10x_4 + 5x_5 + \epsilon \tag{1}$$

with $\epsilon \sim \mathcal{N}(0,1)$. By adding the squared and noised versions of the important variables, the best solution has a root-mean-square error (RMSE) of about 1.8 using a five fold cross-validation.

Friedman 2. The second problem is also from Friedman and Breiman and uses a different approach. The main difference between Friedman 1 and 2 is the larger range of the variables, which will yield a way larger root-mean-square error. There are independent, uniformly distributed variables $x_1, ..., x_4$ with

$$0 \leq x_1 \leq 100 \tag{2}$$
$$40\pi \leq x_2 \leq 560\pi \tag{3}$$
$$0 \leq x_3 \leq 1 \tag{4}$$
$$1 \leq x_4 \leq 11 \tag{5}$$

which form the dependent variable y with

$$y = \sqrt{x_1^2 + (x_2 x_3 - (\frac{1}{x_2 x_4}))^2} + \epsilon, \tag{6}$$

where $\epsilon \sim \mathcal{N}(0, 125)$. The best solution reaches, using five fold cross-validation, a RMSE of around 157.

Friedman 3. In contrast to the wide range of the dependent variable in Friedman 2, the Friedman 3 benchmark limits the dependent variable to $[-\frac{\pi}{2}, \frac{\pi}{2}]$, which will result in a lower RMSE. The variables $x_1, ..., x_4$ are uniformly distributed with

$$0 \leq x_1 \leq 100 \tag{7}$$

$$40\pi \leq x_2 \leq 560\pi \tag{8}$$

$$0 \leq x_3 \leq 1 \tag{9}$$

$$1 \leq x_4 \leq 11 \tag{10}$$

$$y = \text{atan} \left(\frac{x_2 x_3 - (\frac{1}{x_2 x_4})}{x_1} \right) + \epsilon, \tag{11}$$

with $\epsilon \sim \mathcal{N}(0, 0.1)$. The RMSE of the best solution, regarding the cross-validation scheme proposed before, is around 0.19.

2.2 Data from Acoustocerebrography

Acoustocerebrography, or short ACG [3], is a method for monitoring the brain by using molecular acoustics. The raw data are from recordings of active ACG measurements on patients with atrial fibrillation. The measurements were done with the UltraEASY 2 device, which uses multi-frequency ultrasound. Features were forged comparing the received and the sent signal, e.g. the attenuation of the signal. The data set also contains additional medical information about the patients. This paper will focus on the estimated number of white matter lesions, determined by examining MRI images taken prior to the ACG measurements. Further results regarding this study can be found at Dobkowska-Chudon [9] and Olszewski [10]. The data set has in total 114 samples with 1716 continuous features. The task is to predict the approximated number of lesions, which is a continuous feature as well. Since this is a new method, it is unknown which combination of features is useful for regression.

3   Methods

The following methods are all embedded in the same testing environment, which allows to compare and analyse their performance. The RMSE is used as evaluation metric, paired with repeated cross-validation with five folds and ten repeats. That means the evaluation is only performed on the left-out observations, thus there is no bias and by using the same folds for different methods we ensure comparability between these.

3.1 Filtering with Correlation

The squared Pearson correlation r^2 with

$$r = \frac{\sum_{i=1}^{n}(x_i - \bar{x})(y_i - \bar{y})}{\sqrt{\sum_{i=1}^{n}(x_i - \bar{x})^2}\sqrt{\sum_{i=1}^{n}(y_i - \bar{y})^2}} \tag{12}$$

is a direct measure of linear dependency between a feature x and the response y. Since the variables are for a linear regression model this should be a solid choice, although there are some downsides as well, for example outliers have a heavy impact on this. Another possibility is to use Spearman's rank correlation, which is basically the same formula but with ranks instead of the raw values. The downside is that all monotone relationships between some feature x and y result in a good score, even if they are not linear, while the advantage is the robustness to outliers.

The feature selection function f calculates the correlation for every feature with the response and then returns the ten features with the highest r^2. This method is very easy to implement and the fastest, but it does not consider any interactions between the variables.

3.2 Reduction of Dimensionality with PCA and PLS

The well known principal component analysis (PCA) transforms a data matrix into principal components, which are uncorrelated and arranged in such a way, that the first component has the highest variance. By applying this technique, it is possible to explain almost all of the variance of a data set using only few components. Here, in particular the explanatory power of these components is of interest. The training data are used to calculate the scores and loadings, then the first ten components of the scores are used to build the regression model and afterwards the loadings are used to calculate the scores of the testing data, which are used to evaluate the previously built model. This reduces the feature selection function f to a matrix calculation, $f(X) = XL$, where L are the loadings generated from the training data.

A similar approach is to use partial least squares (PLS) components instead of principal components. The main difference here is that it takes the response into account and transforms it as well. This generates loadings and scores from both the data matrix as well as the response vector and has to be applied both to the test data.

3.3 Wrapper Method: Forward Selection

Instead of selecting variables individually or merging all together, their interaction is considered using leap forward selection. The procedure starts with a fixed set of variables and adds another variable, if the underlying metric is improved. If several variables improve the system, the one giving the best improvement is taken. If no further improvement can be done by selecting variables, the algorithm stops. This method is more natural for the problem setup than the leap backwards selection, since it would start with a overfit on the training data, from which on it would be impossible to improve. To make sure variables are only added to the selection when it is necessary, a four fold cross-validation is used to check if the predictive power, represented by a shrinking RMSE, is increased. Note that this is a very cautious and numerically expensive way of adding one variable. A quicker method would be to abandon the training/testing split and use measures like RMSE and correlation coefficient from the fitted training values and the response or information criteria like AIC and BIC. Using this would reduce the total time by factor four, since only one evaluation of the linear model would be necessary. Even more precaution is introduced by using repeated cross-validation. A more sophisticated approach would be to combine the forward leaping with backward steps, where removing variables after several forward steps is possible. The backwards elimination procedure would be similar to above, except instead of checking all variables in $X \backslash I$ all variables in I are individually removed and the model is tested for improvements.

The numerical expense of this procedure is by far larger than for filter methods, PCA or PLS. In this example, ten variables have to be extracted out of 1000, which costs $\sum_{i=1}^{10} 4 \cdot (1001 - i) = 39820$ evaluations of the linear model. Of course this is the worst case, since the algorithm could terminate with less than ten variables, but it is no match for the correlation or the singular value decomposition required in the previous methods.

3.4 Metaheuristic: Genetic Algorithm

The genetic algorithm (GA) has its roots in the mimicking of natural selection processes including selection, recombination and mutation. In 1975 Holland [11] made the concept widely known, even though the first adaptions were made by Barricelli [12] in 1954.

The adaptation to the feature selection problem is straightforward, the processes of selection, combination and mutation will be briefly explained. Selection is the process of picking the best elements of a population. A population is a set of members, in this case each member of the population is a set of variables. Since the maximum number of variables to pick is restricted, each member is restricted to contain ten variables or less. The best members of the population are selected by evaluating the linear model using cross-validation, the rest is eliminated. To fill up the population, the selected members combine to produce new members. Two sets of variables combine by keeping all variables present in both sets, variables that are present in only one of the sets are added by chance.

Another way to add members to the population is to add randomly picked sets of variables to the population. By combining these two methods, the feature space is searched both wide and deep. The last step is mutation, in this case variables are randomly removed and/or added to members of the new population. The restriction of having a maximum amount of ten variables is enforced by randomly eliminating variables until the member has only ten variables left. After that, the whole procedure starts from the beginning with the selection process.

The costs scale linearly with the population size and the number of generations. In the case of a four fold cross-validation as fitness measure, $4 \cdot n_{pop} \cdot n_{gen}$, n_{pop} being the population size and n_{gen} being the number of generations, evaluations of the linear model are necessary. Early stopping can be introduced by checking the improvement of the new generations and stopping if there have been none in several generations.

3.5   Implementation Details

Every algorithm was implemented in R [16]. The caret package [17] was used for the cross-validation scheme and preprocessing, the pls package [18] for the PLS and the doSNOW [19] and foreach [20] packages for parallelizing. The linear model fitting is from the stats package (R core package). All numerical testing was done on the same system with the same random seeds for each method.

4   Results

As it can be seen in Fig. 1, all RMSE are roughly in the same region. Except for one method in one data set, all methods show a lower RMSE for the training data

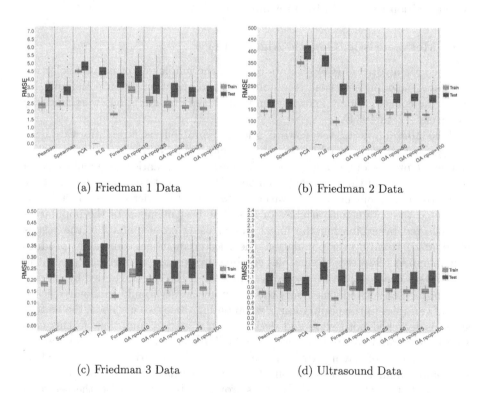

(a) Friedman 1 Data   (b) Friedman 2 Data

(c) Friedman 3 Data   (d) Ultrasound Data

Fig. 1. Training and testing RMSE for all methods and for each data set.

than for the testing data, which means even though precautions were taken, the training data were still overfit. The forward selection has the highest discrepancy between testing and training, which may be explained due to the greedy selection style and the fact that a bad pick of a variable can't be reversed later, when more information are available. This can be improved by doing backward steps, where variables can be eliminated from the selection later on. The two correlation filters show similar behaviour, although the Pearson correlation is slightly better than the Spearman Correlation in the Friedman data sets. The PCA performs very consistent on the ultrasound data set, the mean RMSE is 0.951 with a standard deviation of only 0.0031. This means that the variables with a high variance are not dependent on the fold and that it might be worth to look at these variables. The PLS overfit the training data completely, as it is expected from the way PLS works, since the components are constructed using linear relations between response and data. The resulting components are better than the principal components at predicting for the first two Friedman data sets, level at the third Friedman but outperformed on the ultrasound data. The GA overfits the training data as well, but not as much as the forward selection. The improvement of the testing RMSE does not scale everywhere with an increase in

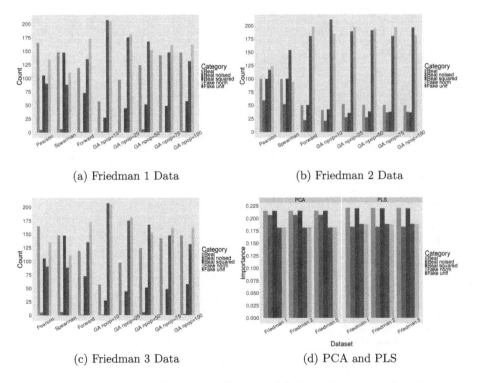

(a) Friedman 1 Data

(b) Friedman 2 Data

(c) Friedman 3 Data

(d) PCA and PLS

Fig. 2. Count of picks for different variable types. (d) shows the averaged and normalized weights for the first 10 components.

population size, while the training RMSE does. To tackle this problem, future evaluations should include a more reliable metric than cross-validation RMSE.

The selection of important features is shown in Fig. 2 for the Friedman data sets. As expected, the Spearman correlation picks more of the squared attributes, while the Pearson correlation picks more random noise variables, which is due to its sensitivity to outliers. Remarkable is also the outstanding performance of the filter methods on the Friedman 2 data set, as they picked the real variables twice as frequently as the other methods. Forward selection and genetic algorithms performed very poorly on this data set, even an increase in the population could not change the outcome significantly. The PCA and PLS components prefer the squared and the noised squared variables equally, but the PCA gives the squared variables more weight. This effect is again based on the composition of the PLS components in comparison to the PCA components. While the PCA only searches for the variance, the PLS is already focused on linear relations between response and data, thus non-linear effects are less likely to be considered. The genetic algorithms preferred the linear and squared real features over the noised as well, these features are only in the Friedman 2 selected, where the metaheuristic was performing very poorly. For the other two, the increase of

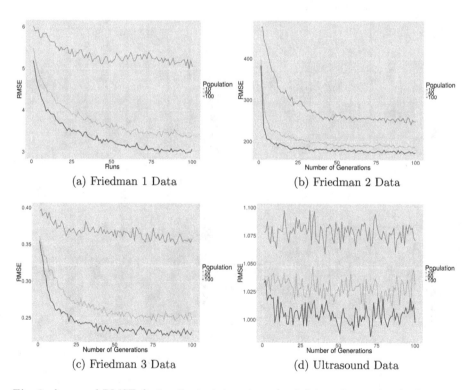

(a) Friedman 1 Data

(b) Friedman 2 Data

(c) Friedman 3 Data

(d) Ultrasound Data

Fig. 3. Averaged RMSE during the training phase (on left out observations) of genetic algorithms.

the population size has a clear effect on the selection of good features, the next paragraph shows the effect of the number of generations.

All simulated data sets perform as it is expected from the genetic algorithm, see Fig. 3. The increase in population size, as well as the increase in the number of generations improves the RMSE. The speed of convergence is quite different, as in the Friedman 2 example the high populations become stationary very quickly, while the other two Friedman data sets converge slowly and further improvement with increased number of generations seems possible.

This differs for the ultrasound data, as the selection seems stationary from the beginning, allowing only small improvements. The underlying optimization problem seems to have several local minima, which are hard to escape from for the GA, unlike the Friedman examples, where there is not much freedom in picking good variables. The increase in population size improves the performance, because the best solution of the first generation is made from purely wide search. This scales the GA down to a Monte Carlo style search, the local search performed by the GA improves the result only slightly.

5 Conclusions

This paper shows a framework for a feature subset selection procedure, which can be easily adapted to other models besides multiple linear regression, for example ridge regression or lasso. The adaptation to classification algorithms is possible as well, although the correlation filter may not be suitable, but instead a filter based on statistical testing like Kruskal-Wallis test is preferred.

The upsides and downfalls of different methods approaching the specific problem setup of having more features than samples have been examined and shown on four different data sets. It turns out, that overall the most simplistic approach, the correlation based filter method, shows already very good results.

The more sophisticated forward selection is overfitting very easily and needs a lot of precaution like the use of a sharp evaluation method for accepting variables into the selection. In our example, we used the RMSE from cross-validation, which could be expanded by using repeated cross-validation, or adding other metrics like correlation or AIC, BIC.

As it comes to data transformation, PLS performed better than PCA on the simulated data sets, while PCA gave better insights and predictions on the ultrasound data set. Both methods produced inferior predictions, compared to the pure selection techniques. An advantage of transformation methods that was not investigated in this paper is that the transformed components can be stacked into the data set along the original variables and improve performance.

The genetic algorithm performed as expected for the simulated data sets, while there was no visible convergence by increasing the number of generations on the ultrasound data set, which may not help finding a good set of variables for regression, but helps gaining insight into the data.

The wrapper and metaheuristics can be improved by adjusting the evaluation function for the acceptance of new variables, respectively the fitness function for ranking the members of the population. Other methods that were not covered in this paper are for example Relief [13], stepwise selection [14] or simulated annealing [15].

Acknowledgements. I want to thank the R&D Team of the Sonovum AG for their constructive comments, open discussions and review of this work and the reviewers for their input and their suggested improvements.

References

1. Everitt, B.S., Skrondal, A.: The Cambridge Dictionary of Statistics. Cambridge University Press, Cambridge (2010)
2. Henderson, H.V., Velleman, P.F.: Building Multiple Regression Models Interactively. Biometrics **37**(2), 391–411 (1981)
3. Bogdan, M., Kolany, A., Weber, U., Elze, R., Wrobel, M.: Computer aided multispectral ultrasound diagnostics brain health monitoring system based on acoustocerebrography. In: Kyriacou, E., Christofides, S., Pattichis, C. (eds.) MEDICO 2016, pp. 983–987. Springer, Cham (2016)
4. Loh, W.-Y.: Variable selection for classification and regression in large p, small n problems. In: Barbour, A., Chan, H.P., Siegmund, D. (eds.) Probability Approximations and Beyond. LNS, vol. 205, pp. 135–159. Springer, Heidelberg (2012). doi:10.1007/978-1-4614-1966-2_10
5. Bernardo, J.M., et al.: Bayesian factor regression models in the "large p, small n" paradigm. Bayesian Stat. **7**, 733–742 (2003)
6. Lee, E.-K., Cook, D.: A projection pursuit index for large p small n data. Stat. Comput. **20**(3), 381–392 (2010)
7. Friedman, J.H.: Multivariate adaptive regression splines. Mach. Learn. **24**, 123–140 (1996)
8. Breiman, L.: Bagging predictors. Ann. Stat. **19**(1), 1–67 (1991)
9. Dobkowska-Chudon, W.: Utilizing comparison magnetic resonance imaging and acoustocerebrography signals in the assessment of focal cerebral microangiopathic lesions in patients with asymptomatic atrial fibrillation (preliminary clinical study results). In: EAA 2016 Congress, Jastrzebia Góra, Poland (2016)
10. Olszewski, R.: The novel non-invasive ultrasound device for detecting early changes of the brain in patients with heart failure. Eur. J. Heart Fail. **18**(Suppl. S1), 307–308 (2016)
11. Holland, J.H.: Adaptation in Natural and Artificial Systems. MIT Press, Cambridge (1975)
12. Barricelli, N.A.: Esempi numerici di processi di evoluzione. Methodos **6**, 45–68 (1954)
13. Kira, K., Rendell, L.: The feature selection problem: traditional methods and a new algorithm. In: Proceedings of the Tenth National Conference on Artificial Intelligence, pp. 129–134. AAAI Press/The MIT Press, Menlo Park (1992)
14. Efroymson, M.A.: Multiple regression analysis. In: Ralston, A., Wilf, H.S. (eds.) Mathematical Methods for Digital Computers, pp. 191–203. Wiley, New York (1960)
15. Debuse, J.C., Rayward-Smith, V.J.: Feature subset selection within a simulated annealing data mining algorithm. J. Intell. Inf. Syst. **9**(1), 57–81 (1997)

16. R Development Core Team: R: A Language and Environment for Statistical Computing. R Foundation for Statistical Computing, Vienna (2008). ISBN 3-900051-07-0, http://www.R-project.org

17. Kuhn, M., Wing, J., Weston, S., Williams, A., Keefer, C., Engelhardt, A.: caret: Classification and Regression Training. R package version 5.15-044 (2012). http://CRAN.R-project.org/package=caret

18. Mevik, B.-H., Wehrens, R., Liland, K.H.: pls: Partial Least Squares and Principal Component Regression. R package version 2.5-0 (2015). https://CRAN.R-project.org/package=pls

19. Revolution Analytics, Weston, S.: doSNOW: Foreach Parallel Adaptor for the 'snow' Package. R package version 1.0.14 (2015). https://CRAN.R-project.org/package=doSNOW

20. Revolution Analytics, Weston, S.: foreach: Provides Foreach Looping Construct for R. R package version 1.4.3 (2015). https://CRAN.R-project.org/package=foreach

21. Leisch, F., Dimitriadou, E.: mlbench: Machine Learning Benchmark Problems. R package version 2.1-1 (2010). https://CRAN.R-project.org/package=mlbench

Colormetric Experiments on Aquatic Organisms

Jan Urban[✉]

Laboratory of Signal and Image Processing, Faculty of Fisheries
and Protection of Waters, South Bohemian Research Center of Aquaculture
and Biodiversity of Hydrocenoses, Institute of Complex Systems,
University of South Bohemia in České Budějovice,
Zámek 136, 373 33 Nové Hrady, Czech Republic
urbanj@frov.jcu.cz

Abstract. The analysis of fish skin color in time lapse experiments is a wide field, since the color changes could be affected by different causes. The classical approach uses single point measurement in CIEL*a*b* color space. On the other hand the image acquisition by digital cameras allows to evaluate spatial statistic, in RGB color space. The reliability of the camera color accuracy as well as the software and method description for the automatic color evaluation, is of the concern of this article. Various fish species were subjected of the test. We performed transformations to other color space and using multivariate data statistics, namely principal component analysis, the relevant color-space is seleted. The presented software solutions are available at http://www.auc.cz/software/.

Keywords: Image · Segmentation · Thresholding · Fish · Skin · Color · Pigmentation

1 Introduction

In this article, the overview of the existing study as well as the latest approach on the time lapse experiments with fish color changes will be presented and discussed to support the increase of image analysis in the color experiments. From the last decade of the 20th century, EU is aiming to the diversification in aquaculture and food production to increase competitiveness among business sphere. The obvious species are represented by carp and tilapia. One of the next promising species, because of its hight grow rate, is the amberjack fish.

These new/emerging species are fast growing and/or large finfishes, marketed at a large size and can be processed into a range of products to provide the consumer with both a greater diversity of fish species and new value-added products. It is of great interest to the aquaculture sector due to its excellent flesh quality, worldwide market availability and high consumer acceptability. Its rapid growth (i.e., short time to market size) and large size makes this species very suitable for product diversification and development of value added products [1].

ⓒ Springer International Publishing AG 2017
I. Rojas and F. Ortuño (Eds.): IWBBIO 2017, Part I, LNBI 10208, pp. 96–107, 2017.
DOI: 10.1007/978-3-319-56148-6_8

With the new species in the aquaculture, it is important to identify and overcome the bottleneck of their own. For example, the smaller scales of the specis might cause other problems, like parasites. Often the emphasis is given to achieving high levels of skin pigmentation, which, together with the size and shape of body and fins, are the most important quality criteria in forming their market value. Also, skin coloration patterns in fish are of great physiological, behavioural and ecological importance. Thus the coloration can be considered as an index of animal welfare in aquaculture as well as an important quality factor in the retail value. The coloration of fish is a result of the overlay and stacking of several types of pigmented cells chromatophores connected to the neural system. The pigment complexes might shrink or stretch according to the neural impulse and fish are able to change their color appearance, e.g. during stress, trying to mimic the background of a darker or lighter environment [2–7]. Final color depends on the depth of imposition and stretch of the pigments in combination with the neural control and environment condition. Skin color of fish depends on the carotenoids present in the diet, mostly as natural phytoplankton or cultured microalgae [8,9]. Usually, when the wild species enters the aquaculture, there are observable changes in the coloration. The color changes are also observable parameter in many time lapse experiments, where is under question the dependency of the fish skin color on the age, season, environment, temperature, and different diet approaches [10–14].

The data analysis comprises pre-processing, data compression and feature extraction and advanced statistic analysis. Altogether, there are plenty of software, packages, toolboxes, applications, programs and recommendations for data processing and/or analysis [15–17]. They all vary in the purpose, used methods, techniques or ideas in the processing steps. The amount of available software solutions is exhausting. The advantage of the situation is, that there is already developed tool to almost any possible evaluation technique. But the major disadvantage is represented by both the complexity and the specificity of the solutions existing. Usually, the biologist working on related topics are highly educated in the sample preparation, instrumental analysis and final interpretation of the results. However, deep induction into the mathematics behind the processing and analysis of the measured datasets is not in their primary focus. Therefore, available amount of software solutions may discourage their enthusiasm as well as confuse the interpretation of the results. Some of the methods are relevant only for particular problems, others are more general. Even then, minimal set of assumptions on data attributes is often necessary to be fulfilled.

There are three typical possible methods of colorimetry. The point measurement by colormeters, spectral measurement a image analysis. The colormeters provides single point or very small are value of the color, usually in CIE L*a*b* color space. The spectral measurements are able to provide the whole spectrum, which could be used to computations of the color values in various color space, however it is still just a point measurement. The image analysis allows to evaluate many points together, estimates the spectra, and use any color space transformation. Under the question was still the color calibration for the image analysis.

In all case, a necessary step is a statistical evaluation of the measured color data, comparison, clustering, etc. [18,19].

2   Methods and Experiments

A digital camera is based on the design of a classic camera; nevertheless the core device is a light sensitive area CCD (Charge Coupled Device - device which collects charge) or CMOS (Complementary Metal-Oxide-Semiconductor).

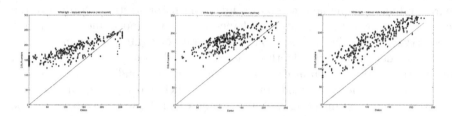

Fig. 1. Comparison of measured (y-axis) and expected (x-axis) RGB (from left to right R, G, B) values using color etalons.

The area of photodiode is sometimes likened to a mosaic [20]. As such, only intensity of the incident light is measured, not direct color. The resulting digital fingerprint of real image is then interpolated and the white balance correction is performed. So called Bayer mask is mounted over the individual chips [21], which is RGB color filter (corresponding to R-red, G-green, and B-blue). Each pixel thus contains information only about one incident color and the other two color components must be interpolated from the neighbouring mosaic stones. Although Bayer interpolation is the most widely used, other designs and patents filters can be used as well. A simple way how to solve bias color is a feature called the white balance. This function takes into account the color temperature of the white point and places the gamut in the right position in the coordinate system. During a series of experiments it is appropriate to always use the same object to balance. Unfortunately, even after the white balance, the *RGB* values from the camera still do not fit exactly to the etalon values (see Fig. 1). This is given by two things: (a) the projection of the measured values from 12 bit acquisition to the 8 bit representation; (b) the color is represented by the *RGB* values ratio, not by the absolute values. The solution for this issue is described in the next subsection.

2.1   Chromatic Colors

Is RGB color space, each pixel has three values – intensity of red, green and blue. To locate fish body in image we just need to select the right thresholds (down and bottom) in these values. This will separate a shape in 3D RGB space that represents the possible values of fish skin color. But, what we see is not only the color, it including brightness also. Two pixels $[r1, g1, b1]$ and $[r2, g2, b2]$ where

$$r1 \neq r2; \quad g1 \neq g2; \quad b1 \neq b2, \tag{1}$$

but the rate

$$\frac{r1}{b1} = \frac{r2}{b2}; \quad \frac{b1}{g1} = \frac{b2}{g2}; \quad \frac{g1}{r1} = \frac{g2}{r2}, \tag{2}$$

have the same color, they differ only in brightness. To eliminate this and cut down the number of dimension in color space is very useful the transformation to chromatic color space [22]:

$$R = \frac{r}{(r + g + b)} \text{ and } B = \frac{b}{(r + g + b)}. \tag{3}$$

The G value is redundant, because R + B + G = 1. Now we have only two values for set the thresholds and we eliminate the problem with brightness. Now we can simply compute for each value for both chromatic color the rate in the picture (or many of them).

2.2 Hue

RGB color space is not the best option, as it does not match the concept of colors as understood by man. In this regard, far more appropriate model is HSV, wherein the value of H (hue-tint) is a parameter corresponding to the color spectrum from the physical point of view, that is the wavelength. The value of S is saturation, chroma, color purity, decrease in white, or ratio of mixing H with gray of the same intensity. V (Value) is a value or intensity, given by the normalized sum of the RGB values (corresponding to the grayscale representation), and it expresses the relative brightness (maximum value) or darkness (minimum value) the color. This model also most closely matches the human (psychophysical) way of perception of color. Therefore, in the HSV color model, the color is represented by the hue. When the camera captured image is converted from the RGB to the HSV color space, as well as the etalon RGB values, the hue values of measured and etalon values fit together (see Fig. 2). Therefore, the image analysis of the white balance calibrated images could provide a reliable outputs, since **the digital camera should be acceptable color measurement device.**

2.3 Image Processing and Analysis

Image processing and analysis generally consists of a number of methods and algorithms, designed to obtain such information from the digitised image, the quantified interpretation of which applies to the real object captured in the image. The vast majority of basic methods for binary images have been developed a long time ago. In the last twenty years a massive adaption of these methods to grayscale (black and white often inaccurately) images have occurred.

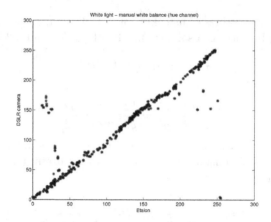

Fig. 2. Comparison of measured (y-axis) and expected (x-axis) hue value using color etalons.

Generalization and methods for color images are still insufficient, the most commonly used treatment is individual for each color channel as grayscale, therefore priori potential emergent properties are ignored. Image processing and analysis provides a useful set of modular tools, whose combination in a suitable sequence of captured images can exploit the information for which the following applies:

- man is able to detect an information value subjectively in qualitative terms, but quantitative expression is time-consuming and inaccurate,
- for a large number of processed data effectiveness and credibility of the subjective assessment of an external observer is declining, but time consumption and error interpretation are growing,
- a part of the information, respectively its presence can not be subjectively recognized.

The objective of image analysis is the development and production of such software tools that facilitate and automate routine performed by man for image evaluation and, moreover, extend them by the mathematical operations of artificial intelligence. Of course there is the possibility of processing of a large number of input data and also a reduction of time-consuming analysis. In principle, each process towards the interpretation of the image can be divided into a few basic steps:

- Pre-processing - this involves noise filtering, transformation of the scene, rotation, translation, scaling, brightness and color standardization.
- Processing - segmentation of the images from the background and segmentation of objects from each other by thresholding, edge or specific objects detection (shape, size, etc.).
- Analysis - quantification of object properties (position, color, number of pixels), comparison of objects between images, identifying symptoms, clustering objects into groups, etc.

2.4   Automatic Thresholding

The gray level thresholding [23] is a nonparametric method of automatic threshold selection for picture segmentation from intensity histogram H(p). The Histogram function H(p) is an intensity function, shows count of pixel f(i, j) with the intensity equal p independently on the position (i, j):

$$H(p) = \sum_{i,j} h(i, j, p) \tag{4}$$

$$h(i, j, p) = 1 \; if \; f(i, j) = p \tag{5}$$

$$= 0 \; if \; f(i, j) \neq p. \tag{6}$$

Firstly, the histogram functions are normalized:

$$o_p = \frac{H(p)}{N}, \tag{7}$$

where N is the total number of pixels in image. For separating histogram into two classes, the probabilities of class occurrence and the class mean are computed:

$$\omega_1 = \sum_{p=1}^{k} o_p; \tag{8}$$

$$\omega_2 = \sum_{p=k+1}^{L} o_p; \tag{9}$$

$$\mu_1 = \frac{\sum_{p=1}^{k} p * o_p}{\omega_1}, \tag{10}$$

$$mu_2 = \frac{\sum_{p=k+1}^{L} p * o_p}{\omega_2}. \tag{11}$$

Also there is necessary to evaluate the total mean level of image:

$$\mu_T = \sum_{p=1}^{L} p * o_p, \tag{12}$$

and between class variance:

$$\sigma_B^2 = \omega_1 * (\mu_1 - \mu_t)^2 + \omega_2 * (\mu_2 - \mu_t)^2. \tag{13}$$

The optimal threshold k^* maximizes σ_B^2 [23].

2.5   Fish Species

Our method of image analysis fish skin color was performed with several fish species: *Carassius auratus, Cyprinus carpio Leuciscus idus, Silurus glanis, Acipenser baeri, Pterophyllum scalare, Hippocampus hippocampus, Amphiprion ocellaris, Amberjakc fish* (see Fig. 3). Changes in fish coloring were a response to the diet enriched with carotenoids from microalgae with their increased content are manifested mainly by a change in xantophores and erytrophores. Microalgae grown in aquaculture have due to their high nutritional value potential to become an important complement to traditional feed products. Due to the content of probiotics, antioxidants and pigments they may positively affect not only the color, but also the health of the fish. The most important carotenoids contained in microalgae include: astaxanthin, lutein, beta-carotene, canthaxanthin, zeaxanthin, violaxanthin and neoxanthin. Selected microalgae can be submitted to fish either by mixing into the feed or indirectly through the food chain. Another possibility is use of dried microalgae, either as powder or pellet.

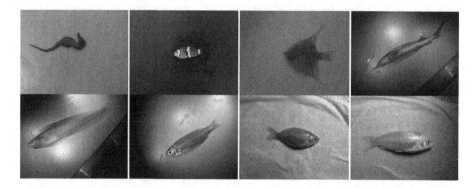

Fig. 3. From left to right, top line: Hippocampus hippocampus, Amphiprion ocellaris, Pterophyllum scalare, Acipenser baeri; bottom line: Silurus glanis, Leuciscus idus, Carassius auratus, Cyprinus carpio. (Color figure online)

2.6   Comparison of Fish Skin Color

Expertomica Fishgui express the changes in the fish skin color. Standardized light conditions and white balance correction on the camera are expected. The fish body in each image is expected to be the dominant object. The background has to be semi-uniform. Expertomica Fishgui is as a standalone Matlab application [24]. The application computes the average (avg) and standard deviation (std) values of the pixels of the fish skin (see Fig. 4). The values are evaluated in the RGB, HSV, and CIE L*a*b color spaces plus the value of the dominant wavelength (lambda). As the output of the processing, two graphs are plotted; the position of the average pixel in the chromaticity diagram and the non-normalised colors distributions across the images in the RGB color space [24–27].

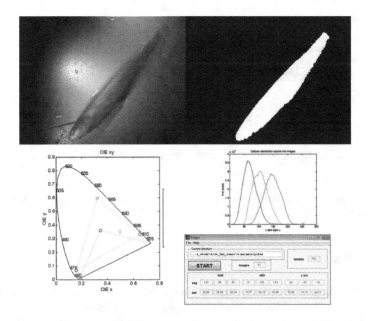

Fig. 4. Example of Fishgui application evaluation of Silurus glanis. (Color figure online)

2.7 Fish Skin Color Evaluation

Software allows to performance evaluation of color from recorded images. Application detect background and analyzed object (fish). Then it computes the values of red, green, blue, brightness, saturation and hue, CIE L*a*b*. The noninvasive method using digital images of the fish body was developed as a standalone application. This application deals with the computation burden and memory consumption of large input files, optimizing piecewise processing and analysis with the memory/computation time ratio. For the comparison of color distributions of various experiments and different color spaces (RGB, CIE L*a*b*) the comparable semi-equidistant binning of multi channels representation is introduced (see Fig. 5). It is derived from the knowledge of quantization levels and Freedman-Diaconis rule. The color calibrations and camera responsibility function were necessary part of the measurement process [28].

2.8 Statistics

The software Fisceapp output is a table of basic statistics of the set, including both positional (mean, median, mode) and scaling (standard deviation) values of RGB, L*a*B*, HSV, L*ch color spaces, and dominant wavelength. Such outputs could be computed for each set (diets, months). To be able to compare the results, we performed also standard MVA principal component analysis with cross correlation of standardized and normalized datasets, to see the 'self-clustering' of the diets and months (Fig. 7), and mainly to select which color parameters are the most relevant (Fig. 6) for the color changes experiments.

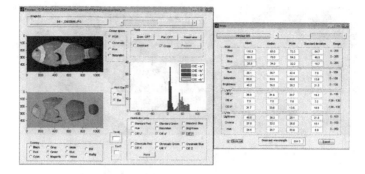

Fig. 5. Example of Fish skin color evaluation application evaluation of Amphiprion ocellaris and basic set statistics. (Color figure online)

3   Results

According to the performed experiments with described fish species and software applications the RGB color space of the digital cameras is not directly acceptable for the color measurements. However the conversion from the RGB to HSV color space represents the color more accurately. The crucial issue is to provide a white balance, before data acquisition. The L*a*b color space is a standard color space for color measurements. The main difference is that the colormetric is single point or small area measurement, while the image could contain the whole object. The segmentation and auto-thresholding of the fish body uses standard Otsu segmentation [23, 24]. The color transformations to any color space, including L*a*b is possible, since the relations are known. The statistic of the image or set represents the estimation of the color distribution across the image or set, respectively. The multivariate data analysis clearly shows the color parameters relevant for the principal components (Fig. 6):

- HSV Hue;
- CIE *a;
- LCH HUE;
- Dominant wavelength.

Since old the feeding experiments used carotenoid enrichment, more color changes is in CIE *a (red-green). Generally it proves that the CIE L*a*b* space is valid for the color evaluation. Moreover both hue values, from HSV or LCH color space are also representative for the color evaluation. Since the hue is little bit easily computed from the RGB values, it should be the correct parameter for the image based color analysis. It is not surprise, that the dominant wavelength is one of the relevant parameters, it is given by its meaning. As was expected the RGB color space do not represent the color information exactly for various reasons mentioned previously.

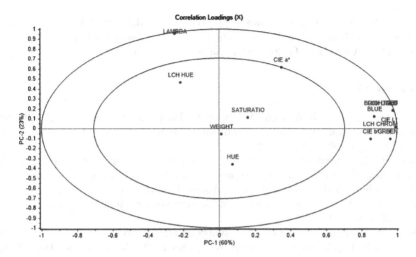

Fig. 6. Loadings of the color parameters as their relevancy for groups/diets classification.

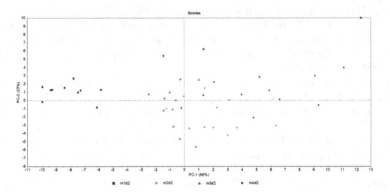

Fig. 7. Evolution of diet month by month according to PC1 and PC2.

4   Conclusion and Discussion

Proposed solution help to increase animal welfare as it represents more simple (in final performance) solution than other invasive methods (histology of dead fishes) with low cost, and evaulated limitations - directly form the obtained datasets. The methodology is based on a much sold and low-priced technology (ordinary digital cameras). The benefit of described methodology for animal health and welfare is immediate, while it cancel the necessity to kill and biochemically analyse the fish. Finally, an adopted application outputs of the running experiments results in the availability of a simple-to-use and usefull research tool for aquaculture research.

Acknowledgement. This work was supported by the Ministry of Education, Youth and Sports of the Czech Republic - projects 'CENAKVA' (No. CZ.1.05/2.1.00/01.0024) and 'CENAKVA II' (No. LO1205 under the NPU I program).

References

1. Saberioon, M.M., Gholizadeh, A., Cisar, P., Pautsina, A., Urban, J.: Application of machine vision systems in aquaculture with emphasis on fish: state-of-the-art and key issues. Rev. Aquac. (2016)
2. Gouveia, L., Choubert, G., Gomes, E., Pereira, N., Santinha, J., Empis, J.: Pigmentation of gilthead sea bream, Sparus aurata (L. 1875), using Chlorella vulgaris (Chlorophyta, Volvocales) microalga. Aquac. Res. **33**, 987–993 (2002)
3. Kelsh, R.N.: Genetics and evolution of pigment patterns in fish. Pigment Cell Res. **17**, 326–336 (2004)
4. Kop, A., Durmaz, Y.: The effect of synthetic and natural pigments on the colour of the cichlids (Cichlasoma severum sp., Heckel 1840). Aquacult. Int. **16**, 117–122 (2008)
5. Oshima, N., Nakamaru, N., Araki, S., Sugimoto, M.: Comparative analyses of the pigment-aggregating and dispersing actions of MCH on fish chromatophores. Comp. Biochem. Phys. C Toxicol. Pharmacol. **129**(2), 75–84 (2001)
6. Pan, C.-H., Chien, Y.-H.: Effects of dietary supplementation of alga Haematococcus pluvialis (Flotow), synthetic astaxanthin and -carotene on survival, growth, and pigment distribution of red devil, Cichlasoma citrinellum (Gunther). Aquac. Res. **40**, 871–879 (2009)
7. Putnam, M.: A review of the nature, function, variability, supply of pigments in salmonid fish. In: de Pauw, N., Joyce, J. (eds.) Aquaculture and the Environment, vol. 16, pp. 245–263. European Aquaculture Society Special Publication, Gent (1991)
8. Gouveia, L., Rema, P.: Effect of microalgal biomass concentration and temperature on ornamental goldfish (Carassius auratus) skin pigmentation. Aquac. Nutr. **11**, 19–23 (2005)
9. Gouveia, L., Rema, P., Pereira, O., Empis, J.: Colouring ornamental fish (Cyprinus carpio and Carassius auratus) with microalgal biomass. Aquac. Nutr. **9**, 123–129 (2003)
10. Ako, H., Tamaru, C.S., Asano, L., Yuen, B., Yamamoto, M.: Achieving natural coloration of fish under culture. UJNR Techn. Rep. **28**, 1–4 (1999)
11. Costa, C., D'Andrea, S., Russo, R., Antonucci, F., Pallottino, F., Menesatti, P.: Application of non-invasive techniques to differentiate sea bass (Dicentrarchus labrax, L. 1758) quality cultured under different conditions. Aquacult. Int. **19**, 765–778 (2011). doi:10.1007/s10499-010-9393-9
12. Costa, C., Antonucci, F., Pallottino, F., Boglione, C., Cataudella, S., Menesatti, P.: Color-warping imaging: a non-destructive technique to evaluate gilthead seabream (Sparus aurata, Linnaeus 1758) freshness. In: CIGR Workshop on Image Analysis in Agriculture, Budapest, pp. 26–27, August 2010
13. Tejera, N., Cejas, J.R., Rodriguez, C., Bjerkeng, B., Jerez, S., Bolanos, A., Lorenzo, A.: Pigmentation, carotenoids, lipid peroxides and lipid composition of skin of red porgy (Pagrus pagrus) fed diets supplemented with different astaxanthin sources. Aquaculture **270**, 218–230 (2007)
14. Torrissen, O.J., Hardy, R.W., Shearer, K.D.: Pigmentation of salmonids Carotenoid deposition and metabolism. CRC Crit. Rev. Aquat. Sci. **1**, 209–225 (1989)

15. Hancz, C., Magyary, I., Molnar, T., Sato, S., Horn, P., Taniguchi, N.: Evaluation of color intensity enhanced by paprika as feed additive in goldfish and koi carp using computer-assisted image analysis. Fish. Sci. **69**, 1158–1161 (2003)

16. Marty-Mahe, P., Loisel, P., Fauconneau, B., Haffray, P., Brossard, D., Davenel, A.: Quality traits of brown trout (Salmo trutta) cutlets described by automated color image analysis. Aquaculture **232**, 225–240 (2004)

17. Stien, L.H., Manne, F., Ruohonen, K., Kause, A., Rungruangsak-Torrissen, K., Kiessling, A.: Automated image analysis as a tool to quantify the colour and composition of rainbow trout (Oncorhynchus mykiss W.) cutlets. Aquaculture **261**, 695–705 (2006)

18. Novelo, N.D., Gomelsky, B.: Comparison of two methods for measurement of red-area coverage in white-red fish for analysis of color variability and inheritance in ornamental (koi) carp Cyprinus carpio. Aquat. Living Resour. **22**, 113–116 (2009)

19. Pavlidis, M., Papandroulakis, N., Divanach, P.: A method for the comparison of chromaticity parameters in fish skin: preliminary results for coloration pattern of red skin Sparidae. Aquaculture **258**, 211–219 (2006)

20. Sonka, M., Hlavac, V., Boyle, R.: Image Processing, Analysis and Machine Vision. Brooks/Cole Publishing Company, Pacific Grove (1999)

21. Bayer, B.E.: Color imaging array. US Patent No. 3971065 (1977)

22. Yang, J., Lu, W., Waibel, A.: Skin-color modeling and adaptation. In: Chin, R., Pong, T.-C. (eds.) ACCV 1998. LNCS, vol. 1352, pp. 687–694. Springer, Heidelberg (1997). doi:10.1007/3-540-63931-4_278

23. Otsu, N.: A threshold selection method from gray-level histogram. IEEE Trans. Syst. Man Cybern. **SMC–9**, 62–66 (1979)

24. Urban, J., Stys, D., Sergejevova, M., Masojidek, J.: Comparison of fish skin colour. J. Appl. Ichthyol. **29**, 172–180 (2013)

25. Zatkova, I., Sergejevova, M., Urban, J., Vachta, R., Stys, D., Masojidek, J.: Albinic form of wels catfish (Silurus glanis). Aquac. Nutr. **17**, 278–286 (2011)

26. Urban J., Sergejevova M., Slepickova I., Kouba A., Stys D., Masojidek J.: Hodnoceni zmen vybarveni okrasnych ryb. Edice Metodik (Technologicka rada), FROV JU, Vodnany, c. 131 (2012). ISBN: 978-80-87437-55-1

27. Kouba, A., Sales, J., Sergejevov, M., Kozak, P., Masojidek, J.: Colour intensity in angelfish (Pterophyllum scalare) as influenced by dietary microalgae addition. J. Appl. Ichthyol. **29**, 193 (2013)

28. Urban, J., Botella, A.S., Robaina, L.E., Barta, A., Soucek, P., Cisar, P., Papacek, S., Domnguez, L.M.: FISCEAPP: fish skin color evaluation application. In: 17th International Conference on Digital Image Processing, Dubai, UAE (2015)

Predicting Comprehensive Drug-Drug Interactions for New Drugs via Triple Matrix Factorization

Jian-Yu Shi[1(⊠)], Hua Huang[2], Jia-Xin Li[1], Peng Lei[3],
Yan-Ning Zhang[4], and Siu-Ming Yiu[5]

[1] School of Life Sciences, Northwestern Polytechnic University, Xi'an, China
jianyushi@nwpu.edu.cn, lijiaxin0932@mail.nwpu.edu.cn
[2] School of Software and Microelectronics,
Northwestern Polytechnic University, Xi'an, China
1363351294@qq.com
[3] Department of Chinese Medicine,
Shaanxi Provincial People's Hospital, Xi'an, China
leipengml@163.com
[4] School of Computer Science,
Northwestern Polytechnic University, Xi'an, China
ynzhang@nwpu.edu.cn
[5] Department of Computer Science,
The University of Hong Kong, Hong Kong, China
smyiu@cs.hku.hk

Abstract. There is an urgent need to discover or deduce drug-drug interactions (DDIs), which would cause serious adverse drug reactions. However, preclinical detection of DDIs bears a high cost. Machine learning-based computational approaches can be the assistance of experimental approaches. Utilizing pre-market drug properties (e.g. side effects), they are able to predict DDIs on a large scale before drugs enter the market. However, none of them can predict comprehensive DDIs, including enhancive and degressive DDIs, though it is important to know whether the interaction increases or decreases the behavior of the interacting drugs before making a co-prescription. Furthermore, existing computational approaches focus on predicting DDIs for new drugs that have none of existing interactions. However, none of them can predict DDIs among those new drugs. To address these issues, we first build a comprehensive dataset of DDIs, which contains both enhancive and degressive DDIs, and the side effects of the involving drugs in DDIs. Then we propose an algorithm of Triple Matrix Factorization and design a Unified Framework of DDI prediction based on it (TMFUF). The proposed approach is able to predict not only conventional binary DDIs but also comprehensive DDIs. Moreover, it provides a unified solution for the scenario that predicting potential DDIs for newly given drugs (having no known interaction at all), as well as the scenario that predicting potential DDIs among these new drugs. Finally, the experiments demonstrate that TMFUF is significantly superior to three state-of-the-art approaches in the conventional binary DDI prediction and also shows an acceptable performance in the comprehensive DDI prediction.

© Springer International Publishing AG 2017
I. Rojas and F. Ortuño (Eds.): IWBBIO 2017, Part I, LNBI 10208, pp. 108–117, 2017.
DOI: 10.1007/978-3-319-56148-6_9

Keywords: Drug-drug interaction · Side effects · Matrix Factorization · Prediction · Regression

1 Introduction

Drug–drug interactions (DDIs) may occur accidentally when drugs are co-prescribed. During drug development (usually the clinical trial phase) [1], a few of DDIs could be identified. However, most of the DDIs among drugs are reported after the drugs are approved due to the very limited number of participants and the great number of drug combinations under screening in clinical trials. DDIs would cause serious adverse drug reactions, such as the reduction in efficacy or the increment on unexpected toxicity among the co-prescribed drugs. Thus, DDIs lead patients, who are typically treated with numerous medications, to be in the unsafe treatment of medication errors [2]. As the number of approved drugs increases, the number of potential DDIs is rapidly rising and the adverse effect of DDI is broadcasting wider and wider. Therefore, there is an urgent need to discover or predict DDIs before clinical medications are administered. However, experimental approaches for predicting DDI (e.g. testing cytochrome P450 [3] or transporter-associated interactions [4]), bear high cost, long duration as well as animal welfare considerations [5]. Computational approaches are able to perform DDI prediction on a large scale such that they provide tools for helping screen potential DDIs. They win many concerns from both academy and industry recently.

Data-mining based computational approaches were developed for detecting DDIs from different sources [5], including scientific literature, electronic medical records, insurance claim databases, and the Adverse Event Reporting System of FDA. However, relying on the clinical evidence in post-market, these approaches cannot provide an alert of potentially dangerous DDIs for clinical medications to be administered soon. Unlike those post-marketed sources, drug features or similarities (e.g. chemical structures [6], targets [7], hierarchical classification codes [8] and side effects [5]) can be obtained before drugs enter into the market. Thus, machine learning-based computational approaches, such as naïve similarity-based approach [6], network recommendation-based [5], classification-based [8], are able to predict potential DDIs in advance, so as that drug safety professionals can screen DDIs rapidly and make appropriate clinical medications. In general, existing approaches were designed for only binary prediction, which only indicates how likely a pair of drugs is a DDI.

However, two interacting drugs may increase or decrease the behavior or effect (e.g. serum concentration) of each other in vivo. Determination of serum concentration (referred to as the amount of a drug in the circulation) is crucial when making optimal patient care, establishing the dosage of a drug, designing prophylactic drug therapy and finding the resistance to therapy with a drug [9]. When interacting with other drugs, a drug would increase or decrease the level of its own and its partners' serum concentration. For example, the serum concentration of Dofetilide (whose DrugBank Id is DB00204) decreases when it is combined with Dabrafenib (DB08912), whereas its serum concentration increases when combined with Dalfopristin (DB01764). Here, we may call the pair of Dofetilide and Dabrafenib as a degressive DDI and the pair of Dofetilide and Dalfopristin as an enhancive DDI.

Consequently, it is very important to know not only whether two drugs interact with each other when combining them together, but also to know exactly whether the interaction increases or decreases their behaviors before making a co-prescription. Nevertheless, to the best of our knowledge, existing approaches only consider the first need, but none of them can meet the second need which asks for a prediction of comprehensive DDIs, including enhancive and degressive DDIs. In addition, existing computational approaches mainly focus on predicting DDIs for new drugs that have none of existing interactions, but none of them can predict DDIs among those new drugs.

In this paper, to address abovementioned issues, we first collect a dataset of DDIs from DrugBank [10], which contains both enhancive and degressive DDIs, as well as the side effects of the drugs from Offsides [11]. Then, proposing a Triple Matrix Factorization, we design a Unified Framework of DDI Prediction (TMFUF). This approach can predict not only conventional binary DDIs but also enhancive and degressive DDIs. Moreover, it is able to predict potential DDIs for newly given drugs that have no known interactions, and further able to predict potential DDIs among new drugs. Experiments on the dataset demonstrate that TMFUF is significantly superior to two state-of-the-art approaches in the conventional binary DDI prediction and also shows an acceptable performance in the comprehensive DDI prediction.

2   Materials and Method

2.1   Dataset and Problem Formulation

We collected 2,329 approved drugs from DrugBank [10] and selected out 603 drugs, which have both DDIs recorded in DrugBank and off-label side effects recorded in Offsides [11]. Totally, 9,149 side effects are involved. Thus, each of 603 drugs is represented as a $1 \times 9{,}149$ dimensional binary feature vector, of which each bit indicates whether or not the drug shows a specific one among the side effects. Meanwhile, we also counted the numbers of enhancive DDIs and degressive DDIs respectively. The details of our DDI dataset is listed in Table 1.

Table 1. Details of comprehensive DDI dataset

# drug	# DDI	# all pairs	# enhancive DDI	# degressive DDI	# side effect
603	24,114	181,503	18,710	5,404	9,149

Without loss of generality, let $\mathbf{D} = \{d_i\}, i = 1, 2, \ldots, m$ be a set of m approved drugs. Their interactions can be organized into an $m \times m$ symmetric interaction matrix $\mathbf{A}_{m \times m} = \{a_{ij}\}$, in which $a_{ij} \in \{-1, 0, +1\}$. If d_i and d_j do not interact with each other, $a_{ij} = 0$. When there is an enhancive DDI between d_i and d_j, $a_{ij} = +1$. Otherwise, $a_{ij} = -1$, when a degressive DDI exists. The conventional binary DDIs $\mathbf{A}_b$ is just the special case, which can be obtained by $\mathbf{A}_b = Binary(\mathbf{A})$. In addition, each drug d_i in $\mathbf{D}$ is represented as a p-dimensional feature vector $\mathbf{f}_i = [f_1, f_2, \ldots, f_k, \ldots, f_p]$, where $f_k = 1$ denotes the k-th specific side effect has been found when taking d_i in clinic and $f_k = 0$ otherwise. All the drugs in $\mathbf{D}$ are sequentially organized into an $m \times p$ feature matrix $\mathbf{F}_{m \times p}$.

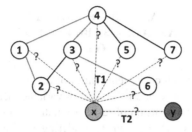

Fig. 1. Illustration of predicting DDIs for a newly given drug and among two new drugs. Nodes represent drugs. The hollow nodes are known drugs (numbered from 1 to 7) and the solid lines between them denote their interactions. Red lines are enhancive interactions and blue lines are degressive interactions. The nodes filled with yellow and green respectively are two newly given drugs, tagged as X and Y. Our problem is to determine which known drugs could interact with the new drug X, how likely X interacts with Y, and what type these potential interactions are. Two kinds of tasks are labeled with 'T1' and 'T2'. (Color figure online)

Two predicting tasks will be handled. The first (T_1) is to predict how likely a newly given drug d_x, having no interaction with any in **D**, interacts with one or more drugs in **D**, and how likely a DDI of d_x is an enhancive DDI or a degressive DDI. The second (T_2) is to predict how likely d_x interacts with another newly given drug d_y and how likely the potential interaction between d_x and d_y is an enhancive DDI or a degressive DDI. Figure 1 illustrates these two tasks.

We believe that drugs can be mapped from their own feature spaces into the common **interaction space** and their inner products are correlated with their interactivity. The drugs in the pair corresponding to an interaction are near to each other in such a space, otherwise are far from each other. Thus, the DDI matrix can be represented as a triple matrix factorization (TMF),

$$\mathbf{A} = \mathbf{F\Theta F'} \tag{1}$$

where Θ is the symmetric projection matrix, in which each entry indicates the importance of the pairs between drug features among enhancive, degressive interactions as well as non-interactions. It builds the bridge between the features of drugs and the interactions between them. Unfortunately, it cannot be directly solved by $\Theta = (\mathbf{F})^{-1}\mathbf{A}(\mathbf{F'})^{-1}$ because $p \gg m$, the multicollinearity between the columns in **F** exists and **A** is not of full rank in general.

2.2 Predicting Model

The main idea to solve Θ is to build a linear bi-regression between **A** and the pair of drugs. We regard **F** as **the observed inputs**, and **A** as the **observed bi-response** of the interaction space. After decomposing **A** into a $m \times r$ drug response matrix $\mathbf{A_d}$ and its transpose matrix $\mathbf{A'_d}$, we solve the bi-regression model as follows,

$$\mathbf{A_d} = \mathbf{FB} + \mathbf{E} = \tilde{\mathbf{A}}_\mathbf{d} + \mathbf{E}$$
$$\mathbf{A} = \mathbf{A_d A'_d} \approx \mathbf{FBB'F'} = \mathbf{F\tilde{\Theta}F'} \tag{2}$$
$$\mathbf{\Theta} \approx \tilde{\mathbf{\Theta}} = \mathbf{BB'}$$

where $\mathbf{B}$ is the $p \times r$ regression coefficient matrix of drugs, $\tilde{\mathbf{A}}_\mathbf{d}$ indicates the regression item of the response matrix of $\mathbf{A}$, and $\mathbf{E}$ is the error variable matrix. In addition, since $\mathbf{A}$ may not have full rank, the dimension of response matrices (r) is just its rank, which is related to the complexity of DDI graph. Considering the equivalent and symmetric roles of the two response matrices of $\mathbf{A}$, we obtain them by singular value decomposition (SVD) as follows,

$$\mathbf{A} = \mathbf{U\Sigma V'} = \mathbf{U}\sqrt{\Sigma}\sqrt{\Sigma}'\mathbf{V'} = \mathbf{A_d A'_d} \tag{3}$$

In fact, each row in $\mathbf{A_d}$ accounts for the topological properties of a drug, because $\mathbf{A}$ can be treated as the adjacent matrix of DDI graph. The joint of the observed inputs and the observed response reflects the underlying **interaction space**.

Based on TMF, we develop two predicting models for T_1 and T_2 in a unified form (TMFUF) as follows.

$$\mathbf{A_{x,D}} = \mathbf{F_x}\tilde{\Theta}\mathbf{F'}, \ \mathbf{A_{x,y}} = \mathbf{F_x}\tilde{\Theta}\mathbf{F'_y} \tag{4}$$

where $\mathbf{F}$ is the feature matrix of approved drugs having known DDIs, $\mathbf{F_x}$ and $\mathbf{F_y}$ are the feature vectors of two newly coming drugs d_x and d_y having no DDI, $\mathbf{A_{x,D}}$ contains the confidence scores which indicate how likely d_x interacting with the drugs in D, and $\mathbf{A_{x,y}}$ is a confidence score that denotes how likely d_x interacts with d_y. Positive and negative confidence scores of drug pairs account for potential enhancive and degressive DDIs respectively. The closer the scores are to zero, the more possibly they are NOT an interaction. Vice versa.

2.3   Partial Least Square Regression

In terms of regression, for m observations, the task is to find the linear relations between their predictor matrix $\mathbf{X}_{m \times p}$ and their response matrix $\mathbf{Y}_{m \times r}$. In this context, $\mathbf{F}$ is the predictor matrix of m drugs $\mathbf{D}$ and $\mathbf{A_d}$ is their response matrix. However, when the number of dimensions of predictors is greater than the number of observations ($p > m$) or when there is multicollinearity among $\mathbf{X}$ values, standard regression will fail. These cases occur in our problem.

For example, drugs can be represented by a list of binary (1/0) bits, such as the profile of OFFSIDES [11] which records ~ 10 thousand adverse events (known as side effects as well) of approved drugs. Each entry of the OFFSIDES profile represents a Boolean determination of the presence of a side effect (e.g. AFIB, aching joints). By contrast, the number of approved drugs in the given dataset is possibly smaller the number of side effect entries. For example, DrugBank has only 2,012 approved small molecule drugs. In addition, there may have a dependency between the entries. Thus, there is a multicollinearity among the side effect entries.

Considering both the number of features possibly greater than the number of drugs and the multicollinearity among features, we utilize multivariate Partial Least-Squares Regression (PLSR) to build the regression model in Formula 2. It finds a linear regression model by projecting the predicted variables and the response variables to a new space, which can be regarded as the Interaction Space. The general underlying model of multivariate PLSR can be solved by SIMPLS algorithm as follows (Dejong 1993).

Pseudo-codes of SIMPLS

INPUT:

X ($m \times p$, column normalized): matrix of predictors

Y ($m \times r$, column normalized): matrix of response variables

$S = Y^T X$: cross-product

K : number of latent factors

Algorithm:

$S_1 = Y^T X$, initial value of cross-product

For i=1 to K

   q_i = dominant eigenvector of $S_i S_i^T$

   $w_i = S_i^T q_i$, gets X weights

   $t_i = X w_i$, gets X latent factors

   $normt \leftarrow t_i^T t_i$, compute norm

   $t_i \leftarrow t_i$ / $normt$, normalizes X latent factor t_i.

   $w_i \leftarrow w_i$ / $normt$, updates weights w_i accordingly.

   $p_i \leftarrow X_i^T t_i$, gets X loadings.

   $c_i \leftarrow Y^T t_i$, gets Y weights.

   $u_i \leftarrow Y^T c_i$, gets Y latent factor.

   If i=1

      $v_i = X^T t_i$, orthogonal loadings for deflation of S

   Else

      $v_i = X^T t_i - V(V^T X^T t_i)$, makes v_i orthogonal to previous loadings.

   End

   $v_i = v_i$ / $sqrt(v_i^T S_i)$, normalizes loadings.

   $S_{i+1} = S_i - v_i(v_i^T S_i)$, deflates S w.r.t current loadings.

   Stores q_i, t_i, w_i, p_i, c_i, u_i, v_i into Q, T, W, P, C, U, V respectively.

End

OUTPUT:

$B = WC^T$, regression coefficients.

3 Experiments

3.1 Preparation

K-fold Cross-validation (CV) is the well-established approach to validate the power of generalization of algorithms in machine learning. To reflect the fact that new drugs have no interaction, we adopted two 10-fold CV schemes (CV1 and CV2) to evaluate the performance of DDI prediction in the two task, T_1 and T_2, accordingly.

In detail, for the given drugs having NO known interaction, CV1 tries to assess the task of predicting new potential interactions between them and those drugs having known interactions, while CV2 attempts to assess the task of predicting new potential interactions among these new drugs.

Different tasks require technically different sets of both training samples and testing samples as follows.

- In CV1, we randomly removed 1/10 drugs out of all the given drugs as the testing drugs and selected the remaining drugs as the training drugs. The pairs among the training drugs were selected as the training samples. Regarding the testing drugs as new drugs, we only selected the pairs between the testing drugs and the training drugs as the testing samples, which are blind to the training. Finally, the above procedures were repeated 10 times and the average of predicting performance in all rounds of CV was taken as the final performance.
- In CV2, same to CV1, drugs were randomly divided into 10 equal-size subsets of drugs. One out of 10 drug subsets was selected as the testing drugs and the union of remaining drug subsets was used as the training drugs. We still labeled the drug pairs among the training drugs as the training samples. Distinctively, we discarded the pairs between the testing drugs and the training drugs in both training and testing, but only kept the pairs among the testing drugs as the testing samples, which are unseen to the training as well. Because of the aim of predicting the interaction between two new drugs having no existing interaction yet, two same sets, of which each contained above 10 subsets, were used to strictly reflect the situation. Consequently, the number of possible pairs of drug testing subsets was 10×10. After considering the symmetry of drug pairs, we finally generated $10 + 10 \times (10 - 1)/2 = 55$ rounds of CV during CV2.

Note that there is a significant difference between CV1 and CV2. CV1 simulates the scenario of the task that has the need to determine whether the given new drugs (having no known interactions) would interact with one or more drugs in the training set. Whereas CV2 imitates the scenario of another task that requires the determination of whether some of the new drugs would interact with each other. The task corresponding to CV1 is useful before one tries to extend existing co-prescriptions by adding new drugs. The task corresponding to CV2 is useful when one tries to make novel co-prescriptions and is also one of the most important steps towards to discover the mechanism of forming DDIs.

To obtain the robust predicting performance, both CV1 and CV2 was repeated 50 times under different random seeds. Their final performance was reported by the average performance over 50 repetitions of CV.

Two measures were adopted to assess the predicting performance, including the area under receiver operating characteristic curve (AUC) and the area under precision-recall curve (AUPR). In the prediction of conventional binary DDIs, since positive and negative samples are interactions and non-interactions, both AUC and AUPR can be calculated by comparing their predicted scores. In the prediction of comprehensive DDIs, to adopt the same way of calculating AUC and AUPR, both enhancive and degressive DDIs first are labeled as positive samples and non-interactions as negative samples. Then, the union of the predicted score of enhancive DDIs and the minus of the predicted score of degressive DDIs is compared with those scores of non-interactions in the same way as that in measuring conventional prediction.

3.2   Prediction

Since TMFUF has one tunable parameter (the number of latent factors) in PLSR, we first investigated how it influences the prediction. In detail, we selected its value from the list {1, 5, 10, 20, 30, 40, 50, 60, 70, 80, 90, 100, 150} and picked out the value (60) corresponding to the best prediction conventional binary DDIs in the task T_1 (Fig. 2) to perform sequential experiments. To perform the conventional prediction, we simply turn those entries equal to -1 into $+1$ in the interaction matrix **A**, so as to obtain the conventional drug-drug interaction matrix.

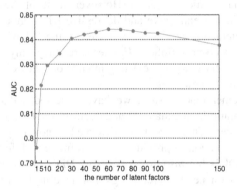

Fig. 2. Illustration of determining the best value for the number of latent factors.

Then, we run TMFUF for four scenarios with respect to two kinds of DDI prediction and two kinds of prediction tasks (Table 2). In T_1, we compared TMFUF with two state-of-the-art methods, Naïve similarity-based approach [6] and label propagation-based approach [5] when predicting conventional DDIs because these methods can only work for this scenario. For a newly given drug, the naïve similarity-based approach directly calculates the summation of its similarities to the drugs interacting with a specific drug to indicate how likely it interacts with the concerned drug. While the latter approach models a set of DDIs and non-DDIs as a binary label network and directly applies label propagation algorithm to infer potential DDIs.

The results show that TMFUF is significantly superior to them. More importantly, two remarkable conclusions can be drawn, (1) comprehensive prediction is more difficult than conventional prediction; (2) DDI prediction in task T_2 is more difficult than that in T_1.

Table 2. Comparison with state-of-the-art methods

Method	Prediction type	Predicting task	AUC	AUPR
Naïve similarity [6]	Conventional	T1	0.759 ± 0.001	0.302 ± 0.002
Label propagation [5]	Conventional	T1	0.774 ± 0.001	0.326 ± 0.002
TMFUF	Conventional	T1	$\mathbf{0.842} \pm \mathbf{0.002}$	$\mathbf{0.526} \pm \mathbf{0.006}$
TMFUF	Comprehensive	T1	0.733 ± 0.004	0.483 ± 0.007
TMFUF	Conventional	T2	0.702 ± 0.004	0.303 ± 0.005
TMFUF	Comprehensive	T2	0.577 ± 0.005	0.246 ± 0.005

4   Conclusion

Discovering and predicting potential DDIs will not only prevent life-threatening consequence in clinical practice, but also prompt safe drug co-prescriptions for better treatments. Computational approaches are able to screen potential DDIs on a large scale before drugs enter into medicine market. However, none of them can predict comprehensive DDIs, including enhancive and degressive DDIs, though it is important to know whether the interaction increases or decreases the behavior of the interacting drugs before making a co-prescription. Furthermore, they only focus on predicting DDIs for new drugs that have none of existing interactions, however, none of them can predict DDIs among those new drugs.

To address abovementioned issues, we have collected a dataset of DDIs from DrugBank, which contains both enhancive and degressive DDIs, and the side effects of the drugs from Offsides. Then, we have designed a novel approach TMFUF for comprehensive DDI prediction. It provides a unified framework for predicting not only conventional binary DDIs but also comprehensive DDIs. Moreover, it enables us to predict potential DDIs for newly given drugs that have no known interaction and to predict potential DDIs among new drugs within the same framework. Experiments on the dataset demonstrate that TMFUF is significantly superior to three state-of-the-art approaches in the conventional binary DDI prediction and also shows an acceptable performance in the comprehensive DDI prediction. In the future, to achieve improved DDI prediction and understand the underlying mechanism of forming DDIs better, more features, (e.g. drug targets, the labels of Anatomical Therapeutic Chemical Classification and chemical structures), should be integrated.

Acknowledgement. This work was supported by RGC Collaborative Research Fund (CRF) of Hong Kong (C1008-16G), National High Technology Research and Development Program of China (No. 2015AA016008), the Fundamental Research Funds for the Central Universities of China (No. 3102015ZY081) and the Program of Peak Experience of NWPU (2016).

References

1. Wienkers, L.C., Heath, T.G.: Predicting in vivo drug interactions from in vitro drug discovery data. Nat. Rev. Drug Discov. **4**, 825–833 (2005)
2. Leape, L.L., Bates, D.W., Cullen, D.J., Cooper, J., Demonaco, H.J., Gallivan, T., Hallisey, R., Ives, J., Laird, N., Laffel, G., et al.: Systems analysis of adverse drug events. ADE Prevention Study Group. JAMA **274**, 35–43 (1995)
3. Veith, H., Southall, N., Huang, R., James, T., Fayne, D., Artemenko, N., Shen, M., Inglese, J., Austin, C.P., Lloyd, D.G., Auld, D.S.: Comprehensive characterization of cytochrome P450 isozyme selectivity across chemical libraries. Nat. Biotechnol. **27**, 1050–1055 (2009)
4. Huang, S.M., Temple, R., Throckmorton, D.C., Lesko, L.J.: Drug interaction studies: study design, data analysis, and implications for dosing and labeling. Clin. Pharmacol. Ther. **81**, 298–304 (2007)
5. Zhang, P., Wang, F., Hu, J., Sorrentino, R.: Label propagation prediction of drug-drug interactions based on clinical side effects. Sci. Rep. **5**, 12339 (2015)
6. Vilar, S., Uriarte, E., Santana, L., Lorberbaum, T., Hripcsak, G., Friedman, C., Tatonetti, N. P.: Similarity-based modeling in large-scale prediction of drug-drug interactions. Nat. Protoc. **9**, 2147–2163 (2014)
7. Luo, H., Zhang, P., Huang, H., Huang, J., Kao, E., Shi, L., He, L., Yang, L.: DDI-CPI, a server that predicts drug-drug interactions through implementing the chemical-protein interactome. Nucleic Acids Res. **42**, 46–52 (2014)
8. Cheng, F., Zhao, Z.: Machine learning-based prediction of drug-drug interactions by integrating drug phenotypic, therapeutic, chemical, and genomic properties. J. Am. Med. Inform. Assoc.: JAMIA **21**, e278–e286 (2014)
9. Koch-Weser, J.: Serum drug concentrations in clinical perspective. Ther. Drug Monit. **3**, 3–16 (1981)
10. Law, V., Knox, C., Djoumbou, Y., Jewison, T., Guo, A.C., Liu, Y., Maciejewski, A., Arndt, D., Wilson, M., Neveu, V., Tang, A., Gabriel, G., Ly, C., Adamjee, S., Dame, Z.T., Han, B., Zhou, Y., Wishart, D.S.: DrugBank 4.0: shedding new light on drug metabolism. Nucleic Acids Res. **42**, D1091–D1097 (2014)
11. Tatonetti, N.P., Ye, P.P., Daneshjou, R., Altman, R.B.: Data-driven prediction of drug effects and interactions. Sci. Transl. Med. **4**, 125ra131 (2012)

Diagnosis of Auditory Pathologies with Hidden Markov Models

Lilia Lazli[1,2,3(✉)], Mounir Boukadoum[2], Mohamed-Tayeb Laskri[3], and Otmane Aït-Mohamed[4]

[1] Department of Electrical Engineering, ÉTS, Montreal, Canada
lilia.lazli.l@ens.etsmtl.ca
[2] COFAMIC Laboratory, Department of Computer Science, UQAM, Montreal, Canada
[3] Department of Computer Science, UBMA, Annaba, Algeria
[4] Department of Electrical Engineering and Computer Science, Concordia University, Montreal, Canada

Abstract. Since about twenty years, the otoneurology functional exploration possesses auditory tool to analyze objectively the state of the nervous conduction of additive pathway. In this paper, we present a new classification approach based on the Hidden Markov Models (HMM) which used to design a Computer aided medical diagnostic (CAMD) tool that asserts auditory pathologies based on Brain-stem Evoked Response Auditory based biomedical test, which provides an effective measure of the integrity of the auditory pathway. Case study, experimental results and comparison with a conventional neural networks models have been reported and discussed.

Keywords: Auditory evoked potentials · Computer aided medical diagnosis · Bayesian models · Vector quantization

1 Introduction

Dealing with expert (human) knowledge consideration, intelligent diagnosis systems or Computer-Aided Diagnosis (CAD) dilemma is one of the most interesting, but also one of the most difficult problems. Among difficulties contributing to challenging nature of this problem, one can mention the need of several knowledge representations, fine classification and decision-making with a certain degree of reliability.

In many applications of interest, it is desirable for the system to not only identify the possible causes of the problem, but also to suggest suitable remedies (systems capable of advising) or to give a reliability rate of the identification of possible causes.

Recently, several decision support systems and intelligent systems have been developed [9, 10] and the diagnosis approaches based on such intelligent systems have been developed for biomedicine applications [11–15]. Indeed, several approaches have been developed to analyze and classify biomedicine signals: electroencephalography signals [12], electrocardiogram signals [13], and particularly signals based on Auditory Brainstem Response (ABR) test, which is a test for hearing and brain (neurological) functioning [11, 16–18].

© Springer International Publishing AG 2017
I. Rojas and F. Ortuño (Eds.): IWBBIO 2017, Part I, LNBI 10208, pp. 118–133, 2017.
DOI: 10.1007/978-3-319-56148-6_10

The analysis and recognition of ABR signals is a medical problem of great importance, since it is the best known technique of the auditory organs evaluation. The task of construction of fully automatic method of ABR recognition present considerable technical difficulties, because the signals are in general hardly readable, and in particular the evaluation of the data part obtained for low intensities of the audio stimulus is especially difficult. It can be assumed that the methods of analysis and recognition of ABR signals can be of some interest to other investigators, not necessarily directly interested in audiology, but trying to cope with the difficulties of interpretation and recognition of totally different signals.

The aim of this work is absolutely not to replace specialized human but to suggest a decision support system with a satisfactory reliability degree for CAD systems. We present in this paper, an original approach which is suggested for CAD systems and applied in biomedicine to auditory diagnosis, based on ABR test.

We propose to use the Bayesian models for classification of electrical signals, with come from a medical test, these are called Auditory Evoked Potentials (AEP). AEP are scalp-recorded electrical responses of the brain elicited by acoustical stimuli. Indeed, since about twenty years, the otoneurology functional exploration possesses a tool to analyze objectively the state of the nervous conduction of additive pathway. The AEP's classification is a first step in the development of a diagnosis tool assisting the medical expert. The classification of these signals presents some problems, because of the difficulty to distinguish one class of signal from the others. The results can be different for different test session for the same patient. Today, taking into account the progress accomplished in the area of intelligent computation or artificial intelligence, it becomes conceivable to develop a diagnosis tool assisting the medical expert. One of the first steps in the development of such tool is the AEP signal classification.

Then we proposed to use the Hidden Markov Models (HMM) and we attempt to illustrate some applications of this theory to real problems to match complex patterns problems as those related to AEP biomedical diagnosis or those linked to social behavior modeling. We focus also on the K-Means clustering algorithm which it is one of the most used iterative partitioned clustering algorithms based in vector quantization. In particularly, we review the theory of discrete HMM and show how the concept of hidden states, where the observation sequences provided using the k-means algorithm, can be used effectively for AEP classification.

In the pattern recognition domain, HMM techniques hold an important place, there are two reasons why the HMM has occurred. First the models are very tick in mathematical structure and hence can form the theoretical basis for use in a wide range of applications. Second the models, when applied properly, work very well in practice for several important applications. Nowadays, HMM are considered as a specific form of dynamic Bayesian networks based on the theory of Bayes [22]. They are a dominant technique for sequence analysis and they owe their success to the existence of many efficient and reliable algorithms.

HMM are used in many areas in modern sciences or engineering applications, e.g. in temporal pattern recognition such as speech, handwriting, gesture recognition, part-of-speech tagging, musical score following, partial discharges. Other areas where the use of HMM and derivatives becomes more and more interesting are biosciences, bioinformatics and genetics [19–25].

The organization of this paper is as follows. In the Sects. 2 and 3, we introduce the theory and the foundation of HMM and vector quantization. In the Sect. 4, we present the AEP signals and we describe our biomedical pattern classifier, implemented with HMM and vector quantization ideas. In the same section, we present the classification results obtained by using a database of 213 AEP like waveforms. A comparison to alternative implementations using neural networks methods is presented. Finally in Sect. 5 we summarize the ideas, discus the presented technique's potential to deal with social behavior modeling and give the prospects that follow from our work.

2 Foundation of HMM

An HMM system is typically characterized by the following quantities [6, 7].

2.1 Elements of an HMM

We define the following notation for an HMM:

1. A set of states S_i, $i \in [1, N]$ that are unobservable though there is often a physical meaning attached to them.
2. A set of M observations. In a discrete HMM [5]. M is the number of codebook vectors, or the number of all possible observations. This implies that any observation, v_t, is quantized into the set $\{x_1, x_2, \ldots x_m\}$ where x_m is the mth codebook vector (to see Sect. 3).
3. A set of state "transition" probabilities represented by matrix $A = [a_{ij}]$ where $a_{ij} = P(q_t = S_j | q_{t-1} = S_i, \lambda)$ with q_t being the state visited at time t, S_i is state i and λ is the model defined by the object class and the corresponding training data.
4. A set of observation probabilities represented by matrix $B = [b_i(x_k)]$ where $b_i(x_k) = P(x_k | q_t = S_i, \lambda)$ is the "emission" probability of the kth quantized observation, x_k, at time t from state S_i if the emission processes are assemply reduces to $P(x_k | S_i, \lambda)$.
5. An initial state distribution or the probability of starting in a given state, i.e., $\pi_j = P(q_1 = S_j | \lambda)$.

Given the number of states N, and the number of observations M, the parameters A, B and π represent the model λ. There are three main issues [5] in order to maximize the performance of the HMM and identify the model in practical applications. These are briefly mentioned in the following. An in depth discussion on these topics can be found in [5–7].

2.2 The Three Basic Problems of HMM

Given the form of HMM of the previous section, there are three basic problems of interest that must be solved for the model to be useful in real-word applications. These problems are the following:

Problem 1. Given the observation sequence $O = o_1 o_2 \dots o_T$, and a model $\lambda = (A, B, \pi)$, how do we efficiently compute $P(O|\lambda)$, the probability of the observation sequence, given the model?

Problem 2. Given the observation sequence $O = o_1 o_2 \dots o_T$, and the model λ, how do we choose a corresponding state sequence $Q = q_1 q_2 \dots q_T$ which is optimal in some meaningful sense (i.e., best "explains" the observations)?

Problem 3. How do we adjust the model parameters $\lambda = (A, B, \pi)$ to maximize P $(O|\lambda)$?

2.3 Solutions to the Three Problems

Solution to Problem 1: Computing Model Probability. The answer to the first problem is the forward-backward procedure [5]. From this procedure we can find the forward variable $\alpha_t(i)$ and it is defined as $\alpha_t(i) = P(O_1 O_2 \dots O_t, q_t = S_i | \lambda)$. This is the probability of the partial observation sequence, $O_1 O_2 \dots O_t$, (until time t) and state S_i at time t given the model λ. The forward-backward procedure also provides us with a backward variable $\beta_t(j) = P(O_1 O_2 \dots O_t | q_t = S_i, \lambda)$ which gives the probability of the · partial observation sequence from t + 1 to the end, given state Si at time t and the model λ. Even though the backward variable is not needed for the first problem it becomes useful when solving problem 3 [5].

Solution to Problem 2: Optimal State Sequence. Problem number 2 is a matter of finding the best state sequence that best fits with the observation. The Viterbi algorithm manages this and finds the single best state sequence, $Q = (q_1 q_2 \dots q_t)$, for the given observation $O = (O_1 O_2 \dots O_t)$ [5, 6].

Solution to Problem 3: Maximization of P(O/λ). The third problem is to adjust the HMM parameters to maximize the probability of the observation sequence. If given an finite observation sequence there is no optimal way of estimating the models parameters. However, it is possible to chose $\lambda = (A, B, \pi)$ such that it is locally maximized for $P(O|\lambda)$ by using the Baum-Welch algorithm [5]. The Baum-Welch algorithm works by assigning initial probabilities to all the parameters. Then, until the training converges, it adjusts the probabilities of the parameters so as to increase the probability the model assigns to the training set [7].

The Baum-Welch algorithm (or Baum-Welch expectation maximization algorithm) makes use of both the forward variable $\alpha_t(i)$ and the backward variable $\beta_t(j)$ when it determines updated parameters for the HMM. Because of this the Baum-Welch algorithm is also known as the Forward-Backward algorithm.

To properly estimate the local maximum for $P(O|\lambda)$ the Baum-Welch algorithm needs several iterations. The algorithm will either be repeated a predetermined number of times, or until the local maximum is found. The local maximum is found when the difference between $P(O|\lambda_{new})$ and $P(O|\lambda_{old})$ reaches a certain value [5].

3 Vector Quantization

Several approaches to find groups in a given database have been developed in literature, but we focus on the K-Means algorithm (vector quantization) [2] as it is one of the
most used iterative partitional clustering algorithms and because it may also be used to
initialize more expensive clustering algorithms (e.g., the EM algorithm).

k-means is one of the simplest unsupervised learning algorithms that solve the
well-known clustering problem. The procedure follows a simple and easy way to classify
a given data set through a certain number of clusters (assume k clusters) fixed *a priori*.

As can be seen in Fig. 1 where the pseudo-code is presented, the k-means algorithm
is provided somehow with an initial partition of the database and the centroids of these
initial clusters are calculated. Then, the instances of the database are relocated to the
cluster represented by the nearest centroid in an attempt to reduce the square-error. This
relocation step (step 3) changes its cluster membership, and then the centroids of the
clusters C_s and C_t and the square-error should be recomputed. This process is repeated
until convergence, that is, until the square-error cannot be further reduced which means
no instance changes its cluster membership [1].

Step 1. Select somehow an initial partition of the
database in K clusters $\{C_1, ..., C_K\}$.

Step 2. Calculate cluster centroid $\overline{w}_i = \dfrac{1}{k_i}\sum_{j=1}^{k_i} w_{ij}$, $i = 1,...K$

Step 3. for every w_i in the database and following the
instance order do
Step 3.1. Reassign instance w_i to its closest cluster
centroid, $w_i \in C_s$ is moved from C_s to C_t if
$\left\|w_i - \overline{w}_t\right\| \le \left\|w_i - \overline{w}_j\right\|$ for all $j = 1,...,K$, $j \ne s$
Step 3.2. Recalculate centroids for clusters C_s and C_t

Step 4. If cluster membership is stabilized then stop
            else go to step 3.

Fig. 1. The pseudo-code of the k-means algorithm.

For the case in which we wish to use an HMM with a discrete observation symbol
density, rather than the continuous vectors above, a vector quantized VQ is required to
map each continuous observation vector into a discrete codebook index. Once the
codebook of vectors has been obtained, the mapping between continuous vectors and
codebook indices becomes a simple nearest neighbor computation, i.e., the continuous
vector is assigned the index of the nearest codebook vector. Thus the major issue in VQ
is the design of an appropriate codebook for quantization.

Fortunately a great deal of work has gone into devising an excellent iterative
procedure for designing codebooks based on having a representative training sequence
of vectors [2]. The procedure basically partitions the training vectors into *M* disjoint

sets (where M is the size of the codebook), represents each such set by a single vector $(v_m, 1 \leq m \leq M)$, which is generally the centroid of the vectors in the training set assigned to the mth region, and then iteratively optimizes the partition and the codebook (i.e., the centroids of each partition). Associated with VQ is a distortion penalty since we are representing an entire region of the vector space by a single vector. Clearly it is advantageous to keep the distortion penalty as small as possible.

4 Validation on Biomedical Classification Paradigm

For our experience, consider using HMM to build a biomedical classifier or a CAMD tool. Assume we have a vocabulary of 3 classes to be recognized: Normal class – Endocochlear class – Retrocochlear class, and that each category is modeled by a discrete HMM. For this type of HMM, there exists a limited number of observations (in our case, number of clusters) which can be made.

Further assume that for each category in the vocabulary, we have a training set of k occurrences (instances) of each category where each instance of the class constitutes an observation sequence, where the observation are some appropriate representation of the characteristics of the class. In order to build a CAMD tool, we perform the following operations:

(1) For each class v in the vocabulary (3 classes in our work), we must build an HMM λ^v, i.e., we must estimate the model parameters (A, B, π) that optimize the likelihood of the training set observation vectors for the vth class.

(2) For each unknown class which is to be recognized, the processing of Fig. 2 must be carried out namely measurement of the observation sequence $O = \{o_1, o_2, \dots, o_T\}$, via a feature analysis of the signal corresponding to the class, followed by calculation of model likelihoods for all possible models, $P(O/\lambda^v)$, $1 \leq v \leq V$, followed by selection of the class whose model likelihood is highest, i.e.,
$$v^* = \arg\max_{1 \leq v \leq V}[P(O/\lambda^v)]$$

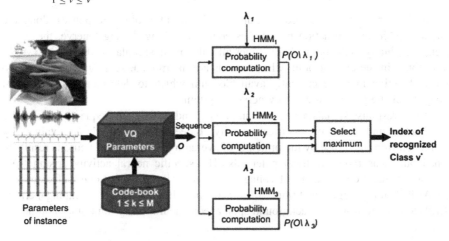

Fig. 2. Block diagram of a biomedical database HMM recognizer.

The probability computation step is generally performed using the Viterbi algorithm (i.e., the maximum likelihood path is used) and requires on the order $V.N^2.T$ computations.

4.1 Background of Brainstem AEP Clinical Test

When a sense organ is stimulated, it generates a string of complex neurophysiology processes. Brainstem auditory evoked potentials (BAEP) are electrical response caused by the brief stimulation of a sense system. The stimulus gives rise to the start of a string of action's potentials that can be recorded on the nerve's course, or from a distance of the activated structures. ABR comprise the early portion (0–12 m-s) of AEP are composed of several waves or peaks. BAEP are generated as follows (see Fig. 3): the patient hears clicking noise or tone bursts through earphones. The use of auditory stimuli evokes an electrical response.

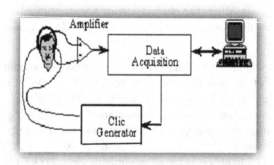

Fig. 3. Brainstem auditory evoked potentials clinical test. (Source: Ref. [3], p. 120)

In fact, the stimulus triggers a number of neurophysiology responses along the auditory pathway. An action potential is conducted along the eight nerve, the brainstem, and finally to the brain. A few times after the initial stimulation, the signal evokes a response in the area of brain where sounds are interpreted. AEP are considered the most objective measure currently available with which to determine the functional integrity of the peripheral auditory nervous system.

These response signals have small amplitude, and so they are frequently masked by the background noise of electrical activity. Indeed, the response is obtained by extraction from the noise by the principle of averaging. The firing of neurons results in small but measurable electrical potentials. The specific neural activity arising from acoustic stimulation, a pattern of voltage fluctuations lasting about one half second, is an AEP. With enough repetitions of an acoustic stimulus, signal averaging permits AEPs to emerge from the background spontaneous neural firing (and other non-neural interferences such as muscle activity and external electromagnetic generators), and they may be visualized in a time-voltage waveform.

Depending upon the type and placement of the recording electrodes, the amount of amplification, the selected filters, and the post-stimulus timeframe, it is possible to detect neural activity arising from structures spanning the auditory nerve to the cortex. Estimating hearing threshold from BAEP signals is a time consuming and labor intensive procedure, and therefore one which recommends itself to computerized automation. The important step is the classification of the signals into Response (R) and No Response (NR) classes, the main difficulties being a poor signal-to-noise ratio and the differentiation of response peaks from artifacts.

The ABR waves or peaks, labelled using Roman numerals I–VII as shown in the Fig. 4, are typically 1 ms apart and have amplitudes of about 100–500 nanovolts. Waves I, III and V are generally considered major peaks, generated by the synchronous electrical activity of the auditory nerve, caudal and rostral auditory brainstem struc-tures, respectively, in response to onset of auditory stimuli. This test provides an effective measure of the integrity of the auditory pathway up to the upper potential level.

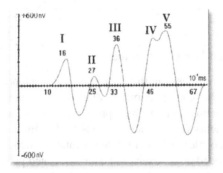

Fig. 4. Perfect AEP (Source: Ref. [3], p. 85)

A technique of extraction, presented in [5] allows us, following 800 acquisitions such as describe before, the visualization of the AEP estimation on averages of 16 acquisitions. Thus, a surface of 50 estimations called Temporal Dynamic of the Cerebral trunk (TDC) can be visualized. The software developed for the acquisition and the processing of the signals is called ELAUDY. It allows us to obtain the average signal, which corresponds to the average of the 800 acquisitions, and the TDC surface. Figure 5 (extracted from [5]) shows two typical surfaces, one for a patient with a normal audition (2-A) and the other one for patient who suffers from an auditory disorder (2-B). This figure shows the large variety of AEP signals even for a same patient. Moreover, this software automatically determinates, from the average signal, the five significant peaks

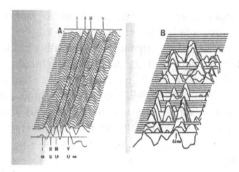

Fig. 5. TDC surfaces (A- Normal patient, B- Patient with auditory disorder) (Source: Ref. [3], p. 85)

and gives the latency of these waves. It also allows us to record a file for each patient, which contains administrative information (address, age,...), the results of the tests and the doctor's conclusions (pathology, cause, confidence's index of the pathology...).

AEP signal and TDC technique are important to diagnosis auditory pathologies. However, medical experts have still to visualize all auditory tests' results before making a diagnosis.

4.2   Classification and Decision-Making

At first, through some examples, an important problem is emphasized to illustrate the problem difficulty of the classification in diagnosis systems. In the biomedicine application described in this section, three patient classes are studied: Retro-cochlear auditory disorder's patients (Retro-cochlear Class: RC), Endo-cochlear auditory disorder's patients (Endocochlear Class: EC), and healthy patients (Normal Class: NC). The AEP signals descended of the exam and their associated pathology are defined in a data base containing the files of 11 185 patients. We chose 3 categories of patients (3 classes) according to the type of their trouble. The categories of patients are the next one:

(1) Normal: the patients of this category have a normal audition (normal class).
(2) Endocochlear: these patients are reached of unrest that touches the part of the ear situated before the cochlea (class Endocochlear).
(3) Retrocochlear: these patients are reached of unrest that touches the part of the ear situated to the level of the cochlea or after the cochlea. (class retrocochlear).

We selected 213 signals (correspondents to patients). So that every process (signal) contains 128 parameters, we were force to respect the values of parameters used in the work describes in the following articles [3, 4] using the LVQ and RBF neural structure respectively. 92 among the 213 signals belong to the normal class, 83, to the class Endocochlear: and 38, to the class retrocochlear. Figure 6 shows two examples of

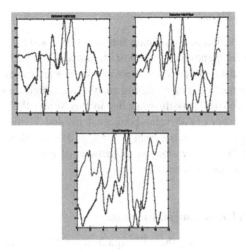

Fig. 6. Two examples of signal representations for RC patients, EC patients, and NC patients.

signal knowledge representations for six patients: RC, EC, and NC. Also, Fig. 7 shows image knowledge representations for the same six patients. These figures illustrate the fact that, signal or image representations could be very similar for patients belonging to different classes, and they could be very different for patients belonging to a same class, demonstrating the difficulty of their classification.

Fig. 7. Two examples of image representations for RC patients, EC patients, and NC patients.

4.3   Analysis of the Signal

Raw BAEPs were amplified and bandpass filtered (100–3000 Hz) to remove the EEG component and high frequency noise. A post stimulus signal of 12.8 ms was sampled at 40 kHz to give 512 data points. Since these raw signals are extremely noisy, standard procedure was to coherently average 1024 of such signals to give a single BAEP signal. This signal can be used for classification but in this study, the signals were further reduced by sampling every eighth value between 1 ms and 11 ms The resulting signal of 50 data points was normalized between 0 and 1. A data set of 321 such input signals was obtained, which included various combinations of hearing impaired and normal subjects and varying stimulus intensities.

4.4   Case Study and Experimental Results

The aim of our work is to classify the AEP signals using HMM models. In our case, the components of input vectors are the samples of the BERA average signals and the output vectors correspond to 3 different possible classes.

To construct our basis of training, we chose the signals corresponding to pathologies indicated like being certain by the physician. All AEP signals come from the same experimental system. In order to value the realized work and for ends of performance comparison with the work of the group describes in the following article [3, 4] and that uses a neural network structure basis of RBF and LVQ networks The basis of training contains 141 signals, of which 25 correspondent to the class retro-cochlear, 55 to the class endocochlear and 61 to the normal class. The ratio of class sizes (the R:NR ratio) in the training set was chosen as 3:1, reflecting the approximate ratio in a clinical setting.

The test set consisted of 213 signals with the same ratio of three R signals to one NR signal. No signals from any of the same subjects used in the training set were included, which added considerably to the difficulty of the learning task.

After the phase of training, when the non-learned signals are presented to the HMM, the corresponding class must be designated.

The convergence of the k-means process has been obtained during the 11th iteration. Figures 8 and 9 illustrate the tradeoff of quantization distortion versus M (on a long scale). Although the distortion steadily decreases as M increases, it can be seen from Fig. 9 that only small decreases in distortion accrue beyond a value of M = 26. Hence HMM with codebook sizes of from M = 26 to 64 vectors have been used in biomedical database recognition experiments using HMM.

Table 1 presents a sample of clustering for two instances taken randomly, after the iteration of convergence. Noting that each instance has been divided in 16 windows that each has 8 parameters. When creating HMMs during this project, we used a HMM implementation for Matlab called "Hidden Markov Model (HMM) Toolbox" for Matlab.

We have three HMM representing respectively: HMM1: normal class; HMM2: endo class; HMM3: retro class. The final parameters of the model that represents the class of normal patients are as follows:

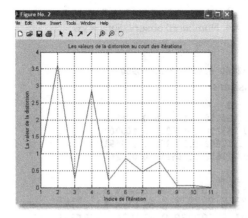

Fig. 8. Process of convergence of AEP basic using the k-means algorithm.

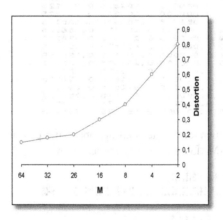

Fig. 9. Curve showing tradeoff of VQ average distortion as a function of the size of the VQ, M (shown of a log scale).

Table 1. The result of clustering of two instances after the iteration of convergence.

Windows	w_1	w_2	w_3	w_4	w_5	w_6	w_7	w_8	w_9	w_{10}	w_{11}	w_{12}	w_{13}	w_{14}	w_{15}	w_{16}
Instance1	14	7	7	14	7	3	7	7	14	7	12	12	12	12	12	14
Instance2	3	3	3	14	3	14	7	7	7	7	12	12	12	8	14	3

Final parameters of HMM1: "Normal class"
Number of states = 4. Number of observations = 26.
Initial Density

1	2	3	4
0.004180	0.482090	0.112750	0.400980

Probability of transitions

	1	2	3	4
1	0.203875	0.496125	0.103875	0.196125
2	0.655800	0.144200	0.100000	0.100000
3	0.274293	0.417642	0.143372	0.164693
4	0.225189	0.371652	0.389914	0.013245

Probability of emission

C01	0.016430	0.107239	0.103475	0.696225
C02	0.013061	0.000879	0.282874	0.726155
C03	0.056308	0.000000	0.243875	0.797125
C04	0.066501	0.000000	0.214875	0.716125
C05	0.000000	0.262201	0.903875	0.000024
C06	0.018004	0.000000	0.402875	0.798125
C07	0.016239	0.000000	0.243875	0.156125
C08	0.014209	0.000000	0.278875	0.790055
C09	0.011219	0.099780	0.313875	0.717128
C10	0.015577	0.000000	0.200475	0.096125
C11	0.000000	0.000073	0.414975	0.792873
C12	0.086889	0.000000	0.247875	0.836125
C13	0.053704	0.000000	0.202875	0.796125
C14	0.127663	0.000000	0.103875	0.792825
C15	0.000000	0.081718	0.201465	0.438125
C16	0.047621	0.001531	0.183475	0.866125
C17	0.024579	0.000000	0.243875	0.986125
C18	0.111468	0.000000	0.682575	0.481125
C19	0.145314	0.000000	0.103875	0.796859
C20	0.123250	0.000000	0.202875	0.835125
C21	0.003183	0.097260	0.549875	0.856125
C22	0.032434	0.000000	0.271875	0.856125
C23	0.000000	0.000000	0.472875	0.296125
C24	0.007458	0.000000	0.421975	0.286125
C25	0.004652	0.000241	0.252175	0.176125
C26	0.004236	0.000000	0.483875	0.795325

The most likely sequence of states and the observing probability of "Instance1" to the HMM1, HMM2, HMM3 respectively are as follows:

HMM1 [1,1,4,2,2,2,2,3,4,4,4,3,3,1,1,1]
P(Obs [14,7,7,14,3,7,7,14,7,12,12,12,,12,14])= 3.527192359999999995e-05

HMM2 [1,1,2,3,4,4,4,3,2,1,2,2,4,4,4,3]
P(Obs [14,7,7,14,3,7,7,14,7,12,12,12,,12,14])= 2.55189279999999995e-05

HMM3 [1,1,2,3,2,2,4,3,3,4,2,2,1,1,1]
P(Obs [14,7,7,14,3,7,7,14,7,12,12,12,,12,14])= 3.48723459999999995e-05

Deduction: "Normal patient"

For the phase of generalization, with application of HMM, the basis of training has been learned correctly at the time of the phase of generalization with 100% rate. The results gotten for the recognition system of the test biomedical basis with using the HMM, are presented like follows.

(1) 98.38% for the normal class.
(2) 58.93% for the class Endocochlear:
(3) 96.15% for the class retrocochlear:

Table 2. Results of classification of the AEP

Evaluation set				Average rate %
Recognizer type	Normal %	Endocochlear %	Retrocochlear %	
RBF	68	58	61	62.5
LVQ	57	62	72	63.7
HMM	**98.38**	**58.93**	**96.15**	**84.5**

Thus, the average rate of success is of **84.48%** for the totality of the data base in relation to the rate of classification of **63.7%** **and 62.5%** of the systems using the LVQ and RBF neural structure respectively describes in the following articles [3, 4]. So the results gotten for every class with the system proposed in this paper in comparison with those of the neural networks structure, is presented in the following table (to see Table 2).

The results with a number of 26 clusters which is the number that we had to get quite satisfactory classification results are presented in the Table 2. The first observation for the results is that the application of HMM for this biomedical BDD is has proven to be very effective for both normal and retro classes with a classification rate of 98.38 and 96.15 respectively and a performance degradation for the endo class with a rate of 58.93. In fact, the signals on which we work can be very different within the endo class compared to other classes. They could be very different for patients belonging to a same class, demonstrating the difficulty of their classification. One can notice, by comparing these results with those obtained with the LVQ and RBF connectionist approaches that include an improvement of the overall performance.

5   Conclusion and Perspective

In this paper, Bayesian approaches is suggested for CAD systems in a biomedicine application: auditory diagnosis based on ABR test. In fact, the aim is then to achieve an efficient and reliable CAD system for three classes: two auditory pathologies RC and EC and normal auditory NC. Implementation and experimental results are presented and discussed.

The setting of this experimentation is off line and it remains again many to make. The results that we got only concern a qualitative approach in a static context. Nevertheless, this preliminary survey will allow us to propose other models that will allow palliating the insufficiencies of HMM.

The main idea is to define a fusion scheme: cooperation of HMM with the multi-network structure in order to succeed to a hybrid model by those providing more effective results than those proposed in this paper. On the other hand, we have studied the potentiality of HMM modeling's application in estimating artificial bots' behavior state in a social negotiation context in the framework of a European Sistine project [8].

The both state in that case may be neutral, aggressive or conciliate. These both is used in the development of innovative training practices for the teaching of negotiation, leading to the development of new teaching and evaluation methodologies. If such technique show a number of strong theoretical advantages, unfortunately, its implementation in a real time interactive multi-user tool still remains inappropriate.

Acknowledgments. This research was supported by grants from "UNESCO for women in Science" and from "ReSMiQ" of Quebec.

References

1. Bradeley, P.S., Fayyad, U.M.: Refining initial points for k-means clustering. In: Proceedings of the Fifteenth International Conference on Machine Learning, pp. 91–99. Morgan Kaufmann Publishers, Inc., San Francisco (1998)
2. Cheeseman, P., Stutz, J.: Bayesian classification (AutoClass): theory and results. In: Advances in Knowledge Discovery and Data Mining, pp. 153–198. AAAI Press, Menlo Park (1996)
3. Dujardin, A.S.: Pertinence d'une approche hybride multi-neuronale dans la résolution de problèmes liés au diagnostic industriel ou médical", Ph.D. thesis, I^2S laboratory, IUT of "Sénart Fontainebleau", University of Paris XII, Avenue Pierre Point, 77127 Lieusaint, France (2003)
4. Dujardin, A.-S., Amarger, V., Madani, K., Adam, O., Motsch, J.-F.: Multi-neural network approach for classification of brainstem evoked response auditory. In: Mira, J., Sánchez-Andrés, J.V. (eds.) IWANN 1999. LNCS, vol. 1607, pp. 255–264. Springer, Heidelberg (1999). doi:10.1007/BFb0100492
5. Motsh, J.F.: La dynamique temporelle du tronc' cérébral: Recueil, extraction et analyse optimale des potentiels évoqués auditifs du tronc cérébral. Thesis, University of Créteil, Paris XII (1987)
6. Rabiner, L.R.: A tutorial on hidden Markov models and selected applications in speech recognition. Proc. IEEE **77**(2), 257–286 (1989)
7. Robenson, M., Azimi-Sadjadi, M.R., Salazar, J.: Multi-aspect target discrimination using hidden Markov models and neural networks. IEEE Trans. Neural Networks **16**(2), 447–459 (2005)
8. Sistine website www.sistine.net
9. Turban, E., Aronson, J.E.: Decision support systems and intelligent systems, Int edn. Prentice-Hall, Upper Saddle River (2001)
10. Karray, F.O., De Silva, C.: Soft Computing and Intelligent Systems Design, Theory. Tools and Applications. Addison Wesley, Boston (2004). Pearson Ed. Limited. ISBN 0-321-11617-8
11. Piater, J.H., Stuchlik, F., von Specht, H., Piater, R.: An adaptable algorithm modeling human procedure in BAEP analysis. Comput. Biomed. Res. **335–353**, 28 (1995)
12. Vuckovic, A., Radivojevic, V., Chen, A.C.N., Popovic, D.: Automatic recognition of alertness and drowsiness from EEG by an artificial neural network. Med. Eng. Phys. **24**(5), 349–360 (2002)
13. Wolf, A., Barbosa, C.H., Monteiro, E.C., Vellasco, M.: Multiple MLP neural networks applied on the determination of segment limits in ECG Signals. In: Mira, J., Álvarez, José R. (eds.) IWANN 2003. LNCS, vol. 2687, pp. 607–614. Springer, Heidelberg (2003). doi:10. 1007/3-540-44869-1_77
14. Chohra, A., Kanaoui, N., Amarger, V.: A soft computing based approach using signal to-image conversion for computer aided medical diagnosis (CAMD). In: Saeed, K., Pejas, J. (eds.) Information Processing and Security Systems, pp. 365–374. Springer, Heidelbreg (2005)

15. Yan, H., Jiang, Y., Zheng, J., Peng, C., Li, Q.: A multilayer perceptron-based medical support system for heart disease diagnosis. Expert Syst. Appl. **30**, 272–281 (2005)
16. Don, M., Masuda, A., Nelson, R., Brackmann, D.: Successful detection of small acoustic tumors using the stacked derived-band auditory brain stem response amplitude. Am. J. Otol. **18**(5), 608–621 (1997). Elsevier
17. Vannier, E., Adam, O., Motsch, J.F.: Objective detection of brainstem auditory evoked potentials with a priori information from higher presentation levels. Artif. Intell. Med. **25**, 283–301 (2002)
18. Bradley, A.P., Wilson, W.J.: On wavelet analysis of auditory evoked potentials. Clin. Neurophysiol. **115**, 1114–1128 (2004)
19. Sha, F., Saul, L.: Comparison of large margin training to other discriminative method for phonetic recognition by hidden Markov models. In: Proceedings of ICASSP 2007, Honolulu, Hawaii (2007)
20. Sha, F., Saul, L.: Comparison of large margin training to other discriminative method for phonetic recognition by hidden Markov models. In: Proceedings of the ICASSP 2007, Honolulu, Hawaii, pp. IV.313–IV.316 (2007). doi:10.1109/ICASSP.2007.366912
21. Al-Ani, T., Hamam, Y.: A low complexity simulated annealing approach for training hidden Markov models. Int. J. Oper. Res. **8**(4), 483–510 (2010)
22. Cheng, C.-C.; Sha, F. & Saul, L.K.: A fast online algorithm for large margin training of continuous-density hidden Markov models. In Proceedings of the Tenth Annual Conference of the International Speech Communication Association (Interspeech-2009), pp. 668–671. Brighton (2009)
23. Ince, H. T., Weber, G.W.: Analysis of Bauspar System and Model Based Clustering with Hidden Markov Models Term Project in MSc Program Financial Mathematics – Life Insurance, Institute of Applied Mathematics METU (2005)
24. Kouemou, G. et al.: radar target classification in littoral environment with HMMs combined with a track based classifier. In: Radar Conference, Adelaide, Australia (2008)
25. Karmakar, T., Khandoker, C.K., Palaniswami, A.H.: Automatic recognition of obstructive sleep apnoea syndrome using power spectral analysis of electrocardiogram and hidden Markov models. In: International Conference on Intelligent Sensors, Sensor Networks and Information Processing (ISSNIP), Sydney, NSW, 15–18 December, pp. 285–290 (2008)

Biclustering Based on Collinear Patterns

Leon Bobrowski[✉]

Faculty of Computer Science, Białystok University of Technology,
Białystok, Poland
l.bobrowski@pb.edu.pl

Abstract. Data mining technique based on minimization of the convex and piecewise linear (*CPL*) criterion functions can be used to extract collinear (flat) patterns from large, multidimensional data sets. Flat patterns consist of data vectors located on planes in a multidimensional feature space. Data subsets located on such planes can represent linear interactions between multiple variables (features). New method of collinear biclustering can also be developed through this technique.

Keywords: Data mining · Flat patterns · *CPL* criterion functions · Biclustering · Linear interactions

1 Introduction

Patterns (regularities) can be discovered in large data sets by using various tools of pattern recognition [1] and data mining [2]. The overall goal of the data mining process is to obtain useful information contained in a given data set on the basis of extracted patterns.

We assume that data sets are composed of feature vectors which represent particular objects (events, phenomena) in a standard way. Different patterns composed of numerous feature vectors can be extracted from the same data set by using different computational tools serving for example for cluster analysis, anomaly detection or interaction mining. Biclustering techniques are currently being developed to explore genomic data [3]. Biclustering procedures are aimed to extract subsets of feature vectors from data sets, and simultaneously to extract subsets of features characteristic for a particular pattern.

Collinear (flat) patterns can be described as numerous subsets of feature vectors situated on planes in vertexical feature subspaces of different dimensionality [4, 5]. Vertexical planes are linked to degenerated vertices in parameter space. Flat patterns can be extracted from large data sets through minimization of a special type of the convex and piecewise linear (*CPL*) criterion function [6]. The technique of the flat patterns extraction based on minimization of the *CPL* criterion function can be compared with the techniques based on the Hough transform [7]. The collinear patterns extraction on the basis of the *CPL* criterion function minimization also provides the opportunity to develop a new method of collinear biclustering.

Theoretical properties of the proposed method of collineear biclustering are analysed in the presented paper.

I. Rojas and F. Ortuño (Eds.): IWBBIO 2017, Part I, LNBI 10208, pp. 134–144, 2017.
DOI: 10.1007/978-3-319-56148-6_11

2 Dual Planes and Vertices in the Parameter Space

Let us assume that each of m objects O_j from a given database is represented by the n-dimensional feature vector $\mathbf{x}_j = [x_{j,1}, \ldots, x_{j,n}]^T$ belonging to the feature space $F[n]$ ($\mathbf{x}_j \in F[n]$). The learning set C consists of m feature vectors $\mathbf{x}_j$:

$$C = \{\mathbf{x}_j\}, \; where \; j = 1, \ldots, m \tag{1}$$

Components $x_{j,i}$ of the feature vector $\mathbf{x}_j$ can be treated as numerical results of n standardized examinations of the j-th object O_j, where $x_{j,i} \in R$ or $x_{j,i} \in \{0, 1\}$.

Feature vectors $\mathbf{x}_j$ from the set C (1) allow to define the following dual hyperplanes h_j^1 in the parameter space R^n ($\mathbf{w} \in R^n$):

$$(\forall \mathbf{x}_j \in C) \quad h_j^1 = \left\{ \mathbf{w} : \mathbf{x}_j^T \mathbf{w} = 1 \right\} \tag{2}$$

where $\mathbf{w}_j = [w_1, \ldots, w_n]^T$ is the parameter (*weight*) vector ($\mathbf{w} \in R^n$).

Each of n unit vectors $\mathbf{e}_i = [0, \ldots, 1, \ldots, 0]^T$ defines the following hyperplane h_i^0 in the n-dimensional parameter space R^n:

$$(\forall i \in \{1, \ldots, n\}) \quad h_i^0 = \left\{ \mathbf{w} : \mathbf{e}_i^T \mathbf{w} = 0 \right\} = \{ \mathbf{w} : w_i = 0 \} \tag{3}$$

Let us consider the set S_k of r_k linearly independent feature vectors $\mathbf{x}_j$ ($j \in J_k$) and $n - r_k$ unit vectors $\mathbf{e}_j$ ($i \in I_k$):

$$S_k = \left\{ \mathbf{x}_j : j \in J_k \right\} \cup \{ \mathbf{e}_i : i \in I_k \} \tag{4}$$

The intersection point of the r_k hyperplanes h_j (2) defined by the feature vectors $\mathbf{x}_j$ ($j \in J_k$) and the $n - r_k$ hyperplanes h_i^0 (3) defined by the unit vectors $\mathbf{e}_j$ ($i \in I_k$) from the set S_k (4) is called the k-th *vertex* $\mathbf{w}_k$ in the parameter space R^n. The vertex $\mathbf{w}_k$ can be defined by the following linear equations:

$$(\forall j \in J_k) \quad \mathbf{w}_k^T \mathbf{x}_j = 1 \tag{5}$$

and

$$(\forall i \in I_k) \quad \mathbf{w}_k^T \mathbf{e}_i = 0 \tag{6}$$

The Eqs. (5) and (6) can be represented in the below matrix form:

$$\mathbf{B}_k \mathbf{w}_k = \mathbf{1}' = [1, \ldots, 1, 0, \ldots, 0]^T \tag{7}$$

where $\mathbf{B}_k$ is the square matrix constituting the k-th *basis* linked to the vertex $\mathbf{w}_k$:

$$\mathbf{B}_k = \left[\mathbf{x}_{j(1)}, \ldots, \mathbf{x}_{j(rk)}, \ \mathbf{e}_{i(rk+1)}, \ldots, \ \mathbf{e}_{i(n)} \right]^T \tag{8}$$

and

$$\mathbf{w}_k = \mathbf{B}_k^{-1} \mathbf{1}' \tag{9}$$

Definition 1: The *rank* r_k ($1 \leq r_k \leq n$) of the k-th vertex $(\mathbf{w}_k [w_{k,1}, \ldots, w_{k,n}]^T$ (9) is defined as the number of the non-zero components $w_{k,i}$ ($w_{k,i} \neq 0$).

Definition 2: The *degree of degeneration* d_k of the vertex $\mathbf{w}_k$ (9) of the rank r_k is defined as the number $d_k = m_k - r_k$, where m_k is the number of such feature vectors $\mathbf{x}_j$ from the set C (1), which define the hyperplanes h_j^1 (2) passing through this vertex $(\mathbf{w}_k^T \mathbf{x}_j = 1)$. The vertex $\mathbf{w}_k$ (9) is *degenerated* if the degree of degeneration d_k is greater than zero ($d_k > 0$).

Definition 3: The basis $\mathbf{B}_k$ (8) is *regular* if and only if the number $n - r_k$ of the unit vectors $\mathbf{e}_i$ ($i \in I_k$ (6)) in the matrix $\mathbf{B}_k$ is equal to the number of such components $w_{k,i}$ of the vector $\mathbf{w}_k = [w_{k,1}, \ldots, w_{k,n}]^T$ (9) which are equal to the zero ($w_{k,i} = 0$).

3   Vertexical Planes in Feature Space

The hyperplane $H(\mathbf{w}, \theta)$ in the n - dimensional feature space $F[n]$ can be defined in the following manner [1]:

$$H(\mathbf{w}, \theta) = \left\{ \mathbf{x} : \mathbf{w}^T \mathbf{x} = \theta \right\} \tag{10}$$

where $\mathbf{w}$ is the *weight vector* ($\mathbf{w} \in R^n$) and θ is the *threshold* ($\theta \in R^1$).

Remark 1: If the threshold θ is different from zero ($\theta \neq 0$), then the hyperplane $H(\mathbf{w}, \theta)$ (10) can be represented as the hyperplane $H(\mathbf{w}', 1)$ with the weight vector $\mathbf{w}' = \mathbf{w}/\theta$ and the threshold θ equal to one ($\theta = 1$).

The ($r_k - 1$) - dimensional *vertexical plane* $P_k(\mathbf{x}_{j(1)}, \ldots, \mathbf{x}_{j(rk)})$ is defined in the feature space $F[n]$ as the standardized linear combination of the r_k ($r_k > 1$) *supporting vectors* $\mathbf{x}_{j(i)}$ ($j \in J_k$) (4) belonging to the basis $\mathbf{B}_k$ (8) [4]:

$$P_k \left(\mathbf{x}_{j(1)}, \ldots, \mathbf{x}_{j(rk)} \right) = \left\{ \mathbf{x} : \mathbf{x} = \alpha_1 \mathbf{x}_{j(1)} + \ldots + \alpha_{rk} \mathbf{x}_{j(rk)} \right\} \tag{11}$$

where $j(i) \in J_k$ (4) and the parameters α_i ($\alpha_i \in R^1$) fulfill the below standardization:

$$\alpha_1 + \ldots + \alpha_{rk} = 1 \tag{12}$$

Two linearly independent vectors $\mathbf{x}_{j(1)}$ and $\mathbf{x}_{j(2)}$ from the set C (1) support the below straight line $l(\mathbf{x}_{j(1)}, \mathbf{x}_{j(2)})$ in the feature space $F[n]$ ($\mathbf{x} \in F[n]$):

$$l(\mathbf{x}_{j(1)}, \mathbf{x}_{j(2)}) = \{\mathbf{x} : \mathbf{x} = \mathbf{x}_{j(1)} + \alpha(\mathbf{x}_{j(2)} - \mathbf{x}_{j(1)})\} = \{\mathbf{x} : \mathbf{x} = (1 - \alpha)\mathbf{x}_{j(1)} + \alpha \mathbf{x}_{j(2)}\}$$
$$(13)$$

where $(\alpha \in R^1)$.

The straight line $l(\mathbf{x}_{j(1)}, \mathbf{x}_{j(2)})$ (13) can be treated as the vertexical plane $P_k(\mathbf{x}_{j(1)}, \mathbf{x}_{j(2)})$ (11) spanned by two supporting vectors $\mathbf{x}_{j(1)}$ and $\mathbf{x}_{j(2)}$ with $\alpha_1 = 1 - \alpha$ and $\alpha_2 = \alpha$. In this case, the regular basis $\mathbf{B}_k$ (8) contains two feature vectors $\mathbf{x}_{j(1)}$ and $\mathbf{x}_{j(2)}$ ($r_k = 2$) and $n - 2$ unit vectors $\mathbf{e}_i (i \in I_k)$. As a result, the vertex $\mathbf{w}_k = [\mathbf{w}_{k,1}, \ldots \mathbf{w}_{k,n}]^T$ (9) contains only two nonzero components $\mathbf{w}_{k,i} (\mathbf{w}_{k,i} \neq 0)$.

Lemma 1: The vertexical plane $P_k(\mathbf{x}_{j(1)}, \ldots, \mathbf{x}_{j(rk)})$ (11) based on the vertex $\mathbf{w}_k$ (9) with the rank r_k greater than 1 ($r_k > 1$) is equal to the hyperplane $H(\mathbf{w}_k, 1)$ (10) in the n - dimensional feature space $F[n]$ defined by this vertex.

Proof: Each feature vector $\mathbf{x}$ can be represented as the linear combination of the basis vectors $\mathbf{x}_{j(i)}$ and $\mathbf{e}_{i(l)}$ (8):

$$\mathbf{x} = \alpha_1 \mathbf{x}_{j(1)} + \ldots + \alpha_{rk} \mathbf{x}_{j(rk)} + \alpha_{rk+1} \mathbf{e}_{i(rk+1)} + \ldots + \alpha_n \mathbf{e}_{i(n)} \qquad (14)$$

where $\alpha \in R^1$.

Let us assume that the feature vector $\mathbf{x}$ is located on the vertexical plane $P_k(\mathbf{x}_{j(1)}, \mathbf{x}_{j(rk)})$ (11) defined by the vertex $\mathbf{w}_k$ (9) with ($r_k > 1$). Then

$$\begin{aligned} \mathbf{w}_k^T \mathbf{x} &= \mathbf{w}_k^T (\alpha_1 \mathbf{x}_{j(1)} + \ldots + \alpha_{rk} \mathbf{x}_{j(rk)} + \alpha_{rk+1} \mathbf{e}_{i(rk+1)} + \ldots + \alpha_n \mathbf{e}_{i(n)}) \\ &= \mathbf{w}_k^T (\alpha_1 \mathbf{x}_{j(1)} + \ldots + \alpha_{rk} \mathbf{x}_{j(rk)}) = \alpha_1 + \ldots + \alpha_{rk} = 1 \end{aligned} \qquad (15)$$

The equality (15) results from the properties (5) and (6) of the vertex $\mathbf{w}_k$ (9). This equality means that the feature vector $\mathbf{x}$ located on the vertexical plane $P_k(\mathbf{x}_{j(1)}, \mathbf{x}_{j(rk)})$ (11) is also located on the hyperplane $H(\mathbf{w}_k, 1)$ (10) in the n - dimensional feature space $F[n]$.

Let us assume now that the feature vector $\mathbf{x}$ is located on the hyperplane $H(\mathbf{w}_k, 1)$ (10). This means that $\mathbf{w}_k^T \mathbf{x} = 1$. Taking into account the Eqs. (5) and (15) we infer that the vector $\mathbf{x}$ is also located on the k-th vertexical plane $P_k(\mathbf{x}_{j(1)}, \mathbf{x}_{j(rk)})$ (11). $\quad\square$

Theorem 1: The j-th feature vector $\mathbf{x}_j (\mathbf{x}_j \in C(1))$ is located on the vertexical plane $P_k(\mathbf{x}_{j(1)}, \mathbf{x}_{j(rk)})$ (11) if and only if the j-th dual hyperplane h_j^1 (2) passes through the vertex $\mathbf{w}_k$ (9) of the rank r_k.

Proof: If the vector $\mathbf{x}_j$ is situated on the vertexical plane $P_k(\mathbf{x}_{j(1)}, \ldots, \mathbf{x}_{j(rk)})$ (11), then the equation $\mathbf{w}_k^T \mathbf{x} = 1$ (15) holds. This means that the hyperplane h_j^1 (2) defined by the j-th feature vector $\mathbf{x}_j$ passes through the vertex $\mathbf{w}_k$ (9).

If the hyperplane h_j (2) passes through the vertex $\mathbf{w}_k$ (9), then $\mathbf{w}_k^T \mathbf{x}_j = 1$. The j-th feature vector $\mathbf{x}_j$ can be represented as the linear combination of the basis vectors $\mathbf{x}_{j(i)}$ and $\mathbf{e}_{i(l)}$ (14):

$$\mathbf{x}_j = \alpha_{j,1}\mathbf{x}_{j(1)} + \ldots + \alpha_{j,rk}\mathbf{x}_{j(rk)} + \alpha_{j,rk+1}\mathbf{e}_{i(rk+1)} + \ldots + \alpha_{j,n}\mathbf{e}_{i(n)} \tag{16}$$

Taking into account the Eqs. (5) and (6) we infer that the j-th vector $\mathbf{x}_j$ is located on the vertexical plane $P_k(\mathbf{x}_{j(1)},\ldots,\mathbf{x}_{j(rk)})$ (11).   □

4 Vertexical Subspaces and Collinear Biclusters

The collinear (*flat*) pattern F_k in the feature space $F[n]$ is formed by a large number m_k of feature vectors $\mathbf{x}_j$ located on the *vertexical plane* $P_k(\mathbf{x}_{j(1)},\ldots, \mathbf{x}_{j(rk)})$ (11). As it results from the *Theorem* 1, the flat patterns F_k is composed of such m_k feature vectors $\mathbf{x}_j$ that the dual hyperplanes h_j^1 (2) pass through the vertex $\mathbf{w}_k$ (9) of the rank r_k:

$$F_k = \left\{ \mathbf{x}_j : \mathbf{w}_k^T \mathbf{x}_j = 1 \right\} \tag{17}$$

Let us introduce a family of the vertexical feature subspaces $F_k[r_k] (F_k[r_k] \subset F_k[n])$ linked to particular vertices $\mathbf{w}_k$ (9). The k-th vertexical feature subspace $F_k[r_k]$ is obtained as a result of the feature space $F[n]$ reduction on the basis of the coordinates $w_{k,i}$ of the vector $\mathbf{w}_k = [w_{k,1},\ldots,w_{k,n}]^T$ (9) [9]:

$$(\forall i \in \{1,\ldots,n\}) \quad \textbf{if } w_{k,i} = 0, \textbf{then} \text{ the } i\text{-th feature } x_i \text{ is not included} \tag{18}$$
$$\text{in the } k\text{-th vertexical feature subspace } F_k[r_k]$$

The k-th vertexical feature subspace $F_k[r_k] (F_k[r_k] \subset F_k[n])$ is composed of such r_k features x_i which are linked to the nonzero weights $w_{k,i}$ ($w_{k,i} \neq 0$). We can remark that the reduced features x_i are linked to the unit vectors $\mathbf{e}_i$ in the regular basis $\mathbf{B}_k$ (8). The vertexical feature subspace $F_k[r_k]$ is composed from such r_k features x_i with such indices i which do not define the unit vectors $\mathbf{e}_i$ in the regular basis $\mathbf{B}_k$ ($i \notin I_k$) (8).

The reduced feature vectors $\mathbf{y}_j = [y_{j,1},\ldots,y_{j,rk}]^T$ belongs to the k-th vertexical feature subspace $F_k[r_k]$ ($\mathbf{y}_j \in F_k[r_k]$) and are obtained from the feature vectors $\mathbf{x}_j = [x_{j,1},\ldots,x_{j,n}]^T$ ($\mathbf{x}_j \in C$ (1)) by neglecting (18) such $n - r_k$ components $x_{j,i}$ which are linked to the unit vectors $\mathbf{e}_i$ in the regular basis $\mathbf{B}_k$ (8). Similarly, the reduced weight vector $\mathbf{v}_k = [v_{k,1},\ldots, v_{k,rk}]^T$ is obtained from the k-th vertex $\mathbf{w}_k = [w_{k,1},\ldots,w_{k,n}]^T$ (9) of the rank r_k by neglecting the $n - r_k$ components $w_{k,i}$ equal to zero ($w_{k,i} = 0$).

Definition 5: The collinear bicluster $B_k(m_k, r_k)$ based on the k-th degenerated vertex $\mathbf{w}_k$ (9) of the rank r_k is defined as the set of such m_k ($m_k > r_k$) reduced vectors $\mathbf{y}_j(\mathbf{y}_j \in F_k)$ which fulfill the equation $\mathbf{v}_k^T \mathbf{y}_j = 1$ with the reduced weight vector $\mathbf{v}_k$ (21):

$$B_k(m_k, r_k) = \left\{ \mathbf{y}_j : \mathbf{v}_k^T \mathbf{y}_j = 1 \right\} \tag{19}$$

Remark 1: The below equalities hold in the k-th vertex $\mathbf{w}_k$ (9):

$$(\forall j \in \{1,\ldots,m\}) \quad \mathbf{w}_k^T \mathbf{x}_j = \mathbf{v}_k^T \mathbf{y}_j \tag{20}$$

The above equations results from the fact that the vector $\mathbf{v}_k$ is obtained from the k-th vertex $\mathbf{w}_k = [w_{k,1},\ldots,w_{k,n}]^T$ (9) through neglecting $n - r_k$ components $w_{k,i}$ (18) equal to zero ($w_{k,i} = 0$).

Remark 2: The collinear bicluster $\boldsymbol{B}_k(m_k, r_k)$ (19) is the set of such m_k reduced feature vectors $\mathbf{y}_j$ ($\mathbf{y}_j \in F_k[r_k]$) which are located on the hyperplane $H(\mathbf{v}_k,1)$ (10) defined in the k-th vertexical feature subspace $F_k[r_k]$ by the reduced weight vector $\mathbf{v}_k = [v_{k,1},\ldots,v_{k,rk}]^T$ with the all r_k components $v_{k,i}$ different from zero ($v_{k,i} \neq 0$):

$$H(\mathbf{v}_k, 1) = \{\mathbf{y} : \mathbf{v}_k^T \mathbf{y} = 1\} \tag{21}$$

We can also remark that m_k feature vectors $\mathbf{x}_j = [x_{j,1},\ldots,x_{j,n}]^T$ linked to the collinear bicluster $\boldsymbol{B}_k(m_k, r_k)$ (19) through the reduced feature vectors $\mathbf{y}_j$ are located on the vertexical plane $P_k(\mathbf{x}_{j(1)},\ldots,\mathbf{x}_{j(rk)})$ (11) of the rank r_k.

$$\begin{gathered}(\forall j \in \{1,\ldots,m\}) \\ if\,\big(\mathbf{y}_j \in \boldsymbol{B}_k(m_k, r_k)\big), then\,\big(\mathbf{x}_j \in P_k\big(\mathbf{x}_{j(1)}, \ldots, \mathbf{x}_{j(rk)}\big)\big)\end{gathered} \tag{22}$$

The k-th collinear (*flat*) bicluster $\boldsymbol{B}_k(m_k, r_k)$ (19) based on the degenerated vertex $\mathbf{w}_k$ (9) has been characterized by the two numbers r_k and m_k, where r_k is the number of features x_i in the vertexical subspace $F_k[r_k]$ (19), and m_k is the number of such feature vectors $\mathbf{x}_j$ ($\mathbf{x}_j \in F[n]$) which define the dual hyperplanes h_j^1 (2) passing through this vertex ($\mathbf{w}_k^T \mathbf{x}_j = 1$). The collinear biclusters $\boldsymbol{B}_k(m_k, r_k)$ (19) should be based on highly degenerated vertices $\mathbf{w}_k$ (9) with a large number m_k.

Let us consider as an example the collinear bicluster $\boldsymbol{B}_k(m_k, r_k)$ (19) with $r_k = 2$. Elements $\mathbf{y}_j$ of such bicluster are based on the straight line $l(\mathbf{y}_{j(1)}, \mathbf{y}_{j(2)})$ (13) in the k-th vertexical feature subspace $F_k[r_k](F_k[r_k] \subset F_k[n])$:

$$\big(\forall \mathbf{y}_j \in \boldsymbol{B}_k(m_k, r_k)\big) \quad \mathbf{y}_j = \alpha_j \mathbf{y}_{j(1)} + (1 - \alpha_j)\mathbf{y}_{j(2)} \tag{23}$$

The line $l(\mathbf{y}_{j(1)}, \mathbf{y}_{j(2)})$ is supported by the two linearly independent vectors $\mathbf{y}_{j(1)}$ and $\mathbf{y}_{j(2)}$.

5 Convex and Piecewise Linear (*CPL*) Collinearity Functions

Feature vectors $\mathbf{x}_j$ from the data set C (1) allow to define the below collinearity penalty functions $\varphi_j^1(\mathbf{w})$ linked to the dual hyperplanes h_j (2) [4]:

$$(\forall \mathbf{x}_j \in C)$$

$$\varphi_j^1(\mathbf{w}) = \left| 1 - \mathbf{x}_j^T \mathbf{w} \right| = \begin{array}{l} 1 - \mathbf{x}_j^T \mathbf{w} \ \ if \ \ \mathbf{x}_j^T \mathbf{w} \leq 1 \\ \mathbf{x}_j^T \mathbf{w} - 1 \ \ if \ \ \mathbf{x}_j^T \mathbf{w} > 1 \end{array} \tag{24}$$

Each of n unit vectors $\mathbf{e}_i$ allows to define the i-th cost function $\varphi_j^1(\mathbf{w})$ in the n-dimensional parameter space R^n, where $\mathbf{w} = [w_1, \dots, w_n]$:

$$(\forall i \in \{1, \dots, n\})$$

$$\varphi_i^0(\mathbf{w}) = \left| \mathbf{e}_i^T \mathbf{w} \right| = |w_i| = \begin{array}{l} -w_i \ \ if \ \ w_i \leq 0 \\ w_i \ \ if \ \ w_i > 0 \end{array} \tag{25}$$

The cost functions $\varphi_i^0(\mathbf{w})$ (25) are linked to the hyperplanes h_i^0 (3). The cost functions $\varphi_i^0(\mathbf{w})$ (24) like the collinearity penalty functions $\varphi_j^1(\mathbf{w})$ (24) are convex and piecewise linear (*CPL*) (Fig. 1).

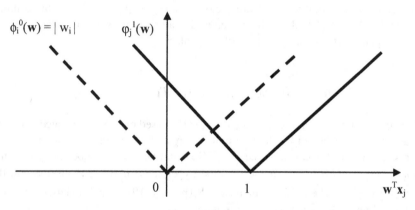

Fig. 1. The penalty functions $\varphi_j^1(\mathbf{w})$ (24) and $\varphi_i^0(\mathbf{w})$ (25)

The k-th collinearity criterion function $\Phi_k(\mathbf{w})$ is defined as the weighted sum of the penalty functions $\varphi_j^1(\mathbf{w})$ (23) linked the feature vectors $\mathbf{x}_j$ from the given subset C_k:

$$\Phi_k(\mathbf{w}) = \sum_{i \in Jk} \beta_j \varphi_j^1(\mathbf{w}) \tag{26}$$

where $J_k = \{j: \mathbf{x}_j \in C_k \subset C \ (1)\}$ and the positive parameters β_j ($\beta_j > 0$) can be treated as the *prices* of particular feature vectors $\mathbf{x}_j$. The standard choice of the parameters β_j values is one $((\forall j \in \{1, \dots, m\}) \ \beta_j = 1.0)$.

It can be proved that the minimal value Φ_k^* of the convex and piecewise linear criterion function $\Phi_k(\mathbf{w})$ (26) can be found in one of the vertices $\mathbf{w}_k$ (9):

$$(\exists \mathbf{w}_k^*)(\forall \mathbf{w}) \Phi_k(\mathbf{w}) \geq \Phi_k(\mathbf{w}_k^*) = \Phi_k^* \geq 0 \tag{27}$$

The basis exchange algorithms which are similar to the linear programming allow to find efficiently the minimal value $\Phi_k(\mathbf{w}_k^*)$ (27) of the criterion functions $\Phi_k(\mathbf{w})$ (26) even in the case of large, multidimensional data subsets C_k [8].

Theorem 3: The minimal value $\Phi_k(\mathbf{w}_k^*)$ (27) of the criterion function $\Phi_k(\mathbf{w})$ (26) defined on elements $\mathbf{x}_j$ of the data subset C_k is equal to zero $\Phi_k(\mathbf{w}_k^*) = 0$), if and only if all the feature vectors $\mathbf{x}_j$ from the subset C_k can be located on some hyperplane $H(\mathbf{w}, \theta)$ (10) with $\theta \neq 0$.

Proof: Let us suppose that all the feature vectors $\mathbf{x}_j$ from the subset C_k are situated on the hyperplane $H(\mathbf{w}', q')$ (10) with $\theta' \neq 0$:

$$(\forall \mathbf{x}_j \in C_k) \quad (\mathbf{w}')^T \mathbf{x}_j = \theta' \tag{28}$$

From this

$$(\forall \mathbf{x}_j \in C_k) \quad (\mathbf{w}'/\theta')^T \mathbf{x}_j = 1 \tag{29}$$

The above equations mean that all the functions $\varphi_j^1(\mathbf{w}'/\theta')$ (24) are equal to the zero in the point $(\mathbf{w}' / \theta')$:

$$(\forall \mathbf{x}_j \in C_k) \quad \varphi_j^1(\mathbf{w}'/\theta') = 0 \tag{30}$$

so (25)

$$\Phi_k(\mathbf{w}'/\theta') = 0 \tag{31}$$

On the other hand, if the criterion function $\Phi_k(\mathbf{w}')$ (26) is equal to the zero in some point $\mathbf{w}'$, then each of the penalty functions $\varphi_j^1(\mathbf{w}')$ (24) has to be equal to zero:

$$(\forall \mathbf{x}_j \in C_k) \quad \varphi_j^1(\mathbf{w}') = 0 \tag{32}$$

or

$$(\forall \mathbf{x}_j \in C_K) \quad (\mathbf{w}')^T \mathbf{x}_j = 1 \tag{33}$$

The above equations mean that each feature vector $\mathbf{x}_j$ from the subset C_k is located on the hyperplane $H(\mathbf{w}', 1)$ (11) in the n - dimensional feature space $F[n]$. $\square$

Remark 3: (the *negative monotonocity property*): The omission of any of the component $x_{j,i}$ (feature x_i) in all the m feature vectors $\mathbf{x}_j = [x_{j,1}, \ldots, x_{j,n}]^T$ (1) cannot reduce the value of Φ^* (26) but the value Φ^* may *increase*.

Remark 3 can be justified by the fact that neglecting some feature x_i from the given feature space $F[n]$ is equivalent to imposing additional constraint "$w_i = 0$" in the parameter space R^n. The minimal value Φ^* (27) of the criterion function $\Phi(\mathbf{v})$ (25) "$w_i = 0$" must not be less.

For the purpose of the feature selection the collinear criterion function $\Phi_k(\mathbf{w})$ (25) can be regularized by adding the cost functions $\varphi_i^0(\mathbf{w})$ (25) [9]:

$$\Psi_\lambda(\mathbf{w}) = \Phi_K(\mathbf{w}) + \lambda \sum_{i \in I} \gamma_i \varphi_i^0(\mathbf{w}) \tag{34}$$

where γ_i ($\gamma_i > 0$) are the *costs* γ_i ($\gamma_i > 0$) of particular features x_i, and λ ($\lambda \geq 0$) is the *cost level* and $I = \{1,\ldots\ldots,n\}$ [9].

The regularized criterion function $\Psi_\lambda(\mathbf{w})$ (34) is convex and piecewise-linear (*CPL*), so their minimum $\Psi_\lambda(\mathbf{w}_k^*)$ can be found in one of vertices $\mathbf{w}_k^*$ (9) by using the basis exchange algorithm [8].

6 Extraction of Collinear Biclusters

The proposed method of the collinear bicluster $B_k(m_k, r_k)$ (19) extraction involves the gradual omitting of selected feature vectors $\mathbf{x}_j$ from the data set C (1).

Remark 4: If the data set *C* (1) which consists of m feature vector $\mathbf{x}_j$ (1) contains the flat pattern F_k (17) composed of m_k feature vectors $\mathbf{x}_j$ ($F_k \subset C$) then the minimal value $\Phi_k(\mathbf{w}_k^*)$ (27) of the criterion function $\Phi_k(\mathbf{w})$ (26) defined on all elements $\mathbf{x}_j$ of the set *C* ($C_k = C$) can be reduced to zero ($\Phi_k(\mathbf{w}_k^*) = 0$) by the exclusion of $m - m_k$ feature vectors $\mathbf{x}_j$ from the set *C*.

The procedure *Vertex* sketched below gives the possibility of extracting the flat pattern F_k (17) from the data set *C* (1) of the n-dimensional feature vectors $\mathbf{x}_j$ [4]:

$$\text{Procedure } Vertex \tag{35}$$

Start: $k = 1$, $C_k = C$ (1).

Step 1. The function $\Phi_k(\mathbf{w})$ (26) is defined on the elements $\mathbf{x}_j$ of the data set C_k.

Step 2. The optimal vertex $\mathbf{w}_k^*$ constituting the minimal value $\Phi_k(\mathbf{w}_k^*)$ (27) of the criterion function $\Phi_k(\mathbf{w})$ (26) is computed

Step 3. If $\Phi_k(\mathbf{w}_k^*) = 0$, then the procedure is **stopped** in the vertex $\mathbf{w}_k^*$, else

Step 4. The data set C_k is reduced by such feature vectors $\mathbf{x}_{j'}$ which have the highest value $\varphi_{j'}(\mathbf{w}_k^*)$ of the penalty functions $\varphi_j(\mathbf{w})$ (23) in the optimal vertex $\mathbf{w}_k^*$ (25)

$$(\forall \mathbf{x}_j \in C_k) \quad \varphi_j(\mathbf{w}_k^*) \leq \varphi_{j'}(\mathbf{w}_k^*) \tag{36}$$

Step 5. The counter k is increased by one ($k \rightarrow k + 1$) and the return to the step *Step* 1 is followed.

The reduction of the data set C_k by the vector $\mathbf{x}_{j'}$ with the highest value $\varphi_{j'}(\mathbf{w}_k^*)$ (*Step* 4) decreases the minimal value $\Phi_{k'}\left(\mathbf{w}_{k'}^*\right)$ (27) of the criterion function $\Phi_{k'}(\mathbf{w})$ (26) defined on the reduced data set $C_k / \{\mathbf{x}_{j'}\}$:

$$\Phi_K(\mathbf{W}_{k'}^*) \leq \Phi_k(\mathbf{W}_k^*) - \varphi_{j'}(\mathbf{W}_k^*) \tag{37}$$

Based on the inequality (37) we can conclude that the minimal value $\Phi_k\left(\mathbf{w}_k^*\right)$ (27) of the criterion function $\Phi_{k'}(\mathbf{w})$ (26) can be efficiently reduced to zero by the exclusion of such feature vectors $\mathbf{x}_{j'}$ which have the highest values $\varphi_{j'}\left(\mathbf{w}_{k'}^*\right)$ (36).

In accordance with the *Theorem* 3, the minimal value $\Phi_k\left(\mathbf{w}_k^*\right)$ (27) of the criterion function $\Phi_k(\mathbf{w})$ (26) defined the data subset C_k ($C_k \subset C(1)$) is equal to zero, if all m_k feature vectors $\mathbf{x}_j$ from the subset C_k are located on the hyperplane $H(\mathbf{w}_k^*, 1$ (10), (21) defined in the n - dimensional feature space $F[n]$ by the optimal vertex $\mathbf{w}_k^*$ (27). If the number m_k of such vectors $\mathbf{x}_j$ is sufficiently large, then the reduced vectors $\mathbf{y}_j$ ($\mathbf{y}_j \in F_k[r_k]$) constitute the collinear bicluster $B_k(m_k, r_k)$ (19).

Data set C (1) created by a large number m of high-dimensional feature vectors $\mathbf{x}_j$ may contain many collinear biclusters $B_k(m_k, r_k)$ (19). The procedure *Vertex* (35) can be modified for the purpose of successive extraction of different collinear biclusters $B_k(m_k, r_k)$ (19). The procedure of successive extraction of collinear biclusters $B_k(m_k, r_k)$ from the given data set C (1) may include multiple minimization of the regularized criterion function $\Psi_\lambda(\mathbf{w})$ (34) with different values of the cost λ. Increasing the value of the parameter λ in the criterion function $\Psi_\lambda(\mathbf{w})$ (34) leads to the reduction of feature subspace $F_k[r_k]$, similarly as in the *RLS* method of feature subsets selection [9].

7 Concluding Remarks

Collinear biclustering has been analyzed in the presented paper. The proposed method of the collinear biclustering is based on multiple minimizations of the *CPL* criterion functions $\Phi_k(\mathbf{w})$ (26) combined with the data subsets C_k (1) reduction (35). New theoretical properties of the proposed method of biclustering have been formulated and proved in the presented paper.

The minimization of the collinear criterion function $\Phi_k(\mathbf{w})$ (26) defined on the set C_k (1) of a large number m of the n - dimensional feature vectors $\mathbf{x}_j$ ($\mathbf{x}_j \in F[n]$) may allow for discovering many collinear biclusters $B_k(m_k, r_k)$ (19) in this set. The collinear bicluster $B_k(m_k, r_k)$ (19) is composed of m_k ($m_k \leq m$) reduced feature vectors $\mathbf{y}_j$ belonging to the k-th vertexical feature subspace $F_k[r_k]$ ($\mathbf{y}_j \in F_k[r_k]$) (18). In accordance with the rule (18) the vertexical feature subspace $F_k[r_k]$ is obtained in a unique manner from the feature space $F[n]$ by neglecting these features x_i which are linked to the weights $w_{k,i}$ equal to zero ($w_{k,i} = 0$) in the k-th optimal vertex $\mathbf{w}_k^*$ (27) of the rank r_k.

The dimensionality r_k of the k-th vertexical feature subspace $F_k[r_k]$ depends on the structure of the flat patterns F_k (17). If the feature vectors $\mathbf{x}_j$ belonging to the flat patterns F_k (17) are located on the line $l(\mathbf{x}_{j(1)}, \mathbf{x}_{j(2)})$ (13) in the feature space $F[n]$, then $r_k = 2$. Similarly, if the feature vectors $\mathbf{x}_j$ belonging to the flat patterns F_k (17) are located on the two-dimensional vertexical plane $P_k(\mathbf{x}_{j(1)}, \mathbf{x}_{j(2)}, \mathbf{x}_{j(3)})$ (12), then $r_k = 3$, and so on.

Perhaps one of the most interesting applications of the collinear biclusters $\boldsymbol{B}_k(m_k, r_k)$ (19) would be in modeling multiple interactions between variables (features). Modeling of multiple interactions can be based on two complementary representations (19) and (22) of the collinear biclusters $\boldsymbol{B}_k(m_k, r_k)$ extracted from the data set C (1).

Acknowledgments. The presented study was supported by the grant S/WI/2/2013 from Bialystok University of Technology and funded from the resources for research by Polish Ministry of Science and Higher Education.

References

1. Duda, O.R., Hart, P.E., Stork, D.G.: Pattern Classification. Wiley, New York (2001)
2. Hand, D., Smyth, P., Mannila, H.: Principles of Data Mining. MIT Press, Cambridge (2001)
3. Madeira, S.C., Oliveira, S.L.: Biclustering algorithms for biological data analysis: a survey. IEEE Trans. Comput. Biol. Bioinform. **1**(1), 24–45 (2004)
4. Bobrowski, L.: Discovering main vertexical planes in a multivariate data space by using CPL functions. In: Perner, P. (ed.) ICDM 2014. LNCS (LNAI), vol. 8557, pp. 200–213. Springer, Cham (2014). doi:10.1007/978-3-319-08976-8_15
5. Bobrowski, L.: Decision rules with collinearity models. In: Czarnowski, I., Caballero, A.M., Howlett, Robert J., Jain, Lakhmi C. (eds.) Intelligent Decision Technologies 2016. SIST, vol. 56, pp. 293–304. Springer, Cham (2016). doi:10.1007/978-3-319-39630-9_24
6. Bobrowski, L., Zabielski, P.: Flat patterns extraction with collinearity models. In: 9th EUROSIM Congress on Modelling and Simulation, EUROSIM 2016, 12–16 September 2016, Oulu Finland. IEEE Conference Publishing Services (CPS)
7. Duda, O.R., Hart, P.E.: Use of the hough transformation to detect lines and curves. Pictures Commun. Assoc. Comput. Mach. **15**(1), 11–15 (1972)
8. Bobrowski, L.: Design of piecewise linear classifiers from formal neurons by some basis exchange technique. Pattern Recogn. **24**(9), 863–870 (1991)
9. Bobrowski, L., Łukaszuk, T.: Relaxed linear separability (RLS) approach to feature (Gene) subset selection. In: Xia, X. (ed.) Selected Works in Bioinformatics, pp. 103–118. INTECH (2011)

A Data- and Model-Driven Analysis Reveals the Multi-omic Landscape of Ageing

Elisabeth Yaneske$^{(\boxtimes)}$ and Claudio Angione

Department of Computer Science and Information Systems,
Teesside University, Middlesbrough, UK
e.yaneske@tees.ac.uk

Abstract. Altered expression of a number of genes has been correlated to biological ageing in humans. The biological age predicted from
gene expression levels is known as transcriptomic age. This differs from
chronological age which is measured as the time that an individual has
lived since their date of birth. Transcriptomic age can be older or younger
than an individual's chronological age. At present, studies have focused
on using transcriptomic data to predict transcriptomic age. However,
this approach largely does not consider the effect that genes have on the
metabolic network and therefore on the observable cellular phenotype.
This research takes the current understanding of transcriptomic ageing
a step further by generating and investigating genome-scale metabolic
models of ageing, using machine learning methods and a multi-omic
approach based on constraint-based modelling. We combine these models with a transcriptomic age predictor and gene expression data from
CD4 T-Cells from human peripheral blood mononuclear cells in healthy
individuals. We show that metabolic models augmented with transcriptomics data of ageing can generate greater metabolic insights into the
differences between chronological and transcriptomic age. Compared to
standard transcriptomic-only approaches, our method provides a more
comprehensive analysis of transcriptomic ageing and paves the way for
a multi-omic understanding of ageing mechanisms in human cells.

Keywords: Ageing · Metabolomic age · Transcriptomic age · Flux
balance analysis · Multi-omics · CD4 T-cells

1 Introduction

Ageing is associated with phenotypes such as wrinkles, greying hair, hair loss
and frailty as well as diseases such as diabetes, osteoporosis and cardiovascular
disease. Underlying these characteristics of ageing are age-related changes to
our metabolic processes [1]. Changes to mitochondrial function, which is vital
for energy metabolism and homeostasis, have been linked to the ageing process
[2,3]. Metabolic dysregulation leads to the build-up of fat stores in the abdominal
cavity [4]. This type of fat is called visceral fat. Higher visceral fat levels mean
a greater risk of insulin resistance, type 2 diabetes, cardiovascular disease and

© Springer International Publishing AG 2017
I. Rojas and F. Ortuño (Eds.): IWBBIO 2017, Part I, LNBI 10208, pp. 145–154, 2017.
DOI: 10.1007/978-3-319-56148-6_12

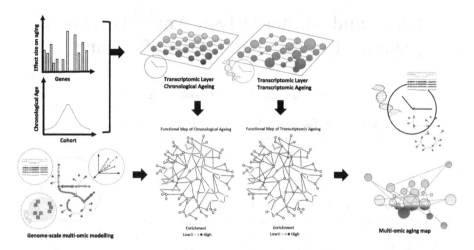

Fig. 1. Multi-omic ageing pipeline. We use the chronological data and corresponding predictors to obtain the effect of both types of ageing on the transcriptomic layer. We then combine the latter with the functional biological network data determined by the metabolism and multi-omic model to obtain the multi-omic ageing map.

cancer [5–8]. However, a recent study by Chee et al. [9] suggests that some age-related metabolic changes may not be inevitable but rather due to lifestyle. A group of young (mean age 21.5) and older (mean age 69.7) participants who were matched for physical activity levels and body composition were found to have no significant difference in insulin resistance and lipid accumulation in their muscles.

The term 'inflammaging' was proposed by Franceschi et al. [10] to describe the imbalance between inflammatory and anti-inflammatory networks which contribute to the chronic diseases of ageing. The immune response declines dramatically with ageing leading to increased frailty e.g. susceptibility to pathogens such as influenza [11]. CD4 cells have recently been shown to be linked to the ageing process. A study comparing gene expression from the same CD4 cells in newborns, middle-aged and long-lived participants found significant expression changes in the transcriptomes across the age groups [12]. The action of CD4 cells is impaired with age which contributes to the deterioration of the immune response [13,14].

Although genes are often regarded as the main players in deciding cell behaviour, it is difficult to predict how they affect the phenotype. Metabolic models can therefore be used to predict the metabotype of a cell, and provide a close-to-phenotype prediction of cell behaviour and fate. Furthermore, metabolic models can be constrained using gene expression data [15].

The idea of this paper is that modelling how the gene expression alterations change metabolic processes provides greater understanding of the ageing process and a more accurate prediction of biological age. This data can then be used to investigate how adding the metabolic information alters and improves transcriptomics-only classifications of patients. Using transcriptomic data from patients, we here generate patient-specific genome-scale metabolic models. We then investigate such models in the context of ageing, therefore obtaining metabolic ageing biomarkers. More specifically, the gene expression data from CD4 cells is mapped to both the chronological age and the transcriptomic age of the individuals in the cohort. This provided us with a baseline of the current understanding of transcriptomic ageing in the transcriptomic layer, and allowed us to investigate the metabolic processes that contribute to both chronological and transcriptomic ageing.

2   Methods

2.1   Transcriptomic Age Predictor

Throughout our pipeline, we use a meta-analysis of CD4 T-Cell gene expression from human peripheral blood mononuclear cells isolated from healthy individuals in the Boston area. The CD4 cell gene expression data was profiled on Affymetrix Human Gene 1.0 ST microarrays [16]. The raw data is accessible through GEO accession number GSE56033. The dataset contains 499 individual patients. Using 1497 age-associated genes, i.e. those found to be differentially expressed with chronological age, the effect size of each individual age-associated gene expression level on chronological age was then calculated [17]. The sum of the effect sizes of the age-associated genes on chronological age was used to calculate a general transcriptomic predictor Z, defined as a linear combination:

$$Z = \sum_i b_i x_i, \tag{1}$$

where x_i is the gene expression level of the ith probe, and b_i is the effect size for the ith probe.

Starting from these predictors, we calculated the transcriptomic age of each individual. To this end, the general transcriptomic predictor was scaled using the mean and standard deviation of the chronological age and the mean and standard deviation of the general predictor [17]. The transcriptomic age of an individual is:

$$SZ = \mu_{age} + (Z - \mu_z)\frac{\sigma_{age}}{\sigma_z}, \tag{2}$$

where μ_{age} and σ_{age} are the mean and the standard deviation of the chronological age, while μ_z and σ_z are the mean and the standard deviation of the predictor Z.

Using (2), we calculated the transcriptomic ages for the 499 individuals. The chronological data was included for each sample. The data was normalised using RMA (Robust Multi-array Average) [18] followed by dividing the gene expression values by the mean value for each probe. The effect size b_i is defined on a gene by gene basis. In cases where a probe represented more than one gene the effect sizes for those genes were averaged to give an on overall average effect size for that probe.

To progress to a multi-omic understanding of transcriptomic ageing we used a constraint-based model of the CD4 cell [19] augmented with transcriptomics through METRADE [20]. This enabled us to create personalised metabolic models of the samples within the cohort. We then mapped the metabolic flux rate to the chronological age of the individuals within the cohort and repeated this pipeline using metabolic flux rate and transcriptomic age.

2.2   Constraint-Based Modelling

Given the matrix S of all known metabolic biochemical reactions and their stoichiometry, and given the vector v of flux rates in a given growth condition [21], constraint-based modelling and flux balance analysis (FBA) allow the prediction of the distribution of flux rates in a given condition The metabolic network is solved by maximising a cellular objective (usually the biomass), with constraints deriving from the steady-state condition $Sv = 0$ and additional constraints v^{min} and v^{max} on lower bound and upper bounds of v.

Using METRADE [20] coupled with the transcriptomic data from each patient, we modify the upper- and lower- limits of reactions as a function of the expression levels of genes involved in the reaction. For each patient, to predict the cellular flux distribution when multiple objectives have to be taken into account, we use the following bilevel linear program:

$$\max g^\mathsf{T} v$$
$$\text{such that} \qquad \max f^\mathsf{T} v, \quad Sv = 0, \qquad (3)$$
$$v^{min}\varphi(\varTheta) \leq v \leq v^{max}\varphi(\varTheta).$$

The vectors f and g are weights to select (or combine) the objectives to be maximised from the vector v. The vector $\varTheta$ converts the gene expression values into coefficients for the bounds of reactions activated by those genes. This is achieved through the function φ, which acts on $\varTheta$, the "expression" of a biochemical reaction, is defined from the patient-specific expression levels of its genes, with a rule depending on the type of enzyme (single gene, isozyme, or enzymatic complex).

2.3   Cluster Analysis

In order to compare the transcriptomic and metabolomic landscapes of ageing, Agglomerative Hierarchical Clustering (AHC) was used to cluster both the transcriptomic and fluxomic data. The transcriptomic data was RMA normalised followed by dividing the gene expression values by the mean value for each probe prior to clustering. AHC requires the distance between every pair of objects in the dataset to be calculated prior to being able link objects together into an hierarchical cluster tree based on proximity. AHC dissimilarity testing was carried out to find the distance and linkage combination that gave the highest Cophenetic Correlation Coefficient (CHC). For the transcriptomic data the linkage method that performed best was Average followed by Weighted, Complete,

Ward then Single. Further testing showed that Euclidean distance and Average linkage gave the best CHC of 0.67. For the fluxomic data the best performing linkage method was Average followed by Single, Weighted, Ward then Complete. The highest CHC of 0.54 was obtained using Average linkage and Squared Euclidean distance.

The average linkage method used the Unweighted Pair-Group Method with Arithmetic Mean (UPGMA) [22] algorithm. UPGMA joins the pair of objects with the smallest distance, then calculates the average between this pair and all the other objects in the dataset, and repeats this cycle until all data has been grouped into one cluster. The distance is computed as:

$$D_{(ij),k} = \left(\frac{n_i}{n_i + n_j} \right) D_{ik} + \left(\frac{n_i}{n_i + n_j} \right) D_{jk}, \tag{4}$$

where D_{ij} is group (ij) which has $n(ij) = n_i + n_j$ members and k is a new cluster.

3 Results

In order to determine the optimal number of natural clusters the pairwise distance data for both the transcriptomic and fluxomic data were visualised using scatter plots (Fig. 2). The most distinct categorisation occurred at four clusters for transcriptomic data. Figure 2 clearly shows the majority of the transcriptomic data forming two large clusters. For the fluxomic data the most distinct categorisation occurred at eight clusters. Figure 2 shows the fluxomic data falling into four main clusters, though less distinct than the transcriptomic data.

Having determined the optimal number of natural clusters for both transcriptomic and fluxomic data the data was further analysed to determine whether the natural clusters correlated with chronological age, transcriptomic age, or both. This analysis builds on recent research which has shown a correlation between CD4 cells and ageing [23]. Scatter plots of transcriptomic age against cluster number were produced for both transcriptomic and fluxomic data. These plots were annotated by chronological age using both colour and size. The results of the clustering with age are displayed in Fig. 3 for the transcriptomic data and for the fluxomic data.

From Fig. 3 it can be seen that although the transcriptomic data formed into two distinct clusters there is very little differentiation in chronological age between the two clusters. There is, however, some differentiation in the transcriptomic age with cluster one containing most of the older patients (approximately 38 and over) and cluster two containing most of the younger patients (approximately 29 and under). These results support the assertion that gene expression levels are related more to biological ageing than chronological ageing. However, the clustering has a low level of granularity with not much differentiation between age groups.

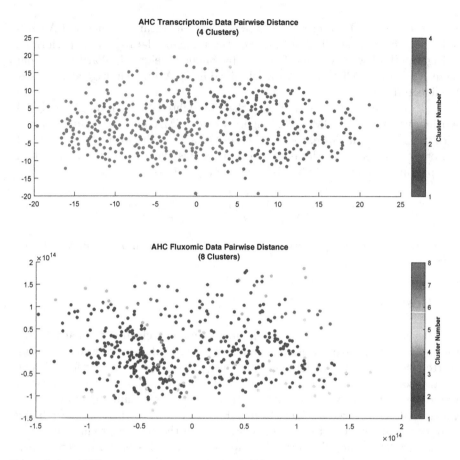

Fig. 2. The AHC pairwise distance matrix data. The pairwise distance plots were obtained after multidimensional scaling. (Top) The top scatter plot shows the pairwise Euclidean distance matrix for the transcriptomic data when classified into four clusters. Here we can see the majority of the data has formed into two distinct clusters. (Bottom) The bottom scatter plot shows the pairwise Squared Euclidean distance matrix for the fluxomic data when classified into eight clusters. The fluxomic data shows a more complex picture with most of the data forming into four less distinct clusters.

The fluxomic data in Fig. 3 shows a more complicated picture. Most of the data is contained in cluster two for both chronological and transcriptomic age. There is some differentiation in chronological ageing with cluster three appearing to capture mainly the 20–30 year age range and cluster five appearing to capture 30s to early 40s. There is also some differentiation in transcriptomic ageing with cluster 3 capturing 30s to mid 40s, cluster five capturing late 30s and early 40s and cluster six capturing early 30s and late 40s. The last two clusters (seven and eight) contain non-overlapping subsets of patients in their 20s and 30s respectively. This appears to show greater differentiation than the transcriptomic data between both chronological and transcriptomic ageing which suggests that further analysis could provide deeper insights into the metabolism of ageing.

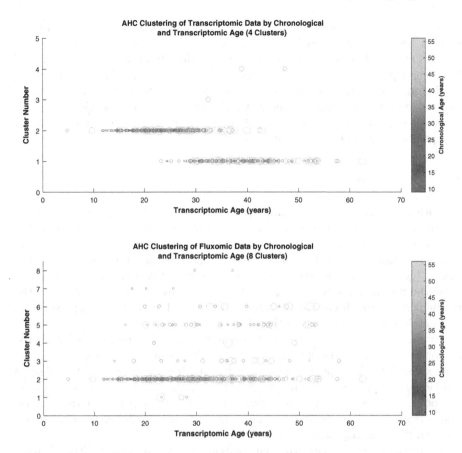

Fig. 3. AHC clustering of transcriptomic and fluxomic data with transcriptomic age annotated by chronological age. Chronological age is annotated using both colour and size. (Top) The top scatter plot shows clustering of the transcriptomic dataset. Cluster 1 and cluster 2 show some differentiation with transcriptomic age. Cluster 1 contains a greater proportion of the older transcriptomic age range from approximately 38 years and older. Cluster 2 contains most of the patients under 29. Very little differentiation can be seen according to chronological age. (Bottom) The bottom scatter plot shows clustering of fluxomic data. Although most of the data is captured in cluster 2, the fluxomic data shows some differentiation in both chronological and transcriptomic age. For chronological age, cluster 3 captures mainly the 20–30 year age range while cluster 5 appears to capture 30s to early 40s. For transcriptomic age, cluster 3 captures 30s to mid 40s, cluster 5 captures 20s, late 30s and early 40s and cluster 6 captures early 30s and late 40s. Cluster 7 and 8 contain small subsets of patients in the 20s and 30s respectively. (Color figure online)

The standard deviation was calculated to determine the within cluster variation of the chronological ages. Fluxomic data had the lowest average standard deviation overall of 5.6776 for 4 clusters. The lowest average standard deviation for the transcriptomic data was 6.0498 for 2 clusters. To determine the between cluster variation an F-test was performed. The fluxomic results for 4 clusters showed the difference of standard deviation between cluster 7 and the other clusters was statistically significant ($p < 0.05$). No significant results were found for transcriptomic data. These results show that, for this dataset, clustering by fluxomic improves the clustering by transcriptomics. Using a multi-variate method such as Principal Component Analysis for further investigation could allow the identification of the principal metabolic fluxes involved in different age groups and hence in the ageing process.

4   Conclusion

As with many diseases and health conditions, transcriptomics only is unlikely to reveal the full picture of ageing [24–26]. Multi-omic stoichiometric modelling can be used to predict the effects of gene expression on the metabolism and vice versa. More specifically, where a set of genes are known to be involved in a disease, genome-scale models can be integrated with transcriptomic data. Using a multi-omic model of ageing, we here achieve a greater understanding of the mechanisms of the disease by modelling the effect of altered gene expression on the metabolism of an organism (Fig. 1).

Interestingly, we are able to obtain a metabolomic network of patients, as opposed to the transcriptomic-only network commonly obtained from gene expression data. We applied agglomerative hierarchical clustering to visualize functional mapping of local enrichment of the metabolic flux rate response to the chronological and transcriptomic ageing processes. We report that the correlation between the chronological age and transcriptomic age shows the relationship between the transcriptomic layer and the metabolomic layer.

Future analyses, such as principal component analysis of the multi-omic data will enable us to show which metabolic pathways correlate with ageing. We will therefore be able to identify biomarkers that, taken together, can define a *metabolomic age predictor*. This can, for example, allow the classification of patients into different phenotypes depending on the metabolic presentation of the disease. Our model could also be used as a diagnostic/prognostic tool e.g. for early chronological ageing.

We finally showed that, using machine learning approaches, we can classify patients according to their metabolic phenotype of ageing by detecting global similarity between patients across multiple omics. As a result, we reach a multi-omic and genome-scale viewpoint that significantly extends the state-of-the-art, but transcriptomic-only, understanding of ageing.

References

1. Barzilai, N., Huffman, D.M., Muzumdar, R.H., Bartke, A.: The critical role of metabolic pathways in aging. Diabetes **61**(6), 1315–1322 (2012)
2. Bratic, I., Trifunovic, A.: Mitochondrial energy metabolism and ageing. Biochim. Biophys. Acta (BBA)-Bioenerg. **1797**(6), 961–967 (2010)

3. Harman, D.: The biologic clock: the mitochondria? J. Am. Geriatr. Soc. **20**(4), 145–147 (1972)
4. Chumlea, W.C., Rhyne, R.L., Garry, P.J., Hunt, W.C.: Changes in anthropometric indices of body composition with age in a healthy elderly population. Am. J. Hum. Biol. **1**(4), 457–462 (1989)
5. Okosun, I.S., Chandra, K.M., Choi, S., Christman, J., Dever, G.E., Prewitt, T.E.: Hypertension and type 2 diabetes comorbidity in adults in the united states: risk of overall and regional adiposity. Obesity **9**(1), 1–9 (2001)
6. Folsom, A.R., Kaye, S.A., Sellers, T.A., Hong, C.-P., Cerhan, J.R., Potter, J.D., Prineas, R.J.: Body fat distribution and 5-year risk of death in older women. Jama **269**(4), 483–487 (1993)
7. Muzumdar, R., Allison, D.B., Huffman, D.M., Ma, X., Atzmon, G., Einstein, F.H., Fishman, S., Poduval, A.D., McVei, T., Keith, S.W., et al.: Visceral adipose tissue modulates mammalian longevity. Aging cell **7**(3), 438–440 (2008)
8. Calle, E.E., Rodriguez, C., Walker-Thurmond, K., Thun, M.J.: Overweight, obesity, and mortality from cancer in a prospectively studied cohort of us adults. N. Engl. J. Med. **348**(17), 1625–1638 (2003)
9. Chee, C., Shannon, C.E., Burns, A., Selby, A.L., Wilkinson, D., Smith, K., Greenhaff, P.L., Stephens, F.B.: Relative contribution of intramyocellular lipid to wholebody fat oxidation is reduced with age but subsarcolemmal lipid accumulation and insulin resistance are only associated with overweight individuals. Diabetes **65**(4), 840–850 (2016)
10. Franceschi, C., Capri, M., Monti, D., Giunta, S., Olivieri, F., Sevini, F., Panourgia, M.P., Invidia, L., Celani, L., Scurti, M., et al.: Inflammaging and antiinflammaging: a systemic perspective on aging and longevity emerged from studies in humans. Mech. Ageing Dev. **128**(1), 92–105 (2007)
11. Blomberg, B.B., Frasca, D.: Quantity, not quality, of antibody response decreased in the elderly. J. Clin. Investig. **121**(8), 2981–2983 (2011)
12. Zhao, M., Qin, J., Yin, H., Tan, Y., Liao, W., Liu, Q., Luo, S., He, M., Liang, G., Shi, Y., et al.: Distinct epigenomes in CD4+ T cells of newborns, middle-ages and centenarians. Sci. Rep. **6**, 38411 (2016)
13. Marco, M.-G., Rebeca, A.-A.: When aging reaches CD4+ T-cells: phenotypic and functional changes. Front. Immunol. **4**, 107 (2013)
14. Lefebvre, J.S., Haynes, L.: Aging of the CD4 T cell compartment. Open Longev. Sci. **6**, 83 (2012)
15. Angione, C., Conway, M., Lió, P.: Multiplex methods provide effective integration of multi-omic data in genome-scale models. BMC Bioinform. **17**(4), 83 (2016)
16. Raj, T., Rothamel, K., Mostafavi, S., Ye, C., Lee, M.N., Replogle, J.M., Feng, T., Lee, M., Asinovski, N., Frohlich, I., et al.: Polarization of the effects of autoimmune and neurodegenerative risk alleles in leukocytes. Science **344**(6183), 519–523 (2014)
17. Peters, M.J., Joehanes, R., Pilling, L.C., Schurmann, C., Conneely, K.N., Powell, J., Reinmaa, E., Sutphin, G.L., Zhernakova, A., Schramm, K., et al.: The transcriptional landscape of age in human peripheral blood. Nat. Commun. **6** (2015). Article no. 8570
18. Irizarry, R.A., Hobbs, B., Collin, F., Beazer-Barclay, Y.D., Antonellis, K.J., Scherf, U., Speed, T.P.: Exploration, normalization, and summaries of high density oligonucleotide array probe level data. Biostatistics **4**(2), 249–264 (2003)
19. Han, F., Li, G., Dai, S., Huang, J.: Genome-wide metabolic model to improve understanding of CD4+ T cell metabolism, immunometabolism and application in drug design. Mol. BioSyst. **12**(2), 431–443 (2016)

20. Angione, C., Lió, P.: Predictive analytics of environmental adaptability in multi-omic network models. Sci. Rep. **5** (2015). Article no. 15147
21. Palsson, B.Ø.: Systems Biology Constraint-Based Reconstruction and Analysis. Cambridge University Press, Cambridge (2015)
22. Michener, C.D., Sokal, R.R.: A quantitative approach to a problem in classification. Evolution **11**, 130–162 (1957)
23. Vasson, M.-P., Farges, M.-C., Goncalves-Mendes, N., Talvas, J., Ribalta, J., Winklhofer-Roob, B., Rock, E., Rossary, A.: Does aging affect the immune status? A comparative analysis in 300 healthy volunteers from France, Austria and Spain. Immun. Ageing **10**(1), 38 (2013)
24. Mamas, M., Dunn, W.B., Neyses, L., Goodacre, R.: The role of metabolites and metabolomics in clinically applicable biomarkers of disease. Arch. Toxicol. **85**(1), 5–17 (2011)
25. Ramana, P., Adams, E., Augustijns, P., Van Schepdael, A.: Metabonomics and drug development. Metabonomics: Methods Protoc. **1277**, 195–207 (2015)
26. Nebert, D.W., Vesell, E.S.: Can personalized drug therapy be achieved? A closer look at pharmaco-metabonomics. Trends Pharmacol. Sci. **27**(11), 580–586 (2006)

Experimental Investigation of Frequency Chaos Game Representation for in Silico and Accurate Classification of Viral Pathogens from Genomic Sequences

Emmanuel Adetiba[1,4(✉)], Joke A. Badejo[1], Surendra Thakur[3],
Victor O. Matthews[1], Marion O. Adebiyi[2,4], and Ezekiel F. Adebiyi[2,4]

[1] Department of Electrical and Information Engineering, College of Engineering,
Covenant University, Ota, Nigeria
emmanueladetiba@gmail.com
[2] Department of Computer and Information Science, College of Science
and Technology, Covenant University, Ota, Nigeria
[3] KZN e-Skills CoLab, Durban University of Technology, Durban, South Africa
[4] Covenant University Bioinformatics Research (CUBRe), Ota, Nigeria
emmanuel.adetiba@covenantuniversity.edu.ng

Abstract. This paper presents an experimental investigation to determine the efficacy and the appropriate order of Frequency Chaos Game Representation (FCGR) for accurate and *in silico* classification of pathogenic viruses. For this study, we curated genomic sequences of selected viral pathogens from the virus pathogen database and analysis resource corpus. The viral genomes were encoded using the first to seventh order FCGRs so as to produce training and testing genomic data features. Thereafter, four different kernels of naïve Bayes classifier were experimentally trained and tested with the generated FCGR genomic features. The performance result with the highest average classification accuracy of 98% was returned by the third and fourth order FCGRs. However, due to consideration for memory utilization, computational efficiency vis-à-vis classification accuracy, the third order FCGR is deemed suitable for accurate classification of viral pathogens from genome sequences. This provides a promising foundation for developing genomic based diagnostic toolkit that could be used to promptly address the global incidence of epidemics from pathogenic viruses.

Keywords: Classification · FCGR · Genome · GSP · Naïve Bayes · Pathogens · Sequences · Virus

1 Introduction

Automatic detection of diverse species of viral pathogens associated with emerging deadly ailments within human populations cannot be over-emphasized as they remain a big threat to both personal and public health. Recent advances in molecular biology, next generation sequencing and online bioinformatics platforms offer a vast computational ecosystem for accurate identification of causative viral pathogens associated with the deadly human diseases. While allowing for extensive analysis, the rapidly

© Springer International Publishing AG 2017
I. Rojas and F. Ortuño (Eds.): IWBBIO 2017, Part I, LNBI 10208, pp. 155–164, 2017.
DOI: 10.1007/978-3-319-56148-6_13

growing databases of genomic sequences also provide an avalanche of resources for improved epidemic surveillance, diagnostics and therapeutics towards promoting healthy living. Furthermore, newer digital signal processing-based bioinformatics methods utilize numerical and/or visual encoding of nucleotide sequences collected from laboratory and environmental surveillances for effective non-alignment analysis [1, 2].

The application of digital signal processing techniques to genomic analysis, coined Genomics Signal Processing (GSP), requires that the nucleotide sequences be encoded numerically or graphically for alignment-free sequence comparison [1, 3, 4]. Next, discriminatory genomic features are extracted from the numeric genome representations to improve on species- or genome-level classification, usually based on a machine learning technique [5]. GSP-based techniques provide alignment-free analyses of the genomes to address the problems of unequal lengths of the sequences, the computational speed and large memory requirements encountered during alignment-based analysis [6, 7]. However for an accurate detection, it is necessary to ensure that the numeric or visual encoding of the nucleotide sequences represents the unique and salient characteristic of the genome as desirable.

Unlike other methods, the Chaos Game Representation (CGR) visually expresses the local patterns of the nucleotide sequences and hence the global structure of the genome in a two-dimensional graphical form [2, 8]. CGR is a scale independent representation developed by Jeffrey [9]. It was derived from the chaos theory, which allows the illustration of frequencies of oligonucleotides in the form of images. With CGR, the oligonucleotides of a genome exhibit the main physiognomies of the whole genome [7]. However, in the original form, CGR is not convenient for processing with a computer, hence, another form of CGR named Frequency Chaos Game Representation (FCGR) was introduced [7, 8, 10]. The CGR pattern of the nucleotide sequences of the same genome are found to be similar but differs quantitatively from the CGR patterns of the genome from another specie. This biological attribute makes the unique genomic signature of CGR and the subsequent features extracted from it, an accurate representation for alignment-free analysis suitable for classification, clustering and identification as proposed in many researches reported recently.

Karamichalis et al. [5] investigated the intra-specie and inter-specie variations of the genomic signatures generated by CGR patterns, using six different distance measures. The study validated the hypothesis that the CGR patterns of the nucleotide sequences of the same genome are similar but differs quantitatively from the CGR patterns of the genome from another specie. The CGR-based genomic signatures also accurately classified the genomic DNA sequences of Homo sapiens and Mus musculus genomes at lower taxonomic levels – class and order.

Messaoudi et al. [11] encoded the genomic sequence of Caenorhabditis elegans (C. elegans) with frequency of CGR patterns, otherwise called Frequency Chaos Game Representation (FCGR), for a time-frequency investigation using the Continuous Wavelet Transform (CWT). The complex Morlet wavelet based CWT revealed significant biological characteristics from the genomic signature of the FCGR patterns.

Kari et al. [12] proposed a molecular distance map developed with the unique genomic signature of CGR suitable for defining relationships between species to identify species, clarify taxonomies and related evolutionary history. Multi-Dimensional Scaling

(MDS) was applied to the distance metrics computed based on the Structural Dissimilarity Index (DSSIM) to produce the map. The map successfully characterized organisms into several taxonomy levels within the Euclidean space that showed the spatial proximity between the nucleotide sequences.

Tanchotsrinon et al. [13] adopted the CGR and Singular Value Decomposition (SVD) for Human Papillomavirus (HPV) genotyping as an approach to fight cervical cancer. Two classes of features were obtained from the SVD-reduced matrices of the original CGR: ChaosCentroid, which captured the structure of the sequences and ChaosFrequency, which represented relevant statistical distribution of nucleotides in the sequences. Their study demonstrated comparative results with no significant difference between their proposed method and the NCBI viral genotyping tool irrespective of the four classification techniques used i.e. Multi-layer Perceptron, Radial Basis Function, K-Nearest Neighbor and Fuzzy K-Nearest Neighbor.

In the current study, we experimentally explored the applicability of FCGR and its appropriate order for classification of viral pathogens from genomic sequences into the right species. This endeavor is aimed at laying a foundation for the development of an alternative, accurate and in silico genomic viral diagnostic tool, which could help in rapid medical interventions in the event of viral pathogens epidemic.

2   Materials and Methods

2.1   Dataset

As shown in Table 1, we extracted the genome sequences of Ebola virus (N = 249), Enterovirus (N = 632), Dengue virus (N = 390), HepatitisC virus (N = 567) and Zika virus (N = 351) from the Virus Pathogen Database and Analysis Resource (ViPR) corpus. This corpus was developed to provide free access to genomic and proteomic sequences of viral pathogens for research and development of vaccines, therapies and diagnostic tools. The Universal Resource Locator (URL) for ViPR as at the time this study was carried out is https://www.viprbrc.org/brc/home.spg?decorator=vipr. The total sample size of the dataset extracted for this study is 2,189. Although, there is a huge collection of pathogenic viral datasets on the corpus, the five viruses were selected due to their prominence as causative agents of diseases that is currently of concern among researchers on a global scale. These viruses are also specifically featured on the home page of the ViPR corpus and the structural diversity of their genomes provide a good basis to investigate the efficacy of FCGR for viral species classification.

Table 1. Extracted dataset for five pathogenic viruses

S/N	Viral species	Number of unique samples
1	Ebola virus	249
2	Enterovirus	632
3	Dengue virus	390
4	HepatitisC virus	567
5	Zika virus	351
	Total	2,189

2.2   FCGR Computation at Different Orders and Naïve Bayes Classifier

FCGR is a numerical matrix in contrast to CGR, which is a graphical representation. Instead of plotting a CGR first and converting it to a FCGR [7, 8]. Wang et al. [10] posits that FCGR can be derived directly from a sequence. Furthermore, Wang et al. [10] introduced the concept of FCGR order, which provides variants in matrix dimensions when FCGR are derived directly from the sequences. For instance, given a sequence S, with f_w representing the frequency of the oligonucleotide w, the matrix structure of a first order FCGR is given as Eq. (1) [10].

$$FCGR_1(S) = \begin{pmatrix} f_C & f_G \\ f_A & f_T \end{pmatrix} \tag{1}$$

The FCGR of (k + 1)th order can be computed by substituting each element f_x in a kth order FCGR with the four elements

$$\begin{pmatrix} f_{CX} & f_{GX} \\ f_{AX} & f_{TX} \end{pmatrix}. \tag{2}$$

Therefore, the matrix structure of a second order FCGR is as shown in Eq. (2) and higher order FCGR can be sequentially computed.

$$FCGR_2(S) = \begin{pmatrix} f_{CC} & f_{GC} & f_{CG} & f_{GG} \\ f_{AC} & f_{TC} & f_{AG} & f_{TG} \\ f_{CA} & f_{GA} & f_{CT} & f_{GT} \\ f_{AA} & f_{TA} & f_{AT} & f_{TT} \end{pmatrix} \tag{3}$$

From Eqs. (1) and (3), it can be seen that a k-th order FCGR is a 2^k x 2^k matrix and it contains 4^k occurrences of the k length oligonucleotides [10, 14]. The direct correspondence of CGR and FCGR, in which a kth order FCGR is equivalent to a CGR of resolution $1/2^k$ was also reported [10]. This makes it possible to observe the major features that are inherent in higher order FCGR (which ordinarily is incomprehensible because of the size) by visual observation of the equivalent CGR. Researchers have also opined that CGR images and correspondingly the FCGR obtained from subsequence of a genome present similar structure as the whole genome [7]. This implies that the CGR image or FCGR of a subsequence is a sufficient genomic signature for species classification rather than the CGR image or FCGR of the whole genome [7, 10]. Therefore, in this study, we ventured to experimentally investigate the efficacy of FCGR at different resolutions or orders ($1 \leq k \leq 7$) for pathogenic virus species classification. We stopped at 7th order because the huge dimension of the matrix elements at 8th order and beyond is computationally expensive without providing any benefits with respect to classification accuracy.

The accurate classification of the viral species from the FCGR-encoded nucleotide sequences was carried out with the Naïve Bayes (NB) classifier, which is a very popular classifier in bioinformatics [15, 16]. NB classifier utilizes a key statistical assumption of conditional independence of the FCGR features within the same class, to assign a class label to each sequence [15]. The class label in this context refers to the viral species, drawn from Table 1. The NB classifier can be trained using different kernel functions such as uniform, epanechnikov, normal and triangular to detect through classification the most probable viral specie from each encoded sequence.

2.3 Experiments

Experiments were performed in this study to determine the efficacy as well as the appropriate order of FCGR for classifying viral pathogens from genomic sequences. The curated sequences for the five viruses were first converted to their numeric equivalents with 1st to 7th order of FCGR. For each of these orders, the FCGR encoded viral sequences were transmitted to train Naïve Bayes classifier using four different kernel functions, namely; uniform, epanechnikov, normal and triangular. Both the FCGR algorithm and naïve Bayes classifier were implemented in MATLAB R2015a, which also provided the in silico platform to perform all the experiments in this study. The PC on which the experiments were performed contains an Intel Core i5-4210U CPU operating at 2.40 GHz speed, with 8.00 GB RAM and runs 64-bit Windows 8 operating system.

3   Results and Discussion

Figure 1 shows the plot of the number of elements in the computed FCGR matrices against the FCGR order. As illustrated on the graph, a first order FCGR matrix contains 4 elements, a second order contains 16 elements, third order contains 64 elements,

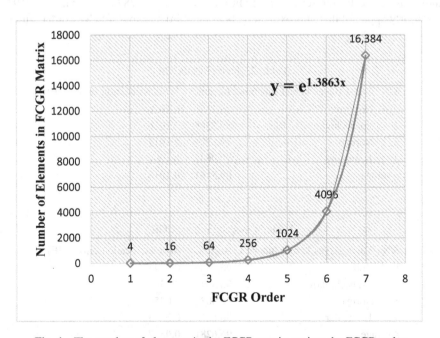

Fig. 1. The number of elements in the FCGR matrix against the FCGR order

fourth order contains 256 elements, fifth order contains 1024 elements, sixth order contains 4096 elements and seventh order contains 16,384 elements. The relationship between the number of elements in the FCGR matrix and the number of FCGR order is clearly an exponential growth which is represented as:

$$y = e^{1.3863x} \qquad (4)$$

where y is the number of elements in the FCGR matrix and x is the FCGR order. The results of the experiments, which are hereafter reported provide an insight on the effects of FCGR order vis-à-vis the number of elements in the corresponding FCGR matrices on pathogenic viral classification accuracy.

Tables 2, 3, 4, 5, 6, 7, and 8 show the results we obtained when the first to seventh order FCGR matrices were respectively utilized to encode the viral genomic sequences. We deemed it expedient to compute the average classification accuracies and average misclassification errors for the four different kernels across the FCGR orders. This provides a compact scheme for the comparison of our results based on the FCGR orders. The average classification results for the first order FCGR is shown in Table 2 (Accuracy = 89.8161%, ME = 0.1018). About 7% increase in performance was obtained for the second order FCGR (Accuracy = 96.6281%, ME = 0.0337) over the first order. Furthermore, increase in performance continued with the third order (Accuracy = 98.3651%, ME = 0.0163) up to the fourth order (Accuracy = 98.1607%, ME = 0.0184). There is however a drastic reduction in the performance results for the fifth order (Accuracy = 94.2098%, ME = 0.0579), which continued for the sixth order (Accuracy = 85.7857%, ME = 0.1422) and the lowest performance results in this study was posted for the seventh order FCGR (Accuracy = 78.0654%, ME = 0.2194). It is clearly apparent that approximately, both the third and fourth FCGR order with 64

Table 2. First order FCGR

S/N	Naïve Bayes kernel function	Accuracy	Misclassification error (ME)
1	Uniform	89.7366	0.1026
2	Epanechnikov	89.7366	0.1026
3	Normal	89.8274	0.1017
4	Triangular	89.9637	0.1004
Average		**89.8161**	**0.1018**

Table 3. Second order FCGR

S/N	Naïve Bayes kernel function	Accuracy	Misclassification error (ME)
1	Uniform	96.5486	0.0345
2	Epanechnikov	96.5940	0.0341
3	Normal	96.5032	0.0350
4	Triangular	96.8665	0.0313
Average		**96.6281**	**0.0337**

Table 4. Third order FCGR

S/N	Naïve Bayes kernel function	Accuracy	Misclassification error (ME)
1	Uniform	98.4105	0.0159
2	Epanechnikov	98.4559	0.0154
3	Normal	98.1381	0.0186
4	Triangular	98.4559	0.0154
Average		**98.3651**	**0.0163**

Table 5. Fourth order FCGR

S/N	Naïve Bayes kernel function	Accuracy	Misclassification error (ME)
1	Uniform	98.7284	0.0127
2	Epanechnikov	98.4559	0.0154
3	Normal	97.0027	0.0300
4	Triangular	98.4559	0.0154
Average		**98.1607**	**0.0184**

Table 6. Fifth order FCGR

S/N	Naïve Bayes kernel function	Accuracy	Misclassification error (ME)
1	Uniform	95.5495	0.0445
2	Epanechnikov	96.9119	0.0309
3	Normal	87.5568	0.1244
4	Triangular	96.8211	0.0318
Average		**94.2098**	**0.0579**

Table 7. Sixth order FCGR

S/N	Naïve Bayes Kernel function	Accuracy	Misclassification error (ME)
1	Uniform	85.6494	0.1435
2	Epanechnikov	93.2788	0.0672
3	Normal	71.4805	0.2852
4	Triangular	92.7339	0.0727
Average		**85.7857**	**0.1422**

Table 8. Seventh order FCGR

S/N	Naïve Bayes kernel function	Accuracy	Misclassification error (ME)
1	Uniform	82.8792	0.1712
2	Epanechnikov	80.2906	0.1971
3	Normal	66.6213	0.3338
4	Triangular	82.4705	0.1753
Average		**78.0654**	**0.2194**

and 256 elements respectively, gave the highest performance results (Approximate Accuracy = 98%, Approximate ME = 0.02). The summary of the entire result is graphically represented in Fig. 2.

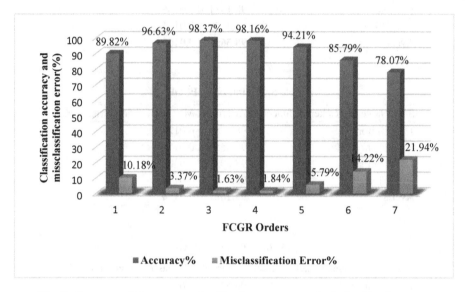

Fig. 2. Summary of the average classification accuracy and misclassification error.

The results in this study clearly agree with the curse of dimensionality philosophy in machine learning, in which too small training data features (first and second order FCGR in this context) may hinder the creation of a reliable classification model for assigning a class to all possible objects in the dataset. Conversely, high dimensions in the training features (sixth and seventh order FCGR in this context) tend to make the contiguity among data points more identical and often lead to lower classification accuracy. Training data with high features has also being reported to lead to high computational cost and memory usage [17], which is the case with the eight order FCGR in this study that motivated its exclusion from the experiments.

Our literature search while undertaking this research yielded few studies that have employed FCGR and other schemes for identification of species ([1, 18–21]. The study by Vijayan et al. [18] utilized a third order FCGR (64 element vector) to encode Eukaryotic organisms with Probabilistic Neural Network (PNN) as a classifier to obtain a classification accuracy of 92.3%. In codicil, the study reported in [1] where a 15-element real Genomic Cepstral Coefficients (GCC) with Radial Basis Function Neural Network (RBFNN) were utilized for identification of four pathogenic viruses gave an accuracy of 97.3%. Obviously, the current result of 98% classification accuracy for five pathogenic viruses with 64 (third order) and 256 (fourth order) elements FCGR and naïve Bayes classifier is comparable to the highlighted similar results in the literature. However, based on the experimental results obtained in this study, the third order FCGR is recommended as an appropriate genomic feature for pathogenic viral classification. This will appositely culminate in economy of memory space, computational efficiency and acceptable accuracy for viral pathogens classification, which is an important contribution albeit moderate, to GSP and bioinformatics body of knowledge.

4 Conclusion

Thus far, we have been able to achieve the objectives of the current study, which are to determine the efficacy of FCGR and its appropriate order for accurate classification of pathogenic virus from genomic sequences. The 98% classification accuracy obtained with the third order FCGR is clearly promising for developing in silico and accurate diagnostic tool for viral pathogens classification using next generation genomic sequences. In the future, we hope to substantially extend this study by increasing the viral pathogens coverage and further experiment with state-of-the-art machine learning methods like deep learning and hierarchical classifiers.

Acknowledgement. The publication of this study is supported and funded by the Covenant University Centre for Research, Innovation and Development (CUCRID), Covenant University, Canaanland, Ota, Ogun State, Nigeria.

References

1. Adetiba, E., Olugbara, O.O., Taiwo, T.B.: Identification of pathogenic viruses using genomic cepstral coefficients with radial basis function neural network. In: Pillay, N., Engelbrecht, A.P., Abraham, A., du Plessis, M.C., Snášel, V., Muda, A.K. (eds.) Advances in Nature and Biologically Inspired Computing. AISC, vol. 419, pp. 281–291. Springer, Cham (2016). doi:10.1007/978-3-319-27400-3_25
2. Hoang, T., Yin, C., Yau, S.S.T.: Numerical encoding of DNA sequences by chaos game representation with application in similarity comparison. Genomics 108(3), 134–142 (2016)
3. Huang, G., Zhou, H., Li, Y., Xu, L.: Alignment-free comparison of genome sequences by a new numerical characterization. J. Theor. Biol. 281(1), 107–112 (2011)
4. Qi, Z.H., Du, M.H., Qi, X.Q., Zheng, L.J.: Gene comparison based on the repetition of single-nucleotide structure patterns. Comput. Biol. Med. 42(10), 975–981 (2012)
5. Karamichalis, R., Kari, L., Konstantinidis, S., Kopecki, S.: An investigation into inter-and intragenomic variations of graphic genomic signatures. BMC Bioinform. 16(1), 1 (2015)
6. Swain, M.T.: Fast comparison of microbial genomes using the Chaos games representation for metagenomic applications. Procedia Comput. Sci. 18, 1372–1381 (2013)
7. Deschavanne, P.J., Giron, A., Vilain, J., Fagot, G., Fertil, B.: Genomic signature: characterization and classification of species assessed by chaos game representation of sequences. Mol. Biol. Evol. 16(10), 1391–1399 (1999)
8. Almeida, J.S., Carrico, J.A., Maretzek, A., Noble, P.A., Fletcher, M.: Analysis of genomic sequences by chaos game representation. Bioinformatics 17(5), 429–437 (2001)
9. Jeffrey, H.J.: Chaos game representation of gene structure. Nucleic Acids Res. 18, 2163–2170 (1990)

10. Wang, Y., Hill, K., Singh, S., Kari, L.: The spectrum of genomic signatures: from dinucleotides to chaos game representation. Gene **14**(346), 173–178 (2005)
11. Messaoudi, I., Oueslati, A.E., Lachiri, Z.: Wavelet analysis of frequency chaos game signal: a time-frequency signature of the C. elegans DNA. EURASIP J. Bioinform. Syst. Biol. **2014**(1), 1 (2014)
12. Kari, L., Hill, K.A., Sayem, A.S., Karamichalis, R., Bryans, N., Davis, K., Dattani, N.S.: Mapping the space of genomic signatures. PLoS one **10**(5), e0119815 (2015)
13. Tanchotsrinon, W., Lursinsap, C., Poovorawan, Y.: A high performance prediction of HPV genotypes by chaos game representation and singular value decomposition. BMC Bioinform. **16**(1), 1 (2015)
14. Stan, C., Cristescu, C.P., Scarlat, E.I.: Similarity analysis for DNA sequences based on chaos game representation. Case study: the albumin. J. Theoret. Biol. **267**(4), 513–518 (2010)
15. Sandberg, R., Winberg, G., Bränden, C.I., Kaske, A., Ernberg, I., Cöster, J.: Capturing whole-genome characteristics in short sequences using a naive Bayesian classifier. Genome Res. **11**(8), 1404–1409 (2001)
16. Wang, Q., Garrity, G.M., Tiedje, J.M., Cole, J.R.: Naive Bayesian classifier for rapid assignment of rRNA sequences into the new bacterial taxonomy. Appl. Environ. Microbiol. **73**(16), 5261–5267 (2007)
17. Janecek, A., Gansterer, W.N., Demel, M., Ecker, G.: On the relationship between feature selection and classification accuracy. In: FSDM, pp. 90–105, 15 September 2008
18. Vijayan, K., Nair, V.V., Gopinath, D.P.: Classification of organisms using frequency-chaos game representation of genomic sequences and ANN. In: 10th National Conference on Technological Trends (NCTT 2009), pp. 6–7, November 2009
19. Nair, V.V., Nair, A.S.: Combined classifier for unknown genome classification using chaos game representation features. In: Proceedings of the International Symposium on Biocomputing, p. 35. ACM (2010)
20. Yang, L., Tan, Z., Wang, D., Xue, L., Guan, M.X., Huang, T., Li, R.: Species identification through mitochondrial rRNA genetic analysis. Sci. Rep. **4**(4089), 1–11 (2014)
21. Adetiba, E., Olugbara, O.O.: Classification of eukaryotic organisms through cepstral analysis of mitochondrial DNA. In: Mansouri, A., Nouboud, F., Chalifour, A., Mammass, D., Meunier, J., ElMoataz, A. (eds.) ICISP 2016. LNCS, vol. 9680, pp. 243–252. Springer, Cham (2016). doi:10.1007/978-3-319-33618-3_25

Gamified Mobile Blood Donation Applications

Lamyae Sardi[1(⊠)], Ali Idri[1], and José Luis Fernández-Alemán[2]

[1] Software Project Management Research Team, ENSIAS,
Mohammed V University, Rabat, Morocco
lamyasardi@gmail.com, ali.idri@um5.ac.ma
[2] Faculty of Computer Science, Department of Computing,
University of Murcia, Murcia, Spain
aleman@um.es

Abstract. Unpaid blood donation is a selfless act of citizenship and the usage of gamification elements in blood donation apps can enhance the donors' experience, especially among youth. This paper analyses the functionalities and explores gamification elements of the existing blood donation apps in the mobile market. A search in Google Play, Apple Apps store, Blackberry App World, and Windows Mobile App store was performed to select 10 gamified BD apps with three duplicates out of 801 pinpointed. The results show that the majority of the blood donation apps selected do not support multiple languages and that the predominant authentication methods are traditional and social logins. Moreover, all the apps were intended for more than one purpose among helping users to find donors and blood centres, track their records and check their eligibility to donate. Most apps installed include notification features and built-in geolocation services to instantly inform the users of donation need in nearby locations. Badges and redeemable points were the most recurrent gamification elements in the blood donation apps selected. There is a need for better incentives in order to not only retain the potential donors but also to recruit non-willing ones.

Keywords: Blood donation · Mobile app · Gamification · Review · m-Health

1 Introduction

Blood donation (BD) is one of the most valuable contribution that an individual can make towards the society. In spite of the significant advances in medical area, no real progress has been made to develop a substitute for blood [1]. This induces the importance of donating blood. In fact, there is a surge demand for blood supplies and blood products [2] each year which are used in different situations ranging from severe childhood anaemia to cancer therapy depending on the country's incomes [3]. Despite the people's awareness about the need for blood, 90% of people who are eligible to donate blood are not currently doing so [4]. Non-donors provide different reasons for their non-willingness such as busy schedule, needle phobia and fear of catching disease [5]. Given this critical situation, many studies investigated how mobile technology can boost potential donors' motivation through a multitude released BD apps. Social networks and text messages have played a leading role in the search and recruitment of blood donors, respectively [6]. However, it has been affirmed that relying on social

© Springer International Publishing AG 2017
I. Rojas and F. Ortuño (Eds.): IWBBIO 2017, Part I, LNBI 10208, pp. 165–176, 2017.
DOI: 10.1007/978-3-319-56148-6_14

networks and text messages to broadcast the need of blood donation could be inefficient if the information is irrelevant to potential donors [7]. Therefore, the BD apps must be developed with in-built geolocation services and push notifications to instantly inform the donor of blood need in nearby locations [7]. One of the critical issues that is encountered in blood donation is the ageing population of most countries which implies the blood centres to mobilize in order to recruit more blood donors among youth as well as improving their retention [8]. In this regards, as the young interaction with mobile devices and game world is significantly high [9], the usage of gamification techniques in BD apps could be of great help. Actually, gamification lies in applying game mechanics and dynamics in non-game contexts [10]. A myriad of domains such as marketing and recruitment have adopted gamification as a strategy to enhance user's loyalty and improve user's motivation [11]. In blood donation area, gamification can enhance the retention rate of donors and motivate the non-donors to start donate through offering proper incentives [12]. Rewarding the blood donors takes several forms. A donor can earn points for each blood unit donated which can be redeemed for material rewards (e.g. gift cards, T-shirt, keychain). Considering that the individual is mostly driven by its intrinsic motivations [13], levelling and acquiring status can have a better influence on donors' engagement and loyalty.

The goal of this paper is to study the functionalities and identify the gamification techniques employed in the existing gamified BD apps. The search for these apps was performed in the four app repositories namely Google Play, Apple store, Blackberry World and Windows Mobile store. Throughout the analysis process, a systematic review approach was used to ensure the accuracy and the objectivity of the process. Of a total of 801 apps, 10 gamified apps were selected and investigated after examining the compliance of their descriptions and screenshots to gamification aspects. This review aims, primarily to study the availability of the BD apps in the app market along with their types. Moreover, this review summarizes the key features and gamification techniques proposed in the BD apps that are likely to boost the blood donation experience.

The remainder of this paper is structured as follows: Sect. 2 describes the research method employed to select the gamified BD apps. Section 3 synthesizes the results obtained and discusses the main findings of this study. Limitations of this study are presented in Sect. 4. Finally, Sect. 5 presents the conclusions and future work.

2   Method

This section describes the methodology used to search for, select and analyse the gamified BD apps.

2.1   Review and Protocol

A set of recommendations set out by PRISMA [14] has been used to address the search for gamified BD apps. Prior to the beginning of the search, a review protocol was developed which sets out the methods of the review including research questions, eligibility criteria and data extraction.

2.2   Research Questions

In order to guide this review, three research questions were formulated according to a previous study on BD apps [15].
Table 1 shows the research questions addressed in this study and their motivations.

Table 1. Research questions

ID	Research question	Motivation
RQ1	Which BD apps are currently available in app stores and what types of BD apps have been selected?	To identify the BD apps that are available in app repositories and the areas of BD that are covered by the selected apps
RQ2	What are the functionalities and features proposed in the installed BD apps?	To study the common functionalities and characteristics as regards BD apps
RQ3	How gamification has been employed in the selected apps?	To analyse which gamification strategies are mostly used in the BD apps

2.3   Eligibility Criteria

The following inclusion criteria (IC) have been used to select the BD apps of this study:

- Free or Paid BD apps available in app repositories.
- Apps for human Blood Donation.
- Apps that are gamified.

IC1 selects the BD apps that are available under the four mobile platforms (iOS, Android, Windows Mobile and Blackberry). IC2 selects the apps that focus only on human blood donation, hence discarding those for pets or those that focus on general health. IC3 assesses the presence of gamification elements in the apps. In this respect, applications were evaluated through their descriptions and screenshots.

2.4   Search Strategy

The study sample was identified through systematic searches in the four app repositories. Apps were identified in June 2016. Search terms were defined using PICO (population, intervention, comparison and outcome) [16] criteria. The population considered is that of blood donors; the intervention consisted on gamified apps for blood donation and the outcomes encompasses all existing outcomes regarding gamified BD apps. The 'comparison' criterion was overlooked since this study does not aim to prove evidence regarding gamified BD apps nor providing alternatives. The search string is defined as follows: *Blood AND (donat* OR give OR bank OR network OR help OR need*).*

2.5   Selection Process

The selection process went through the following main steps:

1. Entering the search terms in the mobile app stores to identify candidate apps.
2. Screening and assessment of the apps using eligibility criteria as shown in Fig. 1.
3. Installing the selected apps in appropriate devices. Since the apps designed for iOS were also compatible with both iPhone and iPad, an iPad 2 was used to explore them. Whereas the apps for android were installed to a Samsung GALAXY Note 2 and the windows app was installed on Nokia Lumia 635 smartphone.

The activities described above were carried out between June and July 2016 by one author. Any discrepancies were resolved through discussion by the rest of authors.

Figure 1 illustrates the selection process of the apps for each app repository. Ten gamified BD apps were selected from a total of 801 apps identified. Although two apps were available for both Android and iOS platforms; they were considered separately.

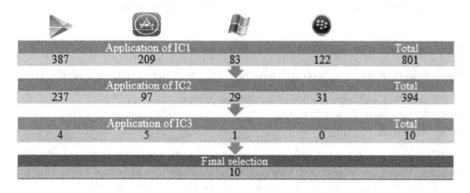

Fig. 1. PRISMA diagram

2.6   Data Extraction

Data extraction process was performed by the first author through a designed form and was reviewed by the two other authors. A set of data items was extracted for each selected BD app. These data items were analysed to show:

- ID. The first letter refers to the source of the app; G represents an Android app, the letter A refers to an iOS app while Windows apps are pointed out with the letter W.
- Name of the App.
- Type of BD App. The following eight types were identified [17].
  - Find donors. Apps which help the user find potential blood donors.
  - Find centres. Apps which help the user find nearby centres to donate blood.
  - Eligibility. Apps which determine whether or not the user is eligible to donate based on their health condition and the date of their last blood donation.
  - Blood Types. Apps which provide information on the different blood groups.

– Blood Calculation. Apps that estimate the user's blood type based on their parental blood types.
– Records. Apps which record and list out the user's donation history.
– Related to a centre. Apps that offer to the user useful information related to a medical institution (e.g. hospital, laboratory …).
– General Information. Apps that give information on the blood donation process.

• App repository link to the App.
• Developer and Country of origin.
• Category.
• Latest update.
• Number of installs.
• Number of raters.
• User rating. It is scored out of 5.

A list of 8 functionality items was also extracted and evaluated for each of the gamified BD apps. These characteristic questions along with their rationales are presented in Table A.1 of Appendix A. The way in which this questionnaire was written was inspired by previous studies [15, 18]. The extracted data and functionality items corresponding to each selected app are listed in Tables B.1, B.2 and B.3 of Appendix B. All appendices can be found at the following link: http://www.um.es/giisw/gbda/appendices.pdf.

3 Results and Discussion

This section describes and discusses the results related to the data extraction process. The extracted data along with the characteristics of the BD apps selected are presented in Appendix B. The search for the gamified BD apps was performed in June 2016. A total of 10 gamified BD apps were selected from the 801 apps identified. As depicted in Fig. 1, eligibility criteria were applied separately to apps for each app repository. In the screening phase, 407 of 801 apps were discarded as they were not intended for human Blood Donation. Only 10 apps appeared to be eligible in terms of incorporating gamification elements and were, consequently, selected for further evaluation in this study.

RQ1: Which BD apps are currently available in app stores and what types of BD apps have been selected?

Among the four app repositories, Apple App store was the app store that offered most BD apps (5 apps) followed by Google Play Store (4 apps). In contrast, the number of BD apps available for Windows and Blackberry devices was less significant. Only one BD app was identified in Windows Phone Store and no app was found relevant in Blackberry app store. This could be owing to the fact that Android and iOS are the most used operating system worldwide [19]. The total of these apps contains duplicates. Actually, three apps are available in more than OS, Android/iOS platform and Android/Windows respectively and were hence included because the features they offer may differ between the corresponding mobile platforms. No restriction was made on the category in which the app figures within the app store in order to include most available gamified BD apps. The small number of the apps selected could be explained by the fact that gamification is a novel discipline in health area in general and blood donation in particular. Hence, incorporating game mechanics in BD apps has not been put heavily into practice. Around 70% of the apps selected have been updated between 2014 and 2016. The G2 and A4 apps have been ultimately updated in 2012 whether because the app is stable and has a good functionality compliance or the app is no longer serving. Table B.2 shows the latest date of update and the current version of each selected gamified BD mobile app. Among the eight aforementioned types of the BD apps, finding centres is the most recurrent type identified in all the selected apps. They help users to find nearby blood drives and donation centres, thus, bridging the gap between blood donors and patients in-need. Five apps of those considered help users to find donors. These apps help meet the constant need of blood by providing an additional support to laboratories and blood banks, especially in emergency situations [20]. The BD eligibility apps selected in this study calculate the eligibility date of the next donation and provide a set of criteria to help the users find out whether they are eligible to donate or not. Considering the World Health Organization (WHO) guidelines on donor suitability for blood donation [21] the main eligibility criteria focus on the lower and upper age limit (18, 65 years old), weight limits (at least 45–50 kg) and vital signs (Pulse, body temperature and body pressure). As depicted in Table B.1, all the apps selected cover more than one type to achieve the utmost efficiency. These apps were counted separately for each type. Around 30% of the apps that combine different types offer the users general information about the process of donating blood. Only Blood donor+ and Blood donor by the American Red Cross apps in their both iOS and Android versions allow the users to maintain a personal record of their blood donations. Nonetheless, no app from those selected was attributed to Blood types and Blood calculation types. This is likely due to the fact that these activities do not need further incentives as they are considered personal and are not a matter of general interest. Fig. C.1 of Appendix C shows the types of the gamified BD apps.

RQ2: What are the functionalities and features proposed in the installed BD apps?

This question was discussed by considering the results obtained when executing the 10 selected apps and answering the questions of Appendix A. The 8 aforementioned characteristics were evaluated for each selected BD app. The results are presented in the following subsections and summarised in Appendix B. The reader may note that installation problems were not taken into account in order to yield an overview of the existing set of gamified BD apps in repositories. However, no significant issue was encountered during the installation of the apps.

Authentication Method. Around 90% of the apps require the user to provide credentials to identify them which can be accomplished in many ways. Table 2 presents the authentication methods employed in the BD apps installed. The Daruj krev s VZP is the only app that provided open access and did not require any login. In the G1, A1 and W1 apps, the user is required to create an account using an email address whereas a social network login particularly Facebook or Twitter is requested in the G3, A2 and A4 apps. The possibility of choosing between a normal login and a social network

Table 2. Authentication methods

Authentication method	App ID
Login	G1 A1 W1
Social network login	G3 A2 A4
Login or social network login	G4 A5
Login + center ID	A3

login is provided by two apps namely G4 and A5. Using a social login in the majority of the apps installed facilitate the access to the app. In order to access the content of Central Blood Bank app, the user can enter the app as a guest although most of the features would not be usable. To harness the app to the utmost, a specific centre ID is needed to create a login account when registering for the first time. This condition shrinks the number of the app's users as it is only accessible for the adherents of the centre. However, it helps to lessen making random and anonym accounts by non-good willing people who may access sensitive details of the users [22]. Thanks to technological advances, the authentication methods are becoming more sophisticated and not easily broken which can crowd out the social and traditional login, still however widely used in the mobile apps.

Multiple Languages. Perhaps one of the actual challenges of the apps is the globalization. Building a mobile app in multiple languages promotes its integration in the international market. However, English remains the most recurrent language supported by the majority of the apps as it is considered to be a universal language [23] mainly because it is the most spoken language worldwide owing to the British and American world economic, political and military domination aspects. In fact, it is supported in 9 of the apps selected beside Dutch, French, German, Romanian, Russian, Spanish and Traditional Chinese in A1 and G1 apps. Although donating blood is frequently made in the surrounding areas [24], developing a BD mobile app with a multiple languages support could be necessary to widen the target users of the app living in the country or area in question. The other language that was identified in the installed apps is Czech available in the Daruj krev s VZP app.

Geographical Restriction. All of the installed apps are dedicated to a specific country. Although there is no restriction during the installation nor the authentication, some of the features such as finding blood banks or donors could be without interest if the user is not located in the targeted area of the app. Four of the apps selected are intended to the people of USA, while five are for countries of Asia namely Singapore, India and Nepal. The remaining app is dedicated to people from Czech Republic. Considering the types of the BD apps selected, their geographical limitation for usage is evident as it would not be plausible to provide blood donation services at a worldwide scale.

User Communication with Medical Institutions. Among those selected, five apps are related to medical institutions. G1 and A1 apps which are the same app developed for Android and iOS smartphones by the American Red Cross, provide details about nearby Red Cross blood drives to schedule appointment. Similarly, the A5 app is linked to the Red Cross of Singapore and gives information about convenient blood banks' location that are ready to accept the donation, among others. However, both apps do not inform the user of any personal blood-related detail after donating. The A3 and A4 are the only apps that are directly connected with a third party. The Central Blood bank app can only be accessed using a particular ID that can be requested at the blood bank which means that the user could receive a feedback after the donation. The A4 app is connected to the blood centre of New Orleans in USA, the user could receive BD test from a laboratory in USA after donating blood, stating whether the user is eligible or not to give blood in the future. The gap between the clinical information of patients and blood banks should be systemically bridged in order to better exploit the usage of blood according the diagnoses of the patient which can be retrieved from different sources (e.g. electronic medical record) [2, 12].

Integration with Social Networking Portals. All the apps selected allow the user to share information via social networks except the G2 App. Facebook is the most predominant social network that is incorporated into the BD apps followed by Twitter. Apart from the G3 and A2 apps which are the same app developed for Android and iOS platforms, the BD apps are integrated with more than one networking site such as Twitter, Instagram, YouTube or others. The American Red Cross Apps for BD allow the users to share achievements through the available social platforms on their devices. Sharing BD information in social networks can be of great help in recruiting new donors, thus increasing the number of voluntary donors among youth in an efficient and economical way [25]. Also, there is evidence that people tend to give a huge importance to their relationships, consequently, they become more motivated to act in a way that maintains their social status [26]. Owing to the fact that social influence shapes the one's behaviour [12], a deeper investigation on the factors impacting donors' behaviour change has to be carried out to increase the recruitment and retention of blood donors.

Geolocation. Geolocation is a paramount feature that should be integrated in mobile apps as it helps the users search for precise locations using maps that illustrate the most direct routes and other navigational data [27]. Nowadays, GPS services for emergency or Location-Based Services are one of the advanced handsets that any smartphone has included which makes the delivery of information goes seamlessly and rapidly. Around 90% of the mobile applications examined are developed with a built-in geolocation feature. Donors and centres' locations are displayed on a map that can be visualized and optimized based on the real-time location data of the user to identify the nearest centres or donors. Since GPS services operates with users' personal information, malicious implications as well as privacy and anonymization concerns can be major issues that need to be addressed to prevent the misuse of data [22].

Notifications. Apart from the G2 App, all the apps installed provide the users with notification feature. The user can be notified in different ways. Among those investigated, a notification can be an instant alert that the user receives about blood shortage in nearby locations or a reminder of a donation appointment or the date on which the next donation is possible. Users can also receive push notifications for upcoming nearby campaigns and receive instant invitation to donation events. Notifications are a powerful tool that gets users' attention in a disruptive manner. However, these notifications may become annoying and intrusive if they convey irrelevant and interesting messages, thus leading to negative perceptions about the app.

BD Recommendations. To ensure the safety of blood donation for both donors and recipients, volunteer donors must be evaluated to check their eligibility according to the guidelines determined by the WHO [21]. Countries may adapt these guidelines to the characteristics of their population. All the mobile applications installed provide blood donation recommendations to the users except three apps; G3, A3 and W1. These recommendations mostly focus on explaining the BD process and eligibility requirements. In addition, users are given tips and guidelines on how to stay comfortable and safe throughout the donation process. In order to decrease wait times at blood drives and streamline the donation experience, the users of some of the apps installed such as A1, A3 and G1 are requested to complete an online health history questionnaire prior to arriving at the donation centre. Users should complete the questionnaire by visiting the official websites of the apps [28, 29] from their personal computer and present it at the blood drive whether printed or scanned to be evaluated.

RQ3: How gamification has been employed in the selected apps?

Gamification stands on applying game mechanics and dynamics in non-game settings. There is evidence that this technique is able to engender motivational drivers of human behaviour in non-game applications [30] by leveraging the users' natural desire for competition, status and achievement, among others [31]. This is accomplished through providing non-monetary incentives and rewards upon performing some actions within the application. These intrinsic rewards mostly consist on social recognition, appreciation and accomplishment feelings [32]. In fact, gamification seizes its essence through using a mixture of motivational affordances such as points, badges and leaderboards. Almost 80% of the mobile applications considered in this paper focus on badges and rewards as gamification elements to incentivise blood donation. The users of A1 and G1 apps are able to acquire badges depending on their donations' type and frequency. The Blood donor app allow the donors to create or join teams and track its ranking on a national leaderboard. Additionally, through the award-winning engine WeWin, partnered retailers offer the users of this app diverse array of extrinsic rewards (e.g. coupons, promo codes) to boost their donations. In the android app Daruj krev s VZP, users can visualize their donation and sharing's mark upon which they can receive badges (silver, gold, diamond). This is also the case of the Android (G3) and iOS (A2) versions of Blood Donor+ app that award various badges to blood donors (Silver, Gold, Diamond, Platinum) alongside the champion and lifesaver badges. Nevertheless, no further information on the rewards is provided within this application. In both android (G5) and windows (W1) versions of the Bloodstore app, users are awarded points so as to recognize the most devoted donors. While in the A3 app, the points are redeemed for gift cards or items that are available in the online store of the blood bank such as T-Shirt and prepaid cards. The total of the awarded points differs according to the donation type. Donors can choose not to redeem their points for external rewards and have the option to donate them to other partnered non-profit-organisations. Alike the American Red Cross app, the red cross of Singapore through its app (A5) rewards the donors an augmented reality video thanking them for their selfless act. In addition, the loyal donors can be rewarded with special offers and discounts from the partnered merchants. Besides, the app confers status to users such as Heroglobin and pacemaster according to the number of lives they have helped save through their donation. The A4 offers external items such as exclusive T-shirts and Free tickets to attend upcoming events in the current month. Additional gifts (e.g. smart TV) could be awarded to users by drawing lots. Given that the secret of behaviour's change using gamification lies in intrinsic incentives, financial or extrinsic rewards could generate a straightforward users' engagement and loyalty towards the application. However, research [33] shows that excessive cash-rewards are not correlated with progress and performance and hence can deflect users' motivation.

4   Limitations of the Study

There are four notable limitations to this study. Firstly, the search string used to pinpoint the gamified BD apps may not have covered all the terms relevant to the study. To alleviate this threat, PICO criteria were used however no gamification-related term was used in order to avoid obtaining games in the results. Secondly, the app repositories do not allow advanced searches due to their limited functionalities thus widening ineffectively the number of search results. Thirdly, as the apps were finally selected after reviewing their screenshots and description, relevant apps could have been missed. Lastly, the steady proliferation of mobile apps is such that the results of this study will no longer portrays the current availability of the apps.

5   Conclusion and Future Work

Despite positive medical progress, there is an increasing need for blood supply because this vital substance which is massively demanded has a short shelf-life. This entails the importance of volunteer blood donors' recruitment. One of the ingenious way to engage volunteer to donate blood consists on gamifying blood donation mobile apps. By means of game design elements, users of BD apps become more willing to give blood as long as they are awarded. This paper has studied the functionalities and gamification aspects of ten gamified BD apps selected from a total of 801 apps available in the four mobile app repositories. The majority of apps installed include mobile features that could enhance the donation experience of the donors. Notifications, Geolocation and social networking integration are the most interesting functionalities in blood donation apps. Gamification was applied in various ways combining points with badges or badges with leaderboards, to cite but a few. In addition, some of the selected apps have opted for financial rewards (e.g. coupons, gift cards) to enhance the retention rate of blood donors. Given that gamification is still in its infancy in eHealth realm, developers may consider these findings to build more compelling BD apps. The impact of gamification techniques on blood donors' behaviour will be investigated in future work. Moreover, we plan to examine the usability compliance of these apps in order to elicit the requirements needed for an effective gamified BD app.

Acknowledgments. This work was conducted within the research project MPHR-PPR1-2015-2018. The authors would like to thank the Moroccan MESRSFC and CNRST for their support.

References

1. Sarkar, S.: Artificial blood. Indian J. Crit. Care Med. **12**(3), 140–144 (2008)
2. Williamson, L.M., Devine, D.V.: Challenges in the management of the blood supply. Lancet **381**(9880), 1866–1875 (2013)
3. Who.int, WHO: More voluntary blood donors needed (2013). https://goo.gl/GCXAHl. Accessed 1 Sept 2016
4. Euro.who.int. Blood safety, Data and statistics (2011). https://goo.gl/Z7dHyB. Accessed 1 Sept 2016
5. Williams, M.S.: Gamification in Blood donation (2012). https://goo.gl/m2iwjo. Accessed 2 Sept 2016
6. Reich, P., Roberts, P., Laabs, N., Chinn, A., McEvoy, P., Hirschler, N., Murphy, E.L.: A randomized trial of blood donor recruitment strategies. Transfusion **46**, 1090–1096 (2006)
7. Setiawan, M.A., Putra, H.H.: Bloodhub: a context aware system to increase voluntary blood donors' participation. In: International Conference on Science and Technology, pp. 231–235. IEEE, RMUTT (2015)
8. Foth, M., Satchell, C., Seeburger, J., Russel-Bennett, R.: Social and mobile interaction design to increase the loyalty rates of young blood donors. In: Proceedings of the 6th International Conference on Communities and Technologies, pp. 64–73. ACM, New York (2013)
9. Smith, A.: Smartphone Ownership–2013 Update, vol. 12. Pew Research Center, Washington DC (2013)
10. Deterding, S., Dixon, D., Khaled, R., Nacke, L.: From game design elements to gamefulness: defining gamification. In: Proceedings of the 15th International Academic MindTrek Conference: Envisioning Future Media Environments, pp. 9–15. ACM, Tampere (2011)
11. Huotari, K., Hamari, J.: Defining gamification: a service marketing perspective. In: Proceeding of the 16th International Academic MindTrek Conference, pp. 17–22. ACM, New York (2012)
12. Fotopoulos, I., Palaiologou, R., Kouris, I., Koutsouris, D.: Cloud-based information system for blood donation. In: Kyriacou, E., Christofides, S., Pattichis, C.S. (eds.) Proceedings of the 14th Mediterranean Conference on Medical and Biological Engineering and Computing, pp. 802–807. Springer, Paphos (2016)
13. Eklund, R.C., Tenenbaum, G.: Encyclopedia of Sport and Exercise Psychology. SAGE Publications, Thousand Oaks (2014)
14. Liberati, A., Altman, D.G., Tetzlaff, J., Mulrow, C., Gøtzsche, P.C., Ioannidis, J.P., Clarke, M., Devereaux, P., Kleijnen, J., Moher, D.: The PRISMA statement for reporting systematic reviews and meta-analyses of studies that evaluate health care interventions: explanation and elaboration. Ann. Intern. Med. **151**(4), 65–94 (2009)
15. Ouhbi, S., Fernández-Alemán, J.L., Toval, A., Idri, A., Pozo, J.R.: Free blood donation mobile application. J. Med. Syst. **39**(5), 1–20 (2015)
16. Stone, P.W.: Popping the (PICO) question in research and evidence-based practice. Appl. Nurs. Res. **15**(3), 197–198 (2002)

17. Ouhbi, S., Fernández-Alemán, J.L., Pozo, J.R., El Bajta, M., Toval, A., Idri, A.: Compliance of blood donation apps with mobile OS usability guidelines. J. Med. Syst. **39**(6), 63:1–63:21 (2015)
18. Bachiri, M., Idri, A., Fernandez-Aleman, J.L., Toval, A.: Mobile personal health records for pregnancy monitoring functionalities: analysis and potential. J. Comput. Methods Programs Biomed. **134**, 121–135 (2016)
19. Statistica. US mobile smartphone OS market share 2012–2016 (2016). https://goo.gl/zOV5lw. Accessed 10 Aug 2016
20. Martínez, M.: Contingency planning for natural disasters. Int. Soc. Blood Transfus. Sci. Ser. **6**(1), 212–215 (2011)
21. Who.int. Blood Donor Selection: Guidelines on Assessing Donor Suitability for Blood Donation. World Health Organization (2012). https://goo.gl/0PpBhG. Accessed 4 Sept 2016
22. McMillan, J.E.R., Glisson, W.B., Bromby, M.: Investigating the increase in mobile phone evidence in criminal activities. In: 46th Hawaii International Conference on System Sciences, HICSS, pp. 4900–4909. IEEE, Hawaii (2013)
23. Pennycook, A.: The cultural politics of English as an international language. Teach. Talk. Teach. **III**(3), 21–23 (1995)
24. James, A.B., Josephson, C.D., Shaz, B.H., Schreiber, G.B.,Hillyer, C.D., Roback, J.D.: The value of area-based analyses of donation patterns for recruitment strategies. Transfusion, **54**(12), 3015–3060 (2014, in press)
25. Lemmens, K.P.H., Abraham, C., Ruiter, R.A.C., Veldhuizen, I.J.T., Bos, A.E.R., Schaalma, H.P.: Identifying blood donors willing to help with recruitment. Vox Sang. **95**, 211–217 (2008)
26. Smith, A., Matthews, R., Fiddler, J.: Recruitment and retention of blood donors in four Canadian cities: an analysis of the role of community and social networks. Transfusion **53**, 180S–184S (2013)
27. Boulos, M.N.K., Yang, S.P.: Exergames for health and fitness: the roles of GPS and geosocial apps. Int. J. Health Geogr. **12**(1), 18 (2013)
28. Redcross.org. Red Cross launches RapidPass Online Health History System (2015). https://goo.gl/PyyUVQ. Accessed 3 Oct 2016
29. Centralbloodbank.org. DonorPass (2014). https://goo.gl/1lljtk. Accessed 3 Oct 2016
30. Deterding, S.: Situated motivational affordances of game elements: a conceptual model. In: A Workshop at CHI. ACM, Vancouver (2011)
31. Sundarde, C., Jain, S., Shraikh, E.: Advancement of blood donation application (2015). https://goo.gl/Fae1pR. Accessed 20 Oct 2016
32. Ibrar, M., Khan, O.: The impact of reward of employee performance (a case study of Malakand Private School). Int. Lett. Soc. Humanist. Sci. **52**, 95–103 (2015)
33. Rimon, G.: It's not about cash: research-based facts on employee engagement (2015). https://goo.gl/cAxut2. Accessed 5 Oct 2016

Medical Entity Recognition and Negation Extraction: Assessment of NegEx on Health Records in Spanish

Sara Santiso[✉], Arantza Casillas, Alicia Pérez, and Maite Oronoz

IXA Group, University of the Basque Country (UPV-EHU),
T649, 20080 Donostia, Spain
{sara.santiso,arantza.casillas,alicia.perez,maite.oronoz}@ehu.eus
http://ixa.si.ehu.eus

Abstract. This work focuses on biomedical text mining. The core of this work is to make a step ahead in the negation detection of biomedical entities on Electronic Health Records (EHRs), where the detection of non-negated entities is as important as the identification of negated entities. For instance, the identification of a negated entity as factual, can produce diagnostic errors in decision support systems.

Negated entity recognition tackles two tasks: (1) entity recognition; (2) entity classification as negated or not. To identify negations, in the literature rule-based and machine-learning techniques have been used. This paper presents an adaptation of the rule-based system NegEx, which uses exact-matching for the aforementioned tasks.

Our contribution consist in assessing the aforementioned two tasks and explored alternatives for each of them, in such a way that the negation detection improves when the entity recognition is able to detect more entities correctly.

The evaluation was carried out within a real domain of 75 EHRs written in Spanish obtaining an f-measure of 76.2 for entity recognition and 73.8 for negation detection.

Keywords: Negation detection · Electronic health records · Text mining · Spanish

1 Introduction

Nowadays, patients' health records are stored electronically, these records can be used in text mining to meet information needs or provide services [2]. The information obtained automatically using text mining techniques can help to improve the patients diagnoses and, as a consequence, it can also improve their health. In addition, a good representation of this information can also help to the personnel of different services within the hospital given that, instead of reading the records, they can obtain the information at a glance. One of the tasks where we can apply text mining is adverse drug reactions extraction, where the recognition, among others, of medical entities and their negations is necessary.

© Springer International Publishing AG 2017
I. Rojas and F. Ortuño (Eds.): IWBBIO 2017, Part I, LNBI 10208, pp. 177–188, 2017.
DOI: 10.1007/978-3-319-56148-6_15

Within the biomedical texts, Nagaz et al. [9] described that "around 13% of sentences found in biomedical research articles contain negation". Ceusters et al. [3] indicated that "A substantial fraction of the observations made by clinicians and entered into patient records are expressed by means of negation or by using terms which contain negative qualifiers". Entities appear frequently negated in the records and their negation can modify the meaning of the extracted information. According to Blanco et al. [1] negation is present in all languages and may be present in all units of language, from words (e.g. "afebrile") to clauses (e.g. "no diabetes mellitus").

In [4] the authors of NegEx describe that for identifying pertinent negatives, a proposition ascribing a clinical condition to a person is identified (e.g. "experiencing nausea") and then the negation of this proposition is determined (e.g. "not experiencing nausea"). Specifically, NegEx preprocesses the text by individual sentences using exact-matching. This means that information across sentences is not used in determining whether a clinical condition is negated. Firstly, it identifies the UMLS terms in the text. Secondly, it searches the trigger words that indicate negation and labels as negated entities those inside a token-window near the negation trigger word.

We chose NegEx because in spite of its simplicity, it gives good results and it has already been used to detect the negation in EHRs. All in all, the main limitation we find to NegEx lies in its Name Entity Recognition (NER) strategy, that simply performs exact-match of entities in the text. We consider negation detection the more robust part in NegEx as although it also uses exact-match, the negation trigger words are shared in different types of texts.

The main contribution of this paper is that we have assessed different entity recognition approaches and their influence on the negation detection. To develop this we have adapted NegEx to make the named entity recognition more flexible.

In brief, this paper tackles clause level negation identification by means of the adaptation of NegEx, which has been assessed in EHRs written in Spanish. We aim at leveraging the flexibility of the first stage that focuses on NER.

The rest of the paper is arranged as follows. First, in Sect. 2 we present the state-of-the-art in negation extraction. Section 3 is devoted to explain our experimentation with the adaptation of the NegEx algorithm. The results for entity recognition and negation extraction are presented in Sect. 4, where we also include an error analysis. Finally, the conclusions drawn from this work together with some ideas for future work are presented in Sect. 5.

2   Related Work

In the literature there are several works that tackle the identification of negated entities within the biomedical domain by means of NegEx, which is the focus of this article.

Chapman et al. [4] developed NegEx to detect the negation of findings and diseases in narrative medical records. This algorithm uses both (i) a list of negation phrases and, (ii) a list of UMLS terms belonging to the finding and diagnosis semantic types, for identifying negatives in textual medical records. These lists have to be modified for each language or domain because NegEx uses exact-matching to index the entities and the negation phrases. After that, by means of regular expressions, the entities inside a window of length 6 near the negation trigger word are assigned the negation status. NegEx was evaluated in a corpus written in English of 1,000 test sentences that contained 1,235 occurrences of UMLS terms with a precision of 84.5 and a recall of 77.8.

Skeppstedt et al. [11] used NegEx in EHRs written in Swedish. To adapt the tool to this language they translated and extended the trigger words. The recognized medical entities were UMLS phrases that belonged to the categories 'Finding', 'Disease' or 'Dysfunction'. The precision was 70.0 and the recall was 81.0 for around 200 sentences with negation trigger words.

Later, Skeppstedt et al. [12] developed an adaptation of the previously mentioned system obtaining the entities from the version in Swedish of SNOMED CT, selecting those terms that belong to the semantic categories 'Finding' or 'Disorder' and using exact-matching. They obtained a precision of 78.0 and a recall of 82.0 when it was tested in 500 sentences with negated phases.

Weegar et al. [13] used the adaptation of NegEx to Swedish in order to find cervical cancer symptoms. The corpus used in the experiments consist of 646 patient records from the *Cervical Cancer Corpus* with 17,263 notes and 776,719 tokens. They tried partial and exact-match for the entity detection and their Clinical Entity Finder extended with NegEx had an average f-measure of 66.7 for the partial-match.

Costumero et al. [5] used NegEx to detect negation in medical documents in Spanish. The authors only changed the negation triggers in a corpus of 500 reports with 422 different sentences and 267 unique clinical conditions. The precision, recall, and f-measure for the negated terms is of 49.5, 55.7, and 52.4 respectively. They find the medical entities replacing the terms of the list used in the original version of NegEx by their translation, some synonyms and terms obtained from the manual annotation of medical texts in Spanish. By contrast, we propose the use of NER tools different from the exact-match approach for the medical entity recognition.

As we can see, NegEx algorithm has been applied to texts written in different languages such as English, Swedish or Spanish inside the medical domain.

3   Adaptation of the NegEx Tool

To detect whether an entity appears negated or not NegEx (1) identifies medical concepts described with terms present in the EHRs and, (2) classifies the recognized entities as negated or not. NegEx develops these tasks by means of exact-match with respect to two lists, one for gathering the medical terms from UMLS, and another one, for listing the negation trigger words.

The use of lists implies that these have to be modified when we want to add new entities or we use NegEx in other domains and languages. Otherwise, some entities are not recognised and, therefore, we can not detect their negations.

For this work we have made an adaptation that let us apply different techniques for the entity recognition not restricted to the simply exact-matching. The aim is to find which of them help us to improve the detection of negated entities when we use NegEx. The different NER systems assessed are:

NER 1: NegEx NER with dictionary list

We use the strategy of NegEx to identify the entities, applying exact-matching to find in the document the words given by a list. This can be considered an approximation to the original NegEx strategy for Spanish given that the list consists of a dictionary created with the medical entities (drugs and diseases) obtained from different sources.

The diseases of the list are the diseases and symptoms of the ICD-10 (International Statistical Classification of Diseases and Related Health Problems in its 10th version). The drugs of the list are the drug families of the ATC (Anatomical Therapeutic Chemical) classification, the drugs of BotPLUS and the active ingredients of ICD-10.

NER 2: NegEx NER with manual annotations list

We also use the strategy of NegEx to identify the entities. However, in this case the list consists fn the medical entities annotated by experts in EHRs.

NER 3: Conditional Random Fields (CRFs)

The entities are recognized using the Conditional Random Fields algorithm. It is a probabilistic framework for labeling and segmenting sequential data. The CRFs construct a conditional model $p(Y|X)$, where X are observation sequences and Y their corresponding label sequences, besides X and Y are jointly distributed [7].

For the creation of a basic CRF we have transformed all the terms to lower case and we have used as features the prefixes and suffixes with consist of the 4 first and last characters of the word. The reason for use this is that some disease are characterized by their affixes, for example, the prefix "*-tis*" indicates inflammation.

NER 4: NegEx NER with manual annotations list+CRFs

It consist of the union of the entities detected by NegEx NER with manual annotations and the entities detected by CRFs.

NER 5: Oracle

The recognized entities are those labeled by the experts. This can be considered as a perfect NER and shall provide us a lower threshold on the error propagation to the negation detection stage.

After the entity recognition, each medical entity (e.g. drugs and diseases) are replaced by a reserved word that correspond to the class or type of the entities. An example of this modification is showed in the following sentence: "*no haber presentado hipoglucemias hasta el momento actual*" (meaning "it has not presented hypoglycemias until the current moment"). In the mentioned generalization, the input of NegEx would be: "*no haber presentado Grp_Enfermedad hasta el momento actual*", where "*Grp_Enfermedad*" is the entity type corresponding to the diseases.

Still, we can turn to NegEx by simply replacing the list of entities by a sort list of three or four entity-types. This makes NegEx easier to scale when we have a high number of entities with different names.

Regarding the negation detection, we continue using the regular expressions with respect to the list of negation trigger words used in the original version of NegEx. Usually, the phrases used to indicate negation are similar across different texts, then it is less probable to loss negated entities.

In this work we have used the list of negation trigger words created by [5]. This list consists of the translation from English into Spanish of the 86 trigger words used in the original NegEx lexicon. We also included in the list 41 of the trigger words used in the Swedish NegEx version by [12].

All in all, we accomplish a list of 121 different trigger words. Some words that appear in both list are, for example, *"sin"* ("without") or *"ausencia de"* ("absence of"), that is, expression that usually indicate the presence of a negation in any language.

Tables 1a and b show the most frequent entities and negation trigger words in the same set of EHRs, respectively. The majority of the negation triggers that appear in the records correspond to 3 words (*"no"*, *"ni"*, *"sin"*). This confirms the idea of robustness of the negation detection stage, given that with only 3 words we detect approximately the 98% of the negations.

Table 1. Lists of the most frequent entities and negation trigger words together with their frequency in the EHRs.

	entity		freq
1	*"hta"*	"ht"	40
2	*"fiebre"*	"fever"	25
3	*"disnea"*	"dyspnea"	24
4	*"dislipemia"*	"dyslipemia"	12
5	*"cardiomegalia"*	"cardiomegaly"	12
6	*"tos"*	"cough"	11
7	*"dolor"*	"pain"	11
8	*"mareo"*	"sickness"	10
...	-	-	-

(a) Entities.

	negation trigger word		freq
1	*"no"*	"no", "not"	215
2	*"ni"*	"nor"	141
3	*"sin"*	"without"	83
4	*"niega"*	"denies"	3
5	*"ausencia de"*	"absence of"	3
6	*"ningún"*	"neither"	1
...	-	-	-

(b) Negation trigger words.

Figure 1 shows the workflow of the original version of NegEx (Fig. 1a) and the workflow of our adaptation (Fig. 1b). In our adaptation the entities can be recognised in a previous step using different NER approaches, and they are replaced with the corresponding entity type. After that, the NegEx algorithm is used as in the original version, but with a short list of entity-types instead of a list of entities. Finally, those entities that appear negated in the records are obtained.

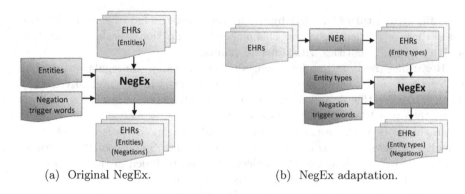

(a) Original NegEx.   (b) NegEx adaptation.

Fig. 1. Workflow of NegEx before and after our adaptation.

4   Results

4.1   Corpora

The corpora used in this work consist of a set of EHRs from Osakidetza (the Basque health system also attached to the Spanish public health system) written in Spanish. These EHRs, as in other languages, contain some misspellings and both standard and non-standard abbreviations. These issues make real EHR processing and entity recognition difficult [2], particularly for languages with fewer resources than English (as it is our case). Despite of being a task of much interest, little progress is being made on EHR processing, possibly due to the fact that EHRs are subject to strict confidentiality regulations, and so happens in this work.

In this work, two corpora have been explored, referred to as Gold-Standard (GS) and Silver-Standard (SS) respectively:

- **Gold-Standard**: It was manually annotated by two experts and they were asked to reach consensus. The IAA (Inter Annotator Agreement) for entities is of 90.53% [10].
- **Silver-Standard**: It contains the GS and a set of documents that was manually annotated by four experts without overlapping in the annotations.

To make the experiments, both corpora were randomly divided in three subsets: *train*, *dev* and *test*. The quantitative description of the corpus is given in Table 2.

The experts annotated the diseases and drugs present in the EHRs. In this work we focus on the diseases, some of them are discontinuous or formed by strings situated in different intervals of the text. For example, in the sentence *"No déficit motor, sensitivo, ni de campo visual"* the entity *"déficit de campo visual"* (meaning "visual field deficit") is composed of two substrings that do not appear together: *"déficit"* and *"de campo visual"*.

Table 2. Quantitative description of the GS and SS. For each subset the number of documents, words (words of the documents, including the repetitions) and vocabulary (words of the document, without repetitions) are given, together with the number of non-negated ($\oplus$Ent) and negated entities ($\ominus$Ent).

Set	Subset	Doc	Word	Voc	$\oplus$Ent	$\ominus$Ent
GS	train	41	20,689	4,934	942	399
	dev	17	11,246	3,430	523	214
	test	17	9,698	2,889	479	150
SS	train	185	96,203	12,548	4,319	614
	dev	67	37,517	7,197	1,796	277
	test	67	33,284	6,409	1,538	212

4.2 Evaluation

To evaluate the NER approaches used in the first stage of the system it has been applied the hold-out evaluation in two ways:

- **Train vs Dev**: The system is trained with the *train* set and evaluated on the *dev* set. In the case of the lists, in the *dev* set we use the list of entities of the *train* set.
- **TrainDev vs Test**: The system is trained with the *train and dev* sets and evaluated on the *test* set. In the case of the lists, in the *test* set we use the list of entities of the *train and dev* sets.

The performance of the system was evaluated using the regular metrics: precision, recall and f-measure and it was assessed at two levels:

- **Exact-match**: The entity found by the system is the same that the entity annotated by the experts. The comparison is made using the offsets, that is to say, the position of the first and last characters of the entity in the text.
- **Partial-match**: The entity found by the system and the entity of the manual annotation overlap, that is to say, the initial offset of one of the entities is between the offsets of the other entity.

Let us pose these two evaluations through an example: given that the system identified the word *"palidez"* (meaning "paleness") with the offsets '1712-1719' and the experts had labeled the word *"palidez C.M."* with the offsets '1712-1724' it is considered as true positive in the partial-match but not in the exact-match.

In brief, these two levels give us complimentary insights into the performance of the system. The evaluation of the entities have been developed using the software of the SemEval task 'Analysis of Clinical Text' [8].

With this evaluation framework, in the following sections we will show the performance of the system in terms of entity recognition and negation detection for the *TrainDev vs Test* evaluation.

4.3   Entity Recognition

Table 3 shows the results achieved in terms of entity recognition. For this work, we have focused on the detection of diseases.

We can see that the approach that offers worse results to recognize the entities is *NegEx NER with dictionary list*. The main reason would be that the entities are not described by the experts in the same way that they appear in these dictionaries, then only few entities are found.

We can also appreciate that if we use he list of entities annotated by the experts more entities are detected, but the precision is still low.

Moreover, the results obtained with the *CRFs* outperforms those obtained with *NegEx NER with manual annotation list*. In general, with the CRFs algorithm the entities are detected with more precision.

We have also developed the evaluation of the approach that combines *NegEx NER with manual annotation list* and *CRFs*. At first sight, the performance of this system should be better. However, the *CRFs* continue obtaining better results because *NegEx NER with manual annotation list* increases the number of False Positives (FPs) and, as a consequence, the precision get worse.

Finally, the results of the *Oracle* approach for the entity detection are showed. As it was expected, all the entities has been found correctly given that this NER approach simply uses the entities labeled by the experts.

In Table 3 we can also appreciate that the results obtained with the GS are better than with the SS. This made us reflect on the importance of getting a stable corpus to gain in credibility on the results attained by the system.

Comparing these results to those presented in Sect. 2, we can see that our partial-match outperforms the results reported by Skeppstedt et al. [12] using exact-match (58 for precision and 23 for recall). With respect to Costumero et al. [5], they obtained better precision and recall for the entity detection (a precision of 86 and a recall of 95). Probably because their list of entities is long enough to cover the majority of the entities.

Table 3. Precision (P), Recall (R) and F-measure (F) for the GS and the SS in entity recognition.

	GoldStandard						SilverStandard					
	Exact			Partial			Exact			Partial		
	P	R	F	P	R	F	P	R	F	P	R	F
NER 1	45.3	4.7	8.5	96.9	10.0	18.2	44.5	6.2	10.8	80.3	11.1	19.5
NER 2	36.5	39.5	37.9	69.4	74.1	71.7	30.7	41.8	35.4	52.7	70.9	60.5
NER 3	60.8	41.0	49.0	91.8	63.8	75.3	48.2	39.4	43.4	72.3	59.8	65.4
NER 4	36.4	44.7	40.1	70.2	83.3	76.2	30.2	46.9	36.8	52.6	80.3	63.6
NER 5	100.0	100.0	100.0	100.0	100.0	100.0	100.0	100.0	100.0	100.0	100.0	100.0

4.4 Negation Detection

After having identified the entities, the adapted NegEx has to detect the negation using a list of negation trigger words. Inherent to the cascade approach the errors from the entity extraction are propagated to the negation detection. To understand the evaluation it is important to explain that we consider a True Positive (TP) to an entity correctly identified as negated by the system.

Table 4 assesses the negation detection ability. According to this, the negation detection using the entities obtained with *NegEx NER with dictionary list* has the worst results. With this approach only has been detected few entities, then, it is not possible to obtain a high recall for the negation detection.

Moreover, the negation detection using the entities of *CRFs* improves the results of the previous systems. The idea was that the performance of NegEx will be better with these entities.

By contrast, the *NegEx NER with manual annotation list* approach is better than the previous one because it is able to recognize more diseases. Approximately the 60% of the FNs correspond to entities that have not been previously identified, that is, although the use of CRFs gives good results mainly for its high precision, its lower recall produces the loss of some negated entities. Discarding these entities the partial-match f-measure is 71.7 for the GS and 49.1 for the SS.

Next, we have evaluated the results obtained with the union of *NegEx NER with manual annotation list* and *CRFs* for the entity recognition. This approach helps to improve the results obtained with *CRFs* but not with *NegEx NER with manual annotation list* because the number of negated entities detected correctly has increased less than the number of those detected incorrectly.

Finally, Table 4 shows how would be the performance of NegEx if all the entities labeled by experts would be recognized. With exact-match the results are better than the offered by the rest of approaches. Nevertheless, it does not happen for the partial-match given that, for example, the system could recognise two entities that experts have labeled as one.

Comparing these results with the related works (mentioned in Sect. 2), the partial-match of the GS lead us to a precision and recall better than those

Table 4. Precision (P), Recall (R) and F-measure (F) for the GS and the SS in negation detection.

	GoldStandard						SilverStandard					
	Exact			Partial			Exact			Partial		
	P	R	F	P	R	F	P	R	F	P	R	F
NER 1	45.5	3.5	6.5	81.8	6.3	11.8	28.9	5.7	9.5	42.1	8.3	13.9
NER 2	42.1	35.9	38.8	80.2	68.3	73.8	21.6	46.6	29.6	33.4	72.0	45.6
NER 3	59.8	34.5	43.8	78.0	45.1	57.1	22.2	18.1	19.9	34.1	28.0	30.7
NER 4	39.2	35.9	37.5	67.7	62.0	64.7	21.3	47.7	29.5	32.9	73.5	45.5
NER 5	87.4	53.5	66.4	87.4	53.5	66.4	85.0	64.8	73.5	85.0	64.8	73.5

obtained by Costumero et al. [5] (precision of 49.0 and recall of 55.0). Considering the partial-match of the GS our results outperform those presented by Weegar et al. [13], as in their work the f-measure is 67.0. Besides, the precision is better than the obtained by Skeppstedt et al. [12] (78.0 for precision and 82.0 for recall).

4.5   Error Analysis

In order to understand the source of errors, the False Negatives and False Positives for each of the two stages (entity recognition and negation detection) were manually examined. To this end, we focused on the errors made in the GS as it is more consistent than the SS given its IAA.

Entity Recognition Errors

– **False Negatives (FNs)**: the CRFs entity recognition system made 279 FNs in the *dev* set and 216 FNs in the *test* set. We found as the main causes the following types of errors:
  - The negation is implicit in the word as a morpheme (typically a prefix): For example, for the word *"afebril"* (meaning "afebrile"), the negation stands on the prefix *"a"*.
– **False Positives (FPs)**: the CRFs entity recognition system made 18 FPs in the *dev* set and 15 FPs in the *test* set. The main causes are these types of errors:
  - Entities that have been found in the document, but the experts have not taken into account as a disease in the given context. Some entities correspond to substances of the clinical analysis, which are diseases depending on the obtained value. In this case it seems that the error is in the manual annotation and not in the entity detection.

Negation Detection Errors

– **False Negatives (FNs)**: the exact-match negation detection system made 69 FNs in the *dev* set and 45 FNs in the *test* set. The main causes are these types of errors:
  - Absence of the negation trigger word in the list.
  - For the discontinuous entities that were matched partially the negation is not often detected by NegEx: For example, the sentence *"orientada en espacio y personas, no en tiempo"* (meaning "she has awareness of place and people, but not of time") contains the negated entity *"no orientada en tiempo"*. Nevertheless, the system only found *"orientada"*. Since the trigger word *"no"* is not in the context, the negation is not found.
– **False Positives (FPs)**: the exact-match negation detection system made 18 FPs in the *dev* set and 24 FPs in the *test* set. The main causes are these types of errors:
  - NegEx annotated negations that the experts skipped. NegEx finds the negations systematically while, at times, the criteria of the experts is to omit them due to the fact that in the given context it might be understood as irrelevant information.

- Close entities: NegEx often fails since it tends to label as negated all the entities in the surrounding of the negation trigger word.

5 Conclusions and Future Work

In this work we present an study about the influence of the entity recognition in the negation detection made by NegEx, which has been used in other works with promising results despite its simplicity. NegEx originally tackles entity recognition by means of exact-match. Nevertheless, exact-match lacks of sparsity of data and it requires a big effort to keep the lists updated, particularly to deal with spontaneous EHRs. We have made an adaptation that enables the use of any context-based entity recognition techniques or other classifiers that are able to generalize and, hence, recognise misspelled entities. Regarding the negation extraction, we have used the same exact-match approach because the negation trigger words are shared in different domains, hence, the adaptation is straightforward.

The different approaches used for this work were assessed with real records written in Spanish. We have evaluated both the ability to recognize entities and the accomplishment on the negation detection using exact and partial-match. According to the results, the entity recognition system that offers better results is the CRFs classifier, which recognize the entities with higher precision. However, the negation detection works better with the *NegEx NER with manual annotation list* because it commits less errors among the identified negations than the union of both. Among the related works, the one developed by Costumero et al. [5] is the most comparable with our work given that they also use documents written in Spanish. For our set of EHRs, the system created in this work obtains an f-measure of 73.8 for the negation, whereas the system created in [5] obtains an f-measure of 52.4 for their set of patients records.

A manual inspection of the errors showed that discontinuous entities are difficult to recognize automatically. Moreover, the presence of discontinuous entities also makes it challenging the detection of negated entities because sometimes the negation trigger word is not in the span considered by NegEx. These challenges are left open for future work. Yet, we find that inferred entity recognition techniques together with an extension of NegEx might result in helpful. In addition, we learned that we could extend NegEx in another direction: that is, it could find the negations implicit to several morphemes (usually prefixes in our case). This would extend NegEx to word level negation detection.

All in all, our final motivation behind the use of this tool is to extract information: not only entities but also events. We are focusing on the extraction of adverse drug reactions and allergies. In other works, such as the developed by Henriksson et al. [6], the negation of the entity was used as feature for adverse drug reaction (ADR) extraction, concluding that it was a key information for this task. Then for future work, we will use the information obtained by NegEx in our ADR extraction system. This information not only can help to the staff of the hospitals to make their work in an easier and faster way, but also helps to improve the diagnoses of the patients.

Acknowledgments. The authors would like to thank the personnel of Pharmacy and Pharmacovigilance services of the Galdakao-Usansolo Hospital. This work was partially funded by the Spanish Ministry of Science and Innovation (EXTRECM: TIN2013-46616-C2-1-R, TADEEP: TIN2015-70214-P) and the Basque Government (DETEAMI: Ministry of Health 2014111003, Predoctoral Grant: PRE 2015 1 0211).

References

1. Blanco, E., Moldovan, D.I.: Some issues on detecting negation from text. In: FLAIRS (2011)
2. Bretonnel, K., Demmer-Fushman, D.: Biomedical Natural Language Processing, vol. 11. John Benjamins Publishing Company, Amsterdam (2014)
3. Ceusters, W., Elkin, P., Smith, B.: Negative findings in electronic health records and biomedical ontologies: a realist approach. Int. J. Med. Inform. **76**, 326–333 (2017)
4. Chapman, W.W., Bridewell, W., Hanbury, P., Cooper, G.F., Buchanan, B.G.: A simple algorithm for identifying negated findings and diseases in discharge summaries. J. Biomed. inform. **34**(5), 301–310 (2001)
5. Costumero, R., Lopez, F., Gonzalo-Martín, C., Millan, M., Menasalvas, E.: An approach to detect negation on medical documents in Spanish. In: Ślęzak, D., Tan, A.-H., Peters, J.F., Schwabe, L. (eds.) BIH 2014. LNCS (LNAI), vol. 8609, pp. 366–375. Springer, Heidelberg (2014). doi:10.1007/978-3-319-09891-3_34
6. Henriksson, A., Kvist, M., Dalianis, H., Duneld, M.: Identifying adverse drug event information in clinical notes with distributional semantic representations of context. J. Biomed. Inform. **57**, 333–349 (2015)
7. Lafferty, J., McCallum, A., Pereira, F.: Conditional random fields: probabilistic models for segmenting and labeling sequence data. In: Proceedings of the Eighteenth International Conference on Machine Learning, ICML, vol. 1, pp. 282–289 (2001)
8. Nakov, P., Zesch, T. (eds.): Proceedings of the 8th International Workshop on Semantic Evaluation (SemEval 2014). Association for Computational Linguistics and Dublin City University, Dublin, Ireland (2014)
9. Nawaz, R., Thompson, P., Ananiadou, S.: Negated bio-events: analysis and identification. BMC Bioinform. **14**, 14 (2013)
10. Oronoz, M., Gojenola, K., Pérez, A., de Ilarraza, A.D., Casillas, A.: On the creation of a clinical gold standard corpus in Spanish: mining adverse drug reactions. J. Biomed. Inform. **56**, 318–332 (2015)
11. Skeppstedt, M.: Negation detection in swedish clinical text. In: Proceedings of the NAACL HLT 2010 Second Louhi Workshop on Text and Data Mining of Health Documents, pp. 15–21. Association for Computational Linguistics (2010)
12. Skeppstedt, M., Dalianis, H., Nilsson, G.H.: Retrieving disorders and findings: results using SNOMED CT and NegEx adapted for swedish. In: Third International Workshop on Health Document Text Mining and Information AnalysisBled, Slovenia, 6 July 2011, Bled Slovenia, Collocated with AIME 2011, pp. 11–17 (2011)
13. Weegar, R., Kvist, M., Sundström, K., Brunak, S., Dalianis, H.: Finding cervical cancer symptoms in swedish clinical text using a machine learning approach and NegEx. In: AMIA Annual Symposium Proceedings. vol. 2015, p. 1296. American Medical Informatics Association (2015)

RISK: A Random Optimization Interactive System Based on Kernel Learning for Predicting Breast Cancer Disease Progression

Fiorella Guadagni[1,2], Fabio Massimo Zanzotto[3], Noemi Scarpato[1(✉)],
Alessandro Rullo[4], Silvia Riondino[2,3], Patrizia Ferroni[1,2],
and Mario Roselli[3]

[1] San Raffaele Rome University, Rome, Italy
{fiorella.guadagni,noemi.scarpato,
patrizia.ferroni}@unisanraffaele.gov.it
[2] IRCCS San Raffaele Pisana, Rome, Italy
silvia.riondino@sanraffaele.it
[3] University of Rome Tor Vergata, Rome, Italy
{fabio.massimo.zanzotto,mario.roselli}@uniroma2.it
[4] Neatec S.p.A., Pozzuoli (Naples), Italy
A.rullo@neatec.it

Abstract. Evaluating disease progression risk is a key issue in medicine that has been revolutionized by the advent of machine learning approaches and the wide availability of medical data in electronic form. It is time to provide physicians with near-to-the-clinical-practice and effective tools to spread this important technological innovation. In this paper, we describe RISK, a web service that implements a multiple kernel learning approach for predicting breast cancer disease progression. We report on the experience of the BIBIOFAR project where RISK Web Predictor has been developed and tested. Results of our system demonstrate that this kind of approaches can effectively support physicians in the evaluation of risk.

Keywords: Multiple kernel · Risk prediction · Breast cancer

1 Introduction

Breast cancer (BC) is the most common type of cancer in women, in whom it represents the second leading cause of cancer deaths. In recent years mortality from BC has declined, possibly as a result of more extensive screening resulting in earlier detection as well as advances in the adjuvant treatment. Accordingly, there has been an extensive search for prognostic factors that may aid in selecting patients most likely to recur, who might potentially benefit from adjuvant therapy. On the other hand, predictive factors may play an important role in a context of precision medicine, as they may help identify the appropriate therapy for an individual patient. However, BC survivability prediction is still a challenging task and growing emphasis is being put on personalized

© Springer International Publishing AG 2017
I. Rojas and F. Ortuño (Eds.): IWBBIO 2017, Part I, LNBI 10208, pp. 189–196, 2017.
DOI: 10.1007/978-3-319-56148-6_16

predictive models based on machine learning techniques, which may allow many BC patients to avoid unnecessary treatments and high medical costs.

To date, few attempts have been pursued to develop predictive models of progression of disease using data mining techniques, which use complex-to-determine clinical variables such as genetic markers. Most of them have been performed on derivative datasets from the SEER (Surveillance Program, Epidemiology, and End Results) that do not include among the patients' attributes some important prognostic parameters such as the St. Gallen criteria (i.e., hormone receptor status, of HER2/NEU expression or Ki67 proliferation index) [1]. Other studies have been performed on hybrid models containing microarray data in addition to the St. Gallen criteria [2]. More recently, SVM or semi-supervised learning techniques have been applied in oncology for predicting survival of breast cancer [3, 4]. Once again, the datasets used were derived from the SEER program, with the limitations referred to above regarding the completeness of prognostic parameters.

In the context of developing predictive models, we have recently demonstrated the potential of routinely-collected clinical data when combined with a machine learning approach based on Multiple Kernel Learning (MKL) [5]. Our MKL combines SVM [6] algorithm and Random Optimization (RO) [7], which is capable of exploiting significant patterns in routinely collected demographic, clinical and biochemical data. Using this approach, we were able to design a clinical decision support system for venous thromboembolism risk stratification prior to chemotherapy start that could be easily adapted to different local situations or medical problems, such as the risk of BC progression.

Here we report the experience of RISK: a Random optimization Interactive System based on Kernel learning for predicting breast cancer disease progression developed in the BIBIOFAR project. Our MKL model has been adapted to estimate the risk of disease progression in an oncological setting of BC patients. To achieve this objective we analyzed a preliminary dataset attained by joint efforts between the PTV Bio.Ca.Re. (Policlinico Tor Vergata Biospecimen Cancer Repository) and the BioBIM (InterInstitutional Multidisciplinary Biobank, IRCCS San Raffaele Pisana). The dataset consisted of 321 consecutive ambulatory BC patients, prospectively followed under the appropriate Institutional ethics approval and in accordance with the principles embodied in the Declaration of Helsinki. All patients provided IRB-approved written informed consent.

BC was pathologically staged according to the TNM classification. Two hundred and eighty women (87%) had primary BC and underwent radical surgery (23% mastectomy, 77% lumpectomy). The remaining 13% of patients had relapsing/metastatic disease and entered the study prior to the start of chemotherapy. Neoadjuvant chemotherapy regimens were instituted in 16% of cases. Adjuvant chemotherapy regimens – both anthracycline and non-anthracycline containing – were instituted in 72% of women. Fifty-seven women (18%) with node–negative disease underwent adjuvant endocrine therapy only (tamoxifen or aromatase inhibitor). Patients with HER2/NEU positivity were all treated with trastuzumab-containing regimens. First-line chemotherapy was instituted in all patients with metastatic disease.

Prognostic routinely-collected factors such as BC stage, menopausal status, pathological grading as well as the St. Gallen criteria (e.g., estrogen (ER) and progesterone (PgR) receptors, HER2/NEU expression and the proliferation index Ki-67) were analyzed for each patient. Furthermore, considering the increasingly awareness that metabolic features might represent an important contributor to BC progression and that insulin may represent a biomarker of adverse prognosis in non-diabetic BC patients [8], we introduced in the model fasting blood glucose and insulin, glycosylated hemoglobin and the presence of an insulin resistance (assessed by the HOMA index), alongside other anthropometric and biochemical indices of the presence of a possible metabolic syndrome. A detailed list of all the features that have been applied to construct the predictor is reported in Table 1.

Table 1. Features included in the model

Patient-related	Tumor-related	Metabolic
Age	Molecular type	Type 2 diabetes
Menopausal status	Histological diagnosis	Body mass index
	Grading	Fasting glycemia
	TNM stage	Fasting insulinemia
	Estrogen receptors	Glycosylated hemoglobin
	Progesterone receptors	HOMA index (insulin resistance)
	HER2/NEU	Total bilirubin
	Ki67 proliferation index	Creatinine

Bayesian analysis was performed, and positive (+LR) and negative (−LR) likelihood ratios were used to estimate the probability of having or not progression, using a free web-based application (http://statpages.org/). Progression free survival (PFS) was calculated from the date of enrollment until relapse or progression of disease. If a patient had not progressed or died, PFS was censored at the time of the last follow-up. PFS curves were calculated by the Kaplan–Meier method and the significance level was assessed according to the log-rank test using a computer software package (Statistica 8.0, StatSoft Inc., Tulsa, OK). The results obtained this approach emphasize that a Multiple Kernel Learning algorithm provides a useful risk predictor for disease progression in BC patients. The details of the *RISK Web Predictor* developed in the context of the BIBIOFAR project are reported below.

2 RISK: The Multiple Kernel Model

In this section we describe the risk predictor multiple kernel approach. As mentioned above we have adopted a multiple kernel learning approach combining SVM and Random optimization.

First of all we identified a set of predictors through the Random Optimization algorithm, then we selected the best of them. A predictor is a model configured with a specific set of weights.

Usually the selection of the best predictor depends on the values of precision, recall and f-measure. To devise the web interface, we selected the two best predictors out of a range of twenty runs. The two predictors were not only the best in term of prediction capacity, but they had also a complementary configuration of weights.

In particular, the first predictor ML + RO-1 had the highest weights for:

- Stage of disease
- Proliferation index Ki67
- Glucose and insulin resistance

ML + RO-2 instead had the highest weights for:

- BC molecular type
- Pathological grading
- Body Mass Index (BMI)
- Glycosylated hemoglobin

The adoption of a couple of predictors imply that the evaluation of the risk isn't a binary value (i.e. YES/NO) but is represented by three levels of risk:

- High risk, if both predictors estimate the patient at risk;
- Intermediate risk, if only one predictor estimates the patient at risk;
- Low risk, if both predictors don't estimate the patient at risk;

This configuration of the model reduces the number of false negative and false positive, but introduce some (approximately 13%) of intermediate risk patients.

As shown in Fig. 1, the combine use of both predictors (both positive – high risk of progression) translates in a positive predictive value (PPV) of 54% and a negative predictive value (NPV) of 87%, which results in an 81% accuracy, a Positive Likelihood Ratio (+LR) of 4.58 [95% Confidence Interval (CI): 2.89–7.14] and a Negative LR (−LR) of 0.576 (95% CI: 0.457–0.712) (Two-tailed exact Fisher test: p < 0.0001).

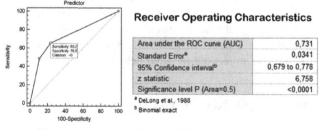

Receiver Operating Characteristics

Area under the ROC curve (AUC)	0,731
Standard Error[a]	0,0341
95% Confidence interval[b]	0,679 to 0,778
z statistic	6,758
Significance level P (Area=0.5)	<0,0001

[a] DeLong et al., 1988
[b] Binomial exact

Criterion values and coordinates of the ROC curve [Hide]

Criterion	Sensitivity	95% CI	Specificity	95% CI	+LR	-LR
>=0	100,00	94,6 - 100,0	0,00	0,0 - 1,4	1,00	
>0 *	65,15	52,4 - 76,5	76,47	70,8 - 81,5	2,77	0,46
>1	48,48	36,0 - 61,1	89,41	85,0 - 92,9	4,58	0,58
>2	0,00	0,0 - 5,4	100,00	98,6 - 100,0		1,00

* Criterion corresponding with highest Youden index

Fig. 1. Performance of risk predictor tool

But the most significant data, confirming the validity of the approach used, were obtained by comparing the survival curves of BC patients stratified in the three risk classes (see Fig. 2), which are distinctly differentiated as a function of the estimated risk. The categorization of patients according to the pair of predictors, in fact, was able to identify patients at increased risk of progressive disease with a hazard ratio (HR) of 5.4 (95% CI: 3.33–8.78; p < 0.0001). In particular, as shown by the Kaplan-Meier curves reported in Fig. 2, patients at high-risk of progression according to selected predictors were characterized by a disease-free survival at 5 years much lower (27%) compared to those classified as intermediate risk (71%) or not at risk (88%, chi-square = 51.6; p < 0.0001).

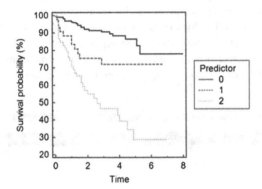

Fig. 2. Survival curves of BC patients

These data are in agreement and extend previous data on the feasibility of machine learning for risk prediction of BC progression. For the first time to our knowledge, we used a random optimization approach to weight the relative importance of attributes.

Moreover, the model developed in this project, although obtained from a dataset less populated than previously used, examined a number of prognostic factors not analyzed in previous models, adding to the risk estimate some "metabolic" parameters that can be available only in the context of a biological bank/clinical database program. As a result, the accuracy obtained with our model (81%) was higher than that reported by other researchers (76% in the study by Kim & Shin, and 71% in that of Park et al.).

Another advantage of the model we propose is that, since all the variables are usually included in the workout routine of BC patient, the risk calculation is practically at no cost to the NHS. Our next step will be to perform a multicenter analysis to clinically validate the model implemented using the web tool developed to aid oncologists in determining progression risk prediction for each BC patient.

3   RISK Web Predictor

RISK Web Predictor is a web service able to provide physicians with a graphical interface of risk model, which is based on routinely-collected clinical data. This service has two main components: a web interface (client) and a decision server.

The web interface allows users (physicians, researchers and so on) to insert values for the evaluation of the risk and shows them a message with the prevision. The decision server is the reasoning component that implements the kernel function to estimate the risk of disease progression, the estimated risk is represented by a ternary value (low, medium, high). The system was developed as a REST (REpresentational State Transfer) service and the internal data representation is in JSON that allows a very fast management of data.

The web interface of RISK is the way the core component can be used in the clinical practice and it is designed to help clinicians to input routinely-collected data in the system. Hence, the client web forms allow users to insert data with a smart interface that provides automatic suggestions and lists of option for the fields (see Fig. 3). Moreover, the form calculates some values by using some other values provided by users. When users insert all values or press the send button, the servlet contacts the server that computes the risk and shows the reply to users in the client form (see Fig. 4). The decision server provides a risk factor even if some clinical values are left empty since our MKL allows some empty values in input data.

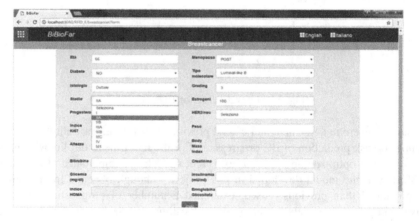

Fig. 3. Client form

An important feature of **RISK Web Predictor** is that all new cases analyzed are stored into a database. This data set will be used to verify the system. Furthermore, all data stored in the database could be exploited to train again the system and improve its performances in term of estimation of the disease progression risk.

Our MKL is open to on-line learning models [9, 10]. So the initial dataset can be extended with the new inserted patients and the model can be refined considering this additional data.

The collection of well-structured medical data, that are mandatory to train machine learning systems, isn't a trivial task. The creation of a graphical interface easy to use for physicians aims to simplify the collection of data.

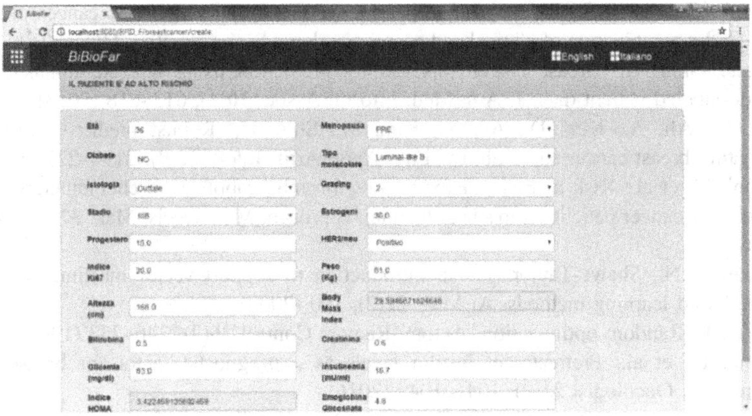

Fig. 4. Client form response

4 Conclusions

There are, of course, some limitations to acknowledge. The most obvious resides in the fact that the model here reported was designed and validated on a dataset derived from a single center. However, primary aim of this study was to test the feasibility of a ML approach in BC prognostic assessment.

Here, we demonstrate that a combined approach of ML algorithms and RO models might be useful in developing local classifiers able to predict the risk of BC progression, in a context of precision medicine. Indeed, the method here described has the unquestionable advantages of selecting the best predictors on training data and to determine the relative weights of clinical attributes.

These results, however, should be regarded with caution and external validation is needed before the clinical value of the RISK Web Predictor in breast cancer can be established. For this reason, the web site has not yet been made publicly accessible, as a multicenter protocol is currently ongoing. Researchers willing to collaborate in the validation phase are welcome to contact the authors to obtain login credentials. Nonetheless, the results here reported add further evidence to the rising idea that locally trained models may provide important information in risk stratification and clinical management of BC patients.

Acknowledgements. This work was partially supported by the European Social Fund, under the Italian Ministry of Education, University and Research PON03PE_00146_1/10 BIBIOFAR (CUP B88F12000730005) to Guadagni F.

References

1. Delen, D., Walker, G., Kadam, A.: Predicting breast cancer survivability: a comparison of three data mining methods. Artif. Intell. Med. **34**(2), 113–127 (2005)

2. Sun, Y., Goodison, S., Li, J., Liu, L., Farmerie, W.: Improved breast cancer prognosis through the combination of clinical and genetic markers. Bioinformatics **23**(1), 30–37 (2007)
3. Kim, J., Shin, H.: Breast cancer survivability prediction using labeled, unlabeled, and pseudo-labeled patient data. J. Am. Med. Inform. Assoc. **20**(4), 613–618 (2013)
4. Park, K., Ali, A., Kim, D., An, Y., Kim, M., Shin, H.: Robust predictive model for evaluating breast cancer survivability. Eng. Appl. Artif. Intell. **26**(9), 2194–2205 (2013)
5. Ferroni, P., et al.: Risk assessment for venous thromboembolism in chemotherapy-treated ambulatory cancer patients: a machine learning approach. Med. Decis. Mak. **37**(2), 234–242 (2016)
6. Cristianini, N., Shawe-Taylor, J.: An introduction to support vector machines and other kernel based learning methods. Ai Mag. **22**(2), 190 (2000)
7. Matyas, J.: Random optimization. Autom. Remote Control **26**(2), 246–253 (1965)
8. Ferroni, P., et al.: Pretreatment insulin levels as a prognostic factor for breast cancer progression. Oncologist **21**(9), 1041–1049 (2016)
9. Filice, S., Croce, D., Basili, R., Zanzotto, F.M.: Linear online learning over structured data with distributed tree kernels. In: Proceedings of the 2013 12th International Conference on Machine Learning and Applications, ICMLA 2013, vol. 1, pp. 123–128 (2013)
10. Filice, S., Castellucci, G., Croce, D., Basili, R.: KeLP: a kernel-based learning platform for natural language processing. In: Proceedings of ACL-IJCNLP 2015 System Demonstrations (2015)

Revealing the Relationship Between Human Genome Regions and Pathological Phenotypes Through Network Analysis

Elena Rojano[1]([✉]), Pedro Seoane[1], Anibal Bueno-Amoros[1],
James Richard Perkins[2], and Juan Antonio Garcia-Ranea[1,3]

[1] Department of Molecular Biology and Biochemistry,
University of Malaga (UMA), 29010 Malaga, Spain
{elenarojano,seoanezonjic,ranea}@uma.es
[2] Research Laboratory, Regional University Hospital of Malaga (IBIMA),
29009 Malaga, Spain
[3] CIBER de Enfermedades Raras, 28029 Madrid, Spain

Abstract. Recent advances in sequencing technologies allow researchers to investigate diseases resulting of genomic variation. This allows us to further develop the concept of precision medicine and determine the best treatment for each patient. We have focused on developing tools for studying genomic loci associated to pathological traits from the perspective of network analysis. We have obtained from DECIPHER database patient information which includes their affected genomic regions by Copy Number Variations (CNV) and their pathologies described as Human Phenotype Ontology phenotypes. We have used different metrics for calculating association values between phenotypes and affected genomic regions to determine which method fits better to our data. The results obtained in this work, can be used in prediction systems for determining and ranking which genomic regions are associated to a concrete phenotype, in order to help clinicians with their diagnosis.

Keywords: Network analysis · Pathological phenotypes · Precision medicine · Rare diseases · CNV

1 Introduction

Many human diseases are due to changes in the genome that affect functional elements such as genes or regulatory elements. Copy Number Variations (CNVs) represent an important class of genetic variation that can affect large areas of the genome. They are caused by duplication or deletion of large genomic regions [1]. There are now many research groups investigating these variants and looking at their association with pathological traits and patient phenotypes, with the aim of better understanding disease and developing personalised therapies [2]. Their study can be aided by the use of systems biology, by creating networks and studying the relationships between their elements [3]. This approach has

© Springer International Publishing AG 2017
I. Rojas and F. Ortuño (Eds.): IWBBIO 2017, Part I, LNBI 10208, pp. 197–207, 2017.
DOI: 10.1007/978-3-319-56148-6_17

been applied to the study of common diseases such as cancer [4], as well as rare diseases, defined as those that occur in less than 5 out of 10 000 people [5].

A key project for the study of CNVs and their role in rare diseases is the DatabasE of genomiC varIation and Phenotype in Humans using ENSEMBL Resources (DECIPHER) (http://decipher.sanger.ac.uk/) [6,7]. This resource includes a database that contains genomic and phenotypic information from more than 22 400 patients throughout the world (December 2016, release v9.12), corroborated by clinicians [6]. It includes data on the type and position of mutation, method of inheritance and pathological phenotypes. The latter are assigned using the Human Phenotype Ontology (HPO, www.human-phenotype-ontology.org) [8], which consists of an standard vocabulary of human pathological terms, organized hierarchically.

Here we present NetAnalyzer, a network analysis tool written in Ruby, and use it to analyse associations between pathological phenotypes and genomic regions. This library can (1) analyse any type of unweighted network, regardless of the number of layers, (2) calculate the relationship between different layers, using various association indices and (3) validate the results. We used this tool to associate genomic regions with phenotypes using DECIPHER data. We also compared the different association methods using 10-fold cross-validation.

2   Methods

In this study, we have developed a pipeline for downloading data from DECIPHER database and processing it to build a tripartite network composed of phenotype (HPO), patient and Genomic Variant Region (GVR) layers. We then used several association indices to measure the connection strength between phenotype and GVR nodes indirectly via the patient layer.

2.1   Data Selection from DECIPHER

We downloaded CNV and HPO phenotype data from the DECIPHER database (15th of Janurary 2017, *"decipher-cnvs-grch37-2017-01-15.txt"*), in accordance with the DECIPHER Data Access Agreement. From this we obtained mutation genomic coordinates, mutation type, inheritance mode and clinically observed phenotypes (if available) for 20 520 patients. Of these, we selected 13 915 patients that have HPO annotations. We filtered these based on inheritance mode, in order to select patients with *de novo* mutations only, under the assumption that these variants are the cause of the pathological phenotype. Under these criteria we finally obtained 4 490 patients. We enriched the phenotypes of these patients using HPO data, by including parent terms from the hierarchy, in order to ensure that patients with different phenotypes can be compared via connections at less specific levels in the hierarchy.

2.2   Tripartite Network Creation

We created a tripartite network by combining GVR-patient and patient-phenotype connections. The GVR-patient connections were obtained through the identification of common and specific regions between DECIPHER patients, as described in Fig. 1. We performed a two-part overlap analysis: firstly we determined the putative overlapping regions using the mutation start and stop positions for each patient (see Fig. 1A, numbered regions). Once these had been determined, we counted how many patients had mutations in these regions (Fig. 1B and classed them as common regions if more than one patient shared this region, or specific regions if only one patient had a mutation in this region. The patient-phenotype connections were directly obtained from DECIPHER. The tripartite network was formed by combining these connection layers (Fig. 1C).

2.3   Establishing Phenotype-GVR Associations

We measured the association between GVRs and phenotypes using networks association methods proposed by Bass *et al.*, [9]. These methods measure the connection between a pair of nodes (in our case, phenotype and GVR nodes) through their shared nodes (in our case, the patient nodes). In order to apply this method to our tripartite network it was necessary to first transform it into a bipartite one. This was done by first combining the phenotype and GVR nodes

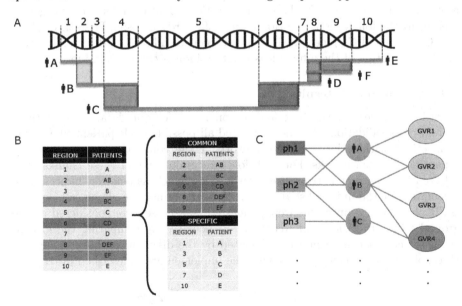

Fig. 1. Genomic Variation Regions (GVR) determination scheme and tripartite network creation. (A) Common and specific GVRs were determined from DECIPHER data. (B) We used both types of regions for the creation of the GVR layer. (C) Tripartite network structure, with 3 layers: phenotype nodes (tagged as ph1–3); patients (A–C); and GVRs (1–4).

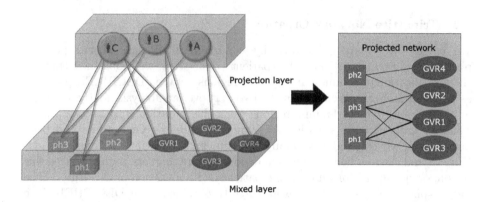

Fig. 2. Creation of the bipartite network from the tripartite network to calculate association indices. From our tripartite network, we created a mixed layer combining both types of nodes, phenotypes (ph1–3) and Genomic Variant Regions (GVR 1–4). Patient nodes (A–C) formed another layer. Connections between both layers were used to create the projected network with a bipartite structure connecting phenotypes and GVRs.

to produce a mixed layer, creating a bipartite network consisting of a patient layer and the GVR-phenotype mixed layer (Fig. 2, 3D representation). From this bipartite network, we projected the patient layer onto the mixed layer in order to establish indirect connections between phenotype and GVR nodes, producing a projected network (Fig. 2, plain representation). This final network was generated using the following association indices: Jaccard, Simpson, Pearson Correlation Coefficient (PCC), geometric, hypergeometric and cosine. More information on the calculation of the different association indices can be found in [9].

2.4   Validation of Results

In order to ascertain the best association index for our data, we performed 10-fold cross-validation. We first generated all possible GVR-patient-phenotype connections within the tripartite network. These connections were randomly separated into ten chunks. For each fold execution, nine of these chunks were selected to construct a tripartite network, for which association indices were calculated for the indices listed above. From the remaining chunk, all possible GVR-phenotype connections were generated, and these were considered as the ground truth. Precision and Recall were calculated by comparing the ground truth to the association indices calculated by the different methods, using 100 thresholds for each method, following the implementation described in [10]:

$$Precision = \frac{\sum\limits_{i=1}^{K} k_i}{\sum\limits_{i=1}^{K} m_i}; Recall = \frac{\sum\limits_{i=1}^{K} k_i}{\sum\limits_{i=1}^{K} n_i}$$

Here, K is the total number of phenotypes with known GVR associations from the ground truth. For each phenotype i, m_i is the number of GVR associations predicted by the association method, n_i is the known GVR associations in the ground truth and k_i is the number of GVR associations common to the predicted and known set GVR associations (the correct predictions). The Precision and Recall measures are averaged across the 10 folds and a Precision-Recall (PR) curve is built for all the association methods in order to determine which has the best performance. For the calculation of the association indices of the tripartite network and the PR values, we developed our Ruby gem NetAnalyzer (https://rubygems. org/gems/NetAnalyzer). The six association index calculations and 10-fold cross-validation procedure was automatised using Autoflow [11], a workflow manager that allowed us to perform all these tasks in an easy and reproducible way.

3   Results

3.1   Coverage Analysis of DECIPHER Data

We performed coverage analysis of the DECIPHER data in order to determine whether there was sufficient information to analyse mutations across the whole genome given our data selection strategy. This was performed using all 13 915 patients with HPO phenotypes, and for the 4 490 of these patients with *de novo* mutations (Fig. 3).

In these graphs we can see peaks that represent a large number of patients in DECIPHER with the same affected genomic region, for example at the start and end of chromosomes 1, 3, 4 and 18. Moreover, there are strong peaks at 2.5×10^7 base pairs of chromosome 15 and 22, and another peak at 2.6×10^7 base pairs of chromosome 16. There are also regions with no coverage, for example the first 2×10^7 base pairs in chromosomes 13, 14, 15 and 22. In general, most of genome is covered by DECIPHER data. We also calculated the average patient coverage, defined as the average number of patients with a mutation at a given nucleotide, which is 16.14 when using all DECIPHER patients (with HPO phenotypes) and 6.93 in the case of *de novo* mutations. As we will describe below, this information will be useful for defining the minimum number of patients necessary to make reliable connections between phenotypes and Genomic Variant Regions (GVR) in our tripartite network.

3.2   Association Index Calculation

We then created our tripartite network GVR-patient-phenotype from DECIPHER data and we calculated the six different association indices described in "Methods". We performed 10-fold cross-validation to obtain Precision-Recall (PR) curves. We accomplished this procedure for networks built with and without HPO enrichment, in order to see if enrichment improved results. We show these PR curves in Fig. 4. We observed better PR performance in the case of enriched data. In addition, we see that the hypergeometric method outperforms the other methods for both, enriched and non-enriched data (Fig. 4, green curve), followed by Simpson method (Fig. 4, pink curve).

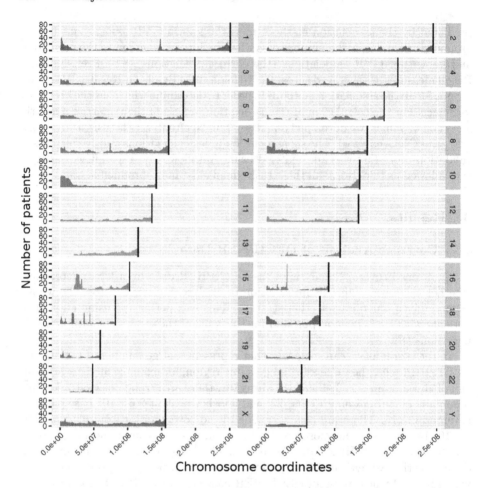

Fig. 3. Coverage analysis performed for *de novo* mutations selected from DECIPHER data. Coverage is represented by a different colour for each chromosome; black vertical lines symbolize the chromosome end. (Color figure online)

Briefly, the hypergeometric index calculates the log transformed probability of finding an equal or greater interaction overlap than the one observed between a given phenotype and GVR [9]:

$$H_{AB} = -log \sum_{i=|N(A) \cap N(B)|}^{min(|N(A)|,|N(B)|)} \frac{\binom{|N(A)|}{i} \cdot \binom{n_y-|N(A)|}{|N(B)|-i}}{\binom{n_y}{|N(B)|}}$$

where $N(A)$ is the number of phenotype nodes, $N(B)$ is the number of genomic variation regions nodes, n_y corresponds to the total number of nodes in the patients layer of the network and i is the number of shared patient nodes between $N(A)$ and $N(B)$.

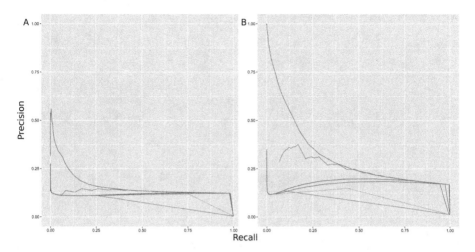

Fig. 4. Precision-Recall curves for each association index method, represented with different colours (green: hypergeometric, red: cosine, pink: Simpson, blue: PCC, light blue: Jaccard and olive green: geometric). In (A) there is the results for without HPO phenotypes enrichment and in (B) the results with HPO phenotypes enrichment. (Color figure online)

Finally, we used the average patient coverage to establish a range of putative values for the minimum number of patients required to keep a phenotype-GVR connection in the network. Given the average patient coverage was 6.93, we chose values of 1, 2, 5 and 10. We used HPO enrichment for this task, as it improves PR as shown above.

Figure 5 shows that results progressively improve according to the PR curves when more than one patient is necessary to support a connection. When it is required al least two patients to keep a GVR-phenotype connection, all specific GVR are removed and this increases the overall method performance. However, this procedure has a drawback, specially at higher cut-offs, the network looses information about the GVR-phenotype relations, as is shown in Fig. 6 (orange line). In fact, when the maximum PR performance is observed (Fig. 5D), corresponding to a minimum number of patients of 10, the network connections are reduced to 16.3% of their original number in the enriched version. Moreover, this percentage is even lower when it is analysed using the non-enriched version (Fig. 6, blue line), keeping only 3.7% of its original connections. This is because HPO enrichment allows very specific phenotypes to be connected at more general levels of the HPO hierarchy, as these more general connections will receive enough patients to be included. Furthermore, it is surprising that Simpson method performance (Fig. 5, pink line) is greatly increased at highest cut-offs although it not reach the high precision values of hypergeometric method on more relaxed cut-offs.

Taking into account the performance improvement and network size reduction, we suggest that the HPO enriched network filtered with a cut-off of two patients is the best option to use for further analysis, as it keeps 66.71% of the original network and moreover it has acceptable PR results, as shown in Fig. 5B.

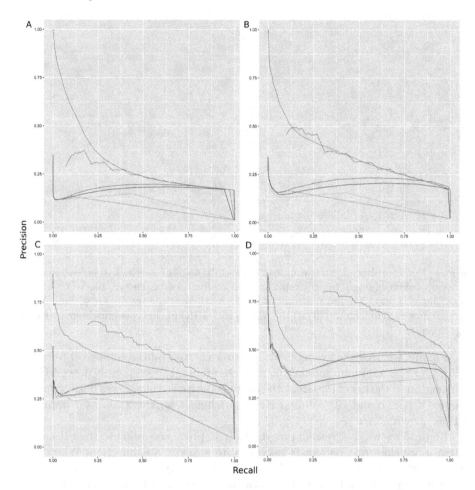

Fig. 5. Precision-Recall curves for each association index method, with supported connections phenotype-GVR by (A) one, (B) two, (C) five and (D) ten patients. The different curves of used methods are represented with different colours (green: hypergeometric, red: cosine, pink: Simpson, blue: PCC, light blue: Jaccard and olive green: geometric). (Color figure online)

4   Discussion

We have analysed the impact of different factors on the analysis of a network linking Genomic Variant Regions (GVRs), patients and their disease related phenotypes, which we have named the GVR-patient-phenotype network. More specifically, we have looked at the minimum number of patients sharing both a phenotype and a GVR needed to reliably support a connection between the two. A GVR-phenotype relationship that is connected by only one patient may be less reliable, as a connection might occur spuriously, or even be due to technology-related problems or erroneous clinical characterisation. Based on our validation

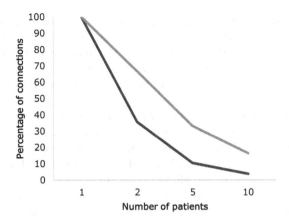

Fig. 6. Network connection reduction on the basis of the minimum number of patients required to keep a connection. Orange line corresponds to the network enriched with HPO phenotypes data, blue line corresponds to non-enriched data. The percentage of connections for 2, 5 and 10 patients was calculated with respect to the 1 patient data (100%). (Color figure online)

results, we see a clear improvement in precision when we increase the minimum number of patients required to connect a GVR and phenotype, suggesting that the criteria of filtering based on the number of patients is well-founded. However, it should be made clear that this criteria reduces the number of connections in the network, potentially reducing the number of detectable associations. It is therefore desirable to establish a cut-off value based on the minimum number of patients needed to support reliable connections.

We also investigated the use of the HPO hierarchy to improve the association results by adding parent terms to the phenotype layer of the network. Let us consider the case that two patients are annotated with similar, but different terms: through enrichment with parental terms of the HPO ontology, we can connect these patients as they will share a parent term. In this manner, the parental nodes can potentially produce stronger connections between patients and GVRs. However, this enrichment can potentially lead to problems as patients with dissimilar phenotypes may end up with the same annotation via a more general parental node. This is likely to affect the different association index calculation methods in different ways. For example, in the case of the hypergeometric method, which penalizes highly connected nodes, the HPO phenotypes enrichment may be accounted for implicitly in the calculation. In contrast, the Simpson method might be more sensitive to the increase of the intersection of nodes [9]. In any case, using the parental terms might be expected to result in higher sensitivity, as the use of additional HPO phenotypes can be associated to GVRs, giving us sufficient information to validate the connections [12].

Given these findings, we assessed various association indices using the network analysis tool we developed, NetAnalyzer. Using a cross validation approach, for which we considered a subset of the GVR-patient-phenotype connections as the ground truth, we found the hypergeometric test outperformed the other methods in terms of the precision-recall curve generated. In contrast to the rest of association methods, which calculate a similarity (Jaccard, Simpson, cosine and geometric) or a correlation (PCC) index, the hypergeometric index is based on a probabilistic model. This method calculate the specificity between associated GVR-phenotype nodes, and it can be used to determine the significance of the association [9].

One potential limitation of this work is the requirement to transform the tripartite GVR-patient-phenotype network into a bipartite network, since methods that we use for the association index calculation were based on bipartite network structure [9]. Future work will use other algorithms for bipartite network analysis [13], as well as new association algorithms based on tripartite network structure [14], and will compare these novel results to those obtained using the original association methods, in order to compare if they explicitly exploit the original. We also plan to explore how the number of patients linking a GVR and phenotype can affect predictive accuracy, as well as other factors such as the phenotype specificity and its impact on the associations values.

Potential applications of NetAnalyzer include patient diagnosis and to help guide the choice of genomic analysis technique to confirm underlying genetic variants for a given disease. As precision medicine and personalised genomics become more commonplace, it could also be used to rank putative associated phenotypes for a patient and thus help guide diagnosis.

We are currently collaborating with clinical research groups in order to carry out experimental validation of our results. We also plan to use annotation tools to help like the genomic regions associated with a pathogen to genes and other functional genomic elements. Finally, although the tools presented here are available for download we are also planning to implement them as a web server, to facilitate their adoption and usage by the research community.

Acknowledgements. This work was supported by the "Plan Nacional" funding from the Spanish Ministry of Economy and Competitiveness (Ref SAF2016-78041-C2-1-R), the Andalusian Government Excellence Research Projects funding (Ref CTS-486) and the European Commission, EU-FP7-Systems Microscopy Network of Excellence (Ref 258068). The CIBERER is an initiative from "Carlos III". Elena Rojano is a researcher from the Plan de Formacion de Personal Investigador (FPI) supported by the Andalusian Government with European Regional Development Fund. James Richard Perkins is a researcher from the Sara Borrell Program (Ref CD14/00242) of the "Carlos III" National Health Institute, Spanish Ministry of Economy and Competitiveness.

The authors thank the Supercomputing and Bioinnovation Center (SCBI) of the University of Malaga for their provision of computational resources and technical support (www.scbi.uma.es/site).

This study makes use of data generated by the DECIPHER community. A full list of centres who contributed to the generation of the data is available from http://decipher.sanger.ac.uk and via email from decipher@sanger.ac.uk. Funding for the project was provided by the Wellcome Trust. Those who carried out the original analysis and collection of the data bear no responsibility for the further analysis or interpretation of it by the Recipient or its Registered Users.

References

1. Doelken, S.C., Köhler, S., Mungall, C.J., Gkoutos, G.V., Ruef, B.J., Smith, C., Smedley, D., Bauer, S., Klopocki, E., Schofield, P.N., Westerfield, M., Robinson, P.N., Lewis, S.E.: Phenotypic overlap in the contribution of individual genes to CNV pathogenicity revealed by cross-species computational analysis of single-gene mutations in humans, mice and zebrafish. Dis. Model. Mech. **6**(2), 358–372 (2013)
2. Robinson, P.N.: Deep phenotyping for precision medicine. Hum. Mutat. **33**(5), 777–780 (2012)
3. Civelek, M., Lusis, A.J.: Systems genetics approaches to understand complex traits. Nature Rev. Genet. **15**(1), 34–48 (2014)
4. Dong, L., Wang, W., Li, A., Kansal, R., Chen, Y., Chen, H., Li, X.: Clinical next generation sequencing for precision medicine in cancer. Curr. Genomics **16**(4), 253–263 (2015)
5. Melnikova, I.: Rare diseases and orphan drugs. Nat. Rev. Drug Discov. **11**, 267 (2012)
6. Firth, H.V., Richards, S.M., Bevan, A.P., Clayton, S., Corpas, M., Rajan, D., Vooren, S.V., Moreau, Y., Pettett, R.M., Carter, N.P.: DECIPHER: database of chromosomal imbalance and phenotype in humans using ensembl resources. Am. J. Hum. Genet. **84**(4), 524–533 (2009)
7. Corpas, M., Bragin, E., Clayton, S., Bevan, P., Firth, H.V.: Interpretation of genomic copy number variants using DECIPHER. Curr. Protoc. Hum. Genet. **72**(8), 14:8.14.1–14:8.14.17 (2012)
8. Robinson, P.N., Mundlos, S.: The human phenotype ontology. Clin. Genet. **77**(6), 525–534 (2010)
9. Fuxman Bass, J.I., Diallo, A., Nelson, J., Soto, J.M., Myers, C.L., Walhout, A.J.M.: Using networks to measure similarity between genes: association index selection. Nat. Methods **10**(12), 1169–1176 (2013)
10. Pandey, G., Steinbach, M., Gupta, R., Garg, T., Kumar, V.: Association analysis-based transformations for protein interaction networks: a function prediction case study. In: Proceedings of the ACM SIGKDD International Conference on Knowledge Discovery and Data Mining, pp. 540–549 (2007)
11. Seoane, P., Ocaña, S., Carmona, R., Bautista, R., Madrid, E., Torrres, A., Claros, G.: AutoFlow, a versatile workflow engine illustrated by assembling an optimised de novo transcriptome for a non-model species, such as Faba Bean (Vicia faba). Curr. Bioinform. **11**, 440–450 (2016)
12. Reyes-Palomares, A., Bueno, A., Rodríguez-López, R., Medina, M.Á., Sánchez-Jiménez, F., Corpas, M., Ranea, J.A.G.: Systematic identification of phenotypically enriched loci using a patient network of genomic disorders. BMC Genom. **17**(1), 232 (2016)
13. Wang, L., Oehlers, S.H., Espenschied, S.T., Rawls, J.F., Tobin, D.M., Ko, D.C.: CPAG: software for leveraging pleiotropy in GWAS to reveal similarity between human traits links plasma fatty acids and intestinal inflammation. Genome Biol. **16**, 190 (2015)
14. Alaimo, S., Giugno, R., Pulvirenti, A.: ncPred: ncRNA-disease association prediction through tripartite network-based inference. Front. Bioeng. Biotechnol. **2**, 71 (2014)

Variety Behavior in the Piece-Wise Linear Model of the p53-Regulatory Module

Magdalena Ochab$^{(\boxtimes)}$, Krzysztof Puszynski, Andrzej Swierniak, and Jerzy Klamka

Silesian University of Technology, Institute of Automatic Control, Akademicka 2A, 44-100 Gliwice, Poland
{Magdalena.Ochab,Krzysztof.Puszynski,Andrzej.Swierniak, Jerzy.Klamka}@polsl.pl

Abstract. Biological processes have very complicated nonlinear dynamics thus generally they are mathematically modeled using nonlinear functions. High nonlinearity in the models cause many difficulties in its analysis. We propose to use piece-wise linear differential equations model instead which is easier for analysis. In this approach we created such model of the regulatory module p53. Protein p53 plays crucial role in the cell response after stress stimuli and enables elimination of the carcinogenic mutated cells. The results shows that after stress stimuli, the number, type and localization of the stationary points are changed, which corresponds to different cell behavior observed in the biological experiments.

Keywords: Biological model · Switches · Piece-wise linear models · Protein p53 · Apoptosis

1 Introduction

1.1 Protein P53 and Its Regulatory Core

Protein p53 is one of the main players in the cellular response for the different intrinsic and extrinsic stimuli. The main activity of the p53 is regulation, activation or repression of the transcription of many genes and consequently the main pool of the protein p53 is the nucleus. In the case of the damages in the DNA structure, the protein p53 activates process of the cell cycle blockade and the DNA repair [1]. If the damages are irreparable, the protein p53 activates transcription of proteins responsible for the programmed cell death, to avoid the multiplication of the improper cells. The dysfunction of the p53 regulatory module leads to the proliferation of the mutated cells and can be a direct cause of the tumor [2]. The proper level of the protein p53 in the cell is assured by the two feedback loops: one positive and one negative. The negative feedback loop in which p53 produces its own inhibitor Mdm2 is responsible for maintenance low p53 level in the properly functioning cells and the oscillatory response of p53 after DNA damage [4]. The positive feedback loop exists by the double

© Springer International Publishing AG 2017
I. Rojas and F. Ortuño (Eds.): IWBBIO 2017, Part I, LNBI 10208, pp. 208–219, 2017.
DOI: 10.1007/978-3-319-56148-6_18

negation; the p53 induces expression of the protein PTEN which, by PIP3 and Akt, inhibits MDM2 translocation to the nucleus. One of the effects observed in the cell after DNA damage is rapid degradation of the MDM2, which leads to increase of the mean p53 level [3]. Kracikova et al. [5] suggest that the p53 concentration is directly related to the cell fate: mean medium p53 level is observed in cell with cell cycle arrest and high p53 level is observed in apoptotic cells.

1.2   Systems with Switchings

Biological and biochemical processes which occurs in cells or tissues are not spontaneous but they are regulated by variety of different molecules such as enzyme in enzymatic reactions or transcription factors which activate genes expression. The biological processes rate dependence on the concentration of the regulatory molecules is highly nonlinear: generally for low concentration the reaction rate is near to zero, with the increase of the concentration the rapid increase of the process rate is observed and with further concentration increase the rate is saturated [6]. The overwhelming majority of the published models are created with the highly nonlinear functions, such as Hill function [7], which can approximate the sigmoidal character of processes (e.g. [8]). The biggest advantage of the nonlinear models is possibility to use variety of nonlinear functions, which can be useful to provide the dynamics nearing the behavior observed in biological experiments. However the parameters of such functions are very difficult to estimate. The highly nonlinear models are very difficult to analyze, even with such basic issues like stability and sensitivity. Moreover finding the analytical solution is almost always impossible, so the results are achieved only by the numerical solutions.

In this work we present completely different approach to modeling the biological processes. Switching control systems are a commonly well known in the control and system engineering (see [9]). As shown by Mestl et al. [11] on some theoretical example, piece-wise linear differential equations (PLDE) models may be successfully applied for biological processes. As we shown in [10] it may be applied also for real biological systems with complicated dynamics. For easier presentation PLDE model can be described by the diagram of the subsequent linear models, and the localization of the stationary points [11–13]. The advantage of presented method is that the linear parts of the model may be easily analyzed, even by finding analytical solutions, however the global properties of such systems are much more difficult to investigate, due to the nonlinearity and discontinuity. In our approach the model consists on the set of linear differential functions and the set of rules to define, which subset of the functions is used for specified state. So basically the model structure and parameters may change depending on its state. The general system equation can be defined as:

$$\frac{dx_i}{dt} = \alpha_i(X) - \gamma_i(X)x_i, \quad i = 1, \dots n. \tag{1}$$

where x_i is the protein level, α_i is the protein production rate, γ_i is protein degradation rate and X is the set of Boolean function functions which values (0 or 1) depends on the state of x.

1.3 Proposed Model

Proposed model is based on our previous work [10]. It describes the main core of the p53 regulatory module by interplay of 4 molecules, which correspond to 4 variables in our model: protein p53 (P), cytoplasmic MDM2 (C), nuclear MDM2 (N) and PTEN (T). Protein p53 activates transcription of the MDM2 and PTEN. Nuclear MDM2 induces the p53 degradation, however its nuclear transport can be blocked by PTEN. The external stress (IR which damage DNA) is denoted by the symbol R, and is responsible for change of the parameter corresponding to MDM2 degradation (both forms). The relationship between proteins is presented on the Fig. 1A.

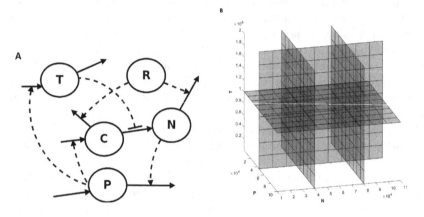

Fig. 1. A. The scheme of the p53 regulatory core. B. State space divided into domains.

The general model is presented by the set of equations (Eq. 2). The parameters marked by upper index K can have different values according to the state.

$$\frac{dP(t)}{dt} = p_1 - d_1^K(N(t)) \cdot P(t),$$

$$\frac{dC(t)}{dt} = p_2^K(P(t)) - k_1^K(T(t)) \cdot C(t) - d_2 \cdot (1 + R) \cdot C(t),$$

$$\frac{dN(t)}{dt} = k_1^K(T(t)) \cdot C(t) - d_2 \cdot (1 + R) \cdot N(t),$$

$$\frac{dT(t)}{dt} = p_3^K(P(t)) - d_3 \cdot T(t). \tag{2}$$

2 Methods

2.1 Threshold

The phase space is divided into regulatory domains by introduced thresholds values for chosen variables. Generally in the piece-wise linear models, the switching function is defined as the step function. Consequently on the threshold values the system is not continuous and the derivative of the function in the switching point cannot be determined but is required for further analysis. To avoid this problem we use the approach presented by Mestl et al. in 1995 [11]. We replace the step function by a continous sigmoid function with the limit $[0, 1]$. Thus the switching function is a monotonic mollifier and is define as $Z_{ij} = Z(x_i, \theta_{ij}, \delta) = 0$ for $x_i \leq \theta_{ij} - \delta$, Z_{ij} increases monotonously form 0 to 1 for $x_i \in \langle \theta_{ij} - \delta, \theta_{ij} + \delta \rangle$ and $Z_{ij} = 1$ for $x_i \geq \theta_{ij} + \delta$. Parameter δ is the distance from the threshold value. When δ tend to zero, this monotonic mollifier approaches the step function. We introduce 4 threshold: one for p53 (θ_P), one for PTEN (θ_T) and two for MDM2 (θ_{N1} and θ_{N2}) which reflects to 4 switching functions denoted by Z_P, Z_T, Z_{N1} and Z_{N2} (see Fig. 1B). Following the Plahte et al. [12] the region where at least one variable lies on its threshold values is called a Δ-region, and according to the number of variables lying on their threshold the Δ-regions can be one-, two-, three-dimensional or more.

2.2   Model Definition

The proposed p53 model with switching function is presented below:

$$\frac{dP(t)}{dt} = p_1 - (d_{10} + d_{11} \cdot Z_{N1}^+ + d_{12} \cdot Z_{N2}^+) \cdot P(t),$$

$$\frac{dC(t)}{dt} = p_{20} + p_{21} \cdot Z_P^+$$
$$- (k_{10} - k_{11} \cdot Z_T^+ + d_2 \cdot (1 + R)) \cdot C(t),$$

$$\frac{dN(t)}{dt} = (k_{10} - k_{11} \cdot Z_T^+) \cdot C(t) - d_2 \cdot (1 + R) \cdot N(t),$$

$$\frac{dT(t)}{dt} = p_{30} + p_{31} \cdot Z_P^+ - d_3 \cdot T(t). \tag{3}$$

We examine how the change of parameter R will influence the model results. All the parameter values and the thresholds are presented in Table 1.

2.3   Model Analysis

The model analysis base on calculation of the stationary points and localization of the cycles. The literature distinguish two types of stationary point - regular, when they lies inside the regulatory domain (RSP) and singular - when at least one of the variables lies on its threshold (SSP) [11]. RSPs are calculated by application of the steady state condition ($dx_i/dt = 0$) for every regulatory domain. If the calculated steady point lies inside the considered domain, this point is called RSP and is asymptotically stable. The SSPs are more difficult to localize. The one-dimensional Δ-region called a wall can contain an asymptotically stable stationary point. The asymptotically stable SSP exist on the wall for $x_i = \theta_{ij}$ only if the derivative $\partial F_i/dZ_{ij} < 0$, where the F_i denotes the time derivative

Table 1. The values of the model parameters

Parameter	Description	Value	Unit
p_1	p53 production rate	8.8	1/sec
p_{20}	Spontaneous MDM2 production rate	2.4	1/sec
p_{21}	p53-induced MDM2 production rate	21.6	1/sec
p_{30}	Spontaneous PTEN production rate	0.5172	1/sec
p_{31}	p53-induced PTEN production rate	3.6204	1/sec
d_{10}	Spontaneous p53 degradation rate	$9.8395 * 10^{-5}$	1/sec
d_{11}	1st nuclear MDM2-induced p53 degradation rate	$6.5435 * 10^{-5}$	1/sec
d_{12}	2nd nuclear MDM2-induced p53 degradation rate	$1.6283 * 10^{-4}$	1/sec
d_2	MDM2 degradation rate	$1.375 * 10^{-5}$	1/sec
d_3	PTEN degradation rate	$3 * 10^{-5}$	1/sec
k_{10}	Spontaneous MDM2 transport to nucleus rate	$1.5 * 10^{-4}$	1/sec
k_{11}	PTEN-inhibited MDM2 transport rate	$1.4713 * 10^{-4}$	1/sec
R	External stress	0, 4, 9	
Thresholds	Description	Value	Unit
θ_P	p53 threshold value	$4.5 * 10^4$	molecules
θ_{N1}	1st nuclear MDM2 threshold value	$4 * 10^4$	molecules
θ_{N2}	2nd nuclear MDM2 threshold value	$8 * 10^4$	molecules
θ_T	PTEN threshold value	$1 * 10^5$	molecules

dx_i/dt. If these condition is fulfilled and the exact localization of the stationary point is inside the determined part of the state space, the point is asymptotically stable SSP. On the two and more dimensional Δ-regions stationary point may exist but they are not asymptotically stable. Such points exist on the crossing of the thresholds values if between domains exist the closed temporal sequence. However such stationary point is stable if its exact localization is contained in the determined part of the state space [12].

We perform the stability analysis for created model. At the beginning we check if the derivative $\partial F_i/dZ_{ij}$ is not equal to zero. Secondly we localize the RSP and build the state transition diagram. Based on the diagram we localize the closed sequences of the transitions between domains and calculate the localization of the stationary points. The stable SSP exists if the following assumptions are fulfilled: values of all switching functions $Z_{ij} \in \langle 0, 1 \rangle$ and all not determined variables lie within the range of the Δ-region. If the SSP exists on the two- or more dimensional Δ-region, the stable limit cycle exists in the system, which results in the oscillations in the model response.

3   Results

In our model we divide the state space into 12 regulatory domains, specified by a vector $\mathbf{B} = [\mathbf{P}, \mathbf{N}, \mathbf{T}]^{\mathbf{T}}$ where P, N, T are the Boolean-like states of denoting if the corresponding variable level is bellow (0) or above (1 in case of P and T and 1 or 2 in case of N) given threshold, for example the region for $P < \theta_P$, $\theta_{N1} \leq N < \theta_{N2}$ and $T \geq \theta_P$ is denoted by the domain $D[011]$ given by the vector $\mathbf{B} = [0, 1, 1]$ and the region $P \geq \theta_P$, $N \geq \theta_{N2}$ and $T \geq \theta_P$ is the domain $D[121]$.

In the p53 model none of the variables depends directly on its own thresholds, so the derivatives $\partial F_i/dZ_{ij}$ for all variables P, C, N, T are equal to zero. Consequently this model does not contain any asymptotically stable SSP on the one-dimensional Δ-region called wall.

3.1 Without Additional Stress ($R = 0$)

For the original parameter values the system has only one RSP in the domain $D[020]$. Based on the state transition diagram (Fig. 2) there is a one closed sequence of the states, so there are two regions which can include the SSP points, the first is two-dimensional $\Delta(\theta_P, \theta_{N2}, T > \theta_T)$ and the second is tree-dimensional $\Delta(\theta_P, \theta_{N2}, \theta_T)$.

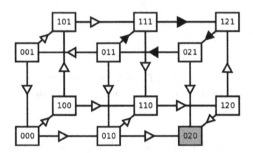

Fig. 2. State transition diagram for the model without external stress. The gray domain includes the RSP. The black arrows emphasize the closed sequence of the states.

Region $\boldsymbol{\Delta(\theta_P, \theta_{N2}, T > \theta_T)}$. To examined the existence of the SSP in this region we use the approach presented by Plathe et al. [12]. We create the equations for this region, taking into account that the variables P and N are equal to its threshold values, correspondingly θ_P and θ_{N2}, and the variable T should be higher than θ_T. Assuming the steady state condition, we can calculate the values of the switching variables Z_P and Z_{N2}, which should be included in the interval $[0,1]$. If these conditions are fulfilled, we calculate exact localization of the point SSP, and check if the assumption for the T value is fulfilled.

The steady state of this region is defined by the following equations:

$$0 = p_1 - (d_{10} + d_{11} + d_{12} \cdot Z_{N2}^+) \cdot \theta_P,$$
$$0 = p_{20} + p_{21} \cdot Z_P^+ - (k_{10} - k_{11} + d_2) \cdot C,$$
$$0 = (k_{10} - k_{11}) \cdot C - d_2 \cdot \theta_{N2},$$
$$0 = p_{30} + p_{31} \cdot Z_P^+ - d_3 \cdot T. \tag{4}$$

The values of the switching variables are $Z_P^+ = 0.0555$ and $Z_{N2}^+ = 0.1936$ so they are satisfying the requirements for $Z_i \in \langle 0, 1 \rangle$. However the T value for this stationary point is $T = 2.3940 * 10^4$ which is much more smaller then θ_T, so this region does not contain the SSP.

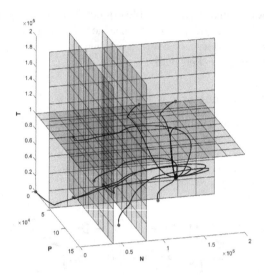

Fig. 3. Time courses in the state space for different initial conditions in system without external stress. Diamond indicates the RSP and gray dots indicate initial points.

Region $\Delta(\theta_P, \theta_{N2}, \theta_T)$. To determine SSP existence in this region we use the same method as presented in previous section. All variables lie on their thresholds, so the model equations in the steady state have the following form:

$$0 = p_1 - (d_{10} + d_{11} + d_{12} \cdot Z_{N2}^+) \cdot \theta_P,$$
$$0 = p_{20} + p_{21} \cdot Z_P^+ - (k_{10} - k_{11} \cdot Z_T^+ + d_2) \cdot C,$$
$$0 = (k_{10} - k_{11} \cdot Z_T^+) \cdot C - d_2 \cdot \theta_{N2},$$
$$0 = p_{30} + p_{31} \cdot Z_P^+ - d_3 \cdot \theta_T. \tag{5}$$

The values of the switching variables are $Z_P^+ = 0.6858$, $Z_{N2}^+ = 0.1936$ and $Z_T^+ = 1.0259$. The Z_T is higher than 1, so this region does not contain SSP. The time courses for this case, for different initial conditions in the state space are presented on the Fig. 3.

3.2   Exposition to the Small Stress (R = 4)

External stress such as DNA damage increases degradation rate of the MDM2, as a result the dynamics of the system response changes. For the low stress

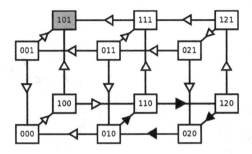

Fig. 4. State transition diagram for the model after stress. The gray domain includes the RSP. The black arrows emphasize the closed sequence of the transitions.

the stationary point in the domain D[020] disappears and the RSP arise in the domain D[101]. Comparing with the normal state, the transition between domains are changed what is easily seen on the diagram in Fig. 4. In this system, the closed sequence is localized between domains D[010], D[110], D[120] and D[020] so SSPs can be found in two regions: $\Delta(\theta_P, \theta_{N2}, T < \theta_T)$ and $\Delta(\theta_P, \theta_{N2}, \theta_T)$.

$\boldsymbol{\Delta(\theta_P, \theta_{N2}, T < \theta_T)}$. The equations describing the steady state in these region are presented below:

$$0 = p_1 - (d_{10} + d_{11} + d_{12} \cdot Z_{N2}^+) \cdot \theta_P,$$
$$0 = p_{20} + p_{21} \cdot Z_P^+ - (k_{10} + d_2 \cdot (1 + R)) \cdot C,$$
$$0 = k_{10} \cdot C - d_2 \cdot (1 + R) \cdot \theta_{N2},$$
$$0 = p_{30} + p_{31} \cdot Z_P^+ - d_3 \cdot T. \tag{6}$$

The switching variables have got values: $Z_P^+ = 0.2602$ and $Z_{N2}^+ = 0.1936$. The variable T in the stationary point is equal to $4.8644 * 10^4$ and satisfy the condition $T_z < \theta_T$ and consequently SSP exists in this region.

$\boldsymbol{\Delta(\theta_P, \theta_{N2}, \theta_T)}$. The equations describing the steady state in these region are presented below:

$$0 = p_1 - (d_{10} + d_{11} + d_{12} \cdot Z_{N2}^+) \cdot \theta_P,$$
$$0 = p_{20} + p_{21} \cdot Z_P^+ - (k_{10} - k_{11} \cdot Z_T^+ + d_2 \cdot (1 + R)) \cdot C,$$
$$0 = (k_{10} - k_{11} \cdot Z_T^+) \cdot C - d_2 \cdot (1 + R) \cdot \theta_{N2},$$
$$0 = p_{30} + p_{31} \cdot Z_P^+ - d_3 \cdot \theta_T. \tag{7}$$

The switching variables have got values: $Z_P^+ = 0.6858$, $Z_{N2}^+ = 0.1936$ and $Z_T^+ = -1.3397$. The variable Z_T is smaller than zero and consequently in this region the stationary point SSP does not exist. In the case of exposition to low stress, two stationary points exist in the system, one regular RSP and one singular SSP. Existing of the SSP results in limit cycle, which cause the oscillations

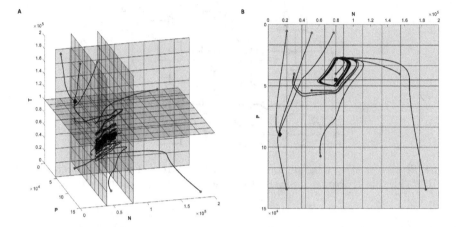

Fig. 5. Time courses in the state space for different initial conditions in system with small external stress (R = 4). Diamond indicates the RSP, square indicates the SSP and gray dots indicate initial points. A. 3D view. B. 2D view on the variables P and N.

in the model response. The behaviors of system for different initial conditions are presented in the Fig. 5.

3.3   Exposition to the Big Stress (R = 9)

The exposition to stress denoted by 9 do not change the state transition diagram comparing to the dose 4 (Fig. 4). There still exist one RSP in the domain D[101] and one closed sequence between domains, so we need to examine two domains for SSP existence.

$\Delta(\theta_P, \theta_{N2}, T < \theta_T)$. The equations describing the steady state in these region are the same as for the $R = 4$ (Eq. 6), where only the value of R is changed. As a result the switching control values are equal: $Z_P^+ = 0.8650$ and $Z_{N2}^+ = 0.1936$. In the stationary point the variable T is equal to $1.2162 * 10^5$ which is bigger than θ_T and consequently in this region SSP does not exist.

$\Delta(\theta_P, \theta_{N2}, \theta_T)$. The equations describing the steady state in these region are the same as for $R = 4$ (Eq. 7). The values of the switching variables are $Z_P^+ = 0.6858$, $Z_{N2}^+ = 0.1936$ and $Z_T = -2.1873$. Due to variable Z_T, which is smaller than 0, in these point the SSP does not exist.

In this system we have only one stable stationary point RSP, so all the time courses will be leading to the domain D[101]. However on the Fig. 6 we can see, that depending on the initial conditions the oscillation of the protein level before the system reach its steady point can occur.

The localizations of all the stationary point are presented in the Table 2.

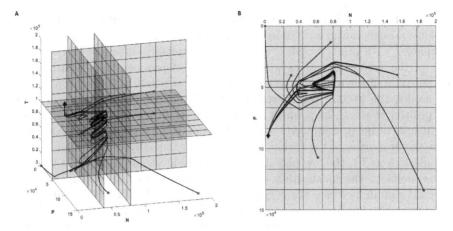

Fig. 6. Time courses in the state space for different initial conditions in system with big external stress (R = 9). Diamond indicates the RSP and gray dots indicate initial points. A. 3D view. B. 2D view on the variables P and N.

Table 2. The exact localization of the stationary points

Stress (R)	Domain	Type	P_s	C_s	N_s	T_s
0	D[020]	RSP	26 857	14 656	159 888	17240
4	D[101]	RSP	89 435	335 101	13 988	137 920
4	$\Delta(\theta_P, \theta_{N2}, T < \theta_T)$	SSP	45 000	36 667	80 000	48 644
9	D[101]	RSP	89 435	170 976	3 568	137 920

4 Discussion

In general biological processes have got very complicated, nonlinear dynamics. This causes problems not only with analyzing of the nonlinear models, but also with determination of the parameter values. Fortunately, majority of biological processes have dynamic described by sigmoid function, which can be approximated by the step function in piece-wise linear differential equations models.

In the [10], based on p53 regulatory module model, we presented that even complicated dynamics of the system can be achieved by linear model with switchings. The problem of stability analysis of the nonlinear systems with switchings can be performed only locally such as in [14]. The piece-wise linear models allow to global analysis of the system stability. Our model is not the first attempt to model the biological system as a piece-wise linear models. A methodology to analyzing such systems were presented in series of papers published by Plathe, Mestl and coworkers [11–13], however in these papers differential equations models consist only on two processes - production and degradation. Transitions from one variable to another or transport between compartments were not considered. In biological system molecules in different compartments (cytoplasm, nucleus, mitochondria, etc.) or in different state (phosphorylated, ubiquitylated, complexed with other proteins) often have completely different activity. Great part of the system dynamics will be neglected if processes like transport and modification of the molecules are not taken into account. Even in this quite simple model, the MDM2 transport from cytoplasm to nucleus enables modeling the positive feedback loop which require delay resulting from that transition.

In spite of being simplified, presented model of the p53 regulatory core maintains the properties of the real biological system. In the system without external stimuli such as IR, the low p53 and high MDM2 levels are observed. Such state corresponds to the localization of the RSP in the domain D[020]. The high external stimuli increases the degradation of the MDM2. As a result the RSP in the domain D[020] disappears and new one in the domain D[101] arises. These reflects to high p53 and PTEN levels with low MDM2 level. Such proteins levels are observed in the biological system after high stress dose, which induces the apoptotic reaction in the cell. The medium dose of the stress results with another stationary point SSP. This point is localized on the thresholds θ_P and θ_{N2}, and is not asymptotically stable. Between the domains D[010], D[110], D[120] and D[020] a limit cycle exists, which results with undamped oscillations of proteins levels. This reflect the state where cells block its cell cycle and tries to repair DNA damage. Moreover for the medium stress two stationary point exist, which corresponds with the biological results. The cells population is divided into two subpopulation: the one contains apoptotic cells and the second one contains the cells with cell cycle blockade [5]. Our results shows that cell fate depends on the initial conditions, which determines which stationary point will be reached.

In this work we analyzed only quite simple regulatory unit of the known biological system. However this model can be extended by a variable for DNA damages, which will enable modeling the process of its generation and repair. Another interesting aspect to consider is the existence of the different number and types of stationary points depending on the parameter d_2. The complex bifurcation analysis may give an interesting insights into the dynamics of the analyzed system.

Acknowledgments. The research presented here was partially supported by the National Science Centre in Poland granted with decision number DEC-2013/11/B/ST7/01713 (for KP), DEC-2014/13/B/ST7/00755 (for AS and JK) and BKM/506/RAU1/2016/15 (MO).

References

1. Vousden, K.H., Lu, X.: Lice or let die: the cell's response to p53. Nature **2**, 594–604 (2002)
2. Schmitt, C.A., Lowe, S.W.: Apoptosis and therapy. J. Pathol. **187**, 127–137 (1999)
3. Jonak, K., Kurpas, M., Szoltysek, K., Janus, P., Abramowicz, A., Puszynski, K.: A novel mathematical model of ATM/p53/NF-B pathways points to the importance of the DDR switch-off mechanisms. BMC Syst. Biol. **10**, 75 (2016)

4. Geva-Zatorsky, N., Rosenfeld, N., Itzkowitz, S., Milo, R., Sigal, A., Dekel, E., Yarnitzky, T., Liron, Y., Polak, P., Lahav, G., Alon, U.: Oscillations and variability in the p53 system. Mol. Syst. Biol. **2**: 2006.0033 (2006)
5. Kracikova, M., Akiri, G., George, A., Sachidanandam, R., Aaronson, S.A.: A threshold mechanism mediates p53 cell fate decision beteen growth arrest and apoptosis. Cell Death Differ. **20**, 576–588 (2013)
6. Alberts, B., Johnson, A., Lewis, J., Raff, M., Roberts, K., Walter, P.: Molecular Biology of the Cell, 4th edn. Garland Science, New York (2002)
7. Hill, A.V.: The possible effects of the aggregation of the molecules of hemoglobin on its dissociation curves. J. Physiol. **40**, 4–8 (1910)
8. Griffith, J.S.: Mathematics of cellular control processes. I. Negative feedback to one gene. J. Theoret. Biol. **20**, 20220 (1968)
9. Liberzon, D.: Switchings in Systems and Control. University of Illionis at Urbana-Campaign, Campaign (2003)
10. Ochab, M., Puszynski, K., Swierniak, A.: Application of the piece-wise linear models for description of nonlinear biological systems based on p53 regulatory unit. In: Proceedings of the XXI National Conference on Applications of Mathematics in Biology and Medicine, Wydawnictwo Uniwersytetu Jana Kochanowskiego w Kielcach, pp. 85–90 (2016)
11. Mestl, T., Plahte, E., Omholt, S.W.: A mathematical framework for describing and analysing gene regulatory networks. J. Theoret. Biol. **176**, 291–300 (1995)
12. Plahte, E., Mestl, T., Omholt, S.W.: Global analysis of steady points for systems of differential equations with sigmoid interactions. Dyn. Stab. Syst. **9**(4), 275–291 (1994)
13. Plahte, E., Mestl, T., Omholt, S.W.: A methodological basis for description and analysis of the systems with complex switch-like interactions. J. Math. Biol. **36**, 321–348 (1998)
14. Ochab, M., Puszynski, K., Swierniak, A.: Structural stability of biological models with switchings. In: 2016 21st International Conference on Methods and Models in Automation and Robotics (MMAR), pp. 333–338 (2016)

Multi-omic Data Integration Elucidates *Synechococcus* Adaptation Mechanisms to Fluctuations in Light Intensity and Salinity

Supreeta Vijayakumar and Claudio Angione[✉]

Department of Computer Science and Information Systems,
Teesside University, Middlesbrough, UK
{s.vijayakumar,c.angione}@tees.ac.uk

Abstract. *Synechococcus* sp. PCC 7002 is a fast-growing cyanobacterium which flourishes in freshwater and marine environments, owing to its ability to tolerate high light intensity and a wide range of salinities. Harnessing the properties of cyanobacteria and understanding their metabolic efficiency has become an imperative goal in recent years owing to their potential to serve as biocatalysts for the production of renewable biofuels. To improve characterisation of metabolic networks, genome-scale models of metabolism can be integrated with multi-omic data to provide a more accurate representation of metabolic capability and refine phenotypic predictions. In this work, a heuristic pipeline is constructed for analysing a genome-scale metabolic model of *Synechococcus* sp. PCC 7002, which utilises flux balance analysis across multiple layers to observe flux response between conditions across four key pathways. Across various conditions, the detection of significant patterns and mechanisms to cope with fluctuations in light intensity and salinity provides insights into the maintenance of metabolic efficiency.

Keywords: Multi-omics · Synechococcus · Stress conditions · Adaptation · Light and salinity · Phototrophic growth

1 Introduction

Metabolism is among the most important biological processes as balancing the production and consumption of metabolites is essential for maintaining life. Furthermore, it is currently the only biological layer that can be modelled genome-wide. Throughout the field of systems biology, there are a number of approaches which endeavour to capture the enormous complexity of biological systems by utilising mathematical modelling and computation to amalgamate the information required to build and refine predictive models of metabolism. The challenges presented by such an undertaking are numerous and persistent owing to the size, format, scale and variation of the disparate data types.

Constraint-based reconstruction and analysis (COBRA) methods are commonly used to express metabolic flux through biochemical pathways based on

© Springer International Publishing AG 2017
I. Rojas and F. Ortuño (Eds.): IWBBIO 2017, Part I, LNBI 10208, pp. 220–229, 2017.
DOI: 10.1007/978-3-319-56148-6_19

knowledge of reaction stoichiometry [1]. During flux balance analysis (FBA), a pseudo-steady state is assumed to calculate all fluxes under time-invariance and spatial homogeneity for purposes of mass conservation. Mass-balance constraints are imposed on the system to identify a range of points representing all feasible flux distributions. A feasible phenotypic state in the solution space is then computed using linear programming with a set of values indicating the optimal conditions required to achieve a given objective function [2]. FBA is particularly suitable for modeling genome scale metabolic networks as the definition of kinetic parameters and metabolite concentrations is not a key requisite. To improve the characterisation of metabolic networks at the whole genome scale, genome-scale models of metabolism can be integrated with heterogeneous multi-omic data to provide a more accurate representation of metabolic capability. This is useful in refining phenotypic predictions across various environmental conditions.

In recent years, genome-scale metabolic models (GSMMs) have been integrated with multiple heterogeneous omic data types in a number of studies. This serves to exploit the large volume of experimental data being generated from high-throughput omics technologies, in order to improve the characterisation of metabolic networks at the whole genome scale. In doing so, additional constraints can be applied during flux balance analysis in order to shrink the solution space [3], thus providing a more accurate representation of metabolic capability as a greater number of factors can be considered to explain cellular behaviour. This can prove useful in refining phenotypic predictions across various environmental conditions or engineering an organism in a way that optimises the production of a certain metabolite, which is highly applicable to fields such as industrial biotechnology and pharmacology.

Cyanobacteria are a group of photosynthetic prokaryotes for which it is imperative to adapt to constant fluctuations in temperature, salinity, light intensity (or irradiance), and nutrient availability, amongst other factors [4]. *Synechococcus* sp. PCC 7002 is a fast-growing cyanobacterium which flourishes in both freshwater and marine environments, owing to its ability to tolerate high light intensity and a wide range of salinities. Harnessing the properties of cyanobacteria has become an imperative goal in recent years owing to their potential to serve as biocatalysts for the production of renewable biofuels [5]. In an industrial setting, *Synechococcus* sp. PCC 7002 has been chosen as a model organism owing to its ease of genetic manipulation as well a tolerance for high salinity and slightly higher temperatures; these are highly desirable traits in micro-algae as this enables cultures to maintain a rapid growth rate in open raceway ponds as well as in photobioreactors, which operate at higher temperatures [6]. Recent studies have examined temporal variations in response to varying light intensity and associated conditional dependencies [7,8]. These need to be accounted for as constraints in genome-scale metabolic models designed to simulate the phototrophic growth in cyanobacteria over diurnal cycles and tackle issues associated with resource allocation.

2   Methods

In this work, we present a heuristic pipeline for analysing a genome-scale metabolic model of the cyanobacterium *Synechococcus* sp. PCC 7002, which is detailed in Fig. 1.

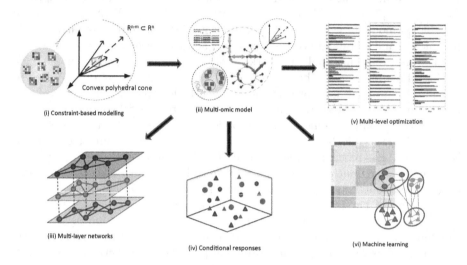

Fig. 1. Pipeline for prediction and classification of *Synechococcus* growth conditions. (i) Stoichiometric coefficients obtained from FBA used to map flux distribution as a convex polyhedral cone; (ii) Multi-omic model of *Synechococcus* sp. PCC 7002 produced by combining the genome-scale metabolic reconstruction with (iii) multi-layer networks of transcriptomic and fluxomic data; (iv) Phenotypic space depicting bacterial responses to varied conditions; (v) Utilisation of multiple objective functions by the model (vi) Unsupervised learning (e.g. PCA) can finally be used to detect latent patterns in unsupervised data by reducing dimensionality and identifying key contributions to variance in datasets.

We initiated our pipeline by mapping the flux distribution for phototrophic growth in *Synechococcus* sp. PCC 7002 using multi-omics flux balance analysis [9] and building condition-specific flux profiles using METRADE [10] and starting from a model recently published by Hendry et al. [5].

Transcriptomic data was acquired in the form of RNA sequencing data from a series of studies previously conducted by Ludwig and Bryant [4,11,12]. Such data were compiled in an online repository known as Cyanomics (available at http:// lag.ihb.ac.cn/cyanomics/) [13], an integrated omics analysis database containing omic data specific to *Synechococcus* sp. PCC 7002. The authors converted sequence data for various culture conditions to fastq format and used a Perl script to filter out low quality reads. A Python script had also been used to calculate the reads assigned per kilobase of target per million mapped reads (RPKMs) as a measure of relative transcript abundance [13]. The reads had been mapped against the genome of *Synechococcus* sp. PCC 7002 using the Burrows-Wheeler algorithm [11]. We calculated fold change values to be centered around 1 by dividing the RPKM values under these conditions by the average expression of three standard control replicates for that gene.

These condition-specific expression profiles were loaded into the model using METRADE [10], for which FBA was carried out using the COBRA Toolbox in MATLAB. For the standard growth conditions, A+ medium was utilised as the culture medium [14] with a temperature of 38 °C, continuous illumination at 250 μmol photons $m^{-2}s^{-1}$, and sparging with 1% (v/v) CO_2 in air; harvestation was performed when the optical density (OD) 730 nm reached 0.7. Specific growth conditions which deviate from the standard conditions are recorded in Table 1, along with a reference to the original paper and the predicted biomass output.

Following this, three reactions involved in energy metabolism (ATP maintenance, photosystem I and photosystem II) were selected to serve as secondary objectives for a bi-level optimisation problem, which was formulated as follows:

$$\begin{aligned} \max \quad & g^{\mathsf{T}}v \\ \text{such that} \quad & \max f^{\mathsf{T}}v, \quad Sv = 0, \\ & v^{\min}\varphi(\Theta) \leq v \leq v^{\max}\varphi(\Theta), \end{aligned} \tag{1}$$

where f is the primary objective function (biomass) and g is the secondary objective function. f and g are Boolean vectors of weights selecting the reactions in v whose flux rate will be considered as the objective. $v^{\min}$ and $v^{\max}$ are vectors which represent the lower and upper limits for the flux rates in v for the unconstrained model. Gene set expression of the reactions associated with the fluxes in v are represented by the vector Θ. φ is a function which maps the expression level of each gene set to a coefficient for the lower and upper limits of the corresponding reaction, and is defined as follows:

$$\varphi(\Theta) = [1 + \gamma |log(\Theta)|]^{\text{sgn}(\Theta - 1)}. \tag{2}$$

The flux distributions calculated for four primary reactions (ATP, Photosystem II, Photosystem I and Biomass) under three pairs of objectives are detailed in Fig. 2. In order to better visualise the differences in flux between conditions, flux values were normalised by dividing by the maximal flux for that reaction across all conditions.

3 Results and Discussion

The multi-omic data used in the multi-layer network consists of gene expression profiles in the transcriptomic layer and steady state flux distributions in the fluxomic layer. In Fig. 1, the nodes in each layer represent environmental conditions such as light intensity and salinity whereas the dashed lines represent interactions between layers of data.

From the transcriptomic studies listed, there are a number of genes which were not transcribed in the controls but specifically transcribed under perturbed

Table 1. Growth and stress conditions for *Synechococcus*sp. PCC 7002. We map the flux distribution for phototrophic growth in *Synechococcus* sp. PCC 7002 using multi-omics flux balance analysis [9]. We then build condition-specific models starting from a model recently published by Hendry et al. [5], and predict the biomass (growth rate, h^{-1}) in all experimental conditions. The unconstrained model representing the optimal growth condition produced 44.04 h^{-1} of biomass.

ID	Condition	Condition specifics	Ref.	Biomass
1	Dark oxic	Incubated in darkness prior to harvest, sparged in N_2	[11]	14.04
2	Dark anoxic	Incubated in darkness prior to harvest	[11]	0.00
3	High light	Illuminated at 900 μmol photons m^{-2} s^{-1} prior to harvest	[11]	37.25
4	OD 0.4	Harvested at OD 730 nm = 0.4	[11]	34.54
5	OD 1.0	Harvested at OD 730 nm = 1.0	[11]	30.35
6	OD 3.0	Harvested at OD 730 nm = 3.0	[11]	28.18
7	OD 5.0	Harvested at OD 730 nm = 3.0	[11]	30.73
8	Low O_2	Sparged in N_2	[11]	42.64
9	Low CO_2	Sparged with air [0.035% (v/v) CO_2]	[12]	17.85
10	N-limited	Cells washed in medium A (lacking NO_3-) and resuspended	[12]	9.59
11	S-limited	Cells washed with $MgCl_2$	[12]	9.53
12	PO_4-limited	Cells washed w/o (PO_4^{3-}), allowed to grow to OD = 0.7 until harvestation	[12]	0.00
13	Fe-limited	Cells washed in medium A with 720 μM deferoxamine mesylate B added at OD 0.35	[12]	21.65
14	NO_3-	Standard growth in medium A (lacking $NaNO_3$) with 25 mM HEPES, 1 μM $NiSO4$, 12 mM $NaNO_3$	[12]	20.36
15	NH_3	Standard growth in medium A (lacking $NaNO_3$) with 25 mM HEPES, 1 μM $NiSO_4$ and 10 mM NH_4Cl	[12]	23.91
16	$CO(NH_2)_2$	Standard growth in medium A (lacking $NaNO_3$) with 25 mM HEPES, 1 μM $NiSO_4$ and 10 mM $CO(NH_2)_2$	[12]	31.04
17	Heat Shock	1h heat shock at 47 °C	[4]	0.00
18	22 °C	Standard growth at 22 °C	[4]	35.10
19	30 °C	Standard growth at 30 °C	[4]	36.26
20	Oxidative stress	5 μM methyl viologen added for 30 min prior to harvestation	[4]	33.55
21	Mixotrophic	Medium A+ supplemented with 10 mM glycerol	[4]	0.00
22	Low salt	Medium A+ containing 3 mM NaCl and 0.08 mM KCl	[4]	0.00
23	High salt	Medium A+ containing 1.5 M NaCl and 40 mM KCl	[4]	33.64

conditions. Many of these genes have yet to be assigned a particular functional category or encode hypothetical proteins, but many more have been linked to specific pathways and compounds and some have been associated with the adaptation of *Synechococcus* sp. PCC 7002 to atypical environmental or growth conditions. The unconstrained model (simulating an optimal growth condition) produces 44.04 h^{-1} of biomass. Therefore, FBA correctly predicts reduced growth in sub-optimal conditions. Some growth is still maintained in some harsh conditions, although severely impaired or null under the dark anoxic, heat shock, phosphate-limited, mixotrophic and low salt conditions. More specifically, inducing a heat-shock response triggers transcriptional activation of heat shock proteins and alternative mechanisms to ensure survival when the main biomass production pathways are inhibited [15].

Although a large number of studies express the maximisation of biomass as the only objective when performing FBA, it is imperative to recognise that in reality most organisms have multiple objectives to satisfy. The addition of multi-level optimisation to our pipeline enables the consideration of more than one objective function and expands the phenotypic solution space so that there are a greater number of feasible optimal points. Specifically, when calculating the flux distribution across conditions, biomass was chosen as the primary objective and the secondary objective set to ATP maintenance, photosystem I and photosystem II. Biomass was chosen to represent the maximisation of growth rate and cellular yields [16], which is a critical consideration for the production of biofuels by cyanobacteria as this informs the substrate uptake rates and maintenance requirements indicating the fundamental growth requirements of the cell. The secondary objectives are key pathways involved in energy metabolism during photosynthesis. Simulating the cost of ATP maintenance can help to examine the energy required for sustaining metabolic activity even in the absence of growth. Incorporation of the photoexcitation reactions occurring within photosystems I and II can characterise how flux under various conditions reflects the light harvestation and energy transfer via photon absorption through these complexes. Thus, solving the linear programming problem between multiple pairs of objectives helps to resolve trade-offs by considering the conditions and constraints affecting each of these objectives.

Figure 2 predicts that the biomass fluxes through Conditions 1–23 are always lower than in Condition 24, the standard control. In Fig. 2, we show that when ATP is set as the secondary objective, the highest fluxes through the ATP maintenance reaction are among conditions which limit growth such as dark anoxic (608.5), mixotrophic (1000), low salinity (1000) or nutrient limitations (ranging from 323.95 to 1000 for nitrogen, sulfur, phosphate and iron-limited conditions). It is likely that this is in order to maintain minimal cellular function when there is no growth (no flux through the biomass pathway) or energy transfer through the photosystems. Lack of light is likely to be a greater contributory factor to decrease in growth as low oxygen concentration does not seem to stunt growth, seeing as the proportional decrease in biomass is low relative to the standard control conditions. On the other hand, there appears to be little to no flux for either

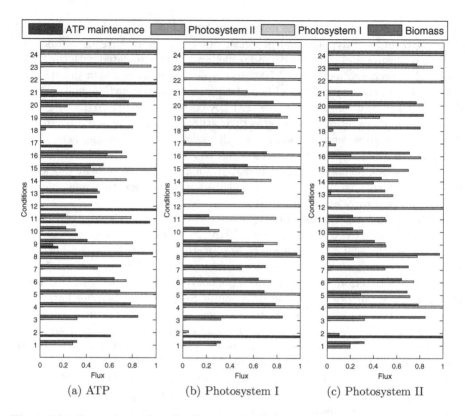

Fig. 2. This figure shows flux distributions for four key reactions: ATP maintenance, photosystem I, photosystem II and biomass when running FBA using three different pairs of objectives (where f = biomass, g = ATP/Photosystem I/Photosystem II). Conditions 1–23 correspond to those detailed in Table 1; Condition 24 is the standard control. Flux values were normalised by dividing by the maximal flux in each reaction across all conditions.

the biomass or photosynthetic pathways in the dark anoxic condition. This is supported by Vu et al. [17], which states that lower yields under dark conditions may be due to the limited generation of energy (ATP) and reductant (NADPH) from glycogen in the absence of photoautotrophic growth. Equal reduction in transcript levels for photosynthetic apparatus was previously observed in all macronutrient-limited conditions studied [12]. We surmise that phosphate limitation has the greatest effect as there is no biomass production across any of our objectives for this condition. This is in line with the findings of Ludwig and Bryant [12], where perturbations caused by phosphate limitation had a greater impact on the global transcription pattern than observed for high irradiance or dark treatments of *Synechococcus* sp. PCC 7002.

Synechococcus sp. PCC 7002 is known to possess one of the greatest tolerances for high light intensity among cyanobacteria (with an upper limit of approximately 2000 μmol photons m^{-2} s^{-1}) [18]. This is evident from our predictions for all three pairs of objectives, where flux through the biomass pathway during high light intensity is only slightly lower than the control condition (37.25 compared to 44.04), even though we observed flux under light intensity of 900 μmol photons m^{-2} s^{-1}. Although fluxes through photosystem II are disrupted, fluxes through photosystem I are still maintained (80.11 in the high light intensity condition as opposed to 250 in the control). Heat shock results in weak fluxes through both photosystems for all three objective pairs and triggers flux through the ATP maintenance reaction when it is set as an objective, presumably to retain cellular function. It was reported by Ludwig and Bryant [4] that transcript levels for genes encoding photosystem I decreased slightly in cells grown at high salinity and remained constant at low salinity; on the other hand, it was found that transcript levels for genes encoding photosystem II did not change in response to fluctuations in salinity. For the high salinity condition, we have observed that fluxes through biomass and photosystem I remain high for our three objective pairs, whereas flux is only maintained in the low salinity condition for the reaction set as the secondary objective g. However, flux through photosystem II for the low salinity condition (250.00) is much higher than flux for the high salinity condition (24.06) when the reaction for photosystem II is set as the secondary objective, and there is no flux through the photosystem II reaction in the control condition.

4   Conclusions

Complex metabolic and phenotypic outcomes as a result of adaptation to a changing environment are difficult to predict from gene expression only. The unified measure of bacterial responses computed by the condition-specific models allows for the detection of coordinated responses shared between different data types as well as the variation in responses across differing growth conditions.

In this work, a heuristic pipeline was constructed for analysing a genome-scale metabolic model of the cyanobacterium *Synechococcus* sp. PCC 7002, which utilises flux balance analysis to obtain flux distributions with multi-level optimisation using linear programming. The four reaction fluxes obtained show clear differences in pathway activity across the various conditions and also between the three pairs of objectives used. This enables the detection of latent, biologically significant patterns and adaptive mechanisms to fluctuations in light intensity and salinity. The aim is to elucidate how *Synechococcus* maintains metabolic efficiency at the cellular level whilst assessing multiple cellular objectives. Considering the vast dimensionality of multi-omic models, the identification of biologically meaningful information can prove to be challenging. As a non-parametric statistical technique, principal components analysis (PCA) can be incorporated into our workflow for identifying patterns in metabolic fluxes within multi-omic models [19].

In this regard, this study integrates transcriptomics with metabolomics and elucidates the unique mechanisms utilised by *Synechococcus* sp. PCC 7002 to adapt to changes in light intensity and salinity to maintain metabolic efficiency for phototrophic growth and light-dependent photosynthesis in a multi-omic fashion. As a result, by predicting and classifying its metabolic profiles in such growth conditions, our approach sheds light on the adaptation process undergone by the cyanobacterium to enable its survival across a wide range of environments and stress conditions.

References

1. Hyduke, D., Schellenberger, J., Que, R., Fleming, R., Thiele, I., Orth, J., Feist, A., Zielinski, D., Bordbar, A., Lewis, N., Rahmanian, S., Kang, J., Palsson, B.Ø.: Cobra toolbox 2.0. Protoc. Exch., 1–35 (2011)
2. Ebrahim, A., Brunk, E., Tan, J., O'brien, E.J., Kim, D., Szubin, R., Lerman, J.A., Lechner, A., Sastry, A., Bordbar, A., Feist, A., Palsson, B.Ø.: Multi-omic data integration enables discovery of hidden biological regularities. Nat. Commun. **7**, 13091 (2016)
3. Reed, J.L.: Shrinking the metabolic solution space using experimental datasets. PLoS Comput. Biol. **8**(8), e1002662 (2012)
4. Ludwig, M., Bryant, D.A.: Synechococcus sp. strain PCC 7002 transcriptome: acclimation to temperature, salinity, oxidative stress, and mixotrophic growth conditions. Front. Microbiol. **3**, 354 (2012)
5. Hendry, J.I., Prasannan, C.B., Joshi, A., Dasgupta, S., Wangikar, P.P.: Metabolic model of synechococcus sp. PCC 7002: prediction of flux distribution and network modification for enhanced biofuel production. Bioresour. Technol. **213**, 190–197 (2016)
6. Ruffing, A.M., Jensen, T.J., Strickland, L.M.: Genetic tools for advancement of synechococcus sp. PCC 7002 as a cyanobacterial chassis. Microb. Cell Fact. **15**(1), 190 (2016)
7. Rügen, M., Bockmayr, A., Steuer, R.: Elucidating temporal resource allocation and diurnal dynamics in phototrophic metabolism using conditional FBA. Sci. Rep. **5**, 15247 (2015)
8. Reimers, A.-M., Knoop, H., Bockmayr, A., Steuer, R.: Evaluating the stoichiometric and energetic constraints of cyanobacterial diurnal growth. arXiv preprint arXiv:1610.06859 (2016)
9. Angione, C., Conway, M., Lió, P.: Multiplex methods provide effective integration of multi-omic data in genome-scale models. BMC Bioinform. **17**(4), 257 (2016)
10. Angione, C., Lió, P.: Predictive analytics of environmental adaptability in multi-omic network models. Sci. Rep. **5**, 15147 (2015)
11. Ludwig, M., Bryant, D.A.: Transcription profiling of the model cyanobacterium synechococcus sp. strain PCC 7002 by next-gen (SOLiD) sequencing of CDNA. Front. Microbiol. **2**(41), 41 (2011)
12. Ludwig, M., Bryant, D.A.: Acclimation of the global transcriptome of the cyanobacterium synechococcus sp. strain PCC 7002 to nutrient limitations and different nitrogen sources. Front. Microbiol. **3**, 145 (2012)
13. Yang, Y., Feng, J., Li, T., Ge, F., Zhao, J.: CyanOmics: an integrated database of omics for the model cyanobacterium synechococcus sp. PCC 7002. Database **2015**, bau127 (2015)
14. Stevens, S.E., Porter, R.D.: Transformation in agmenellum quadruplicatum. Proc. Natl. Acad. Sci. **77**(10), 6052–6056 (1980)

15. Rajaram, H., Chaurasia, A.K., Apte, S.K.: Cyanobacterial heat-shock response: role and regulation of molecular chaperones. Microbiology **160**(4), 647–658 (2014)
16. Feist, A.M., Palsson, B.O.: The biomass objective function. Curr. Opin. Microbiol. **13**(3), 344–349 (2010)
17. Vu, T.T., Hill, E.A., Kucek, L.A., Konopka, A.E., Beliaev, A.S., Reed, J.L.: Computational evaluation of synechococcus sp. PCC 7002 metabolism for chemical production. Biotechnol. J. **8**(5), 619–630 (2013)
18. Xiong, Q., Feng, J., Li, S.T., Zhang, G.Y., Qiao, Z.X., Chen, Z., Wu, Y., Lin, Y., Li, T., Ge, F., Zhao, J.D.: Integrated transcriptomic and proteomic analysis of the global response of synechococcus to high light stress. Mol. Cell. Proteomics **14**(4), 1038–1053 (2015)
19. Brunk, E., George, K.W., Alonso-Gutierrez, J., Thompson, M., Baidoo, E., Wang, G., Petzold, C.J., McCloskey, D., Monk, J., Yang, L., O'Brien, E.J., Batth, T.S., Martin, H.G., Feist, A., Adams, P.D., Keasling, J.D., Palsson, B.Ø., Lee, T.S.: Characterizing strain variation in engineered E. coli using a multi-omics-based workflow. Cell Syst. **2**(5), 335–346 (2016)

MASS Studio: A Novel Software Utility to Simplify LC-MS Analyses of Large Sets of Samples for Metabolomics

Germán Martínez, Víctor González-Menéndez, Jesús Martín,
Fernando Reyes, Olga Genilloud, and José R. Tormo[(⊠)]

Fundación MEDINA, Parque Tecnológico de la Salud (PTS),
Avenida Conocimiento 34, 18016 Armilla, Granada, Spain
jose.tormo@medinaandalucia.es

Abstract. The success of metabolomic analyses relies on the detection method used to analyze the samples and the management of the chemical data. To interrogate the information with biological hypotheses, scientists require a user friendly and manageable way of processing Big Data. Microbial Natural Products Drug Discovery is getting benefits from these techniques that can be applied for a detailed evaluation of the changes in the chemical diversity of the metabolites that a different treatment can induce in a given producer strain. Liquid Chromatography in tandem with Mass Spectrometry (LC-MS) is considered the best cost/effective technique to analyze biological samples that contain metabolites. Simplifying the complexity of these LC-MS sources in a user friendly way can help with the interrogation of the information for a correct use of statistics and scientific hypothesis testing. We describe herein MASS Studio 1.0, a new generation software utility that simplifies LC-MS traces to allow metabolomics analysis on large sets of microbial Natural Products samples in a Drug Discovery Project environment.

Keywords: Metabolomics · Microbial natural products · Mass spectrometry

1 Introduction

Tools are needed to extract useful information from Big Data sources in order to determine trends or isolate populations of samples that perform differently against a given treatment of methodology. Liquid Chromatography coupled with Electrospray Mass Spectrometry (LC-ESI/MS) is the technique of choice in metabolomics because of the large amount on information it generates in a very fast and cost effective way. Natural Products, secondary metabolites produced in nature by animals, plants or microorganism, are a rich source for new therapeutics, with 50% of drugs in development being natural compounds or derived from natural sources. The use of metabolomics in natural products research has increased exponentially in the last years thanks to the advances in LC-MS technologies and Big Data. Frequently, the generation of these large sets of data overflows the discovery processes and require to look for methods of reducing this complexity in a useful way, without losing, a priory, any key information.

© Springer International Publishing AG 2017
I. Rojas and F. Ortuño (Eds.): IWBBIO 2017, Part I, LNBI 10208, pp. 230–244, 2017.
DOI: 10.1007/978-3-319-56148-6_20

We have developed a new tool allowing the metabolomics evaluation of the chemical diversity of our Collections of microbial natural products, used as the source for finding new therapies on diverse medical needs. Starting from previous experience in the design of software tools for metabolomic analyses based on UHPLC-UV data (HPLC Studio) [1–4], we decided to expand our analytic capabilities using low resolution mass spectroscopy (LRMS) detection for generating large sets of data in a fast and robust way, and focus on developing a bioinformatics solution that could produce simplified datasets in a friendly environment for their further use in already available commercial statistics solutions (i.e.: Excel® or BioNumerics®). Once the metabolites of interest or trends of interest could be highlighted, high resolution mass spectrometry (HRMS) techniques developed in house [4] are applied for identifying all the metabolites of interest.

2 Materials and Methods

2.1 Chemical Analyses

Extracts of microbial natural products were analyzed by high pressure liquid chromatography (LC), coupled to low resolution mass spectrometry (LRMS), in reverse-phrase gradient runs of 10 min with an Agilent^{TM} 1100 MSD Mass Spectrometer.

An internal automation application developed *in-house* created a sequence injection file compatible (csv) with the LC software (ChemStation®; Agilent^{TM}) by connecting to our internal database of samples after user selection of the batch to analyze (Nautilus^{TM} 9.0 Thermo-Scientific^{TM} LIMS on an Oracle® DB).

As a result, apart from the indexing sequence file, two sets of m/z data ranging from 150 to 1500 Da (1350 categories, one per m/z), and 0 to 10 min (300 categories, 2 s per scan) were generated for each sample, employing both, negative and positive ionization modes [4].

Metabolite identification, when needed, was performed by low resolution (LC-LRMS) analysis and comparison with a proprietary database of more than 950 known microbial standards. Alternatively, high resolution mass spectrometry database searching (from a Bruker^{TM} maXis® Spectrometer) was performed against a second MEDINA proprietary database of more than 835 microbial metabolites, and, if necessary, the predicted molecular formula of the metabolite of interest was compared with the commercial Chapman & Hall Dictionary of Natural Products database (v25.1) [5].

2.2 Software Development

The MASS Studio 1.0 software utility was developed by using Visual Studio 2013® as the main programming language and designing tool (under Windows7 Operative System; Microsoft Corp.). The Diablo LC DLL® (Diablo Analyticals, Inc) was used for exporting raw data from the Spectrometer as an individual 'csv' file per analyzed sample. Finally, an Excel® Interop® extension (Microsoft Corp.) was used for data file exporting purposes.

Process of analytical raw data was divided into four main modules: Input area, Data Management Area, Process Area and Output area.

2.3   Input Area

MASS Studio 1.0 captures the raw data from batches of LC-MS analytical runs. An Agilent[TM] driver exports the raw data from any analytical run into a 'csv' file where m/z (charge to mas ratio characteristic of the molecules), retention time (in analytical HPLC chromatography, also characteristic of the molecules) and mass intensity (signal intensity detected) are depicted. It can save all runs from an analytical batch (typically two plates/containers, with 160 samples in total) plus an index sequence file into a single folder inside a root network directory of user choice.

Then, Mass Studio 1.0 maps that root directory and can identify all internal files and folders with the LC-MS raw files and their sequence index files, discriminating if the content belongs to a single plate (rack/container) or a set of plates, and the position of each sample inside a given plate. The tool asks the user for the sub-folder that needs to be processed, and if all plates from that batch, or if only a specific plate, needs to be analyzed. It also prompts the user to specify which algorithm must be used to simplify the LC-MS information (the aggregation or the maximum algorithms, or both) (Fig. 1A).

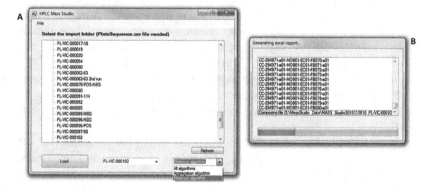

Fig. 1. MASS Studio 1.0 user front-end interfaces: **A.** The software prompts the user which folder inside the configured root pathway contains the raw data that need to be processed, which subset of samples needs to be processed (plate/container or whole batch), and which algorithm to use for simplifying the time dimension of the LC-MS raw data; **B.** The software informs the user on runtime about the individual samples already processed, the file that will contain the composition of the simplified LC-MS data, and the percentage of tasks already performed.

2.4   Data Management Area

Once the LC-MS raw data for the set of samples of interest is selected, the tool proceeds with a series of sequential steps while the user is informed of the tasks performed and the percentage for completion (Fig. 1B).

MASS Studio data management workflow performs sequentially the following tasks:

1. The raw data is loaded into memory.
2. A 90° rotation of the electronic information is performed (matrix transformation).
3. The 'time dimension' of each LC-MS individual analysis is collapsed/simplified.
4. The simplified information for all samples analyzed is collected and combined.

Although each one of these 4 steps (Fig. 2) is essential for the correct management of large sets of samples, the third step is the key point where information is simplified.

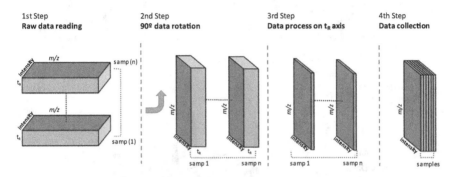

Fig. 2. Data management in MASS Studio 1.0. Batches of raw data files are imported in the application and processed to generate a single output simplified file.

2.5   Algorithms and Process Area

In a detailed chemical analysis of a given sample by LC-MS methodologies, the raw data presents a three-dimensional setup, the LC dimension in seconds/minutes (10 min in our case, one scan every 2 s, 300 entrances in the data matrix), *versus* the MS dimension in ions observed (150 to 1500 *m/z* low resolution ions in our case, 1350 categories in the data matrix), *versus* the intensity detected for each ion (a relative number, inherent to each analytical batch, that is expressed in electric counts detected by the spectrometer, that can go from zero to even the low hundreds of millions of counts).

Some MS spectrometers can detect positive and negative ionization modes simultaneously; others need a specific hardware setup for changing from one mode to the other. Our system can switch between positive and negative modes during runtime for a given sample. So, the typical raw data, before any data management, consisted in two 'csv' files containing each one a [300 × 1350 × intensity] matrix, one in positive and one in negative detection mode.

Classical visualization and process software in LC-MS spectrometers depicts per sample an accumulation of all m/z ions for each given retention time of the chromatographic run (LC), showing a two-dimensional trace of total aggregated intensities across the chromatography (Fig. 3A1). The embedded data in the raw matrix can be typically analyzed in detail by showing specific m/z ions overlapped in the screen in what apparently resembles a side view of the matrix from the chromatographic point of view keeping m/z traces un combined (Fig. 3A2).

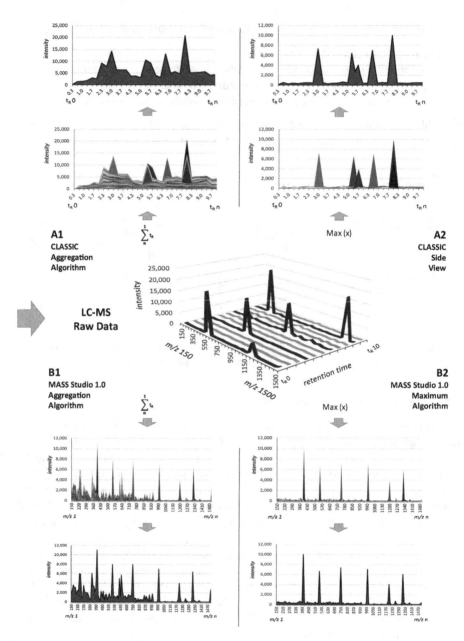

Fig. 3. Two different simplification procedures are available for each individual LC-MS analysis: (**A1**) The classical 'aggregation algorithm' collapsing the m/z dimension; (**B1**) the classical 'aggregation algorithm', included for development comparisons, but collapsing the time dimension; (**A2**) the classical 'side-view' of the individual m/z traces across the time dimension, that is used as an algorithm for comparison purposes although its low resolution/number of categories; and (**B2**) the MASS Studio 1.0 proposed 'maximum algorithm' that keeps the maximum intensity per m/z independently of the retention time where it was detected and presents much larger number of categories. Result matrices scaled for comparison purposes as real raw data number of categories available [4.5:1].

MASS Studio 1.0 simplifies traces by two alternative algorithms, the classical aggregation one (Fig. 3B1; introduced as a development comparison option) and a proposed side-view alternative that maintains the maximum intensities detected for every m/z. But, what is more important due to the mathematical dimensions of the raw data matrix, does it by collapsing the time dimension and not the m/z dimension (Fig. 3B2).

This alternative methodology results in a more convenient simplification of the three-dimensional LC-MS matrices into bi-dimensional ones because: (a) the m/z dimension has much more entries/categories than the time dimension (4.5× in our raw data), keeping when collapsed a far more detailed/resolved bi-dimensional processed result, and (b) it applies a 'side-view' alternative that combines the intensities but avoiding the cumulative side effects that a aggregation would have on resulting signal to noise ratio, that is, typically, very 'noisy' because of the 0 to hundreds of millions of counts that the intensity may present (see Fig. 3 schema where comparisons of the classical and alternative methodologies have been depicted maintaining the [4.5:1] scales of the [m/z:time] categories of the processed matrices).

2.6 Output Area

The Excel® datasheet software (from Microsoft Corp.) was the output format selected for the result matrices. This Excel® format allows almost any kind of future integration with most of the available commercial statistics packages and has, itself, tools to directly re-process or visualize that data.

The detailed output composition report, consisted on a sheet where all m/z intensities for each ion (in rows) are correlated per samples analyzed (one per column). In this sense, all 3D matrix information available per sample has been collapsed into a 2D column of simplified data. In addition to this collapsing of information, each row presents the following metadata: m/z ID, retention time of maximum detection, and maximum value of mass intensity for that m/z within the batch of samples; in the same way, each column presents the following metadata: sample complete ID, Batch, plate and position of the sample in the plate (Fig. 4).

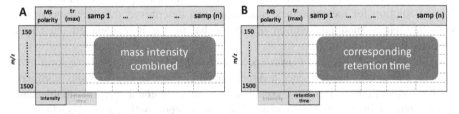

Fig. 4. MASS Studio 1.0 report interfaces: **A.** The software creates an Excel® sheet with the simplified mass spectrometry data (m/z intensities detected) per sample (in columns); **B.** The software also creates a mirror sheet where the retention times for each specified intensity are depicted. This double sheet report can be generated for the aggregation algorithm or the maximum algorithm, in two independent Excel® files.

Two different Excel® files can be generated by MASS Studio, one where the information has been simplified by collapsing the time dimension based on aggregating the data, and/or a second file where the simplification has been accomplished by getting the maximum of intensity for each m/z along all the retention times per sample. In addition, and within each Excel® file, two sheets are created, a main one for mass intensities (Fig. 4A), and an informative second one where the retention time for each detected intensity is specified if needed (Fig. 4B).

2.7 Further Process of Data

For the comparison of large sets of samples, the resulting data matrix generated with MASS Studio 1.0 can then be analyzed with any multivariate statistical analysis software (public as R or commercial). In our case we used commercial BioNumerics® 6.6 software (Applied Maths™) through the generation of a similarity matrix according to the Pearson correlation coefficient and the UPGMA statistics methodology for chemotaxonomy applications (see Sect. 4.3).

In the case of metabolomics analyses, intended for the evaluation of the differential compounds produced in microbial extracts generated in different fermentation conditions, Volcano-plots methodology was applied [6] by using the Excel® commercial software (Microsoft Crop.) [7] (see Sect. 4.1).

3 Results and Discussion

The use of chromatographic runs for comparing samples from different sources has been automated in the past for several uses. When creating HPLC Studio [1–3], a software utility developed by our group that compares ultraviolet (UV) 210 nm profiles (characteristic for Natural Products), it was patent the lack of time-resolution among the 25 min HPLC runs due to the typically broad UV peaks (this was solved, in part, in a second internal version of the tool by using UHPLC runs of 10 min), but specially, to the high limitations that the UV detection presented both, when de-convoluting detected peaks, and when comparing profiles for large sets of samples. As a result, HPLC Studio, although proven to be very useful for selecting the best media conditions to obtain the maximum chemical diversity of a given strain [1–3], resulted in a very limited solution for metabolomics approaches and determined the development of a high throughput modular solution that could handle the more detailed mass spectrometry data.

There are several types of mass spectrometry (MS) analyses available for a given sample. From the simplest one, a direct injection without any previous chromatographic separation, to the most complex one, an orthogonal coupled double chromatography with a MS/MS double detection system with sub-fragmentation of generated ions. Metabolomics studies do often couple one liquid chromatography (LC) to a MS/MS detector. Typical metabolomics approaches (also in our research group [8]) use a LC-MS/MS system where a single chromatographic run is coupled to a MS detection system with an additional fragmentation of the ions detected in a second

MS dimension, producing four-dimensional matrices. This is done on purpose for discriminating the different ions in an untargeted detection mode. This methodology, although very informative, forces the use of high resolution data (HRMS) because it pretends, in the same analysis, to deconvolute the complexity of each sample and to differentiate its components. Thus, the 4 dimensional matrices are very big in number of categories, reaching the low millions of possible m/z due to the high-resolution level of detection needed for the discrimination of the components when the second MS dimension is obtained.

Although low resolution mass spectrometry cannot differentiate two components that present the same molecular mass, in our methodology, we decided to uncouple the metabolomics untargeted detection from the metabolites differentiation, by a simple, but more robust, low resolution (LRMS) step followed by a subsequent identification of the ion/s of interest in a second-high resolution LC-HRMS/MS dereplication step already setup in out lab [5]. This way of dividing the analytical process in two steps had multiple advantages: (a) low resolution spectrometers give more accurate signal to noise ratio and are more accurate when the production of a given metabolite needs to be quantified across many samples; (b) LRMS systems are far more robust than HRMS, allowing the analyses of more complex samples (with less chemical sample preparation/modification) and larger batches of hundreds of samples analyzed within much lower deviation of the hardware detection, requiring less software normalization of the resulting raw data across large batches of samples; (c) the up to four orders of magnitude that HRMS presents in the m/z (ion mass/ion charge ratio) dimension, forces the biostatistic use of 'bucketing' algorithms for allowing the classification of close m/z as the same compound, that may result in the same real metabolite indicated as different due to detection thresholds [9] (Approach not needed in low resolution as in many detections the ion signals go from one to two Daltons, depending on its concentration in the sample due to the isotopic pattern abundances in nature of the different atoms within each metabolite); (d) the additional the double 'bucketing' forced by the use of the time dimension when comparing 4 dimension matrices, that creates multiple false different identifications (up to 9) for the same real compound [9]; and finally, (e) the large amount of hardware calculation power and software development that requires to generate, data manage, process and compare these 4 dimensional HRMS matrices, that no system is able to manage nowadays without several simplifications, that can be performed with multitude of available algorithms each one with advantages and disadvantages.

As a result, we decided to simplify the complexity of the analyzed system, by separating the untargeted metabolomics procedures from the identification/discrimination of the secondary metabolites, by creating the MASS Studio 1.0 tool, benefiting from the advantages of the far larger resolution of the raw data in the m/z (ion mass/charge ratio) dimension when compared to the time dimension, and even avoiding the numerous limitations on lowering the resolution that the 'bucketing' algorithm generates, previously embedded in our previous HPLC Studio 1.0 tool, [1], by directly collapsing the already very low resolution time dimension.

4 Practical Examples

We have multiple examples on the successful use of the MASS Studio 1.0 simplifi-cation of LC-MS data in a relative short period of time. These examples ranged from: the identification of changes in the metabolites that epigenetic modifiers can induce in the fermentations of several fungal strains [4] to the use of LC-MS metabolomics to identify differential compounds or chemotypes produced by the different species of the genus *Preussia* [10]. There are even other applications that are currently being used internally in our Drug Discovery processes for improving efficiency while performing activity-guided fractionation during the purification of new hits from our screening campaigns. Herein we summarize a few of these examples that highlight the power of this tool when performing metabolomics with microbial natural products.

4.1 Differential Metabolomics

Epigenetic modifiers are molecules that can induce the activation of cryptic pathways of microbial strains when present during their fermentation. For depicting how changes on the metabolomics of a given microorganism fermentation can be induced with the addition of epigenetic modifiers we decided to use the strain CF-282001, a *Loratospora* sp. fungal endophyte from *Planistromellaceae* family, isolated from the plant *Retama sphaerocarpa* collected at the Tabernas desert (Spain)., in the presence of the epige-netic modifier sodium butyrate, in a similar setup to the one previously described for *Dothiora* species [7].

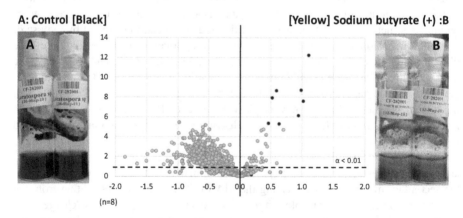

Fig. 5. Volcano-plot comparing changes in the metabolite profiles of the example strain CF-282001, with and without the addition of sodium butyrate. Y-axes correspond to $-\log 10$ of the t-test statistical p-value whereas X-axes correspond to $-\log 2$ of ion masses areas ratios for each *m/z*. Depicted above the dotted line are the differential *m/z* ions with higher production for each condition with statistical confidences above 99% ($n = 8$; $\alpha < 0.01$) [6, 7]. Most differential overproduced *m/z* ions for the condition added with sodium butyrate highlighted in black bold. (Color figure online)

A population of 8 fermentations of CF-282001 in presence of sodium butyrate 100 μM both, during the inocula and the production stage, was compared to a control population of 8 fermentations of CF-282001 without the epigenetic modifier. The MASS Studio 1.0 tool was applied for comparing the 16 LC-MS traces, by creating a single m/z intensity matrix, collapsing the time dimension of each individual analysis, and generating one column per fermentation of each m/z ion. Excel® software was then applied to this matrix for comparing the ratios of production of each ion between fermentation conditions and the significance of the statistical t-test for each population compared in a similar setup to the one used for the differential expression in genomics [6, 7].

Results are depicted in Fig. 5, where ions that present overproduction in one of the conditions are represented at higher (or lower) X-axis and the statistical significances are depicted at higher Y-axis values. There was a clear population of ions overproduced

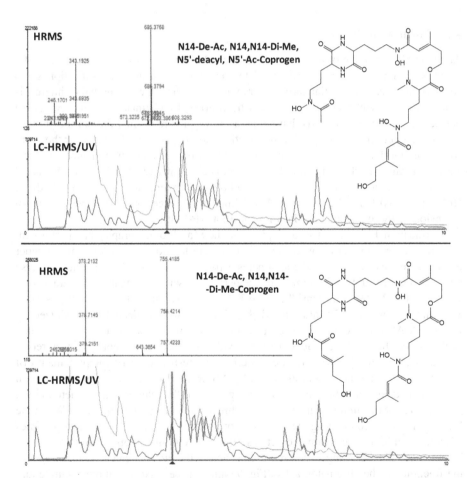

Fig. 6. Identification of the most overproduced m/z ions for the condition added with SBHA on the example strain CF-282001 [5]. LC-HRMS spectra of the two database matches highlighted in yellow in the HPLC trace. Corresponding molecules and HRMS spectra depicted. (Color figure online)

for the strain CF-282001 in presence of the epigenetic modifier sodium butyrate (m/z 756, 755, 686, 685, 379, 378, 343 and 246) highlighted by the metabolomics methodology applied.

When the extracts were analyzed by LC-HRMS looking for these ions, two natural coprogen compounds could be dereplicated (by molecular formula prediction and matching with commercial natural products databases; Fig. 6), as induced by the addition of sodium butyrate. These results support the application of this approach for identifying differential metabolites when fermentation conditions are compared, a method presenting much more suitability for metabolomics than the comparison of HPLC UV-traces (HPLC Studio 1.0, [1]) by getting advantage of the mass dimension during the comparisons without losing the key information while the simplification of the data (MASS Studio 1.0).

4.2   Purification-Plots

An alternative example of the use of the tool can be obtained from the purification of a metabolite of interest from a microbial fermentation. Typically, a purification setup consists on several sequential chromatographic steps where the compound of interest is followed by activity (activity-guided purification) or by chemistry (chemistry-guided purification). In both situations, having an accurate picture of the different metabolites across the fractions, can help with the identification of the one of interest or the other co-eluting interfering ones. In a similar chart to the one obtained with natural products extracts during NP-Plot Screening for small molecules, where presence of each natural products is mapped in function of its m/z and polarity (retention time by a standardized HPLC analysis) [11, 12], we used MASS Studio 1.0 to generate what we call, in an extension of the nomenclature, Purification-Plots (P-Plot).

Purification Plots depict in a single chart how a chromatography has performed, allowing better correlations with the activity of the fractions, and providing a deeper detail on the presence of related components across the fractions. For example, a reverse phase HPLC preparative fractionation was performed with a fermentation extract from the strain CF-282001 (Agilent™ Zorbax® SB-C8, 21.2 × 250 mm, 7 µm; 5–100% acetonitrile in water at 20 mL/min). Eighty fractions were collected and analyzed by LC-MS. MASS Studio 1.0 was used to collapse the time dimension of the individual analyses and the resulting matrix obtained with the 'maximum algorithm' was represented by using the Excel® software. As this could result in a very complex matrix of mixed ions to represent, we introduced in MASS Studio 1.0 the generation of a column of metadata data that indicated for each m/z processed the retention time where the ion was detected at maximum of its intensity within a batch of samples analyzed. When the P-Plot is represented ordering the m/z ions from minimum to maximum retention times of their maximum of detection, a chart describing the chromatographic run in detail can be obtained. This resulting P-Plot represents clearly the m/z that can be detected across the fractions of the chromatography (Fig. 7) and can be used as a more accurate and detailed map for the identification of where the metabolite (or activity of interest) is present after each different purification chromatographic step.

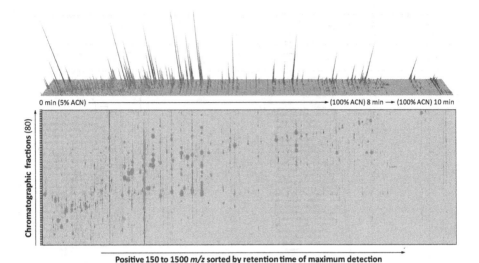

Fig. 7. Purification-Plot for the preparative purification fractionations of the example strain CF-282001 obtained by MASS Studio 1.0 process. Intensities above 15% highlighted in red. Chromatography performance can be followed across the diagonal of the plot. (Color figure online)

4.3 Chemotaxonomy and Chemotype Identification

Recent phylogenic studies on microbial isolates have included the LC-MS analytical profiling of their fermentations to correlate taxonomically close-related strains to differentiate species based of specific produced compounds or chemotypes. Although these type of studies are not very common, they start to appear in microbial taxonomic descriptions as complementary to morphology phenotypes and molecular taxonomy to differentiate among very close strains within the same genus [13–15].

The use of Mass Spectrometry (MS) to measure the mass to charge ratio (m/z) and intensity of ionized molecules in fungal fermentation extracts with the purpose of identifying differential metabolite chemotypes in microbial fermentations is quite recent compared to the classical 210 nm LC-UV profiling, but has typically being applied to the low resolution LC-MS traces (Fig. 3A1). We decided to generate alternative chemotaxonomy comparisons by using MASS Studio 1.0 and applying the 'maximum algorithm' (Fig. 3B2) to compare, as a result, much more detailed and lower signal to noise m/z profiles per each strain.

For validating the methodology, we compared the metabolic profiles of 37 Preussia isolates from plants of the same ecosystem as CF-282001, the Tabernas desert (Spain). Strains were fermented in duplicate in two different production media (MMK2 and YES). MASS Studio 1.0 profiles were compared by classical molecular taxonomy statistical methodology [10]. The resulting dendrograms obtained after the statistical processes with BioNumerics® 6.6 commercial software (one per fermentation medium), depicted the similarity relationships between the different strains and duplicates according to the similarity of the metabolites produced (Fig. 8A–B).

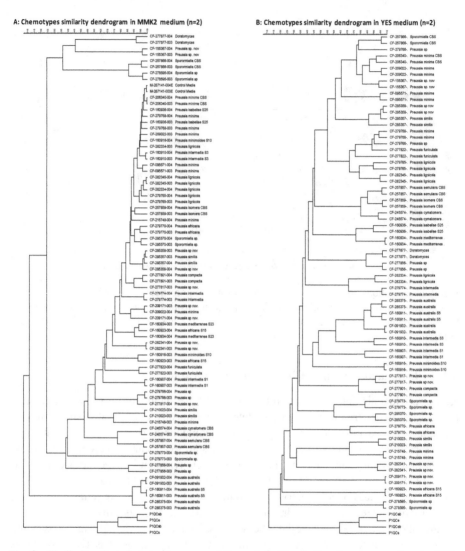

Fig. 8. Similarity clustering dendrograms of the metabolite profiles for an example set of 37 *Preussia* species per duplicated (n = 2), in two different fermentation media (MMK2 and YES). LC/MS control samples (P1QC) and unfermented Control media included.

The closer similarity coefficients obtained for the experimental fermentation duplicates compared to the ones obtained between close related species, not similar by molecular taxonomy data [10], corroborated the accuracy and suitability of this methodology to discriminate microbial metabolite profiles in much more extent than classical LC-UV or even recent LC-MS direct profiling comparisons. Moreover, the MASS Studio 1.0 matrix generated proved to be a very valuable resource for the identification of specific compounds (up to 16) that differentiated chemically several taxonomic clusters among the different species of the genus analyzed [10].

5   Conclusions

MASS Studio 1.0, a software tool for Mass Spectrometry comparison of large sets of samples, has allowed us to perform multiple uses of metabolomics on microbial Natural Products from extracts to fractions. It collapses the time dimension generated during analytical LC-MS runs to simplify their complexity when comparing hundreds of samples, but keeping their high amount of MS information when compared to less informative simpler 'no-LC' MSs runs. As a consequence, this tool improves the selection of fermentation conditions for individual microbial strains by quantifying and highlighting clearly the differential secondary metabolites induced in specific production conditions, for example, during the addition of epigenetic modifiers. It also allows the creation of Purification-Plots for a multidimensional detailed evaluation of a chromatographic fractionation run and allows an improved use of chemical MS information when comparing strains diversity and during the identification of species specific chemotypes.

Acknowledgments. This work was carried out as part of the Master and PhD. Programs from the School of Master Degrees of the University of Granada.

References

1. Garcia, J.B., Tormo, J.R.: HPLC studio: a novel software utility to perform HPLC chromatogram comparison for screening purposes. J. Biomol. Scr. **8**, 305–315 (2003)
2. Tormo, J.R., et al.: A method for the selection of production media for actinomycete strains based on their metabolite HPLC profiles. J. Ind. Microbiol. Biotechnol. **30**, 582–588 (2003)
3. Tormo, J.R., Garcia, J.B.: Automated analyses of HPLC profiles of microbial extracts - a new tool for drug discovery screening. In: Zhang, L., Demain, A.L. (eds.) Nature Product: Drug Discovery and Therapeutic Medicine, pp. 57–75. Humana Press Inc., Totowa (2005)
4. Gonzalez-Menendez, V., et al.: Differential induction of secondary metabolite profiles in endophyte fungi by the addition of epigenetic modifiers. Planta Med. **81**, 924 (2015)
5. Perez-Victoria, I., Martin, J., Reyes, F.: Combined LC/UV/MS and NMR strategies for the dereplication of marine natural products. Planta Med. **82**, 857–871 (2016)
6. Hur, M., Campbell, A.A., Almeida-de-Macedo, M., Li, L., Ransom, N., Jose, A., Wurtele, E.S.: A global approach to analysis and interpretation of metabolic data for plant natural product discovery. Nat. Prod. Rep. **30**, 565–583 (2013)
7. González-Menéndez, V., Pérez-Bonilla, M., Pérez-Victoria, I., Martín, J., Muñoz, F., Reyes, F., Tormo, J.R., Genilloud, O.: Multicomponent analysis of the differential induction of secondary metabolite profiles in fungal endophytes. Molecules **21**, 234–250 (2016)
8. Ríos Peces, S., Díaz Navarro, C., Márquez López, C., Caba, O., Jiménez-Luna, C., Mel-guizo, C., Prados, J.C., Genilloud, O., Vicente Pérez, F., Pérez Del Palacio, J.: Untargeted LC-HRMS-based metabolomics for searching new biomarkers of pancreatic ductal adenocarcinoma: a pilot study. J. Biomol. Screen **21** (2016). doi:10.1177/1087057116671490
9. Ito, T., Odake, T., Katoh, H., Yamaguchi, Y., Aoki, M.: High-throughput profiling of microbial extracts. J. Nat. Prod. **74**(5), 983–988 (2011)

10. Gonzalez-Menendez, V., Martin, J., Siles, J.A., Gonzalez-Tejero, R., Reyes, F., Platas, G., Tormo J.R., Genilloud, O.: Biodiversity and chemotaxonomy of Preussia isolates from the Iberian Peninsula. Two new species discovered. Mycol. Progress (2017, submitted)
11. Osada, H., Nogawa, T.: Systematic isolation of microbial metabolites for natural products depository (NPDepo). Pure Appl. Chem. **84**(6), 1407–1420 (2012)
12. Lim, C.L., Nogawa, T., Uramoto, M., Okano, A., Hongo, Y., Nakamura, T., Koshino, H., Takahashi, S., Ibrahim, D., Osada, H.: RK-1355A and B, novel quinomycin derivatives isolated from a microbial metabolites fraction library based on NP-Plot screening. J. Antibiot. **67**, 323–329 (2014)
13. Stadler, M.: Importance of secondary metabolites in the Xylariaceae as parameters for assessment of their taxonomy, phylogeny, and functional biodiversity. Curr. Res. Environ. Appl. Mycol. J. Fungal Biol. **1**, 75–133 (2011)
14. Stadler, M., Læssøe, T., Fournier, J., Decock, C., Schmieschek, B., Tichy, H.-V., Peršoh, D.: A polyphasic taxonomy of Daldinia (Xylariaceae). Stud. Mycol. **77**(1), 1–143 (2014)
15. Kim, W., Peever, T.L., Park, J.J., Park, C.M., Gang, D.R., Xian, M., Davidson, J.A., Infantino, A., Kaiser, W.J., Chen, W.: Use of metabolomics for the chemotaxonomy of legume-associated Ascochyta and allied genera. Sci. Rep. **6**, 20192 (2016)

Biomedical Engineering

Adsorption of Bilirubin Toxin in Liver by Chitosan Coated Activated Carbon Prepared from Date Pits

Asel Mwafy, Ameereh Seyedzadeh, Waleed Khalil Ahmed,
Basel Alsayyed Ahmad, Betty Mathew, Kamala Pandurangan,
Abdel-Hamid Ismail Mourad, and Ali Hilal-Alnaqbi[✉]

Department of Mechanical Engineering, College of Engineering,
United Arab Emirates University, P.O. Box. 15551, Al-Ain, UAE
{201450073, 201304592, w.ahmed, balsayyed, betmatw,
kamala.p, ahmourad, alihilal}@uaeu.ac.ae

Abstract. The aim of this work was to develop activated carbon (AC) from date pit powder and evaluate its adsorption efficiency of bilirubin toxin. In order to increase the adsorption capacity of bilirubin, an increase in the surface area is necessary. This increase was achieved through pyrolysis technique and to further increase the absorption capacity of AC when coated with chitosan gel, which contains several groups on its chains that act as interaction sites. Results indicated that the presence of the AC lead to a decrease in bilirubin content and the more the AC added to the sample, the faster the rate of adsorption as well as the higher the capacity of adsorption. A 0.3 M AC concentration shows a 0.82 left over bilirubin fraction after 16 h, while a 0.1 M AC concentration shows a 0.9 bilirubin fraction after the same interval of time. Contact time is another factor that also contributed to the increase in adsorption of bilirubin. It was seen that chitosan coated AC shows an increase in adsorption percentage from about 25% to 96% when left for a longer period of time.

Keywords: Activated carbon · Date pit · Adsorption · Chitosan coating · Toxin

1 Introduction

1.1 Liver Failure

Patients with acute liver failure suffer from an increase in toxin levels in the blood stream payable to the liver lacking the ability to remove all the toxins [1]. Most of these toxins bind to protein, usually albumin serum to enable them to travel along the blood stream without poisoning other cells and tissues. Because of this bond, it is not possible to remove toxins, such as bilirubin, using conventional dialysis [2, 3]. Therefore, other methods, using liver support devices are adapted, such as the use of a Molecular Adsorbent Recirculating System (MARS) [1].

MARS is a system used to remove albumin-bound toxins from the blood by selectively removing these toxins using adsorption onto the surfaces of materials through membranes in the machine [3, 4]. Toxins are removed by means of adsorption

© Springer International Publishing AG 2017
I. Rojas and F. Ortuño (Eds.): IWBBIO 2017, Part I, LNBI 10208, pp. 247–260, 2017.
DOI: 10.1007/978-3-319-56148-6_21

onto the surface of activated carbon through a membrane. This method of removing toxins is very important for patients with end-stage liver failure, because so far their only option is liver transplantation, however the shortage of donors and need for the right blood match makes it difficult to find a suitable liver in time [1]. The ability to remove toxins, gives patients, at least, the time they need to find a suitable donor.

Activated carbon (AC) is a form of carbon that is processed to form low-volume pores and large surface area. It is considered as one of the strongest adsorbents and therefore is broadly used in medicine, gas purification, metal extracts, water treatment and industrial applications. AC has a large surface area that can be greater than 1000 m^2/g. Thus, a 3 g sample of AC can have a surface area as sizable as a football pitch [5].

The materials used to prepare AC can be derived from many different raw materials, depending on what raw material is obtainable. Some of these materials comprise wood, coal, coconut, rice husk, shells of plants, stones of fruits, asphalt, metal carbides, carbon blacks, and polymer scraps. Since we are living in the UAE, date palm is one of the major biomass which can be used to form activated carbon. AC produced from date pits has been shown to have mass which varies from 10–15% of total date-fruit mass [6]. The date pits contain high nutritional value because of its reliable source of carbohydrates, fibers, protein, lipids, vitamins, and minerals [7]. Consequently, their utilization is highly wanted by the date processing industries in raising their value-added products.

1.2 Uses of Activated Carbon

Payable to its high quality adsorption properties, AC has recently been utilized in many applications in a variety of fields. This includes its different use in the medical area for adsorbing toxins from the body. AC can effectively avert gastrointestinal absorption of poisons and medicine overdoses if used in the proper manner [16]. AC is managed repetitively over intervals of time until it adsorbs the targeted toxins completely or partially, just enough so that they are not considered toxic anymore.

A very well-known application of AC adsorption is for the removal of contaminants from water. Most water sources contain dissolved micropollutant organic matter that is very hard to remove due to their microscopic sizes. The porous structure and adsorption properties of AC enables the elimination of such matter [17]. Although the effectiveness with which AC removes these wastes highly depends on a variety of factors, such as the concentration of the pollutant, the contact time and the type of AC, it is needless to say that AC plays a crucial role in water purification.

1.3 Tested Bilirubin Adsorption

In previous studies, the effect on bilirubin adsorption due to increases in the albumin concentrations has been analyzed. In order to test for the bilirubin as well as tryptophan toxin adsorption, Annesi et al. [18] used AC with a particle size of 0.3–0.5 mm. Results presented a clear drawback in bilirubin adsorption in the presence of higher albumin concentrations. This shows that devices, such as the MARS have the negative effect of adsorbing albumin over bilirubin. However, the effect of changing the AC concentration on the effectiveness of bilirubin adsorption was not considered. Trying to manipulate the adsorbent using higher concentrations or by coating with high binding affinity solutions, such as chitosan gel can prove to be very effective.

Also derived from preceding literature is that granular AC does not have a large capacity to adsorb bilirubin toxin. Therefore, grinding AC to be used in powder form will significantly increase its adsorption capacity of bilirubin and other toxins in adsorption columns used in liver support devices [19]. Using granular AC does not utilize its full capabilities due to the very small surface area and pore structure as compared with powdered AC. The adsorption properties of AC then highly depend on the particle size and internal surface area.

Nikolav et al. [20] also aimed at adsorbing bilirubin toxin using different types of AC. Nitrogen based granular carbons, AC based on pyrolysis, as well as fibrous AC were used. The tests were carried out for different particle sizes of 7–9 μm for fibrous AC and 0.5–1.0 mm for carbon activated by pyrolysis.

A past study used surface modified chitosan beads to test for their binding affinities for bilirubin in buffer solutions as compared to AC. It was discovered that chitosan beads adsorbed a bilirubin average of 1.18 mg/g of chitosan beads whereas AC adsorbed 0.74 mg/g [21]. From this information, a case of combining powdered AC, having a porous structure and large surface area as well as its high adsorption capacity of small particle sizes, with chitosan's high binding affinities for bilirubin, could potentially provide a larger adsorption capacity for bilirubin than each implemented alone. This method has yet to be tested for the adsorption of protein bound toxins from the liver.

1.4   Activated Carbon Preparation

The methods of preparing AC can be divided into two main kinds; physical activation (PhA) and chemical activation (ChA). In PhA, two main processes are followed; the first is called pyrolysis and is achieved using an inert gas (N_2) to convert the carbonaceous raw material into another more stable and heat resistant compound that consists mainly of carbon (C), while the second process of PhA is the actual activation which occurs in the presence of carbon dioxide (CO_2). In contrast, ChA can be done is one step; pyrolysis and activation are carried out at the same time in the presence of dehydrating agents (e.g. Zn C_{12}, H_2SO_4 or KCl).

AC is going to be used for the adsorption of toxins from the blood stream, therefore, PhA is carried out in order to ensure that no chemicals are involved in such sensitive medical processes. Carrying out ChA would involve using chemicals in the activation of the carbon, which may result in poisoning or contaminating the blood when the AC is used in real life applications to remove toxins from the liver.

Chitosan used to coat the AC contains both amino (-NH$_2$) and hydroxyl (-OH) groups on its chains. These groups can act as interaction sites to increase the adsorption capacity of the AC for albumin-bound toxins [8].

Chitosan is a polymer found in the shells of crustaceans, the shells of insects and the cell walls of fungi and so is a natural source of high abundance in nature [9]. It is grounded into a powder and is used to make chitosan gel, with which the AC will be coated.

In this paper, the tendency of activated carbon (AC) to adsorb bilirubin toxin is studied under different conditions; once with AC, which is achieved through the pyrolysis technique and once with chitosan coated activated carbon. The chitosan coating is expected to increase the adsorption of the AC since chitosan gel should increase the poles on AC providing a greater possibility for bilirubin toxin to bind. The difference in adsorption between AC and chitosan coated AC is observed in the results. Several tests are carried out in order to prove this phenomenon, including FTIR analysis, DSC analysis and scanning electron microscopy. In general, test results show that the presence of AC lead to a decline in bilirubin and further adding AC dropped the levels of bilirubin present. Another observation is related to the contact time between the AC and bilirubin toxin; the longer the AC is left in contact with the toxin, the higher the percentage of adsorption.

2   Materials and Experimental Method

2.1   Materials

Several Materials Were Used for the Preparation and Coating of AC.

The first step taken in the procedure is activating the date pit powder. The powdered date pits are added into a steel roll, which is placed into a furnace through a hole on the right end. In order to control the temperature rises and falls, as well as the optimum temperatures at which activation takes place, a laptop including a specific software (PT software) is connected to the furnace.

In preparing the chitosan gel and coating the AC, a magnetic stirrer is used. A magnetic stirrer works by generating a magnetic field, which interacts with a magnet placed inside the beaker to automatically stir the solution inside. A heating option is also available, where the surface can heat up to warm the beaker and allow the solution to mix and dissolve quicker and more efficiently.

In order to heat beakers of coated AC with the aim of drying and removing moisture and excess chitosan gel, an oven is used at a temperature of around 80 °C. The oven is also used in measuring the moisture content of AC. This is done by weighing a sample of AC before and after drying.

2.2   Setup

The date pit powder used to make the AC is readily purchased having a particle size of 90 μm. physical activation is done over a two-step process. Carbonization first takes place in an inert atmosphere provided by passing nitrogen gas through the sample. This step is initialized by adding the date pit powder into a steel roll, which is then inserted into the furnace. After ensuring that the steel roll is positioned correctly, the furnace is closed on both ends, attaching tubes through which the inert gas will flow.

The nitrogen gas enters and leaves the furnace passing through the powder-filled steel roll. In order to check that approximately the right amount of nitrogen is passing through, a tilted water bottle is connected to the outlet tube and observed for bubbles. The rate of nitrogen flow is controlled from the valve on the nitrogen cylinder.

The inert gas aims to eliminate any impurities, such as oxygen and hydrogen to form a more stable, heat resistant compound that mainly contains carbon. This process is called pyrolysis and primarily involves increasing the temperature of the furnace at a constant rate while passing nitrogen through the powdered date pit sample. The temperature is kept at its peak value for five hours before it is reduced back to room temperature. This is done by connecting the furnace to a laptop with the PT software installed.

The software is left to run, raising the temperature from 0 °C to 900 °C, staying at 900 °C for two hours then dropping back to room temperature at 20 °C. Carbon dioxide was used because it reacts with carbon, converting it into carbon monoxide gas according to the following equation:

$$C + CO_2 \rightarrow 2CO \tag{1}$$

Some of the solid carbon reacts with the CO_2, converting it into carbon monoxide gas, which is removed from the system into the water bottle with the excess gas. This progression is known as gasification, which develops porosity by removing carbon atoms. The pores produced increase the surface area of the remaining carbon making it a better adsorbent.

Next, the chitosan gel is prepared, which will be used to coat the activated carbon in order to increase its adsorption capacity. A diluted acetic acid was first made by mixing 198 ml of water with 2 ml of acetic acid in a reagent bottle. The result is a 1% concentrated acetic acid by volume. 100 ml of the solution was then measured using a measuring cylinder and poured into a beaker with a magnet placed inside and put on a magnetic stirrer to stir and heat the solution first. 0.5 g of chitosan powder is gradually added to the stirring diluted acid, to avoid splashes, clumping and achieve a more even distribution. The magnetic stirrer is set at a temperature around 45 °C during the chitosan gel preparation.

After approximately an hour, when the chitosan gel has reached the desired consistency, 5 g of the activated carbon are weighed and added gradually to the gel after turning off the heater so the mixture is only stirring. The carbon is then left to coat overnight as it stirs (for about 24 h).

The coating procedure is carried out three times for two samples while oven drying the samples between each coating. The end result is two samples of activated carbon coated three times to ensure full coating of the AC.

2.3   Experimental Method (Tests)

The chitosan gel is prepared by mixing chitosan powder with dilute acetic acid over a magnetic stirrer. After some stirring time, the mixture starts turning into a gel form [10]. The AC is then added and left to stir overnight. This process is carried out three times, being left to dry between each coating. Repetition of the coating process is done to ensure that the AC is fully covered with the chitosan gel to maximize its adsorption capacity [11].

For determining the characterization of chitosan coated activated carbon, FTIR, SEM and DSC tests are conducted.

FTIR. Fourier Transform Infrared (FTIR) spectroscopy is used for the characterization of the chitosan coated activated carbon. The main benefits of using an FTIR is that it is quick and takes measurements in a matter of seconds and also has a very high sensitivity due to the extremely delicate detectors employed in the device. FTIR works by emitting infrared radiation of various frequencies through a sample and using a detector to study the frequencies that passed through and those that were absorbed. Infrared spectroscopy gives information about the chemical structure and functional groups of raw materials and the prepared activated carbon. FTIR data gave very important information about the interaction and polymeric association between chitosan and carbon. The Infra-red transmission spectra were recorded with a Perkin Elmer spectrophotometer ranging from 400–4000 cm^{-1} using the KBr technique.

DSC Test. DSC (differential scanning calorimeter) test is done on a sample of AC. The natural gas connected to the DSC is opened and the flow is adjusted, following with turning on the DSC, the cooling system and the PC connected to the DSC. Stabilization is then initiated from the program and the system takes some time to warm up. Meanwhile the sample is prepared by first choosing the suitable aluminum crucible size. The test is carried out using two crucibles; a test and a reference crucible. The crucible and lid are placed on a balance set to mg units and the balance is zeroed before 2–5 mg of the sample are added to the crucible. The lid is added onto the crucible and placed on the dye of the press. The handle is then rotated down to seal the lid onto the sample. The reference and sample are then placed into the DSC and the test is run by initiating it from the computer. This is done after inputting the heating rate, the initial temperature and final temperature.

SEM Test. SEM (Scanning Electron Microscope) test is done to study the surface topography and composition of the activated carbon. To do this, the vacuum pump control is turned on and after pressure equalization, the chamber door is opened. On the tray, the activated carbon sample is placed and the door is closed. Afterwards, the high vacuum pump and electronic control are turned on. The monitor, motorized stage and microscope control are initialed as sufficient vacuum is reached around the samples. After turning on the voltage and cathode, an image appears on the monitor. A tracker ball is used to adjust the position of the image as it is zoomed in (JSM-5600, JeolLtd.,).

Spectrophotometer. A spectrophotometer (Shimadzu UV 1800, Japan) measures the amount of light a sample absorbs. The device basically works by passing a beam of light through a sample and measuring the intensity of the light that passes through using a detector. A double beam photospectrometer is used for tests and it works by comparing the light intensity between two light paths; one having the reference sample and one having the test sample.

In order to start carrying out measurements, a reference sample is prepared. This data can then be used to calculate the transmittance and absorbance of the tested solution.

Bilirubin binds to high and low affinity sites on albumin. Activated carbon adsorbs the bilirubin binded to the low affinity sites. Albumin-bound bilirubin solution is prepared. 30.4 mg of bilirubin is dissolved with 1.15 ml NaOH (0.1 M) and albumin solution is prepared by dissolving it in NaCl (0.5 g Albumin in 100 ml NaCl). The dissolved bilirubin and albumin are mixed together to achieve albumin-bounded bilirubin solution and 26 ml of this solution is used. The albumin-bounded bilirubin solution is then mixed with the PBS solution and a stock solution of 80 µM is prepared and left to sit for 6 h in order to stabilize. This will then be serially diluted into 60 µM and 30 µM (while setting PH at 7.4). From each of the three concentrations; 80 µM, 60 µM and 30 µM, a control is taken, as well as three samples containing different amounts of AC (0.1 g, 0.5 g, and 0.8 g), in amber bottles to avoid photo degradation of the bilirubin. From each concentration, 40 ml were taken for the control and similarly for adding the different amounts of AC. The control and different amounts of carbon containing bottles are placed in a shaking water bath, which is set at 140 rpm and 37 °C. Readings are taken every hour for the first four hours then after 16 h. Between taking the readings, the bottles are placed in the shaking water bath and only removed to take the readings, then put back in. Each time readings are taken for two wavelengths; 416 nm for bilirubin testing and 350 nm for albumin testing. The albumin readings is expected to remain stable and not change with time since activated carbon adsorbs bilirubin and not albumin. Bilirubin on the other hand is expected to decrease with time in the bottles containing activated carbon (Fig. 1).

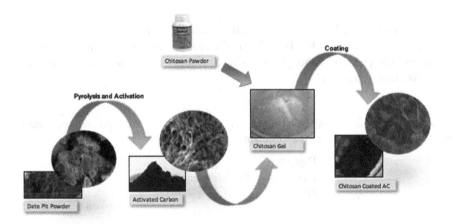

Fig. 1. Graphical abstract

3   Results and Discussion

3.1   FTIR

Fourier Transform Infrared Spectroscopy (FTIR) analysis was carried out in order to identify the functional groups present in activated carbon.

Figure 2(b) shows FTIR spectrum of the date pit activated carbon. In this spectrum an IR band of around 3427 cm^{-1} is assigned to a vibration stretch of hydroxyl group, -OH. C = O group frequency is observed at a wavelength of 1720 cm^{-1}, aliphatic

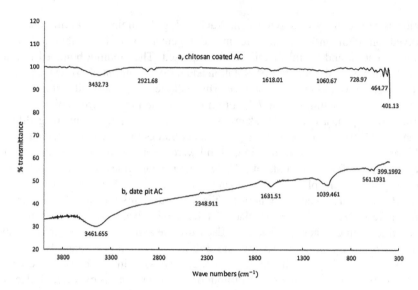

Fig. 2. FTIR spectra for: (a) Chitosan coated date pit activated carbon, and (b) Date pit activated carbon.

groups at around 2923 cm^{-1} and 2855 cm^{-1}, and 1033 cm^{-1} corresponding to C–O stretching, as shown in Table 1(a). The main surface functional groups present were presumed to be phenols, carboxylic acids, and carbonyl groups all of which are typical acidic functional groups [11].

Table 1 Wavelengths of different functional groups for (a) Date pit activated carbon, (b) Chitosan coated date pit activated carbon.

Date Pit AC	
Functional group	Wavelength
OH	3427 cm^{-1}
C=O	1720 cm^{-1}
C-O	1033 cm^{-1}
Aliphatic groups	2923–2855 cm^{-1}
Chitosan coated AC	
Functional group	Wavelength
N-H bending	1630 cm^{-1}
C-N stretching	1157 cm^{-1}
O-H bending	1381 cm^{-1}
C-C-C	1017 cm^{-1}
CH$_3$-C-OH	890 cm^{-1}
OH	647 cm^{-1}
C-C	496 cm^{-1}

In interpreting the FTIR spectra of chitosan coated carbon, the spectrum shows the absorption of both chitosan and carbon. In Fig. 2(a), frequency at 3433 cm^{-1} is observed corresponding to the stretching vibration of O-H and N-H functional groups. The bands at around 2923 cm^{-1} and 2855 cm^{-1} correspond to the asymmetric and symmetric stretching. The peaks in the range of 1630 cm^{-1}, 1157 cm^{-1}, 1381 cm^{-1}, 1017 cm^{-1} and 890 cm^{-1} assigned to N-H bending, C-N stretching, O-H in plane bending, C-C-C Skeletal in the backbone and CH_3-C-OH stretching respectively as recorded in Table 1(b). Bands around 647 cm^{-1} and 496 cm^{-1} indicate the presence of OH and C–C bending vibrations respectively [12].

3.2 DSC Analysis

Figure 3 shows the DSC analysis of chitosan coated activated carbon, which measures the melting point and the glass transition temperature of the chitosan coated AC. This test was carried out by heating the samples from 27 °C to 400 °C at a rate of 5 °C per minute. The date pit AC showed a glass transition temperature of approximately 10 °C and a melting point of around 143 °C. Chitosan coated AC on the other hand showed a glass transition temperature at approximately 40 °C and a phase change (melting point) at approximately 135 °C. While comparing the results of pure chitosan with the present one shows that the thermal behavior of the sample is increased after coating with activated carbon.

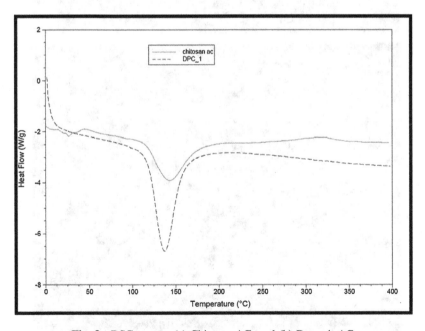

Fig. 3. DSC curves: (a) Chitosan AC, and (b) Date pit AC.

3.3   Scanning Electron Microscopy

Scanning electron microscopy (JSM-5600, Jeol Ltd.) was used to examine the morphology of the surface of the raw date pit, the AC and chitosan coated activated carbon.

It is noted that by activation, Fig. 4(b) shows that the porosity of DP-AC is observed to be significantly higher than the raw material DP. This is expected, as the pores are created during the carbonization step. This is because during the thermal

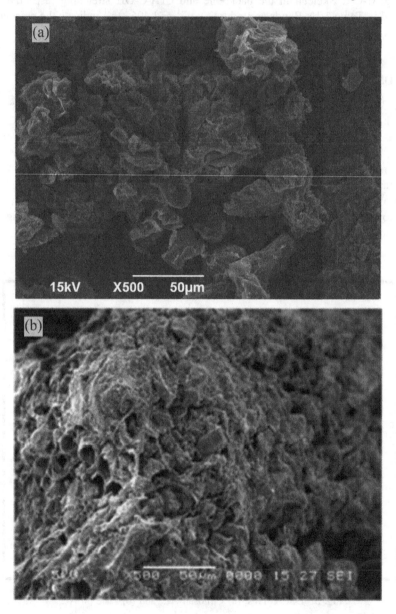

Fig. 4. (a) Chitosan coated activated carbon, (b) DP-AC.

process, the volatile constituents in the DP escape and some macromolecules thermally degrade resulting in the formation of voids in the product. Formation of pores plays prominent role in the adsorption effectiveness of the produced DP-AC, as active sites, are found at the available surface.

Figure 4(a) shows the chitosan coated perfectly on its surface. Chitosan forming a layer which is very rigid to the surface of activated carbon adsorbent surface used for adsorption [13].

3.4 Effect of Contact Time

In order to study the effect of contact time on the bilirubin adsorption using chitosan coated date pit activated carbon (CDPAC), tests were carried out at initial concentration of 80 μM of bilirubin solution, (0.1, 0.5 and 0.8 gm) dosage of CDPAC, pH = 7.4, at 37 °C and the results are shown in Fig. 5. From the figure, it was observed that the adsorption of bilirubin increases with contact time up to 24 h. For the initial stage of the process up to about 4 h, adsorption of bilirubin is slow but then increases in the time period between 4 and 24 h. After a longer period of time, the process of adsorption is expected to reach a constant value (equilibrium) after which, there is no significant change. This phenomenon can be explained during the initial stage of the process, the bilirubin takes some time to react and adsorb to the CDPAC, followed by the intermediate stage with the high availability of active sites on the surface of the adsorbent, which decrease when it reaches the equilibrium, when the active sites are almost all occupied. When comparing the adsorption efficiency, it was found that the efficiency of the chitosan coated activated carbon (CDAC) is comparatively higher than the activated carbon (DP-AC) alone.

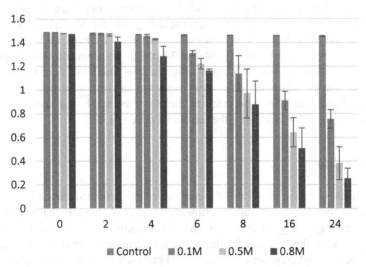

Fig. 5. Effect of contact time for 80 μM, 0.8gm, PH = 7.4.

3.5   Adsorption Percentages

Percentage of Adsorption. The adsorption % is calculated as follows

$$\text{Adsorption } \% = \frac{(C_0 - C_e)}{C_0} \times 100$$

Where C0 and Ce (mg/L) were the initial and equilibrium concentrations of bilirubin in the solution. Figure 6 shows that the percentage of adsorption of chitosan coated date pit activated carbon is more compared with date pit activated carbon.

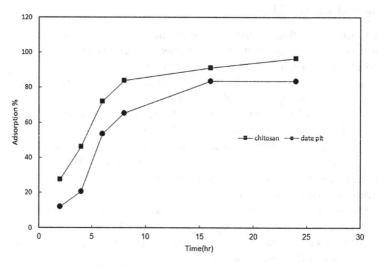

Fig. 6. Adsorption percentage vs. contact time.

Effect of Adsorbent Dose. The amount of the adsorbent is important in the adsorption process. To study the effect of the adsorbent dosage on the adsorption of bilirubin, experiments were conducted using various amounts of CDPAC (0.1, 0.5, 0.8gm), initial concentrations of 80 μM, 60 μM and 30 μM of bilirubin solution at optimized pH = 7.4 and temperature of 37 °C. It is noted that the amount of bilirubin adsorption is higher for a larger adsorbent dose. This explains that some adsorption sites remain unsaturated during the adsorption reaction, therefore, the number of sites available for adsorption increases by increasing the adsorbent dose [14, 15].

4   Conclusions

The present investigation shows that the chitosan coated activated carbon is an effective adsorbent for bilirubin. Several experiments were performed in batch systems at conditions corresponding to pH = 7.4 and temperature of 37 °C, on analyzing the effects of the variables, the results show that the equilibrium state of the interaction between bilirubin and chitosan coated activated carbon surface is reached after 16 h of contact time. The adsorption of bilirubin increases with an increase in chitosan coated carbon dosage. It is also observed that the percentage of adsorption is increased with time and thus, it can be concluded that the present work has demonstrated effectively that the chitosan coated activated carbon has potential as a low cost adsorbent and can be used effectively for the adsorption of bilirubin toxin present in the liver.

Acknowledgement. The authors would like to thank UAEU for their support of the project. This work has been financially funded by the SURE program at UAE University, Al Ain.

References

1. Pless, G.: Artificial and bioartificial liver support. Organogenesis **3**(1), 20–24 (2007)
2. Annesini, M.C., Piemonte, V., Turchetti, L.: Simultaneous removal of albumin-bound toxins in liver support devices: bilirubin and tryptophan adsorption on activated carbon. Chem. Eng. Trans. **32**, 1069–1074 (2013). doi:10.3303/CET1332179
3. Magosso, E., Mauro, U., Luigi, C., Olga, B., Luigi, B., Sergio, S.: A modeling study of bilirubin kinetics during molecular adsorbent recirculating system sessions. Int. Cent. Artif. Organs and Transplant. **30**(4), 285–300 (2006)
4. Puri, P., Anand, A.C.: Liver support devices. Med. Update **22**, 489–493 (2012)
5. Andrade, J.D., Kopp, K., Van Wagenen, R., Chen, C., Kolff, W.J.: Activated Carbon and Blood Perfusion: A Critical Review, Vol 9, No.25 (n.d.), pp. 290–302
6. Mahmoudi, K., Noureddine, H., Ezzeddine, S.: Preparation and characterization of activated carbon from date pits by chemical activation with zinc chloride for methyl orange adsorption. J. Mater. Environ. Sci. **5**(6), 758–769 (2014)
7. Hussain, M.Z., Mostafa, I.W., Vandita, S., Venitia, S., Mohammad, S.R.: Chemical composition of date-pits and its potential for developing value-added product – a review. Pol. J. Food Nutr. Sci. **64**(4), 215–226 (2014)
8. Luk, C.J., Joanne, Y., Chunwah, M.Y., Chiwai, K., Kimhung, L.: A comprehensive study on adsorption behaviour of direct, reactive and acid dyes on crosslinked and non-crosslinked Chitosan Beads. J. Fiber Bioeng. Inf. **7**(1), 35–52 (2014)
9. Kyzas, G.Z., Dimitrios, N.B.: Recent modifications of chitosan for adsorption applications: a critical and systematic review. Marine Drugs **13**, 312–337 (2015)
10. Moussaoui, Y., Mnasri, N., Ben Salem, R., Lagerge, S., De Menorval, L.C.: Preparation of Chitosan Gel, EDP Sciences, pp. 1–8 (2012)
11. Soundarrajan, M., Gomathi, T., Sudha, P.N.: Understanding the adsorption efficiency of chitosan coated carbon on heavy metal removal. Int. J. Sci. Res. Publ. **3**(1), 1–10 (2013)
12. Khaled, M., Noureddine, H., Ezzeddine, S.: Preparation and characterization of activated carbon from date pits by chemical activation with zinc chloride for methyl orange adsorption. J. Mater. Environ. Sci. **5**(6), 1758–1769 (2014)
13. Mirzayant, Y.W.: Experimental study on the use of chitosan coated activated carbon to reduce the content of metal Fe the produced water. ARPN J. Eng. Appl. Sci. **10**(22), 10506–10510 (2015)
14. Kumar, P.S., Kirthika, K.: Equilibrium and kinetic study of adsorption of nickel from aqueous solution onto Bael tree leaf powder. J. Eng. Sci. Technol. **4**(4), 351–363 (2009)
15. Desta, M.B.: Batch sorption experiments: langmuir and freundlich isotherm studies for the adsorption of textile metal ions onto teff straw (eragrostis tef) agricultural waste. J. Thermodyn. **20**(13), 6 (2013). Article ID 375830

16. Neuvonen, P.J., Olkkola, K.T.: Oral activated charcoal in the treatment of intoxications. Med. Toxicol. Adverse Drug Experience **3**(1), 33–58 (1988)

17. Putra, E., Pranowo, R., Sanarso, J., Indraswati, N., Ismadji, S.: Performance of activated carbon and bentonite for adsorption of amoxicillin from wastewater: Mechanisms, isotherms and kinetics. J. Int. Water Assoc. **43**(9), 2419–2430 (2009)

18. Annesini, M., Carlo, C., Piemonte, V., Turchetti, L.: Bilirubin and tryptophan adsorption in albumin-containing solutions I: equilibrium isotherms on activated carbon. Biochem. Eng. J. **40**(2), 205–210 (2008)

19. Ash, S., Sullivan, T., Carr, D.: Sorbent suspension vs. sorbent columns for extracorporeal detoxification in hepatic failure. Ther Apher Sia **10**(2), 145–153 (2006)

20. Nikolaev, V., Sarnatskaya, V., Sigal, V., Klevtsov, V., Makhorin, K., Yushko, L.: High-porosity activated carbon for bilirubin removal. Int. J. Artif. Organs **14**(3), 179–185 (1991)

21. Chandy, T., Sharma, C.: Polylysine-Immobilized Chitosan Beads as adsorbents for bilirubin. Int. J. Artif. Organs **16**(6), 568–576 (1992)

Investigation of the Feasibility of Strain Gages as Pressure Sensors for Force Myography

Him Wai Ng, Xianta Jiang, Lukas-Karim Merhi, and Carlo Menon[(⊠)]

MENRVA Research Group, School of Engineering Science,
Simon Fraser University, Burnaby, BC, Canada
cmenon@sfu.ca

Abstract. Hand gesture recognition is a popular topic of many research studies, and force myography (FMG) has recently emerged for this application. This work investigates a novel sensor system based on electrical resistance strain gages that is fully wearable and easy-to-use. This system consists of eight strain gages embedded in a transparent flexible plastic band, covering the entire wrist. The system was tested with 8 subjects by performing 14 different hand gestures, with an accuracy of 99.2% using support vector machine. The impressive accuracy of the wearable band confirms the capability of strain gages as pressure sensors in force myography in hand gesture recognition applications.

Keywords: Force myography · Wearable technology · Gesture recognition · Electrical resistance strain gages · Machine learning

1   Introduction

The use of hand gestures is a natural way of communication between humans and their environment. Hand gestures need to be recognized and monitored in various applications, such as human-machine interactions (HMI), tele-manipulation of robots, and prosthesis control [1–5]. Researchers have developed various technologies to recognize hand gestures, including computer vision [1, 4], inertial senor [6, 7], data glove [8, 9], and muscle activities sensor [2, 5, 10, 11] based technologies.

Surface electromyography (sEMG) is one of the traditional muscle activity sensor technologies that registers electrical activities generated by the motor units of the skeletal muscles, and is capable of deciphering fine motor movements of the upper extremities, such as grasps and individual finger movements. For example, Ahsan [12] used forearm electromyography to recognize four hand gestures, achieving a test accuracy of 86.8%. However, sEMG systems require a high data sampling rate and high power consumption, which can make most of the systems relatively expensive and not portable. Some other low-cost technologies were proposed, for instance electrical impedance tomography (EIT) [13] which was able to achieve 97% and 87% accuracies in recognizing pinch and hand gestures.

Force myography (FMG) is a non-invasive HMI technique that uses force sensors on the surfaces of muscles to capture their changes in shape, which is also referred to as residual kinetic imaging [14] or muscle pressure mapping [15]. It has been used in monitoring limb and hand movements [16, 17], hand gestures recognition [18, 19], and

© Springer International Publishing AG 2017
I. Rojas and F. Ortuño (Eds.): IWBBIO 2017, Part I, LNBI 10208, pp. 261–270, 2017.
DOI: 10.1007/978-3-319-56148-6_22

prosthesis control [20–23]. Much of the research in FMG has been using force sensing resistors (FSRs) for hand gesture sensing. For example, Amft et al. developed a system to visually distinguish four types of arm gestures on a data plot by using two FSRs [22]. Ogris et al. [23], Wang et al. [24], and Li et al. [21] also demonstrated the possibility to predict different arm and finger gestures by using multiple FSRs pressed against the arm. Xiao and Menon created a novel wearable strap consisting of eight FSRs [16, 17]. The strap was placed on the proximal portion of the forearm and was able to capture the activities of the main muscle groups with eight force input channels. The FSRs within the strap can extract the upper extremities FMG signals from the arm; these signals can then be used to predict different postures of the forearm. Recently, Jiang et al. [18] explored the effect of force exertion during grasping towards the classification accuracy of 16 grasps using a 16 channel FSRs strap on the wrist; they further explored the capability of a FSRs strap in classifying 3 sets of 48 hand gestures on both wrist and forearm, achieving a comparable performance compared to that of using sEMG technology [19].

However, as FSRs are not able to change their electrical resistances rapidly enough when there are fast variations in pressures, and show a time lagging behavior and produce unstable results [11, 25]. This limits its use to large muscle regions with relatively stable pressure variations [11]. In order to monitor rapid pressure variations with higher accuracy, sensors with better sensitivity than FSRs are needed. Electrical resistance strain gages, which are known for its high sensitivity to pressure variations [26], have been studied for such a purpose. The recent work done by Lin et al. demonstrated that strain gages can be used as strain sensors for high accuracy gesture recognition [11]. Our research group has explored the feasibility of using custom-fabricated strain gauge sensors for hand gestures classification, achieved sufficient accuracies of 95% and 80% using cross-validation and cross-trial evaluations [27, 28]. However, these strain gauge sensors were custom fabricated and are not commercially available. Therefore we wanted to examine the use of commercially available strain gauge sensors.

The present work investigated the feasibility of using off the-shelf electrical resistance strain gages as alternative pressure sensors on the wrist for force myography. In comparison to the work done by Lin et al. [11], in which the strain gages were put on the back of palm, which is hard to wear and also limits the movement of the hand. Our choice of putting the strain gages on the wrist helps to minimize the restriction on hand motion. We used eight strain gages to build a wearable band that can be worn on the wrist of a person. The signals generated by the hand while performing different gestures were then analyzed and distinct patterns for each gesture were obtained. Four widely used machine learning algorithms were employed to perform classification on these signals, including Linear Discriminant Analysis (LDA), Support Vector Machine (SVM), Artificial Neural Network (ANN), and K-nearest Neighbor (KNN). This work enables future studies on applying electrical resistance strain gages to hand gesture recognition and other force myography applications.

2   Materials and Methods

2.1   Electrical Resistance Strain Gages Based Wearable Band System

The complete wearable band system is shown in Fig. 1. The entire circuit was built on a small ProtoBoard [29] and a regular breadboard.

We used the KFH-10-120-C1-11L1M2R linear gages [30] as the string gage sensors, which are 10 mm length and allow the coverage of a large portion of skin surfaces. The strain gages were aligned onto the flexible transparent substrate of the band which is made of cellulose acetate, as shown in Fig. 2. Then they were connected

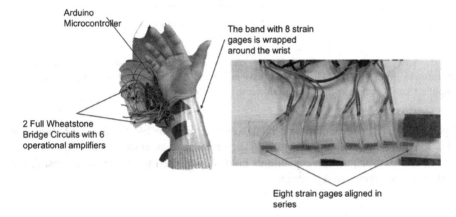

Fig. 1. Prototype of the strain gage based wearable band. The entire system consists of eight strain gages, six operational amplifiers, an Arduino microcontroller and a number of resistors.

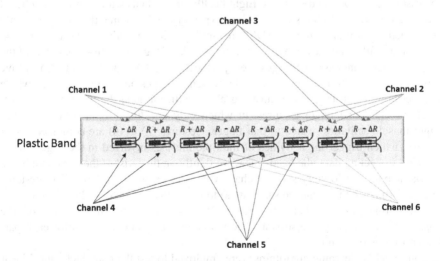

Fig. 2. Schematic of the wearable band showing the arrangement of the strain gages and the location of the corresponding data channels.

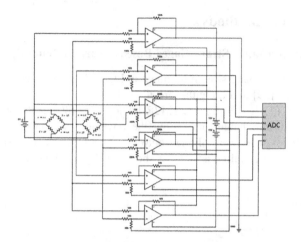

Fig. 3. Schematic of the entire wearable band circuit. The eight strain gages form two Wheatstone bridges and produce six signals in total. Each operational amplifier outputs one amplified signal, which is then passed down to an analog-to-digital converter.

together with six operational amplifiers to implement the electrical circuit shown in Fig. 3. An Arduino microcontroller [31], which has an analog-to-digital converter (ADC) and data acquisition capabilities built-in, was used to convert and transmit the analog output signals to a computer.

2.2   Data Collection and Experimental Methods

An experiment was conducted to evaluate the performance of this strain gage band on the wrist to recognize hand gestures. Eight healthy individuals with both full functional hands were recruited to the study. Various types of hand gestures that are commonly used in activities of daily living (ADL) are employed in the study, as shown in Fig. 4. They contain full hand movements (a–b), digit flexion (d–g), grips (h–i), and part of the American Sign Language (ASL) gestures (j–n) including 2, 9, 10, 'Y', and 'ILY (I love you)' [11, 32, 33]. In the view of muscle movements, gestures (a) and (b) are whole hand gestures which trigger the most muscle activities around the wrist. Gestures (c) to (g) are individual finger gestures, in which only one finger is in action and will trigger muscle activities related to that finger. Gestures (h) to (n) are combined finger gestures, which trigger the activities of multiple muscles involved in these gestures.

When collecting data, the band was worn tightly on the wrist of the dominant hand. The participant comfortably sat in a chair, with arm resting on a table. The gestures were performed in the sequence same as (a) to (n) in Fig. 4, each for 25 s. There were 15 s of resting period in between each gesture. The band was never taken off the subject's wrist for the entire session. The whole data collection process for each participant took about 9 min.

Four machine learning algorithms were employed to test the data, including Linear Discriminant Analysis (LDA), Support Vector Machine (SVM), Neural Network (NN),

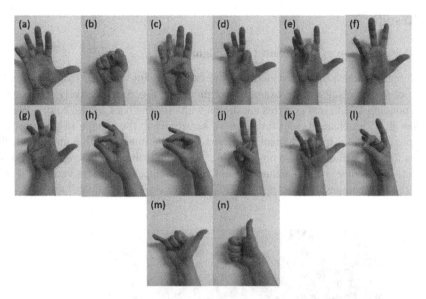

Fig. 4. Fourteen gestures designed to test the performance of the wearable band. Full hand gestures: (a) Fully opened hand (b) Fully closed hand; individual finger movements: (c) Opened hand with thumb closed (d) Opened hand with index finger closed (e) Open hand with middle finger closed (f) Opened hand with ring finger closed (g) Opened hand with pinky finger closed; pinch gestures: (h) Two fingers grabbing (i) Three fingers grabbing; sign language gestures: (j) 2 (k) ILY (I Love You) (l) 9 (m) Y (n) 10

and K-nearest Neighbor (KNN). The LDA classifier used in this study is from MATLAB Statistics and Machine Learning Toolbox [34], which is an extension of Fisher linear discriminant analysis [35]. We used the SVM with a Radial Basis Function (RBF) kernel from Libsvm [36] and the cost and gamma parameters were optimized using a grid search algorithm [17]. We used 1 hidden layer neural network with 20 neurons and 5 nearest neighbors for KNN; both NN and KNN were also from MATLAB Statistics and Machine Learning Toolbox [34] and the parameters (the numbers of hidden layer, neuron, and neighbors) were empirically determined.

The dataset of each participant was tested by above 4 algorithm using 10-fold cross validation. The 10-fold cross-validation is a well-accepted accuracy estimation and model selection method [37], in which the data were first randomized in sample orders and then separated into 10 equal-sized subsets. Of the 10 subsets, one subset was retained as test data for testing the model, and the remaining nine subsets were used for training the model. This cross-validation process was repeated 10 times (folds), with each of the 10 subsets used as the testing data. The results of the 10 repetitions were averaged as a single performance measurement of the participant.

All the methods within this study were in compliance with the declaration of Helsinki and were approved by the Simon Fraser University (SFU) Office of Research Ethics (#2015s0527). All the participants gave informed consent before taking part in the experiment.

3   Results

3.1   Visualization of the Strain Gages Signal

The signals produce distinct patterns to the corresponding gestures from the combination of the six channels; the examples are plotted in Fig. 5, respectively. The signals corresponding to different gestures are shown with different background colors. We can see that different channels vary in levels of sensitivity to the change of gestures. For example, the signal from Channel 3 appears to be the most sensitive among all; the signals for "close thumb", "two finger grab", "ILY", and "thumbs-up" gestures are very distinct. On the other hand, Channel 2 and Channel 5 appear to be less variation across all of the fourteen gestures.

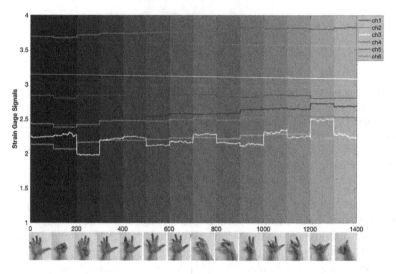

Fig. 5. Example gesture data with 100 data entries per gesture

3.2   Classification Accuracies

The average training and testing accuracies of the 4 classifiers on the dataset are shown in Fig. 6.

The SVM achieved the highest testing accuracy of 99.2% in recognizing the hand gestures and the other 3 models (LDA, NN, and KNN) which achieved accuracies of about 98.5%. These high testing accuracies verified the capability of the strain gages-based wearable being used in FSR applications.

In order to gain better insight into resulting classification accuracies, confusion matrices were generated for all fourteen hand gestures as shown in Fig. 7. As SVM already achieved over 99%, there is almost no confusion in the figure. When the accuracies decrease a little bit for the other three classifiers to about 98.5%, the confusion consistently appears between gesture 6 and 7 (the ring and pinky fingers flexion), which is obviously the most easily misclassified gestures.

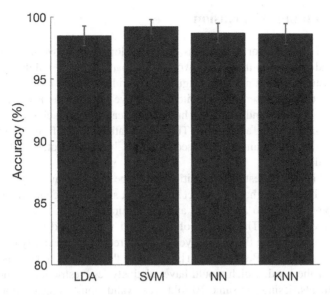

Fig. 6. Testing classification accuracies of the band in performing gesture prediction using LDA in cross-validation scheme

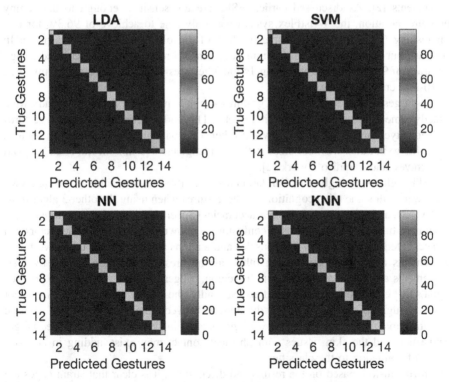

Fig. 7. Confusion matrix for accuracies of the 4 classifiers; the number in x and y-axis is the gesture number as in Fig. 4.

4 Discussion and Conclusion

A novel strain gage sensor system has been developed to be worn on the wrist to recognize hand gestures in the present work. Eight participants tested the system on 14 fine hand gestures including sign language, individual finger movement, and grip gestures. The strain gage sensors exhibited distinctive signal patterns corresponding to different hand gestures and achieved high classification accuracies of >98% using different machine learning algorithms. The classification results from the experiments demonstrated that the strain gages-based wearable band has potential to accurately recognize hand gestures.

BackHand is a hand gesture recognition prototype that used an array of strain gage sensors on the back of the hand to detect 16 different sign language hand gestures [11]. The prototype was able to achieve an average accuracy of 95.8% using an LDA of 10-fold cross validation. The sensors of BackHand were placed on the back of the hand, which could impede the skin movements and reduce the dexterity of the fingers. Compared to the work of BackHand [11], we placed the sensors on the wrist instead of on the back of the hand, which would have relatively less restriction to the hand and finger movements. Using the same 10-fold cross validation testing scheme and LDA algorithm, our system achieved a higher accuracy of 98.5%.

WristFlex is a prototype using 15 FSRs worn on the wrist to monitor finger movements [2]. As discussed earlier, FSRs are not sensitive enough to detect tiny pressure variation, the WristFlex system was only able to achieve of 96.3% for distinguishing 6 pinch gestures, using a SVM of 10-fold cross validation scheme. In contrast, our system only used 8 strain gage sensor on the wrist and achieved a high accuracy of 99.2% for distinguishing 14 hand gestures using the same algorithm SVM in 10-fold cross validation.

Eight gestures (5 sign language gestures + pinch, relax and fist) in our study were exactly same as those in those used in Lin et al. [11] and covered all the 6 hand gestures in Dementyev and Paradiso's [2] work; other 4 individual finger movement gestures (Fig. 4.d–g) in our work were close to the sign language 6–9 in both Lin et al. [11] and Dementyev and Paradiso's work [2].

Although a very high cross validation accuracy (99.2%) was achieved using SVM, there were still some misrecognitions of the gestures when using the other 3 algorithms (LDA, NN, and KNN of about 98.5% accuracies) Specifically, the ring and little fingers flexion gestures more likely caused confusion, as shown in Fig. 7. Multiple factors can be accounted for this. Firstly, the muscles and tendons responsible for these two finger movements are very close to each other, as the pressure variation caused by their movements are likely very similar in the region, triggering the strain gages to produce signals that could confuse the classifiers. Secondly, the size of the strain gages was not small enough to detect very tiny differences between these two gestures' muscle activities, making the differences indistinguishable. Finally, since the band used no gel between it and the skin, issues with physical contact may exist, adding in an extra factor of confusion to the classifier.

In conclusion, based on the results and discussions, it is clear that strain gages are suitable to use as sensors for FMG applications such as hand gesture recognition, and it is able to achieve comparable performance to state-of-art solutions. Future work for the band would be to increase its portability, reduce its power consumption and be more comfortable to users when worn on the wrist, and test the band under more restrict conditions such as real-time, cross-sessions, and cross-subjects.

References

1. Pavlovic, V.I., Sharma, R., Huang, T.S.: Visual interpretation of hand gestures for human-computer interaction: a review. IEEE Trans. Pattern Anal. Mach. Intell. **19**(7), 677–695 (1997)
2. Dementyev, A., Paradiso, J.A.: WristFlex. In: Proceedings of the 27th Annual ACM Symposium on User Interface Software and Technology - UIST 2014, pp. 161–166 (2014)
3. Giuffrida, J.P., Lerner, A., Steiner, R., Daly, J.: Upper-extremity stroke therapy task discrimination using motion sensors and electromyography. Neural systems. IEEE Trans. Neural Syst. Rehabil. Eng. **16**(1), 82–90 (2008)
4. Jiang, H., Duerstock, B.S., Wachs, J.P.: A machine vision-based gestural interface for people with upper extremity physical impairments. IEEE Trans. Syst. Man Cybern.: Syst. **44**(5), 630–641 (2014)
5. Atzori, M., Müller, H.: Control capabilities of myoelectric robotic prostheses by hand amputees: a scientific research and market overview. Front. Syst. Neurosci. **9**, 162 (2015)
6. Georgi, M., Amma, C., Schultz, T.: Recognizing hand and finger gestures with IMU based motion and EMG based muscle activity sensing. In: Proceedings of the International Conference on Bio-inspired Systems and Signal Processing, pp. 99–108 (2015)
7. Xu, R., Zhou, S., Li, W.J.: MEMS accelerometer based nonspecific-user hand gesture recognition. IEEE Sensors J. **12**(5), 1166–1173 (2012)
8. Colasanto, L., Suarez, R., Rosell, J.: Man hybrid mapping assistance of teleoperated grasping tasks. IEEE Trans. Syst. Cybern. Syst. **43**(2), 390–401 (2013)
9. Heumer, G., Ben Amor, H., Jung, B.: Grasp recognition for uncalibrated data gloves: a machine learning approach. Presence Teleoper. Virtual Environ. **17**(2), 121–142 (2008)
10. Al-Timemy, A.H., Khushaba, R.N., Bugmann, G., Escudero, J.: Improving the performance against force variation of EMG controlled multifunctional upper-limb prostheses for transradial amputees. IEEE Trans. Neural Syst. Rehabil. Eng. **24**(6), 650–661 (2016)
11. Lin, J.W., et al.: BackHand: sensing hand gestures via back of the hand. In: Proceedings of UIST 2015, vol. C, pp. 557–564 (2015)
12. Ahsan, M.: Electromygraphy (EMG) signal based hand gesture recognition using artificial neural network (ANN). In: Mechatronics (ICOM 2011), May, pp. 17–19 (2011)
13. Zhang, Y., Harrison, C.: Tomo: wearable, low-cost electrical impedance tomography for hand gesture recognition. In: Proceedings of the 28th Annual ACM Symposium on User Interface Software & Technology, pp. 167–173 (2015)
14. Phillips, S.L., Craelius, W.: Residual kinetic imaging: a versatile interface for prosthetic control. Robotica **23**(3), 277–282 (2005)
15. Radmand, A., Scheme, E., Englehart, K.: High-density force myography: a possible alternative for upper-limb prosthetic control. J. Rehabil. Res. Dev. **53**(4), 443–456 (2016)
16. Xiao, Z.G., Menon, C.: Towards the development of a wearable feedback system for monitoring the activities of the upper-extremities. J. NeuroEng. Rehabil. **11**(2), 13 (2014)
17. Kadkhodayan, A., Jiang, X., Menon, C.: Continuous prediction of finger movements using force myography. J. Med. Biol. Eng. **36**(4), 594–604 (2016)

18. Jiang, X., Merhi, L.-K., Menon, C.: Force exertion affects grasp classification using force myography. IEEE Trans. Human-Machine Syst (2017, accepted)
19. Jiang, X., Merhi, L.-K., Xiao, Z.G., Menon, C.: Exploration of force myography and surface electromyography in hand gesture classification. Med. Eng. Phys. (2017, in Press)
20. Cho, E., Chen, R., Merhi, L., Xiao, Z., Pousett, B., Menon, C.: Force myography to control robotic upper extremity prostheses: a feasibility study. Front. Bioeng. Biotechnol. **4**(March), 1–12 (2016)
21. Li, N., Yang, D., Jiang, L., Liu, H., Cai, H.: Combined use of FSR sensor array and SVM classifier for finger motion recognition based on pressure distribution map. J. Bionic Eng. **9**(1), 39–47 (2012)
22. Amft, O., Troster, G., Lukowicz, P., Schuster, C.: Sensing muscle activities with body-worn sensors. In: International Workshop on Wearable and Implantable Body Sensor Networks (BSN 2006), pp. 138–141 (2006)
23. Ogris, G., Kreil, M., Lukowicz, P.: Using FSR based muscule activity monitoring to recognize manipulative arm gestures. In: 2007 11th IEEE International Symposium on Wearable Computers, pp. 1–4 (2007)
24. Wang, X., Zhao, J., Yang, D., Li, N., Sun, C., Liu, H.: Biomechatronic approach to a multi-fingered hand prosthesis. In: 2010 3rd IEEE RAS & EMBS International Conference on Biomedical Robotics and Biomechatronics, pp. 209–214 (2010)
25. Hollinger, A., Wanderley, M.M.: Evaluation of commercial force-sensing resistors, pp. 1–4 (2006)
26. Hoffmann, K.: An Introduction to Measurements Using Strain Gages. Hottinger Baldwin Messtechnik, Darmstadt (1989)
27. Ferrane, A., Jiang, X., Maiolo, L., Pecora, A., Colace, L., Menon, C.: A fabric-based wearable band for hand gesture recognition based on filament strain sensors: a preliminary investigation. In: 2016 IEEE Healthcare Innovation Point-of-Care Technologies Conference (HI-POCT), pp. 113–116 (2016)
28. Ferrone, A., Maita, F., Maiolo, L., Arquilla, M., Castiello, A., Pecora, A., Jiang, X., Menon, C., Colace, L.: Wearable band for hand gesture recognition based on strain sensors. In: 6th IEEE RAS/EMBS International Conference Biomedical Robotics and Biomechatronics (2016)
29. ProtoBoard. https://www.sparkfun.com/products/retired/8812. Accessed 22 Nov 2016
30. PRE-WIRED STRAIN GAGES, Linear Gages, X-Y Planar Rosettes (Tee Rosette), 0°/45°/90° Planar Rosettes
31. Arduino. https://www.arduino.cc/en/Main/ArduinoBoardUno. Accessed 22 Nov 2016
32. Skirven, T.M., et al.: Rehabilitation of the Hand and Upper Extremity, 6th edn. Elsevier/ Elsevier Mosby, Philadelphia (2011)
33. Cutkosky, M.R.: On grasp choice, grasp models, and the design of hands for manufacturing tasks. IEEE Trans. Robot. Autom. **5**(3), 269–279 (1989)
34. MathWorks, MATLAB - The Language of Technical Computing
35. Xu, Y., Lu, G.: Analysis on fisher discriminant criterion and linear separability of feature space. In: 2006 International Conference on Computational Intelligence Security ICCIAS 2006, vol. 2, pp. 1671–1676 (2007)
36. Chang, C., Lin, C.: LIBSVM. ACM Trans. Intell. Syst. Technol. **2**(3), 1–27 (2011)
37. Kohavi, R., et al.: A study of cross-validation and bootstrap for accuracy estimation and model selection. In: IJCAI, vol. 14, no. 2, pp. 1137–1145 (1995)

Geometric Modelling of the Human Cornea: A New Approach for the Study of Corneal Ectatic Disease. A Pilot Investigation

Francisco Cavas-Martínez[1]([⊠]), Daniel G. Fernández-Pacheco[1],
Dolores Parras[1], Francisco J.F. Cañavate[1], Laurent Bataille[2],
and Jorge L. Alio[3,4,5]

[1] Department of Graphical Expression, Technical University of Cartagena,
Cartagena, Spain
{francisco.cavas,daniel.garcia,dolores.parras,
francisco.canavate}@upct.es
[2] Research and Development Department, Vissum Corporation, Alicante, Spain
lbataille@vissum.com
[3] Division of Ophthalmology, Universidad Miguel Hernández, Alicante, Spain
jlalio@vissum.com
[4] Keratoconus Unit of Vissum Corporation, Alicante, Spain
[5] Department of Refractive Surgery, Vissum Corporation, Alicante, Spain

Abstract. **Purpose:** The aim of this study was to describe the application of a new bioengineering graphical technique based on geometric custom modelling capable to detect and to discriminate small variations in the morphology of the corneal surface.

Methods: A virtual 3D solid custom model of the cornea was obtained employing Computer Aided Geometric Design tools, using raw data from a discrete and finite set of spatial points representative of both sides of the corneal surface provided by a corneal topographer. Geometric reconstruction was performed using B-Spline functions, defining and calculating the representative geometric variables of the corneal morphology of patients under clinical diagnosis of keratoconus.

Results: At least four variables could be used in order to classify corneal abnormalities related to keratoconus disease: anterior corneal surface area (ROC 0.853; $p < 0.0001$), posterior corneal surface area (ROC 0.813; $p < 0.0001$), anterior apex deviation (ROC 0.742; $p < 0.0001$) and posterior apex deviation (ROC 0.899; $p < 0.0001$).

Conclusions: Custom geometric modelling enables an accurate characterization of the human cornea based on untreated raw data from the corneal topographer and the calculation of morphological variables of the cornea, which permits the clinical diagnosis of keratoconus disease.

Keywords: Keratoconus · CAD · Scheimpflug · Surface reconstruction · Virtual model

© Springer International Publishing AG 2017
I. Rojas and F. Ortuño (Eds.): IWBBIO 2017, Part I, LNBI 10208, pp. 271–281, 2017.
DOI: 10.1007/978-3-319-56148-6_23

1 Introduction

The analysis of the geometric characteristics of the cornea has been of interest in ophthalmology given that small variations in its morphology can suppose important changes in visual function of the patient [1]. Thus, the geometric reconstruction of the corneal surface has experienced significant progress in recent years with the development of new technologies [2]. This reconstruction process is currently being developed by the so-called modal methods, based mainly on the development of the Zernike polynomials. However, this approach has limitations as it is inaccurate in abnormal situations: eyes after corneal surgery or eyes with corneal pathology (such as keratoconus) that present significant higher-order aberrations [3–6].

An alternative to these surface reconstruction methods are the zonal methods, mainly based on B-splines, which are stable geometric reconstruction mathematical functions [7, 8]. However, these methods have not been widely used in ophthalmology.

On the other hand, over recent years the development of new computational tools has improved the process of acquisition and image processing [9], enabling to generate three-dimensional shapes [10] and to develop behavioral models which reproduce the geometry of a solid structure more accurately.

One of these tools is the Computer Aided Geometric Design (CAGD), which was created to address the technological needs of companies in automotive and aeronautic sectors. The CAGD allows to study the geometric and computational aspects of any complex physical entity such as areas and volumes to create virtual models [11], which differentiate from physical models for not having a destructive process and reducing costs in terms of production and time [12, 13].

Specifically in the field of Bioengineering, the development of virtual models through CAGD permits the characterization of biological structures, establishing new experimentation procedures in the field of medicine [14]. The development of these three-dimensional models of biological structures [15] allows to optimize the design of prosthesis [16], to perform biomechanical studies using finite elements [17], to obtain virtual representations for educational purposes [18], or simply to generate realistic physical models from a virtual model using 3D printing techniques [19].

Based on all of the above and applied to the field of ophthalmology, this study proposes a new technique for the clinical diagnosis of corneal ectatic diseases. This is a bioengineering graphical technique based on the performance of an integral geometric analysis of a 3D representative model of the corneal structure of a live patient, where the model obtained presents a high sensitivity to small variations that may occur in the morphology of both anterior and posterior corneal surfaces. This accurate characterization of the human cornea enables a new path in the diagnosis of this corneal ectatic disease and offers a new approach that facilitates the follow-up of keratoconus.

2 Materials and Methods

The retrospective study adhered to the tenets of the Declaration of Helsinki and was approved by the local Clinical Research Ethics Committee of Vissum Corporation (Alicante, Spain). Patients examined at the Vissum Corporation (Alicante, Spain) were retrospectively enrolled. Patients were selected from a database of candidates for refractive surgery with normal corneas [20, 21] and also a database of cases diagnosed with keratoconus [20, 21] in both eyes.

All eyes selected (187) underwent a thorough and comprehensive eye and vision examination which included [20, 21] uncorrected distance visual acuity (UDVA), corrected distance visual acuity (CDVA), manifest refraction, Goldmann tonometry, biometry (IOLMaster, Carl Zeiss Meditec AG) and corneal topographic analysis. During this protocol, the Sirius system® (CSO, Florence, Italy) was used, which is a noninvasive system for measuring and characterizing the anterior segment using a rotating Scheimpflug camera that generates images in three dimensions, with a dot matrix fine-meshed in the center due to the rotation. The images taken during the examination are digitalized in the main unit and transferred to a computer and analyzed in detail. Gathered Sirius corneal topographies (data from other topographers such as Pentacam can also be handled [22]) are represented as discrete and finite set of spatial points (point cloud surfaces) in the form of two 31×256 matrices. Both matrices contain the polar coordinates representative of the anterior and posterior corneal surfaces. These data, used only in the first stage of the topographic data acquisition procedure, are called raw data [4, 23]. This warrantied that data were not interpolated or manipulated [22, 23], avoiding any proprietary reconstruction software from topographer's manufacturer. All measurements were performed by the same experienced optometrists, performing three consecutive measurements and taking average values for posterior analysis.

2.1 Diagnosis Procedure

The technique proposed in this article, based on zonal methods, has two well differentiated stages: a first stage where a 3D model of the cornea is reconstructed from the raw data provided by the corneal topographer using CAGD tools, and a second stage where several geometric variables are extracted from the model and analyzed to characterize the cornea.

First Stage: 3D Model Reconstruction
The reconstruction process can be divided into four phases (Fig. 1):

1. Extraction of the point clouds from the Sirius topographer.
2. Generation of both corneal surfaces. The two point clouds were imported to the surface reconstruction CAD software Rhinoceros® V 5.0 (MCNeel & Associates, Seattle, USA). This software provides useful functions, such as B-splines, to generate surfaces with high accuracy. For this study the Rhinoceros' patch surface function was used, obtaining a mean distance error between the 3D point cloud and the solution surface of about $3.60 \times 10^{-4} \pm 6.43 \times 10^{-4}$ mm.
3. Positioning of both corneal surfaces. After the generation of both corneal surfaces, they were engaged by their geometrical center and Z axis.
4. Reconstruction of the 3D model. Once both corneal surfaces were positioned, the peripheral surface (the bonding surface between both sides in the Z-axis direction) was generated and all surfaces were then joined to form a single surface. The surface reconstructed was then exported to the solid modeling software SolidWorks V 2013

(Dassault Systèmes, Vélizy-Villacoublay, France) to generate a 3D model representative of the custom and actual geometry of the cornea.

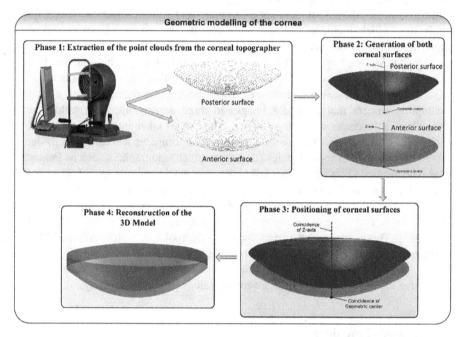

Fig. 1. Geometric modelling process by using DGAO tools.

Second Stage: Characterization of the Cornea

During this stage, some geometric variables representative of the corneal morphology were calculated from the 3D model generated on the previous stage. A detailed description of the geometric variables used during this study can be found in Table 1.

These variables were later statistically analyzed in order to characterize the cornea.

Table 1. Geometric variables analyzed in the study.

Geometric variable	Description
Total corneal volume [mm3]	Volume limited by front, back and peripheral surfaces of the solid model generated
Anterior corneal surface area [mm2]	Area of the front/exterior surface
Posterior corneal surface area [mm2]	Area of the rear/interior surface
Total corneal surface area [mm2]	Sum of anterior, posterior and perimetral corneal surface areas of the solid model generated
Anterior apex deviation [mm]	Average distance from the Z axis to the highest point (apex) of the anterior corneal surfaces
Posterior apex deviation [mm]	Average distance from the Z axis to the highest point (apex) of the posterior corneal surfaces

2.2 Statistical Analysis

A Kolmogorov-Smirnov test was run to assess the data engagement scores. According to this test and thereafter, a Student's t-test or U-Mann Whitney Wilcoxon test was performed (depending on normality), in order to describe differences between normal and keratoconus groups in all the measurements proposed. Additionally, Kruskal Wallis (K-W) and Effect Size (ES) tests were used to compare differences and to quantify the degree of change between groups according to Amsler Krumeich Grading System (AK). ROC curves were established to determine what parameters could be used to classify the diseased corneas by calculating optimal cut-offs, sensitivity and specificity. All the analyses were performed by using Graphpad Prism v6 (GraphPad Software, La Jolla, USA) and SPSS v17.0 software (SPSS, Chicago, USA).

3 Results

From a total of 187 patients, this study included 124 healthy eyes that did not present any ocular pathology [20] and constituted by 69 females (55.6%) and 55 males (44.4%) ranged from 7 to 73 years old, and 63 eyes diagnosed with keratoconus in several grades [21] (53.9% in stage I, 31.7% in stage II, and 14.4% in the most extreme stages, III and IV) and formed by 34 females (53.9%) and 29 males (46.1%) ranged from 14 to 69 years old.

All of the modeled variables showed differences between normal and keratoconic eyes, as seen in Table 2. Total corneal volume presents higher values in healthy eyes (p < 0.0001), while anterior and posterior corneal surface areas are lower in the same subjects (p < 0.0001). This pattern of difference can be seen for most of the variables studied: healthy corneas have anterior and posterior apex deviations lower than keratoconic corneas (p < 0.0001).

Outcomes according to keratoconus severity are shown in Table 3, where comparisons are established according to the AK grading system. Additionally, note that

Table 2. Variables measured in healthy and keratoconic corneas, Mean ± SD (range)

Morphogeometric parameters	Healthy N = 124	Keratoconic N = 63	
	Mean ± SD (range)	Mean ± SD (range)	p value (statistical test)
Total corneal volume [mm^3]	25.90 ± 0.31 (25.59 to 26.21)	23.51 ± 0.48 (23.03 to 23.99)	0.0001 Mann-Whitney
Anterior corneal surface area [mm^2]	43.13 ± 0.06 (43.07 to 43.19)	43.42 ± 0.13 (43.29 to 43.55)	0.0001 Mann-Whitney
Posterior corneal surface area [mm^2]	44.31 ± 0.09 (44.22 to 44.40)	44.81 ± 0.21 (44.6 to 45.02)	0.0001 Mann-Whitney
Total corneal surface area [mm^2]	104.02 ± 0.29 (103.73 to 104.31)	103.68 ± 0.43 (103.25 to 104.11)	0.0001 Mann-Whitney
Anterior apex deviation [mm]	0.0003 ± 0.0002 (0.0001 to 0.0005)	0.0090 ± 0.0035 (0.0055 to 0.0125)	0.0001 Mann-Whitney
Posterior apex deviation [mm]	0.0771 ± 0.0128 (0.0643 to 0.0899)	0.1902 ± 0.029 (0.1603 to 0.2201)	0.0001 Mann-Whitney

Table 3. Variables measured in healthy and keratoconic corneas, Mean ± SD (range)

	Normal	Stage I	Stage II	Stage III-IV	p (KW test)
Total corneal volume [mm³]	25.90 ± 0.31 (25.59 to 26.21)	23.51 ± 0.48 (23.03 to 23.99)	23.09 ± 0.59 (22.5 to 23.68)	20.01 ± 2.88 (17.13 to 22.89)	0.0001
(ES)	–	1.21	1.39	3.41	
Anterior corneal surface area [mm²]	43.13 ± 0.06 (43.07 to 43.19)	43.42 ± 0.13 (43.29 to 43.55)	43.50 ± 0.17 (43.33 to 43.67)	44.31 ± 0.21 (44.1 to 44.52)	0.0001
(ES)	–	−1.16	−1.66	−4.79	
Posterior corneal surface area [mm²]	44.31 ± 0.09 (44.22 to 44.40)	44.81 ± 0.21 (44.6 to 45.02)	44.99 ± 0.22 (44.7 to 45.21)	45.39 ± 0.32 (45.07 to 45.71)	0.0001
(ES)	–	−0.101	−1.49	−4.21	
Total corneal surface area [mm²]	104.02 ± 0.29 (103.73 to 104.31)	103.68 ± 0.43 (103.25 to 104.11)	103.59 ± 0.39 (103.2 to 103.98)	103.53 ± 0.52 (103.01 to 104.05)	0.0001
(ES)	–	0.31	0.20	0.29	
Anterior apex deviation [mm]	0.0001 ± 0.00001 (0.0000 to 0.0002)	0.006 ± 0.0021 (0.0039 to 0.0081)	0.009 ± 0.0035 (0.0055 to 0.0125)	0.012 ± 0.004 (0.008 to 0.016)	0.0001
(ES)	–	−1.11	−1.39	−5.70	
Posterior apex deviation [mm]	0.0771 ± 0.0128 (0.0643 to 0.0899)	0.17 ± 0.029 (0.141 to 0.199)	0.201 ± 0.03 (0.171 to 0.231)	0.237 ± 0.051 (0.186 to 0.288)	0.0001
(ES)	–	−1.19	−1.40	−1.23	

(ES): Effect size

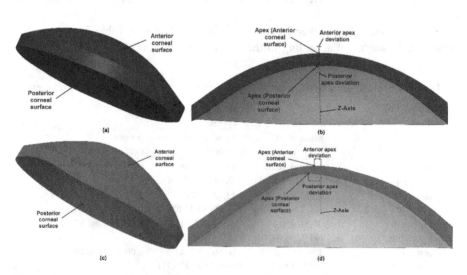

Fig. 2. Geometric variables analyzed during the study that achieved the best results: (a) 3D model of a healthy cornea, (b) Cut of a healthy cornea by a sagittal plane through the corneal apex, (c) 3D model of a diseased cornea, (d) Cut of a diseased cornea by a sagittal plane through the corneal apex.

calculated effect sizes for each disease stage allows quantifying the degree of change, higher for stages III and IV in all of the variables, becoming more evident with the progress of the disease.

The predictive value of the modeled variables was established by a ROC analysis (Fig. 3). From the several geometric variables analyzed during the study, the variables that achieved the best results in the diagnosis of the disease with an area under the ROC curve (AUROC) above 0.7 were the following four: *anterior corneal surface area* (Fig. 2) (area: 0.853, p < 0.0001, std. error: 0.040, 95% CI: 0.762–0.919), with a cutoff value of 43.07 mm², and an associated sensitivity and specificity of 90.27% and 60.01%, respectively; *the posterior corneal surface area* (Fig. 2) (area: 0.813, p < 0.0001, std. error: 0.039, 95% CI: 0.719–0.891), with a cutoff value of 44.18 mm², and an associated sensitivity and specificity of 91.08% and 44.17%, respectively; *anterior apex deviation* (Fig. 2) (area: 0.742, p < 0.0001, std. error: 0.059, 95% CI: 0.641–0.875), with a cutoff value of 0.0013 mm, and an associated sensitivity and specificity of 72.02% and 92.01%, respectively; *posterior apex deviation* (Fig. 2) (area:

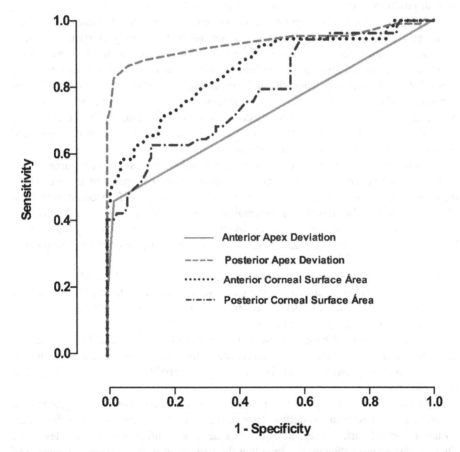

Fig. 3. A ROC analysis modelling the sensitivity versus 1-specificity for variables predicting the existence of keratoconus disease using geometrical custom modelling of the cornea

0.899, p < 0.0001, std. error: 0.041, 95% CI: 0.800–0.964), with a cutoff value of 0.0855 mm, and an associated sensitivity and specificity of 91.28% and 73.07%, respectively. Thus, according to the area under the curve variable calculated for the four variables, it was concluded that the parameter that provides a higher rate of discrimination between normal corneas and corneas with keratoconus is *Posterior apex deviation*. Nevertheless, there are other relevant statistical differences between healthy and diseased eyes, and most of variables studied differ between groups, making it possible to differentiate with high sensitivity and specificity healthy corneas from those patients diagnosed with keratoconus.

4   Discussion

This study has been carried out with the aim of assessing the ability of a specific geometric model to capture diseases on human corneas, accounting only for geometrical data. Specifically, this computational study provides insight into the complex clinical problem of diagnosing corneal ectatic diseases.

It is well known that the mechanical response of any deformable system is affected by its geometry and material properties. When geometry is fully characterized, it is possible to set up a geometric model of the system, which may be used to analyze the geometric response under original conditions. In this case, conditions are defined by the rupture of the geometric balance due to the existence of a biomechanical weakening, as happens in the keratoconus disease. Keratoconus is a disorder characterized by a progressive thinning of the cornea, which is physically presented in its structure as a protrusion or cone type focal curving that entails a redistribution of its pachymetry and some changes in the anatomical morphology of its surfaces.

To date, there are devices that analyze topographically both corneal surfaces and allow an in vivo characterization of curvature changes, corneal pachymetry, etc. These parameters are used in the diagnosis of the disease and therefore could provide a characterization of the underlying morphogeometric alteration. However, the geometrical characterization indices proposed by these devices are not easily compatible between different tomographers, generating confusion in the Ophthalmic Community [24–27].

Other option is the geometric characterization based on raw data, which has been previously used by some authors in the Corneal Biomechanics field [22, 28] and for diagnosis of corneal diseases [29]. However, these studies resort to data interpolation to obtain a specific model for each patient.

Geometric modeling based on raw data that has not been treated by any internal algorithm of the topographer enables an accurate clinical characterization of the human cornea basing on perfectly defined morphological variables in the field of graphic bioengineering.

Furthermore, this study relies on the use of a reduced number of geometrical parameters obtained from modelling tests of the cornea: anterior corneal surface area, posterior corneal surface area, anterior apex deviation and posterior apex deviation. These variables are sufficient to prove that the variability of the geometric response of human corneas is definitely related with disease diagnosis. This method is simplified and more integrative than current diagnostic systems, which analyze separately the anterior and posterior surfaces of the cornea.

This observation has a quite relevant implication in view of the prediction of the response to refractive surgery, i.e. the knowledge of the sole geometry is enough to feed keratoconus diagnosis.

5 Conclusions

The main suggestion derived from this study is to give high priority to the development of non-invasive testing methods that are able to provide through inverse analysis the patient-specific parameters of a sufficiently realistic geometric model of the corneal morphology, which can be obtained with the aid of Computer Aided Geometric Design tools.

This method will allow improving the detection and effects of therapeutic methods used for keratoconus and other corneal ectatic diseases such as post-lasik ectasia. Early studies, currently under publication, have demonstrated the effectiveness of this approach in the early detection of subclinical keratoconus. In a close future, thanks to the analyses of the objective data related to the geometric effect of the intracorneal rings implanted, customized nomograms for the implantation of IntraCorneal Rings will be developed. Later, the analysis of the correlation between the geometric, visual, biomechanical and clinical effects of intracorneal implants in ectatic corneas will allow the development of new therapies and new concepts of corneal implants. The geometric modeling developed will also allow assessing more accurately the outcomes of the corneal crosslinking techniques and its effectiveness in slowing the development of keratoconus.

References

1. Pinero, D.P., Alio, J.L., Barraquer, R.I., Michael, R., Jimenez, R.: Corneal biomechanics, refraction, and corneal aberrometry in keratoconus: an integrated study. Invest. Ophthalmol. Vis. Sci. **51**, 1948–1955 (2010)
2. de Jong, T., Sheehan, M.T., Dubbelman, M., Koopmans, S.A., Jansonius, N.M.: Shape of the anterior cornea: comparison of height data from 4 corneal topographers. J. Cataract Refr. Surg. **39**, 1570–1580 (2013)
3. Klyce, S.D., Karon, M.D., Smolek, M.K.: Advantages and disadvantages of the Zernike expansion for representing wave aberration of the normal and aberrated eye. J. Refract. Surg. **20**, S537–S541 (2004)
4. Ramos-Lopez, D., Martinez-Finkelshtein, A., Castro-Luna, G.M., Pinero, D., Alio, J.L.: Placido-based indices of corneal irregularity. Optom. Vis. Sci. **88**, 1220–1231 (2011)
5. Trevino, J.P., Gómez-Correa, J.E., Iskander, D.R., Chávez-Cerda, S.: Zernike vs. Bessel circular functions in visual optics. Ophthal. Physl. Opt. **33**, 394–402 (2013)
6. Lenarduzzi, L.: Compression of corneal maps of curvature. Appl. Math. Comput. **252**, 77–87 (2015)

7. Ares, M., Royo, S.: Comparison of cubic B-spline and Zernike-fitting techniques in complex wavefront reconstruction. Appl. Opt. **45**, 6954–6964 (2006)
8. Gong, D.W., Chen, J.H., Yuan, C., Ge, R.K., Zhou, M.H.: A new method for reconstruction of corneal topography with Placido disk system. Adv. Mat. Res. **974**, 373–378 (2014)
9. Eklund, A., Dufort, P., Forsberg, D., LaConte, S.M.: Medical image processing on the GPU - past, present and future. Med. Image Anal. **17**, 1073–1094 (2013)
10. Sun, W., Darling, A., Starly, B., Nam, J.: Computer-aided tissue engineering: overview, scope and challenges. Biotechnol. Appl. Biochem. **39**, 29–47 (2004)
11. Farin, G.E., Hoschek, J., Kim, M.-S.: Handbook of Computer Aided Geometric Design. Elsevier, Amsterdam (2002)
12. Pottmann, H., Leopoldseder, S., Hofer, M., Steiner, T., Wang, W.: Industrial geometry: recent advances and applications in CAD. CAD Comput. Aided Des. **37**, 751–766 (2005)
13. Cui, J., Tang, M., Liu, H.: Dynamic shape representation for product modeling in conceptual design. Jisuanji Fuzhu Sheji Yu Tuxingxue Xuebao/J. Comput.-Aided Des. Comput. Graph. **26**, 1879–1885 (2014)
14. Lohfeld, S., Barron, V., McHugh, P.E.: Biomodels of bone: a review. Ann. Biomed. Eng. **33**, 1295–1311 (2005)
15. Almeida, H.A., Bártolo, P.J.: Computational technologies in tissue engineering. WIT Trans. Biomed. and Health **17**, 117–129 (2013)
16. Ovcharenko, E.A., Klyshnikov, K.U., Vlad, A.R., Sizova, I.N., Kokov, A.N., Nushtaev, D. V., Yuzhalin, A.E., Zhuravleva, I.U.: Computer-aided design of the human aortic root. Comput. Biol. Med. **54**, 109–115 (2014)
17. Chiang, I.-C., Shyh-Yuan, L., Ming-Chang, W., Sun, C.W., Jiang, C.P.: Finite element modelling of implant designs and cortical bone thickness on stress distribution in maxillary type IV bone. Comput. Methods Biomech. Biomed. Eng. **17**, 516–526 (2014)
18. Rocha, M., Pereira, J.P., De Castro, A.V.: 3D modeling mechanisms for educational resources in medical and health area. In: Proceedings of the 6th Iberian Conference on Information Systems and Technologies (CISTI 2011) (2011)
19. Schubert, C., van Langeveld, M.C., Donoso, L.A.: Innovations in 3D printing: a 3D overview from optics to organs. Br. J. Ophthalmol. **98**, 159–161 (2014)
20. Montalban, R., Alio, J.L., Javaloy, J., Pinero, D.P.: Correlation of anterior and posterior corneal shape in keratoconus. Cornea **32**, 916–921 (2013)
21. Montalban, R., Pinero, D.P., Javaloy, J., Alio, J.L.: Correlation of the corneal toricity between anterior and posterior corneal surfaces in the normal human eye. Cornea **32**, 791–798 (2013)
22. Ariza-Gracia, M.A., Zurita, J.F., Pinero, D.P., Rodriguez-Matas, J.F., Calvo, B.: Coupled biomechanical response of the cornea assessed by non-contact tonometry. Simulation study. PLoS One **10**, e0121486 (2015)
23. Cavas-Martinez, F., Fernandez-Pacheco, D.G., De la Cruz-Sanchez, E., Nieto Martinez, J., Fernandez Canavate, F.J., Vega-Estrada, A., Plaza-Puche, A.B., Alio, J.L.: Geometrical custom modeling of human cornea in vivo and its use for the diagnosis of corneal ectasia. PLoS ONE **9**, e110249 (2014)
24. Anayol, M.A., Guler, E., Yagci, R., Sekeroglu, M.A., Ylmazoglu, M., Trhs, H., Kulak, A.E., Ylmazbas, P.: Comparison of central corneal thickness, thinnest corneal thickness, anterior chamber depth, and simulated keratometry using galilei, Pentacam, and Sirius devices. Cornea **33**, 582–586 (2014)
25. Hernandez-Camarena, J.C., Chirinos-Saldana, P., Navas, A., Ramirez-Miranda, A., de la Mota, A., Jimenez-Corona, A., Graue-Hernindez, E.O.: Repeatability, reproducibility, and agreement between three different Scheimpflug systems in measuring corneal and anterior segment biometry. J. Refract. Surg. **30**, 616–621 (2014)

26. Savini, G., Carbonelli, M., Sbreglia, A., Barboni, P., Deluigi, G., Hoffer, K.J.: Comparison of anterior segment measurements by 3 Scheimpflug tomographers and 1 Placido corneal topographer. J. Cataract Refract. Surg. **37**, 1679–1685 (2011)
27. Shetty, R., Arora, V., Jayadev, C., Nuijts, R.M., Kumar, M., Puttaiah, N.K., Kummelil, M. K.: Repeatability and agreement of three Scheimpflug-based imaging systems for measuring anterior segment parameters in keratoconus. Invest. Ophthalmol. Vis. Sci. **55**, 5263–5268 (2014)
28. Simonini, I., Pandolfi, A.: Customized finite element modelling of the human cornea. PLoS ONE **10**, e0130426 (2015)
29. Ramos-Lopez, D., Martinez-Finkelshtein, A., Castro-Luna, G.M., Burguera-Gimenez, N., Vega-Estrada, A., Pinero, D., Alio, J.L.: Screening subclinical keratoconus with placido-based corneal indices. Optom. Vis. Sci. **90**, 335–343 (2013)

Virtual Surgical Planning for Mandibular Reconstruction: Improving the Fibula Bone Flap

Dolores Parras[1(✉)], Benito Ramos[2], Juan José Haro[2],
Manuel Acosta[2], Francisco Cavas-Martínez[1],
Francisco J.F. Cañavate[1], and Daniel G. Fernández-Pacheco[1]

[1] Graphic Expression Department,
Universidad Politécnica de Cartagena, Cartagena, Spain
{dolores.parras,francisco.cavas,
francisco.canavate,daniel.garcia}@upct.es
[2] Oral and Maxillofacial Surgery Department,
Hospital Universitario Santa Lucía, Cartagena, Spain
drramosmaxilo@gmail.com, jujohl@hotmail.com,
macostaf@hotmail.es

Abstract. Mandibular reconstruction is still a challenge in modern maxillofacial surgery, using frequently a free fibula flap for its reconstruction. Fibula in-setting has traditionally relied on manual skills, making the results extremely dependent on the surgeon's expertise. To solve this problem, this study proposes the use of virtual surgical planning for mandibular reconstruction from a fibula bone, comprising the following four stages: (i) a scanning phase where both mandible and fibula of patient are scanned by computed axial tomography, (ii) a second phase of study where the osteotomies to be performed both in the mandible and the fibula bone are positioned, and where the guide elements are designed to be used in surgery, (iii) a 3D printing phase where the designed elements will be generated, and (iv) a surgical phase where the surgery is performed. This method permits more accurate and predictable results with an adequate bone-to-bone contact, and reduces costs by shortening the operation, decreasing the ischemia time, and reducing the postoperative complications.

Keywords: 3D reconstruction · Stereolithographic model · Virtual surgical planning

1 Introduction

According to the International Association of Oral & Maxillofacial Surgeons (IAOMS) oral and maxillofacial surgery can be defined as a specialty of medicine dedicated to the diagnosis, surgery and treatment of congenital and acquired diseases of the skull, face, head and neck, oral cavity and jaws (including dentition), which pathology may be tumoral traumatic, degeneration or aging related.

Mandibular reconstruction is still a challenge in modern maxillofacial surgery. Functional and aesthetic requirements are very high and clearly influenced by

© Springer International Publishing AG 2017
I. Rojas and F. Ortuño (Eds.): IWBBIO 2017, Part I, LNBI 10208, pp. 282–291, 2017.
DOI: 10.1007/978-3-319-56148-6_24

reconstructive techniques. In 1989, Hidalgo [1] introduced the free fibula flap for mandibular reconstruction. This flap is especially important for wide mandibular defects because it provides a long bone with a long vascular pedicle and allows incorporation of muscle and skin paddles via the vascular perforators [2].

For this reconstruction procedure, fibula in-setting has traditionally relied on manual skills and surgical expertise to reproduce the curves and shape of the mandible from a straight bone, making the results extremely dependent on the surgeon's skill. This fact has led to apply new improvements in computer software and perform a virtual surgical planning before the surgical process [3, 4].

Virtual surgical planning (VSP) using computed tomographic imaging, computer-aided design and manufacturing technology, allows surgeons to perform virtual surgery, generating templates and cutting guides that will be used for the precise and reliable recreation of the plan in the operating room [5].

The concept of virtual surgery uses surgical simulation rather than relying exclusively on intraoperative manual approximation of mandibular reconstruction. There are studies evaluating the degree to which surgical outcomes in free fibula mandibular reconstructions planned with virtual surgery and carried out with prefabricated surgical plate templates and cutting guides correlates to the virtual surgical plan with good results [6]. Other studies also evaluate the accuracy of mandibular reconstruction and in both virtual planning versus conventional surgery patients.

This study demonstrates that virtual surgical planning permits to achieve a more accurate mandibular reconstruction than conventional surgery. The use of prefabricated cutting guides and plates makes fibula flap moulding and placement easier, minimizing the operating time, and improving clinical outcomes [7].

2 Materials and Methods

VSP is proposed in this study for a mandibular reconstruction from a fibula bone, comprising the following four stages: (i) a scanning phase where both mandible and fibula of patient are scanned by computed axial tomography (CAT), (ii) a second phase of study where the osteotomies to be performed both in the mandible and the fibula bone are positioned, and where the guide elements are designed to be used in surgery, (iii) a 3D printing phase where the designed elements will be generated, and (iv) a surgical phase where the surgery is performed. These steps are described in more detail below.

2.1 Scanning Phase

Imaging diagnosis allows doctors to look inside the human body to search for evidences about a medical condition. There is a great variety of devices and techniques that can create images for the diagnosis of pathologies or as a preventive procedure (Table 1).

Table 1. Some technologies for the acquisition of medical images.

Name	Emitter	Image	Radiation
Radiography	X-rays	2D	Yes
Ecography	Ultrasounds	3D	No
Magnetic Resonance Imaging (MRI)	Magnetic waves	3D	No
Gammagraphy	Radioactive tracer that emits energy (gamma rays) from inside the body	2D	Yes
Positron Emission Tomography (PET)	Radioactive tracer that emits energy (gamma photon) from inside the body	3D	Yes
Photon Emission Computed Tomography (SPECT)	Radioactive tracer that emits energy (gamma rays) from inside the body	3D	Yes
Computed Axial Tomography (CT or CAT)	Multiple X-ray beams released simultaneously from several angles	3D	Yes
Cone Beam Computed Tomography (CBCT)	X-ray cone beam	3D	Yes

In the present study, a computed axial tomography (CAT) was used for the imaging and subsequent three-dimensional reconstruction of the mandible. Specifically, the Somatom Definition AS+ equipment was used for this task. Helical or axial tomography is a medical diagnostic method to get images from inside the human body, obtained through the use of X-rays in the form of transversal millimetric slices, with the aim of studying in detail the entire body, from head to toe. In an X-ray conventional study, the beam of radiation is emitted in a diffuse way, but in the CAT the beam is directed and has a determined thickness that can vary from 0.5 mm to 20 mm, depending on the size of the structure to study. Multiplanar advanced reconstructions and 3D reconstructions can be performed with a high-definition CAT. In addition, the vascularization (arteries and veins) of the organs and their lesions can be evaluated using intravenous contrast.

This system is an excellent non-invasive imaging diagnostic option, where high fidelity images are provided at low cost, the patient is exposed to a minimal radiation, the wide range of gray tones that are handled allows to identify with precision the different tissues and, finally, the dosage of the contrast medium can be controlled by computer.

For the imaging of the leg a computed tomography angiography (CTA) was performed by using the Somatom Definition AS+ equipment (Fig. 1). CTA is used to produce detailed images of blood vessels and tissues in various parts of the body. An iodine-rich contrast material is needed to inject the patient through a small catheter placed in the vein of an arm. Then, a CTA is performed while the contrast is flowing through the blood vessels to the different organs of the body. In the case of the fibula vascularized free flap, it is very important to determine the presence of a normal vascular pattern of the lower extremity. Some variations of the normal pattern, such us bilateral

Fig. 1. Computed tomography angiography of the patient's legs

aplastic tibial anterior artery of peroneal magna artery, if misdiagnosed, may lead to catastrophic consequences if the peroneal artery is harvested with the flap, with ischemia and even necrosis of the lower extremity.

2.2 Study Phase

Once the virtual models of the mandible and fibula reconstructions have been obtained, the study phase is executed and the osteotomies that will be performed in both the mandible and the fibula bone are carefully positioned. This phase is a key point because the guide elements that are designed here will be later used in the surgery, once they have been generated by 3D printing.

The first step is to position the osteotomies in the mandible where the part to be extracted is calculated by designing the guide pieces that will be printed in 3D for their use in the surgery (Fig. 2). Subsequently, a series of cuts are simulated within the 3D model of the fibula. This cuts will allow the construction of a new structure that would occupy the space extracted from the old mandible (Fig. 3).

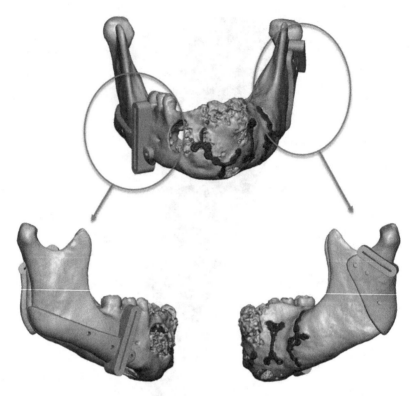

Fig. 2. Virtual surgical planning: osteotomy of the mandible.

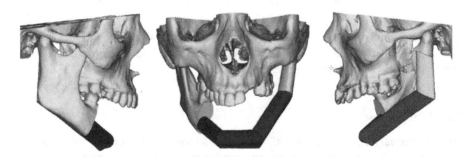

Fig. 3. Virtual surgical planning: reconstruction from fibula bone.

2.3 3D Printing Phase

Once the study of the osteotomies has been performed and the cutting guides that will be used during the surgery have been designed, the next step is to generate these elements by 3D printing. Nowadays, there are different prototyping technologies that allow starting, deforming, melting and joining the base material of a component, and for a few years now, material can also be deposited wherever it is needed.

Current additive manufacturing technologies are based on the principle of dispersion-accumulation. Material and additives attribution processes are those which solidify a material, originally in solid, liquid or powdery state, by successive layers within a predetermined space and by means of electronic procedures. These methods are also known by the acronym of MIM (Material Increase Manufacturing) and can be classified according to two different factors: the starting material and the model obtaining process (Table 2).

Table 2. Additive manufacturing processes.

Name	Tool	Process	Material
Stereolithography (SLA)	Ultraviolet radiation from a laser	Solidification	Liquid
Solid Object Ultraviolet Laser Printer (SOUP)			
Solid Ground Curing (SGC)	Ultraviolet-light lamp		
Inkjet rapid prototyping or Poly jet			
Fused Deposition Modeling (FDM)	Extrusion head	Smelting/solidified	Solid
Headform injection	Injection head		
Laminated Object Manufacturing (LOM) – Selective Deposition Lamination (SDL)	Lamination	Carving and adhesion	
3D printer	Binder	Join	Powder
Selective Laser Sintering (SLS-DMLS)	Laser	Sinter	

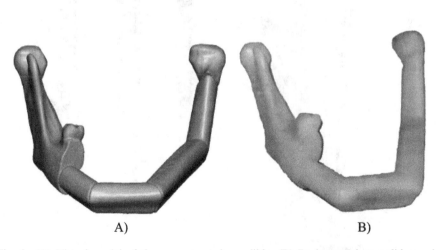

A)   B)

Fig. 4. (A) Virtual model of the reconstructed mandible. (B) Resin model stereolithography used to prebend the reconstruction plate.

Fig. 5. Reconstruction plate of titanium that is blended according to fibula bone segments.

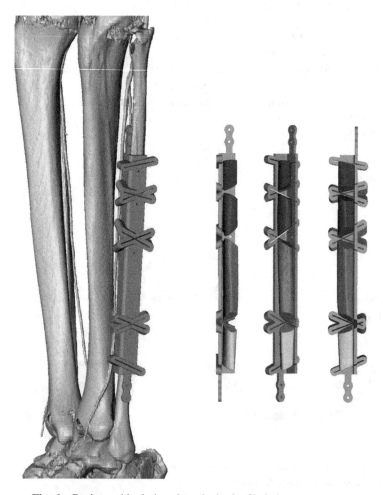

Fig. 6. Cutting guide designed to obtain the fibula bone segments.

The 3D printing technique chosen for this study to manufacture the reconstructed mandibular model has been stereolithography. Data obtained by the CAT are stored in Digital Imaging and Communications in Medicine (DICOM) format and processed by the program Somaris 7 Syngo CT 2012B, VA44A version. The images are converted into STL format where the virtual model becomes a triangulated surface that improves the quality of the image. Stereolithography consists of solidifying the resin, which is in liquid state, with the projection of a laser. This resin is a photopolymer that solidifies when is exposed to ultraviolet radiation. In this way, once a section has been solidified, the platform lowers to solidify the next layer, finally obtaining the complete model.

With the prototype of the mandible reconstructed from the virtual model (Fig. 4), the titanium reconstruction plate (Fig. 5) can be prebent, in advance of surgery, for alignment and fixation of the bone segments that will be extracted from the fibula thanks to the cutting guide previously generated in the study phase (Fig. 6). This process reduces time and the potencial complications that may arise during surgery.

2.4 Surgical Phase

Before the operation, the reconstruction plate was prebent using the stereolithographically reconstructed mandible and sterilized together with the cutting guides. The surgery was performed by two teams. A bilateral cervical approach was used to expose the mandible, which was resected using the mandible cutting guides. They were positioned by screwing to the native mandible, and guided the osteotomies made with a reciprocating saw. The right facial artery and vein were dissected up to the mandibular body to reach the fibula pedicle.

Simultaneously, the left fibula was dissected using a lateral approach, with 6 cm of bone preserved proximally and distally for a harvest of the maximum bone possible and a vascular pedicle 21 cm long, with 1.5 and 3 mm of the caliber for the peroneal artery and the vein, respectively.

With the fibula still attached to the vascular pedicle, its cutting guide was secured with unicortical screws, and the osteotomies were performed, with care taken to protect the vascular pedicle. The prebent reconstruction plate was screwed to the fibula before ligation of the pedicle. The fibula, with the osteotomies and reconstruction plate, was fixed to the native mandible, and the vascular anastomoses to the facial vessels were performed under a microscope.

3 Results

No postoperative complications occurred. The patient had an optimal aesthetic profile that will improve with prosthetic dental rehabilitation. She presented a maximal mouth opening of 32 mm with no temporomandibular symptoms, and an assessment performed 6 months after the surgery confirmed bone union (Fig. 7).

Reproduction of the virtual surgical plan by using the surgical guides, permitted to transfer the VSP to the actual procedure, shortening the operating time while increasing the accuracy of the surgery [6, 8–12]. Differences between the virtual surgical planning and the postoperative values were measured (Table 3).

Fig. 7. TAC of the reconstructed mandible.

Table 3. Differences between virtual surgical planning (VSP) and postoperative values.

	VSP	Postoperative
Length of fibula fragments (mm)	2.70	2.63
	3.72	3.52
	6.70	6.70
	3.42	3.41
Angle mandibule (0)	129	129
Bone-to-bone contact	Excellent	Excellent

4   Conclusions

Mandibular reconstruction today is still one of the largest challenges in cranio maxillofacial surgery. Quality of life can be extremely reduced by tissue defects in the mandible region, whether caused by trauma or tumour. It is a main concern of all cranio maxillofacial surgeons to improve mandibular reconstruction outcomes. Over the last few years, rapid progress in CAD/CAM techniques has become a reality, making possible to perform patient specific reconstructions.

In this aspect, virtual surgical planning permits to obtain stereolithographic models and surgical guides that facilitate moulding and placement of fibula flap, obtaining more accurate and predictable results with an adequate bone-to-bone contact, minimizing the operating time, decreasing the ischemia time, and reducing the postoperative complications.

References

1. Hidalgo, D.A.: Fibula free flap. A new method of mandible reconstruction. Plast. Reconstr. Surg. **84**, 71–79 (1989)
2. Moro, A., Cannas, R., Boniello, R., Gasparini, G., Pelo, S.: Techniques on modeling the vascularized free fibula flap in mandibular reconstruction. J. Craniofac. Surg. **20**, 1571–1573 (2009)
3. Eckardt, A., Swennen, G.R.J.: Virtual planning of composite mandibular reconstruction with free fibula bone graft. J. Craniofac. Surg. **16**, 1137–1140 (2005)
4. Hirsch, D.L., Garfein, E.S., Christensen, A.M., Weimer, K.A., Saddeh, P.B., Levine, J.P.: Use of computer-aided design and computer-aided manufacturing to produce orthognathically ideal surgical outcomes: a paradigm shift in head and neck reconstruction. J. Oral Maxillofac. Surg. **67**, 2115–2122 (2009)
5. Sharaf, B., Levine, J.P., Hirsch, D.L., Bastidas, J.A., Schiff, B.A., Garfein, E.S.: Importance of computer-aided design and manufacturing technology in the multidisciplinary approach to head and neck reconstruction. J. Craniofac. Surg. **21**, 1277–1280 (2010)
6. Roser, S.M., Ramachandra, S., Blair, H., Grist, W., Carlson, G.W., Christensen, A.M., Weimer, K.A., Steed, M.B.: The accuracy of virtual surgical planning in free fibula mandibular reconstruction: comparison of planned and final results. J. Oral Maxillofac. Surg. **68**, 2824–2832 (2010)
7. Wang, Y.Y., Zhang, H.Q., Fan, S., Zhang, D.M., Huang, Z.Q., Chen, W.L., Ye, J.T., Li, J.S.: Mandibular reconstruction with the vascularized fibula flap: comparison of virtual planning surgery and conventional surgery. Int. J. Oral Maxillofac. Surg. **45**, 1400–1405 (2016)
8. Antony, A.K., Chen, W.F., Kolokythas, A., Weimer, K.A., Cohen, M.N.: Use of virtual surgery and stereolithography-guided osteotomy for mandibular reconstruction with the free fibula. Plast. Reconstr. Surg. **128**, 1080–1084 (2011)
9. Foley, B.D., Thayer, W.P., Honeybrook, A., McKenna, S., Press, S.: Mandibular reconstruction using computer-aided design and computer-aided manufacturing: an analysis of surgical results. J. Oral Maxillofac. Surg. **71**, e111–e119 (2013)
10. Shen, Y., Sun, J., Li, J., Ji, T., Li, M.M., Huang, W., Hu, M.: Using computer simulation and stereomodel for accurate mandibular reconstruction with vascularized iliac crest flap. Oral Surg. Oral Med. Oral Pathol. Oral Radiol. **114**, 175–182 (2012)
11. Wang, W.H., Zhu, J., Deng, J.Y., Xia, B., Xu, B.: Three-dimensional virtual technology in reconstruction of mandibular defect including condyle using double-barrel vascularized fibula flap. J. Cranio-Maxillofac. Surg. **41**, 417–422 (2013)
12. Zheng, G.S., Su, Y.X., Liao, G.Q., Chen, Z.F., Wang, L., Jiao, P.F., Liu, H.C., Zhong, Y.Q., Zhang, T.H., Liang, Y.J.: Mandible reconstruction assisted by preoperative virtual surgical simulation. Oral Surg. Oral Med. Oral Pathol. Oral Radiol. **113**, 604–611 (2012)

Elbow Orthosis for Tremor Suppression – A Torque Based Input Case

Gil Herrnstadt and Carlo Menon[(✉)]

MENRVA Lab, School of Engineering Science,
Simon Fraser University, Burnaby, BC, Canada
cmenon@sfu.ca

Abstract. In the endeavor to offer alternative treatments and solutions to patients with pathological tremor, robotic technologies are at the forefront, and represent a noninvasive potential solution to the needs of tremor patients.

We developed a wearable robotic device having one Degree of Freedom, anthropomorphic to the human elbow. The device control architecture involves a speed controlled voluntary driven suppression approach whereby the robotic device tracks the voluntary motion. The tremor motion instead is considered a disturbance and is rejected.

A second motor is attached to the orthosis in order to provide an input simulating the human voluntary and tremor motions. In this work, the human input to the robotic suppression device follows a torque profile. It is demonstrated that the suppression approach can successfully attenuate the tremor while following the voluntary motion component. However, it is also shown that a lag in the voluntary motion ensues.

The experimental testing with the device resulted in a 99.4% reduction of the tremor power and 0.34% reduction to the voluntary component power.

An analysis of the cause of the voluntary component delay is performed and associated hardware limitations are suggested. Finally, we consider whether the lag observed in the experiments, would be expected to also be observed if tests with human subjects are performed, as well as possible future solutions.

1 Introduction

Tremor is a relatively common neurological movement disorder [1]. Yet, perhaps due to its non-life threatening character, tremor seems to be underrepresented in the literature. Parkinson's disease is perhaps the most recognized of tremor related disorders, nevertheless, multiple other tremor conditions exist [2].

The movement disorder society has issued a statement addressing the classification of tremor [3]. Tremor is identified primarily as either rest or action tremor. Each tremor syndrome is typically characterized by a particular tremor frequency band, and activation condition. Most tremor conditions manifest at frequencies distinct from voluntary motion frequencies. Normal motion frequency in activities of daily living is considered to be below 2 Hz, while tremor frequency, for the most part, resides in the 3–12 Hz range. The aforementioned inherent frequency divide is key to the successful

© Springer International Publishing AG 2017
I. Rojas and F. Ortuño (Eds.): IWBBIO 2017, Part I, LNBI 10208, pp. 292–302, 2017.
DOI: 10.1007/978-3-319-56148-6_25

signal processing of movements involving tremor. Furthermore, devices designed to suppress tremor often rely on the notion of the frequency separation to attenuate the tremor while avoiding obstructing the voluntary motion [4, 5].

Several upper limb devices for tremor suppression have been proposed in the literature and tested with a simulated input or tremor subjects. Rosen et al. designed the Controlled Energy Dissipating Orthosis (CEDO) prototype for the suppression of intention tremors [4]. The CEDO was mounted to a chair or table and provided velocity proportional resistance using magnetic particle brakes. A forearm brace interfaced between the user and the CEDO. The device allowed the user's motion in the horizontal plane, suitable for table top activities. A suppression device based on a manually tunable vibration absorber, consisting of a mass-spring-damper system, was proposed by Hashemi et al. [6]. A biomechanical model of the shoulder (flexion/extension only) and elbow joints was developed. Numerical simulation and experimental testing, using a dc motor, were carried out and achieved a tremor attenuation of above 80%. A wrist actuator orthosis design for flexion/extension was developed by Loureiro et al. employing viscous damping with magneto-rheological fluids [7]. Rocon et al. developed the WOTAS, a wearable device that follows the kinematic structure of the arm targeting the elbow, wrist pronation/supination and wrist flexion/extension [5]. The WOTAS utilized angular velocity and force sensors with dc motors to provide the tremor suppression. Testing with tremor subjects was conducted and a tremor reduction between 40–80% was reported for the elbow joint. A more recent system was developed that both simulates the human arm dynamics and the suppressive orthosis for the wrist joint [8]. An impedance suppression approach was implemented which adapted to the tremor fundamental frequency. A tremor attenuation of 98.1% and 74.3% for the fundamental and second tremor harmonic frequencies respectively was reported. Multiple systems utilizing Functional Electrical Stimulation have also been proposed in the literature, typically performing an anti-phase stimulation of the antagonist muscles, with varying degree of effectiveness [9, 10].

When evaluating a potential tremor suppression device, it is advantageous to simulate the human input [6, 8]. Performing simulations can lead to improved system performance prior to human testing.

In this work, tremor suppression results for an elbow joint motion that follows a simulated human torque profile, are presented. This expands on a previous work where the simulated human motion followed a velocity profile. Human motion has been studied in the neuroscience community in an attempt to understand and suggest possible models for human motor control. Despite these efforts, how movement is controlled, from a neurological perspective, in human motion is still not fully understood [11]. There is evidence for the existence of both kinematic and dynamic models in the human motion control [12].

The organization of this manuscript is as follows. Section 2 reviews the developed system, the control structure, and the testing procedure. Sections 3, 4 and 5 present the Results, Discussion and Conclusion respectively.

2   Materials and Methods

2.1   Suppression Orthosis and Control Approach

The elbow orthosis that was developed is composed of a Suppression Motor (SM), a gearbox, a torque sensor and the orthosis frame, as shown in Fig. 1. The orthosis is donned on the upper limb targeting the elbow joint. This work focuses on experimental testing in which the human input was simulated by an additional Driving Motor (DM) with a gearbox. The suppression approach aims to drive the orthosis only with respect to the voluntary motion. Therefore, the voluntary component of the human motion is separated from the total motion. Moreover, the controller, while following the voluntary motion, must reject any disturbances – in this work, the tremor motion is considered a disturbance. Further information and details relevant to the orthosis and control approach implemented are available in [13, 14].

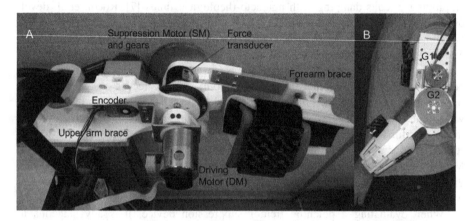

Fig. 1. (A) The suppression system testing assembly with a DM simulating the human input. (B) The suppression orthosis donned.

2.2   Testing Procedure

In this work we provided a torque profile input to the suppression system. Note that in a previous work [14], we demonstrated the suppression device performance when the human input followed a velocity profile. A human user of the proposed system may, when using the system, provide an input that is constructed of either a velocity profile, a torque profile or of a combination of the two kinds of input profiles. We tested the two kind of inputs in order to evaluate the system performance with either input type. As mentioned, the DM was used to simulate the human input. The DM motion profile was constructed based on data collected from tremor patients [15]. It should be noted, for the velocity input case presented in [14], the orthosis forearm gravity component was compensated by the orthosis, whereas in the torque driven input case presented here, the human compensated for the orthosis forearm gravity.

3 Results

A torque (current) driven input is considered as an input provided by the user that follows a torque profile. The torque input was constructed from data of a PD patient (park05). The original raw data and its Power Spectral Density (PSD) plot are shown in Fig. 2. The tremor data from the patient has a first harmonic tremor frequency at 4.5 Hz, which was superimposed with a sinusoidal torque signal having a frequency of 0.8 Hz and amplitude of 0.3 Nm, representing the voluntary motion. The DM (user) performed the gravity compensation for the orthosis arm; it is assumed a user would be able to carry the gravity load imposed by the orthosis arm. The control implemented is as demonstrated in Fig. 3.

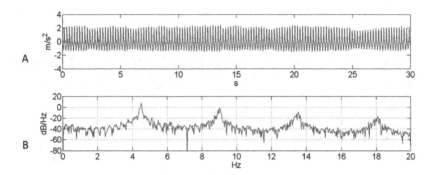

Fig. 2. Raw data from a PD patient (park05). (A) Linear acceleration. (B) PSD

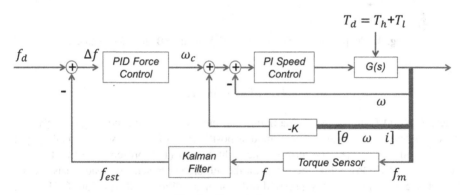

Fig. 3. Control diagram

The intentional velocity associated with the input torque signal was acquired in order to serve as a reference for the suppression-on case. A velocity signal was recorded while the DM was physically disengaged from the SM and with the robot arm attached, representing the unsuppressed human motion, containing the voluntary and tremor components. Consequently, zero phase filtering of the suppression-off velocity resulted in the voluntary velocity component that would serve as the reference signal. When activated, the suppression system should track the filtered velocity signal associated with the suppression-off torque signal.

The velocity tracking and interaction forces obtained are shown in Fig. 4. A delay of approximately 0.5 s is observed in the velocity-tracking signal. The delay is explained by the fact that the torque driven case does not define a specific velocity profile. Rather, the associated velocity profile amplitude and phase depend on the load, in this case imposed by the SM. To demonstrate the aforementioned and to analyze the delay, a model for the system is considered composed of an ideal torque generator, representing the DM, and with inertial and damping loads resisting the motion. The system model is defined as

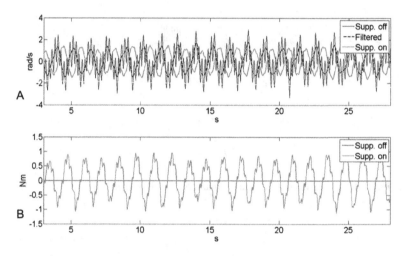

Fig. 4. Suppression system results. (A) Velocity. (B) Interaction force.

$$J\ddot{y} + B\dot{y} = L\sin(\lambda t) \tag{1}$$

where J, B and L are the inertia, damping, and input torque amplitude respectively. y is the dependent variable (in our case angular position) and $\lambda = 2\pi f$ where f is the input signal frequency in Hz. The term on the right side in (1) represents the input torque generated by the DM, while the terms on the left side represent the loads resisting the DM. The expression in (1) is a second order ordinary differential equation (ODE). The general solution to the ODE is composed of complimentary and a particular solutions having the following forms

$$\begin{aligned} y &= y_c + y_p \\ y_c &= C_1 + C_2 e^{-\frac{B}{J}t} \\ y_p &= \alpha\cos(\lambda t) + \beta\sin(\lambda t) \end{aligned} \tag{2}$$

where C_1, C_2 are constants of the complementary solution y_c and t is the independent variable (time in our case). The particular solution y_p is a summation of a sine and cosine where α and β are functions of J, B, λ and L as follows

$$\alpha = -\frac{LB}{\lambda(J^2\lambda^2 + B^2)}$$
$$\beta = \frac{J\lambda}{B}\alpha \tag{3}$$

The particular solution can also be expressed as a sine function with a phase shift as follows

$$\alpha\cos(\lambda t) + \beta\sin(\lambda t) = \gamma\sin(\lambda t + \phi) \tag{4}$$

where γ and ϕ are defined as

$$\gamma^2 = \alpha^2 + \beta^2$$
$$\phi = \tan^{-1}\left(\frac{\alpha}{\beta}\right) \tag{5}$$

Substituting (3) into the expression for ϕ in (5) results in the following

$$\phi = \tan^{-1}\left(\frac{B}{J\lambda}\right) \tag{6}$$

The expression in (6) describes the phase shift as a function of the load inertia and damping as well as of λ. The complimentary solution does not contribute to a phase change; instead the exponential in y_c is responsible to the decay of the signal. The above analysis provides a rationalization for the delay observed in the velocity tracking of the torque driven case. Furthermore, it suggests the delay is not directly related to the suppression system but rather is linked to how the DM is affected by any resistive loads acting on it.

To validate the above model (1), the relationship between J and B From (6) is explicitly derived as a linear relationship where B is a function of J as follows

$$B = J\lambda\tan\phi \tag{7}$$

We can obtain the value for $\lambda = 5.0265$ rad/s from our experiment input frequency ($f = 0.8$ Hz). The time delay, and therefore the phase shift ϕ, can be obtained from the experiment results. Next we find a value for the load inertia applied to the DM by the orthosis system. The expressions for the total inertia, reflected at the DM shaft, for the orthosis suppression-on and -off system configurations are calculated as

$$J_{ON} = J_m + J_{gdm} + \frac{\left(\left(\left(J_m + J_{gsm}\right)K^2_{gsm} + J_{G1}\right)K^2_{ge} + J_{G2} + J_l\right)}{K^2_{gdm}}$$
$$J_{OFF} = J_m + J_{gdm} + \frac{J_l}{K^2_{gdm}} \tag{8}$$

where $J_l, J_m, J_{gsm}, J_{gdm}$ are the orthosis arm, the SM (or DM), SM gearbox, and the DM gearbox moment of inertias respectively. $J_{Gi}, i = 1, 2$ are the external gears moment of inertias, and K_{gsm}, K_{gdm}, K_{ge} are the SM gearbox, DM gearbox, and external gear reductions respectively. Note that the free tremor motion (suppression-off) was recorded while the DM was connected to the robot arm. Substituting the orthosis system parameter values in (8) results in a total moment of inertia of $J_{OFF} = 3.2e - 05$ Kgm^2 and $J_{ON} = 1.136e - 04$ Kgm^2 for the suppression-on and suppression-off cases respectively. For the phase values (ϕ) we look at the phase shift between the input current (A) and the output position (rad). The time delays for the suppression-off and -on cases are 0.76 s and 1.32 s respectively, which converts to $\phi_{OFF} = 3.7824$ and $\phi_{ON} = 6.6978$ (rad).

Consequently, substituting the above calculated inertia values from (8) and the values for λ and ϕ_{OFF}, ϕ_{ON} in (7), we can find the respective damping coefficients $B_{OFF} = 1.2e - 04$ Nm/(rad/s) and $B_{ON} = 2.514e - 04$ Nm/(rad/s).

We can graphically analyse the delay observed in Fig. 4 by reviewing the Bode phase plot between the armature current and output position of the DM. The following transfer function describes the relationship between the DM input current and the output position

$$\frac{\theta(s)}{I(s)} = \frac{K_{gdm}K_t}{s(Js + B)} \qquad (9)$$

where $\theta(s), I(s)$ and K_t are the output rotational position, the input current, and the DM torque constant respectively. The expression in (9) can be written in the time domain as a function of the angular position as follows:

$$J\ddot{\theta} + B\dot{\theta} = K_{gdm}T_m \qquad (10)$$

It is interesting to note the expression in (10) is equivalent to that in (1), with the only difference being the term on the right side. In (1), the right side term is expressed as a sinusoid multiplied by a constant, while in (10) it is expressed simply as the motor torque parameter T_m multiplied by a constant. Furthermore, we consider the following DM transfer function (from DM input voltage to output position)

$$\frac{\theta(s)}{V(s)} = \frac{K_{gdm}K_t}{s\left(JLs^2 + JRs + BR + K_eK_{gdm}^2K_t\right)} \qquad (11)$$

where $V(s), L, R$ and K_e are the armature voltage, armature inductance, armature resistance, and the back EMF constant.

The Bode phase plots of the two DM transfer functions, (9) and (11), are shown in Fig. 5, where the former is represented by the red lines (torque driven case), and the latter is represented by the blue lines (velocity driven input case). Each case is plotted for two values of inertia and damping, namely suppression-off (solid line) and suppression-on (dashed line). In order to show both the suppression-on and -off cases on the Bode plot, the ϕ_{ON} was phase shifted by $-\pi$. Phase shifting does not affect the inertia and damping values calculations, and allows to see both angles within the bode plot phase range of $\pi/2$ rad.

The phase shift, between the suppression-on and -off that was calculated from the experimental results was $(6.6978 - \pi) - 3.7824 = -0.2262$ rad. The same phase shift is seen in Fig. 5 at the 0.8 Hz frequency between the suppression-on phase $(-2.73 + 2\pi = 3.55$ rad) and the suppression-off phase $(-2.5 + 2\pi = 3.78$ rad). It is interesting to note, at the 0.8 Hz frequency, minimal phase change is observed for the velocity driven case between the suppression-on and -off.

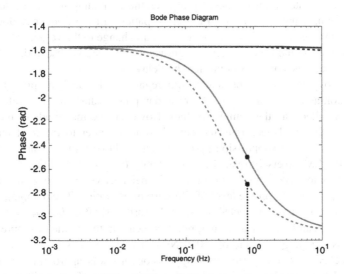

Fig. 5. Phase response of open loop torque (red lines) and velocity (blue lines) driven motor. The marked black squares and the vertical dashed line indicate the 0.8 Hz frequency intersection with the phase lines. (Color figure online)

The PSD of the suppression-off and suppression-on signals were compared and are shown in Fig. 6. A clear power reduction can be observed for the tremor frequencies (above 2 Hz), while in the vicinity of the intentional motion frequency (0.8 Hz) the signals overlap closely, indicating a small impact to the voluntary component power. The tremor power reduction for the 1st and 2nd harmonics was 99.4 and 90.4%

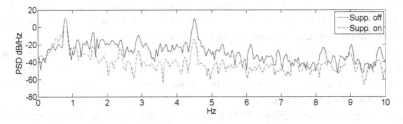

Fig. 6. PSD of suppression-on and -off cases

respectively and 99.4% when combined, while the change to the intentional component power was 0.34%.

The lag in the velocity signal was corrected in order to calculate RMS errors. The calculated RMS values for the velocity tracking error and for the interaction force signals were 0.3 rad/s and 0.55 Nm respectively.

4 Discussion

We defined and tested a torque based profile as the user input to the developed suppression system, using data from a PD tremor patient. In the frequency domain, the result show significant tremor reduction and a small change to the voluntary component power. However, in the time domain, a significant phase lag is observed. In this work we offered an analysis and reasoning for the delay.

Several observations can be made with regard to the analysis provided in the Results section and the value obtained for the damping coefficients. The values for the inertias are smaller than the damping values. However, the magnitude of the acceleration and velocity will determine the contribution of either to the total load force. Additionally, when the suppression system is active, large torques (currents) may be applied by the SM inversely related to the velocity (as revealed in Fig. 4), in a comparable manner to viscous loads. From the DM perspective, the above may represent variable damping, based on the level of SM current activation. The damping value for the suppression-on can be reflected at the DM gearbox shaft as $B_{ONgdm} = 0.069$ Nm/ (rad/s). B_{ONgdm} can be physically interpreted as meaning that while in suppression-on, the system is applying a torque of 0.069 Nm for every 1 rad/s of velocity. The top velocity reached in our experiments (suppression on) was approximately 1.5 rad/s, which would have resulted in a maximum damping force of approximately 0.1 Nm applied to the DM; well within a range that the suppression system is capable of, and in fact, within the measured interaction force values, as seen in Fig. 4 B.

We set out to explore the torque driven case performance and the velocity signal delay observed in Fig. 4 A. We have shown in (1) and in (10) in the Results section, a model to explain the delay. It should be noted, the inertia and damping loads in (1) can exist unrelated to the suppression system (i.e. by any other mechanical element), therefore it is suggested the observed phase shift is attributed primarily to the DM. It is difficult to determine with certainty whether this phenomenon would carry over when testing with humans. It is, however, deemed to be unlikely as, in contrast to the DM (when driven in torque), the human motion involves feedback loops [16, 17] which can help regulate the desired motion. Furthermore, there is evidence that motor control for a body movement from one point to another follows a bell-shaped velocity profile [18, 19], which might suggest the control goal to be velocity related.

A justification for testing in torque mode emerges from the lack of certainty as to how a human motion is generated. It may be that depending on the activity and circumstances, the input from the user will follow a position, velocity, a torque profile or a combination of the above.

Comparing the torque driven PSD results from this work to those obtained in a previous work, involving a velocity driven input [14], a notable difference is observed.

Specifically, a lower attenuation of 90% was achieved for the 2nd tremor harmonic in the torque driven case as opposed to 99.1% in the velocity driven case. It is noted however, that the 2nd tremor harmonic was already relatively reduced in the suppression-off case (peak was approximately 5 dB less than in the velocity driven case) leaving less power to be attenuated. Furthermore, the power contained in the 2nd harmonic formed a relatively small portion of the total tremor power contained in the first two harmonics (0.3% in the torque driven case compared to 0.55% in the velocity driven case). Unlike the velocity driven case, the orthosis arm and mechanical joint were connected to the DM in the torque driven suppression-off case. Consequently, the 2nd harmonic relatively low power may be due to the motion-inhibiting effect of the orthosis arm inertia in combination with the joint friction.

As far as compensating for the phase shift, two approaches are considered, namely compensation by the suppression system or by the DM. The load parameters appear inside the DM transfer functions (velocity or torque driven), which suggests it may not be possible to compensate directly with the DM. Compensation then, should be done by the suppression system such that the DM transfer function parameters (inertia and damping) are unchanged between the unloaded (suppression-off) and loaded (suppression-on) cases. However, to successfully compensate, the SM may need to be controlled in torque. Compensation by the suppression system may be achieved by adding elements that cancel out with the inertia and damping in (1). In practice, this entails using velocity and acceleration feedback loops to provide an equivalent torque command $(-J\ddot{y} - B\dot{y})$ to the left side of (1). In our application however, the suppression system is driven in velocity and thus implementing an accurate torque command for the compensation would not be feasible.

It is interesting to note that a dc motor model, as in (11), has a feedback loop (back EMF) when controlled in voltage, while in the case of the current control, as in (9), it does not. The lack of feedback in the torque driven case may be a contributing factor to the lag that was observed, as is also evident in the Bode plot of Fig. 5.

Finally, in future studies it would be interesting to examine if the delay observed in this work is repeated when testing with tremor participants.

5   Conclusion

This work involved an experimental testing of a suppression system with a dc motor simulating the human input. We explored a human input to the suppression system, corresponding to a torque profile, as an alternative to a velocity profile input that was previously tested [14]. The system achieved significant tremor power reduction levels, similar to those obtained in the velocity input case, and minimal voluntary power attenuation. However, a substantial delay was observed in the measured velocity waveforms. A dynamic model was proposed to explain the delay. The model matches standard models used in the literature (for dc a motor) and the analytical results provide a reasoning for the observed delay. Finally, we discuss how the delay can be accounted for in the system design in future development.

References

1. Wenning, G.K., Kiechl, S., Seppi, K., Müller, J., Högl, B., Saletu, M., Rungger, G., Gasperi, A., Willeit, J., Poewe, W.: Prevalence of movement disorders in men and women aged 50–89 years (Bruneck Study cohort): a population-based study. Lancet Neurol. **4**(12), 815–820 (2005)

2. Elble, R., Deuschl, G.: Milestones in tremor research. Mov. Disord. **26**(6), 1096–1105 (2011)

3. Deuschl, G., Bain, P., Brin, M.: Consensus statement of the movement disorder society on tremor. Mov. Disord. **13**(S3), 2–23 (2008)

4. Rosen, M.J., Arnold, A.S., Baiges, I.J., Aisen, M.L., Eglowstein, S.R.: Design of a controlled-energy-dissipation orthosis (CEDO) for functional suppression of intention tremors. J. Rehabil. Res. Dev. **32**(1), 1–16 (1995)

5. Rocon, E., Belda-Lois, J.M., Ruiz, A.F., Manto, M., Moreno, J.C., Pons, J.L.: Design and validation of a rehabilitation robotic exoskeleton for tremor assessment and suppression. IEEE Trans. Neural Syst. Rehabil. Eng. **15**(3), 367–378 (2007)

6. Hashemi, S.M., Golnaraghi, M.F., Patla, A.E.: Tuned vibration absorber for suppression of rest tremor in Parkinson's disease. Med. Biol. Eng. Comput. **42**(1), 61–70 (2004)

7. Loureiro, R., Belda-Lois, J.: Upper limb tremor suppression in ADL via an orthosis incorporating a controllable double viscous beam actuator. In: ICORR 2005, pp. 119–122 (2005)

8. Taheri, B., Case, D., Richer, E.: Adaptive suppression of severe pathological tremor by torque estimation method. IEEE/ASME Trans. Mechatron. **20**(2), 717–727 (2015)

9. Zhang, D., Poignet, P., Widjaja, F., Tech Ang, W.: Neural oscillator based control for pathological tremor suppression via functional electrical stimulation. Control Eng. Pract. **19** (1), 74–88 (2011)

10. Dosen, S., Muceli, S., Dideriksen, J., Romero, J., Rocon, E., Pons, J., Farina, D.: Online tremor suppression using electromyography and low level electrical stimulation. IEEE Trans. Neural Syst. Rehabil. Eng. **4320**(c), 1–11 (2014)

11. Todorov, E.: Direct cortical control of muscle activation in voluntary arm movements: a model. Nat. Neurosci. **3**, 391–398 (2000)

12. Kawato, M.: Internal models for motor control and trajectory planning. Curr. Opin. Neurobiol. **9**(6), 718–727 (1999)

13. Herrnstadt, G., Menon, C.: Admittance based voluntary driven motion with speed controlled tremor rejection. IEEE/ASME Trans. Mechatron. **21**, 1 (2016)

14. Herrnstadt, G., Menon, C.: Voluntary-driven elbow orthosis with speed-controlled tremor suppression. Front. Bioeng. Biotechnol. **4**(March), 1–10 (2016)

15. Timmer, J., Haussler, S., Lauk, M., Lucking, C.-H.: Pathological tremors: deterministic chaos or nonlinear stochastic oscillators? Chaos **10**(1), 278–288 (2000)

16. Mugge, W., Schuurmans, J., Schouten, A.C., van der Helm, F.C.T.: Sensory weighting of force and position feedback in human motor control tasks. J. Neurosci. **29**(17), 5476–5482 (2009)

17. Shadmehr, R., Smith, M.A., Krakauer, J.W.: Error correction, sensory prediction, and adaptation in motor control. Annu. Rev. Neurosci. **33**, 89–108 (2010)

18. Atkeson, C.G., Hollerbach, J.M.: Kinematic features of unrestrained vertical arm movements. J. Neurosci. **5**(9), 2318–2330 (1985)

19. Rokni, U., Sompolinsky, H.: How the brain generates movement. Neural Comput. **24**(2), 289–331 (2012)

Gremlin Language for Querying the BiographDB Integrated Biological Database

Antonino Fiannaca, Laura La Paglia, Massimo La Rosa$^{(\boxtimes)}$, Antonio Messina,
Riccardo Rizzo, Dario Stabile, and Alfonso Urso

ICAR-CNR, National Research Council of Italy,
via Ugo La Malfa 153, 90146 Palermo, Italy
{antonino.fiannaca,laura.lapaglia,massimo.larosa,antonio.messina,
riccardo.rizzo,dario.stabile,alfonso.urso}@icar.cnr.it

Abstract. In the last decade, biological tasks became much more complex and often their solution requires the simultaneous use of several different resources. Examples of typical bioinformatics scenarios are given by the gene functional studies, the study of microRNA Single Nucleotide Polymorphisms in cancer disease, or the study of protein motifs linked to specific cellular pathways. Available bioinformatics tools give a big contribute in problem solving, but they still require several and time consuming efforts. In this work, we highlight how the Gremlin graph traversal language can be also considered as *lingua franca* and as a valid middleware tool in the implementation, for instance, of some high-level web based tool to solve complex biological tasks. Gremlin queries are tested via a web interface to our previously developed integrated graph database, BioGraphDB.

Keywords: Gremlin language · Graph db · Integrated biological db

1 Introduction

Nowadays, the bioinformatics community can take advantages of a large number of publicly available biological resources. These resources collect both experimentally validated and computationally predicted data, as well as attributes and relationships among biological entities. Unfortunately, since for this kind of data does not exist a standard storage and query format, each resource can require a different approach to be handled properly. In any case, several classes of bio-molecular data, such as transcriptional regulatory networks and protein-protein interaction networks, interact as complex networks and, for this reason, they usually can be modelled as graphs, where nodes (and their attributes) model biological entities and edges contain relationships among these entities. For these reason, biological data and entities are suitable to be modelled as graph databases (graphDB) [1] miRWalk [2] is an example of bioinformatics resource that efficiently exploits a relational DB for integrating biological data such as proteins, genes, biological pathway and microRNA (miRNA). miRWalk database

© Springer International Publishing AG 2017
I. Rojas and F. Ortuño (Eds.): IWBBIO 2017, Part I, LNBI 10208, pp. 303–313, 2017.
DOI: 10.1007/978-3-319-56148-6_26

has been designed in order to be performed with the structured query language (SQL), allowing the user to query a set of pre-defined search methods providing, for example, gene-miRNA-GO annotations or gene-miRNA-pathway relations. By means of a pattern match strategy, it exploits the advantages of the SQL for efficient queries. The main drawback of miRWalk is its lack of flexibility, in fact user can not perform efficiently out-of-the-box SQL queries over this database, because it could imply complex join operations in order to retrieve data [3]. For this reason, a better solution is represented by a graph database-based approach, where a query over biological data can be performed traversing their relations and attributes [1]. Bio4J [4] is an example of this kind of DB developed for bioinformatics. Bio4J defines a framework for the integration of publicly available biological DBs in the field of proteins, gene ontology, biological pathways and enzymes. It is developed in Java language and it also provides support for both Cypher [5] and Gremlin [6,7], that represent two of the most efficient query languages for, respectively, declarative and traversal queries [3]. In this work, we adopt the Gremlin traversal language in order to perform queries over an integrated biological database comprising several biological entities, such as genes, proteins, pathways, functional annotations, microRNA (miRNA). We preferred Gremlin over Cypher because the former is platform-independent, that is it is compliant with different graph db engines, and then, as it will be highlighted later in the paper, it allows tu fully define the exact traversal path. The integrated graph db is called BiographDB, and it has been presented in our previous works [8,9], where we defined the system architecture. Moreover we developed a web application, available at http://tblab.pa.icar.cnr.it/biographdb, in order to submit gremlin queries to BioGraphDB. The rest of the paper is organized as follows. In Sect. 2 we present the Gremlin language, as a well as the main features of the graph database adopted as test case, i.e. the BioGraphDB [8,9], an integrated noSQL graph db collecting heterogeneous biological data from publicly available databases. We also introduce a simple web interface developed in order to let the user to freely query the BioGraphDB database with his own Gremlin query or using some query templates. In Sect. 3 we present two application scenarios in the bioinformatics domain and their implementation using the Gremlin language. Finally in Sect. 4 we make our conclusion.

2   Materials and Methods

In this section, we briefly report the main features of the Gremlin language and how it can be used to proceed at each step in graph traversal. In addition, we introduce implementation details of the proposed BiographDB query console, that allow users to perform Gremlin query over BioGraphDB.

2.1   Gremlin Language

Gremlin [6,7] is a functional, data-flow, graph traversal language designed specifically for the analysis and manipulation of *property graphs*, which are graph data structures distinguished by the following features: (1) both vertices and edges can have any number of properties associated with them; (2) edges in the graph have a directionality; (3) there can be many types of edges and thus, many types of relationships can exist between the vertices.

Every Gremlin[1] traversal is composed of a sequence of steps, able to perform atomic operations on the data stream. Steps, as shown in Table 1, can be categorized as follow: *transform-based* (take an object and emit a transformation of it), *filter-based* (decide whether to allow an object to pass or not), *sideEffect-based* (pass the object, but yield some side effect), and *branch-based* (decide which step to take).

Table 1. Gremlin steps grouped by category.

Transform	*_, both, bothE, bothV, cap, E, gather, id, in, inE, inV, key, label, linkBoth, map, memoize, order, orderMap, out, outE, outV, path, scatter, select, shuffle, transform, V*
Filter	*[i], [i ...j], and, back, dedup, except, filter, has, hasNot, interval, or, random, retain, simplePath*
Side effect	*aggregate, as, groupBy, groupCount, optional, sideEffect, store, table, tree*
Branch	*copySplit, exhaustMerge, fairMerge, ifThenElse, loop*

A Gremlin traversal can be written in a declarative, in an imperative, or in a mixed manner containing both declarative and imperative aspects. With the *declarative* way, you do not tell the traversers the order in which to execute their walk: a traverser is allowed to select a pattern to execute from a collection of other patterns. Instead, an *imperative* Gremlin traversal tells the traversers how to proceed at each step in the traversal. If applied to big integrated bioinformatics graph databases, the latter is suitable to easily and properly state the common or user-defined bioinformatics tasks that a biologist has to address in his daily work.

Gremlin traversals have been tested over BioGraphDB [8,9], an integrated noSQL graph database collecting several and heterogeneous kinds of biological data built on-top of the OrientDB multi-model database system [10]. BioGraphDB integrates the following kinds of biological entities, each one imported from a specific database (shown between parenthesis): microRNA (mirBase [11]), genes (Entrez Gene [12]), proteins (UniprotKB [13]), biological sequences (RefSeq [14]), pathways (Reactome [15]), functional annotations (Gene Ontology - GO [16]), nomenclature (HGNC [17]), miRNA-cancer associations (mirCancer [18]), miRNA-target interactions (mirTarBase [19], miRanda [20], Diana micro-T [21]).

---

[1] The Gremlin version highlighted in this work is from TinkerPop 2.x. TinkerPop is now a part of the Apache Software Foundation and the current release is TinkerPop 3.x, available at http://tinkerpop.incubator.apache.org.

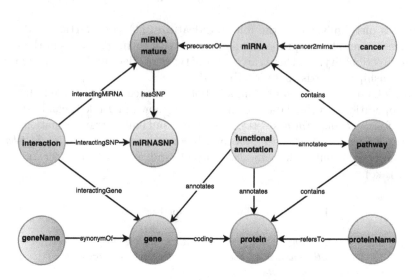

Fig. 1. Entities (vertices) and relationships (edges) Structure of BioGraphDB

According to the graph structure of BioGraphDB, biological entities and their relationships are organized in vertices and edges, as shown in Fig. 1. For example, genes and proteins, as well as their properties, are mapped into a vertices (with attributes) belonging to different classes. A relation between a gene and a protein, such as the "coding" relation, is mapped into an edge belonging to the class *coding*. The import procedure is made through the implementation of ad hoc Extract-Transformer-Loader (ETL) modules, responsible of processing of proper type of biological data.

2.2   BioGraphDB Query Console

Based on the BioGraphDB, we developed, as prototype, a query console web interface, to offer the user a tool to query the database using the Gremlin language. The application is available on http://tblab.pa.icar.cnr.it/biographdb. The query console web interface has a simple interface that is composed of three tabs.

- *DB Schema*: "DB Schema" tab shows a graph of the database schema and a list of informations about database objects vertices, edges and relations. These informations can be used to understand what knowledge can be extracted from the database using Gremlin.
- *Query Console*: "Query Console" tab contains an input area for a Gremlin query.
- *Query Template*: "Query Template" tab contains some pre-formed query, to test the application.

Once a query is submitted, results are shown in a section "Results". Results section is composed of two parts: "Table Results" tab and "Graph Results"

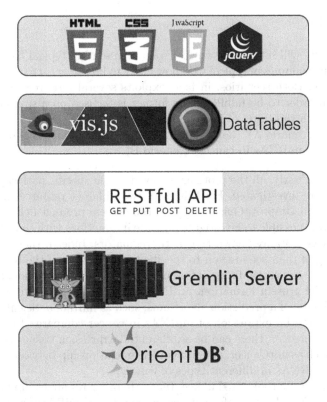

Fig. 2. The stack of components for the implementation of the web interface

tab. "Graph Results" shows data by means of a directed graph, where nodes (biological entities) are highlighted with different colours. Hovering the nodes, user can retrieve the referred information. Hovering the edges, a title shows the relation between the two involved nodes/entities.

The web interface was built on HTML5 and CSS3 graphic template, released under an open source license. The application has a fully responsive design that makes it device-agnostic. For the front-end part, we used javascript, jquery v.1.11.3, dataTables.js v.1.10.13 and vis.js v.4.16.1 library. DataTables is a plug-in for the jQuery Javascript library, highly flexible, that add advanced interaction controls to any HTML table. DataTables is also free open source software, available under the MIT license. Vis.js is a dynamic, browser based visualization library, licensed under both Apache 2.0 and MIT, that is designed to handle large amounts of dynamic data, and to enable data manipulation and interaction. BioGraphApp use vis.js to manipulate and visualize graph data results. The back-end is based on OrientDB v.2.2.13 - a distributed multi-model NoSQL database with a graph database engine. In particular the query console web interface uses OrientDB RESTful services (Rexster 2.6.0) to talk to an OrientDB Server instance using the HTTP protocol and JSON data streams. The set of adopted tools and libraries are summarized in the stack of Fig. 2.

3   Results

Here, we briefly will introduce two scenarios representing typical bioinformatics tasks, used as examples of typical workflow chart to be *traversed* to reach the final study purpose. Both scenarios, in fact, exploits several resources, i.e. different databases, in order to be fulfilled. The former is focused on a specific pathway analysis through the study of its genetic regulators in a specific cellular context as cancer has (scenario A), the latter consists in a target analysis of differentially expressed (DE) miRNAs in cancer (scenario B).

– *Scenario A*: Firstly, all the proteins involved in the specific pathway object of the study are investigated. For this step, two different resources are needed: Reactome and Uniprot. Once evidenced the protein product belonging to the pathway, it is possible to investigate possible interactions with some molecules, i.e., miRNAs, notably known as genetic regulators. Indeed, they act through the binding of their seed region to specific target regions of the RNA messenger (mRNA), causing a disregulation of normal post transcriptional processes that lead the protein formation. miRNA-target interactions are investigated through the use of a prediction algorithms, such as miRanda. Scientific reports show a differential expression of miRNAs in cancer compared to normal tissues, and, moreover, there can be a differential expression tissue-specific. Web services, such as mirCancer, allow to evidence relationship between up or down regulated miRNAs in different types of cancer.

  Together with the tools cited above, there are other resources implicitly used: in this case, NCBI gives informations on genes, and miRbase includes informations about miRNA precursors, sequences and mature forms.

  Because BioGraphDB already integrates all the needed data sources to solve such sort of scenario, we can *translate* the steps above into a sequence of queries and operations on the graph database.

  Let us suppose that the starting pathway object is *Cell Cycle*, which has *R-HSA-1640170* as identifier. To increase the power of the analysis and to address more specifically the study, two filters are also applied. The former is on the free energy associated to the miRNA-target binding site, in order to select miRNAs strongly associated to mRNA target. The latter is on the selection of differentially expressed miRNAs (just up-regulated or down-regulated), to investigate the role of miRNAs as oncogenic or tumour-suppressor molecules, linked to cancer. In this case, the free energy threshold has been set to -32, whereas the miRNAs of interest are the ones *up-regulated*.

– *Scenario B*: Starting from a list of differentially expressed miRNAs linked to a specific disease, we want to verify what are the major target proteins of these miRNAs belonging to particular cellular pathways. For this analysis, four resources are usually needed in sequence: mirCancer, mirBase, miRNA-target interactions, Reactome.

  Supposing to select *acute lymphoblastic leukaemia* as cancer type, we can evidence what are the miRNAs involved in this disease, and what are the pathways most affected by their action. Again, to increase the power of the analysis, a filter on free energy is applied, assigning a threshold value equals to -29.

Now, keeping in mind what is the structure of BioGraphDB in term of classes of vertices and types of edges (Fig. 1), we can go through two different directions depending on the scenario we choose.

Scenario A:

A1. select the desired pathway from the totality of pathways;
A2. find the proteins contained in the pathway;

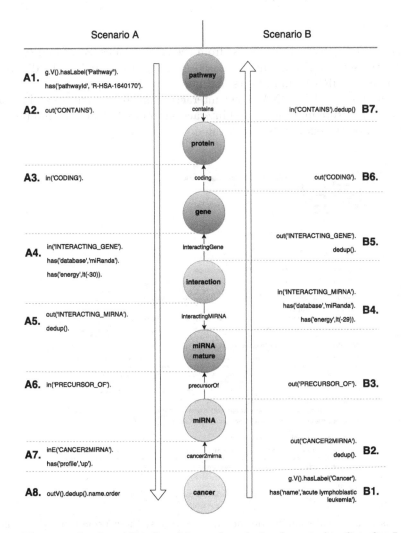

Fig. 3. The scenarios A and B splitted into sub-tasks implemented in Gremlin. Letters refer to the scenario, numbers refer to the sub-tasks ordered by progressive steps.

A3. from each protein, jump to the coding gene;

A4. because a generic miRNA-target interaction involves a miRNA and a gene, for each gene find the related interactions and select just the interactions filtered by the energy threshold value;

A5. starting from each interaction, predicted by miRanda, go to the interacting miRNA and, because a single miRNA can interact with more than one gene, at this step the set of miRNAs may contain duplicates, which are removed;

A6. from each unique miRNA, jump to its precursor;

A7. select just the up-regulated miRNAs;

A8. starting from miRNAs, find the unique cancer diseases and return their names in alphabetical order.

Scenario B:

B1. select the desired cancer disease from the totality of cancers in miRCancer;

B2. find the set of differentially expressed miRNAs associated to the disease;

B3. from each miRNA, jump to its mature;

B4. because a generic miRNA-target interaction involves a mature and a gene, for each mature find the related interactions and select just the interactions filtered by the energy threshold value;

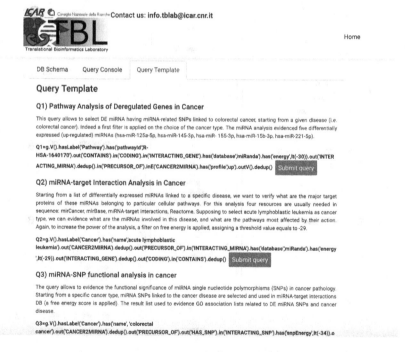

Fig. 4. User can recall some pre-formed queries to test the console using the "Query template". Gremlin queries representing proposed scenarios are respectively Q1 and Q2.

B5. starting from each interaction, go to the interacting gene and, because a single miRNA can interact with more than one gene, at this step the set of genes may contain duplicates, which are removed;

B6. from each unique gene, jump to the coded protein;

B7. find the unique specific pathways that the selected targets belong to.

Figure 3 shows the proposed scenarios A and B, both split in sub-tasks implemented in Gremlin steps.

Similarly for the two previous ones, many other typical bioinformatics scenarios can be solved by a more or less complex Gremlin query. Every scenario can be usually be decomposed in a row of simple sub-tasks, easily translatable into a few of Gremlin code.

Definitively, a scenario can be meant as a graph traversal operation and Gremlin is an ultimate tool to perform such task.

In order to test the proposed scenarios, we used the introduced GraphDB query console. As shown in Fig. 4, both scenarios have been coded as pre-formed gremlin queries, available on "Query Template" tab.

For example, if we execute the query Q1 related to scenario A, we will obtain results in both tabular and graphical format. In particular, the "Table Results"

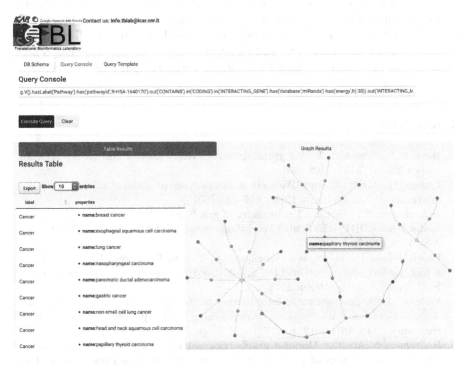

Fig. 5. Query Results in tabular and graph form. In the graph it is possible to follow the whole traversal done by the Gremlin query, i.e. the sequence of entities (nodes) visited in order to obtain the response.

frame will show the query results in a tabular form, where data can be ordered by clicking on any column header, (see the left side of Fig. 5), whereas the 'Graph Results" canvas will show data by means of a directed graph (see the right side of Fig. 5).

4   Conclusion

Nowadays, it is proved that common biological tasks are characterized by remarkable complexity and they also require the combined use of several different resources. It is an undeniable fact that a big integrated bioinformatics database may help a lot in such problems, but it must be properly designed, data must be well related each other, and the queries should be easy to write and fast to execute.

Because bioinformatics scenarios are composed by tasks easily mappable into some graph traversal operations, a graph database is an ideal candidate to integrate data coming from heterogeneous bioinformatics resources and an example of such implementation is given by BioGraphDB.

In this paper, we highlight how the Gremlin graph traversal language could be a good tool to solve biological tasks, even if its level of abstraction is not so high and it is usually a tool closer to computer scientists than to biologists. In order to perform the proposed queries and the user's proper queries, we provided a simple web interface at the following link: http://tblab.pa.icar.cnr.it/biographdb Nonetheless, in a near future, we will adopt Gremlin as the foundation of a high level web-based wizard to allow biologists to easily design, analyse and solve bioinformatics scenarios.

References

1. Have, C.T., Jensen, L.J.: Are graph databases ready for bioinformatics? Bioinformatics **29**(24), 3107–3108 (2013)
2. Dweep, H., Gretz, N.: miRWalk2.0: a comprehensive atlas of microRNA-target interactions. Nat. Methods **12**(8), 697–697 (2015)
3. Holzschuher, F., Peinl, R.: Performance of graph query languages. In: Proceedings of the Joint EDBT/ICDT 2013 Workshops on - EDBT 2013, p. 195. ACM Press, New York (2013)
4. Pareja-Tobes, P., Tobes, R., Manrique, M., Pareja, E., Pareja-Tobes, E.: Bio4j: a high-performance cloud-enabled graph-based data platform. Technical report, Era7 Bioinformatics, March 2015
5. Webber, J.: A programmatic introduction to Neo4j. In: Proceedings of the 3rd Annual Conference on Systems, Programming, and Applications: Software for Humanity - SPLASH 2012, p. 217. ACM Press, New York (2012)
6. Rodriguez, M.A.: The Gremlin graph traversal machine and language (invited talk). In: Proceedings of the 15th Symposium on Database Programming Languages - DBPL 2015, pp. 1–10. ACM Press, New York (2015)
7. Apache Tinkerpop: The Gremlin traversal language

8. Fiannaca, A., La Paglia, L., La Rosa, M., Messina, A., Storniolo, P., Urso, A.: Integrated DB for bioinformatics: a case study on analysis of functional effect of MiRNA SNPs in cancer. In: Renda, M.E., Bursa, M., Holzinger, A., Khuri, S. (eds.) ITBAM 2016. LNCS, vol. 9832, pp. 214–222. Springer, Heidelberg (2016). doi:10. 1007/978-3-319-43949-5_17

9. Fiannaca, A., La Rosa, M., La Paglia, L., Messina, A., Urso, A.: BioGraphDB: a new GraphDB collecting heterogeneous data for bioinformatics analysis. In: BIOTECHNO 2016: The Eighth International Conference on Bioinformatics, Bio-computational Systems and Biotechnolo-gies, pp. 28–34 (2016)

10. Orient Technologies LTD: OrientDB multi-model database engine

11. Kozomara, A., Griffiths-Jones, S.: miRBase: integrating microRNA annotation and deep-sequencing data. Nucleic Acids Res. **39**(Database issue), D-152–D-157 (2011)

12. Schuler, G.D., Epstein, J.A., Ohkawa, H., Kans, J.A.: Entrez: molecular biology database and retrieval system. Methods Enzymol. **266**, 141–162 (1996)

13. The UniProt Consortium: UniProt: a hub for protein information. Nucleic Acids Res. **43**(D1), D204–D212 (2015)

14. Pruitt, K.D., Tatusova, T., Maglott, D.R.: NCBI reference sequences (RefSeq): a curated non-redundant sequence database of genomes, transcripts and proteins. Nucleic Acids Res. **35**(Database), D61–D65 (2007)

15. Croft, D., Mundo, A.F., Haw, R., Milacic, M., Weiser, J., Wu, G., Caudy, M., Gara-pati, P., Gillespie, M., Kamdar, M.R., Jassal, B., Jupe, S., Matthews, L., May, B., Palatnik, S., Rothfels, K., Shamovsky, V., Song, H., Williams, M., Birney, E., Hermjakob, H., Stein, L., D'Eustachio, P.: The Reactome pathway knowledgebase. Nucleic Acids Res. **42**(D1), D472–D477 (2014)

16. The Gene Ontology Consortium: gene ontology consortium: going forward. Nucleic Acids Res. **43**(D1), D1049–D1056 (2015)

17. Gray, K.A., Yates, B., Seal, R.L., Wright, M.W., Bruford, E.A.: Genenames.org: the HGNC resources in 2015. Nucleic Acids Res. **43**(D1), D1079–D1085 (2015)

18. Xie, B., Ding, Q., Han, H., Wu, D.: miRCancer: a microRNA-cancer association database constructed by text mining on literature. Bioinformatics **29**(5), 638–644 (2013)

19. Hsu, S.D., Tseng, Y.T., Shrestha, S., Lin, Y.L., Khaleel, A., Chou, C.H., Chu, C.F., Huang, H.Y., Lin, C.M., Ho, S.Y., Jian, T.Y., Lin, F.M., Chang, T.H., Weng, S.L., Liao, K.W., Liao, I.E., Liu, C.C., Huang, H.D.: miRTarBase update 2014: an infor-mation resource for experimentally validated miRNA-target interactions. Nucleic Acids Res. **42**(D1), D78–D85 (2014)

20. John, B., Enright, A.J., Aravin, A., Tuschl, T., Sander, C., Marks, D.S.: Human microRNA targets. PLoS Biol. **2**(11), e363 (2004)

21. Paraskevopoulou, M.D., Georgakilas, G., Kostoulas, N., Vlachos, I.S., Vergoulis, T., Reczko, M., Filippidis, C., Dalamagas, T., Hatzigeorgiou, A.G.: DIANA-microT web server v5.0: service integration into miRNA functional analysis workflows. Nucleic Acids Res. **41**(W1), W169–W173 (2013)

Modelling of Glucose Dynamics for Diabetes

Tomas Koutny[(✉)]

NTIS - New Technologies for the Information Society,
University of West Bohemia, Univerzitni 8, 306 14 Plzen, Czech Republic
txkoutny@kiv.zcu.cz

Abstract. Diabetes is a heterogeneous group of diseases associated with elevated blood glucose level. Elevated glucose level continuously damages multiple organs, eventually leading to death. Diabetes is one of top 10 leading causes of death. Diabetes does not hurt nor manifests outwardly, until it is too late and the disease has developed. For example, type-1 diabetic patient depends on insulin. The patient takes insulin based on self-monitoring of blood glucose level. Motivated patient takes approximately 3-4 blood samples a day. This cannot capture all important events, e.g., nocturnal hypoglycemia with the risk of death. Therefore, the patient wears a system that continuously measures glucose level in the subcutaneous tissue. Nevertheless, these two glucose levels are not linearly proportional. They can differ considerably. Therefore, a model of glucose dynamics is needed to describe the non-linearity between these two levels. The model is important to estimate blood glucose level from the subcutaneous tissue glucose level and for a development of artificial pancreas. Number of models were developed. In this paper, we present the background, selected models of glucose dynamics, lessons learned and propose what minimal data researchers should share to stimulate further research on this topic.

Keywords: Glucose dynamics · Glucose modelling · Blood glucose · CGMS

1   Introduction

There is a heterogeneous group of diseases and complications frequently associated with elevated blood glucose level, whereas diabetes mellitus is the most common known disease. Eber papyrus dates to 1500 B.C. It describes a condition of "too great emptying of urine". It is considered as a first mention of a diabetes symptom – accompanying symptoms are excessive thirst and severe weight loss. Over 3000 years, symptoms of type 1 diabetes mellitus were known, but the prognosis was always fatal until 1922.

In 1920, Frederick Banting concluded that pancreas secretion regulates sugar in bloodstream. In 1922, the first injection of insulin was given [1]. In 1999, US Food and Drug Administration approved first continuous glucose monitoring system (CGMS). CGMS comprises a sensor installed in subcutaneous tissue and receiver that records the level measured. Researchers use CGMS signal to optimize insulin dosage.

Today, diabetes is 8th most common cause of death [2]. Its prevalence rose from estimated 30 million cases in 1985 to 285 million cases in 2010. The International Diabetes Federation estimates 438 million cases for year 2030. For individuals aged

© Springer International Publishing AG 2017
I. Rojas and F. Ortuño (Eds.): IWBBIO 2017, Part I, LNBI 10208, pp. 314–324, 2017.
DOI: 10.1007/978-3-319-56148-6_27

over 20 years, the prevalence is approximately 11.8% and 10.8% for men and women respectively. For individuals over 65 years, the prevalence is 26.9% [3].

1.1 Glucose Homeostasis

Glucose is primary source of energy for cells of living animals. Glucose and lipids are energy stores for the body. Fructose, glucose and galactose are final products of carbohydrate digestion, whereas glucose has circa 80% share of these. Once absorbed from the intestinal tract, liver rapidly converts fructose and galactose into glucose [4]. Liver cells store excessive glucose as a glycogen. It is as a secondary, long-term energy storage. When blood glucose level is low, liver breaks glycogen down into glucose. Then, glucose gets from liver cells into blood. Most cells are capable of storing at least some glycogen. Brain cannot synthetize nor store glucose for more than few-minutes supply. It is dependent on continuous supply of glucose from blood [5].

Glucose in blood comes from different sources – a dietary source, glycogen breakdown and formation of glucose from its precursors. Eventually, the glucose is removed from the blood. The rate and process of removal is different in different tissues. In general, the glucose is either compacted into an energy store, or the glucose is utilized as a source of energy for cells.

Different compartments may have different glucose levels, and this difference is because capillaries in different compartments have different permeabilities [4]. Subsequently, the cells of different tissues have different glucose utilization rates depending on the tissue's metabolic needs and the level of available glucose.

Insulin promotes uptake of glucose by cells. Pancreas increases insulin secretion with rising blood glucose level. Pancreas releases insulin in bursts [6]. Most cells increase uptake of glucose up to 10-times with insulin. Except liver and brain cells, they would be unable to obtain the minimum required energy without insulin.

Falling blood glucose level causes the pancreas to reduce insulin secretion. Contrary to insulin, pancreas increases secretion of glucagon. Glucagon hormone raises blood glucose level. Lack of insulin with an increased glucagon level activate breakdown of glycogen. About 60% of glucose in the meal is stored in liver and returned into blood later [4]. In addition to glycogen breakdown, glucagon increases gluconeogenesis in the liver. Gluconeogenesis is a process of generating glucose from non-carbohydrate carbon substrates.

A continual exchange of extracellular fluids occurs between blood and the interstitial fluid. The interstitial fluid fills the intercellular spaces. Glucose gets from blood through the capillary membrane into the interstitial fluid. There, cells utilize the glucose or the glucose returns to blood eventually. Lymphatic system represents an accessory exit route into the blood. The extracellular fluids are being continually mixed to maintain almost complete homogeneity throughout the body [4].

1.2 Types of Diabetes Mellitus

Type 1 diabetes mellitus results from autoimmune destruction of insulin producing cells in pancreas. The cause is unknown. The considered causes are combination of genetic makeup, environment, virus-triggered actions, chemicals and drugs that selectively destroy insulin-producing cells. As a result, the blood glucose level is too high, as the missing insulin does not promote glucose utilization. There is a complete or near-total deficiency of insulin [3].

Type 2 diabetes mellitus is a group of heterogeneous disorders characterized by variable degree of insulin resistance, impaired insulin secretion and increased glucose production [3]. Type 2 has strong genetic component. In addition to genetic susceptibility, factors such as obesity, nutrition, and physical activity contribute to the prevalence of type 2. In the beginning, glucose tolerance remains near normal as pancreas increases insulin production. Over the time, insulin-producing cells cannot sustain the increased production and they fail eventually. Meanwhile, diabetes with fasting hyperglycemia develops.

Gestational diabetes mellitus occurs with pregnant women without previously diagnosed elevated blood glucose levels. Insulin resistance is the main symptom. Perhaps, human placental lactogen interferes with insulin receptors. Precise mechanism is unknown.

Stress of illness may elevate blood glucose level. It is a common problem of patients committed to intensive-care unit [7, 8]. For example, sepsis may cause both hypoglycemia and hyperglycemia. Steroid diabetes is another example referring to prolonged hyperglycemia due to glucocorticoid therapy for another medical condition.

2   Commonly Measured Quantities

2.1   Blood Glucose Level

In laboratory experiment, it is possible to obtain blood and CGMS-measured glucose levels at the same time, i.e., to perform frequent glucose level measuring. In such a case, it is possible to smooth the blood-glucose level time-series. Smoothing reduces adverse effects of outliers caused by short-term actions of the metabolism.

In a daily life of type-1 diabetic patient, there is a substantially less number of measured blood glucose levels than measured interstitial fluid glucose levels. As there are several hours between two blood-glucose level measurements, the blood glucose level cannot be approximated nor interpolated and we have to use the measured values only.

When measuring blood glucose level, a blood drop is drawn onto a test strip [9]. Required drop volume varies from 0.3 to 1.0 µl. Despite the relatively small volume, drawing the blood causes an important discomfort to the patient. On the strip, glucose oxidase produces gluconic acid from the blood glucose. The glucometer applies an electric voltage on the terminals of the strip so that an electric current flows through the strip. The electric current reflects the mass of the gluconic acid. Then, the electric current is quantified and converted into a glucose level. There is a key assumption that arterial blood and capillary blood glucose levels are almost identical in concentrations [10].

2.2   Interstitial Fluid Glucose Level

Electrochemical technology is involved in the CGMS sensor measuring [11]. The sensor reads electrical current produced by glucose oxidase reaction. Glucose oxidase is an enzyme that converts glucose to hydrogen peroxide. The peroxide reacts with platinum of the sensor needle, thus producing an electric current. This current is measured, quantified and wirelessly sent to the CGMS system, whereas it can be downloaded into computer. Recently, non-enzymatic optical, impedance and electro-magnetic spectroscopy emerged; see study [11].

In a principle, the number obtained by quantifying the electric current is scaled and reported as a glucose level. Setting the scale is called calibration. For example, if blood glucose level is 10 mmol/l, then CGMS associates the present electric current with 10 mmol/l. If the current would increase by 10%, CGMS would report 11 mmol/l. In the practice, there are more sophisticated mathematical filters that try to remove measurement noise from the obtained signal. The noise could be caused by local effects such as sensor needle design that limits the mass of glucose that can reach the sensor, formulation of glucose oxidase, which catalyzes the reaction that allows the CGMS sensor to measure the glucose, blood clots that formed around the needle, possible effects of medications, etc. Studies [12, 13] discuss CGMS design and algorithms to a greater depth.

The sensor must be calibrated repeatedly, because there is an immune response as the CGMS sensor is a foreigner. Patient's body works towards eliminating and disposing the foreigner as it does with viruses, hostile bacteria and malfunctioning cells. Eventually, the body wins over the special coating of sensor needle and it has to be replaced. Sensor's precision degrades continually, despite a recalibration procedure.

CGMS calibration should only be done in a period, when blood glucose level (and supposedly the CGMS-measured glucose level) is steady. As glucose propagates from blood with a delay, CGMS could easily calculate a wrong scale if at least one of the glucose level is changing. Especially, this might be a problem with the sporadic blood-glucose level measurement in a daily life of diabetic type-1 patient. In the laboratory experiment, this can be avoided with a period of frequent blood glucose level monitoring.

3   Selected Models of Glucose Dynamics

Let us focus on models, which are based on present understanding of glucose home-ostasis. We consider such models to have greatest potential of achieving best accuracy as glucose homeostasis is mostly (except hypothalamus) realized via sensing and responding to various chemicals. Such processes can be modelled with mathematical functions. Hypothalamus exercises limited degree of control over liver and pancreas.

3.1   Sensor Model

Let us consider the Steil-Rebrin model [14–17] that does not calculate with insulin. It rather relates glucose levels in blood and interstitial fluid, where the sensor of continuous monitoring system is. Let us denote blood and interstitial fluid glucose levels with $C_1(t)$ and $C_2(t)$, respectively. Let V_1 and V_2 be volumes of respective compartments and k_{ij} to denote transfer from compartment j to compartment i. k_{02} is a glucose clearance rate from interstitial fluid. According to study [14], we obtain Eq. (1).

$$\frac{dC_2(t)}{dt} = -\frac{1}{\tau} \times C_2(t) + \frac{g}{\tau} \times C_1(t), \quad \text{where } g = \tau \times \frac{k_{21} \times V_1}{V_2} \quad \tau = \frac{1}{k_{02} + k_{12}} \tag{1}$$

Assuming a steady state with no change of interstitial glucose level, we obtain a boundary condition (2).

$$0 = -\frac{1}{\tau} \times C_2(t) + \frac{g}{\tau} \times C_1(t) \tag{2}$$

Equation (2) dictates that $g = 1$; thus τ should be the only parameter to be estimated from data. However, estimating both g and τ parameters improves precision. Study [14] concludes with a need for time-varying modulation.

3.2   Tracer Model

While the Steil-Rebrin model can be satisfied with commonly measured quantities, tracer models are intended to elucidate a particular aspect of plasma-interstitium dynamics. For example, a four-compartment five-rate-constant model was proposed [18]. Equations (3) present this model.

$$\begin{aligned}
\widehat{C_c}(t) &= k_1 \times C_p(t) - (k_2 + k_3) \times C_c(t) + k_4 \times C_e(t) \\
\widehat{C_e}(t) &= k_3 \times C_c(t) - (k_4 + k_5) \times C_e(t) \\
\widehat{C_m}(t) &= k_5 \times C_e(t) \\
C(t) &= (1 - V_b) \times (C_c(t) + C_e(t) + C_m(t)) + V_b \times C_b(t)
\end{aligned} \tag{3}$$

C-s are levels measured in different compartments (whole blood, arterial blood, extracellular fluid, tissue, total activity), k-s are rate constants and V_b is blood volume in the area of interest. A multi-tracer positron emission tomography is needed to obtain all required quantities. The [18F] fluorodeoxglucose analog is used as a tracer. The [11C] glucose would be a better tracer, but it would require yet more complex model calculation with additional metabolic products of glycolysis.

In a different study [19], researchers measured concentrations of three different glucose tracers in both blood plasma and interstitium. Similarly, other studies [20–22] have used an approach based on fluorodeoxyglucose and positron emission tomography.

3.3   Glucose-Insulin Model Combined with the Miachelis-Menten Model

Let us consider a model based on work of Hovorka et al. [23]. The model comprises five subsystems: subcutaneous insulin absorption, plasma insulin kinetics, insulin action, glucose kinetics and gut glucose absorption. The model takes subcutaneous insulin infusion and carbohydrate content of meals on input to produce blood glucose level on output.

The insulin absorption subsystem consists of slow and fast absorption channels. Each channel comprises two-compartment chain.

$$
\begin{aligned}
\widehat{Q}_{is1}(t) &= u_i(t) \times p_i - Q_{is1}(t) \times k_{is1} \\
Q_{is1}(0) &= \frac{u_i(0) \times p_i}{k_{is1}} \\
\widehat{Q}_{is1}(t) &= Q_{is1}(t) \times k_{is1} - Q_{is2}(t) \times k_{is2} \\
Q_{is2}(0) &= \frac{Q_{is1}(0) \times k_{is1}}{k_{is2}} + Q_b \times p_i
\end{aligned}
\tag{4}
$$

With the slow channel (4), $Q_{is1}(t)$ and $Qi_{s2}(t)$ represent insulin masses in first and second compartment. $u_i(t)$ is insulin infusion rate. p_i is portion of subcutaneous insulin absorbed through the slow channel. Q_b is residual insulin from the previous insulin delivery. k_{is1} and k_{is2} are transfer rate parameters.

$$
\begin{aligned}
\widehat{Q}_{if1}(t)(t) &= u_i(t) \times (1 - p_i) - Q_{if1}(t) \times k_{if} \\
Q_{if1}(0) &= \frac{u_i(0) \times (1 - p_i)}{k_{if}} \\
\widehat{Q}_{if2}(t) &= Q_{if1}(t) \times k_{if} - Q_{if1}(t) \times k_{if} \\
Q_{if2}(0) &= Q_{if1}(0) + Q_b \times (1 - p_i)
\end{aligned}
\tag{5}
$$

With the fast channel (5), $Q_{if1}(t)$ and $Q_{if2}(t)$ represent insulin masses in first and second compartment. k_{if} is transfer rate parameter.

One compartment represents insulin kinetics (6), where $Q_i(t)$ is blood insulin level, c_i is background insulin appearance rate. $I_m(t)$ is a function describing time varying components of insulin kinetics.

$$
\begin{aligned}
\widehat{Q}_i(t) &= (Q_{is2}(t) \times k_{is2} + Q_{if2}(t) \times k_{if}) \times I_m(t) - Q_i(t) \times k_e + c_i \\
Q_i(0) &= Q_{i0}
\end{aligned}
\tag{6}
$$

Then, plasma insulin level, $I_p(t)$, is obtained with Eq. (7). $V_i = 190$ ml/kg is the insulin distribution volume and w is subject's weight.

$$
I_p(t) = \frac{Q_i(t)}{V_i \times w} \times 10^6
\tag{7}
$$

Insulin action subsystem (8) is based on experiments with glucose tracers.

$$\widehat{x}_1(t) = -k_{a1} \times x_1(t) + k_{a1} \times S_t \times I_p(t)$$
$$x_1(0) = S_t \times I_p(0)$$
$$\widehat{x}_2(t) = -k_{a2} \times x_2(t) + k_{a2} \times S_d \times I_p(t)$$
$$x_2(0) = S_d \times I_p(0) \tag{8}$$
$$\widehat{x}_3(t) = -k_{a3} \times x_3(t) + k_{a3} \times S_e \times I_p(t)$$
$$x_3(0) = S_e \times I_p(0)$$

$x_1(t)$, $x_2(t)$ and $x_3(t)$ represent the effect of insulin on glucose distribution/transport, disposal and endogenous production. k_a-s are respective rates. S_d and S_t give insulin sensitivity of glucose distribution/transport. S_e is insulin sensitivity of glucose endogenous production.

Glucose kinetics subsystem assumes insulin-independent glucose utilization to be a saturable process accordingly to the Michaelis-Menten model (9).

$$\widehat{Q}_1(t) = -F_{01} \times \frac{Q_1(t)/V}{1 + Q(t)/V} - x_1(t) \times Q_1(t) + k_{12} \times Q_2(t) + EGP_0(1 - x_3(t))$$
$$+ f_g(t) + U_m(t) \times Q_1(0) = Q_{10} \tag{9}$$
$$\widehat{Q}_2(t) = \frac{Q_{10} \times x_1(0)}{x_2(0) + k_{12}}$$

$Q_1(t)$ and $Q_2(t)$ represent glucose levels in accessible (where the measuring is done) and non-accessible compartments. F_{01} is non-insulin dependent glucose utilization. EGP_0 is endogenous insulin production extrapolated to zero insulin level. $EGP_0(1 - x_3(t))$ is zero if $x_3(t) \geq 1$. $U_m(t)$ is meal glucose appearance. $f_g(t)$ represents time-varying characteristics of glucose kinetics.

With $V = 160$ ml/kg as glucose distribution volume, blood glucose level is obtained with Eq. (10).

$$G(t) = \frac{Q_1(t)}{V} \tag{10}$$

Absorption channels represent gut absorption of glucose, each comprising two-compartment chain (11).

$$U_m(t) = (U_{m1}(t) + U_{m2}(t)) \times f_m(t)$$
$$U_{m1}(t) = k_m^2 \times t \times e^{-k_m \times t \times \frac{CHO \times 551}{w} \times p_m}$$
$$U_{m2}(t) = \begin{cases} k_m^2 \times (t - d) \times e^{-k_m \times (t-d) \times \frac{CHO \times 555}{w} \times (1 - p_m)}, & t > d \\ 0, & otherwise \end{cases} \tag{11}$$

$U_{m1}(t)$ and $U_{m2}(t)$ are appearance of first and second absorption channels. k_m is transfer rate. d is delay of the second channel. p_m is portion of meal carbohydrates absorbed through first channel. *CHO* is carbohydrate content of the meal. $f_m(t)$ is time-varying function of absorption profile. It varies considerably between subjects.

3.4 Diffusion Model

While the described models are based on present physiological knowledge, there are still some disadvantages. The Steil-Rebrin model has minimal requirements on input data, but we consider it as too simplified – see study [24]. Tracer models are not intended for everyday use. The glucose-insulin model combined with Michaelis-Menten model considerably increases the level of detail, when modelling glucose homeostasis. It considerably increases the number of model parameters. Then, it can be difficult to determine the parameters correctly. Therefore, we designed a new model that increases the number of parameters modestly, while keeping the minimal requirements on the input data.

Equation (12) relates current blood glucose level (BG) and interstitial fluid glucose level (IG) to future IG (the $b(t)$ and $i(t)$ symbols denote BG and IG at a given time t). Then, we can postulate that changes of current levels determine future rate of change of IG. As a result, the future delay depends on concentration gradient between the current BG and IG. Therefore, Eq. (12) states that the delay is proportional to those IG levels, which are affected by the current BG change [25].

$$\varphi(t) = t + \Delta t + \begin{cases} k \times i(t) \times \frac{i(t) - i(t-h)}{h} & h \neq 0 \\ 0 & h = 0 \end{cases} \tag{12}$$

$$p \times b(t) + cg \times b(t) \times [b(t) - i(t)] + c = i(\varphi(t)) \tag{13}$$

According to study [25]: for the h-long interval, Eq. (12) expresses the effect of a concentration-gradient change rate as it affects IG. The k-parameter is a coefficient of this effect. k and h parameters are less important than p, cg, c and Δt ones [24].

According to study [26]: the p-parameter is an arbitrary constant. It expresses glucose gain from blood, e.g. due to intercellular gaps between the endothelial cells. When compared to Fick's Law of Diffusion [6], $b(t) - i(t)$ is the concentration difference across a membrane, and cg is the surface area, multiplied by membrane permeability, and divided by thickness of the membrane. The c-parameter is an arbitrary glucose level. Besides its physiological relevance, it accommodates calculation error.

4 Comparing the Results

It is difficult to fairly compare two different models of glucose dynamics. There are three obstacles to be solved – model implementation, data set and fitness marker.

4.1 Model Implementation

When a model is published, it is mostly done without a computer program/source code that implements the model. Only the model description is given in the paper. Well-written paper contains enough information to implement the model correctly. One has to be aware that if not all details are given, a seemingly subtle change from the original work can produce a significant deviation in the results. Then, a well-performing model can be rejected because it does not perform as the original paper reported.

4.2 Data Set

Different studies use different experimental setups. Furthermore, the measured data can be pre-processed and/or pre-selected before the models are tested on such data. Therefore, the same data set must be used when comparing two different models of glucose dynamics. Many measured glucose levels will not erase differences due to different experimental setups. For example, a well-compensated diabetic patient is easy to be modelled, when compared to a patient with irregular disturbances in blood glucose level.

Sharing a data set publicly is a subject to law and getting all required approvals. It is considerably easier to obtain data set that will not be shared publicly. However, this hinders the development of glucose models. Despite the diabetes wide spread, a researcher may not get involved just because he/she cannot obtain a representative data set.

While some data sets are shared publicly, the time of their collection matters. As the precision of CGMS sensors tends to increase, such publicly available data sets are getting older. Then, a study based on older data set can be rejected because it no longer corresponds with the current precision of glucose meters.

4.3 Fitness Marker

When a paper devising new model, or a new modification of existing model, is published, it usually claims to outperform another model. Beside using the same data set and having the other model implemented correctly, the chosen fitness marker strongly affects quality of the comparison as well. With diabetes, we can commonly encounter markers such as average and maximum relative error and zone analysis such as Clarke or Parkes' grid. The zone analysis draws a box with measured and calculated quantities on its axes. The box comprises several zones; each zone has a distinct rating that represents severity of the error of the calculated quantity.

We propose to use empirical cumulative distribution function of relative error (ECDF) as the main fitness marker for the following reasons:

1. Giving few markers such as average and maximum relative errors only say nothing about other relative errors.

2. Zone analysis clusters several different relative errors into the same zone, thus effectively making them equal.

With ECDF, we could compare courses of relative error of two different models with a finite number of probabilities (to make ECDF a suitable tool for a paper publication). In addition, ECDF should be given for interstitial fluid glucose level considered as blood glucose level to estimate difficulty of blood glucose level calculation, because different experimental setups will continue to appear.

5 Conclusion and Future Work

In this paper, we presented the background and motivation for the development of models of glucose dynamics. Then, we described selected models to demonstrate various aspects of glucose dynamics modelling. Finally, we have named the main obstacles to fair comparison of different models of glucose dynamics. As the future work, we would like to promote availability of contemporary data sets and usage of common fitness marker – empirical cumulative distribution of relative error. We hope to use the diabetes.zcu.cz portal [27] for this.

Acknowledgement. This publication was supported by the project LO1506 of the Czech Ministry of Education, Youth and Sports.

References

1. Bliss, M.: Rewriting medical history: Charles best and the banting and best myth. J. Hist. Med. Allied Sci. **48**, 253–274 (1993)
2. World Health Organization: The top ten causes of death. Fact sheet No. 310 (2014)
3. Longo, D., Fauci, A., Kasper, A., Hauser, S., Jameson, J., Loscalzo, J.: Harrison's Principles of Internal Medicine. Mc Graw-Hill, New York (2011)
4. Guyton, A.C., Hall, J.E.: Medical Textbook of Physiology. Elsevier Inc., Philadelphia (2006)
5. Poretsky, L.: Principles of Diabetes Mellistus. Springer, New York (2010)
6. Bronzino, J.D.: The Biomedical Engineering Handbook. CRC Press, Boca Raton (2006)
7. Kansagara, D., Fu, R., Freeman, M., Wolf, F., Helfand, M.: Intensive insulin therapy in hospitalized patients: a systematic review. Ann. Intern. Med. **154**, 268–282 (2011)
8. Griesdale, D., de Souza, E., van Dam, R.J., Heyland, D., Cook, K., Malhotra, J., Dhaliwal, A., Henderson, R., Chittock, D., Finfer, D., Talmor, D., Chittock, D.: Intensive insulin therapy and mortality among critically ill patients: a meta-analysis including NICE-SUGAR study data. CMAJ **180**, 821–827 (2009)
9. Clarke, S., Foster, J.: A history of blood glucose meters and their role in self-monitoring of diabetes mellitus. Br. J. Biomed. Sci. **69**, 83–93 (2012)
10. Cengiz, E., Tamborlane, W.W.: A tale of two compartments: interstitial versus blood glucose monitoring. Diabetes Technol. Theret. **11**, S11–S16 (2009)

11. Vaddiraju, S., Lu, D., Burgess, Y., Jain, F.C., Burgess, D.: Technologies for continuous glucose monitoring: current problems and future promises. J. Diabetes Sci. Technol. **4**, 1540–1562 (2010)
12. Gani, A., Gribok, A.V., Lu, Y., Ward, W.K., Vigersky, R.A., Reifman, J.: Universal glucose models for predicting subcutaneous glucose concentration in humans. IEEE Trans. Inf. Technol. Biomed. **14**, 157–165 (2010)
13. Bequette, B.W.: Continuous glucose monitoring: real-time algorithms for calibration, filtering, and alarms. J. Diabetes Sci. Technol. **4**, 404–418 (2010)
14. Guerra, S., Facchinetti, A., Sparacino, G., Nicolao, G., Cobelli, G.D.: Enhancing the accuracy of subcutaneous glucose sensors: a real-time deconvolution-based approach. IEEE Trans. Biomed. Eng. **59**, 1658–1669 (2012)
15. Rebrin, K., Steil, G.: Can interstitial glucose assessment replace blood glucose measurements? Diabetes Technol. Theret. **2**, 461–472 (2000)
16. Steil, G.M., Rebrin, K., Hariri, F., Jinagonda, S., Tadros, S., Darwin, C., Saad, M.F.: Interstitial fluid glucose dynamics during insulin-induced hypoglycaemia. Diabetologia **48**, 1833–1840 (2005)
17. Rebrin, K., Steil, G., van Antwerp, W., Mastrototaro, J.: Subcutaneous glucose predicts plasma glucose independent of insulin: implications for continuous monitoring. Am. J. Physiol. **277**, E561–E571 (1999)
18. Cobelli, C., Dalla Man, C., Sparacino, G., Magni, L., De Nicolao, G., Kovatchev, B.: Diabetes: models, signals, and control. IEEE Rev. Biomed. Eng. **2**, 54–96 (2009)
19. Basu, A., Dube, S., Slama, M., Errazuriz, I., Amezcua, J., Kudva, Y.C., Peyser, T., Carter, R. E., Cobelli, C., Basu, R.: Time lag of glucose from intravascular to interstitial compartment in humans. Diabetes **62**, 4083–4087 (2013)
20. Virtanen, K., Lönnroth, A., Parkkola, R., Peltoniemi, P., Asola, M., Tolvanen, T., Rönnemaa, T., Knuuti, J., Huupponen, R., Nuutila, P.: Glucose uptake and perfusion in subcutaneous and visceral adipose tissue during insulin stimulation in non-obese and obese humans. J. Clin. Endocr. Metab. **87**, 3902–3910 (2002)
21. Bertoldo, A., Peltoniemi, P., Oikonen, V., Knuuti, J., Nuutila, P., Cobelli, C.: Kinetic modeling of [(18)F]FDG in skeletal muscle by PET: a four-compartment five-rate-constant model. Am. J. Physiol. Endocr. Metab. **281**, E524–E536 (2001)
22. Thorn, S.L., de Kemp, R.A.T., Klein, R., Renaud, J.M., Wells, R.G., Gollob, M.H., Beanlands, H., DaSilva, N.: Repeatable noninvasive measurement of mouse myocardial glucose uptake with 18F-FDG: evaluation of tracer kinetics in a type 1 diabetes model. J. Nucl. Med. **54**, 1637–1644 (2013)
23. Haidar, A., Wilinska, M., Graveston, J.A., Hovorka, R.: Stochastic virtual population of subjects with type 1 diabetes for the assessment of closed loop glucose controllers. IEEE Trans. Biomed. Eng. **60**, 3524–3534 (2013)
24. Koutny, T.: Using meta-differential evolution to enhance a calculation of a continuous blood glucose level. Comput. Methods Progr. Biomed. **133**, 45–54 (2016)
25. Koutny, T.: Blood glucose level reconstruction as a function of transcapillary glucose transport. Comput. Biol. Med. **53**, 171–178 (2014)
26. Koutny, T.: Prediction of interstitial glucose level. IEEE Trans. Inf. Technol. Biomed. **16**, 136–142 (2012)
27. Koutny, T., Krcma, M., Kohout, J., Jezek, P., Varnuskova, J., Vcelak, P., Strnadek, J.: On-line blood glucose level calculation. Procedia Comput. Sci. **98**, 228–235 (2016)

Medical Planning: Operating Theatre Design and Its Impact on Cost, Area and Workflow

Khaled S. Ahmed[(✉)]

Electrical Department, Benha University, Banha, Egypt
khaled.sayed@k-space.org

Abstract. Design of Operating rooms department is one of the most complicated tasks of hospital design due to its characteristics and requirements. Patients, staff and tools should have determined passes through operating suite. Many hospitals assume that operating suite is the most important unit in hospital for its high–revenue. Arch Design of these suites is a very critical point to solve an optimized problem in spaces, work flow of clean, dirty and patient in/out in addition to staff together with their relations with adjacent departments. In this study, we will illustrate the most common designs of operating suites and select the most suitable one which satisfy the effectiveness of the operating suite, maximizing throughput, minimizing the costs of necessary, and decreasing the required spaces related to available resources/possibilities. The design should comply with country guidelines, infection control rules, occupational safety and health, and satisfy the maximum befits for patients and staff. A comparative study has been performed on fifteen hospitals and it has been recorded that single input- output technique is the best design.

Keywords: Operating rooms · Design techniques · Workflow · Area and infection control

1 Introduction

Operating suites is a critical point in design any healthcare facilities or hospital due to their needs of definitions and guidelines for sterilization and infection, aseptic practice, equipment and safety, patient and worker safety. Architectural design for this zone plays an important role to satisfy the minimum requirements of those guidelines. Some parameters which should be taken into account in the design process are ceiling, floors, walls, storage area in addition to Heat Ventilation and Air Condition (HVAC). The finished ceiling height of an Operating Room should be a minimum of (~ 3 m) above the floor with selective ceiling materials which may be differentiated in the restricted and semi-restricted areas of the surgical suite (1). Floor materials should be non-porous material, seamless, be readily cleanable which is of primary importance in the surgical suite, as most areas are at high risk for the spread of infections. Also walls should be durable can be disinfected and does not allow bacteria growth. The walls may be anti-bacterial paints, capsules (HPL, Stainless steel, PVC) and glass to have smooth surface against bacteria. Walls should avoid corners and coordinated with casework storage systems (2). Room Finishes, Door with integral sinks should have long-term

© Springer International Publishing AG 2017
I. Rojas and F. Ortuño (Eds.): IWBBIO 2017, Part I, LNBI 10208, pp. 325–332, 2017.
DOI: 10.1007/978-3-319-56148-6_28

durability. Although interior acoustics and/or Noise Control that support speech intelligibility and provide comfort can be difficult to obtain in an operating room where non-porous materials are mandated for infection control requirements, it is important to find ways to control reverberation and noise build-up area (3).

Heat Ventilation and Air Condition is a very important parameter to infection control. It should satisfy specific conditions as temperature, pressure, humidity and if there is a laminar flow. All operating rooms except septic rooms should have positive pressure, total fresh purified air, humidity from 30%–60%, and temperature range 16 C to 20 C. The operating suite department is situated to prevent non-related traffic through the department. It is divided into three distinct areas, which are defined by the physical activities happening through the area and staged in a progressive manner to minimize the potential of cross contamination. The three different areas are the unrestricted area, the semi-restricted area and the restricted area. (1) The unrestricted area includes the central control point that is established to monitor the movement of patients, staff, and equipment. (2) The semi-restricted area ("red line") includes the peripheral support areas of the surgery suite, such as storage areas for clean and sterile supplies, instrument processing areas, scrub sink alcoves, and the corridors leading to the restricted area. (3) Restricted area which access to this area is restricted and appropriate surgical attire as well as coverings for head/facial hair is required (4-8). Many techniques have been used in OR department design, each one has some advantages and dis-advantages based on hospital design (vertical/horizontal), floor, available area, function relations between OR and other department as ICU, ER, others.

1. The first technique is clean core/sterilized zone/sterilized heart system. It has sterilized zone at middle of the operating rooms having all sterilized materials, tools

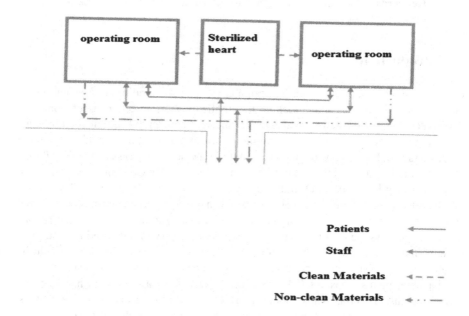

Fig. 1. Sterilized heart technique

and devices. And there is another path to have staff, patients, and un-clean materials as shown in Fig. 1.

2. The second technique is External corridor system. Also it has sterilized zone at middle of the operating rooms having all sterilized materials, tools, devices, and medical staff. And there is another path to have patients, and un-clean materials as shown in Fig. 2.

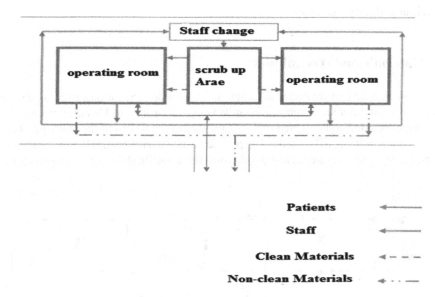

Fig. 2. External corridor system

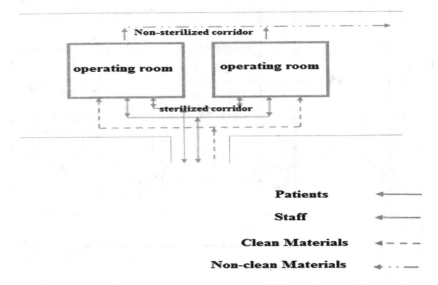

Fig. 3. Two corridor system

3. The third technique is Two corridors system. The first path is sterilized corridor allowing transfer sterilized tools, materials together with medical staff and patients, and there is another corridor (dirty) to have un-clean materials as shown in Fig. 3.
4. The fourth technique is Single corridor. In this system, the operation zone has been designed in one corridor only to have sterilized materials, medical staff and patient taking into account the separation between sterilized and non- sterilized materials (clean and dirty materials).

2 Materials and Techniques

In this paper, based on the previous techniques, fifteen designs for operating zone have been studied in different fifteen general privet hospitals at Egypt. Measured parameters have been recorded and analyzed to determine the best design to fit the maximum benefits. And it is suitable to be used at Egypt. The hospitals have been selected in different cities. These parameters are occupied space area, separated limits between semi-restricted

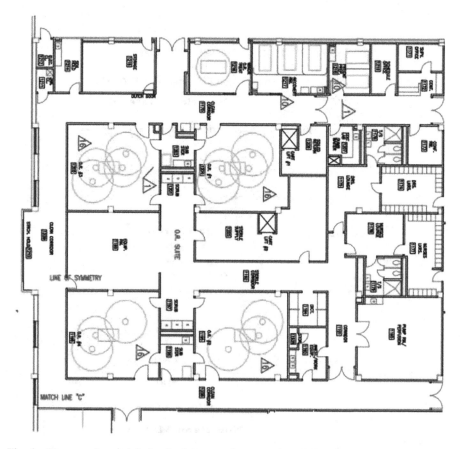

Fig. 4. Clean core hospital design for four operating rooms have HEPA filters and zone separation

and restricted sections, and infection control factors as Air cup sample and wall swab. All conditions inside these zones seem similar from temperature control, positive pressure, HEPA filter and laminar flow. All walls are anti-bacterial paints or fabricated from anti-bacterial materials as (PVC, Stainless steel, glass) (Figs. 4, 5, 6, 7 and 8).

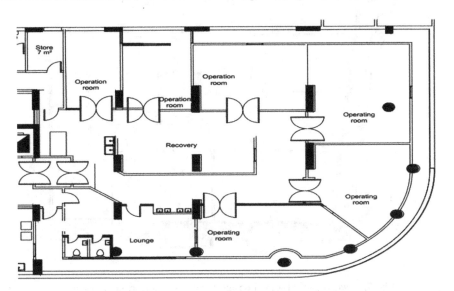

Fig. 5. Two corridor hospital design for five operating rooms

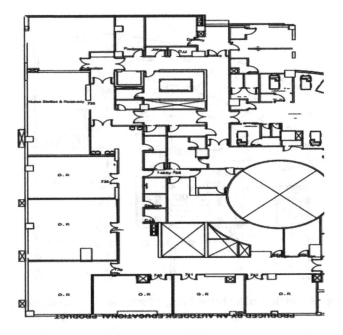

Fig. 6. Single corridor hospital design for six operating rooms

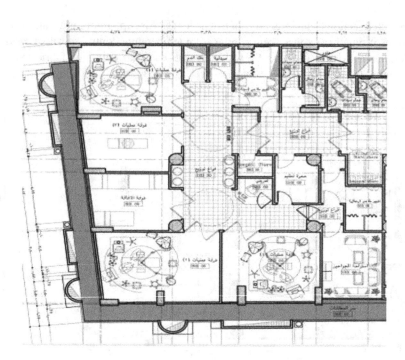

Fig. 7. Two corridor hospital design for three major operating rooms

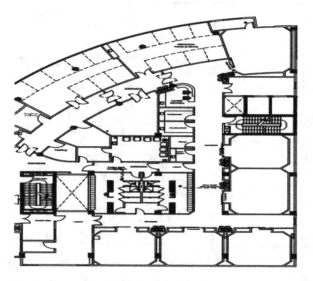

Fig. 8. Single corridor hospital design for six major operating rooms

Hospital location may play an important role against the pollution and infection control especially if it is in an old city filled by factories and high capacity population or traffic. If the hospital is in new city far from smokes and dusts of factories (green/clean area), the hospital will take (N) at location field, otherwise (hospital is in dusty place or occupied zone it will take (O). All hospitals will indicate the status of having HEPA filters in all operating rooms or not (Y/N). The design status will have value 1, 2, 3 and/or 4 based on the used design technique clean core/sterilized heart system, external corridor system, two corridors technique and single corridor technique respectively. Space program for complete operating zone is recognized. Limits separation between semi-restricted area and restricted area should be identified. Finally the air porn sample and swab should indicate the status clean zones or not to satisfy the infection control needs and requirements.

3   Results and Discussion

In this paper, fifteen designs have been selected and studied to determine the best design that fit the minimum requirements of infection control needs and keep the separation among un-restricted, semi restricted and restricted areas. As indicated in Table 1, all hospitals which followed design-3 technique (double/two corridor system) have been failed in all infection control test (air cup sample) especially from the second or dirty corridor in addition to have one hospital has restricted zone infected wall.

Table 1. Hospital number and measured parameters

# Hospital	Location (N/O)	HEPA Filter	Design	Occupied space area (m²)	Limits separation	Infection control	
						Air sample	Wall Swab
Hospital-1	N	Yes	1	1065	Yes	(-ve)	(-ve)
Hospital-2	N	Yes	1	1200	Yes	(-ve)	(-ve)
Hospital-3	O	Yes	1	1300	Yes	(-ve)	(-ve)
Hospital-4	N	Yes	2	990	No	(-ve)	(-ve)
Hospital-5	N	Yes	3	860	Yes	(+ve)	(-ve)
Hospital-6	N	Yes	3	1200	Yes	(+ve)	(-ve)
Hospital-7	O	Yes	3	1160	Yes	(+ve)	(-ve)
Hospital-8	O	Yes	3	1000	Yes	(+ve)	(+ve)
Hospital-9	O	Yes	3	940	No	(+ve)	-ve
Hospital-10	O	Yes	2	1020	Yes	+ve	-ve
Hospital-11	O	Yes	1	1000	No	-ve	-ve
Hospital-12	N	Yes	4	800	Yes	(-ve)	(-ve)
Hospital-13	N	Yes	4	890	Yes	(-ve)	(-ve)
Hospital-14	N	Yes	4	950	Yes	(-ve)	(-ve)
Hospital-15	N	Yes	4	900	Yes	(-ve)	(-ve)

Design 2 or external corridor system has failed in limit separation together with one hospital has infected sample. Design 1 or clean core/sterilized heart designs have been succeed in all samples of labs although it has failed in one limit separation. design 4 single corridor system has been succeed in all samples keeping limit separations and having minimum space areas compared to design-1 which is equal to design-4.

Design-1 space areas are 1000, 1065, 1200 and 1300 m^2 compared to design-4 spaces 800, 890, 900 and 950 m^2 respectively. It has been noted that the maximum area for design-4 (950 m^2) is smaller than the minimum area of design-1 (1000 m^2) keeping the same number of operating rooms (5 operating rooms). Work flow has been satisfied in design-4 in all cases which failed in one hospital at design-1 (limit separation). The finishing cost of dirty corridor or added space area will be increased. Due to the previous notes and records taking into account the type of operation, management, policies and procedures of infection control and quality assurance, the most effective design which satisfy the minimum area and comply with infection control is design-4 or single corridor system.

References

1. American Institute of Architects (AIA): 2006 Guidelines for Design and Construction for Health Care Facilities. AIA Press, Washington (2006)
2. Chefurka, T., Nesdoly, F., Christie, J.: Concepts of flexibility in healthcare facility planning, design and construction, Toronto, Ontario, Canada. http://muhchealing.mcgill.ca/english/Speakers/chefurka_p.html
3. Bartkowski, D.P., Bonter, N.T.: Surgical care in the 21st century. J. Am. Osteopath. Assoc. **105**(12), 545–549 (2005)
4. CDC guidelines and recommendations for ventilation, construction, and renovation of hospitals. http://www.cdc.gov/ncidod/dhqp/gl_construct.html
5. https://www.wbdg.org/ffc/va/space-planning-criteria
6. http://www.cfm.va.gov/til
7. space planning standards for federal healthcare facilities
8. Memarzadeh, F., Manning, A.: Comparison of operating room ventilation systems in the protection of the surgical site. ASHRAE Trans. **108**(Pt. 2) (2002)
9. Memarzadeh, F., Jiang, Z.: Effects of operating room geometry and ventilation system parameter variations on the protection of the surgical site. In: IAQ 2004: Critical Operations: Supporting the Healing Environment through IAQ Performance Standards (2004)

Biomedical Image Analysis

Recognition of Stages in the Belousov-Zhabotinsky Reaction Using Information Entropy: Implications to Cell Biology

Anna Zhyrova$^{(\boxtimes)}$, Renata Rychtáriková, and Dalibor Štys

Faculty of Fisheries and Protection of Waters,
South Bohemian Research Center of Aquaculture and Biodiversity of Hydrocenoses,
Institute of Complex Systems, University of South Bohemia in České Budějovice,
Zámek 136, 373 33 Nové Hrady, Czech Republic
azhyrova@frov.jcu.cz,
http://www.frov.jcu.cz/en/institute-complex-system

Abstract. A common property of a living organism as a non-equilibrium dynamic system is the self-organization including the evolution of this self-organized system through distinct consecutive stages. In this article, the properties of dynamic self-organization is examined on a primitive model of life – the oscillating Belousov-Zhabotinsky (BZ) reaction. This system is sensitive to the changes of external conditions by dynamic reorganization of chemical waves. The generated patterns bring the information on history of the reaction evolution. We performed the pattern classification using calculation of the point information gain entropy density followed by multivariate statistical analysis. It was proved by numerous experiments that each obtained cluster is related to a unique reaction stage with characteristic concentrations of the reactants. The reliability makes this method promising for application to the recognition of stages in variety of complex systems. The results obtained via visual inspection of 6 parallel image series of the BZ reaction together with their statistical analysis approximate cell physiology during development and differentiation of tissues – a small change in the initial conditions leads to a different development of the cell population. This finding also explains a lower reproducibility of measurements of biological systems.

Keywords: Belousov-Zhabotinsky reaction · Information entropy · Pattern recognition · Model of cell biology

1 Introduction

The Belousov-Zhabotinsky (BZ) reaction, which is named after scientists who discovered it [1,2], is a cascade of more than 80 chemical reactions. Starting mechanism and relationship between reaction components remain still unknown.

© Springer International Publishing AG 2017
I. Rojas and F. Ortuño (Eds.): IWBBIO 2017, Part I, LNBI 10208, pp. 335–346, 2017.
DOI: 10.1007/978-3-319-56148-6_29

The hallmark of this chemical process is its ability to change the direction of reaction in the precise frequency which leads to a periodic oscillation. When a thin layer of the reaction mixture is placed onto a vessel, the BZ reaction gives color patterns (chemical waves). These patterns have a complex geometry which changes in time in a distinct order (for the first time analyzed by Winfree [3]) and their shape and duration depend on the actual composition of the reaction mixture and shape of the reaction vessel [4,5]. The time evolution of chemical waves [6,7] proceeds in the sequence from circular target patterns to the variety of spirals (a simply rotating spiral, a meandering spiral, a renascent stable spiral, a convectional unstable spiral, etc.). At the end of the reaction process when the main reagents are exhausted, chemical structures are diluted (waves bleach and disappear at all).

The choice of the BZ reaction as the object of investigation was not accidental, since the observed chemical reaction flow has all initial properties of living objects, such as a dissipative structure [8], ability to self-organization [9], and regime of periodicity [10]. Many processes in the nature, e.g., spiral patterns of *Dictyostelium* molds [11], cardiac muscle [12] or chicken retina [13] can be understood from the point of view of wave transformation.

That make the BZ reaction an extremely valuable pilot object for studying the properties of life [14]. In terms of the bioengineering it means that a causes-effects model explaining the core processes pushing the BZ system to change its states could be applied to the interpretation of the transformation circumstances which underlie more complex forms of life such as cells or biological species.

In this paper, we describe the BZ system using a variable point information gain entropy density [15] which classifies multifractal scaling properties of complex structured systems. The found pattern sequence as well as the duration of individual stages are similar but not identical to properties of living organisms [16].

2 Materials and Methods

2.1 Experiment Setup

The experiments were performed with oscillating bromate-ferroin-bromomalonic acid reaction (a recipe of the Belousov-Zhabotinsky reaction provided by Dr. Jack Cohen [17]). The reaction mixture includes 0.34-M sodium bromate, 0.2-M sulphuric acid, 0.057-M sodium bromide (all from Penta), 0.11-M malonic acid (Sigma-Aldrich) as substrates and a redox indicator and 0.12-M 1, 10-phenanthro-line ferrous complex (Penta) as a catalyst. All reagents were mixed in the above-mentioned sequence. The experiments were performed in a specially constructed thermostat which consists of a Plexiglas aquarium and a low-temperature circulating water bath-chiller. The temperature during the experiments was being kept at 27 °C.

The reaction mixture was placed onto a circular Petri dish with the diameter of 90 mm (this type of vessel was chosen because the majority of experiments described in the literature was performed in it) and mixed counterclockwise using a laboratory three-dimensional orbital shaker TL 10 (Edmund Bühler GmbH, Fisher Scientific) at 14 rpm and angle of tilt $5°$ for 2 min. The experiment was repeated six times.

2.2 Image Processing and Data Calculation Performance

The chemical waves were recorded by a Nikon D90 camera (setting up: Time-lapse shooting 10 s/snapshot, Exposure compensation $+1\frac{1}{2}$ EV, ISO 320, Aperture $\frac{f}{8}$, Shutter speed $\frac{1}{80}$ s). The original 12-bit NEF raw image format was losslessly transformed to an 8-bit PNG format (using a Least Information Loss Converter [18]) and a non-informative image background was cropped using a MATLAB software [19].

So-treated images were further processed using a Image Info Extractor Professional software [20]. The analytic tool for classification of image series was derived from the Rényi entropy

$$I_\alpha = \frac{1}{1-\alpha} \ln \sum_{i=1}^{m} p_i^\alpha, \tag{1}$$

where the parameter α – the Rényi coefficient – $\alpha \geq 0$, $\alpha \neq 1$, indicates the information cost with regards to the examined probability distribution $f(p_i)$. Variable p_i denotes a discrete probability of a given phenomenon i, i.e., in our case, a probability of occurrence of a given intensity in the center of the intersection of the row and column pixel grid of an image (so-called cross type of information surroundings). The total number of intensity levels in the histogram which was created from the intensities on this cross is marked as m. The Rényi entropy provides information characteristics for the multifractal system and is included in the measure of the generalized dimension

$$D_\alpha = \lim_{l \to 0} \frac{I_\alpha(l)}{\ln(l)}, \tag{2}$$

where l is a measure of a spatial element. In the case of an image, the spatial element is a camera pixel and is fixed.

To examine the information contribution of the point (x, y), we calculated the point information gain $\gamma_\alpha(x, y)$ as a difference between the information content of a probability distribution of occurrence of the intensity i in the histogram without ($p_{i,x,y}$) and with (p_i) the examined point:

$$\gamma_{\alpha,x,y} = I_\alpha - I_{\alpha,x,y} = \frac{1}{1-\alpha} \ln \sum_{i=1}^{k} p_{i,x,y}^\alpha - \frac{1}{1-\alpha} \ln \sum_{i=1}^{k} p_i^\alpha = \ln\left(\frac{\sum_{i=1}^{k} p_{i,x,y}^\alpha}{\sum_{i=1}^{k} p_i^\alpha}\right), \tag{3}$$

The relevant probability histograms were created from intensities of pixels around the examined point at the coordinate (x, y). In the Image Info Extractor Professional software, the natural logarithm ln is supplied by the binary logarithm $\log_2$ which gives the information contribution of the given pixel (x, y) in bits.

Unlike the Shannon information entropy (which is a special case of the Rényi entropy for $\alpha = 1$ and is derived for the normal distribution), the α-parameterized Rényi entropy in the computation of $\gamma_{\alpha,x,y}$ enables to specify multifractal structure of the distribution (the intensity histogram) unambiguously via usage of a multiple of the Rényi coefficients α which gives spectra $\gamma_{\alpha,x,y}$ vs. α. By changing the parameter α, we can focus on the probability distribution of one event of interest while suppressing the other. In order to have a general overview on the multifractality of the examined point, we calculated the $\gamma_{\alpha,x,y}$ for thirteen α: 0.1, 0.3, 0.5, 0.7, 1.0 (the Shannon information entropy), 1.3, 1.5, 1.7, 2.0, 2.5, 3.0, 3.5, and 4.0.

To collect the total information of the image, we determined a point information gain entropy H_α as a sum of all $\gamma_{\alpha,x,y}$:

$$H_\alpha = \sum_{x=1}^{s}\sum_{y=1}^{r}\gamma_{\alpha,x,y}, \tag{4}$$

where s and r are dimensions of the image (i.e., numbers of pixels in the rows and columns, respectively). The H_α may be understood as a multiple of the average $\gamma_{\alpha,x,y}$ by the total number of pixels [22]. The full specification of each series image was achieved via calculation of the vectors H_α for each color image channel.

The following cluster analysis (k-means algorithm, squared Euclidian distance) of the obtained spectra H_α vs. α using an Unscrambler X software [21] recognized different evolution states in the chemical system. The optimal number of 7 clusters was defined empirically by the comparison of the cluster dispersion histogram for different amount of clusters with the visual inspection of the image series. The vectors of clusters obtained from six experiments were mutually correlated. To obtain the Pearson's correlation coefficients, we developed an algorithm (written in Matlab) which renumbered the clusters so that the numbers of clusters for each pair of the experiments in Fig. 2 were as much in mutual accordance as possible.

3   Results and Discussions

The course of the reaction was very similar for all experiments:

1. a short latent period, no longer than 400 s, in which no wave exists;
2. a formation of (usually no more than two) wave foci, which further spread over the available reaction surface;
3. a frequent growth of drifting waves;
4. an interaction of wavefronts followed by an initiation of formation of spiral waves in the points of wave breaks;
5. an evolution of the spiral waves, which occupy the two thirds of total reaction time;

6. a damping of chemical oscillation which manifests itself in bleaching of waves and in turning the reaction system into the transparent blue color;
7. the end of the reaction when it is not able to register any change of color of the reaction medium (it is homogeneously blue).

The damping phase seems to be a result of the decay of the chemical energy, while the other phases seem to be independent of the time of chemical decay.

Mainly in Experiment 1 and 4, the waves started to propagate as a concentric structures of irregular geometry at the borders of the dish and further converged into the center of the dish. The border effect also led to the formation of radial patterns with sprained waves at the edges of the vessels followed by their compaction at the center of the dish. The used type of mixing of reagents (Sect. 2) formed a force vector which caused a drift of spiral centers in counterclockwise direction. Also, it should be noted that bleaching of waves started in the center of the reaction space, where the processes such as evaporation of the organic substrate and bromide oxidant (formation of bubbles) was more intensive. This prolonged the attenuation of formation of waves placed near the dish boundaries. Indeed, the reaction process itself represents a transition from one (red reaction surface) to another (complete blue surface) state through many quasistates (target patterns, circular waves, spiral waves, etc.). The samples of photos from 6 parallel experiments are shown in Fig. 1 (the latent reaction period and the end of the BZ reaction is not shown).

Experiment 1: Experiment 1 had the longest latent period from all experiments (200 s). The waves started to be formed at the dish borders (Img. 23) and spread along the perimeter to the dish center (Img. 83). In agreement with literature [6], due to the spatial constrains, the spirals started at the place of interaction of wavefronts (Img. 173). Diffusion processes in the reaction medium promoted rotation of waves, whose centers also drifted anticlockwise through the reaction system under the influence of mixing. After ca. 3000 s, the oscillation damping started (Img. 313) and blue waves broadened till the end of reaction. Even after 5000 s, some weak oscillations were still observed at the dish border (Img. 543). It took the reaction system 2000 s to come to the totally blue stage, where no waves were observed.

Experiment 2: In contrast to Experiment 1, Experiment 2 started after a very short latent period (10 s), when a circular wave center appeared relatively far from the border (Img. 8). The circular wavefronts from this initiation center merged (Img. 83) with those generated at the dish border. In the same ways as in Experiment 1, two waves of different zones of package (more dense and less dense) were formed followed by the phase of spiral waves (Img. 173), the phase of degradation of pattern structures (Img. 313) and the complete attenuation of the reaction due to the exhaustion of the main components (Img. 433). It took 3300 s until the reaction was completely finished.

Experiment 3: As in Experiment 2, the first wave focus appeared far from the dish border (Img. 8), whereas the second one was set at the border. After that, both circular wavefronts spread towards each other. The oscillation center

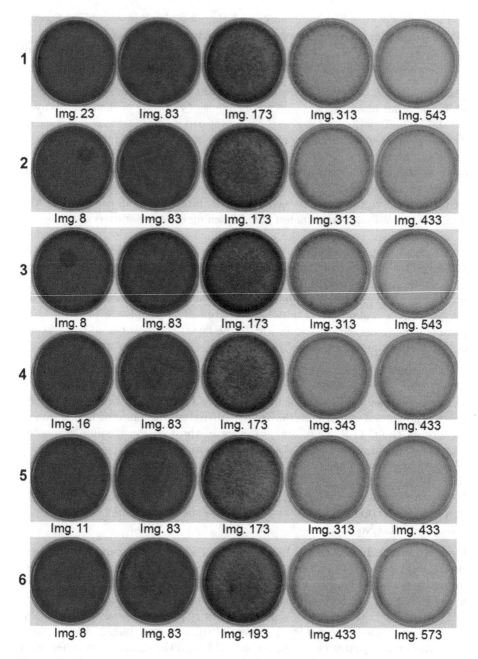

Fig. 1. Key moments in the dynamics of pattern formation in the course of the Belousov-Zhabotinsky reaction (6 experimental repetitions). The serial number of the image indicates the remoteness of the recorded picture on the initial reaction time. All reaction courses exhibit a general law of wave formation from organized circular waves to chaotically oriented spiral waves (*from left to right*). (Color figure online)

was very stable and kept generating the waves (Img. 83) for more than 900 s. More and more wavefronts were evolving and interacted each with other which resulted in their breakage and generation of spiral waves (Img. 173). Spiral waves had more force to shift traveling waves when concurring for the space. The spiral waves evolved to chaotic patterns (Img. 313), which finally ended with the oscillation attenuating (Img. 543). The time of damping phase was equal to Experiment 1 (ca. 200 s).

Experiment 4: As in Experiment 1, Experiment 4 exhibited a long latent period (130 s), when the patterns started to form from the dish border (Img. 16). After that, generated circular waves converged into the center of the dish (Img. 83). As in the previous observations, the spiral waves appeared as a result of the interactions of wavefronts (Img. 173). The following transformation of spirals to the offensive of the chaotic reaction (Img. 343) was finished by the damping of the chemical oscillation (Img. 433). As similar to Experiment 2, the reaction was completely finished after 330 s.

Experiment 5: The course of Experiment 5 almost copied Experiments 2–3: After a 70-s latent period, two wave foci appeared – one focus near the center of the dish and the second focus at the dish border (Img. 11). Such a distribution of the wave centers (Img. 83) led to the wavefronts forming two equivalent spatial clusters with more densely packed waves and less dense ones. The following phase of spiral waves (Img. 173) was replaced by a phase of chaotically oriented patterns (Img. 313). As registered in Experiments 2 and 4, the oscillations finished in 330 s when the first manifestations of the reaction damping was observed (Img. 433).

Experiment 6: The initial period in Experiment 6 took only 30 s, when two wave foci appeared – one of them near the dish border and the second one at the dish border (Img. 8). Due to their short distance, these waves merged soon into a wavefront which was further propagating from one edge of the reaction vessel to the opposite one (Img. 83). Spreading of the joint wavefronts allowed to avoid the breakage of waves for a long time and delayed the occurrence of the phase of spiral waves (Img. 193). Compared with the previous experiments, the onset of the chaotically oriented patterns (Img. 433) as well as the beginning of the oscillation damping (Img. 573) were observed later. This reaction exhibited the shortest period of oscillation damping (1730 s).

In most observed cases (four from six), initial reaction centers were placed far from the dish border. In these cases, after a short latent period, circular waves were smoothly developed. However, if the chemical waves started at the dish border, the course of the BZ reaction underwent significant changes. It thus seems that the formation of structures is to a large extent independent of the chemical composition in the vessel. This conclusion is supported by the observation of (1) re-formation of structures after re-starting a finished experiment by re-shaking the vessel [17] or (2) the change of the course of the BZ reaction in dishes of different shape or size. For instance, compared with experiments in a 200-mm P. dish [25,26] in which waves are generated mainly from the center of the dish, the results described here are more influenced by borders and initial reaction centers are formed more often at the dish border.

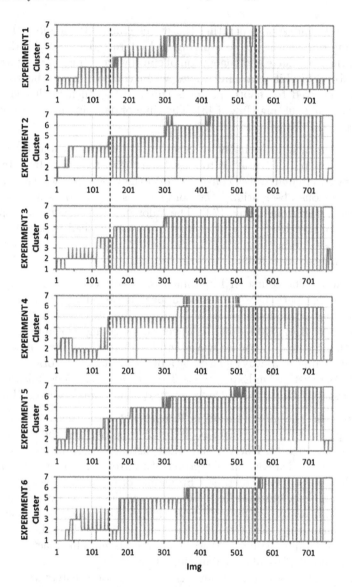

Fig. 2. Time distribution of the reaction phases in the Belousov-Zhabotinsky reaction determined using cluster analysis of vectors of point information gain entropy (6 experimental repetitions). (Color figure online)

The experimental observations of the stages of the BZ reaction are consistent with cluster analysis received from information-entropic data (Fig. 2). The plots of cluster dispersions show clear division of reaction patterns into stages as follows:

1. a wave initiation which is dependent on the length of the latent period (Imgs. 1–30),
2. target waves which completely fill the whole reaction space (Imgs. 31–150),
3. a spiral waves onset which is the same for most of the experiments, regardless of the position of the initial point of reaction (Imgs. 150–200, the beginning of this cluster is marked by a dashed line),
4. spreading of the spiral waves across the reaction space (Imgs. 200–550), and
5. the final stage of reaction oscillation damping, which occurs in the same time for all experiments as well (Imgs. 550–766, the beginning of this cluster is marked by a dashed line).

The distinction of the initial chemical waves from the vessel's border (Fig. 1) is also reflected in the plots of clusters (Fig. 2). When waves started from the dish border (Experiments 1 and 4), the latent period took a longer time, but phases of mixture of spirals and waves came faster – this is presented by two clusters. If the wave foci appeared far from the dish border, the variety of evolved patterns formed several (up to four in Experiment 6) clusters. Anyhow, as notified in the experiments as well as by clustering of experimental data, in all experiments, the stage of spirals and waves began at the same time irrespectively of the position of the reaction initiation centers – in all plots, the beginning of a new cluster took place in Img. 150 (Fig. 2). Cluster analysis allocated evolution of spiral waves to several groups: The first cluster is associated with deployment and followed diffusion of spiral waves. The second cluster begins in the time of switching the phase of spirals and waves to chaotic structures (cf. Imgs. 173 and 313 in Fig. 1). The phase of oscillation damping is the longest stage in all examined experiments. It occupies the biggest cluster in the plots of clusters (Fig. 2) and started approximately at the same time after beginning of the reaction (dashed line through Img. 550 in Fig. 2). The end of the reaction, when no waves were detected, was assigned to the same cluster as the latent period (Experiments 2, 3, 5, and 6). Due to the used calculation approach (i.e., cross point information gain), the distribution of the clusters does not depend on the predominance of blue or red color in the images, but on the presence and shape of the investigated patterns.

The similarities between all six experiments were assessed by correlation analysis of the lengths of each reaction stage that was estimated by the k-mean clustering (Table 1). All experiments showed mutual significant correlation at the significance level $p < 0.05$ with Pearson's coefficients from 0.4585 (experiments 1 and 2) up to 0.9636 (experiments 3 and 5). Obtained correlations as well as the visual inspection of the recorded data confirm our expectation: the way how the chemical waves started to spread change the course of the BZ reaction drastically.

The features detected in the experiments can be interpreted in terms of a proper model of the BZ reaction – an excitable medium of multilevel cellular automaton, which explains qualitatively all observations and simulates an increase/decrease of chemical energy of regular structures [26]. The regular structure is maintained as long as the chemical energy is sufficient and its collapse is to a large extent independent of the operation of the excitable medium.

Table 1. Pearson's correlation coefficients for 6 image series of the Belousov-Zhabotinsky reaction (evaluated for vectors of clusters obtained using vectors of point information gain entropy

Experiment	1	2	3	4	5	6
1	1.0000					
2	0.4585	1.0000				
3	0.6049	0.5615	1.0000			
4	0.6098	0.5742	0.9241	1.0000		
5	0.5412	0.5574	0.9636	0.9051	1.0000	
6	0.6160	0.5284	0.9440	0.9140	0.9538	1.0000

For this reason, the wave damping phase is longer in cases in which the evolution of waves is shorter. In the simulation, the limit set (the last phase) of mixture of spirals and waves is very similar to phase 5 in the experiments. The limit set is a very stable structure with a rather broad zone of attraction.

4   Conclusions

The paper proposed a new approach for a non-invasive detection of developmental phases followed by modeling of the state trajectory of a complex biological object. The transformation of the α-dependent variables point information gain entropy, which correspond to the visible structure of the investigated object, into the space of orthogonal principal components describes macroscopic behavior of a self-organizing system and identifies each development state by a cluster on the charting state trajectory. This mathematical tool for image processing was tested on the series of photographs of the BZ reaction as a primitive simulation of life, with further plans to adopt the method for a wide range of biological phenomena. The obtained models of different modifications of the BZ reaction lead to the conclusion that application of the information-entropic approach to the image processing gives enough specific characteristics for plotting clear mathematical model of the self-developing system and for the automatic recognition of unique states of the investigated process.

The developed procedure with necessary future updates can be used for, e.g., an inexpensive, automatic, contactless recognition of the cell cycle states [23,24, 27,28]. This will bring a major contribution to understanding of the reasons of cell cycle abnormalities that lead to malignization and genesis of cancer, which is a key objective of modern medicine, biology, and related sciences. Under the comparison with other methods, which are traditionally used in cell biology, the data processing using the information-entropic approach is much faster and precise than monitoring of cell growth by a human operator.

Structures which are similar to the final mixture of spirals and waves were observed in many free-evolving self-organized objects. This fact indicates that these structures are results of many – qualitatively similar – processes. When a structure is mechanically constrained, the pattern is changed significantly and the evolution of the structure can be changed up to the level that some stages do not exist. In other words, the properties normally attributed solely to the living systems can be observed in simple chemical mixtures. Based on the correlations of repetitions of the simple self-organizing BZ reaction, the results discussed in the paper also demonstrate a relatively low level of reproducibility of measurements in biological systems.

Acknowledgments. This work was partly supported by the Ministry of Education, Youth and Sports of the Czech Republic – projects CENAKVA (No. CZ.1.05/2.1.00/01.0024) and CENAKVA II (No. LO1205 under the NPU I program).

References

1. Belousov, B.P.: Periodically acting reaction and its mechanism. Collect. Abs. Radiat. Med. **2**, 145–147 (1959)
2. Zhabotinsky, A.M.: Periodic processes of malonic acid oxidation in a liquid phase. Biophysics **9**, 306–311 (1964)
3. Winfree, A.T.: The Geometry of Biological Time, vol. 12. Springer, New York (1980)
4. Zhyrova, A., Štys, D.: Construction of the phenomenological model of Belousov-Zhabotinsky reaction state trajectory. Int. J. Comput. Math. **91**(1), 4–13 (2014)
5. Zhyrova, A., Stys, D., Cisar, P.: Information entropy approach as a method of analysing Belousov-Zhabotinsky reaction wave formation. Chem. Listy **107**(Suppl. 3), S341–S342 (2013)
6. Cross, M.C., Hohenberg, P.C.: Pattern formation outside of equilibrium. Rev. Mod. Phys. **65**, 851–1124 (1993)
7. Jian, L., Zhen-Su, S.: Hierarchical structure description of spatiotemporal chaos (2008). http://arxiv.org/pdf/nlin/0408024.pdf
8. Prigogine, I., Nicolis, G.: Self-Organization in Nonequilibrium Systems: From Dissipative Structures to Order through Fluctuations, 1st edn., p. 512. Wiley-Interscience, New York (1977)
9. Camazine, S., Deneubourg, J.-L., Franks, N.R., Sneyd, J., Theraulaz, G., Bonabeau, E.: Self-Organization in Biological Systems, p. 560. Princeton University Press (2003)
10. Hines, T.M.: Comprehensive review of biorhythm theory. Psychol. Rep. **83**(1), 19–64 (1998)
11. Siegert, F., Weijer, C.J.: Spiral and concentric waves organize multicellular *Dictyostelium* mounds. Curr. Biol. **5**(8), 937–943 (1995)
12. Davidenko, J.M., Pertsov, A.V., Salomonsz, R., Baxter, W., Jalife, J.: Stationary and drifting spiral waves of excitation in isolated cardiac muscle. Nature **355**(6358), 349–351 (1992)
13. Gorelova, N.A., Bures, J.: Spiral waves of spreading depression in the isolated chicken retina. J. Neurobiol. **14**(5), 353–63 (1983)
14. Shanks, N.: Modelling the biological systems: the Belousov-Zhabotinsky reaction. Found. Chem. **3**, 33–53 (2001)

15. Rychtáriková, R.: Clustering of multi-image sets using Rényi information entropy. In: Ortuño, F., Rojas, I. (eds.) IWBBIO 2016. LNCS, vol. 9656, pp. 517–526. Springer, Heidelberg (2016). doi:10.1007/978-3-319-31744-1_46

16. Štys, D., Urban, J., Rychtáriková, R., Zhyrova, A., Císař, P.: Measurement in biological systems from the self-organisation point of view. In: Ortuño, F., Rojas, I. (eds.) IWBBIO 2015. LNCS, vol. 9044, pp. 431–443. Springer, Heidelberg (2015). doi:10.1007/978-3-319-16480-9_43

17. Cohen, J.: Belousov-Zhabotinski Reaction Do-it-Yourself Kit (2009). http://drjackcohen.com/BZ01.html

18. Štys, D., Náhlík, T., Macháček, P., Rychtáriková, R., Saberioon, M.: Least Information Loss (LIL) conversion of digital images and lessons learned for scientific image inspection. In: Ortuño, F., Rojas, I. (eds.) IWBBIO 2016. LNCS, vol. 9656, pp. 527–536. Springer, Heidelberg (2016). doi:10.1007/978-3-319-31744-1_47

19. MATLAB version 7.10.0. Natick, Massachusetts: The MathWorks Inc. (2010)

20. Císař, P., Urban, J., Náhlík, T., Rychtáriková, R., Štys, D.: Image Info Extractor Professional, v. b9 (2015). http://www.auc.cz/software/index5.htm

21. The Unscrambler X version 10.1. Oslo, Norway: CAMO Software (2011)

22. Rychtáriková, R., Korbel, J., Macháček, P., Císař, P., Urban, J., Štys, D.: Point information gain and multidimensional data analysis. Entropy 18(10), 372 (2016)

23. Stys, D., Urban, J., Vanek, J., Cisar, P.: Analysis of biological time-lapse microscopic experiment from the point of view of the information theory. Micron 42, 360–365 (2011)

24. Stys, D., Vanek, J., Nahlik, T., Urban, J., Cisar, P.: The cell monolayer trajectory from the system state point of view. Mol. Biosys. 7, 2824–2833 (2011)

25. Zhyrova, A., Rychtáriková, R., Náhlík, T.: Effect of spatial constrain on the self-organizing behavior of the Belousov-Zhabotinsky reaction. In: IWBBIO 2016, Proceedings Extended Abstracts, Bioinformatics and Biomedical Engineering, pp. 246–258 (2016)

26. Štys, D., Náhlík, T., Zhyrova, A., Rychtáriková, R., Papáček, Š., Císař, P.: Model of the belousov-zhabotinsky reaction. In: Kozubek, T., Blaheta, R., Šístek, J., Rozložník, M., Čermák, M. (eds.) HPCSE 2015. LNCS, vol. 9611, pp. 171–185. Springer, Heidelberg (2016). doi:10.1007/978-3-319-40361-8_13

27. Náhlík, T., Urban, J., Císař, P., Vaněk, J., Štys, D.: Entropy based approximation to cell monolayer development. In: Jobbágy, Á. (ed.) 5th European Conference of the International Federation for Medical and Biological Engineering. (IFMBE), vol. 37. Springer, Heidelberg (2012). doi:10.1007/978-3-642-23508-5_146

28. Rychtáriková, R., Náhlík, T., Smaha, R., Urban, J., Štys Jr., D., Císař, P., Štys, D.: Multifractality in Imaging: application of information entropy for observation of inner dynamics inside of an unlabeled living cell in bright-field microscopy. In: Sanayei, A., Rössler, O.E., Zelinka, I. (eds.) ISCS 2014: Interdisciplinary Symposium on Complex Systems. ECC, vol. 14, pp. 261–267. Springer, Heidelberg (2015). doi:10.1007/978-3-319-10759-2_27

Back Pain During Pregnancy and Its Relationship to Anthropometric Biomechanical Parameters

Julien A. Leboucher[1], Antonio Pinti[2,3(✉)], and Geneviève A. Dumas[4]

[1] LATIM INSERM, CHU Morvan, Brest, France
[2] Université de Valenciennes et du Hainaut-Cambrésis, Valenciennes, France
antonio.pinti@univ-valenciennes.fr
[3] I3MTO, CHR, Orléans, France
[4] Department of Mechanical and Materials Engineering,
Queen's University, Kingston, Canada

Abstract. Numerous studies aiming at testing the relationship between back pain occurrence during pregnancy and demographics, such as parity, age and total body mass, have found conflicting evidence for parity and age, and weak evidence for the total body mass. The aim of this study was to test the possible relationship between anthropometric biomechanical parameters and disability during pregnancy. Anthropometric data were gathered using a stereophotogrammetric method described by Jensen. Data of interest were trunk mass and moments of inertia. The influences of total body mass, age and weight gain on level of disability were also investigated. The latter was quantified using the Oswestry Disability Index (ODI), and painful regions were recorded on a pain drawing. Data were collected at mid- and late-pregnancy, 20 and 34 weeks respectively. Correlations between ODI and other parameters were assessed using the Spearman correlation for ranked data. This test was first performed on the whole mid-pregnancy and late pregnancy samples regardless of pain site. Correlation coefficients were then computed again for sub-samples stemming from the pain drawing data. A positive relationship was found between anthropometric parameters and ODI in the small mid-pregnancy sample, and a negative one has been found in the late pregnancy sample. More specific results appeared for the sub-samples with respect to pain regions. A strong relationship was observed between biomechanical parameters and ODI in the mid-pregnancy low back pain sub-sample. The upper back pain mid-pregnancy sub-sample showed no correlations and the sacroiliac pain group presented rather weak or no relationship with disability. Our findings suggest different etiologies for different kinds of back pain and underline the influence of anthropometric parameters in low back pain observed in pregnancy. Pregnant women suffering from low back pain should be advised to be aware of and control their weight gain as it may have a relationship to lower back pain symptoms. Samples studied here were rather small and further investigation is required to confirm these findings.

Keywords: Biomechanics · Pregnancy · Pain · Anthropometrical parameters · Oswestry score

© Springer International Publishing AG 2017
I. Rojas and F. Ortuño (Eds.): IWBBIO 2017, Part I, LNBI 10208, pp. 347–357, 2017.
DOI: 10.1007/978-3-319-56148-6_30

1   Introduction

The incidence of back pain during pregnancy is above the average for the general population, and is usually reported as over 50% of pregnant women [1–4]. Health care professionals have consequently shown a great interest in back problems during pregnancy. This interest is still growing as assessed by other exhaustive reviews by Wu et al. [5], Bastiaanssen et al. [6], and Timsit [7]. Pain during pregnancy can prevent women from working and it has been cited as a cause of apprehension to become pregnant [8]. The origin of pain in pregnancy is often multifactorial and includes mechanical [9] and hormonal factors [10, 12] as well as poor physical fitness [4]. However, research groups disagree about the relative influence of certain parameters such as age [3, 11, 13, 14], parity [1, 3, 11, 13], and total body mass [15–17]. Moreover, weight gain during pregnancy has been reported as not being related to the incidence of back pain during pregnancy [3, 11, 13], neither have maternal bone density [16, 18] or body height [3, 13, 15, 17]. Few research groups have studied the anthropometric changes of pregnant women [19–21], however Jensen et al. [20, 21] showed significant changes in segment's masses in both upper and lower trunk segments as well as in thighs. As an increase in trunk weight leads to an increase in spine load, the study of such anthropometrical parameters relationship with severity of back pain is of interest. These observations, combined with the fact that biomechanical parameters can be readily modified by specific diet or exercise, motivated our study to focus on possible relationships between trunk biomechanical parameters and different kinds of back pain. This has been achieved through the study of relationship between trunk anthropometric parameters and level of disability due to back pain in working pregnant women.

2   Methods

2.1   Participants

Participants were involved in a larger study led at Queen's University in Kingston (Ontario, Canada) which focused on working pregnant women [22]. To be included in this study women had to be pregnant and employed either full time or part time. Women bearing multiple foetuses or with a pathological pregnancy were excluded in order not to bias results. Data were not collected from women who had stopped working by testing time. The aforementioned study sample consisted of 74 women tested at 20 weeks of pregnancy (n = 18), 34 weeks (n = 20) and at both periods (n = 36). In this study, back pain was assessed by a positive answer to a question asking the pregnant woman if she had back during this pregnancy; over 80% of participants self-reported back pain during their pregnancy. Our study sample included only subjects for which both Oswestry Disability Index (ODI) questionnaire (for women who had back pain) and photographs, for use in Jensen's [23] method, were available. Given the importance of back pain incidence at 20 weeks (16 subjects out of 22, i.e. 72%) and the fact that our complete 34 weeks group complained about back pain, it was decided to restrict our analysis to women who experienced back pain.

Consequently the sample studied consisted of 12 women tested at 20 weeks, 14 tested at 34 weeks and four at both periods, i.e. 30 subjects in all. Mean test dates were (mean (SD)): 20.6 (1.5) weeks and 34.5 (1.5) weeks, for tests at 20 weeks (mid-pregnancy) and 34 weeks (late pregnancy), respectively. This sample's statistics, as well as the ones of the group which they are stemming from, are presented in Table 1.

Table 1. Groups' statistics

Demographics	Larger sample		Study sample		One sample T-test significance
	Mean (SD)	n	Mean (SD)	n	
Age (years)	30.5 (4.2)	72	30.9 (4.2)	30	0.637
Height (cm)	163.7 (6.7)	73	162.8 (5.7)	30	0.421
Weight before pregnancy (kg)	63.2 (11.4)	73	63.6 (11.2)	30	0.876

It is important to notice that the two groups present similar demographic data. This study was approved by Queen's University Ethics Committee and written informed consent was obtained from the subjects prior to testing.

2.2 Anthropometrics

Simultaneous frontal and lateral photographs of the subjects were taken with two Sony Mavica (MVC-FD83) digital cameras, using an experimental set up similar to the one described by Jensen [23] and Jensen and Fletcher [24]. Subjects were standing in a calibrated volume for the photographs. The anthropometrical model used in this study consisted of 15 segments (two feet, two shanks, two thighs, two hands, two forearms, two arms, trunk, neck and head). Their definition was based on Dempster's [25]. Segments were separated from each other using horizontal planes going through joint centres (ankle, knee, wrist and elbow). The thigh segment was separated from the trunk along the inguinal fold in the frontal plane. The neck was separated from the trunk rostral to the collarbones and from the head by a horizontal plane at the level of C1. Figure 1 shows an example of the body segments as used in our study.

It is important to note that the anthropometrical model used in this study merged both trunk segments, leading to a 15 instead of 16 segments model as described by Jensen [23]. This modification was done because, according to Jensen's recommendations, trunk segments are separated by a horizontal plane at the xyphoid process level. This resulted in the breasts being either part of the upper trunk or shared between the lower and upper trunk, depending on the woman's morphology. Merging was achieved by adding segment masses, calculating the coordinates of the new centre of mass and finally adding transferred moments of inertia. Original segment densities, given in the next section were kept to do these computations. Assumption of homogeneous segment density was made for every segment. These densities are shown in Table 2.

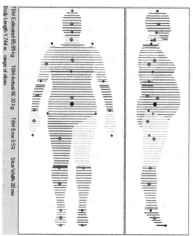

∘ Segment mass center
• Total body mass center
· Proximal and distal reference point

Fig. 1. Body segment separations

Table 2. Homogeneous segments' densities used, from *: Clauser et al. [26], and **: Dempster [25].

Segment	Head	Neck	Upper trunk	Lower trunk	Arm	Forearm	Hand	Thigh	Shank	Foot
Density (g/dm^3)	1070*	1070*	920**	1010**	1060*	1100*	1110*	1040*	1080*	1080*

They correspond to those measured by Clauser et al. [26] for all segments except for trunk segments; for these, Dempster's data [25] were used because they have been considered more appropriate by Jensen et al. [20, 21]. Segments' masses and moments of inertia have been computed according to Jensen's method and using the Slicer software designed at Laurentian University (Sudbury, Ontario, Canada). Sagittal and frontal plane photographs were manually digitized on a PC-type computer. The software then horizontally sliced the segments in constant 20 mm-thick slices. The slices were modelled as elliptical cylinders. Ellipses' axes were measured using slices' length in both views, front and side.

2.3   Questionnaires

The questionnaires, including the Pain Drawing and the modified ODI, were completed by the subjects during testing sessions and aimed at collecting demographic and back pain data. Pain Drawing was inspired by Sturesson et al. [27], and modified ODI questionnaire was inspired by Fairbank et al. [29]. The latter has been modified in order to assess disability due to back pain in retrospect (Fig. 2). The aim of the pain drawing was to distinguish between different locations of back pain (upper back pain, lower back pain and posterior pelvic pain).

2.4   Statistical Analysis

Statistical analysis focused on possible correlations between the Oswestry score (obtained from the ODI questionnaire) and the following variables: age, total body mass (TBM), trunk mass, percentage of mass increase, and moments of inertia with respect to the trunk centre of mass along the anterior-posterior, transverse and longitudinal axes. These tests have been carried out on the whole sample, separately for women tested at 20 and 34 weeks of pregnancy, and then on different sub-samples derived from a classification stemming from the analysis of Pain Drawings. Given the small sample size, Spearman correlation method for ranked data has been chosen because it is more robust than other techniques regarding violation of normal distribution. Significance level has been set at $P < 0.05$ for all computed tests. Statistical analysis has been carried out using commercial package SPSS® version 12.0.

3   Results

3.1   Pain Drawing

Results from the pain drawings, filled in by pregnant women to locate and estimate the severity of back pain symptoms, are presented in Fig. 2A and B, for subjects tested at 20 weeks of pregnancy and for those tested at 34 weeks, respectively.

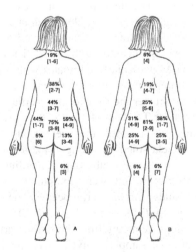

Fig. 2. Pain drawings results at 20 (A) and 34 (B) weeks. Percentages indicate incidence, figures between brackets stand for the range of pain intensity (on a 1 to 10 scale). Blank zones are body parts not recognized as painful.

From Figs. 2A and B, it is important to notice that the most frequently marked zone is the sacroiliac region. This zone presents the highest scores with the adjacent left and right zones. In general, it is the pelvis region in which pain is the most often reported followed by the lumbar region. Less mentioned zones were buttocks at 20 weeks, cervical spine at 34 weeks of pregnancy, and legs for both periods.

3.2   Relationships Between ODI Score and Anthropometric Parameters

In our sample, ODI score ranged from 0.06 to 0.38, and from 0.04 to 0.52 at 20 and 34 weeks of pregnancy, respectively. Mean ODI (SD) were respectively 0.18 (0.10) and 0.22 (0.12) on a scale of 0 to 1. Spearman's coefficients calculated between variables under investigation and ODI scores are shown in Table 3.

Table 3. Spearman's rho coefficients between variables under investigation and ODI score for the complete sample. a: with respect to trunk's centre of mass; *: $P < 0.05$; **: $P < 0.01$

		Age	TBM	Trunk Mass	Percentage of mass increase	Trunk moments of inertia[a]		
						Anterior-posterior Axis	Transverse Axis	Longitu-dinal axis
Spearman's coefficient	20 wks (n = 16)	−0.268 (0.315)	0.517* (0.048)	0.566* (0.022)	0.127 (0.639)	0.616* (0.011)	0.607* (0.013)	0.458 (0.075)
	34 wks (n = 18)	0.136 (0.589)	−0.624** (0.006)	−0.473* (0.048)	0.273 (0.273)	−0.525* (0.025)	−0.524* (0.026)	−0.519* (0.027)

For tests carried out at 20 weeks, TBM, trunk mass and moments of inertia with respect to the anterior-posterior and transverse axes were significantly correlated with ODI score. Positive signs of Spearman's coefficient indicate that ODI scores increased with moment of inertia values. The three other variables (age, percentage of mass increase, moment of inertia with respect to the longitudinal axis) did not significantly correlate with the ODI score. Regarding the sample tested at 34 weeks, all variables under investigation correlated significantly with the ODI score except age and percentage of mass increase. TBM particularly correlates with a significance at the level $P < 0.01$. It is important to notice that, unlike the sample tested at 20 weeks, coefficients are negative, that is the higher the mass or the inertia the lower the ODI score. One should note that, except for TBM at 34 weeks, the coefficients found are rather small and, consequently, none of the correlations were strong. Given the pain drawings results, pregnant women were classified according to the reported location of the pain. The sample was consequently split into subgroups according to location of reported pain. Therefore Spearman's coefficients were computed for the new samples thus created for the rest of the study. Groups have been defined in the following way: (i) upper back pain (UBP) group, (ii) lower back pain (LBP) group, and (iii) posterior pelvic pain (PPP) group. Groups stemming from these new partitions were investigated in the same way as the original samples. Given their small size, n = 3 and n = 4 respectively, UBP and LBP groups for the 34 weeks pregnant subjects were not considered in the analysis. It is important to note that groups complaining about PPP

Table 4. Spearman's rho coefficients between ODI and variables of interest after pain zone sub-sampling. The groups are not exclusive. b: with respect to trunk's centre of mass; UBP Gr.: Upper back pain group; LBP Gr.: Lower back pain group; PPP Gr.: Sacriliac pain group; *: P < 0.05; **: P < 0.01

		Age	TBM	Trunk mass	Percentage of mass increase	Trunk moments of inertia[b]		
						Anterior-posterior axis	Transverse axis	Longitu-dinal axis
Spearman's coefficient (P)	UBP Gr. 20 wks (n = 6)	0.029 (0.956)	0.600 (0.208)	0.429 (0.397)	-0.086 (0.872)	0.486 (0.329)	0.486 (0.329)	0.429 (0.397)
	LBP Gr. 20 wks (n = 7)	−0.805* (0.029)	0.964** (< 0.001)	0.893** (0.007)	0.214 (0.645)	0.929** (0.003)	0.929** (0.003)	0.893** (0.007)
	PPP Gr. 20 wks (n = 13)	−0.118 (0.702)	0.349 (0.243)	0.506 (0.078)	0.041 (0.893)	0.589* (0.034)	0.573* (0.041)	0.351 (0.239)
	PPP Gr. 34 wks (n = 14)	0.090 (0.760)	−0.467 (0.092)	−0.295 (0.305)	0.536* (0.048)	−0.370 (0.192)	−0.410 (0.145)	−0.326 (0.255)

included almost all available subjects and that new sub-samples were not exclusive. Spearman's coefficients are presented in Table 4.

Regarding UBP, correlations found for the whole sample at 20 weeks disappeared. On the contrary, for the LBP-20 weeks sub-sample, correlations were significant at the 0.01 level between anthropometric biomechanical parameters under investigation (TBM, trunk mass, and moments of inertia) and ODI score. It was also interesting to note that this analysis revealed a significant negative correlation between disability intensity and age for this sub-sample only. Concerning the PPP-20 weeks sub-sample, only correlations found with the anterior-posterior and transverse moments of inertia remained significant; a trend was observed with trunk mass. For the 34 weeks-PPP sub-sample, all the correlations found significant for the whole sample were lost, and a positive correlation with the percentage of mass increase reached significance.

4 Discussion

This study aimed at assessing the existence of relations between anthropometrical data and disability. Considering the subjectivity of the reported intensity scores, it was preferred to look at pain drawings only for classification purposes. High occurrence of back pain in Cheng et al. study sample [22] (over 80%), might be explained by several factors: first, over 76% of the subjects reported having already had pain before pregnancy, and this has been pointed as a risk factor by numerous research groups [13, 16, 17, 28]. Second, pain was self-reported and it has been shown by Wu et al. [5] that such questionnaires resulted in a higher occurrence of back pain (around +20%)

than assessed by physicians. Our pain occurrence is beyond the mean reported by these authors. Nevertheless, studies [14, 28, 30], as well as a previous one by Dumas et al. [31], showed similar values. Finally, our study was focusing on back pain and this probably made pregnant women concerned by it more prone to be recruited. Considering trunk segment as a whole anthropometrically was a drawback of our study but was necessary due to women's morphology. This is an issue because, for instance, high values for trunk's mass and moments of inertia will not necessarily lead to an increase of upper spine loading and cause UBP. For this kind of pain, it might be more relevant to look at segments such as and the upper limbs and shoulders. Our pain groups, including both pregnancy test stages, showed the following proportions: PPP 64.5%, LBP 12.9% and mixed pain 22.6%; these figures are very different from those found by Wu et al. [5] in their review (we considered PPP and LBP as the equivalent of their pregnancy-related pelvic girdle pain and pregnancy-related low back pain (PLBP), respectively). Indeed, in their review, pregnancy-related pelvic girdle pain represented around half of the back pain complaints by pregnant women, PLBP around one third, and one sixth for the mixed pain condition. The discrepancy between the proportions may be due to the relatively small number of subjects in this study; however the general trend of the results is in agreement with the literature. Our high incidence of PPP is discussed below. Nevertheless our findings are close to the data from Sturesson et al. [27], whose sample consisted of women between 12 and 40 weeks of pregnancy, except for the sacroiliac region, for which our values are higher (their incidence was 60%), and for the back of the legs (see Fig. 2) for which our values are much lower (their incidence was above 24%). Different results obtained from the same tests performed on different sub-samples at mid-pregnancy suggest that anthropometric parameters are not the only origin of back ache. Indeed, re-sampling pregnant subjects as a function of pain site allowed us to highlight the influence of parameters such as mass and moments of inertia on LBP. Such influence may be overshadowed in studies that do not discriminate over type of pain [32]. Correlation between LBP and moments of inertia might be surprising as the mass increase at this stage of pregnancy is expected to be small. Nevertheless positive correlation has been found [33] between low back pain intensity (measured with a visual analogue scale) and weight before pregnancy. Re-sampling also permitted to demonstrate that mass and moments of inertia were not correlated to ODI for the UBP group. On the contrary, ODI also showed significant correlations with moments of inertia involved in forward and lateral bending of the trunk (i.e. moments along transverse and anterior-posterior axes, respectively). Such correlation between moments of inertia and ODI scores in both LBP and PPP at 20 weeks of pregnancy can be explained by the fact that they lead to greater constraint in the lower spine and surrounding tissues during trunk bending movements. However these findings come from small samples and further research needs to be carried out in order to assess these findings. Concerning the late pregnancy entire sample, the negative correlations found with anthropometrics could be explained by the fact that bigger, or possibly overweight, women may be stronger and proportionally less affected by weight gain and can handle this charge more easily than other women. Reversed coefficients between mid- and late pregnancy might also be explained by the fact that women might have learned to cope with pain through, for instance, a change in spine curvature over the course of pregnancy. Indeed, a correlation was found by

Dumas et al. [31] between both back pain and functional limitations and difference of spine curvature. This correlation was valid when comparing curvatures difference between mid-pregnancy (17–24 weeks) and postpartum with pain during the mid-pregnancy period. In the sample investigated in our study, occurrence of UBP and LBP was twice as high in the 20 weeks group (37.5% and 44%, respectively) as in the 34 weeks group (17% and 22%, respectively). This might have introduced bias in statistics comparing both groups before type of pain sub-sampling. It is possible that UBP and LBP, being a common welfare complaint, lead more often to work stoppage than PPP. As pregnant women had to be working at the period of test, this might also explain the high proportion of PPP in our sample. Considering the negative correlation between pain occurrence and age at mid-pregnancy for the LBP sub-sample, a similar result was found by Östgaard et al. [13], as well as by Wang et al. [14]. However a significant ($P = 0.013$) negative correlation was found between age and TBM, those factors being consequently confounding. Positive correlation found between percentage of mass increase and ODI for the late pregnancy-PPP group indicates that the relative increase of mass seems to be aggravating back pain symptoms even though the mass itself doesn't seem to be involved in ODI score, for this sub-sample. This leads to the conclusion that weight gain should be monitored during pregnancy regarding its possible effects on disability due to back pain. Unfortunately, it has not been possible to carry out tests for UBP and LBP sub-samples for women 34 week pregnant because of the small sample size. In spite of this, it has been shown that correlations found for the whole sample were no longer significant for the PPP sub-sample. It is also important to note that only four women were removed from the whole late pregnancy group to obtain the PPP sub-sample; as the negative correlation disappeared with this removal, further research should be done to confirm this tendency.

5   Conclusion

Our study showed correlations between anthropometric parameters and the ODI. Positive relationships were found at 20 weeks between TBM, trunk mass, and trunk moments of inertia, on one hand, and ODI, on the other hand. These suggest that pregnant women should monitor their weight during pregnancy in order to lessen their impairment. Negative correlations found between the same variables in the late pregnancy sample suggest that adaptation might occur during pregnancy and working pregnant women whose weight tends to be higher appear to be less impaired because of back pain. As these relationships disappeared in the late pregnancy-PPP group further investigation is required. Changes in correlation significance in most of the samples tested after sub-sampling are motivating further investigations as they might explain the lack of agreement, in the literature, between some factors and back pain during pregnancy. The reader is nevertheless advised to consider these results with caution as there were two limitations in this study: only working pregnant women were tested and relationships were assessed on small samples. For these reasons, further investigation is required on larger samples.

References

1. Berg, G., Hammar, M., Möller-Nielsen, J., Lindén, U., Thorblad, J.: Low back pain during pregnancy. Obstet. Gynaecol. **71**, 71–75 (1988)
2. Fast, A., Weiss, L., Ducommun, E.J., Medina, E., Butler, J.G.: Low back pain in pregnancy abdominal muscles, sit-up performance, and back pain. Spine **15**, 28–30 (1990)
3. Mantle, M.J., Greenwood, R.M., Currey, H.L.F.: Backache in pregnancy. Rheumatol. Rehabil. **16**, 95–101 (1977)
4. Östgaard, H.C., Zetherström, G., Roos-Hansson, E., Svanberg, B.: Reduction of back and posterior pelvic pain in pregnancy. Spine **19**, 894–900 (1994)
5. Wu, W.H., Meijer, O.G., Uegaki, K., Mens, J.M.A., van Dieën, J.H., Wuisman, P.I.J.M., Östgaard, H.C.: Pregnancy-related pelvic girdle pain (PPP), I: terminology, clinical presentation, and prevalence. Eur. Spine J. **13**, 575–589 (2004)
6. Bastiaanssen, J.M., de Bie, R.A., Bastiaenen, C.H.G., Essed, G.G.M., van den Brandt, P.A.: A historical perspective on pregnancy-related low back and/or pelvic girdle pain. Eur. J. Obstet. Gynaecol. Reprod. Biol. **120**, 3–14 (2005)
7. Timsit, M.-A.: Pregnancy, low-back pain and pelvic girdle pain. Gynécol. Obstet. Fertil. **32**, 420–426 (2004). (abstract in English)
8. Brynhildsen, J., Hansson, Å., Persson, A., Hammar, M.: Follow-up of patients with low back pain during pregnancy. Obstet. Gynaecol. **91**, 182–186 (1998)
9. Östgaard, H.C., Andersson, G.B.J., Schultz, A.B., Miller, J.A.A.: Influence of some biomechanical factors on low-back pain in pregnancy. Spine **18**, 61–65 (1993)
10. Marnach, M.L., Ramin, K.D., Ramsey, P.S., Song, S.-W., Stensland, J.J., An, K.-N.: Characterization of the relationship between joint laxity and maternal hormones in pregnancy. Obstet. Gynaecol. **101**, 331–335 (2003)
11. Fast, A., Shapiro, D., Ducommun, E.J., Friedmann, L.W., Bouklas, T., Floman, Y.: Low-back pain in pregnancy. Spine **12**, 368–371 (1987)
12. Kristiansson, P., Svärdsudd, K., von Schoultz, B.: Reproductive hormones and aminoterminal propeptide of type III procollagen in serum as early markers of pelvic pain during late pregnancy. Am. J. Obstr. Gynecol. **180**, 128–134 (1999)
13. Östgaard, H.C., Andersson, G.B.J., Karlsson, K.: Prevalence of back pain in pregnancy. Spine **16**, 549–552 (1991)
14. Wang, S.-M., Dezinno, P., Maranets, I., Berman, M.R., Caldwell-Andrews, A.A., Kain, Z. N.: Low back pain during pregnancy: prevalence, risk factors, and outcomes. Obstet. Gynecol. **104**, 65–70 (2004)
15. Damen, L.H., Buyruk, M., Güler-Uysal, F., Lotgering, F.K., Snijders, C.J., Stam, H.J.: Pelvic pain during pregnancy is associated with asymmetric laxity of the sacroiliac joints. Acta Obstet. Gynecol. Scand. **80**, 1019–1024 (2001)
16. Kristiansson, P., Svärdsudd, K., von Schoultz, B.: Back pain during pregnancy. Spine **21**, 702–709 (1996)
17. Orvieto, R., Achiron, A., Ben-Rafael, Z., Gelernter, I., Achiron, R.: Low-back pain of pregnancy. Acta Obstet. Gynecol. Scand. **73**, 209–214 (1994)
18. Björklund, K., Naessén, T., Nordström, M.L., Bergström, S.: Pregnancy-related back and pelvic pain and changes in bone density. Acta Obstet. Gynecol. Scand. **78**, 681–685 (1999)
19. Culver, C.C., Viano, D.C.: Anthropometry of seated women during pregnancy: defining a fetal region for crash protection research. Hum. Factors **32**, 625–636 (1990)
20. Jensen, R.K., Doucet, S., Treitz, T.: Changes in segment mass and mass distribution during pregnancy. J. Biomech. **29**, 251–256 (1996)

21. Jensen, R.K., Treitz, T., Doucet, S.: Prediction of human segment inertias during pregnancy. J. Appl. Biomech. **12**, 15–30 (1996)
22. Cheng, P.L., Dumas, G.A., Smith, J.T., Leger, A., Plamondon, A., McGrath, M.J., Tranmer, J.E.: Analysis of self reported problematic tasks for pregnant women, submitted to Ergonomics (2003)
23. Jensen, R.K.: Estimation of the biomechanical properties of three body types using a photogrammetric method. J. Biomech. **11**, 349–358 (1978)
24. Jensen, R.K., Fletcher, P.: Body segment moments of inertia of the elderly. J. Appl. Biomech. **9**, 287–305 (1993)
25. Dempster, W.T.: Space requirements of the seated operator. WADC Technical report, 55–159, Wright-Patterson Air Force Base, OH (1955)
26. Clauser, C.E., McConville, J.T., Young, J.W.: Weight, volume and center of mass of segments of the human body. AMRL Technical report 69–70. Wright-Patterson Air Force Base, OH (1969)
27. Sturesson, B., Udén, G., Udén, A.: Pain pattern in pregnancy and "catching" of the leg in pregnant women with posterior pelvic pain. Spine **22**, 1880–1884 (1997)
28. To, W.W.K., Wong, M.W.N.: Factors associated with back pain symptoms in pregnancy and the persistence of pain 2 years after pregnancy. Acta Obstet. Gynecol. Scand. **82**, 1086–1091 (2003)
29. Fairbank, J.C.T., Couper, J., Davies, J.B., O'Brien, J.P.: The Oswestry low back pain dissability questionnaire. Physiotherapy **66**, 271–273 (1980)
30. Garshasbi, A., Faghih Zadeh, S.: The effect of exercise on the intensity of low back pain in pregnant women. Int. J. Gynaecol. Obstet. **88**, 271–275 (2005)
31. Dumas, G.A., Reid, J.G., Wolfe, L.A., Griffin, M.P., McGrath, M.J.: Exercise, posture, and back pain during pregnancy (part 2. Exercise and back pain). Clin. Biomech. **10**, 98–103 (1995)
32. Padua, L., Padua, R., Bondì, R., Ceccarelli, E., Caliandro, P., D'Amico, P., Mazza, O., Tonali, P.: Patient-oriented assessment of back pain in pregnancy. Eur. Spine J. **11**(3), 272–275 (2002)
33. Sihvonen, T., Huttunen, M., Makkonen, M., Airaksinen, O.: Functional changes in back muscle activity correlate with pain intensity and prediction of low back pain during pregnancy. Arch. Phys. Med. Rehabil. **79**, 1210–1212 (1998)

Bioimaging - Autothresholding and Segmentation via Neural Networks

Pavla Urbanová[1,2(✉)], Jan Vaněk[2], Pavel Souček[1], Dalibor Štys[1],
Petr Císař[1], and Miloš Železný[2]

[1] Institute of Complex Systems,
South Bohemian Research Center of Aquaculture and Biodiversity of Hydrocenoses,
Faculty of Fisheries and Protection of Waters,
University of South Bohemia in České Budějovice,
Zámek 136, 373 33 Nové Hrady, Czech Republic
losip@frov.jcu.cz
[2] Department of Cybernetics, Faculty of Applied Sciences,
University of West Bohemia in Pilsen, Univerzitní 8, 306 14 Plzeň, Czech Republic

Abstract. Bioimaging, image segmentation, thresholding, and multivariate processing are helpful tools in analysis of series of images from many time lapse experiments. The different methods of mathematic, algorithmization and artificial intelligence could by modified, parametrized or adopted for single purpose case of completely different biological background (namely microorganisms, tissue cultures, aquaculture). However, most of the task is based on initial image segmentation, before features axtraction and comparison tasks are evaluated. In this article, we compare several of classical approaches in bioinformatical and biophysical cases with the neural network approach. The concept of neural network was adopted from the biological neural networks. Th networks need to be trained, however after the learning phase, they should be able to find one solution for various objects. The comparison of the methods is evaluated via error in segmentation according to the human supervisor.

Keywords: Neural networks · Auto-thresholding · Segmentation · Cells · Fish · Images

1 Introduction

Some of the bioinformatics and biophysics experiments requires imaging or bioimaging of the specific biological sample. Especially in the time lapse experiments, where some changes or periods are under investigation. The obtained images has to be processed and analysed. There, are welcome the service of mathematical, informatical and cybernetical methods, namely artificial intelligence, image analysis, and neural networks.

Biological experiments often produce sequences of images. The number of such image sequences is so large that they have to be analysed either automatically or at least semi-automatically. Biological experiments may differ only

© Springer International Publishing AG 2017
I. Rojas and F. Ortuño (Eds.): IWBBIO 2017, Part I, LNBI 10208, pp. 358–368, 2017.
DOI: 10.1007/978-3-319-56148-6_31

slightly from each other; however, even such small changes could lead to significant differences in image processing. Therefore, a re-analysis of the image scene is required with almost every change in experimental condition. Image segmentation algorithms are sensitive to the properties of the image scene. Classical segmentation approaches based on colour space transformations and thresholding methods do not offer an universal solution for variances in the input conditions without the tuning of additional settings. These disadvantages can be easily explained using several examples. Therefore, it is necessary to search for more general segmentation method which could be set by biologically skilled operators. However, these operators would only be able to analyse a few images and not the entire sequence by hand. These hand-segmented images should then be used in the proposed algorithm. The remainder of the sequence processing would run automatically. In this study, a suitable solution based on feed-forward neural networks is tested on three different biological tasks. The number of errors produced by neural networks are significantly lower than that by classical segmentation approaches. Moreover, the results are independent of the used colour space transformation.

Image analysis is often a helpful, perhaps indispensable, tool applied to several issues in biology [1]. The accurate segmentation of biological entities from their background or from each other [2] forms the core of further analysis. The problem is generally independent of the equipment used (e.g. phase contrast microscopy or fluorescence camera) and the scale or object of interest (cells, plant leaves, bugs bodies, etc.). The parameters relevant for segmentation techniques may differ from task to task depending on the properties of the content of the image itself. For example, the image may be transformed to a different colour space to increase object dissimilarity. Some pre-processing steps like filtration and morphological operations [3] are essential for every task to reduce the contribution of noise and to normalize light conditions. Methods for bimodal and multi-modal thresholding are based on different approaches and segmentation criteria [4–8]. Most image processing algorithms were mainly developed for grayscale images. The evolution of algorithms for colour images is often based only on processing independent colour channels as grayscale images [9]. The problem in processing colour images is not only the colour itself but also appropriate colour representation. The source in RGB space is obviously not ideal because then, the images are sensitive to light conditions [10,11]. The results from various colour space thresholding methods may differ from each other, and therefore, the determination of a correct space is non-trivial.

Thresholding algorithms have become important in the image analysis of many biological tasks because of the increasing digitisation of images. Modern measuring devices such as microscopes produce numerous images automatically. The course of an experiment is often observed semi-continuously as an image sequence or a video. This offers the advantage of controlled and relatively constant conditions in image processing during every single run. The result is a constant supply of a large number of data sets exceeding the capacity of analysis by eye and hand. On the other hand, automatic processing requires the initial setting of

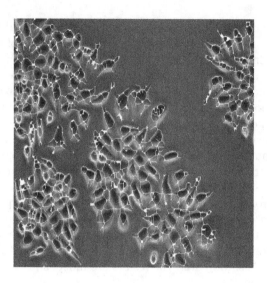

Fig. 1. Example of HeLa cells growth segmentation [12].

parameters or some level of preprocessing, which depends on the task properties, even in the case of non-parametric methods [13]. In life science, the biological conditions of experiments differ only slightly, but from an image processing point of view, the conditions could be completely different from one experiment to another. In such a case, an operator must reinitialize the settings of the automated processing or use other filtration and/or transformation algorithms. This leads to considerable errors in the biological interpretation of the results because the majority of operators are skilled in biological tasks but not in image analysis. Operators only deal with object-to-background segmentation. Such operators require a uniform approach that performs the image segmentation of large sequences automatically together with simple initialization operations. Usually, operators understand the information contained in the images and are able to distinguish the objects present from the background. Therefore, it is simple to demarcate the objects in case of few number of images. With a priori information on the desired results of segmentation, it is possible to prepare the training data for automatic algorithm. After short training of the data, the algorithm may process the entire sequence of images. Classical thresholding methods are not able to use a priori information from training data; therefore, another approach is desirable. The use of feed-forward neural networks (NNs) [14] appears to be a suitable solution for three reasons:

- NNs are classifiers; thus they are able to learn directly from the training data;
- Second, a proper colour space transformation automatically occurs inside an NN during the training process; therefore, NNs are not sensitive to the input colour space;
- NNs are unsupervised, except during training data preparation. No further user intervention is necessary.

In Sect. 2, three different biological tasks are analysed using automatic non-parametric thresholding in three different colour spaces. In Sect. 3, feed-forward neural networks are used for the same biological tasks, and in Sect. 4, the results obtained from the two described approaches are discussed.

2 Classical Thresholding Approach

Simple threshold values are set in many cases only to evaluate object-background dissimilarity. This method is fast and is often sufficiently effective for solving many of segmentation problems. It is based on the search for a local minima or maxima in intensity histograms, which show the thresholds for segmenting an image into the objects that resemble the real objects. It selects parts in the image that correspond to the threshold parameter(s). An increase in the number of threshold parameters may (but not necessarily) improve the segmentation results. The automatic selection of a threshold value(s) becomes more difficult when the situation in an image scene becomes increasingly complicated and includes complexities such as overlapping objects or shadows. Usually, RGB, HSV, or CIELAB colour spaces [15] and their histograms are used as the domains of suitable threshold function(s). For example, Otsu gray-level thresholding is a non-parametric method for automatic threshold selection for image segmentation from an intensity histogram, which maximizes the variance between objects [4].

Three segmentation problems were analysed to compare their characteristic attributes and to find the correct correspondence approach. In each case, the objects differ in shape, colour, number, and background, and so do the observation device and biological purpose. For the satisfactory removal of small blemishes without affecting the overall shapes of the objects, we chose very useful gray-scale morphological operators known as opening-by-reconstruction and closing-by-reconstruction. Opening-by-reconstruction consist of erosion followed by morphological reconstruction, and closing-by-reconstruction is the dual operation of opening-by-reconstruction and dilation, followed by morphological reconstruction [16].

Several fish species such as wells, ide, gold fish, and koi carp (Silurus glanis, Leuciscus idus, Carassius auratus, and Cyprinus carpio; see the first row of Fig. 3) were separated into two groups: experimental and control. In the treatment groups, we used a microalgal biomass as a feed supplement in pelleted feed preparation. We measured all fish individuals several times during the experiment, and hence, we preferred the use of a non-invasive methodology for the colour evaluation of the fish skin. Approximately every 20 days, we took a digital image of each fish. RGB (red, green, and blue) images of the fish species were taken by an ordinary camera Canon PowerShot G5. In general, there is a strict restriction for colour reproduction with respect to the used technology (CCD). The so-called colour gamut is a certain complete subset of colours, which can be accurately represented. Usually, the gamut is a triangle in the CIE colour space (Smith et al. 1932), which represents the entire range of the possible chromaticities and does not cover the entire space of the visible spectrum (See Fig. 2.).

This is a known limitation of the values in the captured images. A device that is able to reproduce the entire visible colour space is an unrealized goal. While modern techniques allow increasingly good approximations, the complexity of these systems often makes them impractical. Therefore, the identification of the colours in an image which are out of the gamut in the target colour space during the processing is critical for ensuring the quality of the final product.

Example (1) for evaluating the changes in the colour of fish skin according to their feeding [17], images of individual fish bodies were taken in a defined time period. A fish was placed approximately in the centre of a scene such that it did not touch the borders. Constituent parts of the fish skin did not share an identical tint. During the experiment, three fish species of different colours were tested. Two of them (Cyprinus carpio and Carassius auratus) were orange, and one (Silurus glanis) was slightly magenta in colour. The most important parameter for the given biological task was average skin saturation. Therefore, it was necessary to capture the entire body of the fish in the image without considering the brightness. The projection of the image into chromatic colour space allows to set segmentation threshold values [10, 18]:

$$r = \frac{R}{R+G+B}, g = \frac{G}{R+G+B}, b = 1 - (r+g). \tag{1}$$

Example (2) ong-run experiments of growing Helacyton gartleri, the genetic chimera of human papillomavirus HPV18, and human cervical cells [19–21] were observed using a phase-contrast microscope. The method of phase contrast allows the observation of soft, colourless, transparent objects, especially living cells. Unfortunately, there are some crucial disadvantages in this method. When the specimen is strongly refracted, a halo effect occurs. This indicates that a very shiny boundary overlaps the real object boundaries. The next limitation is the disappearing of details of saturated objects [22]. The cells themselves change in size and shape during growing and slightly move. This cell movement has the character of vibration rather than translation. The most important parameter from the biological point of view is the evaluation of cell growth that can be expressed as the percent of cells in the observed scene.

Example (3) the growth of Scenedesmus, a genus of colonial green algae with 4 (rarely 2, 8, or 16) cells arranged in a row [23,24], was monitored. This species is a common variety of freshwater plankton, it is used in experimental work on photosynthesis and for artificial cultivation. It has been added to feed (of fish, stock, and humans) because it contains valuable substances (e.g. antioxidants and unsaturated fatty acids). Algae scientists are mainly interested in colonial behaviour and in the influence if cells on each other. The cells were observed using an inverted microscope. Images were taken until the moment of cell splitting. Only one colony was used during a single experiment. RGB, HSV, and chromatic colour spaces were analysed for automatic threshold selection using the Otsu algorithm. The tested images were segmented according to the computed thresholds into two classes: object and background. Reference segmented images were also prepared by hand for every tested image. The 'true' segmented and automatically produced images were then compared. Two separate error rates, for the objects and for the background, were evaluated by the comparison. Detailed results are discussed in Sect. 3.

3   Segmentation via Neural Network

In image segmentation, neural networks are used quite often but mostly as self-organized maps (SOMs) in multi-thresholding tasks [7,9]. However, if training data are available, then the task is similar to a classification problem, and it is preferable to used feed-forward NNs instead. NNs are known as universal approximators and classifiers [14]. The recommended general set-up for non-complicated tasks like thresholding is the following:

- Use of symmetric non-linear activation functions;
- Use of one hidden layer (two hidden layers for more complicated tasks with sufficient training data);
- Setting of the number of neurons in the hidden layer to input dimension (or slightly higher).

The other settings are determined on the basis of the task itself:

- Input dimension of 3 for RGB and similar colour spaces;
- Use of only dimension two for chromatic colour space, because of the third dimension dependency;
- Use of one output neuron.

An example of a feed-forward NN scheme is illustrated in Fig. 1. The NN internal parameters are set by training the data set via a gradient descent algorithm such as back-propagation [14]. The algorithm is iterative and continually improves the initial settings. The number of iterations sufficient for the convergence of the calculation depends on the task and network complexity. The initial parameters setting is generated randomly. Therefore, convergence could yield networks of different quality. It is preferable to run the training process several times and than select the best network. If the classification task is simple and the data are well-distributed in space, the performance of all the networks

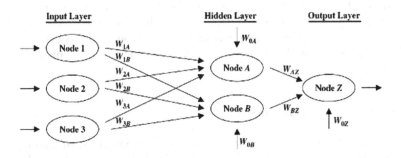

Fig. 2. Example of Feed-forward NN scheme.

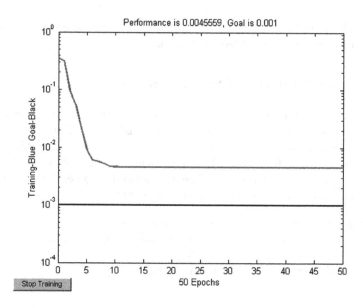

Fig. 3. Example of training process of NN.

would be similar and satisfactory. On the other hand, when the task is more complicated and the data are far from ideal, the training process could fail with some of the initial setting. Network topology complexity influences training stability as well. For an universal binary segmentation algorithm, one network topology was selected after intensive experimentation [25]. It satisfies all the recommendations: It has one hidden layer with three neurons. All the neurons use the same symmetric non-linear activation function called 'tansig'. Matlab neural networks toolbox [26] was used for the training of all the neural networks. Five prepared images were used for training, and another five images were used for testing of each tested domain (three kinds of fish, cell lines, and algae). Because of high memory consumption during training, the selection of pixels for the training set was limited to half a million. The training pixels were selected randomly from the available training data. Hidden layers with different number of neurons were also tested. The best results were obtained for 3–5 neurons; however, the three-neuron topology was expected to demonstrate the best robustness and fastest training. The number of iterations was set to 50. All the networks converged well within this selected number of iterations. An example of the training process is depicted in Fig. 2.

4   Results and Discussion

As expected, the results from automatic non-parametric thresholding differ according to the used colour space (see Table 1). Each task demands a specific colour space. However, this method fails inthe case of more complicated

Table 1. Segmentation error rate of Otsu segmentation method.

Segmentation error rate [%]		RGB			HSV			Chromatic		
		Object	Background	Total	Object	Background	Total	Object	Background	Total
Fish	Cyprinus carpio	44.11	43.05	43.58	1.44	3.74	2.59	7.23	0.00	3.61
	Carassius auratus	73.45	35.01	54.23	0.38	2.20	1.29	1.81	0.00	0.90
	Silurus glanis	46.62	45.78	46.20	25.03	26.90	25.97	1.74	0.00	0.87
Cell line	Helacyton gartleri	48.85	0.50	24.68	51.84	6.79	29.32	78.36	7.65	43.00
Alga	Scenedesmus	42.29	3.62	22.96	47.72	0.13	23.92	39.70	0.15	19.93

Table 2. Segmentation error rate of trained Neural Network.

Segmentation error rate [%]		RGB			HSV			Chromatic		
		Object	Background	Total	Object	Background	Total	Object	Background	Total
Fish	Cyprinus carpio	1.66	0.25	0.95	1.10	0.16	0.63	1.73	0.27	1.00
	Carassius auratus	0.86	0.05	0.45	0.94	0.06	0.50	1.41	0.04	0.72
	Silurus glanis	0.39	0.10	0.24	0.50	0.13	0.31	0.56	0.11	0.33
Cell line	Helacyton gartleri	21.32	1.54	11.43	21.12	1.53	11.33	67.77	1.02	34.40
Alga	Scenedesmus	7.45	0.31	3.88	8.67	0.26	4.47	8.64	0.31	4.48

situations in the observed images, where the boundary between the objects and the background is unclear (Fig. 4).

The input RGB space was not appropriate for fish segmentation because of the various light conditions in the scene. This disadvantage was eliminated using chromatic colours, which produce the lowest error rate across all the three fish species. The thresholding in HSV space for Cyprinus carpio was slightly more successful that in chromatic colour space; however, this method was significantly worse for Silurus glanis. None of the used spaces produced acceptable results for Helacyton gartleri. The object-to-background dissimilarity was difficult to detect even by human eye. Further, the results in all the colour spaces were similar for Scenedesmus algae; here, a high error rate was obtained, especially for the object area. Therefore, the selected thresholding approach was not suitable for transparent and heterogenous objects. It was effective only in the case of fish segmentation with chromatic colour space, provided the properties of the image suited the needs of the task exactly. The transformation to previously used colour spaces was also carried out for comparison with neural networks, both for training and testing data. Commonly, the error rates were almost equal for different input colour spaces (Table 2). The only exception was the segmentation by chromatic colours for Helacyton gartleri, where the removal of brightness information rapidly decreased the quality of object recognition. In other words, segmentation via neural networks is independent of the used input colour space, with the exception of transformations into a space with a lower number of base vectors. In any case, the errors produced by neural networks were less than that by simple thresholding in all the considered cases. Only the preparation of a small training data set (about 5 images for one case, representing 1–3% of entire experiment image sequence) is necessary.

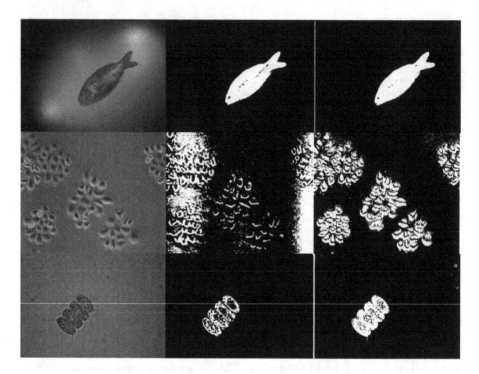

Fig. 4. Illustration of segmentation results. First row is an example of fish, second row is Helacyton Gartleri, third row is Scenedesmus alga. In left column are RGB images, in middle Otsu segmentation using chromatic colors space and in right column are results of neural network in RGB space.

5   Conclusions

In this paper, the differences between classical thresholding segmentation methods and segmentation via feed-forward neural networks were discussed for biological tasks. The proposed approach is suitable for biological tasks that do not have a complicated set-up. It requires the training of only a few images with reference results, which can be prepared by unskilled operators manually or semi-automatically. The approach was tested on three different biological tasks, and it produced significantly lower error rates than classical threshold approaches.

Nowadays, deep neural networks (DNNs) and, especially, convolutional neural networks (CNNs) are very popular in the image analysis domain. Huge amount of data are necessary to successfully training of such large networks. A biologically skilled user is able to annotate only a few images per time-lapse experiment run. It is enough for training of a shallow NN with one or two layers that we used in this paper. In the future work, we would like to use unsupervised and semi-supervised techniques to train DNNs and also CNNs. We also plan to aggregate data from various scale of experiments to support the training of the deep network. The few annotated images may be used to adapt the DNN/CNN to individual experiment conditions.

Acknowledgement. This work was supported by the Ministry of Education, Youth and Sports of the Czech Republic - projects 'CENAKVA' (No. CZ.1.05/2.1.00/01.0024) and 'CENAKVA II' (No. LO1205 under the NPU I program) and FAV.

References

1. Carpenter, A.E., et al.: CellProfiler: image analysis software for identifying and quantifying cell phenotypes. Genome Biol. **7**, R100 (2006)
2. Vanek, J., Urban, J., Gardian, Z.: Automated detection of photosystems II in electron microscope photographs. Tech. Comput. Prague 102 (2006)
3. Beucher, S.: Applications of mathematical morphology in material sciences: a review of recent developments. In: International Metallography Conference, pp. 41–46 (1995)
4. Otsu, N.: A threshold selection method from gray-level histogram. IEEE Trans. Syst. Man Cybern. **SMC–9**, 62–66 (1979)
5. Kapur, J.N., Sahoo, P.K., Wong, A.K.: A new method for gray-level picture thresholding using the entropy of the histogram. Comput. Vis. Graph. Image Process. **29**, 273–285 (1985)
6. Kittler, J., Illingworth, J.: Minimum error thresholding. Pattern Recogn. **19**, 41–47 (1986)
7. Hosseini, H.S., Safabakhsh, R.: Automatic multilevel thresholding for image segmentation by the growing time adaptive self-organizing map. IEEE Trans. Pattern Anal. Mach. Intell. **24**(10), 1388–1393 (2002)
8. Gatos, B., Pratikakis, I., Perantonis, S.J.: Adaptive degraded document image binarization. Pattern Recogn. **39**, 317–327 (2006)
9. Papamarkos, N., Strouthopoulos, C., Andreadis, I.: Multithresholding of color and gray-level images through a neural network technique. Image Vis. Comput. **18**, 213–222 (2000)
10. Yang, J., Lu, W., Waibel, A.: Skin-Color Modeling and Adaptation. Carnegie Mellon University, Pittsburgh (1997)
11. Feris, R.S., de Campos, T.E., Cesar Jr., R.M.: Detection and Tracking of Facial Features in Video Sequences. University of Sao Paulo, Sao Paulo (2000)
12. Urban, J.: Automatic image segmentation of HeLa cells in phase contrast microphotography, Lap LAMBERT Academic Publishing (2012). ISBN-13 978–3846585320
13. Sezgin, M., Sankur, B.: Survey over image thresholding techniques and quantitative performance evaluation. J. Electron. Imaging **13**(1), 146–165 (2004)
14. Ripley, B.D.: Pattern Recognition and Neural Networks. Cambridge University Press, Cambridge (1996). ISBN 978-0521460866
15. Suri, S.J., Setaredan, S.K., Singh, S.: Advanced Algorithmic Approaches to Medical Image Segmentation. Springer, Heidelberg (2002)
16. Vincent, L.: Morphological grayscale reconstruction in image analysis: applications and efficient algorithms. IEEE Trans. Image Process. **2**, 176–201 (1993)
17. Yanong, R.P.E., Curtis, E.W.: Pharmacokinetic studies of florfenicol in koi carp and threespot gourami trichogaster trichopterus after oral and intramuscular treatment. J. Aquat. Anim. Health **17**, 129–137 (2005)

18. Urban, J., Zatkova, I., Vanek, J.: Comparison of the fish skin colour. Tech. Comput. Prague (2006)
19. Van Valen, L., Maiorana, V.C.: HeLa, a new microbial species. Evol. Theory **10**, 71–74 (1991)
20. Bru, A., Albertos, S., Subiza, J.L., Garca-Asenjo, J.L., Br, I.: The universal dynamics of tumor growth. Biophys. J. **85**(5), 2948–2961 (2003)
21. Urban, J., Thornberg, B., Brezina, V., Stys, D.: Evaluation of HeLa cells growth. Tech. Comput. Prague (2007)
22. Jesacher, A., Furhapter, S., Bernet, S., Ritsch-Marte, M.: Shadow effects in spiral phase contrast microscopy. Phys. Rev. Lett. **94** (2005)
23. Reed, H.S.: The growth of scenedesmus acutus. Proc. Natl. Acad. Sci. USA **18**, 23–30 (1932)
24. Gavis, J., Chamberlin, C., Lystad, L.D.: Coenobial cell number in scenedesmus quadrycauda (Chlorophycae) as a function of growth rate in nitrate-limited chemostats. J. Phycol. **15**(s1), 273–275 (1979)
25. Vanek, K., Urban, J., Brezina, V.: Study of scenedesmus alga growth via image analysis. Techn. Comput. Prague (2007)
26. www.mathworks.com/products/neuralnet/

Maximal Oxygen Consumption and Composite Indices of Femoral Neck Strength in a Group of Young Women

Abdel-Jalil Berro[1,2], Nadine Fayad[3], Antonio Pinti[4(✉)],
Georges El Khoury[1,5], Said Ahmaidi[2], Hassane Zouhal[6],
Ghassan Maalouf[7], and Rawad El Hage[1]

[1] Faculty of Arts and Social Sciences, Department of Physical Education,
University of Balmand, El-Koura, Lebanon
[2] EA-3300, APERE, Sport Sciences Department,
University of Picardie Jules Verne, Amiens, France
[3] Faculté de Pédagogie, Université Libanaise, Beirut, Lebanon
[4] I3MTO, EA4708, Université d'Orléans, Orléans, France
antonio.pinti@univ-valenciennes.fr
[5] Laboratoire VIP'S, UFR-APS, Campus la Harpe,
Université Rennes 2, Rennes, France
[6] Movement, Sport and Health Sciences Laboratory (M2S), UFR-APS,
University of Rennes 2-ENS Cachan, Rennes, France
[7] Faculty of Medicine, Bellevue University Medical Center,
Saint Joseph University, Mansourieh, Lebanon

Abstract. The aim of this study was to explore the relationship between maximal oxygen consumption (VO_2 max; mL/mn/kg) and composite indices of femoral neck strength in a group of young women. 41 young women whose ages ranged from 18 to 35 years participated in the present study. Femoral neck bone mineral density was measured by DXA. Composite indices of femoral neck strength (CSI, BSI and ISI) were calculated. VO_2 max (mL/mn/kg) of the participants was measured using a Cosmed Fitmate pro device (version 2.20) while exercising on a bicycle ergometer (Siemens-Elema RE 820; Rodby Elektronik AB, Enhorna, Sweden). VO_2 max (mL/mn/kg) was positively correlated to CSI (r = 0.52; p < 0.001) and ISI (r = 0.45; p < 0.01). The positive associations between VO_2 max (mL/mn/kg) and these two indices (CSI and ISI) remained significant after controlling for body mass index using multiple linear regression models.

Keywords: DXA · VO_2 · Obesity · Bone strength

1 Introduction

Several studies have shown positive associations between anaerobic power and bone strength variables [1–4]. However, little is known concerning the relation between aerobic power and bone strength in young adults [5–10]. Maximal oxygen uptake

© Springer International Publishing AG 2017
I. Rojas and F. Ortuño (Eds.): IWBBIO 2017, Part I, LNBI 10208, pp. 369–375, 2017.
DOI: 10.1007/978-3-319-56148-6_32

(VO$_2$ max) is the maximum rate of oxygen consumption as measured during incremental exercise [5–10]. VO$_2$ max is widely accepted as the single best measure of cardiovascular fitness [5–10]. VO$_2$ max is expressed either as an absolute rate (L/min) or as a relative rate (mL/mn/kg) [5–10]. BMD (Bone Mineral Density) is generally considered as the best determinant of bone strength [11]. However, only 50 to 70 per cent of bone strength variability can be explained by BMD [11–13]. Hip bone strength is also influenced by other factors such as femoral neck width and bending strength [11–13]. Karlamangla et al. [14] have examined the prediction of incident hip fracture risk by composite indices of femoral neck strength (Compression strength index, bending strength index and impact strength index) constructed from DXA scans of the hip. The DXA medical imaging system is a non-invasive system that allows determining body composition and bone mineral density. These indices integrate femoral neck size and body size with bone density [14]. Compression strength index (CSI), bending strength index (BSI) and impact strength index (ISI) reflect the ability of the femoral neck to withstand axial compressive and bending forces and the ability to absorb energy from an impact [15–22]. These indices have been shown to improve hip fracture risk and bone strength assessments in the elderly women [14–16]. In children and young adults, physical activity practice positively influences these indices [18–20]. In young men, obesity is associated with low composite indices of femoral neck strength [22]. In a recent study, we have shown that VO$_2$ max (mL/mn/kg) is a positive determinant of composite indices of femoral neck strength in young men [23]. However, little is known concerning the relationship between maximal oxygen consumption and composite indices of femoral neck strength in young women.

2 Purpose

The aim of the current study was to explore the relationship between maximal oxygen consumption (mL/mn/kg) and composite indices of femoral neck strength in a group of young women.

3 Methods

41 young women (31 normal-weighted (BMI < 25 kg/m2) and 10 overweight (BMI > 25 kg/m2)) whose ages ranged from 18 to 35 years participated in the present study. The 41 participants were recruited from 2 private universities located in North Lebanon. All participants were nonsmokers and had no history of major orthopedic problems or other disorders known to affect bone metabolism or physical tests of the study. Other inclusion criteria included no diagnosis of comorbidities and no history of fracture. An informed written consent was obtained from the participants. The current study was approved by the University of Balamand Ethics Committee. Bone mineral content (BMC) and BMD were determined for each individual by DXA at whole body (WB) (Fig. 2), total hip (TH) and femoral neck (FN) (GE Healthcare, Madison, WI).

Composite indices of femoral neck strength (CSI, BSI and ISI) were calculated as previously described [14].

In Fig. 1, AC is the hip axis length and DE is the femoral neck width. B is the end of the femoral head along the main axis of measurement. The ability of the femoral neck to withstand compressive and bending forces proportional to body weight can be approximated by the following indices:

{1}   Compression Strength Index (CSI) = BMD * femoral neck width (FNW)/weight.
{2}   Bending Strength Index (BSI) = (BMD * FNW^2)/(Hip axis length (HAL) * weight) [15]. Similarly, the ability of the femoral neck to absorb the energy of impact in a fall from standing height is given by:
{3}   Impact Strength Index (ISI) = (BMD*FNW*HAL)/(weight * height) [15].

We directly assessed VO_2 max of the participants using a Cosmed Fitmate pro device (Fig. 3: version 2.20) while exercising on a bicycle ergometer (Siemens-Elema RE 820; Rodby Elektronik AB, Enhorna, Sweden). A progressive 2-min step protocol (20–30 W/step) was used as previously described [23].

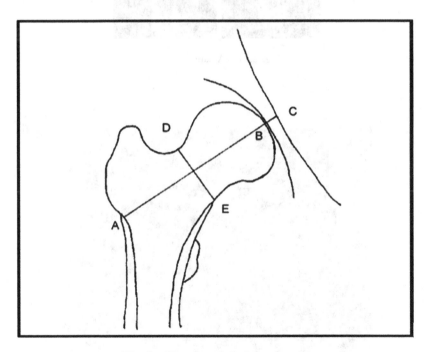

Fig. 1. Geometry of the femoral neck.

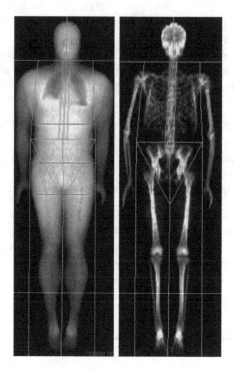

Fig. 2. DXA sample scan.

Fig. 3. Cosemd Fitmate pro device

4 Results

BMI was negatively correlated to CSI (r = −0.48; p < 0.01) and ISI (r = −0.34; p < 0.05). VO_2 max (mL/mn/kg) was positively correlated to CSI (r = 0.52; p < 0.001) (Fig. 4) and ISI (r = 0.45; p < 0.01) (Fig. 5) (Table 1). The positive associations

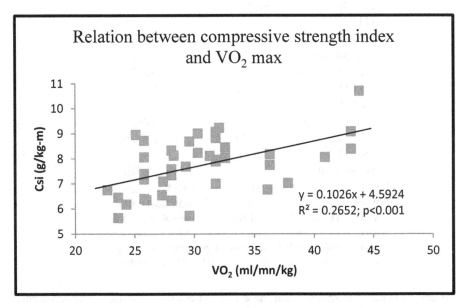

Fig. 4. Relation between compressive strength index and VO_2 max.

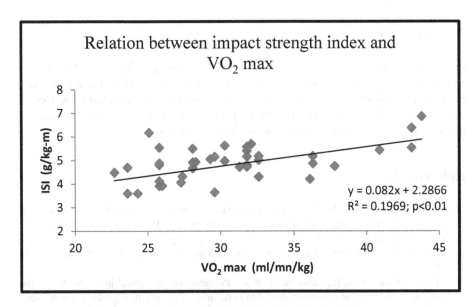

Fig. 5. Relation between impact strength index and VO_2 max.

Table 1. Correlations between clinical characteristics and bone variables

	VO₂ max (ml/min/kg)	p-value
Age (year)	0.200	0.203
Weight (kg)	−0.283	0.0735
Height (m)	0.190	0.235
BMI (kg/m²)	−0.352 *	p < 0.05
FM (kg)	−0.520 **	p < 0.01
LM (kg)	0.106	0.509
WBBMC (g)	0.209	0.189
WBBMD (g/cm²)	0.164	0.305
TH BMD (g/cm²)	0.162	0.319
FN BMD (g/cm²)	0.205	0.197
L1-L4 BMD (g/cm²)	0.265	0.093
CSI (g/kg-m)	0.520 ***	p < 0.001
BSI (g/kg-m)	−0.0461	0.78
ISI (g/kg-m)	0.457 **	p < 0.01

BMI: Body Mass Index; FT: Fat mass; LM: Lean mass; WB: Whole Body; BMC: Bone Mineral Content; BMD: Bone Mineral Density; TH: Total Hip; FN: Femoral Neck; VO₂ max: Maximal Oxygen Consumption; CSI: Compressive strength index; BSI: Binding strength index; ISI: Impact strength index; * p < 0.05; ** p < 0.01; *** p < 0.001.

between VO₂ max (mL/mn/kg) and these two indices (CSI and ISI) remained significant after controlling for body mass index using multiple linear regression models.

5 Conclusion

In conclusion, this study suggests that VO2 max (mL/mn/kg) is a positive determinant of CSI and ISI in young women. To our knowledge, this is the first study to show positive associations between maximal oxygen consumption (mL/mn/kg) and composite indices of femoral neck strength in young women. Enhancing cardiovascular fitness may help to reduce osteoporotic fractures in women.

References

1. Chan, D.C., Lee, W.T., Lo, D.H., et al.: Relationship between grip strength and bone mineral density in healthy Hong Kong adolescents. Osteoporos. Int. **19**(10), 1485–1495 (2008)
2. Zakhem, E., El Hage, R., Bassil, S., et al.: Standing long jump performance is a positive determinant of bone mineral density in young adult women. J. Clin. Densitom. **16**, 129–130 (2012)
3. El Hage, R., Zakhem, E., Zunquin, G., et al.: Performances in vertical jump and horizontal jump tests are positive determinants of hip bone mineral density in a group of young adult men. J. Clin. Densitom. **18**, 136–137 (2013)

4. Vicente-Rodriguez, G., Ara, I., Perez-Gomez, J., et al.: High femoral bone mineral density accretion in prepubertal soccer players. Med. Sci. Sports Exerc. **36**, 1789–1795 (2004)
5. Huuskonen, J., Väisänen, S.B., Kröger, H., et al.: Determinants of bone mineral density in middle aged men: a population-based study. Osteoporos. Int. **11**, 702–708 (2000)
6. Afghani, A., Abbott, A.V., Wiswell, R.A., et al.: Bone mineral density in Hispanic women: role of aerobic capacity, fat-free mass, and adiposity. Int. J. Sports Med. **25**, 384–390 (2004)
7. Huuskonen, J., Väisänen, S.B., Kröger, H., et al.: Regular physical exercise and bone mineral density: a four-year controlled randomized trial in middle-aged men. The DNASCO study. Osteoporos. Int. **12**, 349–355 (2001)
8. Zunquin, G., Berro, A.J., Bouglé, D., et al.: Positive association between maximal oxygen consumption and bone mineral density in growing overweight children. J. Clin. Densitom. (2016, in press)
9. El Khoury, C., Pinti, A., Lespessailles, E., et al.: Physical performance variables and bone mineral density in a group of young overweight and obese men. J. Clin. Densitom. (2016, in press)
10. El Hage, R., Zunquin, G., Zakhem, E., et al.: Maximal oxygen consumption and bone mass in French boys. J. Clin. Densitom. **18**(4), 560–561 (2015)
11. Kanis, J.A., McCloskey, E.V., Johansson, H., et al.: A reference standard for the description of osteoporosis. Bone **42**, 467–475 (2008)
12. Stone, K.L., Seeley, D.G., Lui, L.Y., et al.: BMD at multiple sites and risk of fracture of multiple types: long-term results from the study of osteoporotic fractures. J. Bone Miner. Res. **18**, 1947–1954 (2003)
13. Black, D.M., Bouxsein, M.L., Marshall, L.M., et al.: Proximal femoral structure and the prediction of hip fracture in men: a large prospective study using QCT. J. Bone Miner. Res. **23**, 1326–1333 (2008)
14. Karlamangla, A.S., Barrett-Connor, E., Young, J., et al.: Hip fracture risk assessment using composite indices of femoral neck strength: the rancho bernardo study. Osteoporos. Int. **15**, 62–70 (2004)
15. Yu, N., Liu, Y.J., Pei, Y., et al.: Evaluation of compressive strength index of the femoral neck in caucasians and Chinese. Calcif. Tissue Int. **87**, 324–332 (2010)
16. Ayoub, M.L., Maalouf, G., Bachour, F., et al.: DXA-based variables and osteoporotic fractures in Lebanese postmenopausal women. Orthop. Traumatol. Surg. Res. **100**, 855–858 (2014)
17. Baptista, F., Varela, A., Sardinha, L.B.: Bone mineral mass in males and females with and without down syndrome. Osteoporos. Int. **16**, 380–388 (2005)
18. Sardinha, L.B., Baptista, F., Ekelund, U.: Objectively measured physical activity and bone strength in 9-year-old boys and girls. Pediatrics **122**, e728–e736 (2008)
19. El Hage, R., Zakhem, E., Zunquin, G., et al.: Does soccer practice influence compressive strength, bending strength, and impact strength indices of the femoral neck in young men? J. Clin. Densitom. **17**, 213–214 (2014)
20. El Hage, R.: Composite indices of femoral neck strength in adult female soccer players. J. Clin. Densitom. **17**, 212–213 (2014)
21. El Khoury, C., Toumi, H., Lespessailles, E., et al.: Decreased composite indices of femoral neck strength in young obese men. J .Clin. Densitom. (2016, in press)
22. El Khoury, G., Zouhal, H., Cabagno, G., et al.: Maximal oxygen consumption and composite indices of femoral neck strength in a group of young overweight and obese men. J. Clin. Densitom. (2016, in press)
23. El Hage, R., Zakhem, E., Theunynck, D., et al.: Maximal oxygen consumption and bone mineral density in a group of young Lebanese adults. J. Clin. Densitom. **17**, 320–324 (2014)

Microaneurysm Candidate Extraction Methodology in Retinal Images for the Integration into Classification-Based Detection Systems

Estefanía Cortés-Ancos[1(✉)], Manuel Emilio Gegúndez-Arias[2], and Diego Marin[1]

[1] Department of Electronic, Computer Science and Automatic Engineering,
University of Huelva, Huelva, Spain
{estefania.cortes,diego.marin}@diesia.uhu.es
[2] Department of Mathematics, University of Huelva, Huelva, Spain
gegundez@dmat.uhu.es

Abstract. Diabetic Retinopathy (DR) is one of the most common complications of long-term diabetes. It is a progressive disease that causes retina damage. DR is asymptomatic at the early stages and can lead to blindness if it is not treated in time. Thus, patients with diabetes should be routinely evaluated through systemic screening programs using retinal photography. Automated pre-screening systems, aimed at filtering cases of patients not affected by the disease using retinal images, can reduce the specialist' workload. Since microaneurysms (MAs) appear as a first sign of DR in retina, early detection of this lesion is an essential step in automatic detection of DR. Most of MA detection systems are based on supervised classification and are designed in two stages: MA candidate extraction and further description and classification. This work proposes a method that addresses the first stage. Evaluation of the proposed method on a test dataset of 83 images shows that the method could operate at sensitivities of 74%, 82% and 87% with a number of 92, 140 and 194 false positives per image, respectively. These results show that the methodology detects low contrast MAs with the background and is suitable to be integrated in a complete classification-based MA detection system.

Keywords: Diabetic retinopathy · Microaneurysms · Digital retinal image

1 Introduction

Diabetic retinopathy (DR) is one of the systemic complications of the diabetic mellitus (DM), which is a serious and increasing global health burden – it is estimated that 592 million people suffered from diabetes in 2035 [1].

DR is an important cause of blindness in people of working age in the developed world. It is asymptomatic and it does not interfere with sight until it reaches an advanced stage. People with diabetes are at risk of damage from DR, a condition that can lead to sight loss if it's not treated – 2.6% of global blindness can be attributed to diabetes [2].

© Springer International Publishing AG 2017
I. Rojas and F. Ortuño (Eds.): IWBBIO 2017, Part I, LNBI 10208, pp. 376–384, 2017.
DOI: 10.1007/978-3-319-56148-6_33

Therefore, eye screening for DR is essential for early detection and treatment of this disease. The purpose of screening is to identify asymptomatic individuals who are likely to have DR. In this sense, digital retinal photography is a well-established and cost-effective non-invasive screening method for DR detection [3, 4]. However, increased human resources are required to run these systematic screening programs to accomplish acceptable rates of routine examinations in diabetic patients [5].

This use of digital images of the retinal surface could be exploited for computerized early DR detection. A robust system aimed at filtering cases of patients not affected by the disease using retinal images, would reduce the specialists' workload and increase effectiveness in preventive protocols. Thereby, nowadays, automated systems based on the detection of the earliest visible signs of DR are being developed. One of these signs is the presence of microaneurysms (MA), small red dots on the retina, formed by ballooning out of a weak part of the capillary wall. As a result of this interest, many methods for automated MA detection have been reported over the last years. A review can be found in [6, 7]. They are generally based on supervised classification and are designed in two stages: MA candidate extraction and further description and classification. This work proposes a method that addresses the MA candidate region generation. A proper candidate selection method should reduce false positive rate as well as computing time. Concerning this task, Walter *et al.* [8] proposed a four-step algorithm: pre-processing stage polynomial contrast enhancement applied on green channel image, top-hat transformation applied on the diameter closed image and global thresholding. Fleming *et al.* [9] described contrast normalization, watershed and region growing methods. Quellec *et al.* [10] used optimal wavelet transform, while Hipwell *et al.* [11] applied shade correction to remove changes in the illumination and removed structures greater than MAs. Antal and Hajdu [12] proposed an ensemble-based framework and Oliveira *et al.* [13] explored the Slant Stacking formulation of the Radon transform. Hatanaka *et al.* [14] performed image pre-processing and applied a double-ring filter. Lazar and Hajdu [15] implemented MA candidate detection through the analysis of directional cross-section profiles centred on the local maximum pixels of the preprocessed image. Javidi *et al.* [16] applied a two-dimensional Morlet wavelet transform in all directions and for multiples scales, while Ganjee *et al.* [17] presented a method based on local applying of Markov random field model.

This work proposes a method for MA candidate generation based on the application of a shade-correction algorithm, morphological processing and further dynamic global thresholding. The method has been found adequate to be integrated as the first stage in any classification-based MA detection methodology.

2   Materials

To evaluate the proposed candidate generation method, described in the next section, a test dataset of 83 retinal images was used. These fundus images were digitalized to 1440×960, 2240×1488, 1504×1000 or 1600×1200 pixels, and are 8 bits per colour plane. All fundus images from this database were diagnosed by medical experts with the presence of MAs. These MAs were manually segmented by the experts, which is necessary for methodology evaluation (see Sect. 4). The resulting number of MAs in the whole database is 710.

On the other hand, since database provides no binary masks delineating retina pixels for the images (FOV masks), a FOV mask is generated for each fundus image in the dataset. The use of these masks allows the application of algorithms exclusively within the retina, as well as, to obtain relevant information for the methodology. Thus, fundus image size can be estimated by measuring the diameter of the FOV mask (DFOV). This information is required to adjust all its parameters and confront the methodology to any resolution fundus images. Note that all parameters used were set by experiments carried out on a training database comprising 100 digital retinal images with DFOV 1380 pixels. Therefore, the whole set of the following methodology parameters are referred to 1380 pixel-diameter retinas.

3 Methodology

The MA candidate region generation method works with the green plane (Ig) of the color fundus image RGB. In this channel, MAs present the highest contrast with the background (see Fig. 1 images (a) and (b)). The following process stages can be identified.

First, an averaging filter is applied with the purpose of reducing additive noise from the image:

$$Igm(x,y) = mean\left\{ Ig(i,j) : (i,j) \in S_{x,y}^5 \right\} \tag{1}$$

where Igm denotes the output image and $S_{x,y}^w$ stands for the set of coordinates in a $W \times W$ sized squared window centered on point (x,y).

Second, a shade correction method is performed to enhance the contrast of small and isolated red regions, as well as minimizing the effects of non-uniform illumination throughout the image. The shade correction is obtained as a subtraction of the previous image by its background (Igb). The background is obtained by filtering Igm with a 17×17 median filter:

$$Isc = Igm - Igb \tag{2}$$

with

$$Igb(x,y) = median\{ Igm(i,j) : (i,j) \in S_{x,y}^{17} \} \tag{3}$$

The shade correction algorithm is observed to reduce background intensity variations and enhance contrast in relation to the input mean filtered green channel image (see Fig. 1 images (c) and (d)). Note that the shade-corrected image has been inverted to show the original dark pixels, corresponding to blood vessels and MAs, as bright ones.

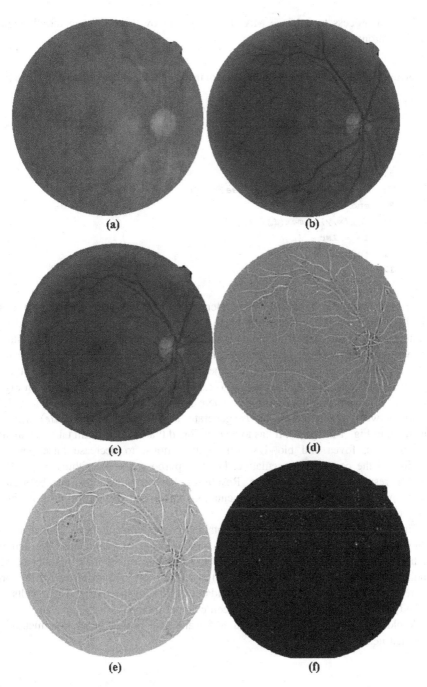

Fig. 1. (a) Original RGB image, (b) green channel (Ig), (c) mean filtered green channel (Igm), (d) shade corrected image, (e) image with suppression of MA (InoMA), (f) MA enhanced image (Ie) (Color figure online)

Then, a process to enhance MAs is carried out by the following morphological transformation:

Algorithm. 1. Pseudocode for the application of the morphological procedure to enhance MA

```
1:   input= Isc
2:   output= Ie
3:   begin
4:   l= 23
5:   for α=0:5:175
7:       I_eroα =imerode(Isc, se(α,l))
8:   end for
9:   I_noMA(x,y) = max{I_eroα(x,y)}
10:  Ie = Isc - I_noMA
11:  return Ie
12:  end
```

where $se\ (\alpha,l)$ is a flat linear structuring element that is symmetric with respect to the neighborhood center, α specifies the angle (in degrees) of the line as measured in a counterclockwise direction from the horizontal axis, and l is approximately the distance between the centers of the structuring element members at opposite ends of the line. For every pixel in the output image, this method selects the maximum value of the set of minimum values generated according to each se rotation. As it can be seen in Fig. 1 image (e), this step produces an image where MAs have been removed (I_{noMA}). Subtraction between this image and the Isc generates a MA-enhanced image (Ie), which is illustrated in Fig. 1 image (f). It has to be mentioned that the algorithm takes advantage of optic disc, fovea and blood vessel segmentations, to decrease false positives and discard the connected vasculature. For this purpose, the algorithms proposed in [18–20], respectively, were applied. Results of these segmentations can be seen in Fig. 1 image (f). Note, that these segmentations have been marked in black in Fig. 1 image (f).

Finally, MA candidate regions are obtained by applying dynamic global thresholding techniques in order to generate a binary image of candidates. We achieve candidate extraction by a bisection algorithm to obtain a predefined number of MA candidate regions. Figure 2 images (b), (c) and (d), show output images of this procedure for 100, 150 and 200 MA candidate regions, respectively. The results are marked in white superimposed to the green channel image.

A block diagram of the proposed MA candidate region generation method is presented next (Fig. 3):

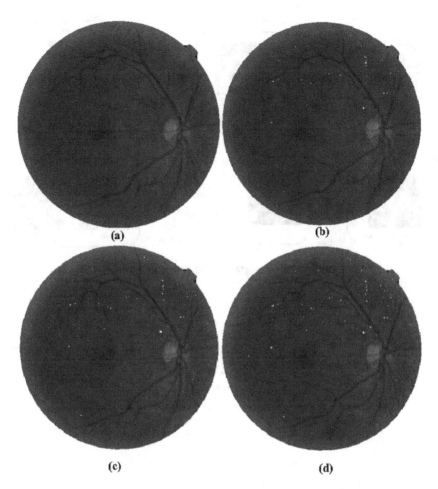

Fig. 2. (a) Green channel image MA candidate images for different numbers of region candidates: (b) 100, (c) 150 and (d) 200

4   Results

The method has been tested in 83 fundus images with presence of MA. These MAs were previously marked and delineated by experts to generate a set of segmented MA regions that serves as ground truth for evaluating the efficiency of the methodology.

Algorithm performance is carried out by calculating the sensitivity of MA candidate detection at different levels of false positive per image (FPR). These false positives were caused mostly by structures that resemble the shape of a MA, like small blood vessels or structures caused by noise.

Thus, the efficiency of the proposed method is measured by the free-response receiver operating characteristic (FROC) curve [21], which plots sensitivity vs. FPR. The FROC curve is showed in Fig. 4. Each point has been generated by selecting a number of regions in our bisection algorithm (see Sect. 3).

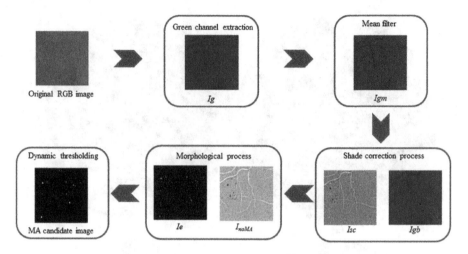

Fig. 3. Block diagram of the proposed method

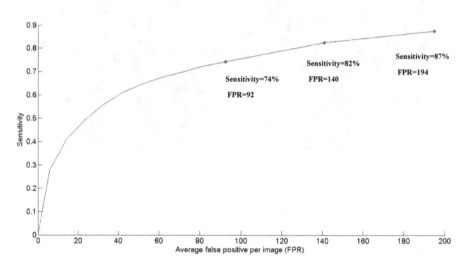

Fig. 4. FROC curve of the proposed method: sensitivity vs. average FP per image with three method operation points.

Possible operating points for the hypothetical practical integration of the method into a complete DR detection system involves sensitivities and FPRs of (74%, 92), (82%, 140) and (87%, 194). These points are indicated in Fig. 4 by red circles. As far as our understanding, no one of the consulted previous works show results of this MA detection stage. These results suggest that the methodology is adequate to be integrated in any automated DR detection system.

5 Conclusions

In this paper we present an automated MA candidate detection methodology based on the application of shade correction, morphological processing and dynamic global thresholding techniques.

The results obtained on a test dataset of 83 fundus images, which were manually marked by ophthalmologists for method evaluation, indicate that the most of the MAs are detected with a reasonable number of false positives. Therefore, the method has been found adequate to be integrated as the first stage in any classification-based MA detection methodology. In any case, it should be underlined that this conclusion is drawn from the results obtained in a small fundus image dataset. Progress in the integration of the proposed method into complete DR systems is definitely associated with the need of carry out a comprehensive system evaluation.

References

1. Guariguata, L., et al.: Global estimates of diabetes prevalence for 2013 and projections for 2035. Diabetes Res. Clin. Pract. **103**(2), 137–149 (2014)
2. Bourne, R.R., et al.: Causes of vision loss worldwide, 1990–2010: a systematic analysis. Lancet Glob. Health **1**, e339–e349 (2013)
3. Abramoff, M.D., et al.: Evaluation of a system for automatic detection of diabetic retinopathy from color fundus photographs in a large population of patients with diabetes. Diabetes Care **31**(2), 193–198 (2008). doi:10.2337/dc07-1312
4. Khan, T., et al.: Preventing diabetes blindness: cost effectiveness of a screening programme using digital non-mydriatic fundus photography for diabetic retinopathy in a primary healt care setting in South Africa. Diabetes Res. Clin. Pract. **101**, 170–176 (2013)
5. Soto-Pedre, E., Navea, A., Millan, S., Hernaez-Ortega, M.C., Morales, J., Desco, M.C., Pérez, P.: Evaluation of automated image analysis software for the detection of diabetic retinopathy to reduce the ophthalmologists' workload. Acta Ophthalmol. **93**, e52–e56 (2015). doi:10.1111/aos.12481
6. Mane, V.M., Jadhav, D.V.: Review: progress towards automated early stage detection of diabetic retinopathy: image analysis systems and potential. J. Med. Biol. Eng. **34**, 520–527 (2014)
7. Mookiah, M.R.K., Acharya, U.R., Chua, C.K., Lim, C.M., Ng, E.Y.K., Laude, A.: Computer-aided diagnosis of diabetic retinopathy: a review. Comput. Biol. Med. **43**(12), 2136–2155 (2013)
8. Walter, T., Massin, P., Erginay, A., Ordonez, R., Jeulin, C., Klein, J.-C.: Automatic detection of microaneurysms in color fundus images. Med. Image Anal. **11**(6), 555–566 (2007)
9. Fleming, A.D., Philip, S., Goatman, K.A., Olson, J.A., Sharp, P.F.: Automated microaneurysm detection using local contrast normalization and local vessel detection. IEEE Trans. Med. Imaging **25**(9), 1223–1232 (2006)
10. Quellec, G., Lamard, M., Josselin, P.M., Cazuguel, G., Cochener, B., Roux, C.: Optimal wavelet transform for the detection of microaneurysms in retina photographs. IEEE Trans. Med. Imaging **27**(9), 1230–1241 (2008)

11. Hipwell, J.H., Strachan, F., Olson, J.A., McHardy, K.C., Sharp, P.F., Forrester, J.V.: Automated detection of microaneurysms in digital red-free photographs: a diabetic retinopathy screening tool. Diabetic Med. **17**(8), 588–594 (2000)

12. Antal, B., Hajdu, A.: An ensemble-based system for microaneurysm detection and diabetic retinopathy grading. IEEE Trans. Biomed. Eng. **59**(6), 1720–1726 (2012)

13. Oliveira, J., Minas, G., Silva, C.: Automatic detection of microaneurysm based on the slant stacking. In: Proceedings of the 26th IEEE International Symposium on Computer-Based Medical Systems, Porto, pp. 308–313 (2013). doi:10.1109/CBMS.2013.6627807

14. Hatanaka, Y., Inoue, T., Okumura, S., Muramatsu, C., Fujita, H.: Automated microaneurysm detection method based on double ring filter and feature analysis in retinal fundus images. IEEE (2012). ISBN: 978-1-4673- 2051-1

15. Lazar, I., Hajdu, A.: Retinal microaneurysm detection through local rotating cross-section profile analysis. IEEE Trans. Med. Imaging. **32**(2), 400–407 (2013)

16. Javidi, M., et al.: Vessel segmentation and microaneurysm detection using discriminative dictionary learning and sparse representation. Comput. Methods Programs Biomed. **139** 93–108 (2016)

17. Ganjee, R., Azmi, R., Ebrahimi Moghadam, M.: J. Med. Syst. **40**, 74 (2016). doi:10.1007/s10916-016-0434-4

18. Marín, D., Aquino, A., Gegúndez-Arias, M.E., Bravo, J.M.: A new supervised method for blood vessel segmentation in retinal images by using gray-level and moment invariants-based features. IEEE Trans. Med. Imaging **30**(1), 146–158 (2011)

19. Marin, D., et al: Obtaining optic disc center and pixel region by automatic thresholding methods on morphologically processed fundus images. Comput. Methods Programs Biomed. **118**(2), 173–185 (2015). http://dx.doi.org/10.1016/j.cmpb.2014.11.003

20. Gegúndez-Arias, M., Marin, D., Bravo, J., Suero, A.: Locating the fovea center position in digital fundus images using thresholding and feature extraction techniques. Comp. Med. Imaging Graph. **37**(5–6), 386–393 (2013)

21. Chakraborty, D.P.: Clinical relevance of the ROC and free response paradigms for comparing imaging system efficacies. Radiat. Prot. Dosimetry **139**(1–3), 37–41 (2010)

A Bio-inspired Algorithm for the Quantitative Analysis of Hind Limb Locomotion Kinematics of Laboratory Rats

Josué González-Sandoval[1], S. Ivvan Valdez-Peña[2], Sergio Dueñas-Jiménez[3], and Gerardo Mendizabal-Ruiz[1(✉)]

[1] Departamento de Ciencias Computacionales, Universidad de Guadalajara, Guadalajara, Jalisco, Mexico
gerardomendizabal@gmail.com
[2] Centro de Investigación en Matemáticas, Guanajuato, Gto., Mexico
[3] Departamento de Neurociencias, Universidad de Guadalajara, Guadalajara, Jalisco, Mexico

Abstract. Rodents are commonly used as biological models in research related to the development of new drugs and treatments. The locomotion analysis of laboratory rats remains a great challenge due to the necessity of time-consuming manual annotations of video data required to obtain quantitative measurements of the kinematics of the rodent extremities. In this article, we address the problem of the synthesis of the hind limb of rats, with the aim of performing an automatic analysis of its kinematics. For this purpose, we recorded video sequences of rats with marks in the posterior hind limbs and we record them while walking in a controlled environment. We perform segmentation of the marks on each frame to generate a binary image corresponding to the marks. A genetic algorithm combined with a local search procedure is used to find an approximation to the optimum hind limb model which best fits the data of the segmented images. The results indicate the feasibility of using the proposed algorithm for automatically detecting the articulation configuration of rats hind limb.

Keywords: Genetic algorithm · Local search · Hind limb · Kinematics analysis

1 Introduction

Rodents are commonly used as biological models for research in several medical-related fields. They play a critical role in the development of new drugs and treatments, for instance, they are used for the assessment of the affectation of diseases, injuries, and medical conditions on the locomotion capacity as well as the locomotor recovery after the application of a treatment [7–9].

Currently, there exist methods used to assess locomotion of rats (e.g., Tarlov scale [4] and Basso, Beattie and Bresnahan (BBB) locomotor scale method [5]).

© Springer International Publishing AG 2017
I. Rojas and F. Ortuño (Eds.): IWBBIO 2017, Part I, LNBI 10208, pp. 385–396, 2017.
DOI: 10.1007/978-3-319-56148-6_34

However, the major drawback of these methods is that they only provide qualitative data. Another approach consists of performing quantitative locomotion analysis on laboratory rats by the use of a transparent tunnel that is designed for video recording the gait of the animals and the subsequent analysis of the acquired images [1–3]. One of the quantitative methods that have been used to assess the kinematics of the rats under different treatments consist of the extraction and characterization of the patterns produced by the angles that are formed between the joints of the extremities of a rat [6]. For this, the posterior hind limbs of the rats are marked to represent the joints of the hind limb from the hip to the ankle (Fig. 1(a)). Then, the curves corresponding to changes of the joint angles with respect to time can be used to compare the gait patterns of different experimental rat groups.

The main limitation of this approach is the significant amount of time required for the manual annotation of the frames of the videos. Moreover, the person who is doing the annotations may perform poorly due to lack of experience or fatigue. Therefore, there is a necessity for automation of the annotation process and the subsequent computation of angle variations with respect to time. In this work, we present advances towards an automatic annotation method that is based on the use of a bio-inspired algorithm that fits a kinematic model of the posterior hind limb of rats to a binary image corresponding to the segmented marker of an image of the rat while walking. The bio-inspired algorithm combines a genetic algorithm for a group of the optimization variables with a local search for a second group of the optimization variables. Our preliminary results indicate the feasibility of employing the proposed approach for the automatic annotation and analysis of the locomotion patterns of the posterior extremities of laboratory rats.

2 Data Acquisition and Hind Limb Markers Segmentation

Laboratory rats were marked on their posterior hind limb using a red water-based non-toxic marker (Fig. 1(a)). Then, the rats were video recorded while walking trough a transparent Plexiglas tunnel at 90 frames per second at a resolution of 640×480.

An observer performed manual segmentation of the colored region from a set of frames from a gait video. As a result, a binary image $S(x, y)$ is obtained where the set of pixels with the value of 1 represents the bones of the hind limb (Fig. 1(b)).

3 Optimization Problem

The proposed method is based on the adjustment of a rat posterior hind limb kinematic model to the binary image representing the markers on the rats leg (Fig. 2). The proposed model include the joints named as $P_1, \ldots, P_5$, the lengths of each element of the leg are denoted by $l_1, \ldots, l_4$ and the angle of each element by $\theta_1, \ldots, \theta_4$.

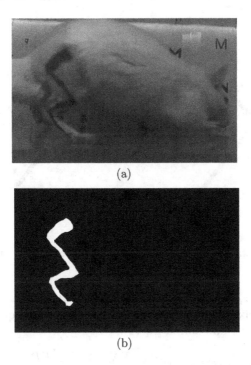

Fig. 1. (a) Example of a frame from the rat's gait video sequence, and (b) its corresponding binary image produced by a manual segmentation $S(x,y)$. (Color figure online)

Given (x_1, y_1) the kinematic model is given by:

$$x_n = x_{n-1} + l_{n-1}(cos(\theta_{n-1})) \tag{1}$$
$$y_n = y_{n-1} + l_{n-1}(sin(\theta_{n-1})) \tag{2}$$

for $n = 2\ldots 5$. Thus, a candidate solution of the problem is given by:

$$X_c = [i_{P_1}, l_1, l_2, l_3, l_4, \theta_1, \theta_2, \theta_3, \theta_4, w_1, w_2, w_3, w_4] \tag{3}$$

where i_{p_1} is an index to get P_1, l_i, for $i = 1\ldots 5$, are lengths of the leg elements, θ_i and w_i are angles and widths of the elements, respectively. The first five variables are optimized via a genetic algorithm (GA), while the remaining are optimized using local search. The first point $P_1 = (x_1, y_1)$ is generated randomly within a small group of pixels G_i; The segmented leg in the image $S(x,y)$ in Fig. 1(b) is divided into ten sets of the same size; the group of pixels G_i used to find the first point is the set located in the uppermost part of this portion.

Given a candidate solution X_c, an artificial image $E(x,y)$, as the one shown in the Fig. 3(a), is generated using the kinematic model in Eq. (2).

The optimization problem is to find a vector of parameters X_c^* which produces the image $E(x,y)$ which best fits the segmented image $S(x,y)$. Figure 3(a) depicts

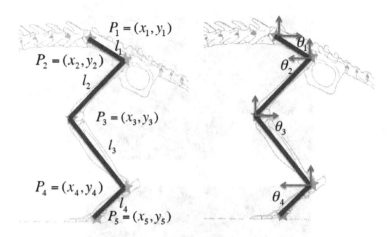

Fig. 2. Kinematic model of the rat posterior hind limb.

an example of $E(x, y)$ superimposed to $S(x, y)$, where FP are true positives, which represents pixels set to 1 in both images. The fitting quality is measured through the objective function in Eq. (4).

$$\max f(X_c) = 1 - \frac{FN}{2\sum_{x,y} S_{xy}} - \frac{FP}{2\sum_{x,y} E_{xy}}. \tag{4}$$

Where FN are false negatives, that is to say, pixels set to 0 in E and 1 in S, FP stands for false positives, that is to say, pixels set to 1 in E and 0 in S, and the denominators are two times the total number of pixels set to 1 in S and E, respectively. In consequence, the maximum and minimum possible values in the second and third term is 0.5 and 0, respectively. Hence, the maximum value of the objective function is 1, when E perfectly fits S, and 0, when the areas covered by them are completely different each other. Notice that the position, length and width of each element impacts the objective function value of a candidate solution.

4   Genetic Algorithm with Local Search for Computing Posterior Hind Limb Model Parameters

We implemented a simple genetic which uses the same operators than the Elitist Non-Dominated Sorting Genetic Algorithm II (NSGA-II) [20], due to it has shown an outstanding performance and is currently one of the most widely used evolutionary algorithms. The genetic algorithm search for the optimal initial point P_1 and lengths l_i of the kinematic model, it is described in Algorithm 1. In line 2 a set of random vectors is generated and stored as the initial population P^0, then, in line 3, each vector is evaluated on the objective function, and the result is stored in F, the evaluation also considers the local search (LS) procedure;

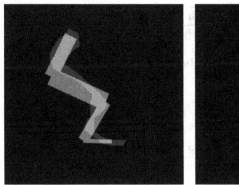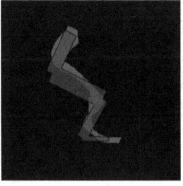

Fig. 3. (a) Example of the estimated image $E(x,y)$ (yellow) overlapped with the segmented image $S(x,y)$ (blue), (b) example of the classification measurements: TP with red, FP with blue and FN with green. (Color figure online)

a vector in P^0 contains the initial point x_1, y_1, and a set of lengths l_i for $i = 1, \ldots, 4$, then, the local search looks for the best angles (θ_i) and widths (w_i) for such lengths (l_i) and initial point (P_1).

After evaluating the population, the best individual is selected and stored, in lines 4 and 10, this part is known as elitism, and the purpose is to identify the individual best adapted to then insert it into the next generation and make it compete with its descendants. The loop stopping criterion evaluates two issues; the first verifies whether the variance of the objective function vector decreases each generation, the second verifies if the objective function value of the best individual increases, the algorithm stops five generations after any of these two conditions is satisfied. The selection in line 7 is named binary tournament, the objective function value of two randomly selected individuals is compared, and the best of both prevails for reproduction. In this case, the reproduction method, in line 8, is called: SBX (Simulated Binary Cross-over) [20]. For the mutation stage, in line 9, we implemented the polynomial mutation proposed by Deb and Goyal [18]. On each iteration, the algorithm generates a sequence of refined estimates, this set of estimations P^t are the same size as the set of the initial population. The behavior expected is that in each new generation, the children are better adapted, and the best approximation to the optimum is improved.

4.1 Local Search

The local search method estimates optimal values of θ_i and w_i using an input vector $X_c = [i_{P_1}, l_1, l_2, l_3, l_4]$ where i_{P_1} is a random index used to get P_1.

The algorithm intends to find the best angle and the best width values for each segment by performing a bisection search. The method starts by generating the initial point $P_1 = [x_1, y_1]$ using the index i_{P_1}, this step is shown in line 2 in Algorithm 2. The local search algorithm seeks for the best angles and widths for

---

Algorithm 1. Genetic algorithm with local search

Data: $S = (x, y)$ Binary Image manually segmented.
$Var_{min} = 1$ (for the stopping criterion)
Result: θ_i, w_i, l_i and $P_1 = (x_1, y_1)$ Initial Points

1 **begin**
2   Initialize Population P^0
3   Evaluate $F^0 = FO(P^0)$ and $LS(P^0)$
4   Elitism $best^0 = Elit(P^0, F^0)$
5      $t = 0$
6   **while** *Termination Condition* ← $tol < 5$ **do**
7   Selection $S^t = BinaryTournament(P^t, f)$
8   Crossover $\hat{D}^t = SBX(S^t)$
9          $D^t = PolinomialMutation(\hat{D}^t)$
10   Evaluate $F^t = FO(P^t)$ and $LS(P^t)$
11   Elitism $best^{t+1} = Elit(P^t, F^t)$
12   Compute Objective Variance $Var^t(F^t)$
13   **if** $Var^t < Var_{min}$ **or** $best^{t+1} > best^t$ **then**
14             $Var_{min} = Var^t$
15             $tol = 0$
16   **else** $tol = tol + 1$

---

each segment, then, in line 4 the segmented image $S(x, y)$ is cropped, for evaluating segment by segment and with the intention of reducing the computational cost. This cropped image $C(x, y)$ has a size of $(2 * l_i, 2 * l_i)$, where the kinematic coordinate (x_i, y_i) is located at the central pixel. For each segment the algorithm uses the same objective function in Eq. 4, denoted by FO, applied on the cropped image. The method consists of three stages of refinement, and in each of these, the search is performed with a higher precision in a reduced range. The first step starts at line 5; with an initial width of $w_0 = 30$, the method performs a search every $15°$ of the angle in which the overlap of the line drawn in the image D_i^I and the image $C(x, y)$ is maximized, according to Eq. 4. The angle found is stored in the variable θ^I. Then, a second search is performed around this last value, every five degrees, this stage is represented by lines 8 through 12. For each angle in the second search, we intend to improve the width of the segment looking for a new width in the range of $w_0 - 5$ to $w_0 + 5$. A third refinement step is performed (from line 13 to 16) for the angle value θ_i^{II}. The width is also varied depending on the sign of the result of $(w^I - w_0)$, where w^I is the best width found in the last stage. The function $sign(a)$ return: 0 if $a = 0$; −1 if $a < 0$ and 1 if $a > 0$.

In the case that the value of the width change in the last stage, the loop (in line 17) varies this value by one pixel at a time until it can not be improved. In line 25, the next coordinate is calculated using the best angle found. Finally, in line 26, the values of the width and the angle are stored; the procedure is repeated until all angles and widths are estimated.

Algorithm 2. Local Search Algorithm

Data: $X_c = [iP_i, l_1, l_2, l_3, l_4]$ and $S(x, y)$
Result: θ_i, w_i
1 **begin**
2   Generate $x_1, y_1 \Leftarrow iP_i$
3   **for** $n \leftarrow 2$ **to** 5 **do**
4   Crop Image $\leftarrow C(x, y) = Crop(S(x, y))$
5   **for** $\theta_i \leftarrow 0$ **to** 180 $steep = 15$ **do**
6   Draw Line from x_{n-1}, y_{n-1} to x_t, y_t with $w_0 = 30$ in D_i^I
7   Find best of $f = FO(D_i^I, C(x, y)) \leftarrow \theta^I$
8   **for** $\theta_{ii} \leftarrow (\theta^I - 10)$ **to** $(\theta^I + 10)$ $steep = 5$ **do**
9   Estimate x_t, y_t with θ_{ii}
10   **for** $w_i \leftarrow (w_0 - 5)$ **to** $(w_0 + 5)$ $steep = 5$ **do**
11   Draw Line from x_{n-1}, y_{n-1} to x_t, y_t with w_i in D_i^{II}
12   Find best of $f = FO(D_i^{II}, C(x, y)) \leftarrow \theta^{II}, w^I$
13   **for** $\theta_{iii} \leftarrow (\theta^{II} - 4)$ **to** $(\theta^{II} + 4)$ $steep = 2$ **do**
14             $w_t = w^I + (sign(w^I - w_0) * 2)$;
15   Draw Line from x_{n-1}, y_{n-1} to x_t, y_t with w_t in D_i^{III}
16   Find best of $f = FO(D_i^{III}, C(x, y)) \leftarrow \theta^{III}, w^{II}$
17   **if** $w^{II} - w^I$ is not 0 **then**
18   **while** $fit_c > fit_b$ **do**
19                 $fit_c = fit_b$;
20                 $w_n = w^{II} + sign(w^{II} - w^I)$;
21   Draw Line from x_{n-1}, y_{n-1} to x_t, y_t with $w_n \leftarrow D_i^{IV}$
22                 $fit_c = FO(D_i^{IV}, C(x, y))$;
23   **if** $fit_c > fit_b$ **then**
24                     $w^{II} = w_n$;
25   Estimate Final Values
26         $x_n = x_{n-1} + l_{n-1} * cos(\theta^{III})$ and $y_n = y_{n-1} + l_{n-1} * sin(\theta^{III})$
27         $w_{n-1} = w^{II}$; $\theta_{n-1} = \theta^{III}$;

---

5 Results

In this section, we describe a methodology used to evaluate the effectiveness of the optimization algorithm. Additionally, we present quantitative and qualitative results to demonstrate the feasibility of the proposed algorithm. The validation process consists of three parts: parameter tuning, quantitative and graphical results, and comparison with the performance of a human observer.

5.1 Parameter Tuning

Table 1 list the objective function values of the best individual in the final generation with respect to the use of different parameters for population size (ps) and

Table 1. Objective function values of the best individual in the final generation with respect to the use of different parameters for population size (ps) and mutation probability (mp).

	$ps = 10$	$ps = 20$	$ps = 50$	$ps = 100$	$ps = 150$
$mp = 2/V$	0.782	0.790	0.817	0.830	0.830
$mp = 1/V$	0.748	0.835	0.815	0.807	0.829
$mp = 1/2V$	0.773	0.796	0.812	0.81	0.804
$mp = 1/4V$	0.789	0.803	0.796	0.806	0.810
$mp = 1/6V$	0.800	0.784	0.807	0.800	0.813

mutation probability (mp). The mutation probability is a function of V, which is the number of variables, in this case, $V = 5$.

5.2   Kinematic Model Adjustment

The parameters with the best results (a population size of 20 and a mutation probability of $\frac{1}{V}$), are kept constant when applying the algorithm to different frames of a video sequence. An example of the qualitative results are depicted in Fig. 4 where a set of three images is shown for each frame. The first column shows the binary image corresponding to the manual segmentation of the marks on the rats posterior hind limb performed by the observer, the second column depicts the resulting image from the genetic algorithm that represents the best individual in the last generation, and the third column depicts the overlap of the manual segmentation and the adjusted kinematic model. In this last column, a different color is assigned to each measurement (TP with red, FP with blue and FN with green).

5.3   Validation

To perform validation of the proposed method, we present a comparison between the values of the mode computed automatically against the values produced by the manual adjustment of the kinematic model on the video frame performed by a human observer. Figure 5, depicts an example of the overlap between the model adjusted by an observer and the binary segmentation image, and by our proposed algorithm with the same segmentation image.

Tables 2 and 3 list the results corresponding to the adjusted kinematic model by the proposed algorithm and a human observer, respectively, for a set of video frames corresponding to a rat gait. Additionally, Table 3 list the difference between the objective function produced by the proposed algorithm and that corresponding to the manual adjustment of the model.

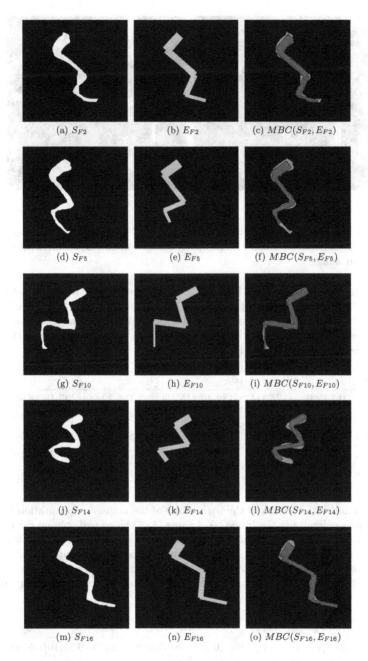

(a) S_{F2} (b) E_{F2} (c) $MBC(S_{F2}, E_{F2})$

(d) S_{F5} (e) E_{F5} (f) $MBC(S_{F5}, E_{F5})$

(g) S_{F10} (h) E_{F10} (i) $MBC(S_{F10}, E_{F10})$

(j) S_{F14} (k) E_{F14} (l) $MBC(S_{F14}, E_{F14})$

(m) S_{F16} (n) E_{F16} (o) $MBC(S_{F16}, E_{F16})$

Fig. 4. Examples of the kinematic model adjustment results: In the first column are shown five random frames manually segmented by expert user, in the second column we place the resulting image from the genetic algorithm in last iteration and the last column depicts the overlap of the manual segmentation and the computational approach. (Color figure online)

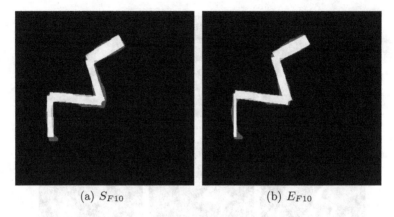

(a) S_{F10} (b) E_{F10}

Fig. 5. Example of the overlap between (a) the model adjusted by an observer and the binary segmentation image, and (b) by our proposed algorithm with the same segmentation image.

Table 2. Quantitative values of the parameters of the kinematic model adjusted image by genetic: The first column list the frame label name (F_i), the second column the objective function value f_g, the third column the position of the first point P_1, the fourth column the angles of the kinematic model θ_i, the fifth column the lengths l_i and the sixth the corresponding widths w_i.

F_i	f_g	P_1	θ_i	l_i	w_i
F_2	0.79596	[175,117]	[132,40,115,15]	[65,125,73,69]	[27,18,18,9]
F_5	0.823432	[264,107]	[145,52,155,63]	[64,112,71,42]	[27,12,17,10]
F_{10}	0.859286	[416,127]	[152,73,187,90]	[73,89,107,82]	[27,14,14,5]
F_{14}	0.7955	[500,108]	[145,57,175,55]	[56,75,100,44]	[27,20,13,18]
F_{16}	0.877426	[559,135]	[123,28,95,12]	[51,116,85,90]	[33,13,20,9]

Table 3. Quantitative values of the parameters of the manual segmentation image by user: The first column list the frame label name (F_i), the second column the objective function value f_u, the third column the position of the first point P_1, the fourth column the angles of the kinematic model θ_i, the fifth column the lengths and the sixth the corresponding widths w_i, and the seventh column the difference of the objective function value between the manual model and the model produced by our proposed algorithm.

F_i	f_u	P_1	θ_i	l_i	w_i	Δf
F_2	0.748737	[174,112]	[128,39,116,27]	[70,122,58,65]	[35,10,15,10]	+ 0.047223
F_5	0.807882	[269,100]	[139,53,154,57]	[73,104,67,59]	[30,10,15,10]	+0.01555
F_{10}	0.820842	[413,125]	[148,72,175,89]	[73,82,104,77]	[20,10,15,10]	+0.038444
F_{14}	0.70746	[501,106]	[150,56,171,40]	[62,75,90,59]	[20,10,15,10]	+0.08804
F_{16}	0.790999	[561,124]	[122,28,96,19]	[62,120,71,76]	[25,10,15,10]	+0.086427

6 Conclusion

We presented a promising method to automatically detect the kinematic variables involved in the analysis of hind limb for laboratory rats. The method is based on a Genetic Algorithm with local search. Our results show that the method is comparable with an expert user. The genetic algorithm globally explores kinematic configurations. Nevertheless, it does not directly deliver the actual positions of the articulations, it is crucial to their determination, because of different lengths initial points produce different angles and widths. Considering the local search, we remark that for a given articulation the local search finds the best angle and width. Our method exhaustively evaluates the whole range of solution; however, it is important to note that our approach avoids a significant computational cost by using several stages for refining the solution. As a result, we can increase the precision and reduce the search interval.

In the argot of evolutionary algorithms, there exist a kind named as memetic algorithms which combine global and local search, although our proposal does the same, it is not a *usual* memetic algorithm, due to memetic algorithms *commonly* use both search processes on the same optimization variables, while our proposal uses the global search for a subset of variables and local search for a different subset, hence, it is a novel way of combining optimization methods.

Finally, our proposal is a first step for the automatic analysis and synthesis of rats kinematics. Future work includes making our method fully automatic, using the kinematics characteristics to classify rat locomotion according to them and to measure the effects of different treatments on the rats.

References

1. Santos, P.M., Williams, S.L., Thomas, S.S.: Neuromuscular evaluation using rat gait analysis. J. Neurosci. Methods **61**(1), 79–84 (1995)
2. Cheng, H., et al.: Gait analysis of adult paraplegic rats after spinal cord repair. Exp. Neurol. **148**(2), 544–557 (1997)
3. Moller, K.Ä., et al.: Gait analysis in rats with single joint inflammation: influence of experimental factors. PLoS One **7**(10), e46129 (2012)
4. Tarlov, I.M., Klinger, H.: Spinal cord compression studies: II. Time limits for recovery after acute compression in dogs. AMA Arch. Neurol. Psychiatry **71**(3), 271–290 (1954)
5. Basso, D., Beattie, M., Bresnahan, J.: A sensitive and reliable locomotor rating scale for open field testing in rats. J. Neurotrauma **12**(1), 1–21 (1995)
6. Osuna-Carrasco, L.P., et al.: Quantitative analysis of hindlimbs locomotion kinematics in spinalized rats treated with Tamoxifen plus treadmill exercise. Neuroscience **333**, 151–161 (2016)
7. Sandi, C., Venero, C., Guaza, C.: Novelty-related rapid locomotor effects of corticosterone in rats. Eur. J. Neurosci. **8**(4), 794–800 (1996)
8. Delfs, J.M., Schreiber, L., Kelley, A.E.: Microinjection of cocaine into the nucleus accumbens elicits locomotor activation in the rat. J. Neurosci. **10**(1), 303–310 (1990)

9. Stephan, F.K., Zucker, I.: Circadian rhythms in drinking behavior and locomotor activity of rats are eliminated by hypothalamic lesions. Proc. Nat. Acad. Sci. **69**(6), 1583–1586 (1972)
10. Kimelberg, H.K., et al.: Acute treatment with Tamoxifen reduces ischemic damage following middle cerebral artery occlusion. Neuroreport **11**(12), 2675–2679 (2000)
11. Franco-Rodriguez, N.E., et al.: Tamoxifen favoured the rat sensorial cortex regeneration after a penetrating brain injury. Brain Res. Bull **98**, 64–75 (2013)
12. Ridler, T.W., Calvard, S.: Picture thresholding using an iterative selection method. IEEE Trans. Syst. Man Cybern. **SMC-8**(8), 630–632 (1978)
13. Bareyre, F.M., et al.: The injured spinal cord spontaneously forms a new intraspinal circuit in adult rats. Nat. Neurosci. **7**(3), 269–277 (2004)
14. Beauparlant, J., et al.: Undirected compensatory plasticity contributes to neuronal dysfunction after severe spinal cord injury. Brain **136**(Pt. 11), 3347–3361 (2013)
15. Edgerton, V.R., et al.: Plasticity of the spinal neural circuitry after injury. Annu. Rev. Neurosci. **27**, 145–167 (2004)
16. Wernig, A., Muller, S.: Laufband locomotion with body weight support improved walking in persons with severe spinal cord injuries. Paraplegia **30**(4), 229–238 (1992)
17. Nocedal, J., Wright, S.J.: Numerical Optimization, 2nd edn. Springer, New York (2006)
18. Deb, K., Goyal, M.: A combined genetic adaptive search (geneas) for engineering desing. Comput. Sci. Inform. **26**, 30–45 (1996)
19. International Commission on Illumination. http://cie.co.at/
20. Deb, K., Agrawal, S., Pratap, A., Meyarivan, T.: A fast elitist non-dominated sorting genetic algorithm for multi-objective optimization: NSGA-II. In: Schoenauer, M., Deb, K., Rudolph, G., Yao, X., Lutton, E., Merelo, J.J., Schwefel, H.-P. (eds.) PPSN 2000. LNCS, vol. 1917, pp. 849–858. Springer, Heidelberg (2000). doi:10.1007/3-540-45356-3_83. http://www.iitk.ac.in/kangal/index.shtml

Quantization and Equalization of Pseudocolor Images in Hand Thermography

Orcan Alpar and Ondrej Krejcar[✉]

Faculty of Informatics and Management,
Center for Basic and Applied Research, University of Hradec Kralove,
Rokitanskeho 62, 500 03 Hradec Kralove, Czech Republic
orcanalpar@hotmail.com, ondrej@krejcar.org

Abstract. Thermal images consist of a single matrix, though several colorization algorithms are mostly applied to highlight the differences in radiation. The colorization is done by built-in colormaps which usually conceal some features, so in case of being already devoted to a specific purpose, hardly useful for other requirements. We therefore proposed a novel quantization and indexing algorithm in hand thermography. The images already extracted by pseudocolor map are segmented by red-channel conversion for turning into grayscale and 4-level quantization. Moreover, through the histograms and cumulative histograms, intermediary indexing is applied by Midway image equalization. The results seem promising for revealing the details with high contrast differences in quantization.

Keywords: Thermal imaging · Hand thermography · Pseudocolor · Quantization · Indexing · Midway image equalization

1 Introduction

Pseudocoloring or pseudocolorization is a color-rendering methodology, applied to grayscale images that have only one layer. Basic difference between false-color and pseudo-color is the number of layers in raw image, since false-coloring can be applied to the multiple matrices of color channels. The main example for the usage of pseudocoloring is thermography, since the thermal infrared cameras have an output of a grayscale image, showing the intensity of the radiation.

Thermography is the methodology for the measurement of temperature by detecting radiation in long-infrared of electromagnetic spectrum. The thermograms give vital information about the slight changes in temperature of the scanned area, which can be related with local physiological process or systemic responses. Thermal imaging technology could contribute to the diagnosis of several disorders and diseases such as, local ischemia [1], peripheral vasoconstriction [2], deep vein thrombosis [3], complex regional pain syndrome [4], breast cancer [5] and similar. Moreover, as a promising diagnostic tool, one of the practice area of thermograms is hand thermography. Hand thermograms are crucial for diagnosis of the Raynaud's syndrome [6], and identification of nerve entrapments [7] in Medicine.

© Springer International Publishing AG 2017
I. Rojas and F. Ortuño (Eds.): IWBBIO 2017, Part I, LNBI 10208, pp. 397–407, 2017.
DOI: 10.1007/978-3-319-56148-6_35

Raynaud's syndrome is a disorder, affecting the extremities caused by reduction of the blood flow. There are two types of this syndrome: primary, triggered by an unknown cause; secondary, associated with a disease or disorder. Each type has common characteristics such as numbness in milder cases and total blockage in severe; however these conditions evolve by time. Therefore, early diagnosis is vital just like in the diagnosis of nerve entrapments in hands. It is possible to find very recently published papers on thermogram analysis in Raynaud's such as: [8, 9].

There are three common types of nerve entrapments affecting the hands: Cubital tunnel syndrome, Carpal tunnel syndrome and Guyon's canal syndrome. Briefly, the Cubital tunnel syndrome is ulnar nerve entrapment in cubital tunnel in the elbows, causing numbness, weakness in grip, incoordination in hands. Carpal tunnel syndrome is the entrapment of the median nerve passing through the wrists, affecting the first three fingers and half of the fourth. The numbness in the other fingers could be triggered by Guyon's canal syndrome, which is the ulnar nerve entrapment in Guyon's canal in the wrists. Unfortunately, very few researches could be found on differential diagnosis of these disorder by thermal imaging [10, 11].

Considering the various requirements for differential diagnosis by the thermal imaging, we propose a novel quantization methodology for pseudocolor thermograms by Midway image equalization. Firstly, we segmented the hands from the thermogram by red channel manipulation for better indexing of the colors. Afterwards, we turned the segmented images into grayscale for calculating histograms and cumulative histograms separately. We indexed the colors from the segmented image by uniform quantization to turn the image into 4-level grayscale and recomputed the histograms. Finally, for intermediary indexing, Midway image equalization is applied to each grayscale image.

The main motivation of this research is to widen the research possibilities of thermal images which are already colorized. The aim and the process of colorization may vary, therefore the images sometimes cannot be a feasible source for research objectives. Quantization is more than just a manipulation of images, it's a first step to preparing for various analysis hereinafter, while equalization is bringing out the hidden features of the images.

In recent years, it is possible to find relevant studies in the literature on RGB image digitization, quantization, sampling and companding. Ozturk et al. [12] presented a new color-based quantization technique for RGB and CIELAB images using artificial bee colony algorithms. Hu et al. [13] introduced a color scheme and a second by K-means algorithm for quantization of RGB images and for compression. Ponti et al. [14] presented several methods of quantization for feature extraction and dimension reduction. Perez-Dalgado [15] proposed an Ant-tree algorithm based color quantization for clustering by color palettes.

This paper starts with introduction of Pseudocolor thermograms and methodology of quantization. Moreover, the bass of midway image equalization is presented. Finally the papers concludes with the experimental results.

2   Pseudocolor Thermograms

Due to built-in colormaps in thermal cameras, the figures are usually extracted by one of each depending on the requirements. Therefore, some important information could be lost along the process and there is no turning back unless there is a reverse code for decoloring. At this point, several grayscaling algorithms are applied to estimate the original, especially when contrast differentiation is needed. On the other hand, the original image is colored as well in most cases, which makes the re-origination nearly impossible. In Fig. 1, several colorization techniques are presented which all are very common in medicine and industry.

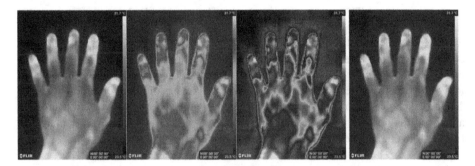

Fig. 1. Pseudocolored hand thermograms (Gray, Rainbow, Rainbow HC, Iron) (Color figure online)

As an outcome, the image similar with the original is on the left which is colored in monochromatic colormap; however if extracted, it would be an RGB image consisting of color layers. The second colormap, named as Rainbow, is the most popular in hand thermograms since it contains clarified information about the temperature. Although the third one is a high contrast version of the rainbow, the main area of utilization is segmentation since the contrast differences are artificially colored. The last one is Iron, which is very suitable for industry to find out the leakages or weak points of pressure or temperature based devices, yet it is useful to recognize the hand vessels through this kind of colormaps.

The major issue is not producing them but using them in the form they were extracted. In other words, taking the images by a thermal camera lets the researchers to process the images, on the contrary, if the images are borrowed from a public dataset or literature, only RGB images could be found. Therefore, it is necessary to index these images for common use by quantization of RGB layers.

3   Quantization

Quantization is a methodology for indexing the multilayer images mostly for compression, dealing with discretization of intensity. Since the images are produced to be views by human-eye, some distortion on the images cannot be recognized. In addition human-eye is an expert of interpolation, therefore the slight differences between uncompressed and compressed images are not so detectable.

Each pixel $p_{i,j,k}$ in an RGB image $I_{i,j,k}$ extracted by Rainbow colormap could be represented as:

$$p_{i,j,k} \in I_{i,j,k}(i = [1:w], \ j = [1:h], \ k = [1:3]) \tag{1}$$

where, w is the width, h is the height of the image and k is the color channel indicator: 1 for red, 2 for green and 3 for blue. Therefore with this notation each pixel in column w and row h in the color channel k could easily be found and identified. Quantification corresponds to a transformation $T(I)$ with the desired level of details in representation by:

$$T(I) : p_{i,j,k} \rightarrow \bar{p}_{i,j} \in Q_{i,j}(i = [1:w], \ j = [1:h]) \tag{2}$$

where $Q_{i,j}$ is the quantized image with the pixel is coordinates $\bar{p}_{i,j}$. Depending on the desired resolution, level of quantization may vary such that it may be 256-level if the colors will be reproduced. On the other hand, quantization is usually irreversible due to loss of information, therefore as the levels increase the loss of the information decrease. The basic choice before quantization is choosing the discrete quantization level to apply so, it is necessary to determine the amount of data to represent whole image. In Fig. 2, 4-level and 8-level quantization examples are given.

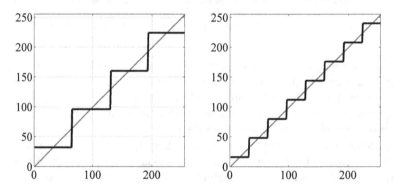

Fig. 2. Examples of quantization. 4-level on the left and 8-level on the right

The images presented in Fig. 2 are the examples of uniform quantization which is the simplest type of zero-memory quantizers. The mapping is done between transition levels t_n and reconstruction levels r_n such that a pixel p is mapped to r_n if between t_n and t_{n+1}. Therefore, given the monochromatic values 0 and 255, for N levels, the uniform quantizer would transform the image by

$$r_n = \frac{255n}{N} - \frac{255}{2N} \tag{3}$$

$$t_n = \frac{255(n-1)}{N} \tag{4}$$

Additionally, quantization of the images produces a natural contours by deleting the continuities. The algorithm also resets the process of blur or gradient effect by disregarding the transitions and approximating the levels with an error range. This process is essential when dealing with indexing of the temperatures in thermograms since the contrast is more crucial than the continuity. An example of this false contour effect is presented in Fig. 3. It is seen that the gradient grayscale image is turned into discrete subregions while preserving the gray intensity values with some approximation errors.

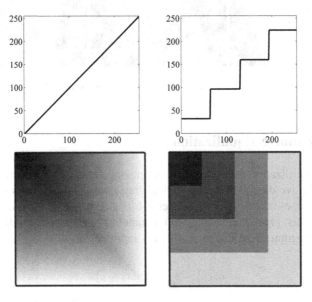

Fig. 3. False contour effect.

However, 256-level image quantization is still possible, so that any multilayer image could be turned into monochromatic image without losing the color data. In addition, if the resolution is satisfactory, there won't be any visible differences between two images. The example is given in Fig. 4 below. Although the first and the last image looks perfectly identical they have structural difference. The grayscale image shows the coordinates of 256 color and when restructured the colors are substituted in corresponding pixel. Therefore any pixel on the last image is consisting of three values and represented on the grayscale image by a number between 0 and 255.

On the contrary, the main focus of quantization methodology in this paper is highlighting the areas of interest to enable reconstruction for different purposes. Therefore 256-level quantization is not necessary, considering temperature levels on a thermogram. Disregarding the color spectrum, there mainly are four key points to be researched, very hot, hot, medium and cold regions; however 4-level quantization is not feasible for raw RGB images. The main solution for this lack of applicability is morphing two indexed monochromatic images by image equalization techniques.

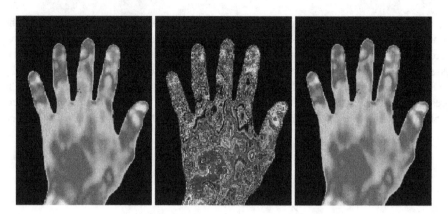

Fig. 4. Original image on the left, 256-level quantized image in the middle, reproduced image on the right.

4   Midway Image Equalization

The Midway equalization deals with pair of grayscale images for find an intermediate histogram while avoiding to erase the original gray levels. Although it is generally used for different images taken in the same scene, it is a perfect tool for indexing a single thermogram. This procedure starts with computing the gray-level histogram and cumulative histogram of an image u, which is represented as

$$h_u(\lambda) = \frac{1}{|\Omega|}\{x; u(x) \leq \lambda\} \tag{5}$$

where h is the discrete histogram of u and $|\Omega| = i.j$ is the number of pixels. If we consider two discrete grayscale images u and v, and the corresponding discrete histograms h_u and h_v, the easiest solution would be the arithmetic mean of the histograms $(h_u + h_u)/2$; however this approach is infeasible due to lack of smoothness. Therefore, in this case, a harmonic mean between two discrete histograms would be more natural, which is called Midway.

$$H_m = \left(\frac{H_u^{-1} + H_u^{-1}}{2}\right)^{-1} \tag{6}$$

The resulting contrast changes δ_{uv} and δ_{vu} are calculated by:

$$\delta_{uv}(x) = \left(\frac{x + H_v^{-1} \circ H_u(x)}{2}\right) \tag{7}$$

$$\delta_{vu}(x) = \left(\frac{x + H_u^{-1} \circ H_v(x)}{2}\right) \tag{8}$$

The equalizer brings two different equalized images, which have different properties, depending on the gray intensity of the original images.

5 Experimental Results

The experiment starts red channel conversion of the rainbow colored thermograms, having a size of 1530×2040 resolution, taken by FLIR thermal camera in 0.96 emissivity, from 30 cm distance in 25 °C room temperature, to find the area of interest I^A on the original image, namely:

$$I^A : I_{i,j,k}(i = [1 : w], \, j = [1 : h], \, k = [1]) \tag{9}$$

The red channel for this experiment represent the heat source, thus the hand, yet it could be altered upon requirements. The red channel is binarized by

$$\hat{I}^A = 1 | I^A / 255 > 0 \tag{10}$$

and multiplied by the original channels to segment the image I^S by dot product

$$I^S_{i,j,k} = \hat{I}^A . I_{i,j,k} \tag{11}$$

The channels are presented in Fig. 5.

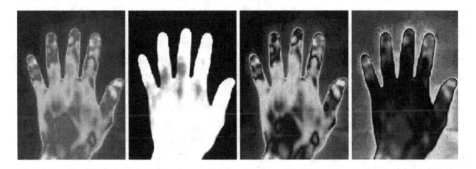

Fig. 5. Original image with red, green and blue channels, from left to right (Color figure online)

For better presentation R $= I^S_{i,j,1}$ G $= I^S_{i,j,2}$ and B $= I^S_{i,j,3}$ abbreviations will be used hereafter. The image is grayscaled to create the first image u, by:

$$u = 0.42(R + G + B) + 0.21R + 0.17(R + G) + 0.125G - 0.04(G + B) - 0.08G \\ - 0.21(R + B) \tag{12}$$

The reason of this grayscaling is the temperature legend of thermograms since the white regions have highest temperature so it will be (R+G+B). On the other hand, coolest place is the violet, so it is represented by $(R + B)$. The second grayscale image v is achieved by the method presented in Sect. 3 and consequently two grayscale images are prepared. Finally the midway algorithm is applied to these grayscale images to find intermediate histograms. The results are presented in Fig. 6.

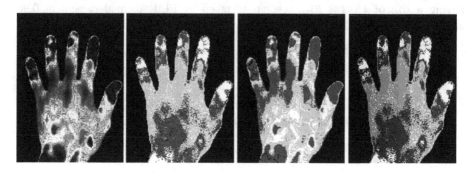

Fig. 6. u, v, δ_{uv}, δ_{vu} from left to right.

The first image u, 256-level grayscale, shows the regions of high radiation by darker colors while the second image v, 4-level grayscale, is produced from the original thermograms so higher radiations are presented by lighter pixels. The equalization brings the images closer yet there are still visible differences between. These results provide the critical contrasts and the unique intensity intervals and could be used by pseudo coloring the indexes.

According to the results, we can see the variance of the temperature in terms of indexes and reveal the contrast and intensity differences to help interpretation of RGB thermal images. AS an important remark, we made the experiments considering that we have no access to the original monochromatic thermogram and we didn't even try to reach it. We repeated this experiment several times with different grayscale images and reached additional results presented in Table 1.

The additional experiments prove that, depending on selected the grayscaling methodology and the requirements, various midway equalized images could be obtained from 4-level quantization. On the other hand, 8-level quantization could be applied when more detail in outputs is necessary. The results of Midway equalization of various grayscale and 8-level quantized thermograms are presented in Table 2.

8-level quantization gives more details as expected, where these details could only be seen in v and δ_{vu}. The grayscaling also changes the color palette therefore every single image in Tables 1 and 2 have their own index.

Table 1. Additional 4-level quantization experiments

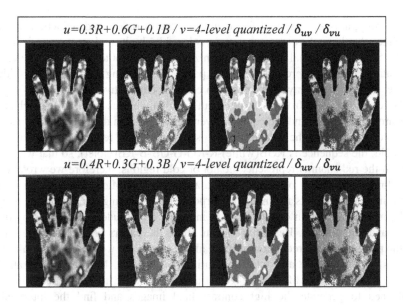

Table 2. Additional 8-level quantization experiments

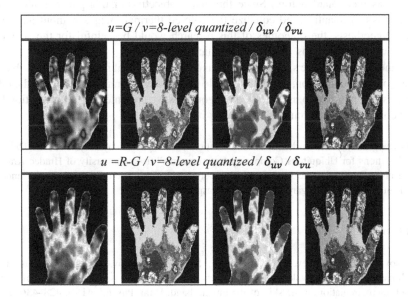

6   Conclusions and Discussions

Hand thermograms are essential for identifying the disorders and reduction in blood flow. Since the thermal images are pseudo colored by built-in algorithms, it is usually hard to achieve the original image to recognize the contrast of temperature. Therefore a quantization method is presented in this paper for indexing. Furthermore, midway equalizer is applied to grayscale images to equalize cumulative histograms.

The results achieved from the experiment are highly promising that we reached two monochromatic images and the equalized versions. The main difficulty in this research is the color sequence in the palette, since it is not ordered by red green and blue. From hot to cold, the sequence is RGB(white)-R-RG-G-GB-B-RB(violet), so that we cannot be sure if the red channel represents the hot or cold regions, therefore we wrote a new grayscaling equation to overcome this drawback. The Midway equalization also brought interesting results for indexing purposes by generating two equalized images.

The main contribution of this paper is indexing the pseudocolor images by quantization and equalization. Regarding quantization, the papers in the literature mostly have the purpose of size reduction of RGB images; however we deal with highlighting the areas of temperature and false contouring. In addition, Midway equalization is generally applied to two different images of a same scene for intensity equalization yet we planned to generate the high-contrast final images and find the intermediate monochromatic thermograms.

Frankly, we came across no important disadvantages or drawbacks worth addressing and yet the main problem is the built-in colorizing algorithms of the thermal cameras, as mentioned before. Since the major objective of this paper is to present several ways to quantize and equalize the colored thermal images without any other specific hypothesis, the outcomes of this research would be helpful for the scientists dealing with thermograms taken from public databases.

On the other hand, we segmented the hands on the thermal images with red-channel conversion and subtraction, since it is simple for the thermograms we had. On the other hand, it is still possible to use one of the algorithms we presented for segmentation [16, 17] or dorsal hand recognition [18].

Acknowledgment. This work and the contribution were supported by project "SP/2102/2017 - Smart Solutions for Ubiquitous Computing Environments" from University of Hradec Kralove. We are also grateful for the support of Ph.D. students of our team (Richard Cimler and Jan Trejbal) in consultations regarding application aspects.

References

1. Volovik, M.G., Kiselev, D.V., Polevaya, S.A., Aleksandrov, N.M., Peretyagin, P.V., Khomiakova, M.I., Kovalchuk, A.V.: Effects of repeated local ischemia on the temperature and microcirculation in the skin of the human hand. Hum. Physiol. **41**(4), 428–436 (2015)
2. Vannetti, F., Matteoli, S., Finocchio, L., Lacarbonara, F., Sodi, A., Menchini, U., Corvi, A.: Relationship between ocular surface temperature and peripheral vasoconstriction in healthy subjects: a thermographic study. Proc. Inst. Mech. Eng. Part H: J. Eng. Med. **228**(3), 297–302 (2014)

3. Deng, F., Tang, Q., Zeng, G., Wu, H., Zhang, N., Zhong, N.: Effectiveness of digital infrared thermal imaging in detecting lower extremity deep venous thrombosis. Med. Phys. **5** (2242–2248), 42 (2015)
4. Choi, E., Lee, P.B., Nahm, F.S.: Interexaminer reliability of infrared thermography for the diagnosis of complex regional pain syndrome. Skin Res. Technol. **19**(2), 189–193 (2013)
5. Dayakshini, S., Kamath, S., Rajagopal, K.V., Prasad, K.: Medical imaging techniques and computer aided diagnostic approaches for the detection of breast cancer with an emphasis on thermography–a review. Int. J. Med. Eng. Inform. **8**(3), 275–299 (2016)
6. Chelicka, I., Matusiak, U., Maj, J., Baran, E., Szepietowski, J.C.: Freezing fingers syndrome, primary and secondary Raynaud's phenomenon: characteristic features with hand thermography. Acta Derm.-Venereol. **93**(4), 428–432 (2013)
7. Ammer, K.: Nerve entrapment and skin temperature of the human hand. In: Infrared Imaging, p. 17-1. IOP Publishing (2015)
8. Horikoshi, M., Inokuma, S., Kijima, Y., Kobuna, M., Miura, Y., Okada, R., Kobayashi, S.: Thermal disparity between fingers after cold-water immersion of hands: a useful indicator of disturbed peripheral circulation in raynaud phenomenon patients. Intern. Med. **5**(461–466), 55 (2016)
9. Kolesov, S.N.: Thermal-vision diagnosis of Raynaud's syndrome and its stages. J. Opt. Technol. **82**(7), 478–486 (2015)
10. Papež, B.J., Palfy, M.: EMG vs. Thermography in Severe Carpal Tunnel Syndrome. INTECH Open Access Publisher (2012)
11. Roldan, K.E., Piedrahita, M.A.O., Benitez, H.D.: Spatial-temporal features of thermal images for Carpal Tunnel Syndrome detection. In: IS&T/SPIE Electronic Imaging. International Society for Optics and Photonics (2014)
12. Ozturk, C., Hancer, E., Karaboga, D.: Color image quantization: a short review and an application with artificial bee colony algorithm. Informatica **25**(3), 485–503 (2014)
13. Hu, Y.C., Lee, M.G., Tsai, P.: Colour palette generation schemes for colour image quantization. Imaging Sci. J. **57**, 46–59 (2013)
14. Ponti, M., Nazaré, T.S., Thumé, G.S.: Image quantization as a dimensionality reduction procedure in color and texture feature extraction. Neurocomputing **173**, 385–396 (2016)
15. Pérez-Delgado, M.L.: Colour quantization with Ant-tree. Appl. Soft Comput. **36**, 656–669 (2015)
16. Alpar, O.: Corona segmentation for nighttime brake light detection. IET Intell. Transport Syst. **10**(2), 97–105 (2015)
17. Alpar, O., Stojic, R.: Intelligent collision warning using license plate segmentation. J. Intell. Transport Syst. **20**(6), 487–499 (2015)
18. Alpar, O., Krejcar, O.: Dorsal hand recognition through adaptive YCbCr imaging technique. In: Nguyen, N.-T., Manolopoulos, Y., Iliadis, L., Trawiński, B. (eds.) ICCCI 2016. LNCS (LNAI), vol. 9876, pp. 262–270. Springer, Cham (2016). doi:10.1007/978-3-319-45246-3_25

Superficial Dorsal Hand Vein Estimation

Orcan Alpar and Ondrej Krejcar[(✉)]

Faculty of Informatics and Management, Center for Basic
and Applied Research, University of Hradec Kralove,
Rokitanskeho 62, 500 03 Hradec Kralove, Czech Republic
orcanalpar@hotmail.com, ondrej@krejcar.org

Abstract. Dorsal hand thermograms give crucial information about the temperature differences and radiation intensity so that it is possible to extract and estimate the hand veins through thermal images depending on how they were colorized. Similar with the finger veins and finger prints, the hand veins represent one of the characteristics of individuals as a unique trait in biometrics and bioinformatics. Although the hand vein identification systems are mostly based on various special equipment and illumination techniques, the dorsal hand veins could easily be recognized by thermal cameras. Therefore in this paper we propose a hand vein estimation and projection methodology for dorsal hand thermograms using anisotropic diffusion and maximum curvature method.

Keywords: Hand thermography · Dorsal hand veins · Feature extraction · Biometrics · Anisotropic diffusion · Maximum curvature

1 Introduction

Biometrics is a term for unique characteristics of the individuals that are nearly impossible to mimic. Along these characteristics, the major types are biological like DNA, physical like fingerprints or habitual like signatures. On the other hand, through emerging technologies, it is now possible to extract more unique features such as, finger and hand veins. However the main difficulty of these tasks is lack of visibility, since it is still hard to determine the veins by normal visible spectrum cameras.

For finger vein detection, the issue mentioned above is solved with LED technologies. Special but not so complicated devices are implemented which have a LED light emitter on the top and visible spectrum camera placed under the fingers. The fingers become slightly permeant under near infrared light, having wavelengths between 700 and 1000 nm, while veins carrying hemoglobin absorbs this spectrum of light. Hand vein recognition systems evolved in a similar way however more light sources were necessary since it is observed that with 20 LEDs, the veins become visible by cameras.

The finger and hand vein recognition systems usually need two special instruments, a light emitter and an image recorder. In addition, the compound device has specific requirements and stationary assembly; however, the thermal infrared cameras are prominent solution for the inflexibility of these kind of devices. Under specified conditions, the hand veins are visible by thermography, depending on the pseudo

I. Rojas and F. Ortuño (Eds.): IWBBIO 2017, Part I, LNBI 10208, pp. 408–418, 2017.
DOI: 10.1007/978-3-319-56148-6_36

colorization method. The main purpose of this research therefore is estimation rather than recognition of the veins, since not all the veins are visible in hand thermography.

We proposed a vein estimation method for dorsal hand thermography starting with thermal imaging of dorsal hands. The thermograms are taken by FLIR thermal camera in 0.96 emissivity, from 30 cm distance in 25 °C room temperature with a resolution of 1530 × 2040. Red channel conversion of Rainbow colored images gives the region of interest and Iron colored images are segmented by dot product of these matrices. Afterwards, anisotropic diffusion filtering is applied to green channel of the segmented images for enhancing the boundaries of the veins. Finally, the superficial veins are estimated and constructed by maximum curvature method on a blank image from scratch. The brief flow chart is presented in Fig. 1.

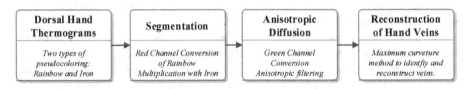

Fig. 1. Workflow of proposed system

The dorsal hand vein recognition systems are widely researched and the methods presented in the literature are so various. Lee et al. [1] proposed 2D Gabor filters for dorsal hand vein recognition and focused on comparison as well as extraction. They used near-infrared camera in a perpendicular position to collect the images of clenched fists on a black background. After segmentation, they extracted the veins using adaptive Gabor filter to create a vector and compare with the database by Hamming distance. Sheetal and Goela [2] dealt with the dorsal side of the fists, as well, however they applied many filtering enhancements and related operations. For noise reduction, the used low-pass and median filters, and enhanced the images by histogram equalization. Finally, through thresholding, they extracted the very low-detail vein patters from the fists.

Janes and Junior [3] prepared a prototype for capturing the dorsal hands by a camera with a removed infrared filter. With the LEDs placed on the bottom of the device, they made the experiments with the modified camera with 850 nm wavelengths. Their results are so promising that the dorsal veins seem to be more visible under the wavelength they had chosen. They also applied curvelet transforms for feature detection and random forest method for classification. Sakthivel [4] presented a prototype device with near-infrared camera as well, but for palm vein recognition. Moreover, Morales-Montiel et al. [5] dealt with dorsal hand vein recognition by infrared imaging. They however focused on the middle regions on the dorsal hand images by Sobel filter, applied median and 2D Wiener filters for noise reduction and finally identified the veins by thresholding. They also presented thermal image enhancing and normalization to find bifurcations on hand veins. Their results for restricted areas of the hands seem exceptional.

Huang et al. [6] also presented a dorsal vein recognition system by fist images and a special device. They focused on key point while detecting the veins and generated the key points by centroid-based circular grid method. Djerouni et al. [7] dealt with contrast enhancements of fist images with NIR camera system. They applied five algorithms to a region on the fists: histogram equalization, adaptive histogram equalization, frangi filter, radon filter and Newtonian operator. On the other hand, the most relevant research we came across so far belongs to Buddharaju and Pavlidis [8] as a chapter in Medical Infrared Imaging book. Although they had very similar papers [9, 10], what we know for certain is: in [8], they dealt with superficial face vein estimation as an addition to their previous research. There are also many more papers worth to mention, published in very recent years, such as [11–17]. The main difference of our paper is that we didn't use any kind of special device to capture the images. We also focused on whole hand snot the fists without searching for region of interest. On the other hand, we used two hind of pseudocolored thermograms for segmentation and vein identification. Comparing the previous results, we can easily state that the veins in thermal images are harder to identify than the NIR images as well.

The main reason of this difficulty is the similarity of the radiation level emitted by the veins and some parts of the skin. In NIR imaging, the wavelength could be calibrated for better discrimination of the veins however it is not possible in thermal images. In addition the clenched fist position makes the veins more visible compared to the normal dorsal hand images. The paper starts with the methodology by introducing the steps separately, with the preliminaries of the methods. The thermal imaging and the segmentation techniques are presented in the beginning. Afterwards, the segmented images are enhanced by anisotropic diffusion for making the dorsal hand veins more visible. Furthermore, using the some parts of maximum curvature method [18], the results are presented. The paper ends with conclusion and discussion.

2 Dorsal Hand Thermography

Thermography is an imaging methodology to reveal the radiation emission in the long-infrared range if there is no visible illumination. In addition, amount of radiation emitted by different sources is various and depending on the temperature of the source, therefore the contrast could also be detected. Thermography is a useful diagnostic tool for medical imaging, bioinformatics, biometrics and biomedicine. In addition, hand thermograms give very important information about the temperature distribution and thus about possible disorders in hands. Some disorders and diseases could easily be diagnosed by the help of thermograms such as Raynaud's and nerve entrapments [19, 20].

The experiment sequence of dorsal hand vein projection system we proposed in this paper starts with thermal imaging. Although there are many more built-in colormaps, we selected rainbow and iron for vein recognition and vein system estimation. The main reason to choose rainbow is the colorization algorithm based on temperature differences so that it is easier to segment the hands by the high radiation channel of red. On the contrary, the Iron colored thermogram gives crucial information about the veins with some error indeed. Each thermogram is generated by built-in FLIR pseudocoloring of the grayscale radiation intensity image, however they have different contrast ratios. The images we used in this process is presented in Fig. 2.

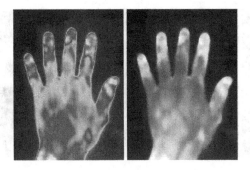

Fig. 2. Colormaps used in this paper: rainbow on the left, Iron on the right. (Color figure online)

Although, the images are the versions of the main thermogram, they have very apparent differences in features. As seen in the Fig. 2, the thresholding is also different as well as the temperature regions: Rainbow tries to reveal the contrast of the slight differences in radiation with 8 major color groups on a red-green-blue(RGB) image: White (RGB) > Red (R) > Red-Green (RG) > Green (G) > Green-Blue (GB) > Blue (B) > Purple/Violet (RB). However the Iron color map has 5 major regons: Hot (White) > Warm (R) > Average (Orange) > Cool (B) > Cold (RB). Given these differences, finding the place of hand is easier with Rainbow while viens are more visible in Iron. Therefore, each pixel $p_{i,j,k}$ on the image of Rainbow colored thermogram $R_{i,j,k}$ could be stated as:

$$p_{i,j,k} \in R_{i,j,k}(i = [1 : w], j = [1 : h], k = [1 : 3]) \tag{1}$$

where w is the width, j is the height and k is the color channel. Since the red channel of $R_{i,j,k}$ always carries the information of radiation differences between the substance and the environment, the raw image is firstly converted to grayscale by red channel conversion.

$$p_{i,j}^r \in R_{i,j}^r(i = [1 : w], j = [1 : h]|k = 1) \tag{2}$$

where r represents the red channel. Moreover, the red channel is binarized by

$$\dot{p}_{i,j}^r \in \dot{R}_{i,j}^r = 1|p_{i,j}^r/255 > 0 \tag{3}$$

However, the final image $\dot{R}_{i,j}^r$ is a matrix of $w.h$ therefore it is not possible to make matrix operations without dividing the Iron image into layers. Any pixel q on an image of Iron colored thermogram is represented by:

$$q_{i,j,k} \in I_{i,j,k}(i = [1 : w], j = [1 : h], k = [1 : 3]) \tag{4}$$

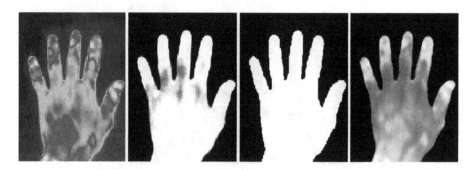

Fig. 3. Segmentation procedure: $R_{i,j,k}$ - $R_{i,j}^r$ - $\dot{R}_{i,j}^r$ - $F_{i,j,k}$ (Color figure online)

So it is necessary to divide into color channels, namely

$$q_{ij}^r \in I_{i,j}^r (i = [1:w], j = [1:h]|k = 1) \tag{5}$$

$$q_{ij}^g \in I_{i,j}^g (i = [1:w], j = [1:h]|k = 2) \tag{6}$$

$$q_{ij}^b \in I_{i,j}^b (i = [1:w], j = [1:h]|k = 3) \tag{7}$$

where the superscripts *r, g, b* represent the color channels: red, green and blue. It is now possible to dot-product the matrices by: $q_{ij}^r \dot{p}_{i,j}^r q_{ij}^r \cdot \dot{p}_{i,j}^r$, $q_{ij}^g \cdot \dot{p}_{i,j}^g$, $q_{ij}^b \cdot \dot{p}_{i,j}^b$ so that the final image $F_{i,j,k}$ consists of three segmented layers. The procedure is shown in Fig. 3 below.

There are some cropping errors indeed, however we concentrated on the hand veins therefore the errors are negligible. The next step is preprocessing of the $F_{i,j,k}$ for vein identification and construction.

3   Anisotropic Diffusion

In thermal imaging of the hand veins, the main vessels could be seen however it is not enough for effective identification. Therefore we applied anisotropic diffusion filter to reduce the noise and to highlight the veins. However, the filtering is applicable for monochromatic images so that we need to turn the segmented image $F_{i,j,k}$ into grayscale without losing the features. The best way for grayscaling the thermograms seemingly is the green channel extraction since the other channels mostly keep the information about the temperature.

$$f_{ij}^g \in F_{i,j}^g (i = [1:w], j = [1:h]|k = 2) \tag{8}$$

The color channels of the $F_{i,j,k}$ are presented in Fig. 4.

Although the veins are slightly visible in $F_{i,j}^g$ as seen in Fig. 4, we applied anisotropic diffusion for sharpening the vein boundaries, by:

$$\frac{\partial F^g(\bar{x}, \Delta t)}{\partial t} = \nabla(c(\bar{x}, \Delta t)\nabla F^g(\bar{x}, t)) \tag{9}$$

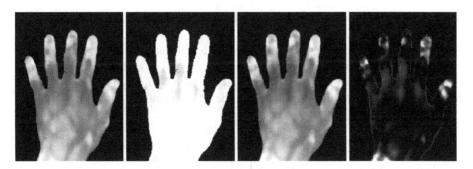

Fig. 4. Color channels: $F_{i,j,k}$ - $F_{i,j}^r$ - $F_{i,j}^g$ - $F_{i,j}^b$ (Color figure online)

where, $c(\bar{x}, t)$ is the diffusion function, $\bar{x}$ represents the spatial dimensions and Δt the time. For each direction d (north, west, south and east) from the starting point (x,y) the diffusion function applied to the image for N times, is:

$$F_{n+1}^g(x,y) = F_n + \frac{c_{d,\Delta t}(x,y)\nabla F_{d,t}^g(x,y)}{4} \tag{10}$$

The coefficient along any direction d is calculated by:

$$c_{d,t}(x,y) = e^{\frac{-\nabla(F_{d,\Delta t}^g)^2}{k}} \tag{11}$$

where

$$\check{F}^g = F_{d,\Delta t}^g = F_{\Delta t}^g(x,y+1) - F_{\Delta t}^g(x,y) \tag{12}$$

and k is the gradient threshold. For the parameters $k = 50$, $\Delta t = 0.2$ for $N = 20$ iterations, we obtained following noise reduced and enhanced image (Fig. 5).

It is seen from the images that not every pixel is darkened yet the vein paths are more visible. Since we deal with the whole hand not with a region, we used some parts of maximum curvature method in the following section.

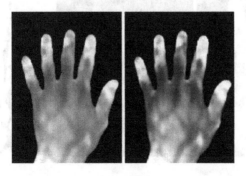

Fig. 5. Green channel F^g on the left and enhanced image $\check{F}^g$ on the right

4 Results of the Estimation by Maximum Curvature

The maximum curvature method is a procedure to calculate the local maximum curvatures on an image. It is usually used for finding the center positions of the veins and connecting them, however we only used the process to find only the curvatures. The curvatures along the vertical direction could be represented as:

$$\kappa(z) = \frac{\frac{d^2 P_f(z)}{dz^2}}{\left(1 + \left(\frac{dP_f(z)}{dz}\right)^2\right)^{\frac{3}{2}}} \tag{13}$$

where $P_f(z)$ is the cross-sectional profile of $\check{F}^g$, z is the position of the profile. A profile would be indexed as concave where $\kappa(z) > 0$ so that the positions of these points are saved as z_i' where $i = 0, 1 \ldots, N - 1$ and N is the number of local maximum points. Afterwards scores are assigned to the center positions as follows:

$$S(z_i') = \kappa\left(z_i'\right) W_r(i) \tag{14}$$

where $W_r(i)$ is the width of the region. The main purpose of this process is to find the large veins for drawing the curves. Every scores obtained are assigned to a blank image V as follows:

$$V\left(x_i', y_i'\right) = V\left(x_i', y_i'\right) + S(z_i') \tag{15}$$

After this process, the estimated vein image V and the complement image $\sim V$ are obtained (Fig. 6).

Although there are some cropping errors, the dorsal hand veins are successfully identified and the vein system could be constructed. The major reason for

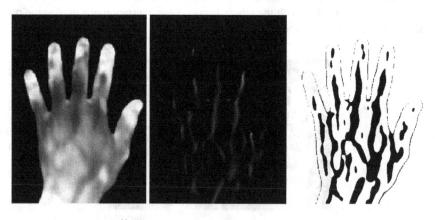

Fig. 6. $\check{F}^g$ on the left, V in the middle, $\sim V$ on the right.

misidentification is the background color that we turned while segmentation. While searching for the veins, our algorithm looks for the darker curves therefore some regions are incorrectly constructed. On the other hand, visible veins were easier to detect however our system successfully estimated the invisible veins in high temperature regions. Considering our vertical approach, one more success in the continuity

Table 1. Additional experiments

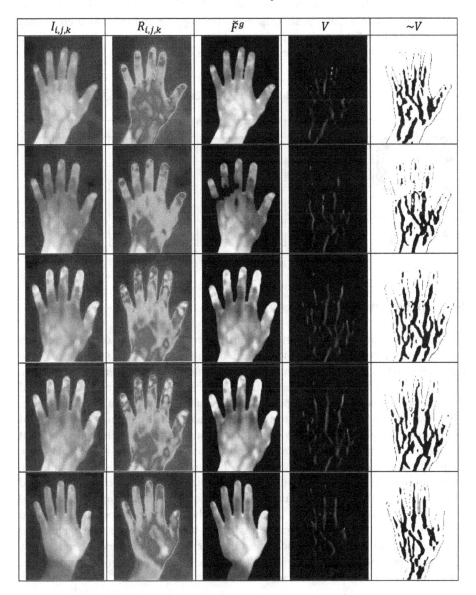

while extracting and constructing the vessel system, yet there are some connection errors in horizontal and diagonal axes.

5 Results of Supplementary Experiments

Even though this paper is dealing with a very preliminary but innovative approach for superficial hand vein estimation, we presented several additional cases in Table 1.

As seen in Table 1, if some regions of the hands are over-heated or discretely cooled down due to indeterminate reasons, some cropping and estimation errors may occur. Generally robust as the experiments show, the estimation of the vein system itself is strictly dependent on environmental factors.

Table 2. An inverse case

$I_{i,j,k}$	$R_{i,j,k}$	$\breve{F}^g$	V	$\sim V$

6 Conclusions and Discussions

Hand thermograms are very important for diagnosis of the possible disorders yet they can be used for dorsal hand vein identification. Therefore we presented a model that can be applied to any hand thermogram to estimate the veins, even the whole vein system is not so clear. The initial results are encouraging since our system could construct the veins on higher temperature regions although there are no veins visible.

As we mentioned before, the main difference and actually the major difficulty of this research is high radiation areas on the dorsal hand images. Therefore, two different algorithms and many conversions were necessary for the recognition of the hand veins. With red channel conversion we segmented the hands and with anisotropic diffusion filters, we reduced the noise and highlighted the veins on the green-channel converted image. Through the maximum curvature algorithm, we only searched the vertical curves and estimated the veins. In addition, throughout our dataset consisting of hand thermographs, we interestingly came across a very high temperature image, presented in Table 2, where the veins are cooler than the other parts of the hand. Therefore it is once again proven that the experiments should be conducted in room temperature, where the veins would be significantly darker.

Among the other papers, the only comparable study belongs to Buddharaju and Pavlidis [8]. Their vein estimation system is applied to the faces on thermograms and the vein systems are successfully determined however very superficially due to lack of visible veins. Extracting the radiation difference as a main feature is what we also dealt with; yet our project has a kernel between recognition and estimation since some veins are totally visible in our dataset.

This paper has very minor but very crucial assumptions: above all, the hands should be around the room temperature and placed vertically. Any artificial hot or cold points would cause the system misidentify the vein system. The veins to be estimated should correspond to darker pixels on hands for successful estimation, and there should be significant difference of the temperature levels between the hands and the background for precise segmentation, as secondary presumptions. The results of this paper could be used for biometric authentication by storing the features extracted from the images. Finally, it is still possible to use one of the algorithms we presented for segmentation [21, 22] or dorsal hand recognition [23], when necessary.

Acknowledgment. This work and the contribution were supported by project "SP/2102/2017 - Smart Solutions for Ubiquitous Computing Environments" from University of Hradec Kralove. We are also grateful for the support of Ph.D. students of our team (Richard Cimler and Jan Trejbal) in consultations regarding application aspects.

References

1. Lee, J.C., Lee, C.H., Hsu, C.B., Kuei, P.Y., Chang, K.C.: Dorsal hand vein recognition based on 2D Gabor filters. Imaging Sci. J. **62**(3), 127–138 (2014)
2. Sheetal, R.P., Goela, K.G.: Image processing in hand vein pattern recognition system. Int. J. Adv. Res. Comput. Sci. Softw. Eng. **4**(6), 427–430 (2014)
3. Janes, R., Júnior, A.F.B.: A low cost system for dorsal hand vein patterns recognition using curvelets. In: Proceedings of the 2014 First International Conference on Systems Informatics, Modelling and Simulation, pp. 47–52. IEEE Computer Society (2014)
4. Sakthivel, G.: Hand vein detection using infrared light for web based account. Int. J. Comput. Appl. **112**(10), 17–21 (2015)
5. Morales-Montiel, I.I., Olvera-López, J.A., Martín-Ortíz, M., Orozco-Guillén, E.E.: Hand vein infrared image segmentation for biometric recognition. Res. Comput. Sci. **80**, 55–66 (2014)
6. Huang, D., Zhang, R., Yin, Y. Wang, Y., Wang, Y.: Local feature approach to dorsal hand vein recognition by centroid-based circular key-point grid and fine-grained matching. Image Vis. Comput. (2016). http://dx.doi.org/10.1016/j.imavis.2016.07.001
7. Djerouni, A., Hamada, H., Loukil, A., Berrached, N.: Dorsal hand vein image contrast enhancement techniques. Int. J. Comput. Sci. **11**(1), 137–142 (2014)
8. Buddharaju, P., Pavlidis, I.T.: Physiology-based face recognition in the thermal infrared spectrum. In: Medical Infrared Imaging: Principles and Practices, pp. 18.1–18.16. CRC Press (2012)
9. Buddharaju, P., Pavlidis, I.T., Tsiamyrtzis, P., Bazakos, M.: Physiology-based face recognition in the thermal infrared spectrum. IEEE Trans. Pattern Anal. Mach. Intell. **29**(4), 613–626 (2007)

10. Buddharaju, P., Pavlidis, I.T., Tsiamyrtzis, P.: Physiology-based face recognition. In: IEEE Conference on Advanced Video and Signal Based Surveillance, AVSS 2005, pp. 354–359. IEEE (2005)

11. Dan, G., Guo, Z., Ding, H., Zhou, Y.: Enhancement of dorsal hand vein image with a low-cost binocular vein viewer system. J. Med. Imaging Health Inform. 5(2), 359–365 (2015)

12. Srivastava, S., Bhardwaj, S., Bhargava, S.: Fusion of palm-phalanges print with palmprint and dorsal hand vein. Appl. Soft Comput. 47, 12–20 (2016)

13. Lee, J.C., Lo, T.M., Chang, C.P.: Dorsal hand vein recognition based on directional filter bank. Signal Image Video Process. 10(1), 145–152 (2016)

14. Wang, Y., Duan, Q., Shark, L.K., Huang, D.: Improving hand vein recognition by score weighted fusion of wavelet-domain multi-radius local binary patterns. Int. J. Comput. Appl. Technol. 54(3), 151–160 (2016)

15. Premalatha, K., Natarajan, A.M.: Hand vein pattern recognition using natural image statistics. Def. Sci. J. 65(2), 150–158 (2015)

16. Wang, J., Wang, G., Li, M., Du, W., Yu, W.: Hand vein images enhancement based on local gray-level information histogram. Int. J. Bioautomation 19(2), 245–258 (2015)

17. Wang, G., Wang, J., Li, M., Zheng, Y., Wang, K.: Hand vein image enhancement based on multi-scale top-hat transform. Cybern. Inf. Technol. 16(2), 125–134 (2016)

18. Miura, N., Nagasaka, A., Miyatake, T.: Extraction of finger-vein patterns using maximum curvature points in image profiles. IEEE Trans. Inf. Syst. 90(8), 1185–1194 (2007)

19. Ammer, K.: Nerve entrapment and skin temperature of the human hand. In: Infrared Imaging, p. 17–1. IOP Publishing (2015)

20. Ammer, K.: The sensitivity of infrared imaging for diagnosing Raynaud's phenomenon is dependent on the method of temperature extraction from thermal images. In: Infrared Imaging, pp. 161–166. IOP Publishing (2015)

21. Alpar, O.: Corona segmentation for nighttime brake light detection. IET Intell. Transp. Syst. 10(2), 97–105 (2015)

22. Alpar, O., Stojic, R.: Intelligent collision warning using license plate segmentation. J. Intell. Transp. Syst. 20(6), 487–499 (2015)

23. Alpar, O., Krejcar, O.: Dorsal hand recognition through adaptive YCbCr imaging technique. In: Nguyen, N.-T., Manolopoulos, Y., Iliadis, L., Trawiński, B. (eds.) ICCCI 2016. LNCS (LNAI), vol. 9876, pp. 262–270. Springer, Cham (2016). doi:10.1007/978-3-319-45246-3_25

Automatic Removal of Mechanical Fixations from CT Imagery with Particle Swarm Optimisation

Mohammad Hashem Ryalat, Stephen Laycock, and Mark Fisher[(⊠)]

University of East Anglia, Norwich Research Park, Norwich NR4 7TJ, UK
{M.Ryalat,S.Laycock,Mark.Fisher}@uea.ac.uk
http://www.uea.ac.uk/computing

Abstract. Fixation devices are used in radiotherapy treatment of head and neck cancers to ensure successive treatment fractions are accurately targeted. Typical fixations usually take the form of a custom made mask that is clamped to the treatment couch and these are evident in many CT data sets as radiotherapy treatment is normally planned with the mask in place. But the fixations can make planning more difficult for certain tumor sites and are often unwanted by third parties wishing to reuse the data. Manually editing the CT images to remove the fixations is time consuming and error prone. This paper presents a fast and automatic approach that removes artifacts due to fixations in CT images without affecting pixel values representing tissue. The algorithm uses particle swarm optimisation to speed up the execution time and presents results from five CT data sets that show it achieves an average specificity of 92.01% and sensitivity of 99.39%.

Keywords: Immobilization mask · CT images · Head and neck cancer

1 Introduction

Head and Neck Cancer (HNC) refers to a group of different malignant tumors that develop in or around the throat, larynx, nose, sinuses, and mouth [1]. Staging of the cancer may be determined by medical imaging, biopsy and blood tests [2]. HNC is the eighth most common cancer in the UK (2014), accounting for 3% of all new cases [3]. Figures published from the United States estimate that 61,760 people developed head and neck cancer in 2015 [4].

A course of Radiotherapy Treatment (RT) typically forms a component in the prescribed treatment of HNC and is delivered in fractions over several weeks. Fixation masks (Fig. 1) are employed [5] to ensure the patient can be consistently positioned for each dose fraction. CT data used for planning HNC radiotherapy treatment contain artifacts due to the mask (Fig. 6a) which in some cases can make planning more difficult and can be troublesome for third parties wishing to reuse the data. This paper proposes an automatic approach to removing them.

© Springer International Publishing AG 2017
I. Rojas and F. Ortuño (Eds.): IWBBIO 2017, Part I, LNBI 10208, pp. 419–431, 2017.
DOI: 10.1007/978-3-319-56148-6_37

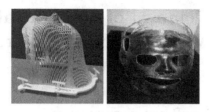

Fig. 1. Immobilization masks (left) thermoplastic (right) polyethylene.

Since Head and Neck CT data are usually acquired prior to radiotherapy treatment many publicly available CT data sets are from patients fitted with immobilization masks or other fixation devices. Segmentation of anatomical structures such as brain, lateral ventricles and skull is required prior to building and rendering 3-D models of these data but artifacts within the CT due to the immobilization mask makes the task more complicated. The removal of artifacts by manually editing individual CT image slices is time consuming and error prone. This is particularly problematic in the regions where the mask contacts the skin [6,7]. Therefore a robust approach to automatically remove the masks from the CT slices represents an appreciable saving in time.

There are numerous studies related to the segmentation and identification of the head/intra-cranial in the CT images [8–11] but in our knowledge, this study is the first to present a fully automatic approach for removing CT image artifacts due the fixation mask. Our algorithm employs an extension of Otsu's method, which classifies pixels as belonging to one of many classes using multi-level thresholding [5]. Exhaustive search for multiple thresholds requires the evaluation of $(n + 1)(D - n + 2)^n$ combinations of thresholds, where n represents the number of thresholds and D represents the absolute difference between the maximum and minimum image pixel value. Since pixel intensities in Dicom images are represented by signed integers the search can be very time consuming. To address this we test three optimisation techniques: Particle Swarm Optimisation (PSO) [12], Darwinian Particle Swarm Optimisation (DPSO) [13] and Fractional Darwinian Particle Swarm Optimisation (FDPSO) [14].

The remainder of this paper is organized as follows. Section 2 presents the proposed approach and Sect. 3 presents the data sets used for evaluation and the experiments that were performed. Section 4 presents the results and finally, Sect. 5 draws conclusions.

2   The Proposed Approach

The basic steps of the proposed approach are presented in Fig. 2. The algorithm applies FDPSO 'slice-by-slice' to segment the image to six different classes under Otsu's criterion [15]. Section 3 explains that we found segmenting the image into six different classes empirically, since in all our experiments this numberclusters all or most of the pixels belonging to the mask as one class. Algorithm 1 further

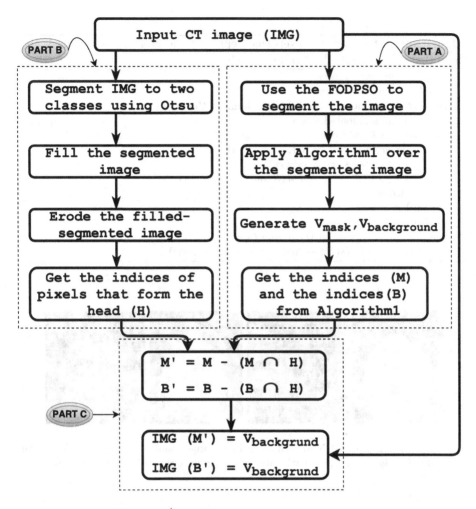

Fig. 2. Overview of the proposed approach.

refines the output of the FODPSO segmentation by employing a heuristic search for pixels in the labelled image that represent the immobilization mask and background.

We assume the top middle pixel represents the image background (air) and search the labeled image until we find a different pixel value (e.g. identified by the red square in Fig. 3). This pixel is assumed to belong to the class labelled mask. Using these pixel labels we identify sets of pixels $\{M\}$ and $\{B\}$ that represent the mask and background respectively.

Figure 4 illustrates that $\{M\}$ and $\{B\}$ sometimes include erroneous pixels because the FODPSO segmentation process groups these as one cluster. The sets $\{M\}$ and $\{B\}$ contain pixels that are misclassified because the FDPSO algorithm only uses intensity to cluster pixels. In our experiments, illustrated in

Algorithm 1. Finding the pixels that represent the immobilization mask and those that represent the background pixels

---
1: $height \leftarrow height\_of\_image$
2: $width \leftarrow width\_of\_image$
3: $mid\_col \leftarrow$ the index of the middle column in the image
4: $s \leftarrow$ Get the segmented image.
5: **for** $row \leftarrow 2, height$ **do**
6:     **if** $s\,(row,\, mid\_col)\, != s(1,\, mid\_col)$ **then**
7:         $v\_mask \leftarrow s\,(row,\, mid\_col)$
8:         $v\_bg\_in\_CT\_img \leftarrow IMG(row - 1,\, mid\_col)$
9:         Break
10:     **end if**
11: **end for**
12: $indices\_of\_mask\_pixels(M) \leftarrow$ find $(s == v\_mask)$
13: $indices\_of\_bg\_pixels(B) \leftarrow$ find $(s == s(1,\, mid\_col))$

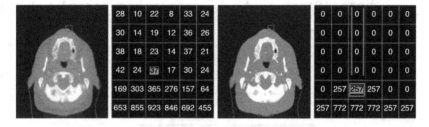

Fig. 3. A CT image in DICOM format and the same image after segmentation using the FODPSO algorithm along with their pixel region tool. (Color figure online)

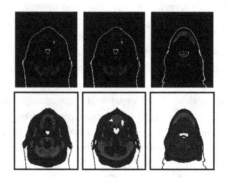

Fig. 4. Examples of pixels mislabeled by FDPSO. Upper row: pixels mislabeled as $\{M\}$; Lower row: pixels mislabeled as $\{B\}$.

Fig. 4 we found that the misclassified pixels are always located inside the skull. We correct this problem by recovering the coordinates of those pixels located within the skull and excluding these from $\{M\}$ by a sequence of operations that split the original CT image to two clusters (i.e. foreground (head) and background (air)), once again by using Otsu's method. We then flood-fill holes that may appear inside the skull using the morphological reconstruction operator described in [16]. Subsequently we proceed by performing an erosion [17] over the filled image. The aim of this process is to erode away the boundaries of the skull so areas of foreground pixels shrink in size. This will guarantee that none of the pixels that belong to the mask will be excluded later and the only pixels that will be excluded are those which are exist inside the skull. The index of those pixels which represent the skull are recovered from the eroded image as $\{H\}$. Equations (1) and (2) identify the sets $\{M'\}$ and $\{B'\}$ that exclude those pixels within $\{M\}$ and $\{B\}$ that are also with the skull $\{H\}$.

$$M' = M - (M \cap H) \tag{1}$$

$$B' = B - (B \cap H) \tag{2}$$

3   Experimental Work

Data Set

Five CT data sets from anonymized patients have been used in this study. The first three data sets ($512 \times 512 \times 155, 512 \times 512 \times 146, 512 \times 512 \times 151$, helical, pixel-spacing 1.367×1.367 mm, slice-thickness 2.5 mm) were acquired at St James's University Hospital NHS Foundation Trust, Leeds, UK and the other two data sets ($512 \times 512 \times 130, 512 \times 512 \times 156$, pixel-spacing 1.08×1.08 mm 0.98×0.98 mm, slice-thickness 3.14 mm) were downloaded from the Cancer Imaging Archive (TCIA)/Head-Neck-Cetuximab [18,19].

Experiments

The accuracy of the PSO, DPSO and FODPSO is evaluated by measuring the fitness (i.e. inter-class variance) for each algorithm and comparing the outputs with the brute-force (BF) method. Table 1 shows the average fitness values generated by the PSO, DPSO and FODPSO algorithms compared with that generated by the BF method. It is clear from Table 1 that the fitness value produced by the FODPSO algorithm is the same or a slightly less than that of the BF method. This indicates that applying the FODPSO algorithm for segmentation leads to a very high accurate outcomes.

Table 2 displays the average CPU processing time that PSO, DPSO, FODPSO and BF methods need to segment all image slices in each data set. The table confirms that the FODPSO algorithm is always slightly faster than the DPSO algorithm and the DPSO algorithm is significantly faster than the PSO algorithm. It is worth noting that the speed of BF search is similar to, but

Table 1. Average fitness values of Brute-Force, PSO, DPSO and FODPSO algorithms for different number of thresholds over different five data sets

Dataset	Thr.	BF	PSO	DPSO	FODPSO
Dataset#1 (155 images)	1	3269.09	**3269.09**	**3269.09**	**3269.09**
	2	3773.38	3773.37	**3773.38**	**3773.38**
	3	3829.39	3829.36	**3829.38**	**3829.38**
	4	3855.43	3854.92	3855.25	**3855.39**
	5	3871.88	3871.81	3871.72	**3871.82**
Dataset#2 (146 images)	1	3488.54	3488.49	**3488.54**	**3488.54**
	2	4067.33	4067.21	**4067.30**	**4067.30**
	3	4142.05	4141.87	4141.91	**4142.02**
	4	4178.69	4178.11	**4178.67**	**4178.67**
	5	4197.88	4197.09	4197.09	**4197.86**
Dataset#3 (151 images)	1	2374.66	2374.39	2374.54	**2374.54**
	2	2635.29	2634.50	2634.88	**2635.26**
	3	2657.87	2654.98	2655.29	**2657.84**
	4	2679.31	2677.82	2679.19	**2679.27**
	5	2688.69	2685.58	2686.44	**2688.64**
Dataset#4 (130 images)	1	3749.77	**3749.77**	**3749.77**	**3749.77**
	2	4264.51	4264.01	4264.32	**4264.51**
	3	4363.87	4363.12	4363.72	**4363.86**
	4	4418.97	4417.99	**4418.96**	**4418.96**
	5	4435.87	4431.89	4435.82	**4435.85**
Dataset#5 (156 images)	1	2870.42	2870.18	**2870.42**	**2870.42**
	2	3386.32	3385.83	3386.12	**3386.32**
	3	3462.11	3460.76	3460.76	**3462.10**
	4	3516.13	3515.73	3516.08	**3516.11**
	5	3535.57	3533.12	3534.84	**3535.53**

less than the speed of FODPSO when the number of thresholds equals one. But as the number of thresholds increases, the difference between the speed of BF and the speed of the other three optimisation algorithms becomes significant. In our case (i.e. removing artifacts due to immobilization masks) we are interested in the case when the number of thresholds equals 5, and then the use of the FODPSO algorithm will make a significant enhancement in terms of speed. The fractional coefficient used in FODPSO allows the convergence rate of the algorithm to be controlled and this explains why FODPSO outperforms the DPSO algorithm.

The standard deviation was used as an evaluation of stability. Table 3 shows that FODPSO produces the most stable results when compared to the PSO and DPSO, and the standard deviation increases as the number of thresholds increase

Table 2. Average execution time (in sec) of the Brute-Force, PSO, DPSO and FODPSO algorithms for different number of thresholds over different five data sets

Dataset	T.holds	Brute-Force	PSO	DPSO	FODPSO
Dataset#1	1	10.86	33.81	12.15	**9.72**
	2	235.52	73.12	58.38	**56.75**
	3	25951	90.85	74.90	**72.53**
	4	2408545	111.01	90.04	**86.82**
	5	>1 week	131.06	103.25	**98.03**
Dataset#2	1	7.28	20.95	8.37	**6.96**
	2	221.46	65.56	53.39	**50.27**
	3	24429	84.90	71.50	**67.20**
	4	1709952	104.41	83.91	**79.08**
	5	>1 week	123.61	97.84	**93.42**
Dataset#3	1	7.25	20.20	7.03	**6.20**
	2	232.58	69.35	56.92	**54.85**
	3	25118	88.02	70.82	**66.54**
	4	1809433	107.73	86.51	**82.36**
	5	>1 week	127.18	100.57	**94.81**
Dataset#4	1	8.06	16.15	8.51	**7.97**
	2	195.84	58.54	47.82	**46.69**
	3	21717	75.35	61.32	**60.80**
	4	1908530	92.85	76.12	**72.91**
	5	>1 week	109.67	86.73	**84.16**
Dataset#5	1	9.26	49.02	8.75	**8.09**
	2	238.19	70.40	57.81	**57.59**
	3	26148	90.99	76.13	**72.12**
	4	2173860	111.82	90.04	**84.29**
	5	>1 week	132.30	104.59	**97.21**

in most cases. Typical results of segmentation using the FODPSO algorithm over one sample image using different number of thresholds is shown in Fig. 5.

Figure 6(a) displays an example of one of the CT slices from the first data set. A previous study [5] evaluated the use of Particle Swarm Optimisation for medical image segmentation and demonstrated the the FODPSO algorithm delivered high accuracy, stability and speed. We found that for our application, segmenting the image to six different classes tends to lead to better results as this number classifies all or most of the pixels belonging to the mask as one class. The FODPSO algorithm delivers significant benefits in terms of execution speed over the BF approach (i.e. exhaustive search) which takes a very long time when the number of clusters equals six. In Sect. 4 we compare the FODPSO algorithm

Table 3. Standard deviation of fitness for PSO, DPSO and FODPSO after running each algorithm 15 times over different five data sets.

Dataset	T.holds	PSO	DPSO	FODPSO
Dataset#1	1	**0**	**0**	**0**
	2	0.0001	**0**	**0**
	3	0.0036	0.0003	**0.0002**
	4	1.2873	0.0114	**0.0105**
	5	0.0206	1.2195	**0.0190**
Dataset#2	1	0.0001	**0**	**0**
	2	0.0009	0.0002	**0.0001**
	3	0.0021	0.0005	**0.0002**
	4	0.0122	0.0120	**0.0113**
	5	0.0787	0.0342	**0.0341**
Dataset#3	1	0.0023	0.0001	**0**
	2	0.0082	0.0009	**0.0001**
	3	0.0810	0.0569	**0.0015**
	4	0.0254	0.0143	**0.0061**
	5	0.5932	0.5437	**0.2903**
Dataset#4	1	0.0011	0.0002	**0**
	2	0.0008	0.0002	**0.0001**
	3	0.0110	0.0089	**0.0073**
	4	0.0196	0.0159	**0.0161**
	5	1.7163	0.0938	**0.0884**
Dataset#5	1	0.0005	0.0003	**0.0001**
	2	0.0012	0.0004	**0.0001**
	3	0.0082	0.0027	**0.0011**
	4	0.0243	0.0199	**0.0190**
	5	1.3081	0.0373	**0.0361**

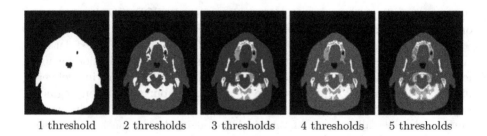

| 1 threshold | 2 thresholds | 3 thresholds | 4 thresholds | 5 thresholds |

Fig. 5. Applying FODPSO using different number of thresholds.

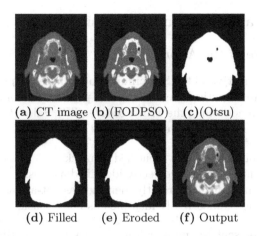

(a) CT image (b)(FODPSO)   (c)(Otsu)

(d) Filled   (e) Eroded   (f) Output

Fig. 6. An example of a CT slice from the first dataset.

with other techniques by tabulating the run time needed to segment a typical HNC CT data set. Figure 6(b) displays the image after it was segmented to six different clusters using the FODPSO algorithm.

In Fig. 6(c–e) we present the output that is generated by part-B of the proposed approach. The image was firstly segmented to two classes (foreground and background) using Otsu's method. It was then filled automatically and eroded as it displayed in Fig. 6(e). Part-A and part-B of the proposed approach produced three data structures of indices $(M, B,$ and $H)$ and those indices were used to form the final output image which is displayed in Fig. 6(f). Finally, Fig. 7 illustrates randomly-selected input images and their outputs after applying the approach.

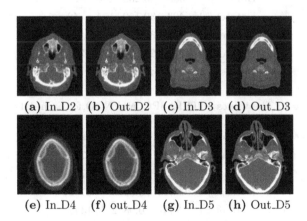

(a) In_D2   (b) Out_D2   (c) In_D3   (d) Out_D3

(e) In_D4   (f) out_D4   (g) In_D5   (h) Out_D5

Fig. 7. One CT slice example from each dataset (input output).

4   Results, Validation and Discussion

The Performance of the Approach

We used the **Sensitivity** and the **Specificity** to evaluate the proposed approach as both of them are statistical measures of the performance of a binary classification test. We have identified the True Positive (TP), False Positive (FP), True Negative (TN) and False Negative (FN) in this context as:

- TP: # of mask pixels correctly identified as mask.
- FP: # of not-a-mask pixels incorrectly identified as mask.
- TN: # of not-a-mask pixels correctly identified as not-a-mask.
- FN: # of mask pixels incorrectly identified as not-a-mask.

The pixels that represent the immobilisation mask were identified by an expert in 25 CT images (5 randomly-selected from each dataset) and compared to the number of pixels identified by the proposed approach. Table 4 displays the average values, rounded to the whole number, of TP, FP, TN and FN for each dataset and Table 5 displays the sensitivity, also called the true positive rate (TPR), specificity (SPC) and the Number-Of-Observations (NOO) for each dataset.

As is shown in Table 5 the average value of the sensitivity (TPR) is 92.01% which indicates to the proportion of positives that are correctly identified (i.e. the percentage of mask pixels which are correctly identified by the proposed approach as mask pixels) and the average value of the specificity (SPC)

Table 4. The average values of TP, FP, TN, and FN for each dataset

Dataset	TP	TN	FP	FN
Dataset#1	389	30,239	100	23
Dataset#2	403	30,152	154	42
Dataset#3	429	30,199	93	30
Dataset#4	1714	29,060	465	203
Dataset#5	841	45,371	148	71

Table 5. The values of TPR, SPC, and NOO for each dataset

Dataset	TPR	SPC	NOO
Dataset#1	0.9441	0.9967	30751
Dataset#2	0.9056	0.9949	30751
Dataset#3	0.9346	0.9969	30751
Dataset#4	0.8941	0.9842	31442
Dataset#5	0.9221	0.9967	46431
Average	**0.9201**	**0.9939**	34,025

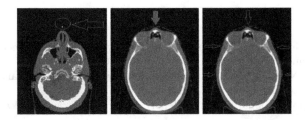

Fig. 8. (**Left**) Example of a CT image includes a noise in the middle column. (**Middle**) A CT image has a disconnected representation of the mask pixels. (**Right**) Defining new start points to seek horizontally.

is 99.39% which points to the proportion of negatives that are correctly identified (i.e. the percentage of not-a-mask pixels who are correctly identified as not-a-mask pixels). The heading 'NOO' in the table indicates to the number of observations which is equivalent to the number of pixels in each image.

Handling Exceptions

We applied our approach over five different data sets (total = 738 images) and noticed that the approach did not work on 13 images of them. This accounts for two reasons (1) some CT images include a noise in the middle column on the top of the mask itself, as displayed in Fig. 8-left (2) some CT images have disconnected representation of the mask pixels, as displayed in Fig. 8-middle. We handled the first exception by applying the median filter over the background area in order to remove the noise from the background area, and we handled the second exception by changing the seeking mechanism in Algorithm 1 by searching the segmented image horizontally and vertically from different five start points as displayed in Fig. 8-right.

5   Conclusion

This paper presented an automatic approach for segmenting immobilization masks in Head-and-Neck CT data sets. The approach identifies the pixels that belong to the immobilization mask and replaces their intensity value with that of air, thereby eliminating the mask from the output image. Five different data sets were tested to evaluate the accuracy of the approach. Sensitivity and specificity were used as statistical measures of the performance of the approach in this study. The evaluation indicates that the proposed approach is robust and of practical use. Some enhancements to speed up the process using Particle Swarm Optimisation were also presented and tested in the paper.

Acknowledgement. We would like to thank Prof. Susan Short and colleagues at St James's University Hospital NHS Foundation Trust for their help and for providing some of the data used in this study.

References

1. Head and Neck Cancer - Overview. http://www.cancer.net/cancer-types/head-and-neck-cancer/overview. Accessed 27 Nov 2016
2. Head and Neck Cancers. https://www.cancer.gov/types/head-and-neck/head-neck-fact-sheet. Accessed 27 Nov 2016
3. The Office for National Statistics Report. https://www.ons.gov.uk/. Accessed 27 Nov 2016
4. Head and Neck Cancer: Statistics. http://www.cancer.net/cancer-types/head-and-neck-cancer/statistics. Accessed 27 Nov 2016
5. Ryalat, M.H., Emmens, D., Hulse, M., Bell, D., Al-Rahamneh, Z., Laycock, S., Fisher, M.: Evaluation of particle swarm optimisation for medical image segmentation. In: Świątek, J., Tomczak, J.M. (eds.) ICSS 2016. AISC, vol. 539, pp. 61–72. Springer, Heidelberg (2017). doi:10.1007/978-3-319-48944-5_6
6. Fisher, M., Applegate, C., Ryalat, M., Laycock, S., Hulse, M., Emmens, D., Bell, D., et al.: Evaluation of 3-D printed immobilisation shells for head and neck IMRT. Open J. Radiol. **4**(04), 322 (2014)
7. Laycock, S.D., Hulse, M., Scrase, C.D., Tam, M.D., Isherwood, S., Mortimore, D.B., Emmens, D., Patman, J., Short, S.C., Bell, G.D.: Towards the production of radiotherapy treatment shells on 3D printers using data derived from dicom CT and MRI: preclinical feasibility studies. J. Radiother. Pract. **14**(01), 92–98 (2015)
8. Wei, K., He, B., Zhang, T., Shen, X.: A novel method for segmentation of CT head images. In: 2007 1st International Conference on Bioinformatics and Biomedical Engineering, pp. 717–720. IEEE (2007)
9. Chen, W., Smith, R., Ji, S.-Y., Najarian, K.: Automated segmentation of lateral ventricles in brain CT images. In: IEEE International Conference on Bioinformatics and Biomeidcine Workshops, BIBMW 2008, pp. 48–55. IEEE (2008)
10. Qian, X., Wang, J., Guo, S., Li, Q.: An active contour model for medical image segmentation with application to brain CT image. Med. phys. **40**(2), 021911 (2013)
11. Zaki, W.M.D.W., Fauzi, M.F.A., Besar, R., Ahmad, W.S.H.M.W.: Qualitative and quantitative comparisons of haemorrhage intracranial segmentation in CT brain images. In: 2011 IEEE Region 10 Conference, TENCON 2011, pp. 369–373. IEEE (2011)
12. Eberhart, R.C., Kennedy, J., et al.: A new optimizer using particle swarm theory. In: Proceedings of the Sixth International Symposium on Micro Machine and Human Science, New York, NY, vol. 1, pp. 39–43 (1995)
13. Tillett, J., Rao, T., Sahin, F., Rao, R.: Darwinian particle swarm optimization. In: Prasad, B. (ed.) Proceedings of the 2nd Indian International Conference on Artificial Intelligence (IICAI 2005), pp. 1474–1487 (2005)
14. Couceiro, M.S., Rocha, R.P., Fonseca Ferreira, N.M., Tenreiro Machado, J.A.: Introducing the fractional-order Darwinian PSO. Signal, Image and Video Processing **6**(3), 343–350 (2012)
15. Otsu, N.: A threshold selection method from gray-level histograms. Automatica **11**(285–296), 23–27 (1975)
16. Soille, P.: Morphological Image Analysis: Principles and Applications. Springer, Heidelberg (1999)
17. Gonzalez, R.C., Woods, R.E.: Image processing. Digit. Image Process. **2** (2007)

18. Clark, K., Vendt, B., Smith, K., Freymann, J., Kirby, J., Koppel, P., Moore, S., Phillips, S., Maffitt, D., Pringle, M., et al.: The cancer imaging archive (TCIA): maintaining and operating a public information repository. J. Digit. Imaging **26**(6), 1045–1057 (2013)
19. The Cancer Imaging Archive. http://doi.org/10.7937/K9/TCIA.2015.7AKG JUPZ/. Bosch, W.R., Straube, W.L., Matthews, J.W., Purdy, J.A.: Data From Head Neck Cetuximab (2015)

Augmented Visualization as Surgical Support in the Treatment of Tumors

Lucio Tommaso De Paolis[✉]

AVR Lab, Department of Engineering for Innovation,
University of Salento, Lecce, Italy
lucio.depaolis@unisalento.it

Abstract. Minimally Invasive Surgery is a surgery technique that provides evident advantages for the patients, but also some difficulties for the surgeons. In medicine, the Augmented Reality technology allows surgeons to have a sort of "X-ray" vision of the patient's body and can help them during the surgical procedures. In this paper is presented an application that could be used as support for a more accurate preoperative surgical planning and also for an image-guided surgery. The Augmented Reality can support the surgeon during the treatment of the liver tumors with the radiofrequency ablation in order to guide the needle and to have an accurate placement of the surgical instrument within the lesion. The augmented visualization can avoid as much as possible to destroy healthy cells of the liver.

Keywords: Augmented reality · Image guided surgery · Computer aided surgery

1 Introduction

The actual trend in surgery is the transition from open procedures to minimally invasive interventions. Using this surgery technique visual feedback to the surgeon is only possible through the laparoscope camera. Furthermore direct palpation of organs is not possible. Minimally Invasive Surgery (MIS), such as laparoscopy or endoscopy, has changed the way to practice the surgery. It is a promising technique and the use of this surgical approach is nowadays widely accepted and adopted as a valid alternative to classical procedures.

The use of MIS offers to surgeons the possibility of reaching the patient's internal anatomy in a less invasive way. This reduces the surgical trauma for the patient. The diseased area is reached by means of small incisions made on the patient body called ports. Specific instruments and a camera are inserted through these ports. The surgeon uses a monitor to see what happens on the surgical field inside the patient body.

With the use of MIS the patient has a shorter hospitalizations, faster bowel function return, fewer wound-related complications and a more rapid return to normal activities. These advantages have contributed to accept these surgical procedures. If the advantages of this surgical method are evident on the patients, these techniques have some limitations for the surgeons. In some systems, the imagery is in 2D. With this limitation the surgeon needs to develop new skills and dexterity in order to estimate the distance

© Springer International Publishing AG 2017
I. Rojas and F. Ortuño (Eds.): IWBBIO 2017, Part I, LNBI 10208, pp. 432–443, 2017.
DOI: 10.1007/978-3-319-56148-6_38

from the anatomical structures. We have to consider that he had to work in a very limited workspace. Modern systems use two cameras to offer a 3D view to the surgeon but this technology is useful when the working volume is not too small.

Medical images (CT or MRI) associated to the latest medical image processing techniques could provide an accurate knowledge of the patient's anatomy and pathologies. This information can be used to guide surgeons during the surgical procedure to improve the patient care.

The Augmented Reality (AR) technology has the potential to improve the surgeon's visualization during the MIS surgery. This technology can "augment" surgeon's perception of the real world with the use of information gathered from patient's medical images. AR technology refers to a perception of a physical real environment whose elements are merged with virtual computer-generated objects in order to create a mixed reality. The merging of virtual and real objects in an AR application has to run in real time. Virtual objects have to be aligned (registered) with real world structures. Both of these requirements guarantee that the dynamics of real world environments remain unchanged after virtual data has been added [1].

The use of AR technology in medicine makes possible to overlay virtual medical images of the organs on the real patient. This allows the surgeon to have a sort of "X-ray vision" of the patient's internal anatomy. The use of AR in surgery produces a better spatial perception and a reduction in the duration of the surgical procedure.

In order to obtain a correct alignment of virtual and real objects is needed a registration procedure that can be accomplished using an optical tracker.

The aim of this paper is to present an AR system that could be used as support for a more accurate surgical preoperative planning and also for image-guided surgery. The application can support the surgeon during the needle insertion in the radiofrequency ablation of the liver tumors. The library on which the application is based has been modified in order to support an optical tracker that could improve the performance of the system.

2 Previous Works

Many research groups are now focusing on the development of systems that assists surgeons during the minimally invasive surgical procedures.

Furtado and Gersak [2] present some examples of how AR can be used to overcome the difficulties inherent to MIS in the cardiac surgery.

Samset et al. [3] present some decision support tools. These tools are based on concepts in visualization, robotics and haptics and provide tailored solutions for a range of clinical applications.

Bichlmeier et al. [4] focus on the problem of misleading perception of depth and spatial layout in medical AR and present a new method for medical in-situ visualization.

Navab et al. [5–7] present a new solution for using 3D virtual data in many AR medical applications. They introduce the concept of a laparoscopic virtual mirror, a virtual reflection plane within the live laparoscopic video.

De Paolis et al. [8] present an AR system that can guide the surgeon in the operating phase. The main goal of the system is to prevent erroneous disruption of some organs during surgical procedures. They provide distance information between the surgical tool and the organs and they use a sliding window in order to obtain a more realistic impression that the virtual organs are inside the patient's body.

Nicolau et al. [9] present a real-time superimposition of virtual models over the patient's abdomen in order to have a three dimensional view of the internal anatomy. The authors have used the developed system in an operating room and to reduce the influence of the liver breathing motion they have tried to simulate the patient's breathing cycle.

LiverPlanner [10, 11] is a virtual liver surgery planning system developed at Graz University of Technology that combines image analysis and computer graphics in order to simplify the clinical process of planning liver tumor resections. The treatment planning stage enables the surgeon to elaborate a detailed strategy for the surgical intervention and the outcome of pre-operative planning can then be used directly for the surgical intervention.

Maier-Hein et al. [12] present a system developed for computer-assisted needle placement that uses a set of fiducial needles to compensate for organ motion in real time; the purpose of this study was to assess the accuracy of the system in vivo.

Stüdeli et al. [13] present a system that provides surgeon, during placement and insertion of RFA needle, with information from pre-operative CT images and real-time tracking data.

De Mauro et al. [14] present the development of an IGS solution for spine surgery based on conventional medical devices and open-source software. The objective is to enhance the surgeon's ability for a better intra-operative orientation by giving him a three-dimensional view and other information necessary for a safe navigation inside the patient. The solution permits to reduce the radiation exposure to the patient and medical staff as well as the operation duration.

Livatino et al. [15] present the advantages of stereoscopic visualization and 3D technologies in medical endoscopic teleoperation; the increased level of depth awareness provided by stereo viewing is remarkable, which facilitates endoscope teleguide, teleexploration and teleintervention, leading to accurate navigation and faster decision-making.

Ricciardi et al [16] present an augmented reality platform for computer assisted surgery in the field of maxillo-facial surgery. A surgical planning module is also integrated in the platform.

3   Augmented Reality in the Treatment of the Liver Tumors

Hepatic cancer is one of the most common solid cancers in the world. Hepatocellular carcinoma (HCC) is the most common primary hepatic cancer. The liver is often the site of metastatic disease, particularly in patients with colorectal adenocarcinoma.

The use of chemotherapy for malignant form of liver cancer rarely led to good results in long-term survival rate. Surgery led to the best results for hepatic cancer care.

Unfortunately only from 5 to 15 per cent of patients with liver cancer can undergo to a potentially curative resection of the liver cancer [17].

Patients with confined disease of the liver could not be candidates to resection because of multifocal disease. Another factor that could preclude surgical treatment is the proximity of tumor to key vascular or biliary structures. This precludes a margin-negative resection potentially unfavorable in case of presence of multiple liver metastases. Very often the tumor is associated to a pre-existent cirrhosis that can further reduce resection margins.

Liver transplant is the only radical therapy that eliminates the risk of recurrence. Unfortunately it can't be always used. So, since most of patients with primary or malignancies confined metastatic at the liver are not candidates for surgical resection, different approaches to control and potentially cure liver diseases were developed.

The Liver Radiofrequency Ablation (RFA) is a minimally invasive treatment for liver cancer used since 1980's. It consists in the placement of a needle inside the liver parenchyma in order to reach the centre of the tumor lesion. When the lesion center is reached, an array of electrodes is extracted from the tip of the needle and it is expanded in the tumor tissue. From these electrodes is injected in the tumor tissue a radiofrequency current that causes tumor cell necrosis for hyperthermia (the local temperature is higher than 60 °C and cancer cells are more sensitive to heat than normal cells).

In Fig. 1 is shown the needle insertion and array expansion of the RFA technique on the liver tumor.

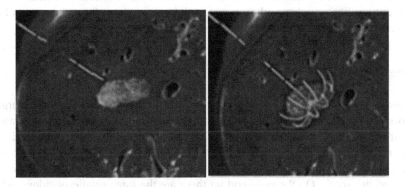

Fig. 1. The effect of the RFA technique at the liver tumor.

One problem in using radiofrequency tumor ablation technique is the correct placement of the needle. To ensure a maximum efficacy of the treatment the needle has to reach the center of the tumor.

Today surgeons use ultrasound, CT or MNR images acquired during the needle placement in order to correctly direct the needle towards the center of the tumor. The use of these two-dimensional images makes the insertion procedure very difficult and requires sometimes more than one insertion. In addition, the surgeon, in order to destroy all tumoral cells, applies the RFA on an extended area of the liver. In this way a large number of healthy cells are also destroyed. This practice can cause to the patient a number of other different consequences. To reduce this problem is of primary importance to reach the center of the tumor.

A guidance system of the needle in tumor ablation procedures can be obtained using the augmented reality technology. With the superimposition of the virtual models of the patient's anatomy (liver, vessels, biliary ducts, cancer, etc.) exactly where are the real ones, it is possible to make the needle placement task less difficult. In this way the surgeon has a sort of x-ray 3D vision of the patient internal anatomy.

The purpose of this AR application is to provide a guidance system that can help the surgeon during the needle insertion in liver RFA. The position and orientation of the ablation tool are measured using some reflective spheres that are detected by an optical tracker in order to measure in real time position and orientation of the real ablation tool.

To achieve a correct augmentation it is necessary to have a perfect correspondence between the virtual organs and the real ones. This is very important in an image-guided surgery application because a very small error in the registration phase can cause serious consequences on the patient due to a bad alignment of the virtual organs on the real ones.

The registration phase is one of the most delicate steps in an AR system for surgery and is obtained using some fiducial points. These points have been defined on the patient body and identified on the patient tomographic images. In this way the application can establish the transformation between the "real" world and "virtual" one. If the patient position can change during surgery it is necessary to fix on the patient body a tracker tool that permits to estimate its new positions and orientations.

4 Developed Application

The developed application is provided of a user interface shown in Fig. 2. The surgeon can load CT images and virtual models and manipulate them in order to understand in the best way the patient anatomy and pathology. The interface has been designed to be simple and functional at the same time; on the left side is placed the application control panel. On the right-top window is shown the 3D models of the organs and the augmented reality scene. On the right-bottom there are the three smaller windows where axial, coronal and sagittal views of CT dataset are placed.

The application offers to surgeon the possibility to study the case study before going in operating room. There is the possibility to apply a clipping modality that permits the surgeon to dissect the model and study its internal structure changing the opacity of the organs (Fig. 3). The dissection could be made along the three principal axes of tomographic images.

The application features described till now are part of what is considered the pre-operative planning task. During this task the surgeon can use the application to study the pathology in a more simple and natural way than that provided by simple CT slice visualization. For the navigation and augmentation task are devoted to the surgery

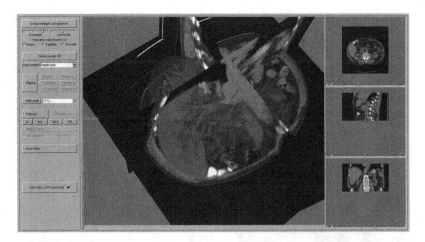

Fig. 2. The application interface.

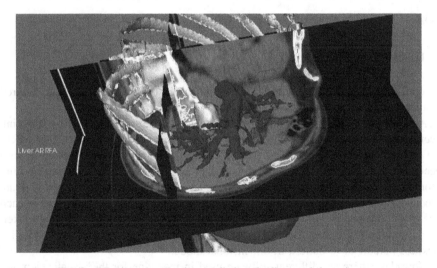

Fig. 3. Clipping visualization applied to the liver and thoracic cage.

room. Here the surgeon needs to use the optical tracker and to carry out the registration task. When the registration process is complete, a virtual ablation tool is shown in 3D view. It is coupled with the real ones and follows its movements.

In Fig. 4 is shown the augmented visualization of 3D virtual model over the patient's body (a dummy) during a preliminary test in the operating room. This visualization should guide the surgeon during the needle insertion in the radiofrequency ablation of the liver tumor.

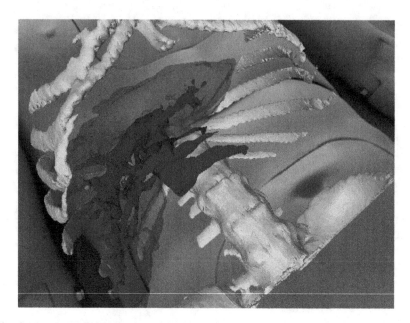

Fig. 4. Augmented visualization of 3D virtual model over the patient's body (a dummy).

5   Used Technologies

A reconstruction of the 3D model of the anatomical structures of the patient is required in order to improve the standard slice view.

An efficient 3D reconstruction of the patient's organs is obtained by applying specific segmentation and classification algorithms to medical images (CT or MRI). The image segmentation consists in the identification of the pixels that belong to a specific anatomical structure. This process can be manual, semi-automatic or fully automatic; the grey levels in the medical images are replaced by colors that are associated to the different anatomical structures [18]. After the segmentation of each slice of the dataset the software can reconstruct the 3D model using a model maker. This algorithm combines the results of segmentation of each slice and the slice thickness parameter to build a tri-dimensional model.

Nowadays there are different software used in medicine for the visualization and the analysis of medical images and the 3D modelling of human organs. Mimics [19], 3D Slicer [20], and OsiriX [21] play an important role among these tools. Some of them are also open-source software.

In the application 3D Slicer, a multi-platform open-source software package for visualization and image analysis, was used for the building of the 3D model of the patient's organ. 3D Slicer has many segmentation algorithms and also a model maker module.

The AR guidance application was developed using Image-Guided Software Toolkit (IGSTK) framework [22]. It is a set of high-level components integrated with low-level open source software libraries and application programming interfaces.

In the developed AR platform is used an optical tracker. This tool is able to detect some retro-reflective spheres intentionally introduced in the surgical scene. These spheres are placed on the surgical tools. They provide within a defined coordinate system the real-time spatial measurements of the location and orientation of the surgical instruments used during the surgical procedure.

In the first prototype of the system the Polaris Vicra optical tracker has been used [23]. After the test in the operating room the tracking system has been changed to ensure surgeons more freedom of movement. For this reason to use the Bonita Vicon tracking system [24] has been decided. This tracking system uses a variable number of separated cameras that can be better positioned in the operating room. A specific library has been developed and tested for interfacing with the Bonita Vicon tracker. This library was integrated inside IGSTK framework.

At the base of the functioning of IGSTK there is the use of a state machine that allows increasing the safety and robustness of the toolkit [25].

The use of a state machine, in fact, allows limiting and controlling the possible behaviours of application in order to ensure that it is always in a formal state planned in the design phase. This guarantees a reproducible and deterministic behaviour that eliminates the risk of application design formal errors.

The separation between public and private interface of an IGSTK component (object) is shown in Fig. 5.

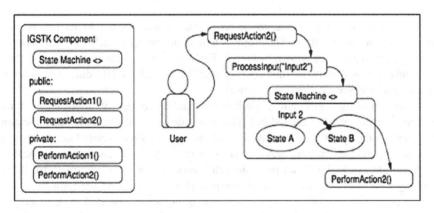

Fig. 5. Working modality of IGSTK library components.

When a user sends a request for an action through a call of a public method, this request is translated into an input for the state machine. The state machine taking into account the current state and the sent input changes its state as expected in the design phase. In any case, the object will always be in a known state because each type of behaviour has been programmed.

The IGSTK Tracker component communicates with the tracker to obtain the position and orientation relative to each tool that is present in the acquisition volume. This information is then transmitted to other IGSTK components that request it.

The two classes of IGSTK framework that provide for trackers management are:

- "igstk::Trackers" that serves to manage the status information of the tracking system;
- "igstk::TrackerTool" that manages the information associated with each tool present on the scene.

The "igstk::Trackers" class uses the interface of the "igstk::Communication" class for the management of communications. This allows preventing blockage of the application in the case of any lack of connection with the tracker. In this case an error event is generated in a non-blocking mode. The tracker acquires data are stored into a memory buffer.

The class "igstk::PulseGenerator" generates the clock used by the application in order to synchronize all events. In this way, reading and updating of the data are asynchronous operations and allow high standards of safety and performance.

To calibrate the instrument used in the surgical application and obtain the transformation elapsing between the tool and the point of interest has been implemented the PivotCalibration algorithm. This algorithm permits to store the position and orientation of the tracker tool holding fixed the position of the instrument tip and moving the tool in the space [26].

6 Application Tests

The first test of application was made on a dummy and some fiducial points were chosen on the dummy surface. The aim was to overlap the 3D models of a real patient's organs on the dummy, as shown in Fig. 4. This test was not successful because there was a misalignment between the dummy and the virtual models due to a different thoracic girth between patient and dummy.

To overcome this problem a preliminary qualitative test of application in the operating theatre has been carried out. This test is designed to evaluate the application uncertainty in the operating room and also all the possible issues related to a live use in the operating room. The test was designed to qualitatively evaluate the application uncertainty and to highlight all possible issues related to a use of the platform in the operating room. The surgeon was very expert in the practice of the radiofrequency ablation procedure and the test was carried out on a patient operated in open surgery with the tumor located on the surface of the liver. The surgeon verified the correct overlapping of tumor and real organs.

The test doesn't add risks for the patient because the application was not used to guide the needle insertion. Anyway the patient was informed on the nature of the experiment and signed an informed consent form.

The registration of the 3D virtual organs on the real ones was obtained just in order to verify if a good overlapping was obtained. The surgeons were satisfied of the obtained result.

7   Conclusions and Future Work

In this paper is presented an augmented visualization system as guided support for needle placement in the treatment of the liver tumor. The application has been tested previously on a dummy and afterwards a first test has been carried out in the operating room during an open surgery procedure for the liver tumor resection.

The laboratory test consists in a simple overlapping of real patient models obtained from a simple anonimized CT dataset on a dummy. The results of this test were not as expected because there was a significally difference in thoracic girth between patient and dummy.

The test in the operating room has also the aim to evaluate all the possible issues of that operating scenario and led to good results: a correct overlap of virtual models on real patient was obtained in this case.

The test in the operating room produced interesting results for all the aspects related to the costraints of the particular environment. Operating rooms are overloaded of systems and devices and an efficient use of space is mandatory. In this first test the correct position of the devices close to the operating table has been defined and a precise definition of the more appropriate fiducial points used for the registration phase has been decided.

Surgeons judged the tracker volume of interaction to be not sufficient. For this reason was decided to change the model of the optical tracker used in this first test to obtain a bigger volume of interaction. A specific library was developed and tested for interfacing with this new tracker. This library was integrated within the IGSTK framework. This new tracker should be tested in the laboratory to evaluate its measure uncertainty and the in the operating room to evaluate if the new volume of inceraction can be sufficient for the surgeons.

Anyway surgeons were excited and think that the AR technology can help them in the needle placement task.

The next test will be done on a pig liver in order to measure the precision of the image-guided application. We're planning also to design a quantitative test in operating room to evaluate what is the application guidance uncertainty.

The possibility to include in the system the simulation of the virtual model deformations due to the breathing of the patient will be also taken into account.

References

1. Maad, S.: Augmented Reality: The Reality of the Global Digital Age. The Horizon of Virtual and Augmented Reality. InTech, Vukovar (2010). ISBN 978-953-7619-69-5
2. Furtado, H., Gersak, B.: Minimally invasive surgery and augmented reality. New Technology Frontiers in Minimally Invasive Therapies, pp. 195–201 (2007)
3. Samset, E., Schmalstieg, D., Vander Sloten, J., Freudenthal, A., Declerck, J., Casciaro, S., Rideng, Ø., Gersak, B.: Augmented reality in surgical procedures. In: SPIE Human Vision and Electronic Imaging XIII (2008)

4. Bichlmeier, C., Wimmer, F., Michael, H.S., Nassir, N.: Contextual anatomic mimesis: hybrid in-situ visualization method for improving multi-sensory depth perception in medical augmented reality. In: Proceedings of Sixth IEEE and ACM International Symposium on Mixed and Augmented Reality (ISMAR 2007), Nara, Japan, pp. 129–138 (2007)

5. Navab, N., Feuerstein, M., Bichlmeier, C.: Laparoscopic virtual mirror - new interaction paradigm for monitor based augmented reality. In: Proceedings of IEEE Virtual Reality Conference 2007 (VR 2007), Charlotte, North Carolina, USA, pp. 10–14 (2007)

6. Bichlmeie, C., Heining, S.M., Rustaee, M., Navab, N.: Laparoscopic virtual mirror for understanding vessel structure: evaluation study by twelve surgeons. In: Proceedings of 6th IEEE International Symposium on Mixed and Augmented Reality, Nara, Japan, pp. 1–4 (2007)

7. Bichlmeie, C., Wimmer, F., Heining, S.M., Navab, N.: Contextual anatomic mimesis: hybrid in-situ visualization method for improving multi-sensory depth perception in medical augmented reality. In: IEEE Proceedings of International Symposium on Mixed and Augmented Reality, Nara, Japan (2007)

8. De Paolis, L.T., Pulimeno, M., Lapresa, M., Perrone, A., Aloisio, G.: Advanced visualization system based on distance measurement for an accurate laparoscopy surgery. In: Proceedings of Joint Virtual Reality Conference of EGVE - ICAT - EuroVR, Lyon, France, pp. 17–18 (2009)

9. Nicolau, S., Garcia, A., Pennec, X., Soler, L., Buy, X., Gangi, A., Ayache, N., Marescaux, J.: An augmented reality system for liver thermal ablation: design and evaluation on clinical cases. Med. Image. Anal. **13**, 494–506 (2009). Elsevier

10. LiverPlanner. http://liverplanner.icg.tu-graz.ac.at

11. Reitinger, B., Bornik, A., Beichel, R., Werkgartner, G., Sorantin, E.: Tools for augmented reality based liver resection planning. In: Proceedings of the SPIE Medical Imaging 2004 on Visualization, Image-Guided Procedures, and Display, pp. 88–99, San Diego, February 2004

12. Maier-Hein, L., Tekbas, A., Seitel, A., et al.: In vivo accuracy assessment of a needle-based navigation system for CT-guided radiofrequency ablation of the liver. Med. Phys. **35**(12), 5385–5396 (2008)

13. Stüdeli, T., Kalkofen, D., Risholm, P., et al.: Visualization tool for improved accuracy in needle placement during percutaneous radio-frequency ablation of liver tumors. In: Medical Imaging 2008 on Visualization, Image-Guided Procedures, and Modeling, Pts 1 and 2, vol. 6918, B9180-B9180, 0277-786X (2008)

14. De Mauro, A., Mazars, J., Manco, L., Mataj, T., Fernandez, A.H., Cortes, C., De Paolis, L. T.: Intraoperative navigation system for image guided surgery. In: 6th International Conference on Complex, Intelligent, and Software Intensive Systems, pp. 486–490 (2012)

15. Livatino, S., De Paolis, L.T., D'Agostino, M., Zocco, A., Agrimi, A., De Santis, A., Bruno, L.V., Lapresa, M.: Stereoscopic visualization and 3-D technologies in medical endoscopic teleoperation. IEEE Trans. Industr. Electron. **62**(1), 525–535 (2015)

16. Ricciardi, F., Copelli, C., Paolis, L.T.: A pre-operative planning module for an augmented reality application in maxillo-facial surgery. In: Paolis, L.T., Mongelli, A. (eds.) AVR 2015. LNCS, vol. 9254, pp. 244–254. Springer, Cham (2015). doi:10.1007/978-3-319-22888-4_18

17. Nagorney, D., Van Heerden, J., Ilstrup, D., et al.: Primary hepatic malignancy: surgical management and determinants of survival. Surgery **106**, 740–748 (1989)

18. Yoo, T.S.: Insight into Images: Principles and Practice for Segmentation, Registration, and Image Analysis. A K Peters Ltd., Boca Raton (2004)

19. Mimics Medical Imaging Software, Materialise Group. http://www.materialise.com

20. 3D Slicer. http://www.slicer.org

21. OsiriX Imaging Software, http://www.osirix-viewer.com

22. Clearya, K., Ibanez, L., Ranjan, S.: IGSTK: a software toolkit for image-guided surgery applications. In: Conference on Computer Aided Radiology and Surgery, Chicago, USA (2004)
23. NDI Polaris Vicra. http://www.ndigital.com
24. Vicon Bonita. http://www.vicon.com/products/bonita.html
25. Cleary, K., Cheng, P., Enquobahrie, A., Yaniv, Z.: IGSTK: The Book. Signature Book Printing, Gaithersburg (2009)
26. Lorsakul, A., Suthakorn J., Sinthanayothin, C.: Point-cloud-to-point-cloud technique on tool calibration for dental implant surgical path tracking. In: Proceedings of SPIE 6918, Medical Imaging 2008 on Visualization, Image-guided Procedures, and Modeling, 17 March 2008

Salient Networks: A Novel Application to Study Brain Connectivity

Nicola Amoroso[1,2], Roberto Bellotti[1,2], Domenico Diacono[2],
Marianna La Rocca[1,2], and Sabina Tangaro[2(✉)]

[1] Dipartimento Interateneo di Fisica "M. Merlin", Università Degli Studi di Bari
"A. Moro", Bari, Italy
[2] Istituto Nazionale di Fisica Nucleare - Sezione di Bari, Bari, Italy
sonia.tangaro@ba.infn.it

Abstract. Extracting meaningful structures and data, thus unveiling
the underlying base of knowledge, is a common challenging task in social,
physical and life sciences. In this paper we apply a novel complex network approach based on the detection of salient links to reveal the effect
of atrophy on brain connectivity. Starting from structural Magnetic Resonance Imaging (MRI) data, we firstly define a complex network model
of brain connectivity, then we show how salient networks extracted from
the original ones can emphasize the presence of the disease significantly
reducing data complexity and computational requirements. As a proof
of concept, we discuss the experimental results on a mixed cohort of
29 normal controls (NC) and 38 Alzheimer disease (AD) patients from
the Alzheimer Disease Neuroimaging Initiative (ADNI). In particular,
the proposed framework can reach state-of-the-art classification performances with an area under the curve AUC = 0.93 ± 0.01 for the NC-AD
classification.

Keywords: Saliency · Complex networks · Alzheimer disease · MRI ·
Machine learning

1 Introduction

Neurodegenerative or inflammatory diseases such as Multiple Sclerosis or
Alzheimer's disease [1,2], can affect brain connectivity. For example, AD onset
is related to cognitive impairment and thinking/behavioral issues. Its spread is
exponentially growing, in fact it is expected that over 115 million people will
develop AD by 2050 [3].

Clinical findings and post mortem studies indicate that AD primarily affects
the hippocampus, the enthorinal cortex and the para-hippocampal gyrus in both
temporal lobes [4–6]: diagnostic features detectable with structural MRI analyses. Besides, diffusion tensor imaging (DTI) measures have shown to be sensitive
to white matter (WM) damage in multiple sclerosis (MS). As a consequence, a
considerable effort has been dedicated to the development of fully automated
whole brain [7,8] and ROI [9,10] segmentation strategies to provide a robust

© Springer International Publishing AG 2017
I. Rojas and F. Ortuño (Eds.): IWBBIO 2017, Part I, LNBI 10208, pp. 444–453, 2017.
DOI: 10.1007/978-3-319-56148-6_39

base of knowledge for building up supervised models. Nonetheless, recent works [11,12] have demonstrated that especially when dealing with multi-center heterogeneous databases, structural features provided by segmentation approaches and the supervised algorithms relying on them to discriminate normal controls and patients hardly succeed in reproducing the performances obtained on different data. Thus, unsupervised approaches, such as complex networks [13,14], have gained popularity to detect anomalies yielded by disease and lessen the role played by ROI detection.

Recently, many investigations have highlighted that scale-free and small-world networks might be important to study for understanding how atrophy affects and modifies human brain networks. However nothing conclusive can be stated yet. In this work, we used MRI T1 brain scans to define a brain connectivity model: brains were automatically parceled in a fixed number of three dimensional patches, representing the network nodes, and the nodes' similarity was measured with Pearson's pairwise correlation.

We present here the first application of the novel salience indicator [15] to reveal the presence of backbone structures in brain networks. In particular, we investigated its application to Alzheimer's disease (AD). The informative power of the salient skeleton was evaluated with the multiplex network framework [16] in order to extract several network features for feeding a supervised machine learning model and detecting AD patterns.

2 Methodology

The proposed approach combines, in a fully automated pipeline, image processing and machine learning techniques within a complex network framework. The algorithmic workflow is schematically depicted in Fig. 1.

```
Algorithm: Salient network application pipeline
1   For all MRI scans do
2       bet -in <output1> -out <output2> -R -B -f 0.4;
3       flirt -in <output2> -out <output3> -dof 12 -interp spline;
4       Supervoxel Correlation Network construction;
5       High Salient Skeletons construction;
6   end
7   Multiplex network construction;
8   Feature extraction;
9   Classification.
```

Fig. 1. The pseudo-code of the proposed methodology.

The proposed approach consists of four main steps:

- Image processing to obtain an intensity and spatial normalization among subjects;
- network building for each subject having patches as nodes and their Pearson's correlations as links;

- identification of the salient links forming the backbone network skeleton;
- construction of the multiplex network of the salient skeletons to extract some features in order to assess informative power of the skeletons using supervised learning techniques.

The following Fig. 2 summarizes the approach.

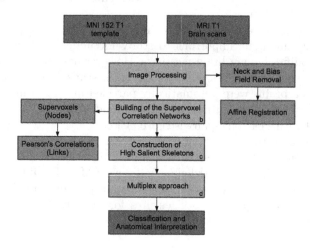

Fig. 2. The figure shows a schematic overview of the proposed methodology which encompasses four principal phases: (a) Image normalization; (b) Brain Network model; (c) High salient skeleton construction; (d) Supervised learning for the method evaluation.

2.1   Study Data

In preparation of this article we used a set of 100 T1 MRI brain scans from the Alzheimer's Disease Neuroimaging Initiative (ADNI). Data were relative to 29 normal controls (NC) and 38 AD patients from a benchmark dataset [17] individuated with the specific purpose of being a small but representative representation of the ADNI database. Demographic information is summarized in the following Table 1. Clinical status, population size, age (average and standard deviation) and gender are provided.

Table 1. This table provides the clinical and demographic description of the set used in this study. Data size, age range, gender and Mini Mental State Examination (MMSE) are shown for each diagnostic group with the relative mean and standard deviation when appropriated.

Clinic	Size	Age	Gender (M/F)	MMSE
NC	29	75 ± 6	16/13	30 ± 1
AD	38	74 ± 8	20/18	23 ± 2

2.2 Image Processing and Network Model

MRI scans were intensity normalized to lessen inter-subject differences, then FSL-BET and FSL-FLIRT, from FMRIB library [8], were used to perform a spatial affine co-registration using the MNI152 T1 template. Once brain scans had been normalized, both in intensity and spatially, we performed a parcellation of the brain in $N = 549$ parallelepiped-shaped patches of volume $3000\,\mathrm{mm}^3$, independently from specific anatomical regions of interest. As demonstrated in previous works [18], the use of this supervoxel size is related to the typical dimension of anatomical brain regions related to AD, such as the hippocampus. The pairwise similarity of patches was measured by means of Pearson's correlation, thus a weighted network, whose nodes were the patches and the weighted undirected edges the correlation measurements, was built.

The image processing and network construction were repeated for each subject, thus a collection of 100 densely connected weighted networks was obtained; to reduce them a wide range of threshold values was explored. A possible constraint to adopt for this search concerned with the desired network topology: especially for biological networks, small-world and scale-free [19] networks should be preferred. Accordingly, we explored how the topology of each network changed when disregarding edges below a particular correlation value r_{thr}. To investigate scale-free topology, the agreement between the resulting degree distribution with a power-law fit in terms of adjusted R-squared statistics, was measured.

The salience indicator has been designed to furnish an overall network description from a node-specific perspective and to define a consensus among nodes on the importance of each edge within the network. Figure 3 shows in detail the procedure adopted to get high salient skeletons from raw data.

Given the set of weights $W = \{w_1, w_2, ..., w_{N'}\}$, with N' being the number of N nodes remaining after the network scale-free transformation, the pairwise distance matrix D is defined. $D_{i,j}$ elements are simply the correlation reciprocal values. Therefore, D captures the intuitive notion for which strongly correlated nodes should be closer than weakly correlated ones. Accordingly, in this way it is also possible to introduce a path length definition: if two patches i and j are connected through a path p consisting of k steps, being them the terminal nodes of p, the length of p is simply the weight sum of the edges belonging to it. It is possible to define several path connecting i and j, in general it will not be true that the paths with minimum number of steps have also minimum length. Thus, throughout this paper the shortest path has to be considered as the path whose weight sum is maximum.

The collection of all links belonging to the shortest paths connecting a generic node n to all the remaining nodes of the network define what is generally called the shortest path tree $T(n)$ which represents, in fact, the most effective routes linking the node n to the network. One can calculate the shortest path trees for all the nodes of the network.

Then for a generic edge (i, j), connecting the nodes i and j, salience $s_{i,j}$ is the indicator accounting how many times the edge belong to a shortest path tree in respect to the total number of shortest path trees. According to this

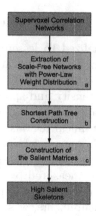

Fig. 3. This Figure shows in detail the procedure adopted to get to high salient skeletons from raw data. (a) Scale free networks with power-law distributed weights were extracted from each initial subject network; (b) Links participating at least once in the shortest paths, starting from a fixed reference node, were recorded in the shortest path tree matrix; (c) Shortest path tree for each reference node were added up to obtain, for each subject, salience matrix, whose values $s_{i,j} = 1$ represent high salient skeleton

definition, $s_{i,j}$ takes into account the fraction of shortest path trees including the edge between i and j.

As $T(n)$ reflects the set of most efficient paths to the rest of the network from the perspective of the reference node n, $s_{i,j}$ is a consensus variable defined by the ensemble of root nodes. If $s_{i,j} = 1$, then link (i, j) is essential for all reference nodes, if $s_{i,j} = 0$, (i, j) is not a fundamental link for the network. We expect that in brain structural networks this approach could emphasize which brain regions are mostly affected by atrophy caused by AD. One of the most important properties salience indicator has is its bimodal distribution with peaks on 0 and 1 values.

2.3 Supervised Learning for the Method Evaluation

Each salient network was used to build a skeleton multiplex network. For each layer, nodes are the patches that have a salient link at least in a subject and links, changing for each layer, are the Pearson's correlation between nodes pairs within the single layers.

From the skeleton multiplex, 4 centrality measures were extracted: the strength and the inverse participation ratio, considering both the single and multiple layer links. Besides, for each quantity the corresponding conditional means.

Overall we obtained a matrix of $8N$ features for M samples, where $N = 549$ is the number of nodes of the SMN and $M = 67$ is the number of subjects or layers of the skeleton multiplex network. To prevent over-training issues yielded by the curse of dimensionality, a feature selection was carried out with a wrapper-based method. The best features were selected, collecting the most important features for each 5–fold Random Forest cross-validation and settling which had a significant probability of occurrence within 1000 rounds. The selected features were used to train a second Random Forest. Forests were grown with 500 trees, a number large enough to reach the typical training plateau for the out-of-bag error. At each split $\sqrt{(N \times 8)}$ features were randomly sampled to grow the trees.

This analysis is certainly of interest for an assessment of the methodology and the feature informative power. In fact, it is well known which there are brain regions related to AD, thus we can use this classification task to observe whether or not high salience skeleton outlines brain regions coherent with the pathology. Skeletons, multiplex networks and the classification process were all built and developed through R free software for statistical computing and graphics [20].

3   Results

3.1   Topology of Brain Networks

There is not an *a priori* reason for which the scale-free topology should have emerged with the same threshold value from different subjects. However, the following Fig. 4 shows how the adjusted R-squared metric averaged over all subjects, adopted to measure the goodness of fit for degree distributions with a power-law function, reaches a high and stable plateau for $r_{thr} > 0.6$.

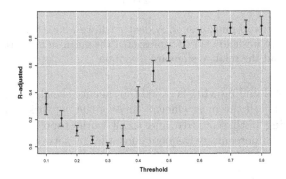

Fig. 4. The brain networks show a consistent power-law degree distribution for thresholds above 0.6. The goodness-of-fit is measured by means of adjusted R-squared coefficient. For each threshold value is represented the mean R-squared coefficient over all subjects and the relative standard deviation.

The $r_{thr} = 0.7$ chosen for subsequent analyses was lying in the middle of the above mentioned plateau. Going back to the original brain network of each subject, the correlations exceeding this threshold value were kept while other edges were disregarded. In this way, for each subject, a scale-free weighted network of highly correlated or anti-correlated nodes was obtained. Moreover, for quantifying and knowing how much the brain network of each subject was small-world as the threshold changes, the mean small-worldness indicator SW averaged over all subjects was studied:

$$SW = \frac{1}{M} \sum_{m=1}^{M} \left(\frac{C_m}{C_m^r} \middle/ \frac{L_m}{L_m^r} \right) \tag{1}$$

where C_m and L_m, and C_m^r and L_m^r are the clustering coefficient and the average shortest path length respectively of each network m to examine, and of the reference random graph with the same node number N and the same link probability p given by the ratio between mean degree $\bar{k}$ and N. It can be noticed as at a threshold of 0.6 in addition to a scale-free topology also a small-world structure emerges, see Fig. 5.

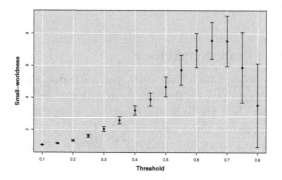

Fig. 5. The brain networks manifest an evident small-worldness behavior for thresholds above 0.6. For each threshold value is represented the mean small-worldness coefficient over all subjects and the relative standard deviation.

This makes natural finding what is called the salience skeleton, *i.e.* the backbone structure including all the most efficient links of the networks, after using a 0.6 threshold granting both scale-free and small world topologies. Figure 6 is an example of salience percentage frequency for a brain network.

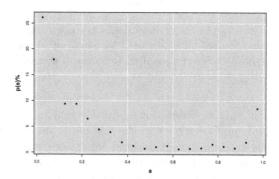

Fig. 6. The salience of a brain network. Saliency values are gathered on 0 and 1 thus it is possible to detect the salient link independently from a particular threshold value. In this network the link fraction contributing to the high salient skeleton is of the 8.46%.

The salience skeleton is by definition the network containing only salient links. As the saliency distribution is bimodal there is no need of finding a particular threshold, but the skeleton stays pretty much unchanged for all threshold values between 0.1 and 0.9, for the present work the threshold 0.5 was used.

3.2   Assessment of Salient Skeleton Based Methodology

We investigated the informative content of the high salient skeletons comparing it with the complete multiplex model. In the NC-AD discrimination, the important features extracted from the skeleton, gave accurate results reported here in terms of area under the receiver-operating-characteristic curve (AUC) and its relative standard error. An AUC of 0.93 ± 0.01 was found for the skeleton multiplex network and an AUC of 0.94 ± 0.01 was reached for the complete multiplex. Corresponding sensitivities and specificities were respectively 0.83 ± 0.02 and 0.91 ± 0.02, for the first multiplex, and 0.85 ± 0.01 and 0.95 ± 0.02, for the second one. Moreover, the two approaches were combined and used to distinguish NC and AD. Results are presented in Fig. 7.

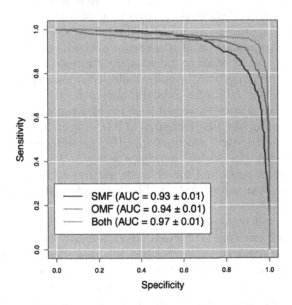

Fig. 7. The receiver-operating-characteristic (ROC) curves and the corresponding areas under the curve (AUC) relative to skeleton (blue curve), original (red curve) and both (green curve) multiplex network features for the NC-AD. (Color figure online)

This study showed that the feature combination makes it possible to achieve an higher classification performance with an AUC of 0.97 ± 0.01 and a sensitivity and specificity respectively of 0.88 ± 0.01 and 0.98 ± 0.01, demonstrating that the skeleton extraction from the network is able to provide additional important information within the multiplex framework. Besides nodes involved in the feature extraction from the skeleton multiplex networks are the middle of the original network nodes providing a relevant computational saving. The total amount of CPU time for the whole analysis pipeline was of 18 h 17 min and 41 s. Therefore each image processing and the relative subject prediction required a CPU time of 16 min 23 s and 2 GB of RAM.

It is worthwhile to note as significant patches covered regions such as hippocampus, amygdala, ventricles, thalamus, brain stem, cerebral sulci, interhemispheric portions and separation areas between gray and white matter. These findings are a further confirm that salient skeletons are an alternative way of detecting AD patterns, able to give supplemental information regarding the disease.

4   Conclusion

In this work, it was proposed an innovative method for reducing complexity of dense brain networks, preserving and especially enhancing, at the same time, their informative content. A study was carried out to bring out the two principal topologies (small-world and scale-free) of the brain networks, able to model and describe the most of the behaviors involved in brain organization. In particular, we studied how a neurodegenerative pathology, such as Alzheimer disease, can modify structural organization of these brain networks extracting high salient skeleton for each one, in order to focus on the strategic hubs and highways of these correlation networks which could be affected in case of atrophy.

We proved, using multiplex network approach, the forcefulness of this methodology, in fact feature extracted from the skeleton multiplex network give in cross-validation, for the discrimination of AD subjects from normal controls, an AUC of 0.93 ± 0.01 which reaches 0.97 ± 0.01 combining these features with those obtained using the original multiplex network. The improvement of the classification performances and the different features extracted from the skeletons demonstrate that salient links highlight an information, at least in part, distinct from that obtained with the initial networks.

Future works will be aiming at validating the methodology on other imaging modalities, for example DTI which could be really helpful to gain further insight in other diseases such as Multiple Sclerosis. Besides, we are going to study mild cognitive impairment subjects, as MCI condition can be in several cases a prodromal stage of AD. Accordingly, the early and accurate detection of impairment could play a pivotal role in the development of drug trials and therapy developments.

References

1. Roosendaal, S., Geurts, J., Vrenken, H., Hulst, H., Cover, K.S., Castelijns, J., Pouwels, P.J., Barkhof, F.: Regional DTI differences in multiple sclerosis patients. Neuroimage 44(4), 1397–1403 (2009)
2. Barnes, D.E., Yaffe, K.: The projected effect of risk factor reduction on Alzheimer's disease prevalence. Lancet Neurol. 10(9), 819–828 (2011)

3. Prince, M., Albanese, E., Guerchet, M., Prina, M.: World Alzheimer Report 2014. Dementia and Risk Reduction: An Analysis of Protective and Modifiable Factors. Alzheimers Disease International, Londres (2014)
4. West, M.J., Coleman, P.D., Flood, D.G., Troncoso, J.C.: Differences in the pattern of hippocampal neuronal loss in normal ageing and Alzheimer's disease. Lancet **344**(8925), 769–772 (1994)
5. Chincarini, A., Bosco, P., Gemme, G., Morbelli, S., Arnaldi, D., Sensi, F., Solano, I., Amoroso, N., Tangaro, S., Longo, R., et al.: Alzheimers disease markers from structural MRI and FDG-PET brain images. Eur. Phys. J. Plus **127**(11), 1–16 (2012)
6. Baron, J., Chetelat, G., Desgranges, B., Perchey, G., Landeau, B., De La Sayette, V., Eustache, F.: In vivo mapping of gray matter loss with voxel-based morphometry in mild Alzheimer's disease. Neuroimage **14**(2), 298–309 (2001)
7. Fischl, B.: FreeSurfer. Neuroimage **62**(2), 774–781 (2012)
8. Jenkinson, M., Beckmann, C.F., Behrens, T.E., Woolrich, M.W., Smith, S.M.: FSL. Neuroimage **62**(2), 782–790 (2012)
9. Amoroso, N., Errico, R., Bruno, S., Chincarini, A., Garuccio, E., Sensi, F., Tangaro, S., Tateo, A., Bellotti, R., Initiative, A.D.N., et al.: Hippocampal unified multi-atlas network (human): protocol and scale validation of a novel segmentation tool. Phys. Med. Biol. **60**(22), 8851 (2015)
10. Chincarini, A., Sensi, F., Rei, L., Gemme, G., Squarcia, S., Longo, R., Brun, F., Tangaro, S., Bellotti, R., Amoroso, N., et al.: Integrating longitudinal information in hippocampal volume measurements for the early detection of Alzheimer's disease. NeuroImage **125**, 834–847 (2016)
11. Bron, E.E., Smits, M., Van Der Flier, W.M., Vrenken, H., Barkhof, F., Scheltens, P., Papma, J.M., Steketee, R.M., Orellana, C.M., Meijboom, R., et al.: Standardized evaluation of algorithms for computer-aided diagnosis of dementia based on structural MRI: the CADDementia challenge. NeuroImage **111**, 562–579 (2015)
12. Allen, G.I., Amoroso, N., Anghel, C., Balagurusamy, V., Bare, C.J., Beaton, D., Bellotti, R., Bennett, D.A., Boehme, K.L., Boutros, P.C., et al.: Crowdsourced estimation of cognitive decline and resilience in Alzheimer's disease. Alzheimer's Dement. **12**(6), 645–653 (2016)
13. Rubinov, M., Sporns, O.: Complex network measures of brain connectivity: uses and interpretations. Neuroimage **52**(3), 1059–1069 (2010)
14. Bullmore, E., Sporns, O.: Complex brain networks: graph theoretical analysis of structural and functional systems. Nat. Rev. Neurosci. **10**(3), 186–198 (2009)
15. Grady, D., Thiemann, C., Brockmann, D.: Robust classification of salient links in complex networks. Nature Commun. **3**, 864 (2012)
16. Menichetti, G., Remondini, D., Panzarasa, P., Mondragón, R.J., Bianconi, G.: Weighted multiplex networks. PloS One **9**(6), e97857 (2014)
17. Boccardi, M., Bocchetta, M., Morency, F.C., Collins, D.L., Nishikawa, M., Ganzola, R., Grothe, M.J., Wolf, D., Redolfi, A., Pievani, M., et al.: Training labels for hippocampal segmentation based on the EADC-ADNI harmonized hippocampal protocol. Alzheimer's Dement. **11**(2), 175–183 (2015)
18. La Rocca, M., et al.: A multiplex network model to characterize brain atrophy in structural MRI. In: Mantica, G., Stoop, R., Stramaglia, S. (eds.) Emergent Complexity from Nonlinearity, in Physics, Engineering and the Life Sciences. Springer Proceedings in Physics, vol. 191. Springer, Heidelberg (2017)
19. Zhang, B., Horvath, S.: A general framework for weighted gene co-expression network analysis. Stat. Appl. Genet. Mol. Biol. **4**(1), 1–17 (2005)
20. R Core Team: R: A Language and Environment for Statistical Computing. R Foundation for Statistical Computing, Vienna (2013)

Clustering of Food Intake Images into Food and Non-food Categories

Abul Doulah and Edward Sazonov[✉]

Department of Electrical and Computer Engineering,
University of Alabama, Tuscaloosa, AL 35487, USA
sayeed.doulah@gmail.com, esazonov@eng.ua.edu

Abstract. For dietary assessment, food images could be useful to identify foods, portion sizes and estimate calorie in meals, either by human nutritionists or through image recognition. Images captured by a wearable camera during eating may include both food and non-food images. To avoid reviewing each image, only informative food images should be included in the analysis and non-food image should be discarded. This study proposed a methodology for clustering of food images into food and non-food groups based on histogram matching, without explicit recognition of the image content. Data was collected from 7 participants wearing an eyeglasses camera. A total of 10 meals were recorded at the sampling rate of 5 s, yielding a total of 1077 images. Each image was labeled by a human rater as a food or non-food image. Histogram matching with Bhattacharyya distance was applied to form a similarity matrix for extracted images from each meal. Both k-means and affinity propagation (AP) algorithms were investigated to cluster the images. Results show that the overall average food image clustering accuracy in respect to human annotation for 10 videos was 0.93 ± 0.04 for AP and 0.90 ± 0.03 for k-means. Similarly, overall average non-food image clustering accuracy was 0.81 ± 0.06 for AP and 0.70 ± 0.09 k-means method.

Keywords: Clustering · Dietary assessment · Histogram matching

1 Introduction

The dietary assessment methods aim to provide an accurate estimation of foods consumed, portion size and calorie intake. To alleviate problems associated with self-reported food intake, various methods of dietary monitoring have been proposed ranging from digital diaries with food images to wearable on body sensors. The technological advances in computer vision allowed the introduction of image/video analysis for diet assessment [1]. Typically, a food imagery based method acquires images before, during and after a meal using a wearable camera or a mobile device. The food images can be analyzed to identify different foods, their portion sizes and nutritional content. In early studies, image analysis was performed by human nutritionists. The study of [2] proposed an approach called the Remote Food Photography Method where the participants submit food images to nutritionist for the analysis of energy intake estimation. Recently, by utilizing crowdsourcing technique, an application called

© Springer International Publishing AG 2017
I. Rojas and F. Ortuño (Eds.): IWBBIO 2017, Part I, LNBI 10208, pp. 454–463, 2017.
DOI: 10.1007/978-3-319-56148-6_40

PlateMate [3] estimated nutrition information from food images provided by users. The crowdsourcing used non-professional nutritionists to assess energy and nutrient content of food from digital images. Recently, automatic image recognition gained popularity in identifying and quantifying food images. Typically, a set of features from the images are extracted and fed to a classifier to recognize various food items. Several methods have been proposed to recognize foods in the image via supervised and unsupervised machine learning techniques. A food recognition application was developed in [4] for the classification of fast-food images into four classes. The study extracted color (normalized RBG values), texture (local entropy, standard deviation, range) and context-based features from images and fed to the artificial neural network for food recognition. The study of [5] applied a support vector machine (SVM) classifier for the recognition of nineteen classes of foods utilizing a set of color (color components and pixel intensities) and texture (responses from Gabor filter) features. In [6], a simple Bayesian probabilistic classifier was used to match images of food items to offline food database containing images of homemade foods, fast-food and fruits. The authors used scale invariant feature transform (SIFT) features and clustered the features into visual words before feeding the classifier. The study of [7] proposed a pairwise classification method that uses the user's speech input to improve the food recognition process. The recognition method incorporated the use of color neighborhood and texton histogram features, Adaboost feature selection and SVM classifier. In [8], the authors created a database of fast-food images and videos in an effort to make a benchmark for food recognition. Based on color histograms and SIFT bag features, the proposed methods were evaluated for seven fast-food classes. In [9], a multiple kernel learning based food recognition method was proposed utilizing the bag of scale invariant feature transform, Gabor filter responses, and color histograms features. Recently, the study of [10] proposed automatic food recognition based on bag of features with support vector machine classifier. A mobile user interface was proposed in [11] that utilizes a client-server image recognition and portion estimation software to estimate calorie content. Some major potential hurdles in automatic approach are to determine exact portion size and accurate food recognition. Apart from that the accuracy of automatic recognition of great variety of foods is still suboptimal. The image capture by mobile devices such as smartphones requires active participation by the person being monitored [2]. Such images are typically taken before and after a meal and are likely to contain clear images of the food and the leftovers. Image captured by wearable cameras does not require active participation of the person being monitored and thus potentially reduces the user burden [12]. However, wearable cameras capture many images that may contain non-food items. Therefore, a method that could separate food images and non-food images, is needed. In this paper, an approach is proposed so that only informative food images can be included in the analysis and non-food images can be discarded. We propose to use the histogram of color textures features directly to find similarities between food images and non-food images. A pairwise matching algorithm compared the meal images and created a similarity matrix for the image collection. Two different unsupervised clustering techniques were investigated to identify groups of food image/non-food images within a meal. The performance demonstrated the potentially of image grouping without recognition.

2 Subject and Methods

Figure 1 shows the flow diagram of the methodology of the proposed system. The following subsections describe each steps in detail.

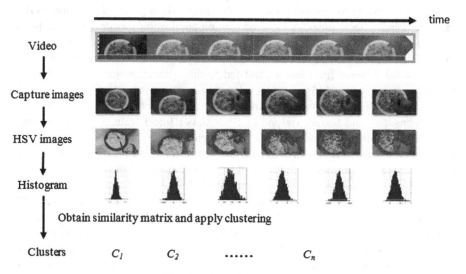

Fig. 1. Flow of processing

2.1 Data Collection and Annotation

The first step involved the recording of meal videos with a wearable camera. The camera was attached to the eyeglasses so that the images would contain most of the food items in food plate/bowl. A total of 7 subjects were recruited for the study. During the visit, subjects were asked to wear the camera during food consumption at breakfast, lunch or dinner. A total of 10 meal videos were recorded at 30 fps and each meal duration was about 10–15 mins. The second step involved image extraction from the recorded videos. A total of 1077 images were obtained from all 10 meal videos at 150 frames (five second) interval. Figure 2 shows examples of selected meal images obtained from each of the captured videos. The images were manually labeled as food or non-food images by a human rater. The food images (class-1) contained the full amount or partial amount of food items. The images that contained no food items, surroundings only, faces of individuals, snapshots of personal gadgets (mobile/laptop) were labeled as non-food images (class-2).

2.2 Histogram Feature and Similarity Matching

Color features of image typically constitute as one of the best descriptor by providing valuable source of information. Among color descriptors, color histograms are the most common and widely used in various object recognition applications. In this work, the

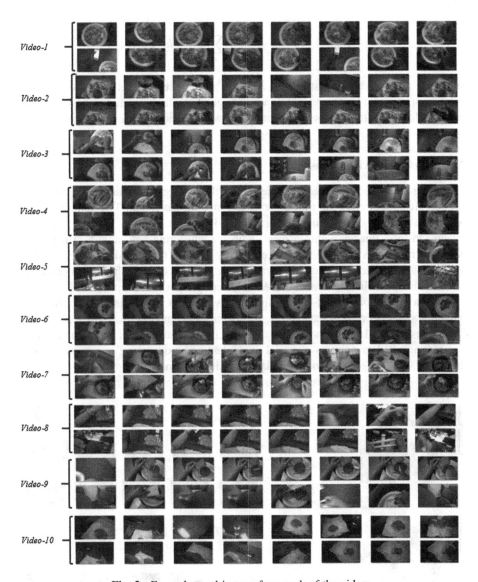

Fig. 2. Example meal images from each of the videos

RGB images in each meal were first transformed into HSV color space images. The color feature for each image was represented using HSV color histogram. A total of 512 bins (8 × 8 × 8) with 8 levels for each color was used. Based on the histogram features of images, a similarity matrix was formed for the image collection. The histogram matching was done utilizing Bhattacharyya distance as,

$$d(h_a, h_b) = \sqrt{1 - \frac{1}{\sqrt{h_a h_b N^2}} \sum_{g \in G} \sqrt{h_a(g) \cdot h_b(g)}} \qquad (1)$$

Where h_a and h_b are two histograms, N and g are number of bins and pixel intensity values respectively. Low scores of d indicated good matches and high scores indicated bad matches. A value of 0 was perfect match while 1 indicated a perfect mismatch.

2.3   Clustering Algorithms

Clustering technique classifies the objects into different groups, partitioning of a data set into clusters of common trait utilizing distance measurement. Two different unsupervised clustering techniques were investigated to cluster groups of food image/non-food images within a meal: K-means clustering and affinity propagation. K-means method is an unsupervised clustering method that classifies the input data objects into multiple classes on the basis of their inherent distance from each other. On the other hand, Affinity propagation (AP) clustering algorithm proposed in [13], takes as input a collection of real-valued similarities between data points and clusters by passing messages between data points. AP identifies exemplars among data points and forms clusters of data points around these exemplars. It operates by simultaneously considering all data point as potential exemplars and exchanging messages between data points until a good set of exemplars and clusters emerges.

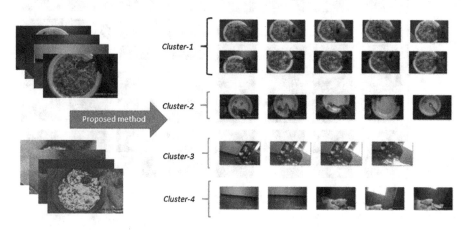

Fig. 3. Clustering by the proposed method

2.4   Performance Evaluation

To provide an objective performance evaluation of the proposed approach, the goodness of image clustering was measured via purity value [14] and overall food image/non-food images clustering accuracy in respect to the human rater.

For a given set of n images labeled as l classes, the images are grouped into m clusters $C_j, j = 1, 2, \ldots, m$, the purity for C_j is defined as,

$$P(C_j) = \frac{1}{|c_j|} \max_l |C_{j,l}| \tag{2}$$

where $C_{j,l}$ consists of images in C_j that belong to class l, and $|C_j|$ represents the size of the set. Each cluster may contain images of different classes. Purity gives the ratio of the dominant class size in the cluster to the cluster size itself. The larger value of purity means that the cluster is a "purer" subset of the dominant class.

The overall food image clustering performance was defined as the percentage of food images clustered together out of originally labeled total number of food images.

$$Acc\_FIC = \frac{\sum_1^x P(C_{j,1}) \cdot |C_{j,1}|}{\sum_1^j |C_{j,1}|} \tag{3}$$

Similarly, the overall non-food image clustering performance was defined as the percentage of non-food images clustered together out of originally labeled total number of non-food images.

$$Acc\_NFIC = \frac{\sum_1^y P(C_{j,2}) \cdot |C_{j,2}|}{\sum_1^j |C_{j,2}|} \tag{4}$$

where x and y are the number of clusters recognized as pure food image clusters and non-food image clusters respectively.

3 Results

The experimental results from two clustering algorithms are provided in Table 1 in terms of purity value of all clusters and overall clustering performance compared to manually labeled images. Figure 3 illustrates the performance of proposed approach for one meal video. The RGB meal images are transformed to HSV before applying to clusters. Top few images for each clusters are shown. Figure 4 demonstrates the comparative performance of both k-means and AP clustering method for all 10 videos. The overall average food image clustering accuracy for 10 videos were 0.93 ± 0.04 for AP and 0.90 ± 0.03 k-means. Similarly, overall average non-food image clustering accuracy are 0.81 ± 0.06 for AP and 0.70 ± 0.09 k-means method. The mean purity of food image clusters for AP is 0.96 and for k-means is 0.93. Also the mean purity of non-food image clustered for both AP and k-means are 0.91 and 0.87 respectively.

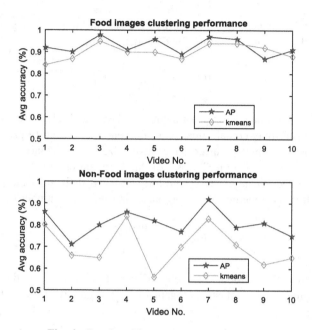

Fig. 4. Results of k-means and AP clustering

4   Discussion

Dietary assessment through food imagery utilizes different image processing techniques. These techniques need to include complex food recognition algorithm which might not be readily available. On the other hand, variety of food items may limit the accuracy of food recognition. Therefore, the method would require human intervention to estimate caloric intake. To assist human nutritionist, clustering meal images could become very useful to reduce the number of food images to be reviewed. The results from the proposed approach suggest that based on similarity between the images, clustering can be done with significant accuracy, without attempting to recognize the food items in the images. Clustering therefore makes it possible to summarize a large collection of images and select just food images to analyze.

The proposed method used HSV color space rather than RGB. HSV images not only include color information but also keep some invariance in intensity and color changes. Since all of the meal videos were recorded in different lighting conditions, HSV image would be more suitable than RGB images.

The experimental results show that the AP clustering algorithm performs better than the k-means for clustering meal images. For few cases, k-means performed comparatively better, however AP exhibited consistency in clustering similar images together in either cases of food images and non-food images. The performance of k-means in clustering of non-food images is significantly less compared to AP. In terms of non-food image clustering, Fig. 4 illustrates a consistent trend for AP but no consistent trend was found in k-means. A potential reason for getting good performance from AP

Table 1. Performance evaluation

No.	Total images	Clustering method	P_{C1} (class)	P_{C2} (class)	P_{C3} (class)	P_{C4} (class)	P_{C5} (class)	P_{C6} (class)	% of food images correctly clustered together	% of non-food images correctly clustered together
1	164	AP	0.97(2)	0.79(2)	1.00(1)	1.00(2)	0.98(1)	0.95(1)	0.92	0.86
		k-means	0.97(2)	1.00(1)	0.79(2)	0.47(2)	0.97(1)	0.96(1)	0.84	0.80
2	89	AP	0.93(1)	0.82(2)	0.90(1)	0.80(2)	1.00(1)	1.00(2)	0.90	0.71
		k-means	0.92(1)	0.74(2)	0.90(1)	1.00(1)	0.80(1)	1.00(2)	0.87	0.66
3	88	AP	1.00(1)	1.00(1)	1.00(1)	0.71(2)	1.00(1)	1.00(1)	0.98	0.80
		k-means	0.50(2)	1.00(1)	1.00(1)	0.71(2)	1.00(1)	1.00(1)	0.95	0.65
4	132	AP	0.79(2)	1.00(2)	1.00(2)	1.00(1)	0.88(1)	0.97(1)	0.91	0.86
		k-means	0.91(2)	1.00(2)	0.79(2)	1.00(1)	0.88(1)	0.97(1)	0.90	0.84
5	113	AP	1.00(1)	1.00(1)	0.97(1)	1.00(2)	1.00(2)	0.77(2)	0.96	0.82
		k-means	0.91(1)	1.00(1)	0.91(1)	0.83(2)	1.00(2)	0.84(2)	0.90	0.56
6	123	AP	0.93(1)	0.82(1)	1.00(1)	1.00(2)	1.00(1)	0.73(2)	0.89	0.77
		k-means	0.81(1)	0.82(1)	1.00(1)	0.73(2)	1.00(1)	1.00(2)	0.87	0.70
7	88	AP	0.97(1)	1.00(1)	1.00(2)	0.90(1)	1.00(1)	1.00(2)	0.97	0.92
		k-means	0.91(1)	1.00(1)	1.00(2)	0.90(1)	1.00(1)	1.00(2)	0.94	0.83
8	140	AP	1.00(1)	1.00(2)	1.00(1)	0.95(1)	1.00(1)	1.00(2)	0.96	0.79
		k-means	1.00(1)	1.00(2)	0.86(1)	0.95(1)	1.00(1)	1.00(2)	0.94	0.71
9	69	AP	0.57(2)	1.00(1)	1.00(1)	0.75(2)	1.00(2)	1.00(2)	0.87	0.81
		k-means	0.67(1)	1.00(1)	1.00(1)	0.75(1)	1.00(2)	1.00(2)	0.92	0.62
10	71	AP	0.89(1)	1.00(1)	0.93(1)	0.90(1)	0.82(1)	1.00(2)	0.91	0.75
		k-means	0.89(1)	0.85(1)	0.93(1)	0.90(1)	0.82(1)	1.00(2)	0.88	0.65
Summary		AP				Average ± Std			0.93 ± 0.04	0.81 ± 0.06
		k-means				Average ± Std			0.90 ± 0.03	0.70 ± 0.09

clustering algorithm is that the AP considers all data points as a potential exemplar at initial iteration and then progress with good match.

For the goodness of clustering analysis, purity value alone may not provide an overall estimate of the system performance even though they provide the quality of image clusters. Because a cluster could contain small number of class images with high purity value. Therefore, one needs to consider both purity value of image clusters and total number of images present in the cluster. For this purpose, overall accuracy measures with respect to all class images were introduced.

It is also observed from both of the clustering algorithm that they create multiple food/non-food image clusters which is expected. During a full meal, the food items are being consumed over time, and therefore yields different portion of foods on the plate constituting different images. Likewise, in the case of non-food images, the subject might take look into different objects and surroundings. Another possible reason could be the surrounding lighting conditions.

The proposed method offers potential use to cluster food and non-food images for a small number of participants. Therefore, further experiments are needed to test the performance of the method on a wider population under diverse lighting conditions.

From the standpoint of generating clusters within the food images, further work needs to be done to separate between full amount of food images and partial amount of foods present in the image. Another potential future direction could be to analyze the time progression of meal from the clusters. As a future work, cluster centers could be identified to rank representative images in a meal.

5   Conclusion

In this paper, a clustering of food image and non-food images method is proposed without food recognition. A pairwise matching algorithm from the histogram of meal images was investigated. Two different clustering approaches were evaluated and compared. The affinity propagation algorithm outperformed k-means clustering method in grouping meal images. The proposed method demonstrates the potential use of minimizing the number of images that need to be analyzed for the estimation of caloric intake. Future works could be done in clustering food images and find out the best few representative images.

References

1. Fontana, J., Sazonov, E.: Detection and characterization of food intake by wearable sensors. In: Wearable Sensors: Fundamentals, Implementation and Applications, pp. 591–616. Academic Press (2014)
2. Martin, C.K., Han, H., Coulon, S.M., Allen, H.R., Champagne, C.M., Anton, S.D.: A novel method to remotely measure food intake of free-living individuals in real time: the remote food photography method. Br. J. Nutr. 101, 446–456 (2009)

3. Noronha, J., Hysen, E., Zhang, H., Gajos, K.Z.: Platemate: crowdsourcing nutritional analysis from food photographs. In: Proceedings of the 24th Annual ACM Symposium on User Interface Software and Technology, pp. 1–12. ACM, New York (2011)
4. Shroff, G., Smailagic, A., Siewiorek, D.P.: Wearable context-aware food recognition for calorie monitoring. In: Proceedings of the 2008 12th IEEE International Symposium on Wearable Computers, pp. 119–120. IEEE Computer Society, Washington, DC (2008)
5. Zhu, F., Bosch, M., Woo, I., Kim, S., Boushey, C.J., Ebert, D.S., Delp, E.J.: The use of mobile devices in aiding dietary assessment and evaluation. IEEE J. Sel. Top. Signal Process. **4**, 756–766 (2010)
6. Kong, F., Tan, J.: DietCam: automatic dietary assessment with mobile camera phones. Pervasive Mob. Comput. **8**, 147–163 (2012)
7. Puri, M., Zhu, Z., Yu, Q., Divakaran, A., Sawhney, H.: Recognition and volume estimation of food intake using a mobile device. In: 2009 Workshop on Applications of Computer Vision (WACV), pp. 1–8 (2009)
8. Chen, M., Dhingra, K., Wu, W., Yang, L., Sukthankar, R., Yang, J.: PFID: Pittsburgh fast-food image dataset. In: 2009 16th IEEE International Conference on Image Processing (ICIP), pp. 289–292 (2009)
9. Joutou, T., Yanai, K.: A food image recognition system with multiple Kernel learning. In: 2009 16th IEEE International Conference on Image Processing (ICIP), pp. 285–288 (2009)
10. Anthimopoulos, M.M., Gianola, L., Scarnato, L., Diem, P., Mougiakakou, S.G.: A food recognition system for diabetic patients based on an optimized bag-of-features model. IEEE J. Biomed. Health Inform. **18**, 1261–1271 (2014)
11. Ahmad, Z., Khanna, N., Kerr, D.A., Boushey, C.J., Delp, E.J.: A mobile phone user interface for image-based dietary assessment (2014)
12. Liu, J., Johns, E., Atallah, L., Pettitt, C., Lo, B., Frost, G., Yang, G.-Z.: An intelligent food-intake monitoring system using wearable sensors. In: 2012 Ninth International Conference on Wearable and Implantable Body Sensor Networks (BSN), pp. 154–160 (2012)
13. Frey, B.J., Dueck, D.: Clustering by passing messages between data points. Science **315**, 972–976 (2007)
14. Chen, Y., Wang, J.Z., Krovetz, R.: Content-based Image retrieval by clustering. In: Proceedings of the 5th ACM SIGMM International Workshop on Multimedia Information Retrieval, pp. 193–200. ACM, New York (2003)

Feature Extraction Using Deep Learning
for Food Type Recognition

Muhammad Farooq and Edward Sazonov[✉]

Department of Electrical and Computer Engineering, University of Alabama,
Tuscaloosa AL 35487, USA
{mfarooq, sazonov}@eng.ua.edu

Abstract. With the widespread use of smartphones, people are taking more and
more images of their foods. These images can be used for automatic recognition
of foods present and potentially providing an indication of eating habits. Traditional methods rely on computing a number of user derived features from
image and then use a classification method to classify food images into different
food categories. Pertained deep neural network architectures can be used for
automatically extracting features from images for different classification tasks.
This work proposes the use of convolutional neural networks (CNN) for feature
extraction from food images. A linear support vector machine classifier was
trained using 3-fold cross-validation scheme on a publically available Pittsburgh
fast-food image dataset. Features from 3 different fully connected layers of CNN
were used for classification. Two classification tasks were defined. The first task
was to classify images into 61 categories and the second task was to classify
images into 7 categories. Best results were obtained using 4096 features with an
accuracy of 70.13% and 94.01% for 61 class and 7 class tasks respectively. This
shows improvement over previously reported results on the same dataset.

Keywords: Deep learning · Transfer learning · Image recognition · Food
recognition · Classification

1 Introduction

In the last few years, recognition of food items from images has become a popular
research topic due to the availability of a large number of images on the internet and
because of the interest of people in social networks. One of the challenging tasks in
image-based food recognition is to determine which food items are present in the pictures. This paper focuses on the task of food item recognition assuming that it is known
that given images contain food and the algorithm is used to determine the food type.

Food type recognition is a hard problem because the shape of different food items is
not well defined and a single image of food can have a variety of ingredients with
varying textures. Color, shape, and texture of a given food type is defined by the
ingredients and the way food is prepared [1]. Even for a given food type, high
intra-class variations in both shape and texture can be observed for example chicken
burgers [1], which can be prepared in a variety of different methods and the final
texture, color, and shape might be different for the same food after preparation.

© Springer International Publishing AG 2017
I. Rojas and F. Ortuño (Eds.): IWBBIO 2017, Part I, LNBI 10208, pp. 464–472, 2017.
DOI: 10.1007/978-3-319-56148-6_41

Researchers have proposed a number of algorithms for recognition of food items in images by employing different feature extraction and classification algorithms. Features computed from images and the choice of classifier plays an important role in food type recognition systems. Yang et al. proposed support vector machine (SVM) based approach with pair-wise statistics of local features such as distance and orientation to differentiate between eight basic food materials [2] on the Pittsburgh fast-food image dataset [1]. On the same dataset, they further classified a given food into one of the 61 food categories with a classification rate of 28.2% [1]. The authors in that work used four different features groups namely color histogram features, bag of scale-invariant feature transform (SIFT) descriptor features, semantic texton forest (STF) features and joint features of orientation and midpoint (OM) for classification of food types. Another work on the same dataset proposed the use of local textural patterns and their global structure using SIFT detector and Local Binary Pattern (LBP) to classify images. Joutou et al. proposed a visual recognition system to classify images of Japanese food into one of the 51 categories [3]. They proposed feature fusion approach where an SIFT-based bag of features, Gabor, and color histogram features were used with multiple kernel learning [4]. Authors in [5] used three image descriptors Bag of Textons, SIFT, and PRICoLBP to classify food images. Another work proposed a combination of bags-of-features along k-means clustering and final classification using linear support vector machine to classify food items into 11 classes with an accuracy of 78% [6]. Random Forest has also been proposed for determining distinctive visual components in food images and to use it for classification of food type [7]. The authors proposed a method to mine parts of images simultaneously for all image classes using random forest models. Other researchers have proposed systems which are able to recognize and segment different food items from images taken by people in real world scenarios using smartphone cameras [8]. In this work, authors proposed a segmentation procedure and used local features computed from segments and each segment was individually classified. The final decision was obtained combining the decision for individual segments.

One of the most critical tasks for any machine learning problem is to extract useful and descriptive features. Feature engineering can be domain-specific and often requires domain knowledge. This can be seen by different feature representations used in literature as reported in previous section. In recent years, Deep Learning algorithms have been successfully applied to a number of image recognition problems [9]. Deep learning architectures have seen an increase in their use in literature because of the availability of large image datasets and the availability of high computing hardware and GPUs. An added advantage of using Deep Learning algorithms is their ability to automatically extract useful representative features during the training phase [10]. A special class of deep learning algorithms called convolutional neural network (CNN) has shown excellent performance on recognition task such as Large Scale Visual Recognition Challenge and is considered as state of the art [11]. Training CNN requires large datasets and are computationally expensive. Therefore, an alternate way is to use a pre-trained CNN model for feature extraction called transfer learning [12] and then use another simpler classifier such as SVM to perform final classification.

The goal of this paper was to explore the use of a pre-trained CNN model for feature extraction for classification of food images into different food categories.

A secondary goal was to explore the classification ability of features extracted from different fully-connected layers of CNN. In this work, SVM classifier was used for classifying food intake using features extracted from the pre-trained CNN models, to perform multi-class classification. This paper further compares the results of the proposed approach with previously reported results on the same image datasets.

2 Methods

2.1 Data

The algorithm designed in this work was tested on the Pittsburgh Fast-food Image Dataset (PFID) [1]. The dataset comprised of images of 61 different fast foods captured in the laboratory. According to the authors, each food item was bought from a fast food chain on 3 different days and on each day, 6 images from different angles with different lightening conditions were taken. The background was kept constant in each image, and the focus was on the food item. The dataset consisted of a total of 1098 images of 61 categories. Details of the dataset are given in [1]. As suggested in [1], in this work, data was divided into 3-folds for each food type, and 3-fold cross validation was performed where 12 images from two days were used for training and the remaining 6 images were used for testing. Figure 1 shows an example of two different food items (burger and salad). First, three rows present images of a chicken burger taken on 3 different days and the last 3 rows show images of salad taken on 3 different days. From this image, variations in shape, texture and color are visible for pictures of the same foods taken after different lighting conditions and different image angles.

Further, authors in [1] proposed to divide foods into seven different categories since different food types might have similar ingredients and similar physical appearance, and the training and validation images were captured on separate days with different view angles. These categories were "(1) sandwiches including subs, wraps; (2) salads, typically consisting of greens topped with some meat; (3) meat preparations such as fried chicken; (4) bagels; (5) donuts; (6) bread/pastries; and (7) miscellaneous category that included variety of other food items such as soup and Mexican-inspired fast food" [1].

This approach resulted in two separate datasets, one with 61 food categories and the second with 7 categories of food items. Two separate classifiers were trained for both problems. For both problems, similar feature computation and classification approaches were used. Traditional classification methods employ user computed features from images and then use classification methods (linear or non-linear classifiers) to classify food images. This work proposes deep learning for feature computation and then use a linear classifier to classify food images. Figure 2 shows the flow of these two methods.

2.2 Feature Extraction: Convolutional Neural Network

Convolutional neural networks (CNN) are state of the art for many image recognition problems. CNN are essentially multi-layer neural networks with multiple convolution and pooling layers. The convolution layer consists of small rectangular patches (filters) smaller than the original image and whose weights are learned during the training

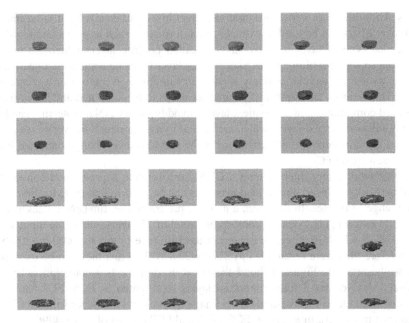

Fig. 1. An example of image categories presents in the PFID food database.

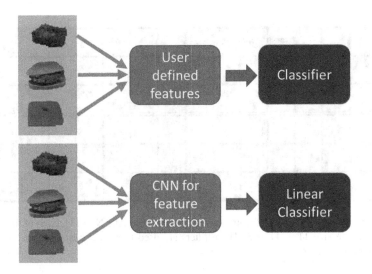

Fig. 2. Feature and classification model flow for user defined features and CNN model extracted features. The first row present conventional machine learning approach where user selected features are used for training of a classifier. The second row shows a deep learning approach where features are extracted automatically by a CNN and then a linear classifier is used for classification.

phases. These filters or kernel are used to extract low-level details from input images. CNN layer filters can be used to extract basic information such as edges and blobs etc. The second type of layer used by CNN is called pooling layers which are used to reduce the spatial size of images at each pooling layer by using some form of activation function over a rectangular window such as maximum or average over a rectangular region. This reduces the number of parameters needs to be computed and hence results in reduced computations at subsequent layers. In addition, a CNN architecture can have multiple fully-connected layers which are similar to regular neural networks, where the layer has full connection to all activations in the previous layer. Fully-connected layers are represented by FC.

In this work rather than training a CNN from scratch, a pre-trained convolution neural network was used. Pre-trained networks can be used to feature extractions from a wide range of images. In this work a network pre-trained on ImageNet dataset called AlexNet was used [11]. AlexNet consists of a total of 23 layers, where the input size is 227-by-227-by-3 (RGB images). Images in the PFID are of the size 600-by-800-by-3 and therefore they were re-sampled to 227-by-227-by-3 so that they can be used as an input to the network. Figure 3 shows the filters used in the first convolution layer of AlexNet. AlexNet has 3 fully connected layers represented as FC6, FC7, and FC8. Fully-connected layers learn higher level image features and are better suited for image recognition tasks [13]. In AlexNet, FC6, FC7, and FC8 consist of 4096, 4096 and 1000 features respectively. In this work, features computed from these three layers were separately used for classification task.

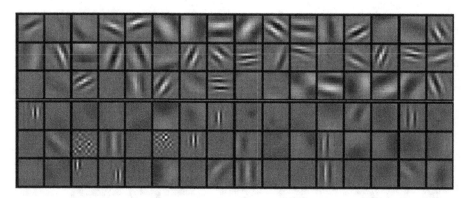

Fig. 3. Example filters used by the first convolution layer in AlexNet [11]. Each of the 96-filters shown is of the size 11 × 11 × 3. These filters are used to extract basic information such as edges, blobs, etc.

2.3   Classification: Support Vector Machine

For the classification task using features computed from deep neural networks, there are two possibilities. One possibility is to use an end-to-end deep network which performs both feature computation as well as classification whereas the second is to use CNN for feature computation and then use a linear classifier for classification. Using end-to-end deep architectures require to tune the parameters of the pre-trained network using the new image dataset and can suffer from overfitting [12]. This procedure also requires the availability of large image dataset. The second procedure using a linear classifier with features extracted by the CNN architecture can help in avoiding overfitting issues since the final classifier is not involved in feature computation [12]. In this work, to perform multiclass classification; linear SVM models were used with features computed by the AlexNet. Training and validation were performed using 3-fold cross validation where for each food type, images taken on two days were used for training and the images taken on the third day were used for validation. This process was repeated three times. Classification accuracies (F-scores) were reported for each food type using confusion matrix. Features from all three fully-connected layers of AlexNet were used for training three separate linear SVM models. These features were used for both 61 class and 7 class multiclass classification problem.

3 Results

Using features extracted from three fully-connected layers of the AlexNet to train linear SVM models resulted in different accuracies for classification of images into 61 categories. Average classification accuracies were 70.13%, 66.39% and 57.2% for features extracted from FC6, FC7 and FC8 layers of the AlexNet.

For 7 classes, the accuracies obtained for features extracted from FC6, FC7, and FC8 layers were 94.01%, 93.06%, and 89.73%, respectively. Tables 1, 2 and 3 show the confusion matrices for seven class classification for features extracted from FC6, FC7 and FC8 fully connected layers of the AlexNet. Confusion matrices for 61 classes/categories are harder to visualize and therefore, are not presented.

Table 1. Confusion matrix; classification into seven food categories based on features extracted from FC6 layer of the AlexNet.

		Predicted Class							
		1	2	3	4	5	6	7	F-score
True Class	1	300	3	1	0	17	0	5	95.24%
	2	0	123	0	0	1	0	0	98.40%
	3	0	0	102	0	0	0	0	97.61%
	4	0	0	0	68	5	0	0	90.67%
	5	4	0	2	9	293	4	9	91.71%
	6	0	0	0	0	0	67	0	97.10%
	7	0	0	2	0	2	0	62	87.32%
								Mean:	94.01%

Table 2. Confusion matrix; classification into seven food categories based on features extracted from FC7 layer of the AlexNet.

		Predicted Class							F-score
		1	2	3	4	5	6	7	
True Class	1	298	4	0	1	16	0	3	95.06%
	2	0	120	0	0	4	0	0	96.39%
	3	0	0	102	0	0	0	0	97.61%
	4	0	1	0	61	3	0	0	85.92%
	5	7	0	5	15	292	4	9	89.85%
	6	0	0	0	0	0	67	0	97.10%
	7	0	0	0	0	3	0	64	89.51%
								Mean:	93.06%

Table 3. Confusion matrix; classification into seven food categories based on features extracted from FC8 layer of the AlexNet.

		Predicted Class							F-score
		1	2	3	4	5	6	7	
True Class	1	296	7	0	1	16	2	10	93.23%
	2	0	114	0	0	3	0	0	94.21%
	3	0	0	101	0	0	0	1	95.73%
	4	0	1	0	62	2	0	0	87.32%
	5	7	3	7	13	294	4	18	88.55%
	6	0	0	0	0	0	65	0	95.59%
	7	0	0	1	1	3	0	47	73.44%
								Mean:	89.73%

4   Discussion and Conclusions

This work presented an approach based on convolution neural network and linear SVM models to differentiate between categories of fast foods from the Pittsburgh dataset. Instead of computing user defined features, AlexNet was used for automatically extracting features from food images. Results suggest that the feature extracted from the FC6 fully-connected layer along with Linear SVM classifier provided the best classification results in both 61 class classification as well as on 7-class classification problem.

The approach presented in this work has improvements over the previously reported results on the same dataset with similar testing conditions. For example, for 61-class problem, previous best results were reported using a combination of Pairwise Rotation Invariant Co-occurrence Local Binary Pattern (PRI-CoLBP$_g$) features with SVM classifier, resulting in classification accuracy of 43.1% [14], whereas the proposed approach in with work resulted in the best accuracy of 70.13%, which shows an improvement of about 27%. On average, the proposed approach consistently performs better than previous approaches, even if features from other two layers are used (accuracies of 66.39% and 57.2%). A possible reason is the ability of CNN to extract local and global features which are more relevant to the classification task.

PFID is a challenging dataset where for each food category, images were taken on 3 different days. On each day images were taken from 6 different viewpoints. Since there are intra-class variations in classes, therefore food types were split into seven major categories i.e. sandwiches, salads/sides, chicken, bread/pastries, donuts, bagels, and tacos. Previous best results for 7 category classification were obtained with a combination of PRI-CoLBP$_g$ features and SVM classifier and resulted in a classification accuracy of 87.3% [14], whereas in this work features extracted from FC6 fully-connected layer with linear SVM obtained classification accuracy of 94.01% with an overall improvement of about 7%. The performance of other classifiers trained with features from FC7 and FC8 layers are also better than previous results.

In this work the image dataset was based on fast food images taken in the laboratory. This work is also important because of the wide use of smartphones for taking images of the foods. The approach presented here can be used to automatically recognize food images and categories similar foods. One limitation of the approach presented here is that images contain only single food items. Future work will focus on images containing multiple food items. Another relevant problem is the use of learning algorithms to differentiate between images of food versus non-food and then use algorithms for recognizing food types. This will be considered in future work.

In the last decade or so, several wearable sensor systems have been proposed for automatic detection of food intake by monitoring of chewing and swallowing such as [15–18]. One such example is a wearable system presented in [16], where a combination of piezoelectric strain sensor and accelerometer were used for detection and recognition of chewing related to eating and, the physical activities performed by the participants. One future direction is to use these systems to automatic detect eating episodes and then automatically trigger a camera to capture images of the food being consumed. As a final step, the approach proposed here can be used to recognize food type and relevant caloric information. There is also a possibility to use food volume estimation method for estimating the volume of the food consumed using 3D models such as the method proposed in [19].

Acknowledgement. Research reported in this publication was supported by the National Institute of Diabetes and Digestive and Kidney Diseases (grants number: R01DK100796). The content is solely the responsibility of the authors and does not necessarily represent the official views of the National Institutes of Health.

References

1. Chen, M., Dhingra, K., Wu, W., Yang, L., Sukthankar, R., Yang, J.: PFID: pittsburgh fast-food image dataset. In: 2009 16th IEEE International Conference on Image Processing (ICIP), pp. 289–292 (2009)
2. Yang, S., Chen, M., Pomerleau, D., Sukthankar, R.: Food recognition using statistics of pairwise local features. In: 2010 IEEE Conference on Computer Vision and Pattern Recognition (CVPR), pp. 2249–2256 (2010)

3. Joutou, T., Yanai, K.: A food image recognition system with multiple kernel learning. In: 2009 16th IEEE International Conference on Image Processing (ICIP), pp. 285–288 (2009)
4. Sonnenburg, S., Rätsch, G., Schäfer, C., Schölkopf, B.: Large scale multiple kernel learning. J. Mach. Learn. Res. **7**, 1531–1565 (2006)
5. Farinella, G.M., Allegra, D., Stanco, F.: A benchmark dataset to study the representation of food images. In: Agapito, L., Bronstein, M.M., Rother, C. (eds.) ECCV 2014. LNCS, vol. 8927, pp. 584–599. Springer, Cham (2015). doi:10.1007/978-3-319-16199-0_41
6. Anthimopoulos, M.M., Gianola, L., Scarnato, L., Diem, P., Mougiakakou, S.G.: A food recognition system for diabetic patients based on an optimized bag-of-features model. IEEE J. Biomed. Health Inform. **18**, 1261–1271 (2014)
7. Food-101 – Mining Discriminative Components with Random Forests. https://www.vision. ee.ethz.ch/datasets_extra/food-101/
8. Zhu, F., Bosch Ruiz, M., Khanna, N., Boushey, C., Delp, E.: Multiple hypotheses image segmentation and classification with application to dietary assessment. IEEE J. Biomed. Health Inform. **19**(1), 377–388 (2015)
9. Lecun, Y., Bottou, L., Bengio, Y., Haffner, P.: Gradient-based learning applied to document recognition. Proc. IEEE **86**, 2278–2324 (1998)
10. Le, Q.V.: Building high-level features using large scale unsupervised learning. In: 2013 IEEE International Conference on Acoustics, Speech and Signal Processing, pp. 8595–8598 (2013)
11. Krizhevsky, A., Sutskever, I., Hinton, G.E.: Imagenet classification with deep convolutional neural networks. In: Advances in Neural Information Processing Systems, pp. 1097–1105 (2012)
12. Pan, S.J., Yang, Q.: A survey on transfer learning. IEEE Trans. Knowl. Data Eng. **22**, 1345–1359 (2010)
13. Donahue, J., Jia, Y., Vinyals, O., Hoffman, J., Zhang, N., Tzeng, E., Darrell, T.: DeCAF: a deep convolutional activation feature for generic visual recognition (2013) ArXiv13101531
14. Qi, X., Xiao, R., Li, C.G., Qiao, Y., Guo, J., Tang, X.: Pairwise rotation invariant co-occurrence local binary pattern. IEEE Trans. Pattern Anal. Mach. Intell. **36**, 2199–2213 (2014)
15. Fontana, J.M., Farooq, M., Sazonov, E.: Automatic ingestion monitor: a novel wearable device for monitoring of ingestive behavior. IEEE Trans. Biomed. Eng. **61**, 1772–1779 (2014)
16. Farooq, M., Sazonov, E.: A novel wearable device for food intake and physical activity recognition. Sensors **16**, 1067 (2016)
17. Farooq, M., Fontana, J.M., Sazonov, E.: A novel approach for food intake detection using electroglottography. Physiol. Meas. **35**, 739 (2014)
18. Farooq, M., Sazonov, E.: Segmentation and characterization of chewing bouts by monitoring temporalis muscle using smart glasses with piezoelectric sensor. IEEE J. Biomed. Health Inform., 1–1 (2016)
19. Chae, J., Woo, I., Kim, S., Maciejewski, R., Zhu, F., Delp, E.J., Boushey, C.J., Ebert, D.S.: Volume estimation using food specific shape templates in mobile image-based dietary assessment. Proc. SPIE. **7873**, 78730K (2011)

Biomedical Signal Analysis

Degrees of Freedom in a Vocal Fold Inverse Problem

Pablo Gómez$^{(\boxtimes)}$ (iD), Stefan Kniesburges, Anne Schützenberger,
Christopher Bohr, and Michael Döllinger

Division of Phoniatrics and Pediatric Audiology,
Department of Otorhinolaryngology Head and Neck Surgery,
University Hospital Erlangen, Waldstraße 1, 91054 Erlangen,
FAU Erlangen-Nürnberg, Germany
pablo.gomez@uk-erlangen.de
http://www.hno-klinik.uk-erlangen.de/phoniatrie/

Abstract. Experimental research on human phonation is negatively affected by the complexity of the process and the limiting anatomy of the larynx. Numerical simulation of the vocal folds and the formulation of an inverse problem is one way to remedy this. Several studies have explored this choosing different degrees of freedom (DOFs) in a biomechanical Two-Mass-Model (2MM). The selection of the DOFs in an inverse problem has a critical impact on the quality of possible solutions, but also affects the complexity of the problem and convergence speed in solving it.

This work compares previous DOF configurations with several new extended configurations in solving the inverse problem for vocal fold recordings of 20 healthy female subjects. Results indicate that, for the 2MM, uncoupled mass and stiffness, and variable collision strength and damping coefficients improve the matching capabilities. They match physiology and lead to up to 50% smaller errors even for a low number of model evaluations.

Keywords: Inverse problem · High-speed videoendoscopy · Vocal fold oscillation · Two-Mass-Model

1 Introduction

The phonatory process in humans has been a topic of continued research for decades due to its complexity and the social impact of voice disorders. Roy et al. [1] report a lifetime prevalence of almost 30% of voice disorders and Wilson et al. [2] found that dysphonia without an obvious laryngeal disease had significant impact on the quality of life. It is therefore desirable to aid diagnosis and further the understanding of phonation and diseases, that impair it.

The human vocal folds oscillate normally with frequencies of up to 350 Hz depending on age, gender and individual [3]. Furthermore the confining anatomy of the larynx limits direct observation and measurement.

© Springer International Publishing AG 2017
I. Rojas and F. Ortuño (Eds.): IWBBIO 2017, Part I, LNBI 10208, pp. 475–484, 2017.
DOI: 10.1007/978-3-319-56148-6_42

One way to account for these complications and gain insight is through the use of numerical models to simulate the vocal fold oscillations. Several models have been suggested, ranging in complexity from one-dimensional lumped-mass models consisting of four masses [4] to models involving complex fluid mechanics [5] or three-dimensional lumped-mass models with 50 masses [6]. As first suggested by Döllinger et al. [7], it is possible to construct an inverse problem to extract the model parameters that lead to a certain oscillation. This is achieved by minimizing the difference between the model and recorded oscillations that were obtained using high-speed videoendoscopy. This idea has been extended upon in several studies [8–11] with considerable success in extracting parameters and describing characteristics like asymmetry.

One question that has not seen substantive research in this field is, which parameters in the model should be considered degrees of freedom (DOFs) for optimal results.

This question is of increasing interest, as with increased computational power, it is feasible to formulate the inverse problem with a higher number of DOFs than it was in the past. The main purpose of this paper is to explore several configurations of DOFs and compare them in regard to their performance and computational demands. This analysis is performed by testing the configurations on the parameter inference of high-speed video recordings of the vocal folds of 20 healthy female subjects during phonation and a subsequent discussion of the results.

2   Methods

2.1   Two-Mass-Model

Originally introduced in 1973 by Ishizaka and Flanagan [4] and later simplified by Steinecke and Herzel [12], the Two-Mass-Model (2MM) remains a popular choice to model the phonatory process. A list of the parameters in the model is given in Table 1 and a schematic depiction in Fig. 1a.

Note, that the model only allows for movement in the lateral direction, which corresponds to the x direction in Fig. 2b. The equations of motion for mass (s, i) where s indicates the side (1 - left, r - right) and i the plane (1 - lower, 2 - upper) are

$$m_{s,i}\ddot{x}_{s,i} = F_{s,i}^a + F_{s,i}^v + F_{s,i}^c + F_{s,i}^d \tag{1}$$

with the forces

$F_{s,i}^a$ - anchor spring force

$F_{s,i}^v$ - vertical coupling force

$F_{s,i}^c$ - force due to collision

$F_{s,i}^d$ - driving force.

Table 1. Description of the parameters in the 2MM

Parameter	Description
$m_{s,i}$	Masses
$k_{s,i}$	Stiffness of the anchor springs
k_s^v	Stiffness of the vertical springs
$k_{s,i}^c$	Stiffness during collision
$r_{s,i}$	Damping coefficients of the anchor springs
l	Glottis length
d_i	Thickness of masses (l, i) and (r, i)
a_i	Area between masses (l, i) and (r, i)
$x_{s,i}^r$	Rest position of the mass (s, i) as distance from the midline
$x_{s,i}$	Position of the mass (s, i) in relation to the rest position
P_s	Subglottal pressure

In the following, the forces $F_{s,i}^a$, $F_{s,i}^v$ and $F_{s,i}^d$ were computed as they were by Döllinger et al. [7], whereas $F_{s,i}^c$ is based on the formulation by Sommer et al. [13].

In the original model by Ishizaka and Flanagan [4] a nonlinear stiffness is proposed. As in the research of Fulcher et al. [14], this is implemented as a variable stiffness $\hat{k}$ so that

$$\hat{k}_{s,i} = k_{s,i}(1 + \eta x_{s,i}^2), \tag{2}$$

where η was set to $100\,\mathrm{cm}^{-2}$. The variable stiffness $\hat{k}_{s,i}$ is then used in the computation of the anchor force $F_{s,i}^a$.

The output of the model are the so-called vocal fold trajectories T_r and T_l, which are defined in the n-th time step as

$$\begin{aligned} T_r[n] &= x_{r,i^*}^r + x_{r,i^*}[n] \\ T_l[n] &= -(x_{l,i^*}^r + x_{l,i^*}[n]), \end{aligned} \tag{3}$$

where

$$i^* = \begin{cases} 1, & \min(a_1[n], a_2[n]) = a_1[n] \\ 2, & \min(a_1[n], a_2[n]) = a_2[n]. \end{cases}$$

Figure 1b shows example trajectories for the default parameters of the model given in appendix Table 3.

The numerical solutions were computed using a Runge-Kutta Algorithm [15] with a time step of $\Delta t = 0.25\,\mathrm{ms}$.

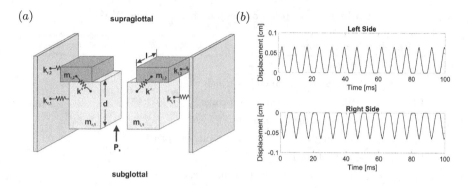

Fig. 1. (a) Depiction of the Two-Mass-Model by Schwarz et al. [10]. (b) Model trajectories for the default parameter values described in the appendix in Table 3. A displacement value of 0 cm indicates a closed glottis, i.e. collision of the vocal folds.

2.2   The Inverse Problem

Experimental measurements of the phonatory process are limited by the human anatomy and the complex processes occurring in the larynx during phonation. One way to still gain insight is by formulating an inverse problem. In an inverse problem, the main interest are the parameters that lead to a certain model output.

Among the numerical models suggested to describe the phonatory process, the 2MM is especially well-suited for the formulation of an inverse problem, as its computational cost is comparatively low, which allows quick model evaluation. Therefore, starting with the works of Döllinger et al. [7] and others [8–10], research was performed on the extraction of mechanical parameters from the 2MM. This process relies on the minimization of the difference between experimental vocal fold trajectories $\tilde{T}_r$ and $\tilde{T}_l$ and the model trajectories T_r and T_l. The experimental trajectories are obtained through high-speed video recordings of the vocal folds similar to the procedure described in previous studies [7,16], the model trajectories by running the 2MM with different initial conditions, i.e. parameters like mass, stiffness or subglottal pressure. The difference between the experimental and model trajectories of length N is described by the objective function

$$\Gamma = \min_{1 \leq j \leq N} \frac{1}{2} \left(L_2^j(\tilde{T}_r, T_r) + L_2^j(\tilde{T}_l, T_l) \right), \tag{4}$$

where

$$L_2^j(\tilde{T}, T) = \sqrt{\frac{\sum_{i=1}^{N}(\tilde{T}[i] - T[i + j \mod N])^2}{\sum_{i=1}^{N}(\tilde{T}[i])^2}}. \tag{5}$$

The *modulo* operation here serves to make Γ phase invariant as a phase shift could be remedied by picking a different starting point of the experimental recording and is therefore negligible.

Given this setup, the inverse problem comes down to deciding, which parameters in the model should be treated as DOFs and then performing a minimization of the function Γ.

For the latter process a Differential Evolution (DE) algorithm implementing the *rand-to-best/1/bin/* strategy, as it was first described by Storn and Price [17], was employed. A detailed overview of the chosen parameters in DE can be found in the appendix in Table 4. The selection of DOFs is the main topic of this work.

2.3   Degrees of Freedom

The number of DOFs in this inverse problem is a critical factor. On the one hand, a lower number might restrict the ability to reproduce certain trajectories with the model. E.g., if the DOFs do not allow for asymmetric model configurations, trajectories with significant left-right asymmetries will not yield good results. On the other hand, a higher number of DOFs leads to a higher dimensionality of the problem. This typically results in a slower convergence and thereby increased computational cost when minimizing Γ.

Previous studies relied on different constellations of DOFs and more are conceivable. Döllinger et al. [7] employed three DOFs, which were used to vary subglottal pressure, mass and stiffness in the model configurations. Furthermore left and right masses and stiffnesses were treated independently, accounting for asymmetries. Schwarz et al. [10] focused on asymmetric configurations in particular and also varied the vertical stiffnesses k_s^v. Symmetric configurations were studied in research by Tao et al. [8], where they employed seven DOFs varying mass, stiffness and damping independently for the lower ($i = 1$) and upper ($i = 2$) planes, and subglottal pressure.

A comparison of the different choices of DOFs using the same objective function Γ is desirable, as it allows to quantify the advantages and impact of certain DOFs on both the range of trajectories that can be replicated, as well as the performance burdens resulting from a higher number of DOFs. To this end, this work compares several combinations of DOFs, shown in Table 2, in-depth.

Table 2. Description of the tested DOF configurations in the 2MM. Stiffness refers only to anchor spring stiffness $k_{s,i}$, damping to damping coefficients $r_{s,i}$. Collision strength is scaled by treating $k_{s,i}^c$ as a DOF.

#DOFs	Description of the DOFs	Source
3	Left/right masses and stiffnesses (coupled), subglottal pressure	[7]
4	As #DOF = 3, additional DOF to scale collision strength	-
6	As #DOF = 4, but mass and stiffness are uncoupled	-
7	Upper/lower plane mass, stiffness and damping, subglottal pressure	[8]
10	Each mass and stiffness is a DOF, subglottal pressure, collision strength	-
14	Each mass, stiffness, damping is a DOF, subglottal pressure, collision strength	-

The DOFs as selected in previous studies [7,8] are compared to other plausible configurations including one that uncouples mass and stiffness (#DOF = 6), but is similar to the one Döllinger et al. [7] used. One configuration, that varies mass and stiffness for each individual mass (#DOF = 10), and one, that builds on the work of Tao et. al [8] and varies mass, stiffness and damping for each individual mass.

2.4   Setup

As test cases for the different DOF configurations, high-speed video recordings of the vocal folds of 20 healthy females (18–24 years old) were analyzed using the in-house developed software tool "Glottis Analysis Tool". Figure 2a depicts the recording process and Fig. 2b is an example image obtained through recording. The trajectories at the 50% anterior-posterior position were used as experimental trajectories $\tilde{T}_r$ and $\tilde{T}_l$. To account for differences in the rest positions $x^r_{s,i}$ between subjects, they were approximated as

$$\hat{x}^r_{s,i} = \frac{\overline{\tilde{T}_s}}{\sqrt{2}}. \tag{6}$$

Excerpts with a length of 100 ms featuring sustained phonation were selected from each recording and a 50 ms interval was cut off from the beginning of the resulting model trajectories as the model starts off in a transient state.

Details of the optimization using DE, such as parameter ranges and algorithm settings are specified in the appendix in Tables 4 and 5. All computations were performed on an Intel® Core™ i5-4590 Processor.

(a)   (b)

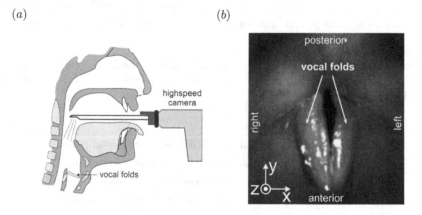

Fig. 2. Depictions of the process of recording experimental trajectories by Semmler et al. [18]. (a) Schematic of the high-speed videoendoscopy. (b) Sample image from a recording of the vocal folds

3   Results

All DOF configurations from the previous section were run three times for several fixed numbers of maximum model evaluations between 1000 and 100000. In total this resulted in 27 runs per recording and DOF configuration. Of the three runs per evaluation limit, only the best was taken into account, as the focus was on determining the best possible results obtainable using a DOF configuration.

Values of Γ in individual runs ranged from 0.11 to 0.6. Averages for each configuration and limit of model evaluations can be seen in Fig. 3. The lowest average Γ value of 0.24 was observed in the runs terminated at 100000 model evaluations with 14 DOFs. In fact, with at least 50000 evaluations, the 14 DOFs configuration outperformed all others. For a lower number of maximum evaluations the configuration with 7 DOFs always performed best.

At 1000 maximum evaluations average Γ values ranged from 0.38 to 0.5, at 100000 they were between 0.24 and 0.40. All configurations besides the 14 DOFs one are not improving significantly anymore at 100000 evaluations, indicating that the matching capabilities of the DOF configurations had been reached. This is expected, as a higher number of DOFs leads to a slower convergence, but should extend matching capabilities.

It also explains why the 14 DOFs configuration starts off worst. The comparatively better performance of the 7 DOFs configuration at a low number of maximum evaluations is probably caused by the fact, that it is the only configuration that neglects asymmetries. The tested recordings contained only very little asymmetries in regard to the amplitude and frequency of the left and right experimental trajectories.

A single optimization run with 3 DOFs and a maximum of 1000 model evaluations required about 2 s and one with 14 DOFs and a maximum of 100000 model evaluations about 130 s.

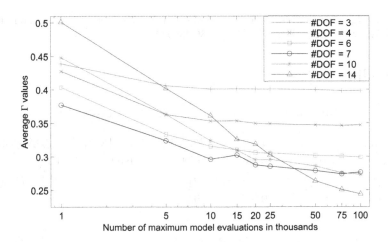

Fig. 3. Average Γ values for runs terminated after a number of model evaluations. For 25000 evaluations or less, the 7 DOFs configuration and after that the 14 DOFs configuration performed best.

4 Discussion and Outlook

The results in this paper give clear indications of several factors that should be considered in the selection of the DOFs in inverse problems involving the 2MM.

First off, note that an increase in DOFs is always a trade-off between a more flexible model and additional computational cost. Our results suggest that a very high number of DOFs is only appropriate, if enough model evaluations can be performed to deal with the slower convergence. The exception here is the configuration with 7 DOFs. It shows that the selection of an appropriate set of DOFs can aid in quick convergence. Assuming symmetric values in the model for the trajectories from healthy phonation likely benefited convergence.

Looking at the benefits of including specific DOFs, the biggest decreases in error stemmed from including a scaling factor for collision strength, and uncoupling mass and stiffness. The improvement through the former can be seen in the comparison of the 3 and 4 DOFs configurations, the latter in the difference between 4 and 6 DOFs configurations. The configuration with 10 DOFs, where each mass and stiffness was a DOF, did not yield significant improvement compared to the 6 and 7 DOFs configurations. Overall, the best results in regard to average Γ values and necessary model evaluations were obtained with the 7 and 14 DOFs configurations, which both included uncoupled mass and stiffness, and DOFs for damping. The differences between the two configurations were the individual treatment of each mass, stiffness and damping as well as the scaling of collision strength with 14 DOFs. Especially the latter is a likely candidate in explaining the differences between them.

Future research will be directed at exploring further improvements to the DOF configurations, analysis of the resulting parameters in comparison to experimental results from other models such as *ex vivo* experiments and a more extensive verification on recordings from male, pathologic or animal subjects.

Acknowledgments. This work was supported by German Research Foundation (DFG) research grants DO1247/8-1 and BO4399/2-1.

Appendix

Table 3. Default parameters of the Two-Mass-Model.

$m_{s,1}\,[g]$	$m_{s,2}\,[g]$	$k_{s,1}\,[N/cm]$	$k_{s,2}\,[N/cm]$	$r_{s,1}\,[Ns/cm]$	$r_{s,2}\,[Ns/cm]$
0.125	0.025	0.08	0.008	0.02	0.02
$k_{s,1}^{c}\,[N/cm]$	$k_{s,2}^{c}\,[N/cm]$	$k_{s}^{v}\,[N/cm]$	$d\,[cm]$	$l\,[cm]$	$P_s\,[cmH_2O]$
$3k_{s,1}$	$3k_{s,2}$	0.025	0.25	0.8	8

Table 4. Parameter settings for DE. Explanations of the parameters can be found in the original paper by Storn and Price [17].

Parameter	Value
Population size	$10 \cdot dimension$
CR	0.5
F	0.8
Reinitialization after # iterations without a change in the population	10

Table 5. Parameter ranges used for the optimization

Parameter	Lower bound	Upper bound
$\tilde{m}_{s,i}$	$0.25 \cdot m_{s,i}$	$3.33 \cdot m_{s,i}$
$\tilde{k}_{s,i}$	$0.3 \cdot k_{s,i}$	$4.0 \cdot k_{s,i}$
$\tilde{r}_{s,i}$	$0.3 \cdot r_{s,i}$	$4.0 \cdot r_{s,i}$
$\tilde{k}^c_{s,i}$	$-1.0 \cdot k^c_{s,i}$	$8.0 \cdot k^c_{s,i}$
$\tilde{P}_s$	$0.1 \cdot P_s$	$10.0 \cdot P_s$

References

1. Roy, N., Merrill, R.M., Gray, S.D., Smith, E.M.: Voice disorders in the general population: prevalence, risk factors, and occupational impact. Laryngoscope **115**(11), 1988–1995 (2005)
2. Wilson, J., Deary, I., Millar, A., Mackenzie, K.: The quality of life impact of dysphonia. Clin. Otolaryngol. Allied Sci. **27**(3), 179–182 (2002)
3. Patel, R.R., Dixon, A., Richmond, A., Donohue, K.D.: Pediatric high speed digital imaging of vocal fold vibration: a normative pilot study of glottal closure and phase closure characteristics. Int. J. Pediatr. Otorhinolaryngol. **76**(7), 954–959 (2012)
4. Ishizaka, K., Flanagan, J.L.: Synthesis of voiced sounds from a two-mass model of the vocal cords. Bell Syst. Tech. J. **51**(6), 1233–1268 (1972)
5. Alipour, F., Berry, D.A., Titze, I.R.: A finite-element model of vocal-fold vibration. J. Acoust.Soc. Am. **108**(6), 3003–3012 (2000)
6. Yang, A., Lohscheller, J., Berry, D.A., Becker, S., Eysholdt, U., Voigt, D., Döllinger, M.: Biomechanical modeling of the three-dimensional aspects of human vocal fold dynamics. J. Acoust. Soc. Am. **127**(2), 1014–1031 (2010)
7. Döllinger, M., Hoppe, U., Hettlich, F., Lohscheller, J., Schuberth, S., Eysholdt, U.: Vibration parameter extraction from endoscopic image series of the vocal folds. IEEE Trans. Biomed. Eng. **49**(8), 773–781 (2002)
8. Tao, C., Zhang, Y., Jiang, J.J.: Extracting physiologically relevant parameters of vocal folds from high-speed video image series. IEEE Trans. Biomed. Eng. **54**(5), 794–801 (2007)
9. Wurzbacher, T., Schwarz, R., Döllinger, M., Hoppe, U., Eysholdt, U., Lohscheller, J.: Model-based classification of nonstationary vocal fold vibrations. J. Acoust. Soc. Am. **120**(2), 1012–1027 (2006)

10. Schwarz, R., Hoppe, U., Schuster, M., Wurzbacher, T., Eysholdt, U., Lohscheller, J.: Classification of unilateral vocal fold paralysis by endoscopic digital high-speed recordings and inversion of a biomechanical model. IEEE Trans. Biomed. Eng. **53**(6), 1099–1108 (2006)
11. Yang, A., Stingl, M., Berry, D.A., Lohscheller, J., Voigt, D., Eysholdt, U., Döllinger, M.: Computation of physiological human vocal fold parameters by mathematical optimization of a biomechanical model. J. Acoust. Soc. Am. **130**(2), 948–964 (2011)
12. Steinecke, I., Herzel, H.: Bifurcations in an asymmetric vocal-fold model. J. Acoust. Soc. Am. **97**(3), 1874–1884 (1995)
13. Sommer, D.E., Erath, B.D., Zanartu, M., Peterson, S.D.: Corrected contact dynamics for the steinecke and herzel asymmetric two-mass model of the vocal folds. J. Acoust. Soc. Am. **132**(4), EL271–EL276 (2012)
14. Fulcher, L.P., Scherer, R.C., Melnykov, A., Gateva, V., Limes, M.E.: Negative coulomb damping, limit cycles, and self-oscillation of the vocal folds. Am. J. Phys. **74**(5), 386–393 (2006)
15. Schwarz, H.R., Köckler, N.: Numerische Mathematik. Springer, Heidelberg (2013)
16. Wittenberg, T., Moser, M., Tigges, M., Eysholdt, U.: Recording, processing, and analysis of digital high-speed sequences in glottography. Mach. Vis. Appl. **8**(6), 399–404 (1995)
17. Storn, R., Price, K.: Differential evolution–a simple and efficient heuristic for global optimization over continuous spaces. J. Glob. Optim. **11**(4), 341–359 (1997)
18. Semmler, M., Kniesburges, S., Birk, V., Ziethe, A., Patel, R., Döllinger, M.: 3D reconstruction of human laryngeal dynamics based on endoscopic high-speed recordings. IEEE Trans. Med. Imaging **35**(7), 1615–1624 (2016)

Information Limits of Optical Microscopy: Application to Fluorescent Labelled Tissue Section

Renata Rychtáriková[1], Georg Steiner[2], Michael B. Fischer[3],
and Dalibor Štys[1(✉)]

[1] Faculty of Fisheries and Protection of Waters,
South Bohemian Research Center of Aquaculture and Biodiversity of Hydrocenoses,
Institute of Complex Systems, University of South Bohemia in České Budějovice,
Zámek 136, 373 33 Nové Hrady, Czech Republic
{rrychtarikova,stys}@frov.jcu.cz
[2] TissueGnostics GmbH, Taborstrasse 10/2/8, 1020 Vienna, Austria
[3] Department of Health Sciences and Biomedicine, Danube University Krems,
Dr.-Karl-Dorrek-Strasse 30, 3500 Krems, Austria
http://www.frov.jcu.cz/cs/ustav-komplexnich-systemu-uks

Abstract. The article demonstrates some less known principles of image build-up in diffractive microscopy and their usage in analysis unravelling the smallest localized information about the original object – an electromagnetic centroid. In fluorescence, the electromagnetic centroid is naturally at the position of the fluorophore. The usage of an information-entropic variable – a point divergence gain – is demonstrated for finding the most localized position of the object's representation, generally of the size of a voxel (3D pixel). These spatial pixels can be qualitatively classified and used for reconstruction of the 3D structures with precision comparable with electron microscopy.

Keywords: Electromagnetic centroid · Point divergence gain · Superresolution · 3D structure reconstruction

1 Introduction

In optical microscopy, features of a light-intensity profile arise essentially by an interplay of two phenomena: by interaction of light with the matter and by modification of the light path by the microscope [1]. In observation of living cells there are mainly two types of responses of the interaction of light with the matter: (i) diffraction of light on particles and (ii) emission of fluorescent light after absorption.

The diffraction of light is a process whose proper description is difficult, in most real cases essentially impossible [2]. A general feature of light behaviour behind a scattering object is the existence of a dark "spire" which is gradually narrowing as the Huygens waves cover the space. The light diffraction on an

© Springer International Publishing AG 2017
I. Rojas and F. Ortuño (Eds.): IWBBIO 2017, Part I, LNBI 10208, pp. 485–496, 2017.
DOI: 10.1007/978-3-319-56148-6_43

object which is circular enough can even produce bright spots on axis, so-called the Arago spots [3]. Biological objects are in most cases formed by dense structured assemblies of proteins, nucleic acids, lipids and other molecules. Objects in the cell interior observable by light diffraction are not merely objects of in-average different refractivity index but they are also internally inhomogeneous. Their detailed scattering properties are beyond a simple description. We are thus content with the information provided by the microscopy experiment, although we have to understand the limitations of this information.

Light fluorescence has been an important examination tool in biological cell for a long time [1]. The fluorescent microscopy of a living cell is in the forefront of the interest due to the efforts to achieve/break the limits of resolution of the observed intracellular objects [4]. All attempts at superresolution were theoretically based on the description of observed phenomena using the Maxwell theory of the electromagnetic field which is a proper description of behaviour of an ensemble of photons [5].

We have recently shown [6] that information projected by light on the screen of a camera can be analyzed down to the level of a single sensor element of the digital camera or, in other words, of a single pixel in a digital image. A single pixel whose size can be determined can be referred to the location of an observed object in the original image. The size of this element does not have any physical limit, i.e., it can be of a-few-nanometer size. This finding contradicts no principles of quantum mechanics, because we do not determine the position of a single photon but the most probable location of a large ensemble of photons whose distribution has a maximum, intensity profile, etc.

Finding the maximum or minimum of an intensity profile does not mean that the object, which gives rise to the response, is located at the particular position in space to which a microscope image seemingly refers. In diffractive imaging, the reason is that both the darkest and smallest point is located outside the object which causes the diffraction. The maximum of fluorescence of a fluorescent object, e.g. a bead, is (in the ideal case) seemingly at a place closest to the surface of the objective lens. The exact maximum or minimum can be defined as a centroid of the outcome of the electromagnetic process which gives rise to the observed light phenomenon.

The attempt to yield a maximal information from the digital image exposes also all non-idealities of the optical path. In case that the resulted image is spectrally resolved, i.e. when a colour digital camera is used, each camera channel typically detects different information [6,9,10]. The difference is generally attributed to the composition of the object which gives rise to the signal, which is ultimately true, but the exact way how this difference is conveyed by the optical path is not easy to unravel. Namely, modern apochromatic lenses utilize combinations of lenses to project all colours at the same place. This assumption is indeed valid only with a finite precision and can be challenged when minute image details are interpreted.

The microscope observation should answer these questions:

1. Where is located the object giving rise to the response?
2. What is the shape of the object?
3. What are the spectral characteristics of the object?

In order to answer these questions, under the microscope, we have analyzed a response of a standard object – a single microparticle – and a section of fluorescently labelled tissue. We introduce a method which systematically determines the electromagnetic centroid of a diffracting object as a centroid of the information in 3D space with the precision of a single voxel (3D pixel).

2   Materials and Methods

2.1   Experiment on Latex Particle

A latex particle of the diameter of 2000 nm was placed on a carbon layer on a electron microscopy copper grid covered by amorphous carbon (prepared at the Institute of Parasitology AS CR, České Budějovice, CZ). The sample was scanned under a optical transmission microscope [6] – nanoscope (Institute of Complex Systems, Nové Hrady, CZ) – equipped by a 12-bit colour Kodak KAI-16000 digital camera with a chip of 4872×3248 resolution (Camera Offset 200, Camera Gain 383, Camera Exposure 2950 ms). A Nikon objective ($60 \times /0.8$, $\infty/0.17$, WD 0.3) which gives the resulted size of the image pixel of 46×46 nm^2 was used. The sample was illuminated by two Luminus 360 LEDs charged by the current of 4000 mA. The standard deviation of the z-step was minimized by the pngparser.exe software [6] which gave a z-stack of 258 images of the average step of 152 nm.

2.2   Experiment on Prostate Cancer Tissue

A fixed sample of prostate cancer tissue was imunnofluorescently labelled by DAPI (4',6-Diamidine-2'-phenylindole) and Anti-Cytokeratin 18. The treated sample was scanned using a TissueFaxs PLUS fluorescent microscope (TissueGnostics GmbH, Vienna, AT) equipped by a 12-bit grayscale camera with a chip of 1560×1960 resolution. A 100× oil objective gave the resulted image pixel of the size of 328×328 nm^2. The microscope step along the z-axis was 100 nm. The full z-stack series contained 82 images.

2.3   Image Processing and Visualization

The 3D reconstructions of the image z-stacks were performed by the method similar to that described in detail in [6]. The modifications were as follows:

1. The segmentation of the object of interest using an algorithm which searches for intensities of unchanged values between two consecutive images was applied only to the z-stack of the bead, not to the z-stack of the dyed tissue.

2. The main difference was in the selection of a focused part of both z-stacks using k-means clustering into two groups using Unscrambler® X software, Norway. Instead of vectors of point information gain, vectors of point divergence gain [7] were computed from intensity histograms of the whole images for a set of $\alpha = \{0.1, 0.3, 0.5, 0.7, 0.99, 1.3, 1.5, 1.7, 2.0, 2.5, 3.0, 3.5, 4.0$ (the bead), 5.0, 6.0, 7.0 (the tissue)$\}$. In case of the 2-μm bead scanned in light transmission, the z-stack which underwent the computation of the point divergence gain vectors, was firstly transformed into 8 bits by the Least Information Lost (LIL) algorithms [9] which enables to yield the maximum of information during the bit-depth reduction and to compare images through the whole stack. The clustering reduced the number of images from 258 to 81 (Imgs. 83–163) and from 82 to 52 (Imgs. 26–77) for the bead and the tissue, respectively.

3. The points of almost unchanged intensities in two consecutive in-focus images were stacked after (i) segmentation of the object demarcated by green intensity 1000 in the image followed by the calculation of point divergence gain for $\alpha = 4.0$ (the bead) and (ii) computation of the point divergence gain for α of the value of 7.0 and 6.0 for DAPI and Anti-Cytokeratin 18, respectively, followed by the automatic selection of two kinds of relevant points as written in Sect. 3.2.

The figures of sections of the point spread function of the microbead in each colour channel as well as of positions of the electromagnetic centroids (Fig. 2) were plotted using Matlab® 2016b (Mathworks, USA) software.

All figures in the article were also visualized using the LIL conversion into 8 bits. The original and processed image sets and relevant Matlab algorithms are available at [8].

3   Results

3.1   Electromagnetic Centroid

Figure 1 shows a typical example of a microscopic image of the simplest object – a latex particle of the diameter of 2 μm – at the position of visually determined focus. For each (red, green, blue) image channel, the response of the object differs in position and in shape of the intensity distribution.

As a consequence of light interactions, one can find two smallest objects on the apparent axis of the response of interactions of the electromagnetic field with a diffracting object in each colour channel: A dark spot, which is an outcome of the destructive light diffraction, and a bright spot, which is surrounded by dark intensities and is called the Arago spot (Fig. 1). In the ideal case, this bright spot appears due to constructive light interference at the optical axis at the position of the center of a circular diffracting object [3].

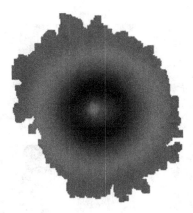

Fig. 1. Image 127 from a z-stack of microscopic images of a 2000-nm latex particle. (Color figure online)

Figure 2a demonstrates xz- and yz-planes of the 3D microscopic image of the latex particle in Fig. 1 including all non-idealities of the microscope optics. The detected intensities are combinations of an interaction of the light electromagnetic field with the sample followed by a transformation of the intensities by the microscope optics. Figure 2b depicts the positions of the dark and bright maxima – the centroids of the electromagnetic field – in each colour channel. The bright green maximum (Fig. 2b, *left*) consists of many points of identical intensities due to the saturation of the 12-bit intensity signal at high intensities.

3.2   Fluorescently Labelled Cells

Figure 3 shows a typical microscopic image of a fluorescently labelled tissue, where the observed intensities correspond to the positions of a fluorophore in the sample. The fluorophore changes its spectral properties and quantum efficiency in response to the environment. The information analysis of the given z-stack of the labelled tissue using the point divergence gain was focused on identification of positions of fluorophores which, according to given rules, differ from other (Fig. 4, *upper*). The given rules are to examine (and display) points whose intensity does not change over two z-levels and which were selected according to

- their proximity, i.e. dense areas are displayed and
- different intensity, i.e. spectrally different object are displayed.

In less favourable cases, any simple rule is not available and the image has to be analyzed from the complete dataset. For instance, (Fig. 4, *lower*) shows an example of the 3D reconstruction which arose from images in which the unchanged points occurred only very sparsely.

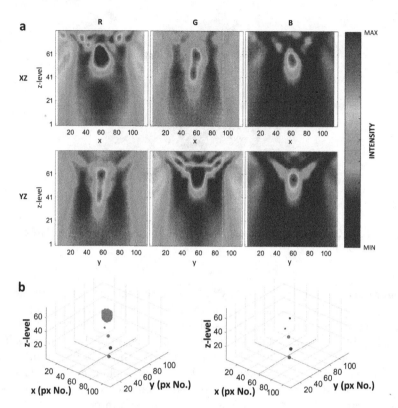

Fig. 2. 3D intensity maps of the 2000-nm latex bead in Fig. 1 in the red, green, and blue image channel. The minimal/maximal intensities are 623/2706 (R), 759/4095 (G), 430/3752 (B). Voxel size is 46 nm (horizontally) and 152 nm (vertically). (a) Sections in the xz- and yz-plane, respectively. (b) Positions of electromagnetic centroids (*colour-coded*). The bigger points and the smaller points correspond to negative and positive light interferences, respectively. The bright spot in the green channel is not sufficiently resolved due to the saturation of the signal. The positions of xz- and yz-planes relevant to a are highlighted by bold lines. (Color figure online)

4 Discussion

This paper tries to answer how to recognize unchanged information between two optical cuts in optical microscopy in order to find the most localized information about the position of an object. It has few technical determinants:

1. First of all, the comparison of the optical cuts is limited by the analog-to-digital (AD) conversion. The AD converters in standard digital cameras provide 12- or 16-bit conversion (4096 or 65536 intensity levels). However, experimenters usually work with 8-bit images with 256 intensity levels. Image distortions which accompany the 12/16-bit to 8-bit conversion was reported earlier [9]. In this paper, we report an analysis of original 12/16-bit datasets.

[0,0]   y

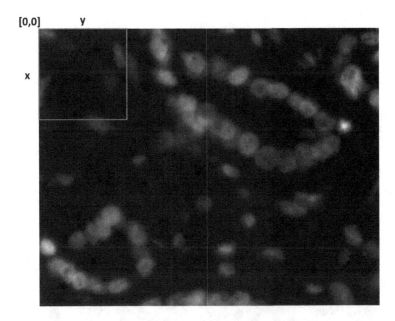

x

Fig. 3. Microscopic image of the section of prostate cancer tissue labelled by DAPI dye targeted to nuclei. Pixel size is $328 \times 328\,\mathrm{nm}^2$. (Color figure online)

2. The second main technical determinant is a size of pixel or voxel (3D pixel) which determines a theoretical size of the object. It is not the goal of this article to discuss limits of the discriminability of objects in terms of the theory of light, we assume hereby that it may be determined experimentally.

3. The third determinant is the exposure time which is closely related to the sensitivity of camera chip. The exposure of time of camera is usually set so that all intensity signals would be captured without any oversaturation. In addition, the linear responses of the chip elements to exposure time are never guaranteed and the calibration is, according to our experience, never correct.

4. The acquisition time of cameras also plays a pivotal role. In case of observation of moving object, e.g., living cells, the acquisition time can be so long that the object changes its position and a false negative signal is obtained.

5. Further, the objective determinants of the precision of finding the most localized information about the position of an object come from various sources of noise. In the examples presented in this article, when the samples were observed under high light intensities (i.e. under a large ensemble of photons), it is unlikely that the quantum noise [11] was the key limit which prevent to find the intensity maximum or minimum. In these examples, the signal distortions originate from numerous sources related to the optical path and the instrument electronics. The noise modulates the signal by the multiplication by a function. Since no precise characteristics of the noise is known, it is most appropriate to assume that it has a multifractal character. This mutifractal intensity response is then discretized in space and in time. A method of calculation of point divergence

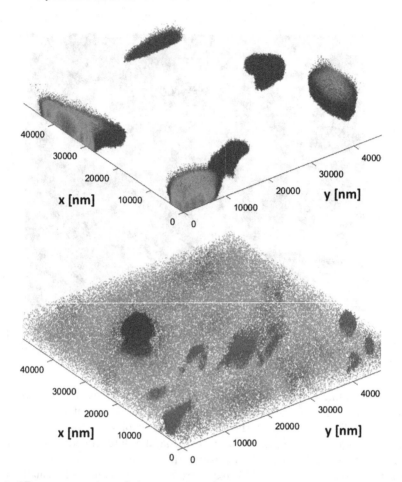

Fig. 4. 3D reconstruction of the positions of points of unchanged intensities in the section of prostate cancer tissue labelled by DAPI dye targeted to nuclei (*upper*) and Anti-Cytokeratin 18 targeted to keratin (*lower*). The section is located in the upper left corner of Fig. 3 (*orange section*). Voxel size is $328 \times 328 \times 100\,\mathrm{nm}^3$. Colorbar is the same as in Fig. 2a but with the minimal/maximal intensities of 684/3495 for DAPI and 248/1522 for Anti-Cytokeratin 18. (Color figure online)

gain [6] enables to account for the whole spectrum of these distortions. The points (pixels) of almost unchanged intensities are grouped with regards to different assumptions about the distributions of occurrence of intensity signals. We used the method of point divergence gain [6] for resolution of the information in z-stacks of images of living cells into elementary information contributions in the diffractive imaging. In the green image channel, there was observed a lower number of identical points between two consecutive z-levels. It is assigned to a broader intensity spectrum, i.e., a broader wavelength range of the green filters

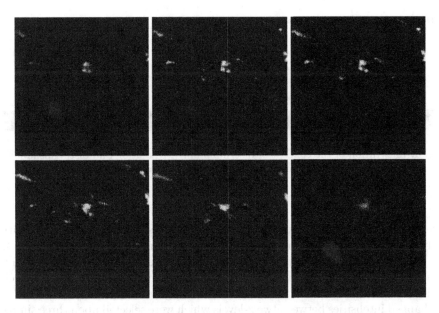

Fig. 5. 7th, 16th, 21st, 26th, 36th, and 46th image (*from upper right to lower left*) of the series from which Fig. 4, (lower), *lower* was constructed. Pixel size is $328 \times 328 \, \text{nm}^2$.

of the Bayer mask [12]. This illustrates a paradox of the information analysis: The broader the information spectrum is, the more the differences are found (Fig. 5).

A latex bead of the 2-μm size is an object of the size which is very relevant to intracellular objects such as mitochondria or other oval organelles. Its size in the visible light spectrum is from 5 light waves (shortwave blue, 400 nm) to less than 3 light waves (far red, 720 nm), which is on the border between macroscopic behaviour and behaviour of, e.g., metal nanobeads of the size of the fraction of the light wavelength. If we stick with the macroscopic description, we can say that the Arago spot [3] is located in the vicinity of the object and is directly adjacent to the dark area of diffraction.

The interpretation of the imaging of a single bead/organelle has relation to two fundamental concepts of the signal analysis [13]: the resolution and the distinguishability. The microscopic resolution is defined as a distance at which two first-order valleys of the Airy waves [14] exactly merge. As seen in Fig. 2, the Airy pattern is not observable in a real point (object) spread function. In contrast, the positions of the centroids of the interaction of light with a particle (i.e. intensity maximum/minimum) can be always found. Thus, searching for centroids of an object's response is a more realistic approach than the the theoretical Airy concept of resolution. Hence we suggested the information resolution concept [6].

The resolution can be also defined by absence or presence of points which do not have the identical intensity at the same position in the next z-level. In other words,

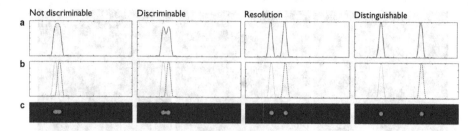

Fig. 6. Illustration of the concept of discriminability, resolution and distinguishability in optical microscopy. Total intensity profiles (a), intensity profiles of each peak (b), and focal plane views (c).

a resolved object has to be surrounded by non-objects. Such an approach is critically dependent on the technical set-up and was discussed previously [6] (Fig. 6).

The fluorescence imaging is interpreted much more simply than the diffractive imaging. Fluorescent molecules are of an infinitely small size (relative to a pixel) and we search for their distribution in a cell. Figure 4, *upper* shows objects of unchanged intensities between two z-levels which were selected upon three simple assumptions:

1. Each voxel contains one or none fluorophore which represents one intensity level.
2. When signals are present in numerous neighbouring pixels, these pixels can be assigned to an object. Unless signals in regions of lower density are of different intensity than the average, they belong to the background.
3. The voxels of different intensities than the average one are occupied by fluorophores of different spectral properties or there are more fluorophores in one voxel, they are considered relevant irrespective to the density of non-zero voxels in the neighbourhood. The detailed analysis showed that there are both (i) high intensity points in the centre of the dense areas – the nucleoli – which most likely represent several fluorophores per voxel and (ii) voxels of different intensities, mostly higher, outside dense regions, which most likely represent specific binding to objects other than nuclei, which is the primary target of the dye.

If an image is full of labelled objects of different intensities (Fig. 4, *lower*), it seems that any simple rule of data analysis cannot be applied. However, the assumption of zero point divergence gain significantly clarifies the dataset and enables a realistic 3D analysis of the observed structures.

5   Conclusion

In this article we have highlighted a few important technical aspects which can limit the complete yield of information from a microscopic image:

1. Usage of 12-bit colour depth of an image must not be sufficient, usage of a higher-bit depth can be necessary.
2. It is legitimate to observe bright spots in difractive images. These spots are not necessarily (auto)fluorescent objects.
3. The information can be localized with an infinite precision. It requests for microscopic cameras with a high number of pixels and usage of a higher magnification.

Since samples are usually expensive and often irreplaceable, maximum of information from optical microscopic experiment should be acquired and analyzed.

Acknowledgments. This work was supported by the Ministry of Education, Youth and Sports of the Czech Republic – projects CENAKVA (No. CZ.1.05/2.1.00/01.0024), CENAKVA II (No. LO1205 under the NPU I program), The CENAKVA Centre Development (No. CZ.1.05/2.1.00/19.0380) – and by the CZ-A AKTION programme.

References

1. Thorn, K.: A quick guide to light microscopy in cell biology. Mol. Biol. Cell **27**(2), 219–222 (2016)
2. Mie, G.: Beiträge zur Optik trüber Medien, speziell kolloidaler Metallösungen. Ann. Phys. (Berl.) **330**(3), 377–445 (1908)
3. Fresnel, A.J.: OEuvres Completes 1. Imprimerie impériale, Paris (1868)
4. Moerner, W.E.: Single-molecule spectroscopy, imaging, and photocontrol: foundations for super-resolution microscopy (nobel lecture). Angew. Chem. Int. Ed. **54**, 8067–8093 (2015)
5. Maxwell, J.C.: A dynamical theory of the electromagnetic field. Phil. Trans. Roy. Soc. Lond. **155**, 459–512 (1865)
6. Rychtáriková, R., Náhlík, T., Shi, K., Malakhova, D., Macháček, P., Smaha, R., Urban, J., Štys, D.: Super-resolved 3-D imaging of live cells organelles from bright-field photon transmission micrographs. (under the revision in Ultramicroscopy) http://arxiv.org/pdf/1608.05962.pdf
7. Rychtáriková, R., Náhlík, T., Smaha, R., Urban, J., Štys Jr., D., Císař, P., Štys, D.: Multifractality in imaging: application of information entropy for observation of inner dynamics inside of an unlabeled living cell in bright-field microscopy. In: Sanayei, A., Rössler, O.E., Zelinka, I. (eds.) ISCS 2014. ECC, pp. 261–267. Springer, Switzerland (2015)
8. ftp://160.217.215.251:21/InfoLimits (user: anonymous; password: anonymous)
9. Štys, D., Náhlík, T., Macháček, P., Rychtáriková, R., Saberioon, M.: Least Information loss (LIL) conversion of digital images and lessons learned for scientific image inspection. In: Ortuño, F., Rojas, I. (eds.) IWBBIO 2016. LNCS, vol. 9656, pp. 527–536. Springer, Cham (2016). doi:10.1007/978-3-319-31744-1_47
10. Císař, P., Náhlík, T., Rychtáriková, R., Macháček, P.: Visual exploration of principles of microscopic image intensities formation using image explorer software. In: Ortuño, F., Rojas, I. (eds.) IWBBIO 2016. LNBI, vol. 9656, pp. 537–544. Springer, Switzerland (2016)
11. Mizushima, Y.: Detectivity limit of very small objects by video-enhanced microscopy. Appl. Opt. **27**(12), 2587–2594 (1988)

12. Bayer, B.E.: Color Imaging Array. Patent US3971065 A (1975)
13. Urban, J., Afseth, N.K., Štys, D.: Fundamental definitions and confusions in mass spectrometry about mass assignment, centroiding and resolution. TrAC-Trend. Anal. Chem. **53**, 126–136 (2014)
14. Airy, G.B.: On the diffraction of an object-glass with circular aperture. Trans. Camb. Phil. Soc. **5**, 283–291 (1835)

Hands-Free EEG-Based Control of a Computer Interface Based on Online Detection of Clenching of Jaw

Mahta Khoshnam, Eunice Kuatsjah, Xin Zhang, and Carlo Menon[✉]

School of Engineering Science, Simon Fraser University, Burnaby, BC, Canada
{mkhoshna,ekuatsja,xzal10,cmenon}@sfu.ca

Abstract. This paper presents a method for characterizing electromyogram (EMG) contamination of electroencephalogram (EEG) signals caused by clenching of jaws. The goal of this study is to use volitional contraction of jaw muscles to generate command signals for triggering an external device. Frequency analysis of the EEG signal acquired from one electrode above the ear shows that clenching is demonstrated by high frequency components not observable during other jaw movements. Based on this analysis, a clenching index is defined and a threshold is chosen such that clenching index values above this threshold show the contraction of jaw muscles. The proposed algorithm is then paired with a virtual reality platform in which the user can navigate by sending commands through clenching her/his teeth. Our experimental results demonstrate that the sensitivity of the proposed method for online detection of clenching is 80% and its specificity is 88%.

Keywords: Computer interfaces · Jaw clenching · EEG signals · Power spectrum analysis

1 Introduction

Assistive technologies can help people with cognitive, sensory or motor impairments by extending their capabilities to interact with their surrounding environment and by assisting them in performing activities of daily living. Advancement in technology has enabled the introduction of a variety of assistive devices to improve quality of life for individuals with impairment, *e.g.*, by assisting with mobility and/or augmenting communication ability. However, how well a new assistive technology will be received by its targeted users depends on its functionality as well as its ease of use, among other factors [1]. To facilitate control and manipulation of assistive technologies, brain-computer interfaces (BCI) are used to translate user's neural signals into control commands for activation and manipulation of external devices [2]. For reliable function of such devices, it is important to identify the user's intentions from his/her brain signals and relate them to corresponding tasks or actions [3]. In this regard, we present a proof-of-concept study in which we use electroencephalogram (EEG) recording device to detect artifacts created by clenching of teeth for activation of an external

© Springer International Publishing AG 2017
I. Rojas and F. Ortuño (Eds.): IWBBIO 2017, Part I, LNBI 10208, pp. 497–507, 2017.
DOI: 10.1007/978-3-319-56148-6_44

platform. EEG signals can be easily and non-invasively collected, and are considered a practical modality for developing BCI technologies [3].

Brain-computer interfaces based on EEG signals use electrodes placed on the scalp to monitor brain activities and thus, provide a hands-free non-invasive technique for controlling assistive devices. Activating a software through detecting user's attention to a specific stimulus [4], steering a wheelchair by interpreting user's thoughts to generate commands for turning left or right [5, 6], and manipulating a robotic arm to perform reaching/grasping tasks through imagining different motor movements [7], are examples of different applications of EEG-based BCIs. The main principle in developing such BCIs is that when the user imagines a mental task associated with an intended movement, such an intention is reflected in the EEG signals and can be detected by implementing proper pattern recognition techniques [3]. Therefore, the performance of BCI systems depends on the selection of signal features as well as the implemented classification algorithm [8].

One of the challenges in designing EEG-based BCIs is electromyogram (EMG) and electrooculography (EOG) contamination, *i.e.*, EEG data are contaminated by artifacts generated from muscle or eye movements [9, 10]. Therefore, users are often asked to refrain from unnecessary movements while performing an experimental task, which limits the performance of EEG-based BCIs in practical applications [11]. Developing automatic techniques to detect and remove user-generated artifacts has been a recent topic of research in the field (*e.g.*, [12–14]). Such techniques generally involve pre-processing of EEG signals to remove artifacts caused by jaw movements, eye blinks, eye movements and similar muscle activities. However, when users are able to perform such muscle movements, these artifacts can be exploited to be part of the BCI [15, 16]. Moreover, since these artifacts are generated by muscle activities, they can be more easily detected and are well-suited for control purposes [9, 10].

In this paper, we focus on online detection of clenching of jaw from EEG signals for activating an external platform. Although jaw movements have generally been treated as undesired artifacts that should be removed to obtain "clean" EEG signals [13, 17], jaw clenching is characterized with distinctive features that can be used as an additional control input to a BCI. In [18], it was shown that tooth clicks, *i.e.*, brief contact between the upper and lower teeth, generate vibration signals that can be detected using accelerometers worn behind the ear. Such a tooth-clicking detection system was used along with a head tracking camera to enable people with severe upper limb paralysis to control a computer. Foldes and Taylor [19] analyzed EEG signal collected from one scalp surface electrode and implemented classification algorithms to distinguish different jaw activities, namely double-jaw-clench pattern, triple-jaw-clench pattern and non-patterned activity. It was shown that jaw muscle contractions are detectable from EEG data and can be used as discrete switch commands to activate assistive devices.

In this paper, we characterize the EEG artifacts generated by jaw movements with the aim of using jaw clenching as a trigger for assistive devices. Therefore, EMG contamination of EEG signals created by jaw muscle contraction is the part of the biosignal that is of interest. Considering that jaw clenching introduces high frequency vibrations into the EEG signals [19], in this paper, we perform a power spectrum analysis on the signal collected from one scalp surface electrode to detect high-frequency components of the signal. The basic idea is close to that implemented in a different context to determine leg and knee tremors from foot acceleration signals [20]. Power spectrum analysis shows that artifacts contaminating EEG signals during different jaw muscle activities have different frequency components. Moreover, during jaw clenching, the majority of power occurs at higher frequencies, *i.e.*, at frequencies higher than a threshold frequency (f_t). Based on this analysis, a clenching index (CI) is defined as the power in the high-frequency band divided by the power in all frequency ranges. At each time interval, if CI is greater than a predefined clenching threshold (CT), jaw clenching is detected. Since each individual has a different clenching pattern, f_t and CT are determined for each person by collecting and analyzing a sample of his/her EEG data. Furthermore, the effect of sampling window size and the amount of overlap between signal windows on the performance of the proposed algorithm are studied. To demonstrate the potential use of the suggested method for activating assistive devices, the algorithm is integrated within a platform which acquires and analyzes EEG signals in real-time to detect jaw clenching to navigate in a virtual reality environment. This platform was tested on five able-bodied participants and experimental results show that the sensitivity and specificity of the proposed algorithm in online detection of jaw clenching are 80% and 88%, respectively, and that it is possible to control a virtual environment using jaw muscle contractions. The system is capable of detecting clenching periods as short as 1.5 s and the detection delay is 1.5 s. These results show that the proposed system can be potentially used for control of assistive devices.

2 Methods

The power spectrum analysis technique proposed in this paper is first developed and evaluated using pre-acquired EEG signals. Parameters required for online processing of signals are also derived. In this section, each of these steps is explained in details.

2.1 Data Acquisition

Continuous EEG was recorded at 128 Hz using a wireless EEG headset (Emotiv Epoc headset, Emotiv, San Francisco, CA) [21] from five able-bodied participants. Two electrodes were placed at temporal sites, T7 and T8, according to the international 10–20 system [22] and signals were referenced to the left and right mastoids. It has been previously shown that the effect of jaw clenching can be observed in EEG signals collected at these electrode locations [19]. Each participant was asked to perform a series of jaw movements, including talking, chewing and clenching, at given cues and to continue the same activity for at least 5 s. The order of activities was random and a zero to 5-second pause was administered between two consecutive cues. Participants were asked not to shake their heads vigorously during the data collection process, but eye movements, such as blinking, were allowed. Total data collection time in each trial was between 20 and 40 s. Two sets of data were acquired from each participant and collected EEG signals were saved for further analysis.

2.2   Power Spectrum Analysis

Figure 1 shows a sample of collected EEG signals and their spectrograms, *i.e.,* variation of frequency components with time. The acquired signal is divided into one-second windows with 50% overlap and a 256-point fast Fourier transform (FFT) is used for calculating the spectrogram. Different jaw movements are also labeled on Fig. 1. It can be observed that jaw clenching is characterized by high frequency components such that the majority of power occurs at frequencies higher than a threshold frequency (f_t). For the sample signal shown in Fig. 1, this threshold frequency is $f_t = 20$ Hz. Hereinafter, the frequency band containing the frequencies higher that f_t is called "clenching band". According to the spectrogram, resting periods of jaw (periods of no activity) mainly contain lower frequency components, while the power during talking and chewing intervals is distributed within the entire frequency range. These characteristics are used to distinguish clenching from other jaw activities: during jaw clenching, the majority of power occurs in the "clenching band".

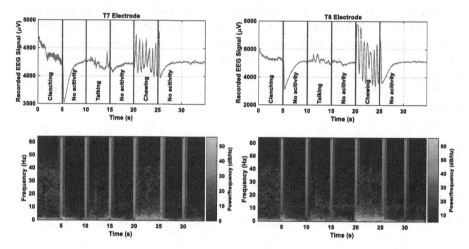

Fig. 1. A sample of collected data: (a) EEG signal collected from T7 and its corresponding spectrogram, (b) EEG signal collected from T7 and its corresponding spectrogram. Different jaw activities are labeled on top panels.

For each collected EEG signal, a sampling window is used to obtain a segment of the signal. The length of the sampling window is l_w and the window is centered at time points, t_{cw}, located at equal time intervals. For each segment, power spectral density (PSD) is calculated using fast Fourier transform (FFT). The clenching index (CI) is then defined as the ratio between the square of the area under the power spectrum of the signal segment in the "clenching band" and the square of the area under the power spectrum of the signal segment in the entire frequency range:

$$CI = \left(\frac{\text{area under the power spectrum curve in the "clenching band"}}{\text{area under the power spectrum curve in the entire frequency range}} \right)^2 \quad (1)$$

This index shows the ratio of the power corresponding to higher frequency components and can distinguish clenching from other jaw activities. If CI is greater than a predefined clenching threshold (CT), then jaw status in the analyzed signal segment will be determined as clenching. Figure 2 illustrates the process of clenching detection for a sample of EEG signal collected at T7 electrode.

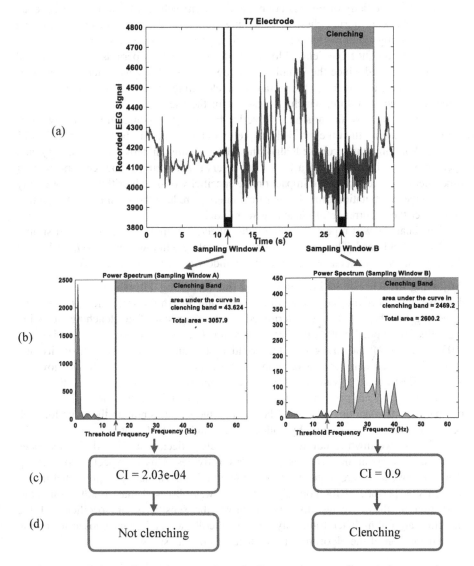

Fig. 2. (a) EEG data collected from T7 electrode. Two random sampling windows are chosen for demonstration, (b) Power spectrum is calculated for each sampling window. The area under the power curve in the clenching band as well as the total area under the power curve are calculated, (c) Clenching index (CI) is obtained based on Eq. (1), (d) When CI is greater than the clenching threshold, clenching is detected. In this example, CT is 0.4.

Effect of Window Length and Window Centers

The effect of the length of the sampling window on the performance of the clenching detection algorithm was studied by varying the window length: $l_w = 0.5, 1, 1.5, 2, 5$ s. The time interval between two consecutive window centers was 0.5 s. Applying the clenching detection algorithm showed that best results for sensitivity and specificity are obtained with choosing $l_w = 1$ s. Shorter sampling window lengths, such as $l_w = 0.5$ s, decrease the specificity of the detection algorithm. In such cases, short high-frequency intervals that occur during chewing or talking are falsely detected as clenching. Using longer sampling window lengths ($l_w = 1.5$, 2, 5 s), the start time of clenching is detected with a noticeable delay. Moreover, shorter clenching periods are not detected if a long l_w is used, since the sampling window acts as a low-pass filter. This implies that with using a longer l_w, the duration of clenching should also be longer to be detected by the algorithm, which is tiresome for the user.

Setting $l_w = 0.5$ s, the amount of overlap between two consecutive windows was varied by changing the time interval between window centers to be 0.1, 0.25, 0.5, 0.75 or 1 s. Applying the proposed algorithm, best results in terms of sensitivity and specificity are obtained using a 0.5 s time interval between window centers. Shorter time intervals have a negative impact on the specificity of the algorithm and noticeably increase the computation time. Longer time intervals reduce the sensitivity of detection and detect the occurrence of clenching with a delay.

This analysis was repeated for the data collected from all participants and similar results were obtained. Thus, the length of the sampling window was set to $l_w = 1$ s and the time interval between two consecutive window centers was set to 0.5 s.

Adjusting the Threshold Frequency and the Clenching Threshold

To take into account individual differences, the threshold frequency, f_t, and the clenching threshold, CT, are determined for each person. The clenching detection algorithm is applied with different values for f_t and CT (5 Hz $< f_t <$ 40 Hz and 0.05 $<$ CT $<$ 0.4) and the values that yield best results in terms of sensitivity and specificity are chosen for that participant. A sample of obtained results is shown in Table 1. Best threshold values for each participant as well as obtained sensitivity and specificity results are given in Table 2. Although f_t and CT values are different from one participant to another, our tests showed that these values remain the same for each person and can be determined by collecting a sample of EEG data.

It is also worthwhile noting that EEG signals collected at T7 and T8 electrodes demonstrate very similar features during jaw activities. This is expected considering that jaw movements are assumed to be symmetric [19]. The analysis presented in this section was performed for both signals and it was observed that simultaneous consideration of these signals does not improve the sensitivity or specificity of the detection algorithm. Therefore, only one signal collected at T7 was chosen for online implementation of the algorithm in the next step of this study.

Table 1. Results of applying the clenching detection algorithm with different values for the threshold frequency (f_t) and the clenching threshold (CT) for sample EEG data collected from Participant 2: (a) sensitivity results, (b) specificity results. For this participant, these values were chosen to be $f_t = 15$ Hz and CT = 0.4.

(a) Sensitivity %								
CT→ f_t (HZ)↓	0.05	0.1	0.15	0.2	0.25	0.3	0.35	0.4
5	100	100	100	100	100	100	100	100
10	100	100	100	100	100	100	100	100
15	100	100	100	100	100	100	100	91.7
20	100	100	100	100	100	83.3	75	66.7
25	100	100	100	83.3	75	58.3	50	33.3
30	100	75	50	16.7	0	0	0	0
35	8.33	0	0	0	0	0	0	0
40	0	0	0	0	0	0	0	0

(b) Specificity %								
CT→ f_t (HZ)↓	0.05	0.1	0.15	0.2	0.25	0.3	0.35	0.4
5	62	68.9	77.6	77.6	81	87.9	89.7	91.4
10	70.7	77.6	79.3	81	86.2	89.7	91.4	93.1
15	75.9	77.6	82.8	84.5	89.6	93.1	94.8	96.6
20	77.6	81	84.5	86.2	93.1	94.8	96.6	96.6
25	79.3	84.5	86.2	91.4	96.6	96.6	96.6	96.6
30	84.5	91.4	96.5	96.6	98.3	100	100	100
35	100	100	100	100	100	100	100	100
40	100	100	100	100	100	100	100	100

Table 2. Threshold frequency (f_t), clenching threshold (CT) and resulting sensitivity and specificity obtained for different participants.

	f_t	CT	Sensitivity %	Specificity %
Participant 1	10	0.05	100	98.7
Participant 2	15	0.4	91.7	96.6
Participant 3	20	0.4	100	98.1
Participant 4	15	0.25	90.9	100
Participant 5	15	0.15	95.2	95.3

3 Experimental Results

To assess the performance of the proposed technique for potential use with assistive technologies, the system was paired with a virtual reality (VR) platform in which the user can move forward using clenching cues. To this end, the algorithm described in the previous section should be implemented online. In this section, the experimental platform and obtained results are presented in details.

Real-Time Data Acquisition and Processing
In this phase of the study, EEG data were recorded at 250 Hz using a wireless EEG acquisition device (BioRadio, Great Lakes NeuroTechnologies, Valley View, OH) [23] from five able-bodied participants. One electrode was placed at T7. Fpz was used for ground and the signals were referenced to the left mastoid. The data streaming from BioRadio were windowed in steps of 125 samples (0.5 s) by one-second windows (250 samples). Each window segment was band-pass filtered and fed into the clenching detection algorithm which calculated the power spectrum of the signal segment and subsequently, determined the clenching index (CI) for that window. If the calculated CI was greater than the pre-defined clenching threshold (CT) for that participant, the algorithm reported the occurrence of clenching by displaying a message on the screen. The code required for receiving EEG data from BioRadio, the clenching detection algorithm and the Graphical User Interface (GUI) were implemented in MATLAB. While the system is running, MATLAB collects data every 80 ms to minimize system overhead.

Virtual Reality Platform
The virtual reality (VR) platform uses Unreal Engine 4 game engine [24] obtained from [25] to create a virtual model of a house in which the user can walk around and also interact with the environment through actions such as opening/closing doors and turning off/on lamps. A screenshot of this environment is shown in Fig. 3. Micro-controllers (Arduino Uno and Micro) are used to translate the result of the detection algorithm into actions in the virtual environment, such that when clenching is detected, a "move forward" command is sent to the VR platform.

Fig. 3. A screenshot of the virtual reality environment.

Experimental Task and Results

To determine the threshold frequency and the clenching threshold, each participant was instructed to clench her/his teeth for three times, each time for at least two seconds at given cues. A rest period of approximately five seconds was administered between two consecutive cues. The detection algorithm was then updated with f_t and CT values determined from this test. In the next step, the VR platform was activated and participants were instructed to move forward in the virtual environment using clenching cues at their own pace. Users were allowed to talk and slightly move their head during the experiments. Each participant completed two series of trials and, in each trial, performed clenching 10 times. The average time for each trial was approximately two minutes. Based on the results obtained from 100 clenching cues from five participants, in this experiment, the sensitivity of the clenching detection algorithm was 80% and its specificity was 88%. Moreover, it was observed that the proposed algorithm detects jaw clenching when its duration is longer than 1.5 s, but can miss shorter clenching intervals. The average delay between the user clenching and detection of its occurrence was approximately 1.5 s. Table 3 shows sources of delay in the system. These delay times were calculated by adding timers in the MATLAB code.

Table 3. Sources of delay in detection of clenching

Process	Delay (s)
Windows refresh rate	0.5
Filtering each window segment	0.02
Processing each window segment	0.1
Buffer check rate	0.08
Communication delay	0.8
Total	1.5

4   Discussion

The study presented in this paper suggests detection of EEG artifacts generated by muscle movements as an additional signal for controlling assistive devices. In this regard, we characterized the effect of various jaw movements on the EEG data collected from one scalp surface electrode located above the ear. The suggested placement for electrodes is non-invasive and allows for designing a compact EEG acquisition device in alignment with recent technological advancements [26]. Moreover, since only one electrode is used, the system can be quickly setup and the computational effort is minimal. It is shown that tooth clicks do not impose excessive fatigue on the user [18], which is an important factor in the design of assistive devices [1]. Participants in this study quickly became adept at controlling the virtual platform and reported its usability, but did not express fatigue or difficulty resulting from clenching their jaw. However, they reported that the delay in detecting the clenching is noticeable and adversely affects the intuitiveness of navigation in the virtual environment. According to Table 3, the major cause of delay is the communication delay between the EEG acquisition device and the computer. Utilizing more advanced communication systems can effectively reduce the observed delay. As mentioned in the previous section, decreasing the time interval between two consecutive window centers, *i.e.*, windows refresh rate, increases the computational duration. Therefore, a trade-off should be made to optimize the performance.

This study was performed with participation of able-bodied adults. Further study is required to evaluate the performance and usability of the proposed system for people with impairment. Nevertheless, this study shows the potential of using EEG artifacts generated from jaw clenching as an additional input for controlling devices.

5 Conclusions

In this paper, we presented a technique for detecting artifacts created by clenching of teeth in EEG signals in order to activate an external platform, namely a virtual reality environment. Our analysis showed that different jaw movements, such as chewing, talking and clenching, are characterized by different frequency components in the EEG signals and that jaw clenching can be accurately distinguished by analyzing the power spectrum of the signal. To show the potential of the proposed technique for control of assistive devices, the suggested algorithm was paired with a virtual reality environment to form a platform in which user can navigate in the virtual environment by clenching their teeth. The system was evaluated with healthy participants and it was shown that the proposed algorithm can detect clenching and activate the VR platform with good sensitivity and specificity. Future steps in this work include reducing the detection delay and evaluating the reliability and usability of the system for individuals with impairment.

References

1. Rupp, R., Kleih, S.C., Leeb, R., R. Millan, J., Kübler, A., Müller-Putz, G.R.: Brain–computer interfaces and assistive technology. In: Grübler, G., Hildt, E. (eds.) Brain-Computer-Interfaces in Their Ethical, Social and Cultural Contexts. TILELT, vol. 12, pp. 7–38. Springer, Dordrecht (2014). doi:10.1007/978-94-017-8996-7_2
2. Collinger, J., Boninger, M., Bruns, T., Curley, K., Wang, W., Weber, D.: Functional priorities, assistive technology, and brain-computer interfaces after spinal cord injury. J. Rehabil. Res. Dev. 50(2), 145–160 (2013)
3. Millán, J.D.R., Rupp, R., Müller-Putz, G.R., Murray-Smith, R., Giugliemma, C., Tangermann, M., Vidaurre, C., Cincotti, F., Kübler, A., Leeb, R.: Combining brain–computer interfaces and assistive technologies: state-of-the-art and challenges. Front. Neurosci. 4(161) (2010)
4. Zickler, C., Riccio, A., Leotta, F., Hillian-Tress, S., Halder, S., Holz, E., Staiger-Sälzer, P., Hoogerwerf, E.J., Desideri, L., Mattia, D., Kübler, A.: A brain-computer interface as input channel for a standard assistive technology software. Clin. EEG Neurosci. 42(4), 236–244 (2011)

5. Tanaka, K., Matsunaga, K., Wang, H.O.: Electroencephalogram-based control of an electric wheelchair. IEEE Trans. Robot. **21**(4), 762–766 (2005)
6. Li, J., Liang, J., Zhao, Q., Li, J., Hong, K., Zhang, L.: Design of assistive wheelchair system directly steered by human thoughts. Int. J. Neural Syst. **23**(03), 1350013 (2013)
7. Onose, G., Grozea, C., Anghelescu, A., Daia, C., Sinescu, C.J., Ciurea, A.V., Spircu, T., Mirea, A., Andone, I., Spânu, A., Popescu, C.: On the feasibility of using motor imagery EEG-based brain–computer interface in chronic tetraplegics for assistive robotic arm control: a clinical test and long-term post-trial follow-up. Spinal Cord **50**(8), 599–608 (2012)
8. Lotte, F., Congedo, M., Lécuyer, A., Lamarche, F., Arnaldi, B.: A review of classification algorithms for EEG-based brain–computer interfaces. J. Neural Eng. **4**(2), R1–R13 (2007)
9. Goncharova, I.I., McFarland, D.J., Vaughan, T.M., Wolpaw, J.R.: EMG contamination of EEG: spectral and topographical characteristics. Clin. Neurophysiol. **114**(9), 1580–1593 (2003)
10. Fatourechi, M., Bashashati, A., Ward, R.K., Birch, G.E.: EMG and EOG artifacts in brain computer interface systems: a survey. Clin. Neurophysiol. **118**(3), 480–494 (2007)
11. Felzer, T., Freisleben, B.: HaWCoS: the hands-free wheelchair control system. In: Proceedings of the Fifth International ACM Conference on Assistive Technologies, pp. 127–134 (2002)
12. Mognon, A., Jovicich, J., Bruzzone, L., Buiatti, M.: ADJUST: an automatic EEG artifact detector based on the joint use of spatial and temporal features. Psychophysiology **48**(2), 229–240 (2011)
13. Lawhern, V., Hairston, W.D., McDowell, K., Westerfield, M., Robbins, K.: Detection and classification of subject-generated artifacts in EEG signals using autoregressive models. J. Neurosci. Methods **208**(2), 181–189 (2012)
14. Mammone, N., La Foresta, F., Morabito, F.C.: Automatic artifact rejection from multichannel scalp EEG by wavelet ICA. IEEE Sens. J. **12**(3), 533–542 (2012)
15. Wolpaw, J.R., Birbaumer, N., McFarland, D.J., Pfurtscheller, G., Vaughan, T.M.: Brain–computer interfaces for communication and control. Clin. Neurophysiol. **113**(6), 767–791 (2002)
16. Dewan, E.M.: Occipital alpha rhythm eye position and lens accommodation. Nature **214**, 975–977 (1967)
17. McMenamin, B.W., Shackman, A.J., Maxwell, J.S., Bachhuber, D.R., Koppenhaver, A.M., Greischar, L.L., Davidson, R.J.: Validation of ICA-based myogenic artifact correction for scalp and source-localized EEG. Neuroimage **49**(3), 2416–2432 (2010)
18. Simpson, T., Broughton, C., Gauthier, M.J., Prochazka, A.: Tooth-click control of a hands-free computer interface. IEEE Trans. Biomed. Eng. **55**(8), 2050–2056 (2008)
19. Foldes, S.T., Taylor, D.M.: Discreet discrete commands for assistive and neuroprosthetic devices. IEEE Trans. Neural Syst. Rehabil. Eng. **18**(3), 236–244 (2010)
20. Moore, S.T., Yungher, D.A., Morris, T.R., Dilda, V., MacDougall, H.G., Shine, J.M., Naismith, S.L., Lewis, S.J.: Autonomous identification of freezing of gait in Parkinson's disease from lower-body segmental accelerometry. J. Neuroeng. Rehabil. **10**(19), 1 (2013)
21. Emotiv Epoc. www.emotiv.com/epoc
22. Aspinall, P., Mavros, P., Coyne, R., Roe, J.: The urban brain: analysing outdoor physical activity with mobile EEG. Br. J. Sports Med. **49**(4), 272–276 (2015)
23. Great Lakes NeuroTechnologies. www.glneurotech.com
24. Unreal Engine. https://www.unrealengine.com/
25. Downtown Visuals. www.downtownvisuals.com/ue4-virtual-tour/
26. In-Ear EEG. http://spectrum.ieee.org/the-human-os/biomedical/devices/in-ear-eeg-makes-unobtrusive-brain-hacking-gadgets-a-real-possibility

Wavelet Decomposition Based Automatic Sleep Stage Classification Using EEG

Nieves Crasto[1] and Richa Upadhyay[2]([⊠])

[1] Universite Grenoble Alpes, Grenoble, France
nievescrasto@gmail.com
[2] NMIMS's Mukesh Patel School of Technology Management and Engineering,
Mumbai, India
richa.upadhyay@nmims.edu

Abstract. The diagnosis of sleep related disorders like sleep apnea, insomnia, restless legs syndrome, begins with grading sleep into stages to analyze the problem. The R&K rules recommend dividing the polysomnographic record of sleep consisting of EEG, EOG and EMG into 30 s epochs and classifying them as Stage 1, 2, 3, 4, Rapid Eye Movement (REM) and Wake state. In this paper, data from a single EEG electrode are decomposed into its wavelet coefficients (Daubechies wavelet from 2 to 6). Instead of using statistical parameters like entropy, energy, etc. of the coefficients as features, the coefficients are directly applied as input to a neural network for classification. Prior to training the neural network, the high dimensional input data are reduced to its principal components. The proposed method helps in isolating Stage 3 and 4, rather than identifying them as a combined Slow Wave Stage (SWS). Best results were obtained using Daubechies 2 wavelet, with an overall accuracy of 86%.

Keywords: EEG · Wavelet · Artificial neural networks · Principal component analysis

1 Introduction

Sleep is a state of natural rest during which the brain is relatively more responsive to internal than external stimuli. Electroencephalography (EEG) signals are classified into five types of brain waves, distinguished on the basis of frequency ranges. These frequency bands from low to high frequencies are called delta (δ), theta (θ), alpha (α), beta (β) and gamma (γ). Table 1 elaborates various brain waves [1].

The interplay of brain circuits during sleep results in Non Rapid Eye Movement (NREM) and Rapid Eye Movement (REM) sleep. NREM is further subdivided into four stages of 1 (drowsiness), 2 (light sleep), 3 (deep sleep) and 4 (very deep sleep). At night in the course of sleep stages 1, 2, 3 and 4 are followed by REM sleep, and a complete cycle from stage 1 to REM takes about one and half hour. A person spends 50% of total sleep time in stage 2 sleep, about 20% in REM sleep, and the remaining 30% in the other stages [2]. NREM sleep

© Springer International Publishing AG 2017
I. Rojas and F. Ortuño (Eds.): IWBBIO 2017, Part I, LNBI 10208, pp. 508–516, 2017.
DOI: 10.1007/978-3-319-56148-6_45

Table 1. Details of brain waves

Wave	Frequency range	Usually associated with
Delta	0.5 to 4 Hz	Deep dreamless sleep
Theta	4 to 8 Hz	Dreams, deep meditation, hypnosis
Alpha	8 to 13 Hz	Relaxed wakefulness
Beta	13 to 22 Hz	Attentiveness, concentration & anticipation
Gamma	22 to 30 Hz	Processing of various stimuli (visual, auditory, etc.)

restores the physiological functions, while REM restores the mental functions. Therefore sleep staging is a significant part of process assessing sleep disorders such as Sleep Apnea Syndrome. The scoring of different sleep stages is done on the basis of Rechtschaffen and Kales standard (R&K) [3]. The process of human scorers manually marking sleep stages by calculating the probabilities of occurrence of R&K events is a tedious job. Thus a number of papers [4] deals with developing a system to classify sleep stages using data from a single EEG electrode.

1.1 Recent Work

Most of the works on automatic sleep stage identification were carried out using time-frequency analysis, mainly wavelet transform [8,9] and different classification procedures like neural networks (NN) [10] or support vector machine (SVM) [5]. Fraiwan et al. [4] in their work achieved a success rate of 83% in classifying six stages using time frequency analysis and entropy measures for feature extraction. Ebrahimi et al. [2] achieved accuracy of 93% to classify four (Awake, Stage1+REM, Stage 2 and Slow Wave Stage) sleep stages using wavelet packet coefficients and neural networks. Similarly, Koley et al. [5] used time domain, frequency domain and non-linear analysis for feature extraction and SVM recursive feature elimination (RFE) technique to find the optimum number of feature subset, which provided 85% classification performance for five different sleep stages. Also, Tagluk et al. [6] used EEG, electromyography (EMG) and electrooculogram (EOG) data, trained a feed forward neural network and achieved a accuracy of 74.7% to classify five sleep stages. 95.55% classification rate was achieved by Sinha et al. [7] in a study which employed the use of wavelet transform and NN to classify three stages of (a) sleep spindles (b) REM sleep and (c) awake stages. Additionally, sleep is mostly classified by combining Stage 3 and Stage 4 in a single stage called Slow Wave Sleep (SWS). In this paper, principal components of the wavelet coefficients are used as features instead of the coefficient's energy and entropy at each decomposition level [2], to isolate the Stage 3 and Stage 4.

The EEG sleep data from Physionet's Sleep-EDF [Expanded] Database has been used in this study. The files are whole-night polysmnographic (PSG) sleep recordings of Caucasian males and female subjects, containing EEG (from Fpz-Cz and Pz-Oz electrode locations) sampled at 100 Hz [11,12]. Data from Fpz-Cz

Table 2. Number of epochs in the dataset

Stage	S1	S2	S3	S4	REM	Wake	Total epoch
Epoch	2583	16164	3140	2253	7018	3487	34645

electrode is considered for sleep stage classification. Epochs of 30 s duration are used to score sleep stages as Wake, REM, Stage 1 (S1), Stage 2 (S2), Stage 3 (S3) and Stage 4 (S4) (Table 2). The EEG signal contains artifacts in the form of EOG signal which are separated out using Independent Component Analysis (ICA). The mixture of EEG and EOG signals, $\mathbf{X}$ is considered as a linear combination of pure EEG and EOG signal, $\mathbf{S}$ [26].

$$\mathbf{X} = \mathbf{AS} \tag{1}$$

where

$$\mathbf{X} = \begin{bmatrix} x_1(n) \\ x_2(n) \end{bmatrix} \qquad \mathbf{S} = \begin{bmatrix} s_1(n) \\ s_2(n) \end{bmatrix}$$

EEG and EOG are assumed to be statistically independent as they are generated by different sources [28]. Therefore they can be recovered from linear mixture $\mathbf{X}$ by finding a transformation in which the transformed signals are as independent as possible. Thus aim is to estimate the unmixing matrix $\mathbf{Z}$ which, in turn enables the estimate of the signal sources $\mathbf{U}$ to be obtained by

$$\mathbf{U} = \mathbf{ZX} \tag{2}$$

The matrix $\mathbf{Z}$ is determined so that the mutual information of the transformed components $\mathbf{U}$ is minimized. The FastICA algorithm has been chosen because the rate at which the error function (mutual information) convergences to a minimum is least quadratic compared to linear convergence rate of stochastic gradient descent methods [27].

2   Feature Extraction

2.1   Wavelet Decomposition

The wavelet is a smooth and quickly vanishing oscillating mathematical function with good localization both in frequency and time. A wavelet family $\psi_{a,b}(t)$ is a set of elementary function generated by dilations and translations of a unique admissible mother wavelet $\psi(t)$

$$\psi_{a,b}(t) = \frac{1}{\sqrt{|a|}} \psi\left(\frac{t-b}{a}\right) \tag{3}$$

where $a, b \epsilon R$, $a \neq 0$, a, b are the scaling (dilation) and translation parameters, respectively [9]. Any general function can be represented as a linear combination

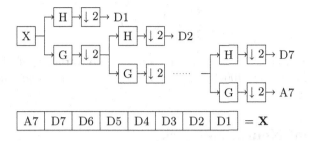

Fig. 1. Wavelet decomposition upto 7 levels

of wavelet functions obtained from the mother wavelet. The decomposition of the signal leads to a set of coefficients called wavelet coefficients (Fig. 1). At each stage in decomposition two sets of coefficients are produced, the approximation coefficients and the detail coefficients. The approximation coefficients are generated by convolving the input signal with a low pass filter, G and down-sampling the results by a factor of two. The detail coefficients are similarly generated, by convolving the input with a high pass filter, H and down-sampling by a factor of two. The approximation coefficients are then treated as new inputs and the above operation is repeated recursively as a binary tree algorithm till the final stage of decomposition is reached. Instead of finding the energy of coefficients at each decomposition level [2], the actual approximation coefficients at final level (A7) along with the detail coefficients at every level (D7, D6, D5, D4, D3, D2, D1) are selected as input vectors to the neural network. Daubechies wavelet from 2 to 6 are considered, and the best accuracy was achieved using Daubechies 2 wavelet.

2.2   Principal Component Analysis

If all the wavelet coefficients are considered as features (approx 3000), training the neural network with such very high dimensional dataset would be time consuming. Dimensionality reduction facilitates the classification, visualization, communication, and storage of high-dimensional data. Principal components analysis (PCA) is used as a method to reduce the data dimensions by finding the directions of greatest variance in the data set and represent each data point by its coordinates along each of these directions [24]. Each row of the training matrix $\mathbf{X}_{m \times n}$ represent a point in an n dimensional Euclidean space. If $\mathbf{E}_{n \times n}$ be the matrix whose columns are the eigenvectors of $\mathbf{X}^T \mathbf{X}$, ordered as largest eigenvalue first, then $\mathbf{XE}$ gives the points of $\mathbf{X}$ transformed into a new coordinate space. In this space, the initial columns which correspond to largest eigenvalues (λ) capture most of the variability of $\mathbf{X}$. Therefore choosing eigenvectors associated with the largest eigenvalues and ignoring the rest, $\mathbf{E}'_{n \times k}$ is multiplied with $\mathbf{X}_{m \times n}$ to transform it to a space with fewer dimensions, $\mathbf{X}'_{m \times k}$ [13,18]. To choose the k features out of the n features, where $k < n$ we use the following condition [18].

$$\frac{\sum_{i=1}^{k} \lambda_i^2}{\sum_{i=1}^{n} \lambda_i^2} \geq 0.9 \tag{4}$$

400 features are chosen which retain 95% of the variance of the data set. Considering more than 400 features did not provide significant improvement in the accuracy.

3   Artificial Neural Network

Gradient-based learning algorithms compute the hypothesis $\mathbf{h}_i = f(\mathbf{x}_i, \mathbf{W})$ where $\mathbf{x}_i$ represents the ith input feature vector and $\mathbf{W}$ represent the adjustable weight matrix [25]. The feed forward neural network has 3 layers: input layer (400 neurons), hidden layer (50 neurons) and output layer (6 neurons for S1-S4, REM and Wake). Based on empirical analysis, 50 neurons in the hidden layer resulted in optimum results. Since the average firing frequency of biological neurons as a function of excitation follows a sigmoidal characteristic [20], the activation function f of layer 2 (hidden layer) by the layer 1 (input layer), for the ith input, $\mathbf{h}_i^{(2)}$ is chosen as (5).

$$\mathbf{h}_i^{(2)} = f(\mathbf{x}_i, \mathbf{W}^{(1)}) = \mathbf{1} \oslash \{\mathbf{1} + exp^{-(\mathbf{W}^{(1)^T}\mathbf{x}_i + \mathbf{b}_0)}\} \tag{5}$$

where $\mathbf{W}^{(1)} \in \mathbb{R}^{400 \times 50}$ is the weight matrix controlling function mapping from layer 1 to 2, $\mathbf{x}_i \in \mathbb{R}^{400 \times 1}$ is the ith input vector, $\mathbf{b}_0 \in \mathbb{R}^{50 \times 1}$ is the bias vector, $\mathbf{1} \in \mathbb{R}^{50 \times 1}$ is the ones vector and $\oslash$ signifies Hadamard division. Similarly, the activation function of layer 3 (output layer) by layer 2 for the ith input, $\mathbf{h}_i^{(3)}$ is

$$\mathbf{h}_i^{(3)} = f(\mathbf{x}_i, \mathbf{W}^{(2)}) = \mathbf{1} \oslash \{\mathbf{1} + exp^{-(\mathbf{W}^{(2)^T}\mathbf{h}_i^{(2)} + \mathbf{b}_0)}\} \tag{6}$$

where $\mathbf{W}^{(2)} \in \mathbb{R}^{50 \times 6}$ is the weight matrix controlling function mapping from layer 2 to 3, $\mathbf{b}_0 \in \mathbb{R}^{6 \times 1}$ is the bias vector and $\mathbf{1} \in \mathbb{R}^{6 \times 1}$ is the ones vector. The error function $J(\mathbf{W}^{(1)}, \mathbf{W}^{(2)})$ measures the difference between the desired output of the ith example, $\mathbf{y}_i \in \mathbb{R}^{6 \times 1}$ and the hypothesis, $\mathbf{h}_i^{(3)} \in \mathbb{R}^{6 \times 1}$ averaged over all the training examples, m [25]. The regularization parameter, λ is used to prevent the neural network from overfitting the data.

$$J(\mathbf{W}^{(1)}, \mathbf{W}^{(2)}) = -\frac{1}{m} \sum_{i=1}^{m} \left[(1 - \mathbf{y}_i^T) \log(1 - \mathbf{h}_i^{(3)}) + \mathbf{y}_i^T \log \mathbf{h}_i^{(3)} \right] +$$

$$\frac{\lambda}{2m} \sum_{i=1}^{2} tr(\mathbf{W}^{(i)^T}\mathbf{W}^{(i)}) \tag{7}$$

where tr stands for trace of the matrix.

The backpropagation algorithm has been used to minimize the error function in weight space using the method of conjugate gradients instead of the gradient descent algorithm. To avoid the issues of determining the optimum step size and the error function reaching a local minima using gradient descent, the Polack-Ribiere flavour of conjugate gradients is used to compute search directions, and a line search using quadratic and cubic polynomial approximations. The Wolfe-Powell stopping criteria is used together with the slope ratio method for guessing initial step sizes [14].

4   Results and Conclusion

The dataset is divided into training (60% of total number of epochs), cross-validation (20% of total number of epochs) and test sets (20% of total number of epochs). Random sub-sampling cross-validation is used to fine tune the neural network's regularization parameter, λ to prevent overfitting. Once the optimum value for λ is obtained, the network is trained using the training plus cross-validation examples (80% of total number of epochs) and then tested of the remaining 20% of total number of epochs. The performance of the proposed method is evaluated using sensitivity (SE), specificity (SP) and accuracy (AC) of each stage and is calculated, as follows [4]:

$$SE = \frac{TP}{TP + FN} \tag{8}$$

$$SP = \frac{TN}{TN + FP} \tag{9}$$

$$AC = \frac{TN + TP}{TN + TP + FN + FP} \tag{10}$$

$$\text{Total AC} = \frac{\sum_{i=1}^{6} TP_i}{\text{Total no. of epochs}} \tag{11}$$

where TP is the number of true positives, TN is the number of true negatives, FP is the number of false positives, and FN is the number of false negatives.

Table 3. Confusion matrix of the epochs in the dataset (each row signifies the true class)

	S1	S2	S3	S4	REM	Wake
S1	268	57	0	0	10	55
S2	138	2999	140	5	45	50
S3	2	108	403	83	1	4
S4	3	8	71	355	0	4
REM	45	58	0	0	1333	12
Wake	56	11	1	1	3	600

Table 4. Sensitivity using Daubechies wavelets 2 to 6 for the dataset

Wavelet	Db 2	Db 3	Db 4	Db 5	Db 6
S1	0.523	0.523	0.494	0.523	0.500
S2	0.925	0.919	0.925	0.925	0.931
S3	0.655	0.654	0.639	0.631	0.652
S4	0.800	0.791	0.784	0.809	0.802
REM	0.958	0.960	0.960	0.962	0.958
Wake	0.828	0.812	0.822	0.819	0.808

Table 5. Specificity using Daubechies wavelets 2 to 6 for the dataset

Wavelet	Db 2	Db 3	Db 4	Db 5	Db 6
S1	0.981	0.977	0.980	0.980	0.979
S2	0.898	0.901	0.892	0.899	0.892
S3	0.969	0.966	0.968	0.969	0.970
S4	0.987	0.986	0.986	0.987	0.988
REM	0.979	0.980	0.980	0.979	0.980
Wake	0.988	0.988	0.988	0.989	0.989

Table 6. Accuracy using Daubechies wavelets 2 to 6 for the dataset

Wavelet	Db 2	Db 3	Db 4	Db 5	Db 6
S1	0.947	0.943	0.944	0.946	0.944
S2	0.910	0.909	0.907	0.910	0.909
S3	0.940	0.938	0.938	0.936	0.941
S4	0.974	0.973	0.973	0.975	0.976
REM	0.974	0.976	0.975	0.975	0.975
Wake	0.971	0.969	0.970	0.971	0.970
Total	**0.860**	**0.855**	**0.855**	**0.857**	**0.858**

Tables 4, 5 and 6, show the results obtained for Daubechies wavelets from 2 to 6. Best accuracy of 86% was obtained using Daubechies wavelet 2. Thus by using wavelet coefficients directly, a higher accuracy was achieved compared to a 74.4% obtained by [6] where only 5 sleep stages were classified using EEG, EMG and EOG signals. It also outperformed the accuracy of 83% achieved by [4] for 5 stages of sleep using random forest classifier. A marginal improvement in accuracy of 87% was achieved by [19] using a recurrent neural network but was again limited to classifying five stages of S1, S2, SWS, REM and Wake. In [2] only four stages were classified with an accuracy of 93% using wavelet sub-band energy as features. However Table 3 shows that the prediction was slightly skewed towards stage 2 because of the fact that 50% of total sleep time is dominated by stage 2. Since a fully connected neural network leads to overfitting of the training data, auto-associative neural networks could be used to help better define decision boundaries.

References

1. Khandpur, R.S.: Handbook of Biomedical Instrumentation. Tata McGraw-Hill Education, New York City (1992)
2. Ebrahimi, F., Mikaeili, M., Estrada, E., Nazeran, H.: Automatic sleep stage classification based on EEG signals by using neural networks and wavelet packet coefficients. In: 30th Annual International IEEE EMBS Conference Canada (2008)
3. Rechtschaffen, A., Kales, A.: A Manual of Standardized Terminology, Techniques, and Scoring System for Sleep Stages of Human Subjects. UCLA, Brain Research Institute/Brain Information Service, Los Angeles (1968)
4. Fraiwan, L., Lweesy, K., Khasawneh, N., Wenz, H., Dickhaus, H.: Automated sleep stage identification system based on time-frequency analysis of a single EEG channel and random forest classifier. Comput. Methods Programs Biomed. **108**, 10–19 (2012)
5. Koley, B., Dey, D.: An ensemble system for automatic sleep stage classification using single channel EEG signal. Comput. Biol. Med. **42**(12), 1186–1195 (2012)
6. Tagluk, M.E., Sezgin, N., Akin, M.: Estimation of sleep stages by an artificial neural network employing EEG, EMG and EOG. J. Med. Syst. **34**(4), 717–725 (2010)
7. Sinha, R.K.: Artificial neural network and wavelet based automated detection of sleep spindles, REM sleep and Wake States. J. Med. Syst. **32**(4), 291–299 (2008)
8. Prochazka, A., Mudrova, M., Vysata, O., Hava, R., Araujo, C.P.S.: Multi-channel EEG signal segmentation and feature extraction. In: 14th International Conference on Intelligent Engineering Systems (INES), pp. 317–320 (2010)
9. Gandhi, T., Panigrahi, B.K., Anand, S.: A comparative study of wavelet families for EEG signal classification. Neurocomputing **74**(17), 3051–3057 (2011)
10. Ronzhina, M., Janousek, O., Kolarova, J., Novakova, M., Honzik, P., Provaznik, I.: Sleep scoring using artificial neural networks. Sleep Med. Rev. **16**(3), 251–263 (2012)
11. Goldberger, A.L., Amaral, L.A.N., Glass, L., Hausdorff, J.M., Ivanov, P., Mark, R.G., Mietus, J.E., Moody, G.B., Peng, C.K., Stanley, H.E.: PhysioBank, PhysioToolkit, and PhysioNet: components of a new research resource for complex physiologic signals. Circulation **101**(23), 215–220 (2000)
12. Kemp, B., Zwinderman, A.H., Tuk, B., Kamphuisen, H.A.C., Oberye, J.J.L.: Analysis of a sleep-dependent neuronal feedback loop: the slow-wave microcontinuity of the EEG. IEEE Trans. Biomed. Eng. **47**(9), 1185–1194 (2000)
13. Ghodsi, A.: Dimensionality reduction: a short tutorial. Technical report, University of Waterloo (2006)
14. Navon, I.M., Legler, D.M.: Conjugate-gradient methods for large-scale minimization in meteorology. Mon. Weather Rev. **115**, 1479–1502 (1987)
15. Susmskova, K.: Human sleep and sleep EEG. Measur. Sci. Rev. **4**(2), 59–74 (2004)
16. Sen, B., Peker, M., Cavusoglu, A., Celebi, F.: A comparative study on classification of sleep stage based on EEG signals using feature selection and classification algorithms. J. Med. Syst. **38**, 18 (2014)

17. Radha, M., Garcia-Molina, G., Poel, M., Tononi, G.: Comparison of feature and classifier algorithms for online automatic sleep staging based on a single EEG Signal. In: 36th Annual International Conference of the IEEE Engineering in Medicine and Biology Society (EMBC), pp. 1876–1880 (2014)
18. Leskovec, J., Rajaraman, A., Ullman, J.D.: Mining of Massive Datasets. Cambridge University Press, Cambridge (2014)
19. Hsu, Y.L., Yang, Y.T., Wang, J.S., Hsu, C.Y.: Automatic sleep stage recurrent neural classifier using energy features of EEG signals. Neurocomputing **104**, 105–114 (2013)
20. Schalkoff, R.J.: Artificial Neural Network. McGraw-Hill, New York City (2013)
21. Rojas, R.: Neural Networks. Springer, Heidelberg (1996)
22. Mallat, S.: A theory for multiresolution signal decomposition: the wavelet representation. IEEE Pattern Anal. Mach. Intell. **11**(7), 674–693 (1989)
23. Addision, P.S.: The Illustrated Wavelet Transfor Handbook Introductor Theory and Appication in Science, Engineering, Medicine and Finance. IOP Publishing Ltd., Bristol (2002)
24. Hinton, G.E., Salakhutdinov, R.R.: Reducing the dimensionality of data with neural networks. Science **313**(5786), 504–507 (2006)
25. LeCun, Y., Bottou, L., Bengio, Y., Haffner, P.: Gradient-based learning applied to document recognition. Proc. IEEE **86**(11), 2278–2324 (1998)
26. Vigon, L., Saatchi, M.R., Mayhew, J.E.W., Fernandes, R.: Quantitative evaluation of techniques for ocular artefact filtering of EEG waveforms. IEEE Proc. Sci. Measur. Technol. **147**(5), 219–228 (2000)
27. Hyvrinen, A., Oja, E.: Independent component analysis: algorithms and applications. Neural Netw. **13**(4–5), 411–430 (2000)
28. Vigario, R., Sarela, J., Jousmiki, V., Hamalainen, M., Oja, E.: Independent component approach to the analysis of EEG and MEG recordings. IEEE Trans. Biomed. Eng. **47**(5), 589–593 (2000)

An Impact of Severe Preeclampsia on Cardiovascular System Adaptation of Newborns in Early Neonatal Period

Olga Kireeva[1,2,3], Eugene Bushmelev[1,2,3], Elena Emelianchik[1,2,3],
Alla Salmina[1,2,3], and Michael Sadovsky[1,2,3(✉)]

[1] Krasnoyarsk State Medical University,
p. Zheleznyaka str., 1, 660022 Krasnoyarsk, Russia
msad@icm.krasn.ru
{lenacor,allasalmina}@mail.ru, eugene.bushmelev@gmail.com
[2] Institute of Computational Modelling of SB RAS,
Akademgorodok, Krasnoyarsk 660036, Russia
[3] Kemerovo Regional Perinatal Center, Oktjabrsky pr. 22, 650000 Kemerovo, Russia
http://icm.krasn.ru

Abstract. Some preliminary results toward the relation of various factors of age adaptation in early neonatal period, and parameters of the development of cardiovascular system are present. It was fond both healthy babies (verification data set) and those born by mothers with severe preeclampsia are stably divided into two classes, by K-means. Functional and clinical differences between the classes still require more detailed investigation.

Keywords: Dynamics · Correlation · Model · Blood pressure

1 Introduction

Prenatal period is the key factor to determine a physiologically normal parturition and postnatal adaptation of a newborn. Good conditions result in the implementation of compensatory reactions with no damage for a fetus development, in early neonatal period. Negative external and inner factors affect significantly decrease adaptive capacity of a fetus, may result in a premature birth and cause the critical condition of a newborn in early neonatal period.

Severe preeclampsia (PE) is one of the extremal factors here. It is well know fact that heavy hypertension makes a core clinical problem of PE. PE increases risk of a stroke of a mother at the third trimester of pregnancy for $3 \div 13$ times; similarly, it increases the risk at the sixth to twelfth month after parturition [1]. Chronical placental insufficiency makes a core impact factor of PE on a fetus. It retards the prenatal development and slow down the newborn adaptation to the external life conditions [2].

Circulatory system of a newborn is the key chain in adaptation. Heart and vessels of a newborn go through radical changes resulted from the start of a

© Springer International Publishing AG 2017
I. Rojas and F. Ortuño (Eds.): IWBBIO 2017, Part I, LNBI 10208, pp. 517–523, 2017.
DOI: 10.1007/978-3-319-56148-6_46

function of pulmonary circulation and decay of a portal bloodstream. Blood pressure growth in a ventricle and systemic circulation vessels are essential for compensatory changes in organism. Thus, arterial pressure could be considered as an integrative figure of a blood circulation state, and the dynamics of that former reflects the adaptation processes running in it.

The violations of placental circulation accompanied with PE cause hypoperfusion and hypoxia of organs and tissues of a fetal, and hypotensive therapy may cause reperfusion injury [3]. Simultaneously, the oscillations of arterial blood tension are a sensitive marker of catecholaminergic organism reactions, as well as the violations in systolic and diastolic heart functions, kidney ischemia [4]. We studied the patterns of adaptation process of newborn children circulation system, who were borne by mothers suffering from severe PE.

2   Materials and Methods

To do the study, we have carried out monitoring of arterial pressure during early neonatal period for 60 children. Also, we observed 40 children whose mothers have not PE, to verify the results.

We used LIFESCOPE I heart monitor produced by NIHON KOHDEN, Japan, model BSM-2301K to trace the arterial presser of newborn; both systolic and diastolic arterial pressure have been determined. To measure the pressure, we used tonometer cuff for newborn #1 to #4, in dependence on the baby weight. The timetable for blood pressure measurements was the following: 12 measurements (every two hours) during the first 24 h, then 8 measurements (every 4 h) on the second and third days, and 4 measurements (every 6 h) on the fourth and fifth days. During the first day of the life, both systolic and diastolic pressure figures have been detected, on the third, sixth, twelfth and twenty third hour of the life. Later, a diurnal average figures have been recorded.

Recorded figures of blood pressure were compared to the normative figures established for premature babies [5], with respect to body mass and gestation age [5].

Comparison of blood pressure figures observed for newborns of extremely low birth weight has been carried out to the normal figures for premature babies [6]. For each newborn, the index of hypertension time has been determined, during the starting fife days of life. That latter was determined as a percentage of the measurements exceeding the normal figures expected for this time period. The index allows to estimate a duration period of the blood pressure growth during the first day of life; the index is also knows as *blood pressure load*. Typical figures for the index is 15–25% for adults. An excess of the index of 25% level should be interpreted as a lability of arterial hypertension, and the case ≥50% is unambiguously interpreted as a stable arterial hypertension.

2.1   Methods in Statistical Images Treatment

The main goal of this study is to retrieve knowledge from the above-mentioned data. All the data for research has been collected from medical records, and that was very typical situation; in practice, this way of the data collection is not free from lacunae occurrence. The curse of dimensionality was another problem in our study. Since there is no clear, concise and comprehensive theory explaining all the relations between preeclampsia and cardiovascular system peculiarities, we tried to gather as much data, as possible. This approach resulted in the curse of dimensionality problem: the number of variables describing a patient is comparable to the number of all the records available for the study.

In such capacity, we mainly focused on the order search, in the available data, and the knowledge retrieval from the data, rather than to pursue standard techniques of statistical analysis. In brief, we used (non-linear) cluster analysis to treat the data, and elastic map is the key tool here [7,8].

To develop an elastic map, we took into consideration two versions of the observed data:

(1) the data on blood pressure gathered during first week of life of a baby (more details on the data collection schedule see above), and
(2) the data characterizing heart function, collected at the same time moments, as pressure data.

Simultaneously, some biochemical data have been collected, including ANP and troponin levels; these data were considered to a kind of attribute. In other words, we did not incorporate them into the data set to develop a clusterization, or elastic map. We checked whether the clusters identified through two sets of data enlisted above yield a correlation with those biochemical indicators.

Elastic map, in brief, is a tool to visualize and cluster multidimensional data. Elastic map implementation consists of several step. At the first step, two main principal components must be found (these are the directions in the multidimensional space yielding the greatest extension of data set); these are determined by the first and the second eigenvectors of covariance matrix developed over the data set. At the second step, one puts on a plane over these two directions, and make a projection of each data point on the plane. As soon as the projections are found, each data point must be connected with the projection point by a (mathematical) spring; all springs are of the same elasticity coefficient λ.

At the third step, one allows the plane to be flexible and expandable; the plane is supposed to be uniform in terms of the flexibility and expansion properties. Releasing the system, one gets a jammed (nonlinear) surface; the surface must meet some constraints, among them a continuity of transformations, and topology (no glues are allowed).

At the fourth step, each data point is redefined over the surface: a projection must be changed with the point on the surface that is the nearest to the given data point. Finally, one should release the (mathematical) springs allowing the surface to became plane back. This method provides an identification and evaluation of cluster structure that may not be revealed by linear statistics methods (see [7,8] for details).

3   Results and Discussion

To begin with, we consider a clusterization of the data due to K-means. We pursued the clusterization for two, three and four classes. Stability of clusterization has been observed for two classes, only. Figure 1 shows the clusterization. The most astonishing thing here was that K-means for two classes failed to discriminate thick patients from the reference healthy ones.

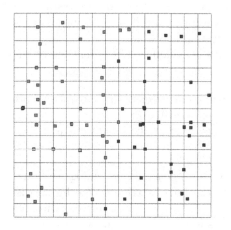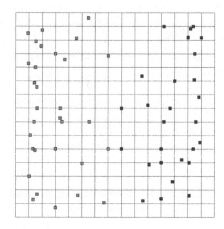

Fig. 1. Elastic map (in inner coordinates) showing the clusterization of the patients over two clusters provided by K-means; the clusters are shown in color. (Color figure online)

Nonetheless, the clusterization is rather stable, in both sets of variables; it proves indirectly the existence of some structure in the data that may not be revealed over the sick and reference patients, but still manifests in the distribution of all the patients in the clusters.

To figure out the clusterization in more detail, we have developed the nonlinear clusterization with elastic map technique, as described above; Fig. 2 shows those clusterizations. The clusters in Fig. 2 are indicated in yellow (for variables from heart data set), and just colored in grey scaling (for the variables from pressure data set). Few words should be said towards the cluster identification in Fig. 2. To do that, we used contrast coloring that reveals the dense areas in data set. The procedure is as following: each point on the elastic map is supplied with a "hat" (that is Gaussian function, in our case). The parameters of each hat are the same. Then the "hat" functions are summarized, and the sum function is shown (in grey level contours) on the map. The "hat" function has unique parameter δ that defines a width of the function; we used $\delta = 0.15$ in our studies.

Let now focus on Table 1. As we said above, no good discretion has been found between sick and healthy (these are reference) patients. In other words, the clusters identified over the sets of points did not show any discretion between

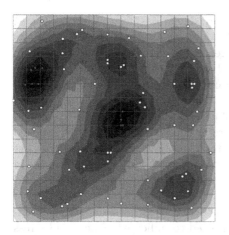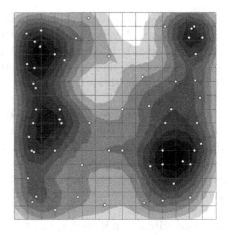

Fig. 2. Elastic map of patients distribution over two clusters (left), and four clusters (right). The left distribution is obtained over the heart characteristics, while the right one over the pressure data. (Color figure online)

Table 1. Levels of troponin I and ANP in the clusters identified with elastic map technique, for the indexed and total data bases (*H1* and *HG*, correspondingly); same for pressure. τ is mean level for troponin, and β is mean level of ANP: σ is standard deviation. c is the correlation between troponin and ANP, in th3e relevant cluster.

Data	Cluster	$\tau_{\text{Trop I}}$	$\sigma_{\text{Trop I}}$	β_{ANP}	σ_{ANP}	c
H1	1	0.270	0.065	366.714	84.817	−0.037
	2	0.345	0.132	320.750	93.600	0.770
	3	0.270	0.120	357.000	121.156	−0.234
HG	1	0.232	0.070	335.400	82.373	−0.033
	2	0.350	0.077	302.500	67.727	0.473
P1	1	0.344	0.086	320.800	62.886	0.457
	2	0.310	0.066	330.000	64.086	0.924
PG	1	0.190	0.079	320.200	114.027	0.835
	2	0.283	0.100	323.667	68.766	0.137
	3	0.304	0.103	364.714	63.827	−0.005
	4	0.273	0.108	421.000	139.056	−0.369

healthy and sick patients, for both data sets. Nonetheless, there is difference between them, and the difference manifests in the correlations between two very important biochemical compounds involved in cardiovascular system regulation and development: these are troponin and ANP-protein. We did not include these variable into the analysis, but stipulated them to be an attribute. In other words, we checked the hypothesis whether the clusters identified through heart characteristics would differ in the content of those biochemical substances.

The table shows mean values and standard deviations, for those compounds, calculated over the clusters. Indeed, cluster 2 for the indexed data set exhibits evidently higher level of troponin, with increased standard deviation of the mean value. Meanwhile, there is nothing unusual in ANP behaviour, nor in troponin behaviour among those clusters; but the correlation coefficient between them quite informative.

Table 1 shows three very high figures of correlation between ANP concentration, and troponin concentration. These are in cluster 1 obtained with indexed data set in heart describing variables, cluster 2 obtained with indexed data set in pressure describing variables, and cluster 1 obtained with general data set in pressure describing variables. These figures differs strongly from all others, and such difference may not be accidental.

On the contrary, an increased correlations between (rather independent) variables describing a population has been reported to indicate an increased stress [9–11]. Thus, one should definitely expect that the correlation indices shown in the table indeed tell us on the existence of a group in stress. It should be stressed that average values of various variable remain stable, and do not report on a stress, at least unless that latter is too strong. This phenomenon requires more detailed investigation.

References

1. Roberts, J.M., Pearson, G.D., Cutler, J.A., Lindheimer, M.D.: Summary of the NHLBI working group on research on hypertension during pregnancy. Hypertens. Pregnancy **22**(2), 109–127 (2003)
2. Chen, Q., De Sousa, J., Snowise, S., Chamley, L., Stone, P.: Reduction in the severity of early onset severe preeclampsia during gestation may be associated with changes in endothelial cell activation: a pathological case report. Hypertens. Pregnancy **35**(1), 32–41 (2016)
3. Canoy, D., Cairns, B.J., Balkwill, A., Wright, F.L., Khalil, A., Beral, V., Green, J., Reeves, G.: Hypertension in pregnancy and risk of coronary heart disease and stroke: a prospective study in a large UK Cohort. Int. J. Cardiol. **222**, 1012–1018 (2016)
4. Wang, Y.A., Chughtai, A.A., Farquhar, C.M., Pollock, W., Lui, K., Sullivan, E.A.: Increased incidence of gestational hypertension and preeclampsia after assisted reproductive technology treatment. Fertil. Steril. **105**(4), 920–926 (2016)
5. Saleh, L., Danser, J.A.H., van den Meiracker, A.H.: Role of endothelin in preeclampsia and hypertension following antiangiogenesis treatment. Curr. Opin. Nephrol. Hypertens. **25**(2), 94–99 (2016)
6. IngelfingerJ, R., Powers, L., Epstein, M.F.: Blood pressure norms in low birth weight infants: birth through 4 weeks. Pediatr. Res. **17**, 319 (1983)
7. Mirkes, E.M., Coats, T.J., Levesley, J., Gorban, A.N.: Handling missing data in large healthcare dataset: a case study of unknown Trauma outcomes. Comput. Biol. Med. **75**, 203–216 (2016)
8. Gorban, A.N., Zinovyev, A.Y.: Fast and user-friendly non-linear principal manifold learning by method of elastic maps. In: Proceedings DSAA 2015 - IEEE International Conference on Data Science and Advanced Analytics, Paris, October 2015

9. Gorban, A.N., Tyukina, T.A., Smirnova, E.V., Pokidysheva, L.I.: Evolution of adaptation mechanisms: adaptation energy, stress, and oscillating death. J. Theor. Biol. **405**, 127–139 (2016)

10. Gorban, A.N., Smirnova, E.V., Tyukina, T.A.: Correlations, risk and crisis: from physiology to finance. Phys. A **389**(16), 3193–3217 (2010)

11. Gorban, A.N., Smirnova, E.V., Tyukina, T.A.: General laws of adaptation to environmental factors: from ecological stress to financial crisis. Math. Model. Nat. Phenom. **4**(6), 1–53 (2009)

Classification Algorithms for Fetal QRS Extraction in Abdominal ECG Signals

Pedro Álvarez, Francisco J. Romero, Antonio García, Luis Parrilla,
Encarnación Castillo$^{(\boxtimes)}$, and Diego P. Morales

Facultad de Ciencias, Department of Electronics and Computer Technology,
University of Granada, Granada, Spain
{pedroalgui7, franromero}@correo.ugr.es,
{grios, luis, encas, diegopm}@ugr.es

Abstract. Fetal heart rate monitoring through non-invasive electrocardiography is of great relevance in clinical practice to supervise the fetal health during pregnancy. However, the analysis of fetal ECG is considered a challenging problem for biomedical and signal processing communities. This is mainly due to the low signal-to-noise ratio of fetal ECG and the difficulties in cancellation of maternal QRS complexes, motion, etc. This paper presents a survey of different unsupervised classification algorithms for the detection of fetal QRS complexes from abdominal ECG signals. Concretely, clustering algorithms are applied to classify signal features into noise, maternal QRS complexes and fetal QRS complexes. Hierarchical, k-means, k-medoids, fuzzy c-means, and dominant sets were the selected algorithms for this work. A MATLAB GUI has been developed to automatically apply the clustering algorithms and display FHR monitoring. Real abdominal ECG signals have been used for this study, which validate the proposed method and show high efficiency.

Keywords: Abdominal ECG · Fetal heart rate · Clustering algorithms · MATLAB · GUI

1 Introduction

One the most important ways to detect cardiac anomalies in early stages of the fetus heart forming and supervise its well-being is the monitoring of the fetal heart activity [1]. The fetal electrocardiogram (FECG) [2] signal may not be directly measurable, and it has to be determined from the measurement of a composite signal as for example the abdominal ECG. Noninvasive fetal electrocardiography consists in the signal recording by using surface electrodes placed on the abdomen of a pregnant woman. It has huge potential applications, but presents some drawbacks mainly due to abdominal recordings of fetal ECG have lower signal-to-noise ratio (SNR) as compared with the invasive procedure. The significant amount of noise comes from fetal brain activity, muscle contractions, mother electromyogram (EMG) and respiration, movement artifacts and etc. Moreover, the considerably higher amplitude of the maternal ECG (MECG) components as compared to the FECG components makes difficult the extraction of fetal information. Discrete Wavelet Transform (DWT) [3] can be used for

© Springer International Publishing AG 2017
I. Rojas and F. Ortuño (Eds.): IWBBIO 2017, Part I, LNBI 10208, pp. 524–535, 2017.
DOI: 10.1007/978-3-319-56148-6_47

the suppression of different types of noise including DC levels and wandering [4, 5]. Different approaches have been also proposed for the extraction of the FECG and the detection of parameters from this signal [6, 7]. However, the AECG signal processing can be oriented to extraction of information such as FHR, but avoiding the processing required by MECG removing. In the present paper, different clustering algorithms are studied to be applied on the denoised AECG signals [5] for fetal QRS detection. From this study, a new approach for FHR extraction using AECG signals is proposed which uses single-channel, does not require the use of a maternal ECG reference signal and does not need removing the maternal components.

2   Clustering Overview

Clustering algorithms are used for data partitioning into a certain number of clusters [8]. Most researchers describe a cluster by considering the internal homogeneity and the external separation [9], *i.e.*, patterns in the same cluster should be similar to each other, while patterns in different clusters should not. Some simple mathematical descriptions of several clustering methods [10] are presented at this review.

2.1   Clustering Procedure

The general scheme of a clustering procedure is shown in Fig. 1.

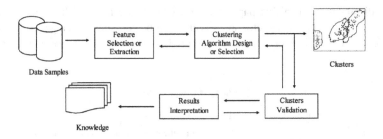

Fig. 1. General scheme of a clustering procedure [8].

- **Feature selection or extraction.** Feature selection chooses distinguishing features from a set of candidates, while feature extraction utilizes some transformations to generate useful and novel features from the original ones [11, 12]. In general, ideal features should be used to distinguish patterns belonging to different clusters, immune to noise, easy to extract and easy to interpret.
- **Clustering algorithm design or selection.** This step is usually combined with the selection of a corresponding proximity measure and the construction of a criterion function. Once a proximity measure is chosen, the construction of a clustering criterion function makes the partition of clusters to be an optimization problem. There is no clustering algorithm which can be universally used to solve all

problems. Therefore, it is important to carefully investigate the characteristics of the problem at hand, in order to select or design an appropriate clustering strategy.

- **Cluster validation.** Given a data set, each clustering algorithm can always generate a division, no matter whether the structure exists or not. Moreover, different approaches usually lead to different clusters; and even for the same algorithm, parameter identification or the presentation order of input patterns may affect the final results. Therefore, effective evaluation standards and criteria are important to provide the users with a degree of confidence in the clustering results.
- **Result interpretation.** The final goal of clustering is to provide users with meaningful insights from the original data, so that they can effectively solve the problems. Experts in the relevant fields will interpret the data partition. Further analysis may be required to guarantee the reliability of extracted knowledge.

2.2 Distance Measures

Given a set of patterns, $X = \{x_1, \ldots, x_j, \ldots, x_N\}$, where $x_j = (x_{j1}, x_{j2}, \ldots, x_{jd})^T \in R^d$ each measure x_{ji} is said to be a feature. The distance or dissimilarity function is the main tool to measure similarity and it is defined to satisfy the following conditions [8]:

1. Symmetry: $D(x_i, x_j) = D(x_j, x_i)$
2. Positivity: $D(x_i, x_j) \geq 0$ for all x_i and x_j.
3. Minkowski inequality: $D(x_i, x_j) \leq D(x_i, x_j) + D(x_k, x_j)$ $\forall x_i, x_j, x_k$
4. Reflexivity: $D(x_i, x_j) = 0$ if $x_i = x_j$.

The distances used in this paper are shown in Table 1. The Minkowski distance [13] is a metric in a normed vector space which can be considered as a generalization of both Euclidean and Cityblock distance. The Euclidean Squared distance metric uses the

Table 1. Measure table

Measure	Mathematical expression	Comments
Minkowski distance	$D_{xy} = (\sum_{l=1}^{d} |x_l - y_l|^n)^{1/n}$	Features with large values and variances tend to dominate over other features
Cityblock distance	$D_{xy} = \sum_{l=1}^{d} |x_l - y_l|$	Special case of Minkowski metric at $n = 1$. Tend to form hyperrectangular clusters
Euclidean distance	$D_{xy} = (\sum_{l=1}^{d} |x_l - y_l|^2)^{1/2}$	The most commonly used metric. Tend to form hyperspherical clusters
Squared euclidean distance	$D_{xy} = \sum_{l=1}^{d} |x_l - y_l|^2$	
Chebychev distance	$D_{xy} = \max|x_l - y_l|$	Special case of Minkowski metric at $n \to \infty$
Mahalanobis distance	$D_{xy} = (x - y)^T S^{-1}(x - y)$ S covariance matrix	Invariant to any nonsingular linear transformation. Tend to form hyperellipsoidal clusters

same equation as the Euclidean distance metric, but does not calculate the square root. As a result, clustering with the Euclidean squared distance metric is faster.

2.3 Clustering Algorithms

Clustering algorithms are divided into two groups: hierarchical and partitional. Hierarchical clustering attempts to construct a tree-like nested structure partition of $X = \{H_1, \ldots, H_Q\}$ $(Q \leq N)$, such that $C_i \in H_m, C_j \in H_l$ y $m > l$ imply $C_i \in C_j$ or $C_i \cap C_j = \emptyset$ for all $i, j \neq i, m, l = 1, \ldots, Q$. Linkage-based clustering starts by assigning each data point to a single point cluster. Then, repeatedly, merge the "closest" clusters of the previous clustering, decreasing the number of clusters with each round. If kept going, this algorithm would eventually results in the trivial clustering in which all of the domain points share one large cluster. Then, to clearly define these algorithms we have first to decide how to measure the distance between clusters, and, second, to determine when to stop merging. The most common ways of extending the distance to a measure of distance between domain clusters are:

- Single Linkage Clustering: the distance between clusters is defined by the minimum distance: $D(C_i, C_j) = \min\{d(x, y) : x \in C_i, y \in C_j\}$.
- Average linkage Clustering: the distance is defined to be the average distance between a point in one of the clusters and a point in the other: $D(C_i, C_j) = \frac{1}{|C_i||C_j|} \sum_{x \in AC_j, y \in C_j} d(x, y)$.
- Max linkage Clustering: the distance is the maximum distance between clusters: $D(C_i, C_j) = \max\{d(x, y) : x \in C_i, y \in C_j\}$.

Without employing a stopping rule, the outcome of such an algorithm can be described by a clustering dendrogram, i.e., a tree of domain subsets, having the singleton sets in its leaves, and the full domain as its root [14].

On the other hand, hard partitional clustering decomposes a data set X into a K disjoint clusters $C = \{C_1, \ldots, C_K\}(K \leq N)$ such that:

(1) $C_{i_K} \neq \emptyset, i = 1, \ldots, K$
(2) $U_{i=1}^K C_i = X$
(3) $C_i \cap C_j = \emptyset, i, j = 1, \ldots, K,$ y $i \neq j$

Contrary to hard partitional clustering (in which each pattern only belongs to one cluster), for fuzzy clustering [15] a pattern may belong to all clusters with a degree of membership $u_{i,j} \in [0, 1]$, which represents the membership coefficient of the j-th object in the i-th cluster and satisfies the following two constraints: $\sum_{i=1}^c u_{i,j} = 1$ and d $\sum_{i=1}^c u_{i,j} < N, \forall j$.

3 Clustering Technique for Fetal QRS Extraction

The proposed clustering technique for the extraction of fetal QRS complexes has been modeled using MATLAB. First, a previously proposed wavelet-based preprocessing is applied to the AECG signal in order to remove wandering and noise [5, 7]. Then, the new clustering technique is used for fetal QRS extraction. Finally, the false positive and false negative procedure proposed in [7] is used to correct false-detected and/or non-detected fetal QRS complexes. This section is devoted to the study of different clustering algorithms for the classification of special features in AECG signals in order to determine the cluster corresponding to fetal QRS complexes.

The first step consists in selecting the signal features allowing us to distinguish patterns belonging to different clusters. After an analysis of the AECG signals, the amplitude difference between a local maximum followed by a local minimum has been selected as the main feature. Extracting this amplitude feature from AECG signals, three different clusters can be obtained. The cluster with a greater value of amplitude indicates that our data corresponds to RS-peaks of maternal ECG complexes. The cluster with an intermediate value corresponds to RS-peaks of a fetal ECG complex. Lastly, the cluster with lower value corresponds to noise or other waves.

3.1   Avoiding Local Minima

Clustering is an optimization problem in which to solve the problem it is necessary to reach an absolute minimum. In spite of the computational cost, we use equivalent algorithms with less complexity. In contrast, like many other types of numerical minimizations, the solution often depends on the starting points. It is possible to reach a local minimum for any clustering algorithm, where reassigning any one point to a new cluster would increase the total sum of point-to-centroid distances, but where a better solution does exist. However, the algorithm can be replicated several times to overcome that problem, without increasing the computational cost. Figure 2 displays an example where there is a local minimum at 5 replicates.

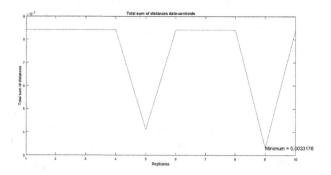

Fig. 2. Total sum of data-centroid distances based on the number of replicates

3.2   Optimal Number of Clusters

One of the most difficult tasks in any clustering problem is to find the best number of clusters, k. For the proposed clustering classifications over AECG signals, the desirable number of clusters is three and it should be also the minimum. Despite this, sometimes the signals will not be correctly classified, needing more than three groups. To solve this problem it is necessary to use different criteria indicating the optimum number of clusters. Basically the minimum average distance to centroid trying different groups is looked for. When the best k is found, the average falls rapidly. There are two criteria which work well with our dataset, Davies-Bouldin criterion [16] and silhouette value [17]. Note that it is necessary to delimit these criteria in order to optimally solve our problem. In conclusion, it is not suitable to use these criteria for $k > 8$.

3.3   Clustering Algorithms for Fetal RS-Peak Detection

The clustering algorithms evaluated for fetal RS-peak detection are k-means, k-medoids, fuzzy C-means and hierarchical. About metrics, we have choosen squared euclidean (sqe) and cityblock (cb) distances. When using an unidimensional amplitude vector, cityblock distances are equal to Euclidean an Mahalanobis.

3.3.1   K-means Algorithm
This algorithm begins with initialization of the centroids. For this task we have selected k-means++ instead of Lloyd's k-means. Using a simulation study for several cluster orientations, Arthur and Vassilvitskii [18] have demonstrated that k-means++ achieves faster convergence to a lower sum of within-cluster, sum-of-squares point-to-cluster-centroid distances than Lloyd's algorithm. Then, the algorithm assigns each data to the cluster that has the closest centroid. When all data have been assigned, it recalculates the positions of the centroids. Assigning data and recalculating centroids are repeated until the centroids no longer move. Figure 3 shows the data classification for this algorithm using abdomen-4 r01 recording from PhysioNet database [19]. The blue points corresponds to maternal QRS complexes, green points to fetal QRS complexes and brown points to noise or other waves. The centroids are shown using red marks.

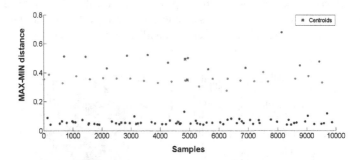

Fig. 3. K-means classification (squared Euclidean distance) for 10-seconds ab4 r01 recording (Color figure online)

Comparing these classification results with database annotations, there is only one false-detected fetal QRS complex (FP) and one non-detected (FN). The FN occurs because the fetal QRS complex is masked by a maternal complex. Remember that false positive and negative corrections proposed in [7] are applied to improve the results.

3.3.2   K-medoids Algorithm

This algorithm is similar to k-means, thus its goal is to divide a set of measurements into k subsets or clusters so that the subsets minimize the sum of distances between a measurement and a center of the measurement's cluster. In the k-means algorithm, the center of the subset is the mean of measurements in the subset, often called a centroid. In the k-medoids algorithm, the center of the subset is a member of the subset, called a medoid [20]. Figure 4 shows k-medoids classification for 10-seconds abdomen-2 r01 recording, where green points corresponds to maternal QRS complexes, brown to fetal QRS complexes and blue to noise or other waves.

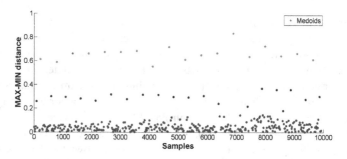

Fig. 4. K-medoids classification (squared Euclidean distance) for 10-seconds ab2 r10 recording (Color figure online)

3.3.3   Fuzzy C-means Algorithm

Fuzzy C-means [21] is a clustering method allowing each data to belong to multiple clusters with varying degrees of membership. For data classification we have chosen the greatest probability of every data, assigning it the group with a higher probability. Figure 5 shows the data set along to their probability mass function.

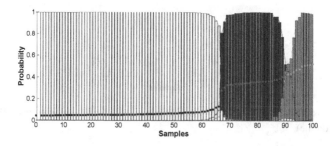

Fig. 5. Fuzzy C-means probability density function for 10-seconds ab4 r01 recording

3.3.4   Hierarchical Algorithm

Hierarchical algorithm groups data over a variety of scales by creating a cluster tree or dendrogram. The tree is not a single set of clusters, but rather a multilevel hierarchy, where clusters at one level are joined as clusters at the next level. Once obtained the dendogram, criteria Davies-Bouldin and silhouette criterion value help us to determine the final number of clusters. Figure 6 shows the dendogram obtained for hierarchical classification of 10-seconds ab4 r01 recording.

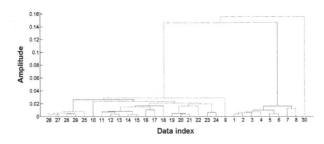

Fig. 6. Dendogram for hierarchical classification of 10-seconds ab4 r01 recording

4   Results

Recordings from the *Abdominal and Direct Fetal Electrocardiogram Database* [19] have been used for the training and the validation of the proposed method. Each recording comprises four differential signals acquired from maternal abdomen and the reference direct fetal electrocardiogram registered from the fetus head. The fetal R-wave locations were automatically determined in the direct FECG signal by means of on-line analysis applied in the KOMPOREL system [19]. The recordings, sampled with 16-bit resolution at 1ksps, are 5-minute long and the signal bandwidth is 1 Hz–150 Hz. Figure 7 shows an example for the application of the proposed method. First subplot shows the original signal, second subplot shows the denoised signal including the detected maternal and fetal R-peaks using k-means algorithm and squared Euclidean distance, and finally, last subplot shows the FHR monitoring.

To assess the performance of the proposed FHR extraction method, the accuracy parameter can be studied [7]:

$$Acc = \frac{(TD)}{(TD + FP + FN)}$$

where TD are the true-detected fetal QRS complexes, FN are false negatives and FP are false positives. Table 2 shows the accuracies obtained for 1-minute for recordings of the database (4th minute s) using the algorithms and distances detailed in Sect. 3. Annotations of database were used in order to detect the FPs and the FNs. Acc_1 indicates the accuracy obtained after clustering classification step while Acc_2 indicates the accuracy of the proposed method, after the FP and FN correction step. The average accuracies manifests that K-means cityblock and K-medoids squared Eucliden and

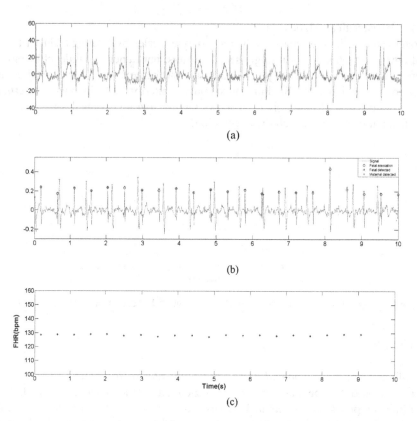

Fig. 7. 10-seconds ab4 r01 recording (a) Original signal (b) Denoised signal including detected maternal and fetal R-peaks (k-means and squared Euclidean distance) (c) FHR monitoring

Table 2. Evaluation results

	K-means sqe		K-means cb		K-medoids sqe		K-medoids cb		Fuzzy c-means		Hierarchical	
Recording	Acc_1	Acc_2	Acc_1	Acc_2	Acc_1	Acc_2	Acc_1	Acc_2	Acc_1	Acc_2	Acc_1	Acc_2
Ab1 r01	88.3	93.8	88.3	93.8	88.3	93.8	88.3	93.8	88.3	88.3	57.2	57.2
Ab4 r01	83.1	88.0	63.1	70.2	86.9	90.8	85.3	90.8	63.5	66.0	63.2	66.0
Ab2 r04	86.2	92.3	87.0	90.8	86.2	90.8	86.9	90.8	86.2	88.0	46.4	46.5
Ab3 r07	87.5	93.7	86.2	91.5	87.5	93.1	86.2	91.5	86.8	92.2	87.5	93.7
Ab4 r07	54.0	58.6	83.8	94.4	83.8	94.4	83.8	94.4	83.8	94.4	83.8	94.4
Ab1 r08	95.3	99.3	95.3	99.3	95.3	99.3	95.3	99.3	95.3	99.3	95.3	99.3
Ab1 r10	90.9	99.4	90.9	99.4	90.9	99.4	90.9	99.4	90.9	90.9	27.9	29.4
Ab4 r10	87.0	94.4	87.0	94.4	87.0	94.4	87.0	94.4	87.0	94.4	87.0	94.4
Average	84.04	89.9	85.2	**94.8**	88.2	**94.5**	88.0	**94.3**	85.2	93.7	68.5	87.8

cityblock achieve the best results. Figures 8 and 9 display the FHR monitoring obtained using k-means algorithm and squared Euclidean distance for fetal QRS extraction in two different recordings.

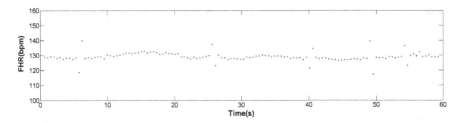

Fig. 8. FHR monitoring for 1-minute ab4 r01 recording

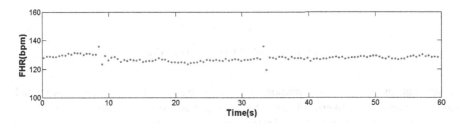

Fig. 9. FHR monitoring for 1-minute ab1 r08 recording

4.1 Graphical User Interface

A graphical user interface (GUI) has been developed as a platform for evaluating the results of the clustering algorithms proposed in this work. Figure 10 shows a capture of this GUI. The user can select between the following input parameters:

- Signal parameters: AECG signal, channel, window length, start and final points.
- Cluster parameters: minimum number of groups ($k = 3$) and maximum (advisable to use $k \leq 8$), to calculate the number of clusters which makes the best classification, according to Davies-Bouldin criterion and silhouette value.

After the classification of the data, the principal table of this GUI (displayed in Fig. 10) shows the following results: used algorithm and metrics, number of clusters, computational cost, *Acc*, *Se* and *PDV* values [7]. *FPs and FNs* section shows the extraction evaluation when the correction over the algorithm with the best *Acc* is computed. *Fetal Heart Rate* section shows the instantaneous FHR after the FP and FN correction step. Finally, *Average results* section displays the average *Acc* of every cluster algorithm. Attending to GUI graphs shown in Fig. 10, upper-left-corner graph represents the AECG signal with the threshold algorithm proposed in [7] including a fix threshold, where fetal QRS complexes marked in pink, and the annotated complexes are marked in black (in every graph). Upper-right-corner graph displays the AECG signal with red marks that corresponds to the fetal QRS complexes extracted from the clustering algorithm with the highest *Acc*. Lower-left-corner graph shows the data

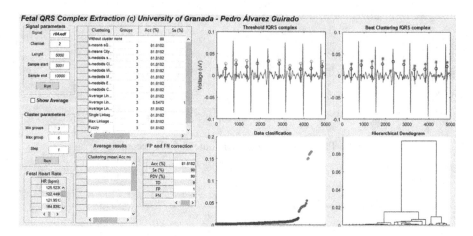

Fig. 10. Capture of the Graphical User Interface (Color figure online)

groups (data are shorted by amplitude) accordingly with the clustering algorithm which has the highest *Acc*. Lower-right-corner graph represents the dendogram of hierarchical clustering with the highest *Acc*. Finally, the GUI also displays an additional graph which is not shown in Fig. 10, with the fetal QRS complexes marked after FPs and FNs corrections.

5   Conclusion

This paper presents a new method for FHR extraction from AECG signals. It introduces a novel clustering procedure for the data classification and subsequent determination of the fetal QRS complexes. It consists of extracting the amplitude of the located maxima followed by minimum as main feature, which helps in the identification of RS-peaks of fetal QRS complexes. The obtained amplitudes are classified into three clusters corresponding to maternal QRS complexes, fetal QRS complexes and noise or other waves. This paper also shows the study of different clustering algorithms and related parameters meeting the best classification and thus the best accuracy in the fetal QRS complex extraction. From the extracted fetal QRS complexes, FHR monitoring is carried out. The proposed clustering-based presented was validated for real AECG signals obtaining high accuracy.

Acknowledgments. This work has been partially funded by Project CEMIX UGR-MADOC 02/16.

References

1. Jenkins, H.M.: Thirty years of electronic intrapartum fetal heart rate monitoring: discussion paper. J. R. Soc. Med. **82**, 210–214 (1989)

2. Taylor, M.J., Thomas, M.J., Smith, M.J., Oseku-Afful, S., Fisk, N.M., Green, A.R., Paterson-Brown, S., Gardiner, H.M.: Non-invasive intrapartum fetal ECG: preliminary report. BJOG: Int. J. Obstet. Gynaecol. **112**(8), 1016–1021 (2005)
3. Mallat, S.G.: A theory for multiresolution signal decomposition: the wavelet representation. IEEE Trans. Pattern Anal. Mach. Intell. Arch. **11**(7), 674–693 (1989)
4. Addison, P.S.: Wavelet transforms and the ECG: a review. Physiol. Meas. **26**(5), R155–R199 (2005)
5. Castillo, E., Morales, D.P., García, A., Parrilla, L., Lopez-Ruiz, N., Palma, A.J.: One-step wavelet-based processing for wandering and noise removing in ECG signals. In: IWBBIO, pp. 491–498 (2013)
6. Nandhini, P., Meeradevi, T.: Literature review of fetal ECG extraction. Bonfring Int. J. Adv. Image Process. **2**(1), 55–62 (2014)
7. Castillo, E., Morales, D.P., Botella, G., García, A., Parrilla, A., Palma, A.J.: Efficient wavelet-based ECG processing for single-lead FHR extraction. Digit. Signal Process. **23**(6), 1897–1909 (2013)
8. Xu, R., Wunsch, D.: Survey of clustering algorithms. IEEE Trans. Neural Netw. **16**(3), 645–678 (2005)
9. Gordon, A.D.: Classification, 2nd edn. Chapman & Hall, London (1999)
10. Hansen, P., Jaumard, B.: Cluster analysis and mathematical programming. Math. Program. **79**, 191–215 (1997)
11. Jain, A., Murty, M., Flynn, P.: Data clustering: a review. ACM Comput. Surv. **31**(3), 264–323 (1999)
12. Bishop, C.: Neural Networks for Pattern Recognition. Oxford University Press, New York (1995)
13. Deza, E., Deza, M.M.: Encyclopedia of Distances, p. 94. Springer, Heidelberg (2009)
14. Shalev-Shwartz, S., Ben-David, S.: Understanding Machine Learning: From Theory to Algorithms. Cambridge University Press, Cambridge (2014)
15. Zadeh, L.A.: Fuzzy sets. Inf. Control **8**(3), 338–353 (1965)
16. Davies, D.L., Bouldin, D.W.: A cluster separation measure. IEEE Trans. Pattern Anal. Mach. Intell. **PAMI-1**(2), 224–227 (1979)
17. Rouseeuw, P.J.: Silhouettes: a graphical aid to the interpretation and validation of cluster analysis. J. Comput. Appl. Math. **20**(1), 53–65 (1987)
18. Arthur, D., Vassilvitskii, S.: K-means++: the advantages of careful seeding. In: SODA 2007: Proceedings of the Eighteenth Annual ACM-SIAM Symposium on Discrete Algorithms, pp. 1027–1035 (2007)
19. Goldberger, A.L., et al.: PhysioBank PhysioToolkit, and PhysioNet: components of a new research resource for complex physiologic signals. Circulation **12**(101): e215–e220 (2000). http://physionet.org/physiobank/database/nifecgdb/
20. Park, H.-S., Jun, C.-H.: A simple and fast algorithm for K-medoids clustering. Expert Syst. Appl. **36**, 3336–3341 (2009)
21. Bezdec, J.C.: Pattern Recognition with Fuzzy Objective Function Algorithms. Plenum Press, New York (1981)

Evaluation of Algorithms for Automatic Classification of Heart Sound Signals

Ricardo Enrique Pérez-Guzmán[1(✉)], Rodolfo García-Bermúdez[2],
Fernando Rojas-Ruiz[3], Ariel Céspedes-Pérez[1], and Yudelkis Ojeda-Riquenes[4]

[1] Department of Computer Sciences, University of Las Tunas, Las Tunas, Cuba
{ricardopg,arces}@ult.edu.cu
[2] Faculty of Computer Science, Technical University of Manabi, Manta, Ecuador
rodolfo.garcia@live.uleam.edu.ec
[3] Department of Computer Architecture and Technology, CITIC,
University of Granada, Granada, Spain
frojas@ugr.es
[4] Teaching Hospital Ernesto Guevara de la Serna, Las Tunas, Cuba
diana@ltu.sld.cu

Abstract. Auscultation is the primary tool for detection and diagnosis of cardiovascular diseases in hospitals and home visits. This fact has led in the recent years to the development of automatic methods for heart sound classification, thus allowing for detecting cardiovascular pathologies in an effective way. The aim of this paper is to review recent methods for automatic classification and to apply several signal processing techniques in order to evaluate them in the PhysioNet/CinC Challenge 2016 results. For this purpose, the records of the open database PysioNet/Computing are modified by segmentation or filtering methods and the results were tested using the challenge best ranked algorithms. Results show that an adequate preprocessing of data and subsequent feature selection may improve the performance of machine learning and classification techniques.

Keywords: Heart sounds · Phonocardiogram · Signal processing · Classification algorithms · Preprocessing

1 Introduction

Cardiovascular diseases (CVD) are the major cause of morbidity and mortality worldwide. An estimated 17.5 million people died from CVD in 2012, accounting for 31% of global deaths [1]. Aging population, low fertility rate and increased life expectancy, are factors which indicate a possible rise in the number of patients who suffer from cardiovascular disease. On the other hand, technologies such as magnetic resonance or ultrasound have displaced auscultation in richer countries. This diagnostic technique is still the most important tool for ambulatory clinic and it is inexpensive when compared with the other methods.

© Springer International Publishing AG 2017
I. Rojas and F. Ortuño (Eds.): IWBBIO 2017, Part I, LNBI 10208, pp. 536–545, 2017.
DOI: 10.1007/978-3-319-56148-6_48

Limitations of the audible frequency range, ambient noise and variations in the signals according to the location where they are recorded are the main drawbacks to an effective diagnosis through auscultation. In some countries (specially in underdeveloped ones), access to specialists show reduced rates of more than 50000 patients per doctor. A potential solution is to provide automatic methods for the classification of heart sound signals.

Automatic CVD classification using heart sound signals has been described for over 50 years [2]. However, although auscultation is a common practice in medical examinations, the use of phonocardiogram (PCG) is associated with lower preference if other tools, such as echocardiogram or electrocardiogram, are available. This is partially caused by the lack of robust algorithms for automatically classifying PCGs [3] and by the difficulty for isolating the heart sound signal from the noise. Also, the practical applications of heart sound highly depends on cognitive skills and expertise of the medical examiners.

For example, Fig. 1 shows the frequency distribution of the main components of heart sound (A represents a healthy record while B represents a CVD diagnosed patient). As can be noted, the components S1, S2, S3 and S4 overlap significantly in the frequency domain. Similarly, murmurs and artifacts from respiration and other non-physiological events also overlap notably. Therefore, separation of the fundamental sounds (S1, S2, S3 and S4), as well as the noise, is very difficult in the frequency domain. In addition, the morphological similarity of the noise in normal or abnormal PCG records, converts the identification of these into an extremely complex task in the time domain [4].

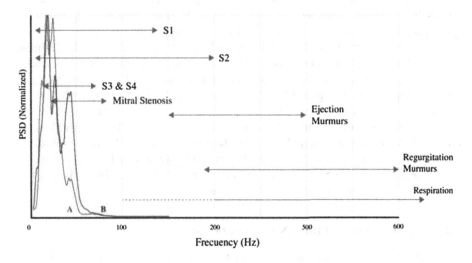

Fig. 1. Spectral region for different cardiac sounds and other physiological sounds. The arrows indicate typical theoretical frequencies. Taken from [4].

Due to the variability of techniques for classification of cardiac sound signals, the lack of a rigorously validated high-quality database and limitations in automatic systems to solve these problems, PhysioNet/CinC Challenge 2016 launched a challenge to the scientific community described in [2,4,5]. Its main goal is to classify each record as normal or abnormal, according to a training data set. In addition, the challengers had to classify those recordings that were too noisy as uncertain. Therefore, no classification technique could be applied to these data.

Typical methods for heart sound classification can be grouped into four categories: Artificial Neural Network (ANN) classification, Support Vector Machines (SVM) based methods, cluster based classification and Hidden Markov Models (HMM). In this sense, ANN is the most commonly used machine learning method, using wavelet attributes [6–8].

In recent years the use of SVM [9,10] for PCG classification has increased as this technique can be combined with other classification techniques and thus the results were improved. For its part, HMM is not only used for segmentation, but also to classify pathologies within phonocardiographic records [11,12]. Finally, cluster-based methods regularly use k-nearest neighbors (kNN) [13,14].

2   Materials and Methods

2.1   Dataset

To achieve the PhysioNet/CinC Challenge 2016 goal, organizers grouped several records around the world in the largest open PCG databases. It is detailed in [4], and includes 4430 records taken from 1072 subjects, for a total of 233512 sounds from healthy and sick patients. Four of the databases were divided into both training and test sets with a 70–30 training-test split.

Up to now, there are 3153 available records which can be used in the training set, while the test set includes a total of 1277 heart sound recordings. The dataset provided by the organizers to evaluate their functions is part of the training set, therefore yileding to the possibility of certain overfitting in the algorithms.

The signal acquisition process was performed from different measurement devices, both in clinical and non-clinical environments (such as home visits). The smallest recording time is 5 s., while the longest takes 120 s. In each recording, a PCG lead was used (except in the training set a which also contains a simultaneously recorded ECG), at nine different locations of the body (aortic, pulmonary, tricuspid and mitral, among others).

2.2   Segmentation Versus Non-segmented Methods

The PCG analysis resides essentially in two categories. The first approach is based on temporal segmentation, i.e. identifying the cardiac cycles and the position of the first and second heart sounds. Length variation in S1 and S2 and their intensities are considered as the conclusive signs of cardiac anomalies [15]. The typical steps in the analysis of PCG in order to diagnose CVD are described in Fig. 2.

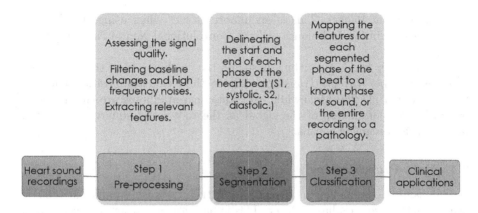

Fig. 2. Typical steps in the analysis of PCG signals using segmentation.

Numerous studies have been developed using PCG segmentation for classification. Most common methods can be grouped into 4 fundamental categories: envelope-based methods [16], feature-based methods [17], machine learning techniques [18] and Hidden Markov Model methods (HMM) [19].

An alternative classification approach consists of not proceeding with signal segmentation [20,21], thus avoiding the segmentation step in the process described in Fig. 2. The advantage of this type of techniques with respect to the first group lies in the elimination of the dependence of a precise segmentation, besides it reduces the computational cost. For example, Zabihi et al. in [15] extracted 18 features from time and frequency domains and they were ranked in the challenge in the second place.

Unfortunately, comparing these two approaches is affected by the lack of a high quality database which it would have been rigorously validated. PhysioNet/CinC Challenge 2016 [5] tries to solve some of these drawbacks by building a well-structured database [4], and allowing researchers to freely choose the alternative they consider most effective, as well as the programming language.

2.3   Feature Selection

A review of the literature shows that a wide variety of features can be used for PCG signal classification. Nonetheless, features can be classified into four main domains: time domain, frequency domain, statistics or a combination of time-frequency [22,23]. The most common time domain features are the standard deviations (STDs) of the main PCG waves (for example, the RR interval), the absolute mean ratio of the amplitude between S1 and S2 [24], kurtosis and amplitude deviation.

Within the frequency domain the most commonly features are those that use the power spectrum of each heart sound state (S1, systole, S2, diastole) using a Hamming window, the Fourier Transform or Mel Frequency Cepstral Coefficents (MFCC) [25]. The use of frequency domain based features allows to treat the data as images. This is an important fact when using classification without segmentation techniques or from Convolutional Neural Networks CNN [26].

On the other hand, algorithms based on the combination of time-frequency domains use the Wavelet Transform as a method of decomposition and feature extraction [27,28]. Meanwhile, the statistical attributes are characterized by extracting parameters of the signal as the root mean square (RMS), QR factor [27] and segment means.

2.4   Data Preprocessing

Default sampling frequency in the database is 2 kHz, so it is resampled at 1000 Hz using a polyphase antialiasing filter. This is because each heart sound was segmented using Springer's improved version of Schmidt's method [29] to identify four states (S1, S2, systole and diastole). Figure 3 shows the procedure followed to evaluate the classification algorithms. In the preprocessing stage, it was verified that the lowest frequency in the database is 20 Hz. To reduce the effect of using multiple recording devices and different environments, all recordings were filtered by a third order bandpass filter of 20–400 Hz before proceeding to signal segmentation.

Noise and artifacts can corrupt the sound recording, and so can also affect the processing data. For instance, in training database *e* completely silent segments in the PCG signal can be observed. The challenge evaluation in this contribution is based on the method described in [28], where if silent segments longer than 25 ms are detected, the longest non-silent region is chosen for further processing and the whole signal is penalized.

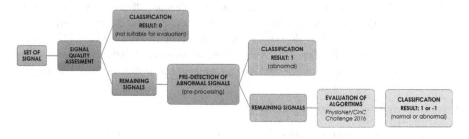

Fig. 3. Flow diagram of the algorithm evaluation process.

The next step was to derive the parameters associated with cardiac registers, such as heart rate or systolic length. For this purpose, the algorithm provided in the sample entry [29] was used with minor modifications. Cardiac cycle and systolic length for the training database are calculated by using the hand-corrections annotation of the records, like the one made by [28]. Heart rate in the training database varied between 35 bpm and 159 bpm, and the systolic interval between 0.172 s and 0.537 s.

Final preprocessing step was the identification of heart sound limits using the original Springer algorithm [29]. As a result, the onset of S1, systole, S3 and diastole were obtained, which will be evaluated in the algorithms developed by PhysioNet/Computing Challenge 2016 participants.

3 Results

3.1 Algorithm Scoring

The final results of the challenge are based on the number of records classified as normal, abnormal or uncertain (too noisy), as shown in the Table 1. The sensitivity (Se) (Eq. 1) and specificity (Sp) (Eq. 2) are defined in [2] and the positions are given according to $MAcc$ as reference Eq. 3. Values wa_1 and wa_2 are the percentages of good signal quality and poor signal quality in all abnormal recordings respectively, and are used as weights for calculating Se. On the other hand, wn_1 and wn_2 are the percentages of good signal quality and poor signal quality in all normal recordings, and are used as weights for calculating Sp.

$$Se = \frac{wa_1 \cdot Aa_1}{Aa_1 + Aq_1 + An_1} + \frac{wa_2 \cdot (Aa_2 + Aq_2)}{Aa_2 + Aq_1 + An_1} \tag{1}$$

$$Sp = \frac{wn_1 \cdot Nn_1}{Na_1 + Nq_1 + Nn_1} + \frac{wn_2 \cdot (Nn_2 + Nq_2)}{Na_2 + Nq_1 + Nn_1} \tag{2}$$

$$MAcc = \frac{Se + Sp}{2} \tag{3}$$

The top ranked algorithm is a combination of the AdaBoost classifier and a Convolutional Neural Network (CNN) [24], based on the segmentation of the PCG signal. In contrast, the second best ranked algorithm [15] combines ANN and SVM to achieve classification, without segmenting the signal. Differences between the scores of the first and the second algorithm are relatively small, as can be seen in Table 1.

Ensemble machine learning algorithms combine the predictions of several learning models into a single "ensemble" model, with the objective of improving their performance. Common approaches to ensemble learning include bagging, boosting, and stacking, among others [30].

Even when segmentation is the most popular method for heart sound classification, it could be verified that there is no marked difference between using segmentation or not for the classification of PCG signals. For example, ranked algorithms 1st and 3rd used segmentation, while algorithms 2nd and 7th do not segment the signal. These algorithms maintain a very similar final score ($MAcc$), with high indexes of sensitivity or specificity. This implies that an appropriate feature selection will be one of the most important steps together with the ability of the algorithm to process those features.

Table 1. Final scores for the top 8 of 48 entrants, the example algorithm provided and a simple voting approach. MFCC (*Mel Frequency Cepstral Coefficients*), NN (*Neural Network*), SVM = (*Support Vector Machine*), CNN (*Convolutional NN*), KNN (*K Nearest Neighbors*). Taken from [2].

Rank	Entrant	Se	Sp	MAcc	Method note
1	Potes et al.	0.9424	0.7781	0.8602	AdaBoost & CNN
2	Zabihi et al.	0.8691	0.8490	0.8590	Ensemble of SVM
3	Kay Agarwal	0.8743	0.8297	0.8520	Regularized neural network
4	Bobillo	0.8639	0.8269	0.8454	MFCCs, wavelets, tensors & KNN
5	Homsi et al.	0.8848	0.8048	0.8448	Random Forest & LogitBoost
6	Maknickas	0.8063	0.8766	0.8415	Unofficial entry - no publication
7	Plesinger	0.7696	0.9125	0.8411	Probability-distribution based
8	Rubin et al.	0.7278	0.9521	0.8399	Convolutional NN with MFCs
17	Shadi et al.	0.7120	0.9015	0.8068	Simple mode
43	Sample entry	0.6545	0.7569	0.7057	Springer algorithm

3.2 Evaluation of Classification Algorithms

The aim of this research is not to modify the models of mathematical classification, or the values trained by the challengers. The main purpose is to modify the search limits of the fundamental components in PCG signals, after the study of the morphological characteristics of data (especially the data that belongs to CVD patients). To this end, the processing technique described in Sect. 2.4 was evaluated in the top ranked algorithms of [5] and the results shown in Table 2 were obtained.

As a result, boundaries to search for the peak were modified, thus corresponding to the heart rate in the autocorrelation function. New boundaries were set 0.4 s and 1.8 s. For systolic peak searching, the upper limit was restricted to a maximum of 550 ms.

It was verified that the number of errors decreases from 112 to 81 in the case of heart rate and from 92 to 47 in the case of systolic length. The estimation is considered correct if it does not differ from the reference by more than 10% in the case of the heart rate and by more than 50 ms in the case of the systolic interval.

As it can be seen in Table 2, the specificity (Sp_{eval}) of algorithms 1st and 3rd is slightly improved and, consequently, the final qualification ($MAcc_{eval}$). Nevertheless, the sensitivity decreases 0.66 units, while the specificity increases around 1.33 points. This increase makes the algorithm more robust compared to the detection and classification of potentially noisy signals.

The methods proposed by Bobillo and Homsi et al. are able to correctly classify 100% of the evaluation set. Instead, when trying to generalize new examples, it does not get as good results as Potes [24] or Zabihi [15]. The reason for this

Table 2. Evaluation of the performance of some algorithms before and after preprocessing. Se_{eval}, Sp_{eval} and $MAcc_{eval}$ are the results obtained with the proposed approach. Best performances are shown in bold.

Rank	Entrant	Se	Se_{eval}	Sp	Sp_{eval}	$MAcc$	$MAcc_{eval}$
1	Potes et al.	**0.9669**	0.9603	0.6867	**0.700**	0.8268	**0.8301**
3	Kay and Agarwal	**0.9338**	0.9272	0.9133	**0.9267**	0.9236	**0.9269**
4	Bobillo	**1.0000**	0.9934	**1.0000**	1.0000	**1.0000**	0.9967
5	Homsi et al.	**1.0000**	0.9934	**1.0000**	1.0000	**1.0000**	0.9967
9	Jiayu	0.9272	**0.9272**	0.6533	**0.6600**	0.7902	**0.7936**
10	Shadi et al.	0.9603	**0.9669**	0.9600	**0.9733**	0.9601	**0.9701**
28	Leal et al.	0.9205	**0.9272**	0.8400	**0.8400**	0.8803	**0.8836**
32	Quintana et al.	0.6358	**0.6755**	0.9133	**0.9200**	0.7745	**0.7977**
37	Tamanna et al.	**0.9205**	0.9139	0.7133	**0.7267**	0.8169	**0.8203**
42	Wang et al.	0.7020	**0.7152**	0.7133	**0.7133**	0.7077	**0.7143**

overfitting behaviour is that the evaluation function is excessively adapted to the training set and it fails when trying to generalize to new samples.

From position 10th, the sensitivity and specificity values of the evaluated methods begin to improve. All algorithms published in [5] which were programmed with Matlab were verified and the main results are shown. It is demonstrated that an adequate data preprocessing can improve the efficiency of techniques for automatic classification of PCG signals. Future research and development could concentrate on the creation of an algorithm that is able to distinguish between the different types of diseases. Noise immunity of the algorithm and its tolarance towards dissimilaties in recording circumstanses should also be improved in the future.

4   Conclusions

This paper presents a machine learning approach for the evaluation of top ranked algorithms for heart sound signals classification. From the results, we can conclude that it is not possible to determine for the moment whether PCG segmentation or not is a better choice. However, the most important stage it is and adequate feature selection and the ability of the algorithm to deal with noise and with the obtained set of features themselves.

It was demonstrated that the modification of the signal limits, the adequate filtering of the records and the derivation of the extracted attributes, improve the classification process. When using the new boundaries, the number of errors decreases from 112 to 81 in the case of heart rate and from 92 to 47 in the case of systolic length. Through these modifications, the results for Se and Sp of the top ranked PhysioNet/CinC Challenge 2016 algorithms were improved. Finally, better ranking results are obtained if more than one automated learning algorithm were used, as the most recent researches show.

Acknowledgments. This work has been partially supported by the project TIN2015-67020-P of the Spanish Ministry of Economy and Competitiveness.

References

1. WHO. 2016 world statistics on cardiovascular disease. Technical report (2016). http://www.who.int/mediacentre/factsheets/fs317/en/
2. Clifford, G.D., Liu, C., Moody, B., Springer, D., Silva, I., Li, Q., Mark, R.G.: Classification of normal/abnormal heart sound recordings: the physionet/computing in cardiology challenge 2016. Comput. Cardiol, pp. 609–612 (2016)
3. Ortiz, J.J.G., Phoo, C.P., Wiens, J.: Heart sound classification based on temporal alignment techniques. In: International Conference on Computing in Cardiology 2016 (2016)
4. Liu, C., Springer, D., Li, Q., Moody, B., Juan, R.A., Chorro, F.J., Castells, F., Roig, J.M., Silva, I., Johnson, A.E., et al.: An open access database for the evaluation of heart sound algorithms. Physiol. Meas. **37**(12), 2181 (2016)
5. Classification of normal/abnormal heart sound recordings: the physionet/cin cardiology challenge 2016 (2016). http://physionet.org/challenge/2016/
6. Akay, Y., Akay, M., Welkowitz, W., Kostis, J.: Noninvasive detection of coronary artery disease. IEEE Eng. Med. Biol. Mag. **13**(5), 761–764 (1994)
7. Liang, H., Nartimo, I.: A feature extraction algorithm based on wavelet packet decomposition for heart sound signals. In: Proceedings of the IEEE-SP International Symposium on Time-Frequency and Time-Scale Analysis, pp. 93–96. IEEE (1998)
8. Uguz, H.: Adaptive neuro-fuzzy inference system for diagnosis of the heart valve diseases using wavelet transform with entropy. Neural Comput. Appl. **21**(7), 1617–1628 (2012)
9. Zheng, Y., Guo, X., Ding, X.: A novel hybrid energy fraction and entropy-based approach for systolic heart murmurs identification. Expert Syst. Appl. **42**(5), 2710–2721 (2015)
10. Ari, S., Hembram, K., Saha, G.: Detection of cardiac abnormality from PCG signal using LMS based least square SVM classifier. Expert Syst. Appl. **37**(12), 8019–8026 (2010)
11. Wang, P., Lim, C.S., Chauhan, S., Foo, J.Y.A., Anantharaman, V.: Phonocardiographic signal analysis method using a modified hidden Markov model. Ann. Biomed. Eng. **35**(3), 367–374 (2007)
12. Saracoglu, R.: Hidden Markov model-based classification of heart valve disease with PCA for dimension reduction. Eng. Appl. Artif. Intell. **25**(7), 1523–1528 (2012)
13. Bentley, P., Grant, P., McDonnell, J.: Time-frequency and time-scale techniques for the classification of native and bioprosthetic heart valve sounds. IEEE Trans. Bio-med. Eng. **45**(1), 125 (1998)
14. Quiceno-Manrique, A., Godino-Llorente, J., Blanco-Velasco, M., Castellanos-Dominguez, G.: Selection of dynamic features based on time-frequency representations for heart murmur detection from phonocardiographic signals. Ann. Biomed. Eng. **38**(1), 118–137 (2010)
15. Zabihi, M., Rad, A.B., Kiranyaz, S., Gabbouj, M., Katsaggelos, A.K.: Heart sound anomaly and quality detection using ensemble of neural networks without segmentation. In: International Conference on Computing in Cardiology 2016 (2016)

16. Moukadem, A., Dieterlen, A., Hueber, N., Brandt, C.: A robust heart sounds segmentation module based on S-transform. Biomed. Signal Process. Control **8**(3), 273–281 (2013)
17. Varghees, V.N., Ramachandran, K.: A novel heart sound activity detection framework for automated heart sound analysis. Biomed. Signal Process. Control **13**, 174–188 (2014)
18. Tang, H., Li, T., Qiu, T., Park, Y.: Segmentation of heart sounds based on dynamic clustering. Biomed. Signal Process. Control **7**(5), 509–516 (2012)
19. Sedighian, P., Subudhi, A.W., Scalzo, F., Asgari, S.: Pediatric heart sound segmentation using hidden Markov model. In: 2014 36th Annual International Conference of the IEEE Engineering in Medicine and Biology Society, pp. 5490–5493. IEEE (2014)
20. Yuenyong, S., Nishihara, A., Kongprawechnon, W., Tungpimolrut, K.: A framework for automatic heart sound analysis without segmentation. Biomed. Eng. Online **10**(1), 1 (2011)
21. Deng, S.W., Han, J.Q.: Towards heart sound classification without segmentation via autocorrelation feature and diffusion maps. Future Gener. Comput. Syst. **60**, 13–21 (2016)
22. Balili, C.C., Sobrepena, M.C.C., Naval, P.C.: Classification of heart sounds using discrete and continuous wavelet transform and random forests. In: 2015 3rd IAPR Asian Conference on Pattern Recognition (ACPR), pp. 655–659. IEEE (2015)
23. Moukadem, A., Dieterlen, A., Brandt, C.: Shannon entropy based on the S-transform spectrogram applied on the classification of heart sounds. In: 2013 IEEE International Conference on Acoustics, Speech and Signal Processing, pp. 704–708. IEEE (2013)
24. Potes, C., Parvaneh, S., Rahman, A., Conroy, B.: Ensemble of feature-based and deep learning-based classifiers for detection of abnormal heart sounds. In: International Conference on Computing in Cardiology 2016 (2016)
25. Godino-Llorente, J.I., Gomez-Vilda, P.: Automatic detection of voice impairments by means of short-term cepstral parameters and neural network based detectors. IEEE Trans. Biomed. Eng. **51**(2), 380–384 (2004)
26. Rubin, J., Abreu, R., Ganguli, A., Nelaturi, S., Matei, I., Sricharan, K.: Classifying heart sound recordings using deep convolutional neural networks and mel-frequency cepstral coefficients. In: International Conference on Computing in Cardiology 2016 (2016)
27. Homsi, M.N., Medina, N., Hernandez, M., Quintero, N., Perpiñan, G., Warrick, A.Q.P.: Automatic heart sound recording classification using a nested set of ensemble algorithms. In: International Conference on Computing in Cardiology 2016 (2016)
28. Goda, M.A., Hajas, P.: Morphological determination of pathological PCG signals by time and frequency domain analysis. In: International Conference on Computing in Cardiology 2016 (2016)
29. Springer, D.B., Tarassenko, L., Clifford, G.D.: Logistic regression-HSMM-based heart sound segmentation. IEEE Trans. Biomed. Eng. **63**(4), 822–832 (2016)
30. Han, J., Pei, J., Kamber, M.: Data Mining: Concepts and Techniques. Elsevier, Amsterdam (2011)

Automatic Glissade Determination Through a Mathematical Model in Electrooculographic Records

Camilo Velázquez-Rodríguez[1](✉), Rodolfo García-Bermúdez[2],
Fernando Rojas-Ruiz[3], Roberto Becerra-García[4], and Luis Velázquez[5]

[1] Grupo de Procesamiento de Datos Biomédicos (GPDB),
Universidad de Holguín, 80100 Holguín, Cuba
`cvelazquezr@uho.edu.cu`
[2] Facultad de Ciencias Informáticas, Universidad Técnica de Manabí,
Manta, Ecuador
`rodgarberm@gmail.com`
[3] Departamento de Arquitectura y Tecnología de Computadores,
ETS Ing. Informática y Telecomunicación, Universidad de Granada,
Granada, Spain
`frojas@ugr.es`
[4] Universidad de Málaga, Málaga, Spain
`idertator@gmail.com`
[5] Centro para la Investigación y Rehabilitacón de Ataxias Hereditarias,
Holguín, Cuba
`cirahsca2@cristal.hlg.sld.cu`

Abstract. The glissadic overshoot is characterized by an unwanted type of movement known as glissades. The glissades are a short ocular movement that describe the failure of the neural programming of saccades to move the eyes in order to reach a specific target. In this paper we develop a procedure to determine if a specific saccade have a glissade appended to the end of it. The use of the third partial sum of the Gauss series as mathematical model, a comparison between some specific parameters and the RMSE error are the steps made to reach this goal. Finally a machine learning algorithm is trained, returning expected responses of the presence or not of this kind of ocular movement.

1 Introduction

Several events are present in electrooculographic (EOG) records, classified as ocular movements according to different criteria. The saccades and fixations are two of the most studied events due to the clinical meaning of those signals. The saccades are fast and accurate ballistic eye movements used in repositioning the fovea to a new location in the visual environment [1] and the fixations are present when an object of interest is held approximately stable on the ocular retina [2].

There are various tests that can be realized in order to evaluate a certain condition of a subject, such as saccadic test, smooth pursuit test, horizontal and

© Springer International Publishing AG 2017
I. Rojas and F. Ortuño (Eds.): IWBBIO 2017, Part I, LNBI 10208, pp. 546–556, 2017.
DOI: 10.1007/978-3-319-56148-6_49

vertical calibration, etc. The saccadic tests are those related with the capture of saccades and fixations as main events. This kind of test is developed in the presence of a stimulus that the subject had to follow with the eyes.

The EOG records have been studied through several years and by many authors. Some characteristics of this kind of biological record that can be observed without much effort are the level of noise that comes with the signal, amplitude and length of the record. Other characteristics are inherent to this signal, so it is necessary to apply different techniques of signal processing to be able of studying it, for instance, the exact points of several events, the velocity in any point of the record and the independent components responsible of the generation of events like the saccades mentioned before.

The exact determination of the events present in EOG records is a complex task due to factors such as the error in the measure of the signal, the presence of unwanted noise captured from several sources and the peculiar dynamics of a very fast kind of movement such as the oculars. However, investigations like [3–6] propose procedures for the identification of saccades and fixations in EOG records.

The work developed by Becerra et al. in [7] reinforce the use of machine learning models that works in a very efficient way in tasks like the determination of events. The investigations mentioned previously allow the extraction of events like saccades and fixations from EOG records, even with the presence of noise. However, in the majority of saccades a phenomenon known as glissadic overshoot happens, which is disregarded and not considered in the segmentation of the signal.

The glissadic overshoot is characterized by an unwanted type of movement known as glissades. The glissades are a short ocular movement that describe the failure of the neural programming of saccades to move the eyes in order to reach a specific target. After the target is passed, comes a rectification of the eyes with the objective to finally reach the desired goal. The steps mentioned above as failure and rectification are summarized by the glissade [8].

One of the first authors that investigated this type of phenomena was Terry Bahill. In [8] he hypothesized that this kind of ocular movement is generated by a mismatch between the neural components that generate the saccades, the pulse and step. Also, in [9] he concludes that from the biological point of view, this movement appended in the end of saccades is caused by fatigue in the saccadic eye movement system. The use of computational models allow the study of overshoot in saccadic eye movements as is exposed in [10]. One of the reasons that provoke the mismatch between the neural signals could be pulse width errors as mentioned in [11].

As the glissadic overshoot is appended after the saccade finalize, sometimes it is considered as part of the saccade movement, which provoke an extended signal larger than the standard saccadic movement. This phenomenon also provokes a shorter fixation caused by the extension of saccade. When a numerical differentiation to the EOG record with glissades is applied, the velocity profile obtained shows a minor movement after the saccade, as shows the Fig. 1.

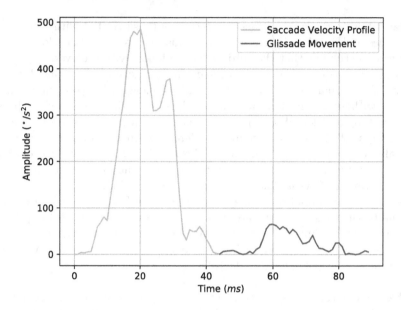

Fig. 1. Numerical differentiation of a saccade with a glissade appended.

In a previous work [12], we describe a mathematical model that make a very good approximation of the velocity profile of a saccade. Through the realization of this work we use a modified version of the previous mathematical model, with the main objective of characterizing the glissadic phenomena.

The knowledge of the meaning of the parameters in the used model, will allow us analyze those related with the glissadic phenomena. The values of these parameters and the presence or not of glissades will let the construction of a dataset in order to train a classification machine learning algorithm. The model trained will classify new saccades with the presence or not of glissades and will give a tool for the researchers to know how to determine a better segmentation of events.

In Sect. 2 we describe how we did the acquisition of the data, as well the mathematical model that we use to describe both, the saccade and glissade. The Sect. 3 is dedicated to the presentation of results and a discussion of it. Finally the conclusions resume all the work done in this investigation.

2   Materials and Methods

The electrooculographic records were obtained by the medical staff of the Centre for the Research and Rehabilitation of the Hereditary Ataxias (CIRAH) at Holguín, Cuba. A two-channel electronystagmograph (Otoscreen, Jaeger-Toennies, Hochberg, Germany) was used to record saccadic ocular movements. The stimulus and patient response data were automatically stored in ASCII files as comma separated values (CSV) by the Otoscreen electronystagmograph, according to its user manual specifications.

Electrooculographic study from several subjects were captured at a sampling rate of 200 Hz with a bandwith of 0.02 to 70 Hz, for a total of 163 study made. Each one of the study have at least tests of $10°, 20°$ and $30°$ of visual stimulation, therefore containing several records captured to the same subject. Typically, saccadic record have at least one horizontal channel and one stimulus signal.

After the records were captured, each one of the tests included in every record was processed. In a first step, the signals recorded were filtered in order to remove the unwanted noise that carry many of these tests. The median filter has proven to be very robust in eliminating high frequency signal noise while preserving sharp edges. A study carried out by Juhola in 1991 demonstrated that this kind of filters is appropriate for eye movements signals [13]. For eliminating non desired noise present in the signals we use a median filter with a window size of 15 samples obtaining good results. This is accomplished using the **medfilt** function of the scientific Python library SciPy [14].

To obtain the velocity profile of the filtered signals, it was applied one of the Lanczos differentiators, specifically Lanczos 11 or Lanczos of 11 points, due that this kind of numerical differentiators use curve fitting instead of interpolation in their procedure [15]. The following equation shows the mathematical description of Lanczos 11:

$$f' \approx \frac{f_1 - f_{-1} + 2(f_2 - f_{-2}) + 3(f_3 - f_{-3}) + 4(f_4 - f_{-4}) + 5(f_5 - f_{-5})}{110\,h} \tag{1}$$

where h is the step of sampling, which in the obtained records has a value of $h = 4.88$ ms.

The velocity profile obtained as a result of the application of Lanczos 11, present positive and negative values representing in a physical way the direction of the saccadic movement. In order to standardize all the velocity profiles obtained, the **absolute value** function provided natively by Python and also included in the numerical library of this programming language named NumPy, was applied to the differentiated signal [16]. One of the velocity profiles obtained is shown in the Fig. 2.

As can be seen in the Fig. 2, there are several velocity profiles in a signal, so it is necessary to split the signals according to the different events that occur in it, in order to get every profile extracted and ready to be modeled. With the goal of separating the velocity profiles we use a **findpeaks** MATLAB similar function in the Python library. The mentioned function is called **peakutils** and is founded in the web repository of Python located in http://pypi.python.org.

The Fig. 3 shows the peaks, marked as red exes, detected through the use of the **peakutils** library mentioned before. To find the start points we use an iterative algorithm that start in the maximum peaks detected and in every iteration go backward, after reaching a defined threshold, we find the minimum in the population of points near to the stopping criteria. The results of this procedure are shown in the Fig. 4.

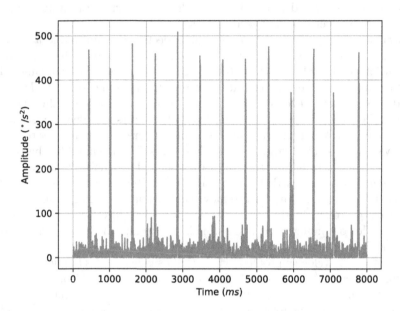

Fig. 2. Velocity profile as result of Lanczos 11 and the **abs** function application.

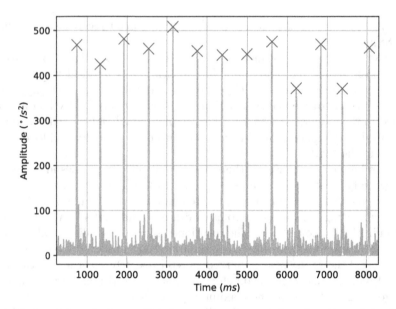

Fig. 3. Peaks of the velocity profiles detected by the use of **peakutils** module. (Color figure online)

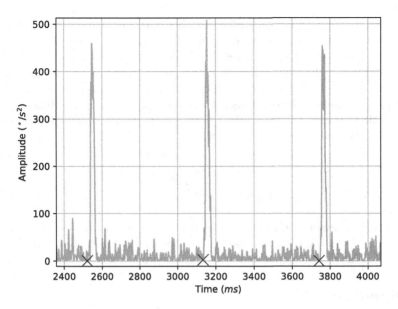

Fig. 4. Start points detected using the backward procedure.

Detected the start points for every velocity profile in the differentiated signal, we can split the latest from start point to the next start point and obtain a part of the signal similar to the one showed in the Fig. 1. The last velocity profile is chosen until the end of the signal.

A velocity profile extracted from the differentiated electrooculographic test, could be modeled for various equations that describe this kind of shape. In a previous work [12] we compare several mathematical models in the task of describing the velocity profile. The best model in that work was the second partial sum of the Gauss series or **gauss2** for short, due to have one of the lower errors in modeling the data, only surpassed by the third partial sum of the Gauss series or **gauss3**.

We choose in [12] gauss2 over gauss3 because of the presence of 3 less parameters to be optimized and similar fitting results. However, in this new scenario, we need to use gauss3 because not only the velocity profile will be fitted, but also the glissade signal appended in the end of the latest. The third partial sum of the Gauss series is described as follows:

$$gauss3(x) = \sum_{i=1}^{3} a_i e^{\left[-\left(\frac{x-b_i}{c_i}\right)^2\right]} \tag{2}$$

where a_i the amplitude, b_i is the location of the centroid and c_i is related to the width of every gaussian peak that is optimized in order to get a good approximation of the model. The most significant parameters for the purpose of this work are b_1, b_2 and b_3. These parameters give the location of centers of each gaussian peak, it is very important to analyze the behavior of this parameters because they can reflect the presence or not of glissades in a velocity profile.

The significance of the parameters involved in the mathematical modeling will present different values according to fitting made. However, if the error in the fitting is large, the meaning of b_i as the parameters taken in consideration, loose validity. The standard error of regression, also known as Root Mean Squared Error (RMSE) is one of the statistical metrics more used in modeling procedures. The Eq. 3 gives an insight of this metric:

$$RMSE = \sqrt{\frac{\sum_{i=1}^{n}(y_i - \hat{y}_i)^2}{n}} \tag{3}$$

where n is the amount of points in the data, y_i represent the ith point of the data and $\hat{y}_i$ is the ith value estimated by the model in evaluation. Values closer to 0 in this metric mean that the fit is more useful for prediction.

3   Results

The 163 study captured, contain 904 electrooculographic records that were used to extract the velocity profiles that conform the available data in this work. After the application of the mentioned procedure in the before section, we extracted 25853 velocity profiles from the EOG records. These data were partitioned, taking 13101 velocity profiles to realize the mathematical curve fitting by the use of gauss3, representing the 50.7% of the available data.

As shows the Eq. 2, the parameters b_i locate the center of each one of the gaussian peaks. We hypothesize that if the error of the curve fitting is low and the saccade present glissade at the end of it, the gauss3 curve will fit this rectification phenomena, as shows the Fig. 5.

The error in the curve fitting that shows the Fig. 5 was 21.8, meaning that it was not an excellent modeling, but good enough to partially describe the glissadic curve as shows the mentioned figure. This visual perception comes more clearly analyzing the chosen parameters b_i with values 48.9, 24.8 and 52.8. The parameters values manifest a separation between one of the values and the other two that are very close to each other.

Making this analysis, a procedure can be formulated for determining the presence or not of glissadic phenomena, due to the values of the selected parameters. This procedure is as follow: analyze the values of the b_i and calculate the absolute values of their differences then, inspect if the three differences are lower than a defined threshold in order to return an unmismatch response, any other case contain the presence of glissades accordingly to the gauss3 mathematical model.

With the data collected by different responses we are able to form a training dataset for a classification algorithm. This dataset will contain the RMSE error result of the fitted procedure, each one of the parameters b_i and the response of the presence or not of glissades as the class of this formed dataset.

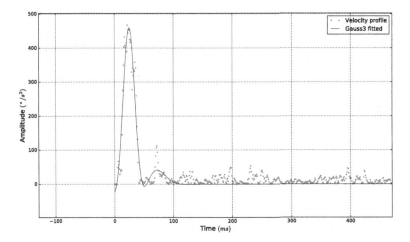

Fig. 5. Velocity profile with glissade, fitted using gauss3 as model.

We use machine learning tools to solve the classification task to determine if a saccade have the presence of glissades or not. Because we are using Python technologies, we selected Scikit-Learn as machine learning library, specifically we evaluate four different models: Support Vector Machines [17], CART decision trees [18], K-Nearest Neighbors [19], and an ensemble method known as Random Forest [20].

Each classification algorithm was trained using a technique known as cross validation, we use the folding equals to 10, meaning that the dataset is sliced in 10 equals parts, one for training and the rest for evaluation, exchanging the trained part in every evaluation. As a final score of the algorithm we find the mean of trained values in each fold, also we find the standard deviation in order to know the total spectrum of the models score. With the goal of observe the behavior of each model, we apply the detailed previous procedure 50 times as can be seen in the Fig. 6.

It is important to mention that we use the classification algorithms with their default values, except from the K-Nearest Neighbors, being modified the number of neighbors in each iteration. Also, it can be seen in the Fig. 6 a better general performance by the ensemble method known as Random Forest, the increment of neighbors in the K-Nearest Neighbors does not increase the performance of this method, on the other hand, make it worse. The best performance of KNN is when the number of neighbors is 4. The CART decision tree, has a good performance also, but not so good like the Random Forest procedure. In the case of the Support Vector Machine algorithm remains invariable in the value of 94.5 of exactitude, due to the invariability of its parameters.

We choose the Random Forest classification trained model due to the presence of the highest scores in the training task, like shows the Fig. 6 some values surpass the 0.96 score, and also the majority of the scores are higher than the

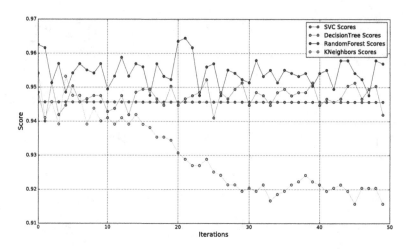

Fig. 6. Classification algorithms training responses.

0.95 value. The model trained can be used to evaluate new velocity profiles that were excluded from the training procedure.

The 49.3% of the available velocity profiles was not considered in the training step in order to realize testing and evaluation of the machine learning model chosen by the authors. This velocity profiles are not classified with the presence or not of glissades, so in order to evaluate the trained model it is necessary to select a portion of the data randomly and manually, classify if it has glissades or not. After obtaining the response of the model, it can be evaluated against the manual classification.

From the total of velocity profiles remaining, 100 records were chosen randomly, following this random process a normal distribution. A visual inspection was made in order to assign a class to the velocity profiles chosen. The objective is to know the presence or not of glissades, so a binary class can be formed assigning the value of 1 to the first statement and 0 to the other.

A curve fitting procedure was applied to the 100 records selected in order to obtain the RMSE metric and the b_i parameters mentioned before. These four collected values will serve as input to the trained Random Forest model. The response of the model was in correspondence with the expected values stated in the manual procedure. Many of the responses were correct, even in some of them, the trained model returned an answer better than the human prediction.

4   Conclusions

This contribution exposes the definition of a natural phenomenon known as glissadic overshoot. This kind of phenomenon is present in most of the saccadic ocular movements, specifically at the end of it, and it is supposed that fatigue is one of the main reasons for the presence of this particular movement.

Also, a computational procedure to automatically determine the glissades in a velocity profile signal is proposed. The computational algorithm involves the mathematical modeling by the use of the third partial sum of the Gauss series, due to the great similarity that this model has with the data analyzed.

The third partial sum of the Gauss series or gauss3 possesses several parameters that describe different parts of the optimized curve. Analyzing the b_i parameters of gauss3, we can define the presence or not of the glissadic phenomena, due to the meaning of this specific parameters that are the centers of the gaussian curves. Also, the RMSE error of each optimization made was measured, in order to know the validity of the values.

An algorithm was developed, that given certain threshold and the values of the b_i, can determine the presence or not of glissades. Obtained the RMSE error, the mentioned parameters and the class of the data, we built a dataset in order to train a machine learning classification algorithm. For this purpose four machine learning paradigms are compared: Support Vector Machines, K-Nearest Neighbors, Random Forest and Classification and Regression Trees, resulting the Random Forest procedure with the better performance.

The trained model can automatically predict with great performance if a determined saccade has appended glissades or not. Many of the responses were in correspondence with the expected values and in others cases improve the human prediction made about certain decisions.

Acknowledgements. This work has been partially supported by the project TIN2015-67020-P of the Spanish Ministry of Economy and Competitiveness.

References

1. Goffart, L.: Saccadic eye movements. In: Encyclopedia of Neuroscience, pp. 437–444. Academic Press, Oxford (2009)
2. Lukander, K.: Measuring gaze point on handheld mobile devices. Ph.D. thesis, Helsinki University of Technology, Helsinki, Finland (2004)
3. Inchingolo, P., Spanio, M.: On the identification and analysis of saccadic eye movements. A quantitative study of the processing procedures. IEEE Trans. Biomed. Eng. **32**(9), 683–695 (1985)
4. Juhola, M., Jäntti, V., Pyykkö, I., Magnusson, M., Schalén, L., Åkesson, M.: Detection of saccadic eye movements using a non-recursive adaptive digital filter. Comput. Methods Programs Biomed. **21**, 81–88 (1985)
5. Wyatt, H.: Detecting saccades with jerk. Vis. Res. **38**(14), 2147–2153 (1998)
6. Salvucci, D.D., Goldberg, J.H.: Identifying fixations and saccades in eye-tracking protocols, Palm Beach Gardens, Florida, United States, pp. 71–78. ACM (2000)
7. Becerra-García, R., García-Bermúdez, R., Joya-Caparrós, G., Fernández-Higuera, A., Velázquez-Rodríguez, C., Velázquez-Mariño, M., Cuevas-Beltrán, F., García-Lagos, F., Rodríguez-Labrada, R.: Non spontaneous saccadic movements identification in clinical electrooculography using machine learning. In: Rojas, I., Joya, G., Catala, A. (eds.) IWANN 2015. LNCS, vol. 9095, pp. 56–68. Springer, Heidelberg (2015). doi:10.1007/978-3-319-19222-2_5

8. Bahill, A.T., Clark, M.R., Stark, L.: Glissades-eye movements generated by mismatched components of the saccadic motoneuronal control signal. Math. Biosci. **26**(3–4), 303–318 (1975)

9. Bahill, A.T., Stark, L.: Overlapping saccades and glissades are produced by fatigue in the saccadic eye movement system. Exp. Neurol. **48**, 95–106 (1975)

10. Bahill, A.T., Clark, M.R., Stark, L.: Computer simulation of overshoot in saccadic eye movements. Comput. Programs Biomed. **4**(4), 230–236 (1975)

11. Bahill, A., Hsu, F., Stark, L.: Glissadic overshoots are due to pulse width errors. Arch. Neurol. **35**(3), 138–142 (1978)

12. García-Bermúdez, R., Velázquez-Rodríguez, C., Rojas, F., Rodríguez, M., Becerra-García, R., Velázquez-Mariño, M., Arteaga-Vera, J., Velázquez, L.: Evaluation of fitting functions for the saccade velocity profile in electrooculographic records. In: Rojas, I., Joya, G., Catala, A. (eds.) IWANN 2015. LNCS, vol. 9095, pp. 592–600. Springer, Heidelberg (2015). doi:10.1007/978-3-319-19222-2_49

13. Juhola, M.: Median filtering is appropriate to signals of saccadic eye movements. Comput. Biol. Med. **21**(1–2), 43–49 (1991)

14. Jones, E., Oliphant, T., Peterson, P.: SciPy: open source scientific tools for Python (2001)

15. Becerra García, R.A.: Plataforma de procesamiento de electrooculogramas. Caso de estudio: pacientes con Ataxia Espinocerebelosa tipo 2. Master Thesis, Universidad de Holguín (2013)

16. Oliphant, T.: Python for scientific computing. Comput. Sci. Eng. **9**(3), 10–20 (2007)

17. Cortes, C., Vapnik, V.: Support-vector networks. Mach. Learn. **20**(3), 273–297 (1995)

18. Breiman, L., Friedman, J., Stone, C.J., Olshen, R.A.: Classification and Regression Trees. CRC Press, Boca Raton (1984)

19. Silverman, B.W., Jones, M.C.: An important contribution to nonparametric discriminant analysis and density estimation. Int. Stat. Rev. **57**(3), 233–238 (1989)

20. Breiman, L.: Random forests. Mach. Learn. **45**(1), 5–32 (2001)

Evaluation of the Differentiation of Noisy Electrooculographic Records Using Continuous Wavelet Transform

Rodolfo Garcia-Bermudez[1]([✉]), Fernando Rojas[2], Gabriel Demera[1],
Christian Torres[1], David Zambrano[1], Gonzalo Joya[3], and Roberto Becerra[3]

[1] Universidad Técnica de Manabí, Portoviejo, Ecuador
{gabriel.demera,christian.torres,david.zambrano}@fci.edu.ec,
rodgarberm@gmail.com
[2] Architecture and Technology, CITIC, University of Granada, Granada, Spain
frojas@ugr.es
[3] Department of Electronic Technology, University of Malaga, Malaga, Spain
gjoya@uma.es, idertator@gmail.com

Abstract. The differentiation of signals in presence of noise results complicated, due to the amplification effect of the traditional methods. In this work we do a preliminary evaluation of the use of the wavelet transform to obtain the first derivative in electrooculographic records generated by means of simulation and strongly contaminated by white and biological noise and interference of 60 Hz of the AC line. The Haar and Gauss1 wavelets produce excellent results when compared with other existing methods, revealing a very promissory field of application of this tool.

Keywords: Wavelet differentiation · Electrooculography · Velocity · Saccades

1 Introduction

In a general sense, noise is one of the main problems to be faced in the processing of biomedical signals [1]. The computation of the derivatives of noisy signals can constitute in many cases a non trivial task, by considering that usually it is impossible to define a derivable analytic function suitable to represent the signal apart from the noise it has incorporated. It is more frequent the use of numerical derivatives by means of some of the methods traditionally used for this purpose. This problem is not completely solved concerning the analysis of electrooculographic records of saccadic movements, which are the rapid displacement of the eyes to change the visual field of attention. These records are formed by discrete voltage values captured by skin electrodes placed around the eyes, representing the angle position of the eyes. However a considerable amount of the algorithms and methods that extract useful information for diagnosis and research are not based on the original position information but on the velocity profile of the signal.

© Springer International Publishing AG 2017
I. Rojas and F. Ortuño (Eds.): IWBBIO 2017, Part I, LNBI 10208, pp. 557–566, 2017.
DOI: 10.1007/978-3-319-56148-6_50

Saccades are a valuable source of information about the condition of patients in a wide group of affections related to the neurology of the human beings, amongst them we could cite the Parkinson disease and hereditary ataxias. Currently there exist an important set of tests specially designed by the specialists to diagnose and evaluate the course of these diseases, including the controlled presentation of visual stimuli conducing to a saccadic response of the patients [2].

On the other hand, these disorders are concurrent with a gradual lost of the motor control of the patient, and they are the cause of involuntary movements and tremors. As a consequence the ocular bio-potentials are very often contaminated with artifacts non directly related to the signal of interest to the analysis of the saccades, and the obtention of the velocity profile becomes more complicated. In our experience the oculomotor records of patients of severe ataxia are very noisy. The saccadic waveform is seriously compromised by the slowness and the late reactions of the subjects, and the presence of a considerable tremor with frequencies between 3 Hz and 7 Hz, added to the usual contamination with the 60 Hz of the AC line and white noise [1,3,4].

In this work, we carry out a preliminary evaluation of the use of wavelets for the differentiation of noisy electrooculographic records. A set of saccadic signal strongly contaminated by noise is generated by means of the simulation, then they are processed to compute the first derivative with Haar and Gauss1 wavelets, and compared to the traditional methods of eight-points central difference and adjacent points difference in order to know its feasibility for this purpose.

1.1 Numerical Differentiation

The numerical differentiation have been probably the most known method to compute the derivatives of discrete signals, amongst them finite differences and polynomial interpolation [5]. The numerical differentiation in general amplifies greatly any noise present in the signal [6].

The finite differences methods can be implemented to carry out the derivatives of first order or higher orders depending on the values of the coefficients, while the accuracy is defined by the number of these coefficients. The points used to the calculation of the derivative can be symmetrical to the point to be differentiated, in this case it is known as central differences, or be to the left or right of the point, and depending on this they are known as backward or forward finite differences. In this work one point forward finite differences and eight points central differences are the selected methods to be compared with wavelet derivatives. The expression for the one point forward finite differences is given in Eq. 1.

$$y' = \frac{y_{+1} - y_0}{t_s} \tag{1}$$

Where:

y: signal to be differentiated
y': first derivative of y
t_s: sampling time

The Eq. 2 is the expression to compute the eight points central differences [7].

$$y' = \frac{a_4(y_{+4} - y_{-4}) + a_3(y_{+3} - y_{-3}) + a_2(y_{+2} - y_{-2}) + a_1(y_{+1} - y_{-1})}{t_s} \quad (2)$$

Where:

$a1 = 0.8024$
$a2 = -0.2022$
$a3 = 0.03904$
$a4 = -0.003732$

1.2 Wavelet Differentiation

The wavelet transform is a tool for the analysis in frequency and time, it has proven to be very valuable for local analysis of nonstationary signals or with fast transient because of its good estimation of time and frequency localizations [8]. The wavelet is defined by the Eq. 3 [8]:

$$C(\tau, a) = \frac{1}{\sqrt{|a|}} \int_{-\infty}^{\infty} x(t)\psi\left(\frac{t - \tau}{a}\right) dt \quad (3)$$

Where:

C: wavelet transform coefficients
ψ: mother wavelet
τ: displacement of the mother wavelet over the signal
a: scale of the wavelet (dilation or contraction)

The mother wavelet ψ is displaced and convoluted with the signal $x(t)$, to obtain the coefficients describing the signal in this level or scale. The scale is defined by the dilation or contraction of mother wavelet, giving the resolution of the analysis of the signal. In the lower scales the wavelet has a good resolution in time, while for higher scales the resolution in frequency is better.

The wavelets have been used before for numeric differentiation, several works have remarked the benefits and limitations of the application of continuous (CWT) and discrete wavelet transform (DWT) [9–13], however most of the references with specific applications we have found are related to the analytical chemistry [14–17].

In a general analysis, covering some of the most known families of the DWT and CWT, these are applied to three basic waveforms, the Gaussian, Lorentzian and Sigmoid functions related to the chromatogram, spectrum and titration in analytical chemistry studies [12]. The wavelet families Daubechies, Symlet, Coiflets, Biorthogonal and Derivative of Gaussians have been proved in this and other works with better results than conventional methods. The wavelet transform combines simultaneously the smoothing with the differentiation, and if an appropriate scale parameter is selected, the signal to noise ratio of the derivatives can be improved without a considerable degrading of the resolution [12].

We have found no references of the differentiation of noisy electrooculographic signals with the application of wavelet transforms. Central point differences methods have been amongst the preferred by the researchers and some implementation of the Savitsky-Golay filters [18–20].

DWT could be not the best option for saccadic signals differentiation because of the short duration of saccades and other impulse waveforms like glissades, if the sampling rate is not adequately high, there will be too few points to perform the successive decimation steps associated to this transform. This is one of the most important considerations to be taken in account to choose the CWT wavelets for the differentiation of these signals.

2   Materials and Methods

2.1   Simulation of the Registers

A set of 30 electrooculographic records was generated, each of the records with 20 saccades separated by fixations of several seconds. This experimental set was uniformly conformed of records with saccadic amplitudes of $10°, 30°$ and $60°$, sampled at a frequency of 1000 samples/s (Fig. 1).

1. A velocity profile was modeled by using the Gamma function described in [21] with a set of pseudo-aleatory parameters generated from a real database.
2. Linear segments were added between saccades in irder to simulate fixations conforming an unique vector.
3. By means of numerical integration, the velocity profile was converted into a signal of position.
4. A mixture of noisy components of amplitude $0.1°$ was added. This combine signal is composed by white noise, colored noise in the interval of 3 Hz to 7 Hz, and interference of 60 Hz.

2.2   Differentiation of the Signal

As it was described above, Haar and derivative of Gauss of order 1 have been used in previous works referenced in the area of analytical chemistry. Taking in account these precedents, we selected these wavelets in order to be compared with finite differences derivatives. The following methods have been used in this work.

1. Haar continuous wavelet transform, with scales from 4 up to 128 increasing in powers of two.
2. Gauss1 (first derivative of Gaussian) continuous wavelet transform, with scales from 4 up to 128 increasing in powers of two.

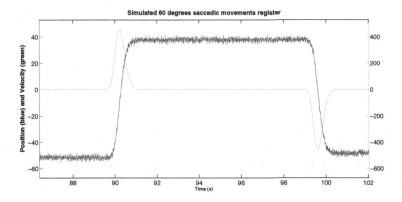

Fig. 1. A section of a simulated saccadic EOG record

3. One point forward finite differences, a median filter with a kernel time of 10 ms was applied to the signal before the calculation of the derivative. The same filter was also used again in order to smooth the differentiated signal.
4. Eight-points central difference, a median filter with a kernel time of 10 ms was applied to the signal before the calculation of the derivative, and the same filter was used again to smoothing the differentiated signal.

2.3   Evaluation of the Differentiation

Root medium square error (RMSE) was the metric used to compute the results of each method. The RMSE was calculated between the velocity calculated by using the derivatives and the original simulated velocity profile. It is known that the maximum value of the velocity profile can be modified due to differentiation. Derivative wavelets at high levels are more affected by this effect as a consequence of the increasing of the smoothing. The same effect was observed in traditional numerical differentiation, caused by low pass filtering.

In order to compute the value of RMSE without the distortion introduced by the displacement of the peak value, a coefficient was calculated in every saccade in order to adjust the derivative waveform with respect to the original simulated velocity profile.

It was implemented by means of the function **fminsearch** from Matlab which was configured to find a value of k to minimize sum of the square of the residuals y as it is expressed in Eq. 4:

$$y = k * vS - vD \tag{4}$$

Where:

vS: Original simulated velocity profile
vD: Velocity profile obtained by using differentiation
k: Coefficient to be calculated

The RMSE of the difference in peak values between the original simulated profile of velocity and the calculated derivative was also computed.

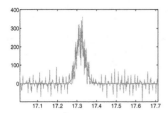

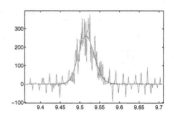

Fig. 2. Differentiation using scale 32 Haar wavelet (blue) and 8-points central difference (red) for 10° (Color figure online)

Fig. 3. Differentiation using scale 8 Gauss1 wavelet (blue) and 8-points central difference (red) for 10° (Color figure online)

All the processing in this work has been made with Matlab R2013 64 bits for Linux. The differentiation by means of the Haar and Gauss1 wavelet was implemented according to the work of Bai et al. [13].

3 Results

After the experiments were performed, a visual inspection of every simulated velocity profile was carried out. Thus, the Haar and Gauss1 wavelets were able to do a better differentiation than both traditional methods of finites differences. However, Haar and Gauss1 wavelets showed the best performance in different scales. Figures 2 and 3 show the derivatives obtained by the Haar and Gauss1 wavelet and the two versions of finite differences.

The presence of considerable levels of noise in the derivatives using finite differences are evident in both figures. However this is completely different in the case of the signals obtained by wavelets derivatives, with a high level of noise cancellation.

Table 1 shows the means (rmseM) and standard deviations (rmseSD) between the differentiated signal with the methods that we are testing and the original

Table 1. RMSE of the differentiation

Amplitude	Wavelet	Param.	Wv4	Wv8	Wv16	Wv32	Wv48	Wv64	Diff1	DifCtrl8	Minimum
10	gauss1	rmseM	15.5	4.6	7.4	22.5	38.6	53.3	38.7	39.4	**Wv8**
10	gauss1	rmseSD	0.8	0.2	0.5	1.3	2.0	2.5	1.7	1.8	
10	haar	rmseM	37.8	24.4	12.2	3.8	6.1	9.8	38.7	39.4	**Wv32**
10	haar	rmseSD	1.8	1.2	0.6	0.2	0.4	0.7	1.7	1.8	
30	gauss1	rmseM	11.3	4.0	8.0	26.3	48.1	69.7	33.0	33.6	**Wv8**
30	gauss1	rmseSD	0.4	0.1	0.2	0.7	1.2	1.6	1.0	0.9	
30	haar	rmseM	29.3	17.8	9.0	3.8	6.4	10.6	33.0	33.6	**Wv32**
30	haar	rmseSD	0.9	0.6	0.3	0.1	0.2	0.3	1.0	0.9	
60	gauss1	rmseM	9.1	3.0	4.9	16.6	32.6	50.0	28.3	28.8	**Wv8**
60	gauss1	rmseSD	0.5	0.1	0.2	0.5	0.9	1.3	1.3	1.3	
60	haar	rmseM	24.4	14.5	7.2	2.6	3.9	6.3	28.3	28.8	**Wv32**
60	haar	rmseSD	1.2	0.8	0.4	0.1	0.1	0.2	1.3	1.3	

Table 2. RMSE of the differentiation peak values

Amplitude	Wavelet	Param.	Wv4	Wv8	Wv16	Wv32	Wv48	Wv64	Diff1	DifCtrl8	Minimum
10	gauss1	rmsePkM	22.6	8.9	22.9	69.3	106.7	133.3	139.1	144.8	Wv8
10	gauss1	rmsePkSD	8.0	6.3	9.4	7.9	7.0	6.5	10.7	11.4	
10	haar	rmsePkM	104.3	45.1	18.3	9.4	17.3	31.1	139.1	144.8	Wv32
10	haar	rmsePkSD	11.3	10.5	6.7	7.4	9.3	9.2	10.7	11.4	
30	gauss1	rmsePkM	189.8	165.9	137.5	57.3	31.7	91.7	314.5	319.4	Wv48
30	gauss1	rmsePkSD	23.0	22.2	21.3	18.8	11.4	14.5	24.0	24.1	
30	haar	rmsePkM	274.4	214.5	184.8	162.6	146.2	125.4	314.5	319.4	Wv64
30	haar	rmsePkSD	23.6	23.9	22.8	22.0	21.5	20.9	24.0	24.1	
60	gauss1	rmsePkM	370.8	347.8	326.6	260.7	180.7	104.3	498.4	499.4	Wv64
60	gauss1	rmsePkSD	41.5	41.2	40.1	37.2	33.5	29.6	37.2	38.7	
60	haar	rmsePkM	456.5	394.8	366.0	344.8	333.6	318.2	498.4	499.4	Wv64
60	haar	rmsePkSD	39.9	41.3	41.5	40.8	40.6	39.7	37.2	38.7	

simulated signal. This table shows a consistent performance for the wavelet differentiation, with significant differences with respect to the traditional methods.

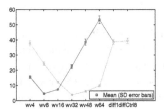

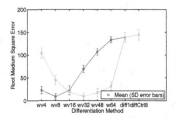

Fig. 4. RMSE for Haar (blue) and Gauss1 (green) wavelet for 10°, the last two points representing traditional differentiation (Color figure online)

Fig. 5. RMSE of the peaks for Haar (blue) and Gauss1 (green) wavelet 10°, the last two points representing traditional differentiation (Color figure online)

For the best scale of the wavelets, the RMSE values and the RMSE of the peaks are visibly lower, in correspondence with the visual inspection of the records. For all the amplitudes of the saccades the scale 8 of Gauss1 and scale 32 of Haar wavelets show the best results in terms of the differentiation RMSE. Average values are significantly lower than both traditional derivatives, while the low values of standard deviation suggest an uniform behavior of the differentiation process.

RMSE of the differences in the peak values of these signals is shown in Table 2, where rmsePkM and rmsePkSD are the means and standard deviation of this peak difference. There exists certain erratic behaviour according to the minimum of this value in the three different amplitudes evaluated.

Figures 4 and 5 show more clearly the behaviour of the differentiators for every level of decomposition of the wavelets for an amplitude of the saccades

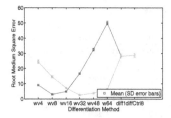

Fig. 6. RMSE for Haar (blue) and Gauss1 (green) wavelet for 60° (Color figure online)

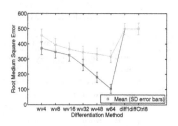

Fig. 7. RMSE of the peaks for Haar (blue) and Gauss1 (green) wavelet for 60° (Color figure online)

of 10°, while Figs. 6 and 7 are for 60°. Although the minimum in each of the mother wavelet is at a different level, in both cases this minimum has a similar low value.

4   Conclusions

The method of eight point central difference in combination with the median filtering has been traditionally used to obtain the velocity profile of electrooculographic records in the research of ocular movements. However our own experience is that this method is very sensible to the noise present in the records from ataxia patients, and it is necessary to apply an aggressive low pass filtering which causes severe deformations of the waveform.

Use of wavelets as a transform to compute derivatives has shown to be effective in the field of analytical chemistry amongst others, but it has not been tested in electrooculographic signals. The results of this work by using simulated saccades strongly contaminated by noise suggest that this could be a successful approach in order to obtain the derivatives of this kind of signals.

A problem that must be solved is the relative decreasing of the peak value of the velocity, although is not so high as with traditional methods, this is a critical parameter in the evaluation of the condition of the patients of neurological diseases. The scale 8 for Haar and 32 for Gauss1 have shown the best performance for the signals with saccadic amplitudes of 10°, 30° and 60°. An important issue to be solved is the determination of the best scale with real records. Further work is necessary with real records of patients of neurological diseases where the presence of biological artifacts like tremor and involuntary movements might hinder the process of differentiation.

It is also necessary to test other mother wavelets and to make comparisons with other differentiation methods as filters of the family of Savitsky-Golay, specially some implementations of Lanczos filters, known by their immunity to noise.

Acknowledgments. This work has been partially supported by the project TIN2015-67020-P of the Spanish Ministry of Economy and Competitiveness.

References

1. Ranjbaran, M., Jalaleddini, K., Lopez, D.G., Kearney, R.E., Galiana, H.L.: Analysis and modeling of noise in biomedical systems. In: 2013 35th Annual International Conference of the IEEE Engineering in Medicine and Biology Society (EMBC), pp. 997–1000, July 2013
2. Chambers, J.M., Prescott, T.J.: Response times for visually guided saccades in persons with Parkinson's disease: a meta-analytic review. Neuropsychologia **48**(4), 887–899 (2010)
3. Garcá-Bermúdez, R., Rojas, F., Becerra García, R.A., Velázquez Pérez, L., Rodríguez, R.: Selection of wavelet decomposition level for electro-oculographic saccadic de-noising. In: Rojas, I., Joya, G., Cabestany, J. (eds.) IWANN 2013. LNCS, vol. 7903, pp. 135–141. Springer, Heidelberg (2013). doi:10.1007/978-3-642-38682-4_16
4. Becerra, R., Joya, G., García Bermúdez, R.V., Velázquez, L., Rodríguez, R., Pino, C.: Saccadic points classification using multilayer perceptron and random forest classifiers in EOG recordings of patients with ataxia SCA2. In: Rojas, I., Joya, G., Cabestany, J. (eds.) IWANN 2013. LNCS, vol. 7903, pp. 115–123. Springer, Heidelberg (2013). doi:10.1007/978-3-642-38682-4_14
5. Li, J.: General explicit difference formulas for numerical differentiation. J. Comput. Appl. Math. **183**, 29–52 (2005)
6. Ahnert, K., Abel, M.: Numerical differentiation of experimental data: local versus global methods. Comput. Phys. Commun. **177**, 764–774 (2007)
7. Inchingolo, P., Spanio, M.: On the identification and analysis of saccadic eye movements-a quantitative study of the processing procedures. IEEE Trans. Biomed. Eng. **BME-32**, 683–695 (1985)
8. Haddad, S.A., Houben, R., Serdijn, W.A.: First derivative Gaussian wavelet function employing dynamic translinear circuits for cardiac signal characterization. In: Proceedings of the ProRISC Workshop on Circuits, Systems and Signal Processing, Veldhoven, The Netherlands, pp. 288–291. Citeseer (2002)
9. Zhang, S., Zheng, J., Gu, W., Zhang, H., Hou, X., Gao, H.: Application of spline wavelet transform in differential of electroanalytical signal. Chin. Sci. Bull. **46**, 550–555 (2001)
10. Chen, Y.-M., Wei, Y.-Q., Liu, D.-Y., Boutat, D., Chen, X.-K.: Variable-order fractional numerical differentiation for noisy signals by wavelet denoising. J. Comput. Phys. **311**, 338–347 (2016)
11. Melcer, T., Danielewska, M.E., Iskander, D.R.: Wavelet representation of the corneal pulse for detecting ocular dicrotism. PloS One **10**, e0124721 (2015)
12. Shao, X., Ma, C.: A general approach to derivative calculation using wavelet transform. Chemometr. Intell. Lab. Syst. **69**, 157–165 (2003)
13. Bai, J., Luo, J., Shao, J.: Application of the wavelet transforms on axial strain calculation in ultrasound elastography. Prog. Nat. Sci. (09), 942–947 (2006)
14. Shao, X., Pang, C., Su, Q.: A novel method to calculate the approximate derivative photoacoustic spectrum using continuous wavelet transform. Fresenius' J. Anal. Chem. **367**, 525–529 (2000)
15. Zhang, X., Jin, J.: Wavelet derivative: application in multicomponent analysis of electrochemical signals. Electroanalysis **16**, 1514–1520 (2004)
16. Nie, L., Wu, S., Lin, X., Zheng, L., Rui, L.: Approximate derivative calculated by using continuous wavelet transform. J. Chem. Inf. Comput. Sci. **42**, 274–283 (2002)

17. Elzanfaly, E.S., Hassan, S.A., Salem, M.Y., El-Zeany, B.A.: Continuous wavelet transform, a powerful alternative to derivative spectrophotometry in analysis of binary and ternary mixtures: a comparative study. Spectrochim. Acta Part A: Mol. Biomol. Spectrosc. **151**, 945–955 (2015)
18. Becerra-García, R., et al.: Non spontaneous saccadic movements identification in clinical electrooculography using machine learning. In: Rojas, I., Joya, G., Catala, A. (eds.) IWANN 2015. LNCS, vol. 9095, pp. 56–68. Springer, Heidelberg (2015). doi:10.1007/978-3-319-19222-2_5
19. Larsson, G.: Evaluation Methodology of Eye Movement Classificatio Algorithms. Skolan för datavetenskap och kommunikation, Kungliga Tekniska höskolan (2010)
20. Yee, R.D., Schiller, V.L., Lim, V., Baloh, F.G., Baloh, R.W., Honrubia, V.: Velocities of vertical saccades with different eye movement recording methods. Invest. Ophthalmol. Vis. Sci. **26**(7), 938–944 (1985)
21. Van Opstal, A.J., Van Gisbergen, J.A.M.: Skewness of saccadic velocity profiles: a unifying parameter for normal and slow saccades. Vis. Res. **27**(5), 731–745 (1987)

Biomedicine

Scores of Intestinal Fibrosis from Wavelet-Based Magnetic Resonance Imaging Models

Ian Morilla[1,3]($\boxtimes$) (iD), Sabrina Doblas[2], Philippe Garteiser[2], Magaly Zappa[2], and Eric Ogier-Denis[3]

[1] Laboratoire Analyse Géométrie et Applications CNRS UMR 7539, Université Paris 13 Sorbonne-Paris-Cité, 93430 Villetaneuse (Paris), France
[2] CRI, U1149, Team "Novel Imaging Biomarkers for Inflammation, Fibrosis and Cancer", Université Paris-Diderot Sorbonne-Paris-Cité, Research Centre of Inflammation, BP 416, 75018 Paris, France
[3] INSERM, UMRS1149, Team "Inflammation Intestinale", Université Paris-Diderot Sorbonne-Paris-Cité, Research Centre of Inflammation, BP 416, 75018 Paris, France
morilla@univ.paris-13.fr
https://www.math.univ-paris13.fr/laga/

Abstract. Intestinal fibrosis is a common complication of inflammatory bowel disease caused by an excessive deposition of extracellular matrix components. Currently, there are no reliable scores to identify early stages of fibrosis prior to clinical symptoms. Potential biomarkers of intestinal fibrosis, including gene variants, serum microRNAs, serum extracellular matrix proteins or circulating cells (i.e. fibrocytes) have shown neither harmonious results nor specific for fibrostenosis on heterogeneous patients' cohorts. In this work, we develop reproducible mathematical models of intestinal fibrosis activity based on continuous wavelet-based analysis of univariate and bivariate time series derived from magnetic resonance imaging on irradiated-rat models of rectitis. This approach enable us to provide clinicians with a non-invasive and reliable score of fibrosis state necessary prior to make any decision on the course of a patient's treatment.

Keywords: Wavelet transform · Intestinal fibrosis · Damage repair · Magnetic resonance imaging · Inflammatory bowel disease

1 Introduction

Severe mucosal tissue damage requiring efficient wound healing is a main feature of inflammatory bowel disease (IBD), in its two disease forms, Crohn's disease and ulcerative colitis (UC). Despite an extensive and detailed investigation on the immunological pathways involved in chronic inflammation in recent years, the physiology and pathophysiology of mucosal wound healing has remained widely unexplored. At this point in time, we are just beginning to understand

© Springer International Publishing AG 2017
I. Rojas and F. Ortuño (Eds.): IWBBIO 2017, Part I, LNBI 10208, pp. 569–578, 2017.
DOI: 10.1007/978-3-319-56148-6_51

the mechanisms that lead from intestinal inflammation to fibrosis. Current concepts view fibrosis as a reactive process. A chronic or recurrent inflammation is considered a necessary precondition for the initiation of intestinal fibrosis. However, there are no reliable scores distinguishing among fibrosis stages. In effect, candidate biodrivers of intestinal fibrosis such as gene variants, microRNAs or fibrocytes display conflicting results on patients with a heterogeneous stratification. On the other hand, magnetic resonance imaging (MRI) has been successfully applied for the diagnosis and characterisation of IBD. Although endoscopy and biopsy are still regarded as standard procedures for IBD evaluation, [1] recent studies indicate that MRI accurately detects and quantifies extent and activity of IBD [2,3]. In the end, MRI transforms a signal from the time or space domain into the frequency domain (see the discussion on natural stimulus statistics in [4]). Therefore, it seems plausible to make use of Fourier (FT) or wavelet transform (WT) in the analysis of such data derived from MRI. Although Fourier techniques have some similar links to wavelet methods, the later ensures that a signal can be spotted in both time and frequency. The name wavelets means small waves (whereas the sinusoids involved in Fourier analysis are "big waves"), and, in short, a wavelet is an oscillation that decays quickly. Upon the prompt progression of the field, wavelets are loosely accepted to be "building blocks that can quickly decorrelate data" [5]. In recent years, wavelets analysis has been applied to a large variety of biomedical signals [6], and there is a growing interest in using wavelets in the analysis of sequence and functional genomics data leading to personalised medicine. In this work, we provide a wavelet-based analysis of magnetic resonance imaging derived data with the goal of creating a reliable score of intestinal fibrosis evolution.

2   Materials and Methods

Briefly, this section provides a fair landscape to understand the analysis procedure showed in Sect. 3.

2.1   Magnetic Resonance Imaging on Irradiated-Rat Models of Rectitis

Magnetic resonance imaging (MRI) is an imaging technique used primarily in medical settings to produce high quality images of the inside of the human body. MRI is based on the principles of nuclear magnetic resonance (NMR), a spectroscopic technique used by scientists to obtain microscopic chemical and physical information about molecules. In this work, we generated 45 rat models of inflammation: 10 samples served as control and the others 35 were irradiated with the goal of measuring fibrosis activity on them upon periods of 2, 3, 4, 6, 9, 11, 12, and 13 weeks. In that way, the parameters captured were associated with the following imaging variables:

1. Flow sensitive alternating inversion recovery (FAIR)
2. Diffusion imaging
3. T_1, T_2 and the spin density ρ
4. Magnetisation transfer contrast in colon

While FAIR amounts to flow sensitive alternating inversion recovery (FAIR)-MRI protocol to measure hemispheric cerebral blood flow in a rat stroke model, diffusion imaging can reveal abnormalities in white matter fiber structure and provide models of brain connectivity. In addition, T_1, T_2 and ρ are the spin-lattice relaxation time, spin-spin relaxation time, and the spin density respectively and are properties of the spins in a tissue. The value of these quantities change from one normal tissue to the next, and from one diseased tissue to the next. They are therefore responsible for contrast between tissues in the various image types. Finally, magnetisation transfer contrast is a new method of increasing the contrast between tissues by physical rather than chemical means.

Scores of Intestinal Fibrosis. Combining the continuous parameters of the MRI variables and the grade of expertise in inflammatory lesions of our practitioners, clinicians based at the "Centre de Recherche sur l'Inflammation" (CRI, U1149); Université Paris-Diderot Sorbonne, tuned a discrete range between 0 and 4 of fibrosis stages. Encouraged by scarcity of reliable evaluating methods in damage repair during the treatment of inflammatory bowel disease, we wanted to extend the usage of these non-invasive scores of inflammation severity by constructing and simulating reproducible mathematical models.

2.2 Wavelet Transform: Mathematical Basics

Our intermediate goal in this stage is to learn parameters from wavelet-based on MRI in order to find mathematical functions reproducing models numerically. To better understand the motivation underlying our choice of wavelet methods, we introduce some basic mathematical concepts about them.

In wavelet theory, a function is represented by an infinite series expansion in terms of dilated and translated version of a basic function ψ called the "mother" wavelet [7–9]. We made used of the continuous wavelet transform (CWT), which for a function $f(t)$ is defined as:

$$CWT(f, \alpha, \beta) = \alpha^{-\frac{1}{2}} \int_{-\infty}^{\infty} f(t)\psi\left(\frac{t-\beta}{\alpha}\right) dt \tag{1}$$

where α (the scale parameter) > 0, β (the translation parameter) $\in \mathbb{R}$. The CWT maps a one-dimensional signal to a two-dimensional time-scale joint representation. It is calculated by continuously shifting a continuously scalable function over a signal and calculating the correlation between the two. We analyse the frequency structure of uni- and bivariate time series derived from MRI parameters using the Morlet wavelet[1]. This continuous, complex-valued wavelet leads

---

[1] Morlet et al., [10,11].

to a continuous, complex-valued wavelet transform of the time series at hand, and is therefore information-preserving with any careful selection of time and frequency resolution parameters. The "mother" Morlet wavelet, in the version implemented in this manuscript is:

$$\psi(t) = \pi^{-\frac{1}{4}} e^{i\omega t} e^{-\frac{t^2}{2}}, \tag{2}$$

We set the rotation rate for the "angular frequency" ω to 6 radians per time unit, which is the agreed value in literature since it makes the Morlet wavelet approximately analytic.

The bivariate analysis allows us to compare the frequency contents of two time series, and projecting conclusions about the series' synchronicity at specific periods and across certain ranges of time. This method is referred to as cross-wavelet analysis. Other measure of our interest implies the time-averaged power calculation since it is useful to investigate the overall strength of periodic phenomena. Additionally, the concept of Fourier coherency measures the cross-correlation between two time series as a function of frequency; an analogous concept in wavelet theory is the notion of wavelet coherency, which, however, requires smoothing of both the cross-wavelet spectrum and the normalising individual wavelet power spectra [12]. Notice all calculations in this stage were executed by the R package "WaveletComp" [13].

2.3   Numerical Methods Reproducing Data from Wavelets Learned Parameters

Upon learning of statistically informative parameters derived from the above described wavelet methods, we perform simulations of autoregressive integrated moving average (ARIMA) model [14] to reproduce data and make inference on our MRI rat models. These models are fitted to time series data either to better understand the data or to predict future points in the series. Seasonal ARIMA models are usually denoted $ARIMA(p, d, q)(P, D, Q)_m$ where parameters p, d, and q, which refer to the non-seasonal part of the model, are non-negative integers, p is the order (number of time lags) of the autoregressive model, d is the degree of differencing (the number of times the data have had past values subtracted), and q is the order of the moving-average model (MA), i.e., a filter applied to white noise whose impulse response (or response to any finite length input) is of finite duration. And m is the number of periods in each season, while uppercase P, D, Q refers to the autoregressive, differencing, and moving average terms respectively for the seasonal part of the ARIMA model. In presence of non-stationarity data, a convenient advantage of these models derived from applying an initial differencing step (corresponding to the "integrated" part of the model) to reduce the non-stationarity [15].

Another approaches to time series such as generalised autoregressive conditional heteroscedasticity (GARCH) [17] – what is an extension of ARIMA if there is reason to believe that, at any point in a series, the terms will have a characteristic size, or variance –, and non-linear methods like NGARCH, Neuronal Networks (NN) [18] or Least-Squares Support Vector Machine (LS-SVM) [19] have also demonstrated themselves as good indicators in time series modelling. However, we opted for the ARIMA models since those have been proven to be successful in a much wider range of experimental results. In any case, the combination of various dissimilar methods in time series modelling, has been already proposed in literature as one possible improvement over the performance of any method individually. This may help to unify time series modelling accuracy [20,21].

3 Results

3.1 Intestinal Fibrosis Wavelet Transformation

Random forest variable importance tests [16] and partial dependence analysis of MRI variables predicting intestinal fibrosis (data not shown) revealed magnetisation transfer contrast as the most informative variable in terms of statistics. We reached up to 80% of fitness along a goodness of 0.05 in the case of functional partial dependence between intestinal fibrosis and magnetisation transfer contrast variables. Afterwards the magnetisation transfer contrast, the variable importance methods set a ranking of MRI variables consisting of the following order: FAIR, T_2, and T_1 density, and diffusion imaging variables. Subsequently, from now onwards, we based our calculations on the instestinal fibrosis and the selected MRI variable dependency. The wavelet power spectrum plot shown in Fig. 1 includes the region of significant periods in $x - axis$ for each $t - period$ released upon 10 (*left panel*) and 100 simulations (*right panel*) respectively. These simulations provide us with p-values to assess whether or not there is periodicity in the series (null hypothesis). We observe how the most significant areas can be characterised by a nonlinear bounded trend in the initial period of 4 weeks.

3.2 Intestinal Fibrosis vs Magnetisation Transfer Contrast Cross-Wavelet Coherency

From the wavelet cross-correlation, i.e., coherency plot (Fig. 2, *left panel*), the period 4 shows joint significance across the entire time interval, but in the middle; according to the construction of magnetisation transfer contrast and intestinal fibrosis series (Fig. 1). The arrows indicate that magnetisation transfer contrast and intestinal fibrosis are in phase on the right and left, with magnetisation transfer contrast leading, whereas (non-significant) phase differences at this period shifting off the middle. In addition, the average power (Fig. 2, *right panel*) decreases for the periods 6, 8, 10, 12 and 13 with shape peaks and low points remaining further separate from each other. This scenario reflects a dense periodicity detection of magnetisation transfer contrast and intestinal fibrosis series.

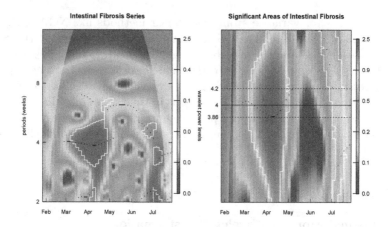

Fig. 1. Wavelet power spectrum of the fibrosis series with nonconstant period. The ridge of the time-period grid appears nonlinear (*left panel*), despite of the period in $x - axis$ increasing linearly. Zoom in to show significant areas of fibrosis series (*right panel*) leads to the central period 4 (*straight line*) in the range [3.86, 4.2] (*dashed lines*) for the logarithmic period scale.

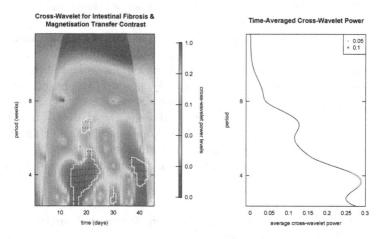

Fig. 2. Cross-wavelet power spectrum for magnetisation transfer contrast and intestinal fibrosis cross-correlation. Significant periods (*highlighted with arrows*) are displayed with time, what stands for the structure of intestinal fibrosis series (*left panel*). The average power (*right panel*) confirming that the version of cross-wavelet power provides sound results for the periods indicated (*solid balls in the right panel*) at significance levels of 10% (blue) and 5% (red). (Color figure online)

3.3   Reconstruction of Fibrosis Series

A natural issue to be solved concerns the wavelet "calibration" in the transformation. In this way, we retrieved the original data from parameters "learned" in the wavelet cross-correlation transform (Fig. 3). As detected in Fig. 2, left and right time regions are on phase ensuring the proper reconstruction of the original fibrosis(magnetisation) series at both areas, whereas the remaining areas show a poorer performance of data retrieval (Fig. 3, *red sinuoids*). From this plot, we can easily validate the "richness" attained by our wavelet cross-correlation based on prior selected periods. This value maybe useful for later filtering purposes.

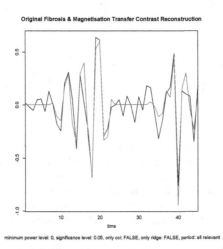

Fig. 3. Reconstruction of the series with non-constant period. The reconstructed magnetisation transfer contrast and intestinal fibrosis original series (*highlighted in red*), based on the selected periods by the wavelet cross-correlation shown in Fig. 2. The $x - axis$ stands for time, $y - axis$ represents scores of fibrosis depending on magnetisation. In black, the original series calculated by cross-wavelet power spectrum. (Color figure online)

3.4 Inferred Mathematical Models Reproduce Scores of Intestinal Fibrosis

We made use of parameters and periods inferred in previous sections to reproduce scores of intestinal fibrosis (Fig. 4). To this end, we fitted data to the best general linear model in ARIMA algorithm. One can observe how the time series reflect slow time-shifting levels, what leads to non-stationarity in mean. Hence, there is the need for stochastic trends correction in $ARMA(p, q)$ within ARIMA model. The simplest method to estimate the overall non/seasonal parts of the ARIMA model is the brute-force scan of the full Cartesian product of all combinations specified by a grid search. The best performance in terms of error bounds for both parts of ARIMA model resulted in the combination between the parameters $(1, 0, 0)$ and $(2, 1, 0)$. Such combination yielded a confidence levels of 95% to predict scores of fibrosis (magnetisation) time period ahead the original series. The selected period identified in Fig. 2 is also included in Fig. 4 (*upper central panel*).

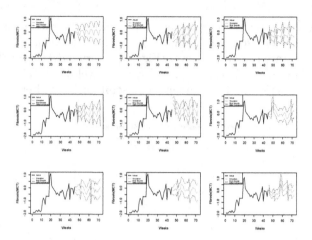

Fig. 4. Numerical simulations of intestinal fibrosis vs. magnetisation transfer contrast series. Reproducible mathematical methods for a wide range of periods (from 3 –closest to selected period– to 13, corresponding to one panel each) including that showed in Fig. 3 (*upper central panel*). These simulations predict scores of intestinal fibrosis depending on the parameters of the MRI variable magnetisation transfer contrast ($x -$ $axis$) with time ($y - axis$). Highlighted in black, the originally calculated fibrosis vs. magnetisation transfer contrast series. In red, predicted scores of fibrosis, whereas the dashed blue lines are the error bounds at 95% of confidence. (Color figure online)

4   Conclusion

Evaluation of intestinal fibrosis is still highly dependent on the subjective opinion of hospital practitioners world-wide. Currently, there is no efficient and reproducible enough methods providing scores of fibrosis in colon, in terms of wound repair activity. Recent studies indicate that MRI bowel-targeted methods can accurately evaluate inflammation during the treatment of a patient having been diagnosed with Crohn's disease or ulcerative colitis. In this study, we have demonstrated the potential ability of wavelet methods in capturing hidden components from MRI data in a cohort of irradiated rat models. Those are reproducible models of inflammation enabling the collection of reliable MRI-based images. The reconstruction of initial scores of intestinal fibrosis based on MRI dependant of magnetisation transfer contrast is used to reinforce their reliability. In addition, mathematical simulation extends these scores to any period of time series. Overall, this methodology would provide clinicians with systematic mechanisms useful to avoid spending a large amount of effort in assessing handpicked therapeutic hypotheses related to wound healing activity in the colon. By means of the extended non-invasive and reliable scores of fibrosis state, provided in here, the diagnosis efficiency and accuracy in preclinical animal studies can be enhanced. Further studies are needed to evaluate the role of MRI in the clinical situation.

Acknowledgments. We acknowledge the financial support by Institut National de la Santé et de la Recherche Médicale (INSERM), Inserm-Transfert, Association François Aupetit (AFA), Université Diderot Paris 7, and the Investissements d'Avenir programme ANR-11-IDEX-0005-02, Sorbonne Paris Cité, Laboratoire d'excellence INFLAMEX.

References

1. Gee, M.S., Harisinghani, M.G.: MRI in patients with inflammatory bowel disease. J. Magn. Reson. Imaging: JMRI **33**(3), 527–534 (2011). doi:10.1002/jmri.22504

2. Nahon, S., Bouhnik, Y., Lavergne-Slove, A., et al.: Colonoscopy accurately predicts the anatomical severity of colonic Crohn's disease attacks: correlation with findings fromcolectomy specimens. Am. J. Gastroenterol. **97**, 3102–3107 (2002)

3. Kettritz, U., Isaacs, K., Warshauer, D.M., Semelka, R.C.: Crohn's disease. Pilot study comparing MRI of the abdomen with clinical evaluation. J. Clin. Gastro. enterol. **3**, 249–253 (1995)

4. Donoho, D.L., Flesia, A.G.: Can recent innovations in harmonic analysis "explain" key findings in natural image statistics? Technical report. Stanford University, Department of Statistics, Stanford, CA (2001)

5. Sweldens, W.: Wavelets: what next? Proc. IEEE **84**, 680–685 (1996)

6. Aldroubi, A., Unser, M. (eds.): Wavelets in Medicine and Biology. CRC Press, Boca Raton (1996)

7. Daubechies, I.: Ten Lectures on Wavelets. SIAM, Philadelphia (1992)

8. Chui, C.K.: An Introduction to Wavelets. Academic Press, New York (1992)

9. Mallat, S.G.: A theory for multiresolution signal decomposition: the wavelet representation. IEEE Trans. Pattern Anal. Mach. Intell. **11**, 674–693 (1989)

10. Morlet, J., Arens, G., Fourgeau, E., Giard, D.: Wave propagation and sampling theory - part I: complex signal and scattering in multilayered media. Geophysics **47**, 203–221 (1982)

11. Morlet, J., Arens, G., Fourgeau, E., Giard, D.: Wave propagation and sampling theory - part II: complex signal and scattering in multilayered media. Geophysics **47**, 203–221 (1982)

12. Liu, P.C.: Wavelet spectrum analysis and ocean wind waves. In: Foufoula-Georgiou, E., Kumar, P. (eds.) Wavelets in Geophysics, pp. 151–166. Academic Press, San Diego (1994)

13. Roesch, A., Schmidbauer, H.: WaveletComp: computational wavelet analysis. R package version 1.0. (2014). http://www.hs-stat.com/projects/WaveletComp/WaveletComp_guided_tour.pdf

14. Hamilton, J.D.: Time Series Analysis. Princeton University Press, Princeton (1994)

15. Kwiatkowski, D., Phillips, P.C.B., Schmidt, P., Shin, Y.: Distribution of the estimators for autoregressive time series with a unit root. J. Econom. **54**, 159–178 (1992)

16. Diaz-Uriarte, R.: GeneSrF and varselRF: a web-based tool and R package for gene selection and classification using random forest. BMC Bioinform. **8**, 328 (2007)

17. Bollerslev, T.: Generalized autoregressive conditional heteroskedasticity. J. Econom. **31**(3), 307–327 (1986). doi:10.1016/0304-4076(86)90063-1

18. Shanmuganathan, S.: Artificial Neural Network Modelling, 1st edn. Springer International Publishing, Heidelberg (2016)

19. Burges, C.J.C.: A tutorial on support vector machines for pattern recognition. Data Min. Knowl. Disc. **2**, 121–167 (1998)
20. Zhang, G.P.: Time series forecasting using a hybrid ARIMA and neural network model. Neurocomputing **50**, 159–175 (2003)
21. Armstrong, J.S.: Findings from evidence-based forecasting: methods for reducing forecast error. Int. J. Forecast. **22**, 583–598 (2006)

L_1-regularization Model Enriched with Biological Knowledge

Daniel Urda[1,3]([✉]), Francisco Aragón[2], Leonardo Franco[2,3],
Francisco J. Veredas[2,3], and Jose M. Jerez[2,3]

[1] Universidad de Málaga, Andalucía Tech, ETSI de Informática, Málaga, Spain
durda@lcc.uma.es
[2] Universidad de Málaga, Departamento de Lenguajes y Ciencias de la Computación,
ETSI de Informática, Málaga, Spain
[3] Instituto de Investigación Biomédica de Málaga (IBIMA),
Inteligencia Computacional en Biomedicina, Málaga, Spain

Abstract. Biomarkers discovery in RNA-Seq gene expression data aims
to identify a subset of genes with high prediction capabilities in complex
human traits. Ideally, robustness of a genetic signature is an a priori
desired property difficult to obtain in practice due to the nature of the
data. In this paper we propose a simple way of including biological knowledge into an l_1-regularization model. In particular, the paper describes
two different approaches (*Gene-specific*, *Gene-disease*) of how to extract
biological information from an online resource (PubTator) based on a
gene citations scheme, and it shows how to use it to estimate a predictive model. The results of the analysis carried out in a public RNA-
Seq dataset of breast invasive cancer (BRCA) shows that our proposal
obtains some improvement in the predictive performance and a substantial improvement in terms of biomarkers stability (genetic signatures
400% times more robust in the worst case scenario).

Keywords: RNA-Seq · Biomarkers selection · Biological knowledge ·
Machine learning · Precision medicine

1 Introduction

Human global health is a research field with unstopping and increasing interest
throughout the years. Over the last decade, a huge investment has been made to
improve high-throughput sequencing technology in order to sequence a sample
of a patient faster, cheaper and with higher quality procedures [1,2]. Analyzing
individual-level data is encouraging since there is a hope that the information
encoded at genes, transcripts and proteins levels may explain differences, in
terms of diagnostic or response to treatments, to identical clinical profiles (which
is known as precision medicine [3]). This information is, however, complex in
nature, as datasets usually consist of a small number of samples (N) in contrast
to the thousand of variables (P) describing each sample, what is commonly
known as large-p-small-n problems [4].

© Springer International Publishing AG 2017
I. Rojas and F. Ortuño (Eds.): IWBBIO 2017, Part I, LNBI 10208, pp. 579–590, 2017.
DOI: 10.1007/978-3-319-56148-6_52

Several machine learning (ML) models and techniques have been traditionally used to develop predictive models in different areas [5–8], providing high performance rates in these large-p-small-n problems [9,10]. In this sense, one of the key procedures in the development of predictive models for complex human traits based on genomic data is, therefore, feature selection (FS). Three different well-known approaches have been proposed and published in the literature to address this issue: filter, wrapper and embedded procedures [11]. Independently of the FS procedure used, the goal is to identify a genetic signature with high prediction capabilities in a totally new and unseen test dataset, different to the one used to build the predictive model.

Nevertheless, biomarkers selection becomes unstable as soon as the number of features gets larger (like in the $P >> N$ scenario), specially due to the existing high correlation among the thousands of genes describing each sample. In fact, Van't Veer and colleagues [12] came up with a genetic signature of 70 genes that allows to predict clinical outcome of breast cancer with a good performance rate, and this signature is actually implemented in a commercial product known as the *MammaPrint test*. Two years later, Wang and colleagues [13] published a genetic signature of 76 genes that performed as well as the one discovered in [12], although only 3 genes were overlapped across both gene signatures. Finally, Venet et al. [14] showed that one can randomly pick any subset of genes that will significantly be associated with breast cancer outcome. These results clearly indicate that it is necessary to impose some constrains to the ML models and FS techniques to overcome the huge variability observed.

In this paper, we propose a simple linear model with l_1-regularization that incorporates prior biological knowledge from a well-known public repository. The proposal aims to quantify the importance of a given gene in the estimation of the predictive model based on the number of citations found in PubTator [15–17] for that particular gene. Based on the results published in thousands of previous studies, this approach makes those genes with a larger number of citations more likely to be selected by the FS procedure and thus being included in the final genetic signature. Furthermore, a hypothetical less important gene will also be part of the genetic signature if this gene adds predictive value. Two different approaches for quantifying the importance of each gene are proposed in this paper (Gene-specific, Gene-disease) and their predictive performance and biomarker stability have been tested on a public RNA-Seq gene expression dataset for breast invasive carcinoma. Additionally, we show the advantages of our methodology in a controlled artificial dataset.

The rest of the paper is organized as follows: Sect. 2 describes the datasets used within the experiments as well as the proposed methodology. Next, the results obtained both in the artificial and BRCA datasets are shown in Sect. 3. Finally, in Sect. 4 we present some conclusions obtained from this work.

2   Materials and Methods

2.1   Dataset

A public RNA-Seq gene expression dataset for breast invasive adenocarcinoma (BRCA), freely available at The Cancer Genome Atlas (TCGA) website[1], was used to carry out the present analysis. This dataset has already been batch-corrected and RSEM normalized [18]. In addition, we first removed those genes that do not show any expression across the samples (they do not add predictive value) and we performed a log_2 transformation of the genes expression level to ensure they closely approximate to a normal distribution. In order to test and show the potential benefits of our methodology, an artificial controlled dataset based on this real RNA-Seq data was generated as follows. First, $K = 100$ random genes out of the total number of $P = 20021$ genes were selected. Then, $K = 100$ β coefficients were randomly chosen from a uniform distribution between $[0, 1]$. And finally, the class label of this artificial dataset was generated by applying the logistic function (see Eq. 2) to the linear combination of the 100 β coefficients and pre-selected genes. Therefore, the ground truth is a priori known within this artificial dataset and the best solution that can be obtained is the identification of those K genes among the initial P ones. In Table 1 the overall details of both datasets used are shown. In both considered cases, the event of interest is the vital status of a given patient ($0 =$ "*alive*", $1 =$ "*dead*") at a fixed time t.

Table 1. Overall description of the datasets: number of samples (N), number of genes (P) and class distribution (*control* $= 0$, *cases* $= 1$).

Name	N	P	Controls	Cases
Artificial	1212	20021	583	629
BRCA	1212	20021	1013	199

2.2   Methodology

In this work, we propose to use an l_1-regularization model with built-in prior biological knowledge from public well-known repositories. L_1-regularization models are simple linear models with a LASSO penalty [19], that works by trying to set as many coefficients as possible to zero unless the data tell us not to do it. LASSO models have been previously shown to work well in the large-p-small-n scenario being able to overcome overfitting issues.

Let us assume that a dataset is represented as $D = \{x_i, y_i\}$, with $i \in \{1 \dots N\}$ samples, x_i representing the vector of P features describing the i-th sample, and y_i being the class label. Under the LASSO approach, an objective function is minimized, as shown in Eq. 1 for a classification setting:

---

[1] https://cancergenome.nih.gov/.

$$\min_{\boldsymbol{\beta}} \sum_{i=1}^{N} (y_i - F_{log}(\boldsymbol{\beta x_i}))^2 + \lambda \sum_{j=1}^{P} |\beta_j| \tag{1}$$

where the function F_{log} corresponds to the logistic function and is defined as follows:

$$F_{log}(x) = \begin{cases} 1, & \text{if } \frac{1}{1+e^{-x}} \geq 0.5 \\ 0, & \text{otherwise} \end{cases} \tag{2}$$

Using this definition of the LASSO problem, we are considering homogeneous priors over the independent variables x_{ij}, where $j \in \{1 \ldots P\}$ genes, i.e., every single gene is equally treated and regularized in the optimization procedure. An extension of this model where heterogeneous priors are considered has been introduced in 2006 and named adaptive-LASSO [20]. This method allows to weight each coefficient differently in the l_1-penalty and it has been shown to perform as well as if the true underlying model is given in advance. Equation 3 shows the minimization problem that the adaptive-LASSO approach tries to optimize for a classification setting.

$$\min_{\boldsymbol{\beta}} \sum_{i=1}^{N} (y_i - F_{log}(\boldsymbol{\beta x_i}))^2 + \lambda \sum_{j=1}^{P} \gamma_j |\beta_j| \tag{3}$$

It is easy to infer that Eq. 3 is equivalent to Eq. 1 when $\gamma_j = 1, \forall j \in \{1 \ldots P\}$. On the other hand, if $\gamma_j = 0$, then Eq. 3 would be equivalent to the objective function of the logistic regression when no regularization is applied. Therefore, adaptive-LASSO can be seen as a model that lies in between simple LASSO and logistic regression, and where $\boldsymbol{\gamma}$ could be interpreted as a vector that somehow measures how important each gene is. In this sense, the lower and closer to zero γ_j is, the more important the j-th gene will be and, thus, the fewer it will be regularized in the optimization procedure. And vice versa, the bigger and closer to one, then the lower important the j-th gene is and the more its β_j coefficient will be regularized in the optimization procedure.

To this end, we propose to enrich the adaptive-LASSO model by building a vector $\boldsymbol{\gamma}$ based on online public repositories. This will allow us to incorporate prior biological knowledge obtained from thousands of previous studies with the hope of getting better and more robust results. In concrete, we propose to come up with a vector of individual-gene importance using PubTator[2]. This online resource incorporates manual literature curation from PubMed citations, in particular storing all the PubMed IDs of articles that have been published together with the gene or genes referenced on each of those articles, as well as the disease or diseases involved in the study.

In this sense, our proposal considers two approaches for the $\boldsymbol{\gamma}$ score construction, both based on a counting citations scheme that uses genes IDs to merge both databases (PubTator and the studied RNA-Seq dataset):

---

[2] https://www.ncbi.nlm.nih.gov/CBBresearch/Lu/Demo/PubTator/.

1. Gene-specific: given the j-th gene, γ_j will be linked to the number of times that the j-th gene is cited within the repository in any article.
2. Gene-disease: given the j-th gene and a disease of interest, e.g. BRCA, then γ_j is linked to the number of times that the j-th gene is cited within the repository in any article linked to breast invasive adenocarcinoma. The advantage of this approach is that it does not take into account references of the j-th gene linked to other diseases. The elimination of these references from the counting scheme assumes that in these cases the j-th gene may not be important to predict the event of interest (BRCA).

Independently of the scheme, the procedure basically starts with $\gamma_j = 1$ and then increments the count by one every time a citation of interest is found within the repository. Once this procedure is carried out for every gene in the dataset, the final γ vector is obtained as shown in Eq. 4:

$$\gamma = (1/\gamma)^\epsilon \tag{4}$$

The parameter ϵ controls the smoothness of the individual-gene regularization for those genes identified as significant according to the information in the repository (genes with more citations), where $\epsilon \in (0, 1]$. A high value of ϵ, e.g. $\epsilon = 1$, implies a smoother individual-gene regularization in those genes. On the other hand, if $\epsilon \approx 0$ then the individual-gene regularization will be more aggressive and closer to the regularization applied to those genes with no citations in the repository. The best value for the ϵ parameter is empirically chosen from a set of sensible values tested within the analysis.

3 Results

In all the experiments, a 10-fold cross-validation strategy where the complete dataset is partitioned in 10 folds of equal sizes was used to estimate the performance of each model. In this sense, models are fitted in 9 folds and tested in the unseen test fold left apart within an iterative procedure that rotates the train and test folds used. Since the BRCA dataset is highly imbalanced (see Table 1), the Area Under the Curve (AUC) is used to measure the goodness of a given model fitted to data. The whole analysis was carried out under the R software using the package "glmnet" [21] which already implements a nested cross-validation to learn the regularization parameter λ.

3.1 Artificial Data

Using the artificial controlled dataset generated as described in Sect. 2.1, a set of different LASSO models were tested mainly differenced in the number of genes used and the way each gene is penalized in the analysis. Table 2 shows the results obtained for each of these models, where:

– $LASSO_{200}$: standard LASSO model with homogeneous priors fitted to the K genes used plus another 100 randomly selected.

Table 2. Average test data results obtained in a controlled artificial dataset using different models. The Area Under the Curve (AUC), average number of selected genes (#genes), and average number of genes within the K genes used to generate the class label (#genes*) are shown.

Model	AUC	#genes	#genes*
$LASSO_{200}$	0.9920 ± 0.00	137.9	88.7
$LASSO_{2000}$	0.9504 ± 0.02	230.7	56.5
$LASSO_{20021}$	0.9325 ± 0.03	286.8	24.8
$LASSO_{19921}$	0.8972 ± 0.03	254.8	0
$E_1\text{-}LASSO_{20021}$	0.9805 ± 0.01	133.7	66.9
$E_2\text{-}LASSO_{20021}$	0.9923 ± 0.01	100	100

- $LASSO_{2000}$: equivalent to the previous one but adding 1900 randomly selected genes to the K genes used.
- $LASSO_{20021}$: equivalent to the first one but now fitted to the entire dataset, thus using the whole 20021 genes.
- $LASSO_{19921}$: standard LASSO model with homogeneous priors fitted to the entire dataset after removing the K genes used to generate the class label.
- $E_1\text{-}LASSO_{20021}$: a LASSO model with heterogeneous priors set in such a way that $\gamma_j = 1$ for the 19921 genes not used to generate the class label in contrast to a random value in $(0, 1)$ assigned to the K genes that were used. The ϵ parameter was fixed to 1.
- $E_2\text{-}LASSO_{20021}$ equivalent to the previous one but now $\gamma_j = 0$ for the K genes used to generate the class label.

The results shown on Table 2 confirm what we would expect to see when using the artificial data set. The first three settings show how the complexity of the analysis increases while more genes are added to the dataset (the AUC drops from 0.9920 to 0.9325). Accordingly, the average number of selected genes within the K genes used to generate the class label (column *#genes**) reflects how unstable is the FS procedure when the aim is to identify the ground truth in wider datasets (*#genes** drops from 88.7 to 24.8 genes). In addition, the fourth setting supports the statement made in [14], where a good performance (AUC $= 0.8972$) can be achieved even if the K genes were not included for the analysis. In this sense, it shows that it is possible to find another genetic signature with high prediction capabilities due to the existing high correlation among genes. Finally, the last two settings show the advantages of incorporating biological knowledge into the l_1-penalty. The fifth setting simulates a possible scenario where the K genes used are less regularized (γ_j set to a random number in $(0, 1)$), supposing that these genes were found to be more important according to an online public resource. Then, the obtained AUC goes up to 0.9805 in contrast to 0.9325 of the setting where homogeneous priors were used in the l_1-penalty. At the same time, the average number of selected genes within the K genes used to generate the class label is now 66.9 instead of 24.8. The last setting shows an ideal scenario, supposing that we can set $\gamma_j = 0$ for the K genes used according to the information within an online public resource (in practice, this may be unachievable). In this case, we would be able to get the genetic signature that is known to be the ground truth ($\#genes^* = 100$) and the best performance (AUC $= 0.9923$).

3.2   BRCA Data

The BRCA RNA-Seq dataset was analyzed in order to predict the vital status of a patient and compare the results from three different settings: (i) standard LASSO with homogeneous priors as baseline model, (ii) LASSO with heterogeneous priors obtained by the *Gene-specific* approach, and (iii) LASSO with heterogeneous priors obtained by the *Gene-disease* approach. Ten different and fixed values of ϵ were considered for (ii) and (iii). As we are not only interested in outperforming our baseline model (LASSO with homogeneous priors) in terms of accuracy but also in terms of stability of the genetic signatures obtained, we propose to use a robustness index (RI) defined as follows:

$$RI = \frac{\#(\cap genes_f)}{\#genes}, f \in [1, 10] \tag{5}$$

where the numerator corresponds to the number of overlapping genes across the 10 folds of the cross-validation and the denominator measures the average number of retained genes across the folds. The higher and closer to 1 the RI is, the more robust the solution, as a larger overlap will be found in the genetic signatures.

Table 3 shows the results for the different models. In terms of AUC and for both approaches proposed in this paper (*Gene-specific, Gene-disease*), it is possible to find a parameterization of the models for which the baseline is outperformed. The *Gene-specific* approach achieves an AUC of 0.7 using $\epsilon = 0.1$ and the *Gene-disease* an AUC of 0.69 setting $\epsilon = 0.2$, in contrast to the 0.66 of a standard LASSO model with homogeneous priors. It may not look an impressive improvement, but 0.04 is still quantitatively a good result taking into account that LASSO is a linear model that assumes a linear relationship between the independent variables and the outcome, thus not capturing possible non-linearities existing in the data. Moreover, these results encourage us to properly tune the hyper-parameter of our model, ϵ, e.g., by learning it from data through nested cross-validation, with the hope of pushing the performance further.

Regarding the stability of the genetic signatures, both of our approaches outperform the baseline model by far. The worst case setting of the *Gene-specific* approach obtains a RI of 0.08 compared to 0.02, thus being 400% times more robust. Further, the *Gene-disease* approach is even more robust, achieving a RI of 0.19 for the worst case setting ($\epsilon = 0.4$) meaning that there is almost a 20% of overlapping genes across genetic signatures. In [12,13], only 3 out of 70–76 genes were respectively overlapped among the genetic signatures provided. That means less than 5% of overlapping. Therefore, our proposal in this paper turns out to be a good choice to improve biomarker stability as well as predictive performance.

Table 3. Average test data results obtained in the BRCA RNA-Seq dataset for the baseline (standard LASSO with homogeneous priors) and our proposed approaches. The Area Under the Curve (AUC), average number of selected genes (#genes), and robustness index (RI) are shown.

Model	ϵ	AUC	#genes	RI
Lasso	-	0.66 ± 0.06	259.5	0.02
Gene-specific	**0.1**	**0.70±0.06**	**232.3**	**0.19**
Gene-specific	0.2	0.67 ± 0.06	262.3	0.18
Gene-specific	0.3	0.66 ± 0.05	288.7	0.17
Gene-specific	0.4	0.64 ± 0.05	252.8	0.08
Gene-specific	0.5	0.62 ± 0.08	250.6	0.08
Gene-specific	0.6	0.64 ± 0.06	229.7	0.09
Gene-specific	0.7	0.64 ± 0.06	183.5	0.10
Gene-specific	0.8	0.64 ± 0.07	128.5	0.18
Gene-specific	0.9	0.61 ± 0.07	192.3	0.18
Gene-specific	1	0.61 ± 0.08	238.8	0.26
Gene-disease	0.1	0.67 ± 0.10	193.2	0.20
Gene-disease	**0.2**	**0.69±0.05**	**212.4**	**0.22**
Gene-disease	0.3	0.67 ± 0.04	234.8	0.21
Gene-disease	0.4	0.65 ± 0.04	202.1	0.19
Gene-disease	0.5	0.66 ± 0.04	170.8	0.29
Gene-disease	0.6	0.67 ± 0.04	148.6	0.39
Gene-disease	0.7	0.67 ± 0.04	149.9	0.42
Gene-disease	0.8	0.68 ± 0.04	148.8	0.46
Gene-disease	0.9	0.67 ± 0.04	147.2	0.48
Gene-disease	1	0.67 ± 0.04	155.4	0.5

Focusing our attention on the predictive performance of the proposed approaches, the *Gene-specific* one has more variability on the results. Averaging the AUC across the values tested for ϵ, it results in 0.643 ± 0.028 and 0.67 ± 0.01 for *Gene-specific* and *Gene-disease,* respectively. Despite the *Gene-specific* providing the best individual setting performance (AUC = 0.7 when $\epsilon = 0.1$), in average it performs worse than the *Gene-disease* approach. A possible explanation for this result would be the way γ is built on the *Gene-specific* approach. Here, all citations found in PubTator of a given gene are taken into account, whether the papers linked to those citations are related to BRCA or not. This definition is logically introducing some noise in γ because some genes may be set not to be regularized (or less regularized) when in reality it has not yet been published any association of those genes to BRCA (but to other diseases). Therefore, this result suggests to consider the *Gene-disease* approach as a preferable choice, although this also opens the question of considering other definitions of γ or even possible combinations of different definitions of γ.

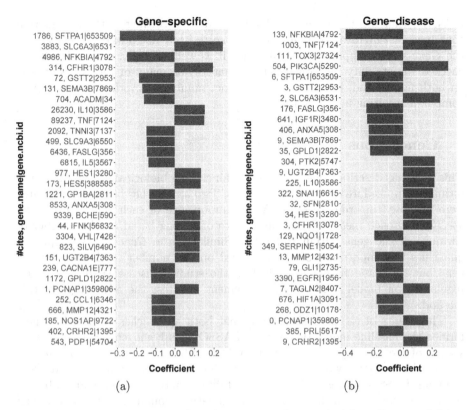

Fig. 1. Summary of the top-30 selected genes according to the $|\beta|$ coefficients of, (a) a genetic signature of 213 genes obtained using a LASSO model enriched with the *Gene-specific* approach, setting $\epsilon = 0.1$, and fitted to the complete dataset; (b) a genetic signature of 295 genes obtained using a LASSO model enriched with the *Gene-disease* approach, setting $\epsilon = 0.2$, and fitted to the complete dataset.

An added value of parametric models and linear models in particular is that they are interpretable models, thus allowing analysts to verify the contribution of each retained gene on the target outcome. Therefore, we chose the best configuration for the *Gene-specific* and *Gene-disease* approaches according to Table 3 and fitted the corresponding models to the complete BRCA data. In this sense, Fig. 1 shows a sorted list of the top-30 genes that contribute more to predict the outcome in both genetic signatures. The higher a gene appears in the figures, the more it contributes to predict the vital status of a patient. In addition, those genes highly expressed with positive coefficients (bars positioned to the right) will increase the chances of not surviving while genes highly expressed with negative coefficients (bars positioned to the left) are protective of not surviving. Discussing the discovery of new associations is beyond the scope of this paper but in principle it can be highlighted how our proposal, as an extension of LASSO, allows to do it. In concrete, the *Gene-disease* approach finds the gene

PCNAP1 that has not been linked yet to BRCA according to the information contained in PubTator (it does have one record but not linked to BRCA and the *Gene-specific* approach actually finds it as well). Finally, it can visually be checked the stability of both genetic signatures as many of the genes appear on both figures and, in particular, the exact number of overlapping genes was computed resulting in 135 genes.

4   Conclusions

This work has described one simple way of including biological knowledge in the estimation procedure of genetic signatures. Two approaches based on PubTator were proposed in order to quantify the importance of each gene. The *Gene-specific* approach measures the importance of a gene based on the number of citations of the gene within PubTator. Similarly, the *Gene-disease* approach quantifies the importance of a gene based on the number of citations of the gene but restricted to papers linked to the studied disease (BRCA). An l_1-regularization model was used to implement and test the advantages of this method.

The results of the analysis showed that both approaches proposed outperformed the baseline model (a standard LASSO with homogeneous priors). One setting of the *Gene-specific* achieved an AUC of 0.7 while for the *Gene-disease* case an AUC of 0.69 was obtained, noting that AUC = 0.66 was obtained using the plain LASSO approach. In terms of stability, both approaches are substantially more robust regarding the genetic signatures found. The baseline model showed a RI of 0.02 while the *Gene-specific* and the *Gene-disease* approaches increased the RI up to 0.08 and 0.19 respectively on their worst case scenario. In other words, the genetic signature found by the *Gene-specific* is 400% more robust than the one obtained by plain LASSO. The best case of the *Gene-disease* approach achieved a RI of 0.5 with no loss of predictive performance compared to the baseline model, meaning that there is a 50% of genes overlapped across the genetic signatures. In a pairwise comparison of the two approaches proposed, the *Gene-disease* approach appeared to be a preferable choice as a better and more robust average performance across the different settings tested was obtained. Furthermore, as these approaches are an extension of the LASSO family, they allow to discover genes that have not been previously cited in the repository when these genes have some predictive value.

The authors believe that this methodology can be further extended and improved. On one hand, there was a clear indication in the results that the method may perform better with a specific tuning of the ϵ values, e.g., by learning them from the data through nested cross-validation. Additionally, not only other ways apart from PubTator could be considered to measure the importance of a gene, but also identifying a possible combination of γ vectors built from different resources may potentially improve the quality of the constrained search procedure and the final results therefore. And finally, it will be more challenging but at the same time desirable to expand this methodology to other non-parametric or more sophisticated ML methods such as Support Vector Machines (SVM), Gaussian Processes (GP), Deep Learning (DL), etc.

Acknowledgements. The authors acknowledge support through grants TIN2014-58516-C2-1-R from MICINN-SPAIN which include FEDER funds, and from ICE Andalucía TECH (Spain) through a postdoctoral fellowship.

References

1. Reuter, J., Spacek, D.V., Snyder, M.: High-throughput sequencing technologies. Mol. Cell **58**(4), 586–597 (2015)
2. Kircher, M., Kelso, J.: High-throughput dna sequencing – concepts and limitations. BioEssays **32**(6), 524–536 (2010)
3. Aronson, S.J., Rehm, H.L.: Building the foundation for genomics in precision medicine. Nature **526**(7573), 336–342 (2015)
4. Johnstone, I.M., Titterington, D.M.: Statistical challenges of high-dimensional data. Philos. Trans. Roy. Soc. Lond. A: Math. Phys. Eng. Sci. **367**(1906), 4237–4253 (2009)
5. Stallkamp, J., Schlipsing, M., Salmen, J., Igel, C.: Man vs. computer: benchmarking machine learning algorithms for traffic sign recognition. Neural Netw. **32**, 323–332 (2012)
6. Perlich, C., Dalessandro, B., Raeder, T., Stitelman, O., Provost, F.: Machine learning for targeted display advertising: transfer learning in action. Mach. Learn. **95**(1), 103–127 (2014)
7. Ghahramani, Z.: Probabilistic machine learning and artificial intelligence. Nature **521**(7553), 452–459 (2015)
8. Deng, L., Hinton, G., Kingsbury, B.: New types of deep neural network learning for speech recognition and related applications: an overview. In: Proceedings of the International Conference Acoustics, Speech and Signal Processing (2013)
9. Fukunaga, K., Hayes, R.R.: Effects of sample size in classifier design. IEEE Trans. Pattern Anal. Mach. Intell. **11**(8), 873–885+ (1989)
10. Brain, D., Webb, G.I.: On the effect of data set size on bias and variance in classification learning. In: Richards, D., Beydoun, G., Hoffmann, A., Compton, P. (eds.) Proceedings of the Fourth Australian Knowledge Acquisition Workshop (AKAW 1999), pp. 117–128. The University of New South Wales, Sydney (1999)
11. Saeys, Y., Inza, I., Larrañaga, P.: A review of feature selection techniques in bioinformatics. Bioinformatics **23**(19), 2507 (2007)
12. van 't Veer, L.J., Dai, H., van de Vijver, M.J., He, Y.D., Hart, A.A.M., Mao, M., Peterse, H.L., van der Kooy, K., Marton, M.J., Witteveen, A.T., Schreiber, G.J., Kerkhoven, R.M., Roberts, C., Linsley, P.S., Bernards, R., Friend, S.H.: Gene expression profiling predicts clinical outcome of breast cancer. Nature **415**(6871), 530–536 (2002)
13. Wang, Y., Klijn, J.G., Zhang, Y., Sieuwerts, A.M., Look, M.P., Yang, F., Talantov, D., Timmermans, M., Meijer-van Gelder, M.E., Yu, J., Jatkoe, T., Berns, E.M., Atkins, D., Foekens, J.A.: Gene-expression profiles to predict distant metastasis of lymph-node-negative primary breast cancer. Lancet **365**(9460), 671–679 (2005)
14. Venet, D., Dumont, J.E., Detours, V.: Most random gene expression signatures are significantly associated with breast cancer outcome. PLOS Comput. Biol. **7**(10), 1–8 (2011)

15. Wei, C.H., Kao, H.Y., Lu, Z.: PubTator: a web-based text mining tool for assisting biocuration. Nucleic Acids Res. **41** (2013)

16. Wei, C.H., Harris, B.R., Li, D., Berardini, T.Z., Huala, E., Kao, H.Y., Lu, Z.: Accelerating literature curation with text-mining tools: a case study of using PubTator to curate genes in PubMed abstracts. Database(Oxford) **18** (2012)

17. Wei, C.H., Kao, H.Y., Lu, Z.: PubTator: a PubMed-like interactive curation system for document triage and literature curation. In: BioCreative 2012 Workshop, vol. 05 (2012)

18. Li, B., Dewey, C.N.: RSEM: accurate transcript quantification from RNA-Seq data with or without a reference genome. BMC Bioinform. **12**(1), 323 (2011)

19. Tibshirani, R.: Regression shrinkage and selection via the lasso: a retrospective. J. Roy. Stat. Soc.: Ser. B (Stat. Methodol.) **58**(1), 267–288 (1996)

20. Zou, H.: The adaptive lasso and its oracle properties. J. Am. Stat. Assoc. **101**(476), 1418–1429 (2006)

21. Friedman, J., Hastie, T., Tibshirani, R.: Regularization paths for generalized linear models via coordinate descent. J. Stat. Softw. **33**(1), 1–22 (2010)

Secret Life of Tiny Blood Vessels: Lactate, Scaffold and Beyond

Vladimir Salmin[1,2], Andrey Morgun[1,2], Elena Khilazheva[1,2],
Natalia Pisareva[1,2], Elizaveta Boitsova[1,2], Pavel Lavrentiev[1,2],
Michael Sadovsky[1,2(✉)], and Alla Salmina[1,2]

[1] Research Institute of Molecular Medicine and Pathobiochemistry,
Krasnoyarsk State Medical University,
p. Zheleznyaka str., 1, 660022 Krasnoyarsk, Russia
vsalmin@gmail.com,msad@icm.krasn.ru,sadovsky.mikhail@gmail.com
[2] Institute of Computational Modelling of SB RAS, Akademgorodok,
660036 Krasnoyarsk, Russia
{441682,elena.hilazheva,allasalmina}@mail.ru
http://icm.krasn.ru, http://krasgmu.ru

Abstract. We studied the model of cerebral angiogenesis *in vitro* using lactate-releasing gelatin bioscaffolds and primary culture of brain endothelial cells. We found that development of microvessels from actively proliferating rat brain microvessels endothelial cells was greatly modified by the presence of lactate at the surface of the scaffold with different lactate-releasing ability. Fractal dimension of newly-established vessel loops allows precise characterizing the local microenvironment supporting cell growth on various types of gelatin scaffolds.

Keywords: Brain microvessel endothelial cells · Angiogenesis *in vitro* · Gelatin scaffold · Fractal dimension · Form factor

1 Introduction

Cerebral angiogenesis (development of new vessels from the pre-existing ones) is a complex process occurring in developing, mature, and aging brain. Establishment of new microvessels is governed by local concentrations of pro-angiogenic factors released by activated neuronal and glial cells. As a result, endothelial cell layer surrounding by pericytes, astroglial endfeet, and neurons, form the neurovascular unit (NVU) and the blood-brain barrier (BBB) [8]. The latter is responsible for selective permeability for huge number of molecules whose activity greatly depends on structural and metabolic plasticity of endothelial cells.

Being activated, tip and stalk endothelial cells respond to the actual levels of locally produced factors (VEGF, angiopoietics, cytokines etc.) and start proliferation, differentiation and migration along the concentration gradient of regulatory stimuli. It is commonly accepted that these events are under the control of VEGF-VEGFR-Delta-like ligand 4 (Dll4)-Jagged-Notch pathway [13].

© Springer International Publishing AG 2017
I. Rojas and F. Ortuño (Eds.): IWBBIO 2017, Part I, LNBI 10208, pp. 591–601, 2017.
DOI: 10.1007/978-3-319-56148-6_53

In addition, endothelial progenitor cells either of bone-marrow origin or local tissue residents with pericyte phenotypic characteristics contribute a lot to the angiogenic events in the active or damaged brain regions [12]. Angiogenesis is stopped by dramatic changes in the concentrations of molecules with pro- or anti-angiogenic activities. Then, maturation of BBB resulting in establishment of fully active tight junctions between the contacting endothelial cells leads to sealing the barrier in the newly-formed microvessel.

Microvascular density in the brain depends on various endogenous (hormones, cytokines etc.) and exogenous (running, cognitive training, brain stimulation etc.) regulatory signals. Deciphering molecular mechanisms of cerebral angiogenesis associated with brain activity or post-injury recovery would give us new approaches to pharmacotherapy of brain disorders and controlling the BBB permeability.

Recently, many systems for *in vitro* assessment of angiogenesis have been developed. They usually consist of a polymer scaffold (collagen, fibronectin, gelatin, matrigel etc.) which provides the substrate for endothelial cell adhesion [6]. However, mechanistic reconstruction of the basement membrane analogue or extracellular matrix composition could not reproduce the complex microenvironment required for controlled endothelial cells proliferation and differentiation. Therefore, there are many attempts to develop biomimetic scaffold fully matching functional and metabolic plasticity of endothelial cells *in vitro* and *in vivo* [1].

There is the accumulating evidence that metabolic plasticity of brain microvessel endothelial cells (BMEC) and perivascular cells of non-endothelial lineage (i.e. astrocytes or pericytes) affects key angiogenic events as is clearly seen within the NVU in (patho)physiological conditions. As an example, highly-glycolytic astroglial cells may not only respond to the metabolic needs of neighboring neurons by releasing lactate which serves as an alternative fuel for mitochondria-enriched neurons, but may also be involved in the regulation of BMEC functional activity, angiogenesis, and BBB maturation [9]. Therefore, we have proposed that changing local concentrations of lactate would control the proliferative potential and angiogenic activity of BMEC grown on bioscaffolds. To prove this, the model of *in vitro* cerebral angiogenesis was established and analyzed.

2   Material and Methods

Experiments have been performed on Wistar rats. All the procedures with experimental animals have been done in a strict accordance to the principles of European Regulations. The study was approved by the Local Ethic Committee of the Prof. V.F. Voino-Yasenetsky Krasnoyarsk State Medical University.

2.1   Preparation of Lactate Gradient Gelatin Scaffolds (LGGS)

Lactate-releasing gelatin scaffolds (LGGS) with the step lactate gradients were prepared from holographic films (VRP-M, Slavich, Russian Federation). The films were treated as follows:

(1) removal of AgCl and sensitizing stain at the fixate solution at 20 °C; upon the exposure for 20 min they were washed in deionized water at 20 °C, for 20 min;
(2) treatment in tanning solution with paraformaldehyde at 20 °C, for 120 min followed by washing in deionized water at 20 °C, for 20 min (for tanned F type of LGGS only);
(3) drying at RT for 1 h;
(4) cutting the LGGS;
(5) coating the half of the LGGS surface with the protecting coverage at the gelatin film side;
(6) saturation of non-coated LGGS side in 40% lactate solution for 45 min;
(7) removal of superficial lactate film with cold water at 15−20 °C, for 3−5 s;
(8) drying at 75 °C for 45 min, and
(9) UV sterilization of LGGS followed by removal of protecting coverage and placement of the LGGS into the wells of the culture plate.

2.2   Cell Culture

BMEC have been isolated from the brain of Wistar rats (postnatal day 10). Isolation and establishment of primary culture of BMEC was performed according to the protocol of Liu et al. [7]. The BMEC obtained were phenotyped with monoclonal antibodies against brain microvessels endothelial marker (adaptor protein ZO1) along the standard immunohistochemistry protocol using primary anti-ZO1 antibodies (Santa Cruz Biotechnology, Inc., sc-8147) and secondary antibodies labeled with Alexa Fluor 488 (Abcam, ab150117) followed by the detection at the luminescent microscope.

For further experiments BMEC have been plating on LGGS (diameter of 16 mm) at the border of lactate-releasing and non-lactate-releasing halves of the scaffold (5×10^4 cells per scaffold) in 12-wells plate (Corning, 3516). The cells were cultured in DMEM/F12, 20% FBS, with glutamine, antibiotics/antimicotic added. To prolong phase preceding cells monolayer establishment, BMEC plates were placed asymmetrically at one pole of the well (Fig. 1; here M is the zone of cell monolayer, A is angiogenesis zone, G&L is lactate-releasing part of gelatin scaffold, G is non-lactate-releasing part of gelatin scaffold. Gated area represents the zone of interest to analyze angiogenic events in the cell culture). When proliferation of BMEC started, monolayer zone (M) expanded, thereby the zone of angiogenesis (A) gradually moved along the border of lactate-releasing half of scaffold (G&L) and non-lactate-releasing half of the scaffold (G).

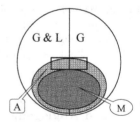

Fig. 1. Zones of brain microvessel endothelial cells growth *in vitro*; see details in text.

2.3   Angiogenesis Assessment

An assessment of angiogenesis *in vitro* was carried out in the gated area of angiogenesis at various time moments (120–390 min) after the cell plating using phase-contrast microscopy (ZOE, Bio-Rad Laboratories, Inc., USA). The number of BMEC-derived loops, the square inside the loops (mm^2), and the length of loop perimeter (mm) have been calculated and analyzed according to the protocol [10]. The size of the area was 0.5×1.0 mm from each side of the border between G&L and G zones. The location of area was chosen as close as possible to the border of BMEC monolayer. Image analysis was performed with the ImageJ 1.43 software (NIH, USA).

2.4   Statistical Analysis

Descriptive statistics (mean and standard deviation) was used. The data are presented as $mean \pm SD$ (standard deviation). Student's t-criterion was used for parametric analysis with $p \leq 0.05$ considered as statistically significant. Regression analysis was applied for data analysis.

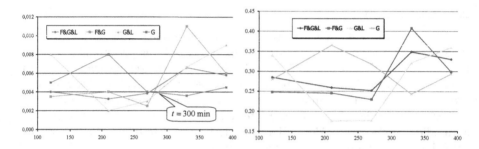

Fig. 2. Dynamic changes of parameters of square (left) and perimeter (right) of BMEC-derived angiogenic loops formed on the scaffolds with the step gradient of lactate with or without preliminary paraformaldehyde tanning. Horizontal axis is time (in minutes), and vertical axis is square (in mm^2, left) and perimeter (in mm, right).

2.5   Description of the Fractal Structure of Blood Vessels and Angiogenic Cell Culture

Blood itself could hardly be considered as an ideal liquid, especially in meso- and microscales. Indeed, it consists of a real solution of salts, large biological moleculae, and a suspension of various biological particles (erythrocytes, etc.). Besides, those particle are permanently involved into a tremendous network of various biological and biochemical reactions running both in a blood, and on a border between blood and other tissues. Hence, it makes blood to be rather viscose substance with a non-trivial circulatory dynamics.

Thus, the blood vessels are highly adapted to the features mentioned above. These peculiarities force to implement some special tools and techniques for the analysis of the patterns of vessels development; fractal dimension of that latter might be valuable.

In general, fractal structure is defined as a self-similarity in a pattern. Skipping a rich mathematical theory of that phenomenon, we focus on the regular manifestation of fractal structuredness: unlike the regular relations between geometric sizes of a figure (say, diameter, or something similar), and volume, a fractal structure exhibits a fractional exponent in such relations [2,11].

For example, a circle area S is related with the radius R of that latter in the following form: $S = \pi R^2$. Changing for logarithms, one gets a linear equation $\ln S = \ln \pi + 2 \ln R$, or in the most general form,

$$y = \alpha + \gamma x \tag{1}$$

with $\gamma = 2$, where y and x stand for logarithms of area and radius, respectively. If shape of an area differs from a circle, both α and γ may change. Here we focus on γ behaviour. Actually, for a number of shapes, γ remains the same; $\gamma \neq 2$ (and, moreover, is not integer) for rather jammed and distorted patterns.

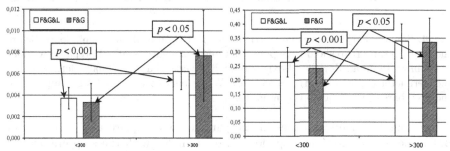

Fig. 3. Square (left) and perimeter (right) of BMEC-derived angiogenic loops before and after 300 min (threshold time point) exposure obtained on gelatin scaffold prepared with preliminary tanning. Horizontal axis is time, and vertical axis is perimeter (left, in mm^2) and square (right, in mm), respectively. Here and further (see Figs. 4 and 7) the inset boxes indicate the data couples yielding a reliable difference observed in the experiments.

Another type of fractal structure may take place with rather smooth and circle-like objects. Here the fractal structure describes not a single profile, but an ensemble of them. The patterns observed over the cultivated cells look like a tiling of a surface with a (considerably dense) set of figures which are pretty close to a circle. Meanwhile, the radii of those tiles are different, and the gaps between the tiles are small enough to neglect them. The pattern, simply speaking, resembles a foam layered over a substrate. Here the fractality of pattern manifests in the equation similar to (1), but y here is logarithm of an average area of cells, and x is logarithm of an average radius of them. Further, we shall follow this definition of fractality, in our studies.

3 Results

We tested various approaches to establish LGGS able to create adequate microenvironment for BMEC proliferation and differentiation evident as angiogenic loop formation and variability of loop fractal variance, respectively. Therefore, we applied various protocols including establishment of lactate-embedded gelatin scaffolds and preliminary gelatin film tanning with paraformaldehyde to control the speed of lactate diffusion from the scaffold to the medium. Final concentration of scaffold-derived lactate in the medium was $2\,\mu M$ that corresponds to the real lactate level in the brain tissue *in vivo*.

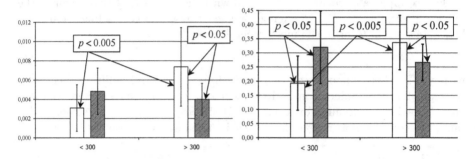

Fig. 4. Square (left) and perimeter (right) of BMEC-derived angiogenic loops before and after 300 min (threshold time point) obtained on gelatin scaffold prepared without preliminary tanning. Horizontal axis is time, and vertical axis is square (left, in mm^2) and perimeter (right, in mm), respectively. White bars correspond G&L composition, and grey bars correspond to G composition.

We found dynamic changes in the parameters of square and perimeter of BMEC-derived angiogenic loops formed on scaffolds with the step gradient of lactate with or without paraformaldehyde tanning (see Fig. 2); here and everywhere in the text F&G&L is tanned lactate-releasing scaffold, F&G is tanned non-lactate-releasing scaffold, G&L is non-tanned lactate-released scaffold, and G is non-tanned non-lactate-releasing scaffold. Minimal deviation of the parameters at all the scaffolds tested was evident at 270–340 min. Then, 300 min time point could be used as a threshold time for further analysis of angiogenesis-associated changes in the cell culture.

Values of mean square and perimeter of angiogenic loops before and after 300 min (threshold time point) obtained on gelatin scaffold prepared with tanning are shown on Fig. 3. Significant changes were evident in the lactate-enriched and lactate-free parts of the scaffold, however, there were no differences between square and perimeter values in both the zones. Also, enlargement of loops is clear in G&L and G parts of the scaffold.

Table 1. Parameters of fractal dimensions and form factors for various types of gelatin scaffolds.

Type of LGGS	D	F
F& G& L	1.424752	−3.64299
F&G	2.224755	−2.55444
G&L	2.084779	−2.72607
G	1.840845	−3.04429

Dynamic parameters of angiogenesis on the scaffold prepared without tanning are demonstrated in Fig. 4. Significant dynamic difference was observed in the part of lactate-releasing scaffold. Lactate-enriched and lactate-free zones of cell growth could be distinguished on both the mentioned parameters after 300 min of cell growth, but perimeter values were significantly different even before the threshold time point. Enlargement of BMEC-derived angiogenic loops was evident in the G&L zone of the scaffold.

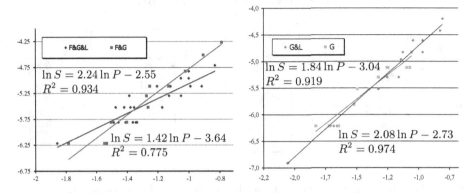

Fig. 5. Fractal dimensions D and form factors F of BMEC-derived angiogenic loops obtained on tanned (left) and non-tanned (right) gelatin scaffolds. Horizontal axis is logarithm of perimeter, and vertical axis is logarithm of square.

Evaluation of loop shape could be done with the parameter of fractal dimension D and geometric form factor F. Therefore, we determined the parameters of trend lines for all the types of scaffolds:

$$\ln(S) = D * \ln(P) + F. \tag{2}$$

Figure 5 demonstrates the parameters of fractal dimension and formfactors of loops obtained on tanned (top) and non-tanned (bottom) scaffolds. P was $< 10^{-5}$ for the difference between fractal dimension and form factors in the zones with or without lactate on both the scaffold types (Table 1).

We found that parameters of fractal dimension and form factors strongly correlated (Fig. 6). Here and everywhere in the text F&G&L is tanned lactate-releasing scaffold, F&G is tanned non-lactate-releasing scaffold, G&L is non-tanned lactate-released scaffold, and G is non-tanned lactate free scaffold.

Thus, the loop geometry on various gelatin scaffolds could be characterized with the single parameter (fractal dimension or form factor). Let us use further fractal dimension only for characterization of a scaffold. Dynamics of fractal dimension for BMEC-derived angiogenic

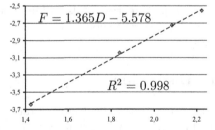

Fig. 6. Correlation of fractal dimension D (horizontal axis) and form factor F (vertical axis) for the data obtained from all the types of gelatin scaffolds (see details in Table 1). Straight line shows the linear trend.

loops appeared on the scaffolds of various types shown on Fig. 7 confirms that this parameter D is an invariant with weak time-dependence but with a sufficient power to distinguish the types of scaffolds.

We follow a hypothesis that cell differentiation potential determined as the ability to form angiogenic loops of various fractal dimension should affect fractal dimension D parameter. Then, the dynamics of cell differentiation could be easily tracked from the ratio of D variance to D itself: $\eta = \sigma_D/D$.

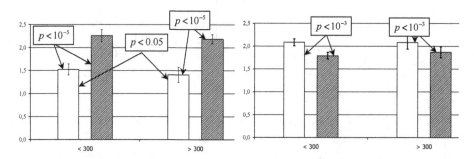

Fig. 7. Fractal dimension D in all the scaffolds types tested (tanned is left, and non-tanned is right); horizontal axis it time (in minutes), and vertical axis is D (in relative units). White bars correspond F&G&L composition, and grey bars correspond to F&G composition.

Figure 8 shows the dynamics of BMEC differentiation potential obtained from D parameter for various scaffolds types. Lactate presence in the scaffold results in growth of D variance, especially for tanned gelatin scaffolds. A variation of σ_D to D ratio in time is shown with four linear trend equations:

$$\eta_{F\&G\&L} = 0.0003t - 0.0014 \tag{3}$$

$$\eta_{G\&L} = 0.0002t - 0.0264 \tag{4}$$

$$\eta_{F\&G} = -0.0001t + 0.0793 \tag{5}$$

$$\eta_{G} = -0.0001t + 0.0734, \tag{6}$$

where Eq. (3) corresponds to blue lines in Fig. 8, Eq. (4) corresponds to green lines, Eq. (5) corresponds to red lines, and finally Eq. (6) corresponds to black ones.

4 Discussion

Analysis and assessment of cerebral angiogenesis *in vitro* is an important tool in cell biology. It is very promising approach to decipher pathogenesis of numerous brain disorders, in neuroscience, and for manipulating BBB permeability, as well as to test drug candidates affecting angiogenic events in cerebral vasculopathy, neurodegeneration, aberrant brain development, brain injury, brain tumors, and neuroinflammation.

Modeling of angiogenesis *in vitro* using BMEC provides several meaningful advantages including precise control of cell growth dynamics, application of variations in matrix and medium components, possibility to use huge number of methods to modify angiogenic properties of endothelial cells (i.e. genetic

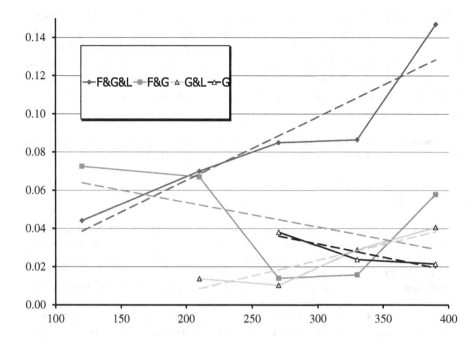

Fig. 8. Variance η of fractal dimension D in various scaffold types as measure of DMEC differentiation potential; horizontal axis is time (in minutes). Linear trends are shown in dashed lines. See details on linear trend explanation in text (Color figure online).

manipulations) or to minimize the effects of exogenous and endogenous factors with the presumptive pro- or anti-angiogenic activity. Recently, novel approaches have been suggested to optimize angiogenesis *in vitro* assays [3,5]. However, it is strongly required that in development of analytic tools to interpret the data on the angiogenic potential of endothelial cells as well as on angiogenesis-supporting properties of scaffolds.

Our study aims to find out some general characteristics of lactate-releasing gelatin scaffolds that are significant for the support of cerebral angiogenesis *in vitro*. Therefore, we tested gelatin scaffolds with different lactate-releasing ability to obtain the data on the angiogenic potential of BMEC *in vitro*. Apparently, high local concentrations of astroglia-derived lactate in the perivascular zone of brain microvessels seems to be a strong pro-angiogenic stimulus for BMEC [9]. Thus, selection of the scaffold (tanned and lactate-embedded) with the lactate-releasing ability close, to some extent, to the perivascular astrocytes in NVU results in better angiogenesis *in vitro*.

Moreover, we find out the parameters (these are fractal dimension D and form factor F) applicable for the precise analysis of proliferation and differentiation potential of BMEC grown *in vitro*. Thus, the fractal structure of newly-formed tubes *in vitro* should be taken into the consideration when analyzing the angiogenic properties of endothelial cells or extracellular matrix. Our data are consistent to some previously reported approaches based on dimensional, topological and fractal properties of capillary-like networks *in vitro* [4], and provide reliable analytical tools for the evaluation of angiogenesis-supporting scaffolds.

In summary, our data reveal and prove the novel possibilities to manage BMEC growth on biomimetic scaffolds to analyze angiogenesis *in vitro* with high degree of accuracy.

Acknowledgement. The study was supported by the grant of the Russian Science Foundation (project N 14-25-00054) (VS, AM, EK, NP, EB, PL, AS).

References

1. Boccardo, S., Gaudiello, E., Melly, L., Cerino, G., Ricci, D., Martin, I., Eckstein, F., Banfi, A., Marsano, A.: Engineered mesenchymal cell-based patches as controlled VEGF delivery systems to induce extrinsic angiogenesis. Acta Biomater. **15**(42), 127–135 (2016). doi:10.1016/j.actbio.2016.07.041
2. Falconer, K.J.: The Geometry of Fractal Sets (Cambridge Tracts in Mathematics). Cambridge Univeristy Press, Cambridge (1986). p. 180
3. Faulkner, A., Purcell, R., Hibbert, A., Latham, S., Thomson, S., Hall, W.L., Wheeler-Jones, C., Bishop-Bailey, D.: A thin layer angiogenesis assay: a modified basement matrix assay for assessment of endothelial cell differentiation. BMC Cell Biol. **15**, 41 (2014). doi:10.1186/s12860-014-0041-5
4. Guidolin, D., Albertin, G., Ribatti, D.: Exploring in vitro angiogenesis by image analysis and mathematical modeling. In: Méndez-Vilas, A., Díaz, J. (eds.) Microscopy: Science, Technology, Applications and Education, pp. 876–884. Badajoz, Formatex (2010)
5. Heiss, M., Hellström, M., Kalén, M., May, T., Weber, H., Hecker, M., Augustin, H.G., Korff, T.: Endothelial cell spheroids as a versatile tool to study angiogenesis in vitro. FASEB J. **29**(7), 3076–3084 (2015). doi:10.1096/fj.14-267633
6. Hielscher, A., Ellis, K., Qiu, C., Porterfield, J., Gerecht, S.: Fibronectin deposition participates in extracellular matrix assembly and vascular morphogenesis. PLoS One **11**(1), e0147600 (2016). doi:10.1371/journal.pone.0147600
7. Liu, Y., Xue, Q., Tang, Q., Hou, M., Qi, H., Chen, G., Chen, W., Zhang, J., Chen, Y., Xu, X.: A simple method for isolating and culturing the rat brain microvascular endothelial cells. Microvasc. Res. **90**, 199–205 (2013). doi:10.1016/j.mvr.2013.08.004
8. Ruhrberg, C., Bautch, V.L.: Neurovascular development and links to disease. Cell Mol. Life Sci. **70**(10), 1675–1684 (2013). doi:10.1007/s00018-013-1277-5
9. Salmina, A.B., Kuvacheva, N.V., Morgun, A.V., Komleva, Y.K., Pozhilenkova, E.A., Lopatina, O.L., Gorina, Y.V., Taranushenko, T.E., Petrova, L.L.: Glycolysis-mediated control of blood-brain barrier development and function. Int. J. Biochem. Cell Biol. **64**, 174–184 (2015). doi:10.1016/j.biocel.2015.04.005
10. Staton, C.A., Reed, M.W.R., Brown, N.J.: A critical analysis of current in vitro and in vivo angiogenesis assays. Int. J. Exp. Pathol. **90**, 195–221 (2009). doi:10.1111/j.1365-2613.2008.00633.x
11. Tricot, C.: Curves and Fractal Dimension, vol. XIV. Springer, New York (1995). p. 324

12. Vallon, M., Chang, J., Zhang, H., Kuo, C.J.: Developmental and pathological angiogenesis in the central nervous system. Cell Mol. Life Sci. **71**(18), 3489–3506 (2014). doi:10.1007/s00018-014-1625-0

13. Wälchli, T., Wacker, A., Frei, K., Regli, L., Schwab, M.E., Hoerstrup, S.P., Gerhardt, H., Engelhardt, B.: Wiring the vascular network with neural cues: a CNS perspective. Neuron **87**(2), 271–296 (2015). doi:10.1016/j.neuron.2015.06.038

Increasing of Data Security and Workflow Optimization in Information and Management System for Laboratory

Pavel Blazek[1,2](✉), Kamil Kuca[1,3](✉), Jiri Krenek[1], and Ondrej Krejcar[1]

[1] Faculty of Informatics and Management, Center for Basic and Applied Research, University of Hradec Kralove, Hradec Kralove, Czech Republic
pavel.blazek@unob.cz, {kamil.kuca, jiri.krenek, ondrej.krejcar}@uhk.cz
[2] Faculty of Military Health Sciences, University of Defence, Hradec Kralove, Czech Republic
[3] Biomedical Research Centre, University Hospital Hradec Kralove, Hradec Kralove, Czech Republic

Abstract. In order to ensure the protection of biomedical data, it is necessary, due to their special characteristics, to approach them differently than the common data. In the biologic laboratory environment, it is possible to understand their protection as a part of consequential and interconnected measures which complement themselves. Even though there are some procedures and potentially there are also some available manuals for individual areas, it is not possible to realize them without consideration or mutual interconnection. The methods for an attack and for protection have been changing in the recent times. Therefore, it is necessary to periodically review and update the given set of measures. Even higher level of security can be reached by optimization of the workflow and by the usage of new technologies. If a more effective identification of persons and of their realized tasks is achieved in the system, it is possible to limit the leakage of sensitive information or dangerous material outside of the laboratory. Moreover, this also increases the effectiveness in generating research results.

Keywords: Data collecting · Security · Information system · Laboratory

1 Introduction

The biomedical laboratory environment and the support of its activities are from the point of the sw equipment constantly developing. However, some old procedures are still being used. The performance of the laboratories increases together with the higher efficiency of processed tasks [1]. A significant part of this development is attributed to the investments into suitable program-equipment, which exactly fulfils the requirements for the support of all activities. The modern laboratories that would like to take part in the top-class projects need to prove their quality by the accredita-tion according to ISO 17025 or by the 21 CFR Part 11 compliant.

© Springer International Publishing AG 2017
I. Rojas and F. Ortuño (Eds.): IWBBIO 2017, Part I, LNBI 10208, pp. 602–613, 2017.
DOI: 10.1007/978-3-319-56148-6_54

The operation of the biomedical research is primarily influenced by the essence of its activity [2]. Precisely this contains the aims of the research, used technologies, used chemical or possibly biological material. Due to this, it is obvious that it needs to be secured on all levels – physical, electronic, functional and informational. [3] The security system has to be global and must be able to detect successful and unsuccess-ful security incidents. The aim is the results protection of multiple years' work of people and teams.

2 Problem Definition

In the common business environment, it is possible to set up applications without the danger of the crash of an operating system due to their influences. The laboratories are from this point of view very unique. It can be easily recognized by concentrating on just the computer-driven laboratory devices, on which even the installation of anti-virus programs can have a negative impact on its operation. However, these devices have to be connected to the network in order to enable data export. To be able to comply with company politics, it is necessary to set up exceptions which can be classified as potential entries for harmful code. Another area is the development and operating of one-purpose applications. On one hand the required results can be obtained in this way. On the other hand, from the working station safety point of view, where the processing of data takes place, such utility files may also cause an interruption in security. Therefore, it is more suitable to concentrate on the Laboratory Information and Management System (LIMS) [4]. These systems are modular and can contain or create required functionalities [5]. The amount of stand-alone applications is in this way considerably reduced.

The LIMS modules can be also used as connectors for laboratory devices and sensors from which the data is recorded into the own database. The significant source of information is presented in the connection to external databases. In this case, it is necessary to find solutions to the connector, the data transformation, data anonymi-sation and interpretation [6]. Such source is for example the connection to PACS data-base of images. The DICOM format contains apart from the displayed parts also metadata which specify the information from the point of the doctor and also the information about the patient. Therefore, the connector has to be able to provide ensured authorized connection, sending of a query in a standardized format and anonymisation of obtained data. Even though, the system has its own security mea-sures, successful attempts to penetrate into dangerous data storages have been recorded.

During models development, the authors focus on existing working procedures and they offer their optimisation [7, 8]. The problem is that not all solutions are in com-pliance with requirements for information system security already at the application layer. If it is taken into consideration that the lab environment is equipped with physical security elements and the system of access security that does not necessarily mean the data are fully secured. Mostly, the situation is resolved by implementation of the autonomous system for physical security of the area, where a person entering the lab identifies himself/herself with an ID card, and on the basis of its code finding in the database, the door lock is released and the person is allowed to enter. Another autonomous system cares for verification of a person who logs into the computer network environment. In the next step, he/she can be asked for verification of the application access. Users naturally try to simplify their working environment as much as possible even by going around the security elements [9].

The aim of the management system module presented hereinafter is to create a proposal environment with balanced security elements. Security of lab data has to be comprehended as a complex of preventive measures which lead to their protection from the view of legislation, intellectual property of research work results [10, 11] and last but not least as potentially dangerous information in the hands of unethically behaving persons.

In comparison with other LIMS, the presented system is more sophisticated. Roles setting in studied systems is similar, however, not sufficient from our point of view. Compared LIMS [12–14] uses the Active Directory or LDAP account verification for logging into the system, but a group division is general (Administrator, Technician, External User, Manager, Chemist). After logging, this model enables viewing of all parts or all experiments. This cannot be classified as a bad solution. In labs, where there is a stable working team and workers are motivated to work in the team, the risk of individual's failure is expected to be lower compared to the labs, where lab technicians' and interns' practical training is carried out.

The suggested system has to correspond to the organizational structure from which the access authorization into individual parts of information system with the possibility of more detailed setting, also for current experiments is derived.

3   Review

The database Web of Science was used as the source of information during the study of problematic of the Laboratory Security of information systems. The main point of interest were articles with the JCR® category ranking in quartiles Q1 and Q2. Out of the 180 results on the query "Laboratory Information and Management System" exist overall 104 sources to the articles published since the year 1983. During the last 20 years, there were 55 articles published. The distribution since 1997 is displayed in the left Fig. 1. The graph on the right side shows the same period, but with 21 filtered articles from magazines from the quartiles Q1 and Q2. During the whole examined period from 1945, there were 58 registered magazines found in the database. From those graphs, it is recognizable that the LIMS problematic is still important to the researchers and these systems are also being further developed. In regard to the content of those articles, which is further described, is the reason for those publications obvious.

The articles in publications with ranking Q1 and Q2 reached 380 citations or 340 when auto-citations were excluded. The average citation per article was 6.55. During the last 20 years, this value is decreased to 2.75. The distribution of articles' citation from quartiles Q1 and Q2 per years is displayed in the Fig. 2. The information systems became more important with the consistent development and the expansion of technologies, mainly connectivity-related. This is reflected by the increase in citations which can be observed since approximately 2004.

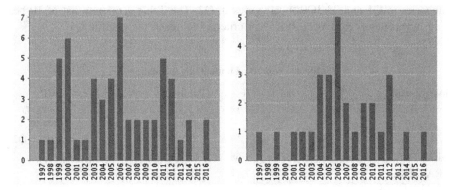

Fig. 1. Published item each year all (left), filtered titles JRC® classification Q1 and Q2 (right)

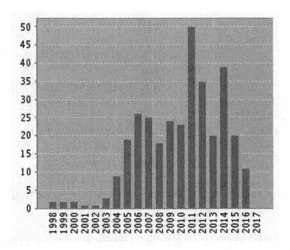

Fig. 2. Citations for individual years for the last 20 years – articles from Q1 and Q2

In the next approach, the focus was put on the selection of specific articles from the 104 mentioned sources. The results analysis discovered that those articles are not only closely connected to the category of Computer Science - this category is represented by only 14.4%. However, the most common group with 43 titles is the Chemistry analytical which consists of 41.3%. Similarly, these data can be analyzed in regards to the number of articles in the individual magazines. The selection was concentrated on those articles that reach the JCR® Q1 or Q2 rating in the given category. CHEMOMETRICS AND INTELLIGENT LABORATORY SYSTEMS clearly leads in these categories with 15 articles.

Overall, 24 magazines were chosen for further processing. From those 24, 14 magazines belong to the Q1 and 10 articles into the Q2. This figure takes into account the highest reached quadrant since it often happens that magazines' content is ranked into more JRC® categories. The 24 mentioned magazines extents into 23 JCR®

categories in Q1 and 23 JCR® categories in Q2. The Table 1 offers a list of 10 best rated magazines according to IF 5 year criteria that reach the quartile Q1.

Table 1. Top 10 of journals by evaluations

Journal name	IF	IF 5 year	Publisher	JCR category quartil	
				Q1	Q2
BIOINFORMATICS	5,766	7,685	OXFORD UNIV PRESS	3	
TRAC-TRENDS IN ANALYTICAL CHEMISTRY	7,487	7,474	ELSEVIER SCIENCE BV	1	
ANALYTICAL CHEMISTRY	5,886	5,922	AMER CHEMICAL SOC	1	
ANALYTICA CHIMICA ACTA	4,712	4,481	ELSEVIER SCIENCE BV	1	
BMC GENOMICS	3,867	4,276	BIOMED CENTRAL LTD	1	1
PROTEOMICS	4,079	3,666	WILEY	1	2
JOURNAL OF PHARMACEUTICAL AND BIOMEDICAL ANALYSIS	3,169	2,904	PERGAMON-ELSEVIER SCIENCE LTD	1	1
JOURNAL OF FOOD ENGINEERING	3,199	3,512	ELSEVIER SCI LTD	2	
CYTOTHERAPY	3,625	3,479	INFORMA HEALTHCARE	2	2
BMC BIOINFORMATICS	2,435	3,435	BIOMED CENTRAL LTD	1	1

After this executed selection of articles, a research was conducted about whether and how the question of data security that are processed and saved into described LIMS is approached.

From the 21 selected articles, a certain chronology can be recognized, where the first article published in 1984 [15] describes the basics of the information systems, including the planning and setting of the infrastructural elements. The newer articles then further develop the workflow specification and its implementation into the laboratory environments [16, 17]. Mostly, all articles mention the need for data centralization, meaning their storing into the central DB in order to share them [18] and more effectively use them, and the possibilities of their validation [19–21]. Even in the newer articles [22, 23] the same drawback can be recognized. The favorite ones are the web-based solutions [18, 24, 25], which are independent of the platform and that enable easier access. Open Source solutions are mentioned [18, 23, 24, 26] as an advantage due to their low cost, and mainly for their ability to being precisely adjusted for the requirements of the given laboratory. This is proved by their implementation in proteomic [23, 25–28] and food laboratories [1, 29]. The specific implementations of LIMS are mentioned even in genetic research [24, 30], mainly then in connection to storage of large amount of data, need for its processing [29, 31] and archiving [28] – in general Big Data. The problematic also concentrates on the previously mentioned proteomic [23, 25]. Some articles mention optimization and effectiveness in a general sense [15–17], others describe the usage of procedures' automation for the research in more detail [23, 25] as a subworkflow. A support can be presented in the form of helping instruments which were already previously used in other branches, such as barcodes and QR codes [22, 32, 33].

The security to the data access is resolved only partially [22], potentially obtaining of the authorization on the level of user account on basis of classification to a group [18] is considered as sufficient. On the other hand, PARPs database [28] offers a higher form of security where the user can access the data only after the authorization in LDAP database and the permission is granted on the level of individual projects.

4   Organization Structure

The described solution comes from the real lab environment. Designing a logical scheme was extended [34] by a security concept limiting unauthorized data leakage. The mentioned modification involves not only application security of a part of data in a production database but also the workplace organization and technical equipment [35]. For comprehension of the whole concept, it is necessary to describe elements, which are involved in the system and affect its function.

The elements can be divided into five groups:

- **Management** – supervisors who lead activities in the lab and assign authorizations and tasks for processing.

- *In silico* also called Dry Lab is the group of computer chemistry which collects information and data to the assigned task on the basis of which molecules models are designed. With the use of computer simulation, behavior and forces of mutual bonds are modelled and analyzed.

- **Synthetic team** deals with physical formation of compounds on the basis of not only *in silico* team results.

- **Biochemical, toxicological and pharmacokinetic team** – groups which carry out tests of the synthetized sample in vivo and in vitro independently on each other, if needed pharmacokinetic tests. The result reveals an effect of the tested substance on living organisms, its toxicity or possibly a capability of living organisms to degrade the substance.

4.1   Initial Function Description

The management group conducts lab operation and cooperation of the teams. The first task is mostly directed to the *In silico* team for provision of theoretical tests. The preparation lies, besides the theoretical plan, in building a local "ligand database" and a "biological target" model. It is not complicated to acquire data for the actual database owing to a possibility to download different molecular "subsets" (e.g. lead-like, fragment-like, drug-like) from the huge on-line "drug" databases (e.g. zinc.docking.org).

For further data processing in the "docking" program, it is necessary to array out a format conversion and preliminary chemical calculations. A similar situation occurs in proteins, the structures of which are available in on-line databases (e.g. www.pdb.org) already "adjusted for molecular docking". The processing requires a sufficient computing power which may offer the environment dedicated in the cloud [36, 37]. Potential biological and physical properties of the studied substances are the results of the mentioned above procedure. The team uses data stored in the database itself or in an external source to design the model. By chemical structures modelling, the team will get a group of potential substances which correspond with the assigned task. Selecting these substances, the team will get the formula of one or more theoretically significant substances. Information is handed over from the *In silico* team in the form of a chemical formula and metadata. Those can be linked to more than one experiment, that is why the modules LabApp and Library are linked with the bond N:M. Information on chemical properties and procedures received from particular experiments are then stored in the LabApp database, it means 1:N, and inserted into the metadata table. Simultaneously, the identification of the person, who inserted the data into the system, is automatically stored here together with the timestamp. The timestamp defines the relevance in case of a conflict with data in subsequent verifying experiments. The record prepared this way facilitates processing of citations sources into reports, scientific articles or for a patent procedure. Laboratory structure and its connections are drawn in Fig. 3.

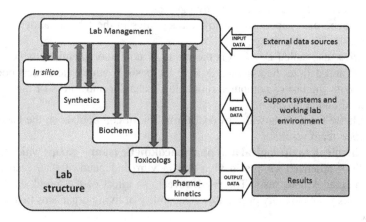

Fig. 3. Logical scheme of laboratory environment

4.2   Modified Model (Logical Description)

The management group is the only leading element. The definition of a potentially interesting substance may come from the *in silico* team. The outputs of theoretical analyses with information on the best assessed model are submitted to the supervisors who will subsequently decide which substance will be synthetized. Information on selection process is not public, it is assigned to a particular person in the synthetic team together with a container marked with the bar code. The person returns the container with the content which corresponds with the assigned task [38]. If the synthesis is successful, the label with the barcode is changed and handed over for in vivo analysis, then in vitro test and finally pharmacokinetic test is carried out in the same way and procedure in which a capability of the living organism to keep or eliminate a potential medicament is tested.

Only the supervisors are acquainted with the results of individual phases of the experiment. The next procedure depends on supervisors' decision. Every time, the tested sample is handed over to the lab's supervisor, who will mark the sample for each phase of processing with a single barcode, for further testing. The complex set of generated codes used in the given experiment is sorted in the database. Together with generation of a new label, the task is defined and assigned to a particular worker.

Since several experiments are usually carried out simultaneously in the lab, it is possible to assure that the substance will be always assigned to a concrete person who is responsible for a particular test but not aware of its origin. This person's task is to provide appropriate testing on the unidentified sample and to elaborate required reports.

The LabIS database is a relation database and comprises all records connected with the experiment. Everybody, who was assigned to the experiment by the supervisors, participates in creation of its content. On the basis of the assigned task, the selected person can have an access to the GUI only to those items which he/she has to know and those which have to be completed with measured figures. In common LIMS, the experiment records can be freely listed. Reading of the results here is in hands of persons who lead the experiments.

From the viewpoint of data structure security, on the application layer has to be ensured, that while assigning the task in a particular phase of the experiment, the authorized person can open only those items which are required for providing the procedure and the box for filling out the results of his/her own contribution in the experiment. This logic is resolved mainly in the graphic LabIS interface which communicates with the database.

The users' accounts of lab workers are linked into groups according to roles. Each experiment represents one line in the table with 37 items divided into four logical parts, whereas 4 cells from the whole are used for storage of service data. From the graphic interface of the Management group workers, tasks can be assigned to persons according to their workload, capabilities and according to their group appointment. Partial and complete results are available in the form of a structured table, for the Management group and through this group for next persons [39] who participate in publication of results.

4.3   Basic Modules Used in LabIS

Basic building elements of the proposed LIMS which are used by all modules of the application layer are concerned.

CAM – the central authentification module is the database of all authorized persons. The Active directory usually forms its ground. It allows verification of user accounts and setting authorization for access to objects. It supports interconnection of user accounts with identification codes e.g. RFID tokens, which are used for identification of person's access in the workplace. The CAM appears to be the real central authority for verification of person's permit to enter the building and there for defining premises as well as for their logging into the LIMS information system with determined authorizations.

DBS – the database store which is primarily used not only for storage of operational data from the LabIS and Library applications but also for storage of records on the audit of incidents connected with activities in LabIS and lab environment. Centralization of these record enables to find easily and fast links to a security incident and consequently also a key person.

LabIS – is the central application involving operations connected with managing lab activities, with collection and evaluation of data from the experiments. In the frame of the application, it enables to define access authorizations to individual tasks at the group and individual levels.

MAUD – the module of incidents audit follows-up directly to LabIS and is actually a part of it. It is presented as an independent module because its inputs are not only incidents from LabIS. It is possible to store the data from other IS components in its database.

Entry registration – is the module closely linked to the MAUD, the purpose of which is to centralize data collection from sensors recording persons' movement. The outputs may be used not only for the security audit but also as the data source for the attendance system.

Store – the module used for monitoring movement and state of the material, mainly consumable. Together with technical gadgets, it offers automated assessment of supplies state and calls for their replenishment before they are entirely exhausted.

Connectors – the module for connection to extern data sources (hospital IS, DICOM) and to devices used in the laboratories.

Library – the module used for electronic storage of publications, comments on them and for their link with running experiments. All persons, who were verified in logging via the CAM, are allowed to insert documents and to make comments.

eLearning – is an optional module primarily designated for laboratories with unstable staff, where training of interns and students practice is carried out. The module is intended for training and knowledge testing of new workers.

Security audit – the module which facilitates a central view on the stored data from the LabIS, on the Entry registration and the Store modules from the time line of incidents perspective. It is possible to read there, on the basis of person's ID and the time period, the sequence of incidents connected with the given person.

4.4   LabIS Infrastructure

The LabIS system is designed as a web-based platform independent application. The interconnection of its modules enables the sharing of required data among the users. However, it prevents the access to information which exposes the sensitive parts of experiments. Backing-up of the data and the data revival clearly belong to the collection of elements for data security. However, these were not mentioned as individual functionalities among LIMS modules due to whole system being run in the cloud environment that is located outside of the laboratory environment [35]. It includes in its configuration the data back-up on the level of virtual devices. Therefore, the requirement, which was not yet mentioned, for the minimization of IT system maintenance by the laboratory workers is satisfied.

5 Conclusion

Nowadays, biomedicine is made a priority. Laboratory experiments are a competitive environment and there is no doubt that it is also a place where there are substances, which in inappropriate hands are changed in weapons of mass destruction. As it was mentioned in the introduction, the lab security is to be thought in the sense of perspective. It is to be considered as the connection building security with personnel and data security. The presenting lab information systems do not provide necessary application security.

Protection of premises and intellectual properties is indisputable requirement. Systems which are used for this purpose, an applicable integrated element is not always easy to find. Environment which is poorly secured is with its actual security close to that which has the security of a higher level but it is decentralized. The reason is that its users are burden with various identification procedures and thus they obey them. The proposed system is, from the view of its service very simple, it eliminates variability of ways of log in and the range of information which the user gets acquainted with. The aim of the system security within acceptable limits of restrictions was achieved.

Acknowledgment. This work and the contribution were supported by project "SP/2102/2017 - Smart Solutions for Ubiquitous Computing Environments" from University of Hradec Kralove.

References

1. Cagindi, O., Otles, S.: Importance of laboratory information management systems (LIMS) software for food processing factories. J. Food Eng. **65**(4), 565–568 (2004). doi:10.1016/j.jfoodeng.2004.02.021. ISSN 0260-8774
2. Blazek, P., Krenek, J., Kuca, K., Jun, D., Krejcar, O.: The system of instant access to the life biomedical data. In: Computational Intelligence and Informatics, pp. 261–265 (2014). doi:10.1109/CINTI.2014.7028686
3. NRC. Prudent Practices in the Laboratory: Handling and Management of Chemical Hazards. Updated ed. National Academies Press, Washington (2011)
4. LIMSwiki. http://www.limswiki.org
5. Quo, C.F., Wu, B., Wang, M.D.: Development of a laboratory information system for cancer collaboration projects. In: 27th Annual Conference on IEEE Engineering in Medicine and Biology (2005). doi:10.1109/IEMBS.2005.1617070

6. Rabinovici-Cohen, S.: Biomedical information integration middleware for clinical genomics. In: Feldman, Yishai A., Kraft, D., Kuflik, T. (eds.) NGITS 2009. LNCS, vol. 5831, pp. 13–25. Springer, Heidelberg (2009). doi:10.1007/978-3-642-04941-5_4

7. Kammergruber, R., Robold, S., Karliç, J., Durner, J.: The future of the laboratory information system – what are the requirements for a powerful system for a laboratory data management? Clin. Chem. Lab. Med. (CCLM) **52**(11) (2014). doi:10.1515/cclm-2014-0276

8. Prasad, P.J., Bodhe, G.L.: Trends in laboratory information management system. Chemom. Intell. Lab. Syst. **118**, 187–192 (2012). doi:10.1016/j.chemolab.2012.07.001

9. Hansen, M., Köhntopp, K., Pfitzmann, A.: The open source approach - opportunities and limitations with respect to security and privacy. Comput. Secur. **21**(5), 461–471 (2002). doi:10.1016/S0167-4048(02)00516-3

10. Ekins, S.: Computer Applications in Pharmaceutical Research and Development, pp. 57–61 Wiley (2006). ISBN 0-471-73779-8

11. Hu, Y.: Development of information management system used in laboratory. Advanced materials research. In: 27th Annual International Conference of the Engineering in Medicine and Biology Society, IEEE-EMBS 2005, Shangai, China, pp 2859-2862 (2005)

12. LABWARE Result Count. http://www.labware.com/

13. Lab Collector. http://labcollector.com/

14. Qi Analytica. LISA.lims, http://lisa.lims.cz/

15. Megargle, R.: Laboratory information management-system. Anal. Chem. **61**(9), 1104–1113 (1989). doi:10.1002/1615-9861(200209)2:9<1104::aid-prot1104>3.0.co;2-q

16. Megargle, R.: A Report on the 3rd International Laboratory Information Management-Systems Conference, Held in Egham UK, 6–8 June 1989. Trac-Trends Anal. Chem. **8**(10), 353–354 (1989). doi:10.1016/0165-9936(89)85070-8

17. Gibbon, G.A.: Trends in laboratory information management-systems. Trac-Trends Anal. Chem. **3**(2), 36–38 (1984). doi:10.1016/0165-9936(84)87049-1

18. Grimes, S.M., Ji, H.P.: MendeLIMS: a web-based laboratory information management system for clinical genome sequencing. BMC Bioinform. **15** (2014). doi:10.1186/1471-2105-15-290

19. McDowall, R.D.: The role of laboratory information management systems (LIMS) in analytical method validation. Anal. Chim. Acta **391**(2), 149–158 (1999). doi:10.1016/s0003-2670(99)00107-5

20. McDowall, R.D.: A matrix for the development of a strategic laboratory information management-system. Anal. Chem. **65**(20), A896–A901 (1993)

21. McDowall, R.D., Mattes, D.C.: Architecture for a comprehensive laboratory information management-system. Anal. Chem. **62**(20), A1069–A1076 (1990)

22. Triplet, T., Butler, G.: The EnzymeTracker: an open-source laboratory information management system for sample tracking. BMC Bioinform. **13** (2012). doi:10.1186/1471-2105-15-290

23. Morisawa, H., Hirota, M., Toda, T.: Development of an open source laboratory information management system for 2-D gel electrophoresis-based proteomics workflow. BMC Bioinform. **7** (2006). doi:10.1186/1471-2105-7-430

24. Voegele, C., Tavtigian, S.V., De Silva, D., Cuber, S., Thomas, A., Le Calvez-Kelm, F.: A laboratory information management system (LIMS) for a high throughput genetic platform aimed at candidate gene mutation screening. Bioinformatics **23**(18), 2504–2506 (2007). doi:10.1093/bioinformatics/btm365

25. Cho, S.Y., Park, K.S., Shim, J.E., Kwon, M.S., et al.: An integrated proteome database for two-dimensional electrophoresis data analysis and laboratory information management system. Proteomics **2**(9), 1104–1113 (2002). doi:10.1002/1615-9861(200209)2:9<1104::aid-prot1104>3.0.co;2-q

26. Helsens, K., Colaert, N., Barsnes, H., Muth, T., et al.: Ms_lims, a simple yet powerful open source laboratory information management system for MS-driven proteomics. Proteomics **10** (6), 1261–1264 (2010). doi:10.1002/pmic.200900409
27. Nebrich, G., Herrmann, M., Hartl, D., Diedrich, M., et al.: PROTEOMER: a workflow-optimized laboratory information management system for 2-D electrophoresis-centered proteomics. Proteomics **9**(7), 1795–1808 (2009). doi:10.1002/pmic.200800522
28. Droit, A., Hunter, J.M., Rouleau, M., Ethier, C., Picard-Cloutier, A., Bourgais, D., Poirier, G.G.: PARPs database: a LIMS systems for protein-protein interaction data mining or laboratory information management system. BMC Bioinform. **8** (2007). doi:10.1186/1471-2105-8-483
29. Sanchez-Villeda, H., Schroeder, S., Polacco, M., McMullen, M., et al.: Development of an integrated laboratory information management system for the maize mapping project. Bioinformatics **19**(16), 2022–2030 (2003). doi:10.1093/bioinformatics/btg274
30. Honore, P., Granjeaud, S., Tagett, R., Deraco, S., Beaudoing, E., Rougemont, J., Debono, S., Hingamp, P.: MicroArray Facility: a laboratory information management system with extended support for Nylon based technologies. BMC Genomics **7** (2006). doi:10.1186/1471-2164-7-240
31. Monnier, S., Cox, D.G., Albion, T., Canzian, F.: TaqMan information management system, tools to organize data flow in a genotyping laboratory. BMC Bioinform. **6** (2005) doi:10.1186/1471-2105-6-246
32. Vu, T.D., Eberhardt, U., Szoke, S., Groenewald, M., Robert, V.: A laboratory information management system for DNA barcoding workflows. Integr. Biol. **4**(7), 744–755 (2012). doi:10.1039/c2ib00146b
33. Wray, B.R.: Bar code sample identification - the key to laboratory information management-systems (LIMS) productivity. Trac-Trends Anal. Chem. **7**(3), 88–93 (1988). doi:10.1016/0165-9936(88)85027-1
34. Penhaker, M., Cerny, M., Rosulek, M.: Sensitivity analysis and application of transducers. Paper Presented at the 5th International Workshop on Wearable and Implantable Body Sensor Networks, BSN 2008, in Conjunction with the 5th International Summer School and Symposium on Medical Devices and Biosensors, ISSS-MDBS 2008, Hong Kong (2008)
35. Dolezal, R., Sobeslav, V., Hornig, O., Balik, L., Korabecny, J., Kuca, K.: HPC cloud technologies for virtual screening in drug discovery. In: Nguyen, N.T., Trawiński, B., Kosala, R. (eds.) ACIIDS 2015. LNCS (LNAI), vol. 9012, pp. 440–449. Springer, Cham (2015). doi:10.1007/978-3-319-15705-4_43
36. Pavlik, J., Komarek, A., Sobeslav, V.: Security information and event management in the cloud computing infrastructure. In: Computational Intelligence and Informatics, pp. 209–214 (2014) doi:10.1109/CINTI.2014.7028677
37. ownCloud. https://owncloud.org/
38. Machina, H.K., Wild, D.J.: Laboratory informatics tools integration strategies for drug discovery: integration of LIMS, ELN, CDS, and SDMS. J. Lab. Autom. **18**(2), 126–136 (2012). doi:10.1177/2211068212454852
39. Sahiti, M., Vimla, L.P.: Organization of biomedical data for collaborative scientific research: a research information management system. Int. J. Inf. Manag. **30**(3), 256–264 (2010). doi:10.1016/j.ijinfomgt.2009.09.005. ISSN 0268-4012

Challenges Representing Large-Scale Biological Data

GIS-Aided Modelling of Two Siberian Reservation Sites

Marina Erunova[1,3(✉)] and Michael Sadovsky[1,2]

[1] SFU, 660041 Krasnoyarsk, Russia
marina@icm.krasn.ru
[2] Institute of Computational Modelling of SB RAS,
Akademgorodok, 660036 Krasnoyarsk, Russia
msad@icm.krasn.ru
[3] Krasnoyarsk State Agrarian University,
Mira prosp., 90, 660049 Krasnoyarsk, Russia
http://icm.krasn.ru

Abstract. Reserved territories seem to be the best reference sites of wildnature, where the long-term observations are carried out. Simulation model of spatially distributed processes of contamination of such reservation is developed, and the dynamics of some pollutants is studied. An issue of the generalized evaluation of an ecological system status is discussed.

Keywords: Pollution · Correlation · Analysis · Modelling · Spatial analysis

1 Introduction

Human activities impact heavily various ecological systems, especially at the regions with severe climate, and the impact may be irreversible. Non-ferrous and heavy metals the number one in global pollution (see [1] for details). Mathematical models are the tool to study a dynamics of deteriorated ecosystems. Studying spatially distributed communities, one faces the problem towards the interplay of geography of a site, and the biological features of the inhabitants. A number of landscape-scale models have been developed in attempt to predict a vegetation pattern or other natural resources [2–7].

Russia has a unique experience in wildnature reservation through the establishment of areas (called *zapovednik*) specially protected by law; special federal agency supervises them. No activity, nut a research is allowed in such areas. The sites play a key rope in understanding and forecasting of the dynamics of global pollution. Here GIS address the complex study of dynamics of physical, geographical and biological processes run at zapovedniks best of all. The studies of protected areas/reservations with GIS tools are not too convenient; some results in that direction see in [8,9]). The studies of fully preserved territories located at Russia, in Central Siberia are presented by Prechtel [10,11]). Here we present the basic solution (developed primarily for *Stolby* zapovednik, see [1,17]) to unify and standard GIS based solution to be applicable for any area.

© Springer International Publishing AG 2017
I. Rojas and F. Ortuño (Eds.): IWBBIO 2017, Part I, LNBI 10208, pp. 617–628, 2017.
DOI: 10.1007/978-3-319-56148-6_55

Here we present some results on the life dynamics observed at *Tzentral'no-sibirskii* zapovednik. Basically, we intend:

(1) to reveal the statistical relationship between a forest type and some physical and geographical features of the landscape;
(2) to reveal the statistical relationship between a forest type and features of landscape;
(3) to reveal and present a pattern of the contamination with some pollutants.

2   Materials and Methods

2.1   Study Area

Tzentral'nosibirskii State Biospheric Wildnature Reservation (*zapovednik*) is located approximately 1000 Km north from Krasnoyarsk, occupying two sites on both banks of the river of Yenisei (see Fig. 1), thus preventing the anthropogenic impact on the river ecosystem. Its total area is 10118.49 square kilometres. This is a huge area; the altitude of zapovednik varies from see level to app. 600 m above. Zapovednik has extended hydrography network containing more than thousand of lakes and watery rivers, and more than 6000 small rivers and springs. It was established in 1985, there are six stationary observatories (they are the check-points, simultaneously), and more than a hundred of temporary observatories located site-wide.

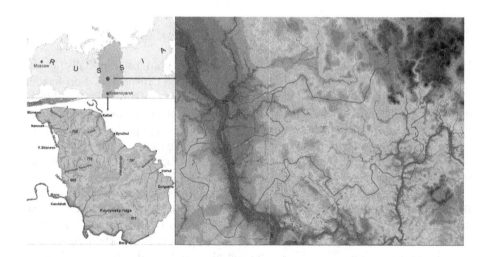

Fig. 1. Location of Siberian *zapovedniks* at Russia.

2.2   Data Collection Description

GIS-database incorporates relief data, soil type distribution data, river network data, road network data, and the detailed forest estimation data. The forest estimation database specifies forty data fields. All these databases are conveyed in the digital form.

Tzentral'nosibirskii zapovednik is the perfect reference site, for the purposes of global monitoring of some pollutants, since it exhibits the least possible influence resulted from a human activity. Currently, winter precipitations are monitored. The observatory network is quite complex, since the territory of zapovednik is extended very much. Previously, few anomalies of heavy metals content have been found; hence, the observatories are located rather sporadically, in the areas with low pollutants level. Reciprocally, the observatories are located in more regular manner, at the sites with increased pollutants content, to verify whether the anomaly comes from geographical features, or it results from a human activity.

We traced the concentrations of Ph, F, S, Cl, Pb, Sr, Al, Cd, Fe, Cu, Zn, Cr, Mn, Ni, Co, K, Na, Ca, and Mg, at winter precipitations. The collected samples were treated as described in our previous paper [1,16,17].

2.3   GIS-Based Method

ArcGIS software package supplied with Spatial Analyst and 3D Analyst modules have been used to debelol digital model of zapovednik, and carry out the spatial data analysis. MapInfo Professional package has been used to convert the data formats. EasyTrace software package was used to digitalize the cartography data on vegetation mapping.

Federal Agency for Cartography and Geodesy provided the primary geographical data of $1 \div 200\,000$ scale; the vegetation map of $1 \div 50\,000$ scale was used. Ten individual map-cases were cohered, and data bases were merged. The vegetation map was processed through the scanning and digitalization, and the attributive data were converted from paper sources, namely from the charts of vegetation type description.

The techniques for spatial analysis [15] were used to study geographical features of the site. The following item have been implemented:

- primarily spatial analysis (positioning, object detection via their attributes, computations involving geometry objects of higher level);
- measurements (linear features of polygons, shape measures, distances);
- classification (thematic cartography, neighbourhoods search, filtering, buffers, slopes and aspects);
- statistically defined surfaces (relief digital models, interpolations, grid models);
- spatial distributions (triangulations, lines and polygons distributions, orientation of linear and polygonal objects);
- overlays (selection, mapping of point, linear and polygonal objects).

Thematic maps for the pollutants mentioned above are developed, currently, for winter precipitations. The heavy metal concentrations, as well as some halogens concentrations were used as the input data for soil pollution maps to study the distribution of metals in various compartments of zapovednik. Spatial interpolation of data was carried out with Inverse Distance Weighting (IDW). Grid size of $15 \times 15\,\text{m}^2$ was implemented using all the input points available with fixed weight of an observation point.

3 Results

3.1 Model of Zapovednik

Tzentral'nosibirskii zapovednik digital model differs, to some extent, from the similar one implemented for *Stolby* zapovednik [1]. First of all, no spatial dynamics based on GIS solution has been developed, yet. The shortage results from a lack of long-time observations, since the zapovednik has been founded at 1985. The simulation model consists of the digital layers, and the data from numerous databases attributed to the points, in those layers. All the points in the layers are categorized for those incorporating the observations properly located at the layers, and interpolated ones, with special respect to relief peculiarities.

The digital model of zapovednik has the following layers: altitude isolines, reservation borders, rock pillars, river network (both polygonal, and linear objects), springs, forest roads, walking paths, check-points, forestry borders, compartment lines, elevation points, forest estimation, soils, and others.

Unlike for the digital model of "Stolby" zapovednik, this model contains four separate layer of the forests; it results from the huge geographical extensions of "Tzentral'nosibirskii" zapovednik. Each layer of the forests corresponds entirely to a forestry. These for layers could not be merged into a single one, since they are too large. Nonetheless, the basic principle of the digital model implementation (that is the basin based implementation) makes no problem with the data exploration over these four separate layers. It should be also said, that the soil layers are obtained from a soil studies of the territory, but from the forest type description; thus, a soil layer is indeed a component of the afforestation inspection of a forest layer. The inspection has been carried out in 1990.

Also, river layer differs from the similar one developed for "Stolby" zapovednik. Since the geographical map and vegetation map exhibit different scale, the river layer was implemented in two copies, differing in details, with two scales: $1 \div 200000$ and $1 \div 50000$. A choice of a specific river network layer is determined by the task to be done.

3.2 Spatial Analysis of Physics Geography Data

The analysis of distribution of forest types with respect to an altitude yields a prevalence of pine vegetation at almost any height (see Fig. 2). The altitude vegetation distribution pattern is rather irregular, exhibiting significant variations in landscape features (slopes, mountain ridges, etc.). Figure 2 shows percentage of area of the reservation in dependence on the height. Thus, Fig. 3 shows the distribution of individual vegetation types with respect to a height indicating the absolute area measure occupied by a specific vegetation type at the given height. For example, pines occur mainly at the places located from 300 m to 700 m in height, with total occupation area up to 25 km^2. The maximal area of pine occupation is observed for the altitude ranging from 500 m to 520 m. Similar pattern of a vegetation type distribution could be obtained for other species.

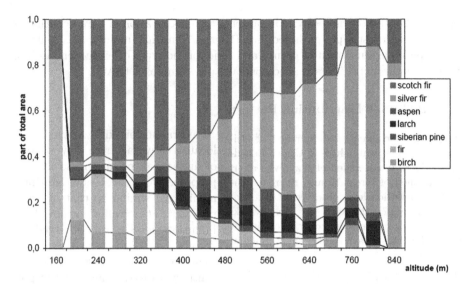

Fig. 2. Distribution of types of forestry over the altitude, at *"Stolby"* *zapovednik*, in terms of a portion of total area.

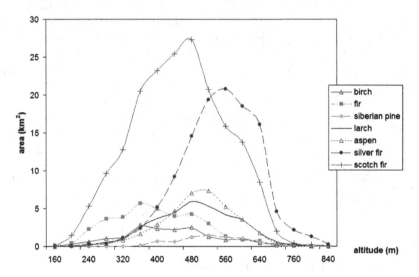

Fig. 3. Distribution of types of forestry over the altitude, at *"Stolby"* *zapovednik*.

Figure 3 represents the area of each vegetation type as a portion of the entire reservation area. This chart reveals the part of each vegetation type occupying the specific height. In contrast to Fig. 2, this figure clearly presents the contribution of various forest species, within a specific height layer of the site. Besides, we have verified the vegetation map of "Stolby" state reservation, with particular respect to the succession of aspen for fir, for the last years. Large scale vegetation map $(1 \div 25\,000)$ of the reservation presents the classification of flora, identifying 70 groups of forest types gathered into 21 series. This map mainly aims to figure out a spatial arrangement of groups and series of the types of biogeocenoses into basic structural entities: so called zone altitude complexes (ZAC).

3.3 Pollutant Distribution Pattern

Table 1. 1: element; 2: watershed of Osinovka river; 3: watershed of Upper Lebedjanka river; 4: watershed of Inzyrevka river; 5: watershed of Great Komsa river; 6: generalized pollution level observed at "Stolby" Reservation; 7: the city of Krasnoyarsk, observations carried out over the district with the most favourable environmental conditions.

1	2	3	4	5	6	7
Ph	5.0	5.6	4.9	4.7	5.5	8.0
F, $\times 10^{-2}$	1	1	1	1	1	28
S	0.10	0.29	0.10	0.10	0.67	4.42
Cl	0.150	0.150	0.150	0.150	0.940	1.890
Pb, $\times 10^{-4}$	7	7	4	9	28	42
Sr, $\times 10^{-3}$	4	7	4	4	4	102.4
Al	0.015	0.030	0.015	0.015	0.015	1.230
Cd	$9 \cdot 10^{-5}$					
Fe, $\times 10^{-1}$	0.10	0.16	0.16	0.24	0.44	6.53
Cu, $\times 10^{-3}$	1.7	2.4	0.9	1.7	2.9	3.4
Zn, $\times 10^{-2}$	0.83	3.28	1.01	0.93	0.89	17.2
Cr, $\times 10^{-5}$	5	5	5	5	60	260
Mn, $\times 10^{-3}$	3.7	5.2	3.9	2.7	N/A	N/A
Ni, $\times 10^{-4}$	4	15	2	4	N/A	N/A
Co, $\times 10^{-4}$	1	2	1	1	N/A	N/A
K	1.1	1.6	0.7	0.9	N/A	N/A
Na	0.5	2.5	0.3	0.4	N/A	N/A
Ca	0.20	0.56	0.29	0.14	N/A	N/A
Mg	0.02	0.07	0.04	0.02	N/A	N/A

"Tzentral'nosibirskii" zapovednik is located very far from any sources of industrial pollution, so it was proposed to cancel environmental monitoring similar to that one carried out at "Stolby" zapovednik. Pollution level data observed at "Stolby" zapovednik had gaps, for some specific pollutants [1]. On the contrary, pollution level of the pollutants to be observed at "Tzentral'nosibirskii" zapovednik was supposed to be equal to the background one. We traced the level of those pollutants at "Tzentral'nosibirskii" zapovednik, trying to heal the gaps mentioned above. In 2006, a pilot screening research has been carried out, at four basins located alongside Yenisei river. The motivation behind such screening was the hypothesis towards a lack of global pollution to be observed at these point.

Surprisingly, the screening observations had shown an excess of some pollutants over the background level (see Table 1).

3.4 Modelling of Pollutant Distribution

A distribution of chemical pollutants, as well as their migration is affected with numerous complex factors. The behaviour of pollutants could hardly be described and present in simple figures; the point is that some pollutants are tended to a reasonably stable bunching, while others are not. Thus, one can quite precisely foresee the behaviour of some pollutants observing some of them considered to be very indicative. Here we figure out two patterns of pollutant distribution: the former is a general one, and the latter is Zn.

It is a common practice to show the results of monitoring of global pollutants in generalized index, in site of the fact that this latter may loose some (maybe, important) details. To address the problem, we developed GIS-based model producing such index:

$$K_c = \frac{C}{C_b}, \tag{1}$$

where K_c is the concentration factor of a chemical element, C is the real content of that latter, and C_b is the background content of the element. K_c indicates an excess of the element content over an averaged background level of that latter determined over a territory. The reference sites are expected to show the pollution figures minimally affected from outside.

Quite often, compartments of an ecosystem are poisoned with several elements. Here the total contamination index could be derived representing an effect of a group of elements. This factor is determined for all elements detected within a sample attributed to a site where the sampling has been done; both the formula for the factor, and its value are established legally. Thus, the total pollution index Z_c is defined as follows:

$$Z_c = \sum_{j=1}^{n} K_c^{(j)} - (n-1), \tag{2}$$

where $K_c^{(j)}$ is the concentration index of j-th pollutant, and n is the number of pollutants.

Basically, the idea to make a generalized index stands of the concept of maximum permissible concentration (MPC). The point is that MPC methodology fails to take into account the specific features of a region, or a specificity of a pollutant impact on environment, etc. This discrepancy was broken through due to the model implementation, that changes a background level values of pollutants in compartments for clarkes. Besides, the background indices for fluorine and heavy metals are not available for the site.

3.5   Spatial Analysis of Zinc Distribution

To begin with, we start from the data on zinc distribution. Zinc is motile element at acid and low-acid soils, with moderate phytotoxicity. The growth of concentration of zinc yields a suppression of vegetation [3]. Zinc is mainly delivered to soil with industrial wastes from non-ferrous metallurgy, paintwork material industry, galvanism sewage and slugs of municipal sewage treatment facility.

The analysis of zinc distribution at the substrate (see Fig. 4, C) shows that zinc content exceeds the background level, in general. There is a single site close to Synzhul check-point where the concentration is less than 40 mg/kg. Zinc contaminates the reservation, at most; the content ranges from 40 to 80 mg per kg, twice exceeding the background level. The area located closely to Kaltat river, Sliznevaja river, Sliznevaja Rassokha river, and Drjannaja river, as well as close to Bykovsky spring are exposed to high zinc content (more than 80 mg/kg).

Zinc distribution in soil (see Fig. 4, D) differs from that former in substrate. Basically, zinc content is close to the background level; this latter is equal to

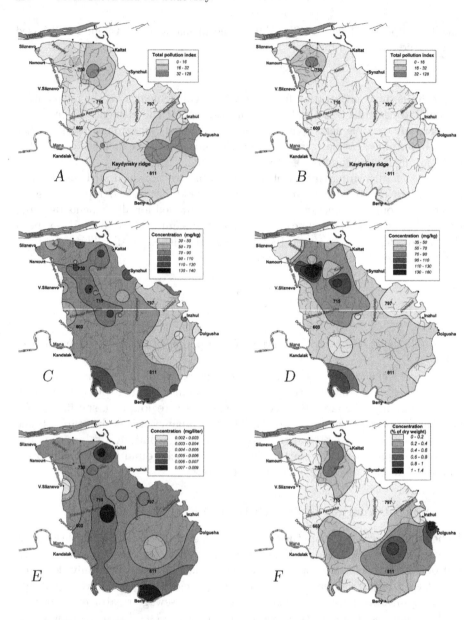

Fig. 4. Pollution patterns for general pollution (in substrate (A) and in soil (B), and zinc pollution (see details in text).

51 mg/kg. Nevertheless, few spots of strong contamination are observed. In general, these spots coincide to the localities of the increased zinc content in the substrate. This fact probably follows from an active migration of zinc in the *substrate – soil* compartment, since one observes an acid medium there.

Correlation coefficient of zinc content in substrate vs. that latter in soil is 0.63. Such correlation reveals vertical migration of zinc at the soil profile.

Increased zinc content in precipitates is observed at the watershed of Medvezhii spring and Pletnjazhnaja river, and at the area immediately neighbouring the suburb of the city of Kranoyarsk (see Fig. 4, E). No congruence of the contamination spots is observed among needles, soil and substrate. Two regions exhibit the increased zinc content in vegetation (see Fig. 4, F). The upper reaches of Namourt and Kaltat, and junction area of Sliznevaja Rassokha with Bol'shaja Sliznevja are these regions. It should be noticed, that the areas of increased zinc content in vegetation are not related to the areas of the increased zinc content in substrate and upper soil level. Thus, zinc content in vegetation is related to the content of that former in rock, rather than to other compartments.

4 Discussion

Modelling of spatially distributed dynamics of various ecological systems still challenges the experts in various fields ranging from pure mathematics to specific issues in biology and computer science. The variety of approaches is great enough to meet all the aspects of these rather complicated studies. Clear, apparent and powerful modelling of the dynamics of compartments of spatially distributed ecological system is still doubtful, for many reasons. Here we present the results of the modelling of environmental processes running at the discrete and spatially extended ecological system of "Stolby" state reservation. We have implemented the simulation model of the site that figures out some typical steady regimes in the distribution of vegetation over the territory of zapovednik, and distribution of global pollutants affecting the ecosystem. GIS technologies and solutions are the key issues of the modelling implementation. Similar approach is present in [14]. They discuss the model for habitat estimation, where this latter is affected by a number of factors. GIS implementation allows to figure out the effect of spatially distributed factors. Similarly, the modelling of generalized pollution pattern at zapovednik allows to evaluate the habitat conditions in that latter, in general. It should be said, that the studies of the environmental processes with the special emphasis to the reserved territories and reference sites with wild nature ecological pattern of habitation draw the attention of researchers (see, e.g., [10,11]). These works present the modelling results of the environmental conditions at Katun State reservation located in Altai, where the impact of relief could hardly be taken into account but the GIS technologies and solutions.

It is a common fact, that environmental conditions impact the plant community and vegetation pattern of a (complex) ecological system, while the detailed pattern of such correspondence is not clear. The altitude of the vegetation place is the key factor here. We found rather stable and distinctive typical communities with prevalence of different forest species occupying the different altitude zones. Figures 3 and 4 show this relation between the altitude and forest species occupancy, in detail. Such complexes of dominating species are argued sometimes, while they are found by different researchers, in different climatic and environmental conditions; see, e.g., [10,11] for Siberia, [7] for mountain China region.

Quite detailed and smart study of such zone structure of forests is studied and discussed by Hörsch [12].

The developed model reveals some statistical relations between all four compartments of the studied ecological system. Relief seems to be the leading factor here. Thus, soil types correlate quite closely to the altitude zones. Four types of soils are roughly identified at zapovednik; the types exhibit the prevalence with the altitude. This is not a point, since the types are mainly identified being based on the altitude zone structure. A relation between the class of soil and the altitude is less obvious. Eight classes of soil are determined at zapovednik; each class preferably occupies a specific altitude belt. A dispersion of the classes among themselves observed at zapovednik results from the highly jagged relief of the reservation.

On the contrary, one sees quite poor correlation between vegetation type, and soil class. One could expect to meet an increased correlation between these two compartments of the ecosystem, while there is observed rather poor correlation between these entities. The point is that soil formation is affected with a number of factors, including relief, rock, climate, season, vegetation, etc. All these factors force the relation simultaneously and, quite often, in opposite directions. Such complex interaction obturates discretion of the effect of a peculiar factor. More close study of the relations between soil class and vegetation reveals a decreased correlation among them. This fact means that the structure of forest associations fails to identify unambiguously the relevant class of soil. Similar observations are discussed in [12].

The efficiency of the approach to the modelling shown above should be used to study the dynamics of various pollutant distributions over an ecosystem. The dynamics of the transfer of such compounds could hardly be taken with the analytical modelling solely. The point is, that the pollutants are transferred both actively, by biotic components of an ecosystem, and passively, through the diffusion and other types of inactive transportation web. Such approach was used in [8], where the smart decision-making support system was implemented for the tasks of forest management and control. The GIS-based modelling is widely used for the studies of the dispersion of contaminations among the compartments of various ecosystems (see, e.g., [13]). The monitoring system supported with various Internet solutions becomes a new interactive scientific tool for a community of experts in various fields; see, e.g. http://info.krasn.ru/stolby/ and http://res.krasu.ru/ses/doc/1_1.shtml or [10].

The results of the study presented here are mainly focused on the development of the simulation model describing the dynamics of vegetation and pollutant distribution for the purposes of environmental monitoring at "Stolby" reservation. The most up-to-date methods of cartography modelling and GIS-based analysis were used due to implement informational technologies of a complex analysis of spatial geographical data and environmental data. The methodology presented above is currently implemented for the studies of the "Tzentral'nosibirskii" State Biospheric Wildnature Reservation located at the southern part of Turukhansk region of Krasnoyarsk krai, and, partially, at

Evenk autochthon area (Baikit region). The area of the reservation is 972017 hectares, exceeding more than 20 times the area of "Stolby" State Reservation. An efficiency of CIS technologies for the purposes of wildnature monitoring is evident; the technologies could be expanded for other territories, which meet a defence by society. A detail study due to GIS-technologies of the state and dynamics of natural processes at the "Tzentral'nosibirskii" State Biospheric Wildnature Reservation reveals the reference pattern of the dynamics of natural processes at the biosphere, since anthropogenic influence on the reservation is very low. The data gathered at the reservation could be used for estimation of the state of any reserved territory, both at Russia, and outside.

References

1. Erunova, M.G.: Implementation of geoinformation system of the "Stolby" state reservation. Zapovednoe delo. (9), pp. 76–80 (2001). Moscow
2. Iverson, L.R., Dale, M.E., Scott, T., Prasad, A.: A GIS-derived integrated moisture index to predict forest composition and productivity of Ohio forests (USA). Landscape Ecol. **12**, 331–348 (1997)
3. Tappeiner, U., Tasser, E., Tappeiner, G.: Modelling vegetation patterns using natural and anthropogenic influence factors: preliminary experience with a GIS based model applied to an Alpine area. Ecol. Model. **113**, 225–237 (1998)
4. Münier, B., Nygaard, B., Ejrnæs, R., Bruun, H.G.: A biotope landscape model for prediction of semi-natural vegetation in Denmark. Ecol. Model. **139**, 221–233 (2001)
5. Dymond, C.C., Johnson, E.A.: Mapping vegetation spatial patterns from modelled water, temperature and solar radiation gradients. Photogramm. Remote Sens. **57**, 69–85 (2002)
6. Pfeffer, K., Pebesma, E.J., Burrough, P.A.: Mapping Alpine vegetation using vegetation observations and topographic attributes. Landscape Ecol. **18**, 759–776 (2003)
7. Zhao, C., Nan, Z., Cheng, G., Zhang, J., Feng. Z.: GIS-assisted modelling of the spatial distribution of Qinghai spruce (Picea crassifolia) in the Qilian Mountains, Northwestern China based on biophysical parameters. Ecol. Model. (Available online 10 August 2005, in press)
8. Ostwald, M.: GIS-based support tool system for decision-making regarding local forest protection: illustrations from Orissa, India. Environ. Manag. **30**(1), 35–45 (2002)
9. Ji, W.W., Leberg, P.A.: GIS-based approach for assessing the regional conservation status of genetic diversity: an example from the southern Appalachians. Environ. Manag. **29**(4), 531–544 (2002)
10. Prechtel, N.: GIS-Aufbau für den Naturschutz im Russischen Altai. In: Geoinformationssysteme – Theorie, Anwendungen, Problemlösungen. Kartographische Bausteine. Bd. 21, pp. 82–100, Institute for Cartography, TU Dresden (2003a)
11. Prechtel, N.: Selected Problems and Solutions for Drainage Modelling and Handling in a GIS. In: Geoinformationssysteme – Theorie, Anwendungen, Problemlösungen. Kartographische Bausteine. Bd. 21, pp. 101–109. Institut für Kartographie, TU Dresden (2003b)

12. Hörsch, B.: Modelling the spatial distribution of montane and subalpine forests in the central Alps using digital elevation models. Ecol. Model. **168**(3), 267–282 (2003)
13. Lee, C.S.L., Li, X., Shi, W., Cheung, S.C.N., Thornton, I.: Metal contamination in urban, suburban, and country park soils of Hong Kong: a study based on GIS and multivariate. statistics. Sci. Total Environ. **356**(1–3), 45–61 (2006)
14. Store, R., Jokimäki, J.: A GIS-based multi-scale approach to habitat suitability modelling. Ecol. Model. **169**(1), 1–15 (2003)
15. Erunova M.G., Yakubailik O.E., Kadachnikov A.A. Geoinformation analysis of nature condition at "Stolby" reservation. Geogr. Nat. Resour. (2), 136–142 (2006)
16. Erunova, M.G., Gosteva, A.A., Yakubailik, O.E.: GIS support for the environmental monitoring of wildnature reservations. SFU J. Ser. Techn. Technol. **1**(4), 366–376 (2008)
17. Erunova, M.G., Sadovsky, M.G., Gosteva, A.A.: GIS-aided simulation of spatially distributed environmental processes at "Stolby" state reservation. Ecol. Model. **195**(3–4), 296–306 (2006)
18. Gavrikov, V.L., Sharafutdinov, R.A., Knorre, A.A., Pakharkova, N.V., Shabalina, O.M., Bezkorovaynaya, I.N., Borisova, I.V., Erunova, M.G., Khlebopros, R.G.: How much carbon can the Siberian boreal taiga store: a case study of partitioning among the above-ground and soil pools. J. For. Res. **27**(4), 907–912 (2016)

What Can the Big Data Eco-System and Data Analytics Do for E-Health? A Smooth Review Study

Sidahmed Benabderrahmane[(✉)]

Paris Dauphine University, Place du Marchal de Lattre de Tassigny,
75016 Paris, France
Sidahmed.benabderrahmane@gmail.com

Abstract. In this paper we present a global overview of the present usage and future trends of the different big data ecosystems in the E-Health's scientific domains. Indeed, bioinformaticians as well as medicine practitioners are actually generating very large amounts of data, and thus storing, managing, and analyzing these large scale data-sets still represent a big challenge. The used Big Data ecosystems are involved at different steps of the production chain, i.e., from the acquisition of both structured and non-structured data, the storage in traditional and/or NoSQL databases, and finally the analytics using the Map Reduce framework. We will discuss in this smooth survey, all these parts of the ecosystem and will give some use cases on real data-sets in the domain.

Keywords: Big data · Data mining · Machine learning · Data science · E-health · Bioinformatics · Medicine

1 Introduction

Nowadays, the size of the data that is being generated and created in different organizations is increasing drastically. Due to this large amount of data, several areas in artificial intelligence and data science have been raised. Hence, after having collected the data from different sources and stored them in various databases, we can proceed to many analysis steps for the aim of extracting hidden information and discriminant patterns [1]. In fact, most organizations do actually a set of techniques to discover intricate relationships, discover complex patterns, and predict trends in the data. These data are modeled and pertinent variables are identified in order to obtain useful insight on the problem and obtain intelligent decision. Figure 1 illustrates an example of what we call the advanced analytics and its components. Indeed it contains descriptive, diagnostic, predictive, and prescriptive analytics. Altogether, theses methods allow us to know what is hidden in our data, what has happened, why it happened, what will happen and what we can do to change what will happen in a way that benefits us [2].

© Springer International Publishing AG 2017
I. Rojas and F. Ortuño (Eds.): IWBBIO 2017, Part I, LNBI 10208, pp. 629–641, 2017.
DOI: 10.1007/978-3-319-56148-6_56

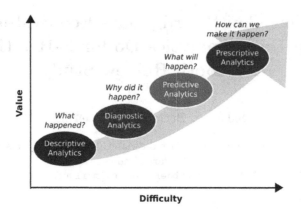

Fig. 1. Data analytics work-flow in data science.

During the descriptive analysis stage, which is the first step of our analysis, a global summary on the data yields to create a historical view on the data and prepare them for future analysis. This analysis uses illustrative figures such as histograms, pie chart, or tables, and the user can easily interpret the hidden information. The diagnostic analytics is the next step of the analysis. It uses methods such as correlations factors for the interpretation of factors that contributed to the outcome. Here, it is important to verify the possible correlation dependencies that may exist between the input variables. It can be calculated with different way, and the less distant to 1 is the result, the more correlated the variables are. The predictive analytics is used for making the prediction of future values and identifying unknown events. It is the most important stage of the advanced analytics, since it involves very sophisticated methods of machine learning and data mining. Finally, Prescriptive data analytics is the last step of an advanced data analysis. It consists of the application of the predictive model to determine the best solution or outcome among various choices, given the known parameters. In this phase, not only is predicted what will happen in the future using our predictive model, but also is shown to the decision maker the implications of each option [3].

Actually, medical laboratories and health-care industries are generating very large amounts of data. Such huge volumes of complex data are used at different steps in data analytics to support decision making, patient monitoring, anomalies detection, disease tracking, drugs discovery, and fundamental scientific research [4].

It is important for these different organizations to extract, store, and manage the data for further analysis, by taking into account the different issues, such as the type of the infrastructure, the data format, and the frequency of acquisition. Researches in E-health, medical informatics, or computational biology attract more and more of data scientists as well as biologists or bioinformaticians. Their works are evolved in conjunction with a high focus on data analytics and data preparation and cleaning. With the emergence of novel technologies such as connected objects or Internet of Things (IOT) [5], the way of data generation in health domains has been drastically transformed. For example, in the case of patients data, Electronic Health Records (EHR) are massively produced thanks to intelligent sensors, pervasive smart phones applications, and social media [6]. By consequence, these technologies have participated in the proliferation of the concept of Big Data that allowed health-care industries and life science laboratories to set up a variety of business and scientific researches such as, improving patient hospitalization and profiling, enhancing chemical drug discovery, supporting clinical tests, accelerating DNA sequencing, ... etc.

We discuss in this paper the different landscape and use cases of Big Data in E-Health domains, by highlighting the real gain when using these technologies to manage and mine medical data. The rest of the paper is organized as following: the next section gives a wide-range state of the art study of the different big data technologies, starting as a chronological way, from parallel computing to distributed systems. Section 3 presents the application of big data to resolve health-related problems. Section 4 discusses the current challenges and the future trends in the domain and concludes the paper.

2 Big Data Eco-System: The Chronology

2.1 High Performance Computing and Grid Computing

The arrival of the big data is the result of several research studies on parallel computing. Historically, works have started earlier in the 90' with the multi-threading programming frameworks [7], with which were proposed programs that run on several processor cores in a parallel manner. After that, technologies such as High Performance Computing (HPC) systems and Grid Computing (GP) raised, allowing the use of multiple machines for a common computation task [8]. These distributed processing systems, are based on the connection between several computers (nodes in a cluster) on which the large data-sets are distributed to solve a problem. The shared resources in the clusters are memories, hard disks, and CPUs [9]. The issue with HPC and GP is that they are very expansive due to their high technology multi-core CPUs. Indeed, the currently used high quality processors handle the most computationally heavy operations such as physical simulations, offline rendering, or forecasting. Some of these heavy calculations are nevertheless easily parallelizable and can therefore benefit from an architecture designed for parallel computation. GPU computing has come on the market as an alternative, since the GPU (Graphical processing unit) components are cheaper than CPUs, and offering competitive characteristics in large-scale data processing. However, the problem with GPU high-throughput computing is the difficulty of communicating over a network. Moreover GPUs cannot handle virtualization of resources [10].

2.2 Google Distributed File System

Research works on big data were initiated earlier in past decade by Google scientists, for the aim of the storage of a wide range of complex and heterogeneous data that they have been producing (query logs, maps, videos, images, ...). The proposed GFS (Google File System) is a new way to store and process data in a paralleled manner on a set of connected machines [11]. The idea was to create a distributed, scalable, and fault-tolerant storage system using commodity hardware, alternatively to HPC and GPU that are expensive. The proliferation of big databases in the different domains such as computational biology, health, economy, retails, ..., helped in the emergence of novel industrial opportunities and research challenges for both data storage and analysis. The Big-Data terminology has been used to describe large volume of databases, which have 4 important properties, also known as 4V: Volume, Variety, Velocity and Veracity [12]. The volume property means that a big data system should be capable to support storing and analyzing big data-sets (from some Tera to Zeta-bytes). The variety illustrates how a big data framework can analyze both structured or unstructured data, textual, images, video, batch or streaming data. The velocity represents the growing speed with which the data are changing during time. With GFS, the size of disk blocks units were expanded (up to 64 MB), hence it was possible to deal with big data, since large files were splitted in small disk units. Moreover, GFS offers the possibility to replicate data on different machines, hence decreasing the probability of loosing data [13].

2.3   Hadoop Framework: The Great Twist of Map Reduce

As reported above, the history in data processing shows that initially the computing applications were performed on a machine with a single processor and a main memory. Parallel computing was introduced for large databases analytics using very expansive and not publicly accessible solutions such as HPC. With the emergence of Big Data concepts, and the wide use of Internet with different materials (smart-phones, laptops, connected objects,...), the need of large computing solutions with several thousands of nodes became evident. To that aim, Doug Cutting started working from 2004 on a software called Hadoop [12,14], which is a framework that can be deployed on a cluster of commodity machines. The great advantages of Hadoop are: it is open source[1], it permits large scale distributed data analysis, and it also provides a robust and fault-tolerant Hadoop Distributed File System (HDFS). This new technology has been inspired from Google's file system [11], and was published starting from 2008. The main added value in Hadoop is a Java-based API, which enables parallel processing via the nodes of the cluster using the MapReduce paradigm (MR) [13,15,16].

Similarly to GFS, Hadoop HDFS uses large disk blocks, and it saves each file in a configurable number n of replication on different blocks of the nodes of the cluster, hence it ensures fault-tolerance. HDFS is composed of a master machine called NameNode (NN) that is responsible of saving namespace of the file system (FS), and manage acces to data. It also communicates with slave machines called DataNodes (DN) that are data storage units. Datanodes send heartbeat signals at different time stamps to NameNode, and it is with this mechanism that the

---

[1] hadoop.apache.org.

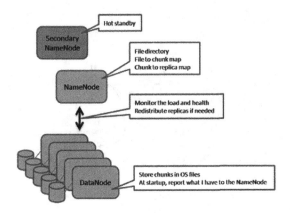

Fig. 2. The Hadoop HDFS file system. A cluster is composed by a NameNode, a lot of DataNodes and a secondary NameNode. Data are replicated in many DataNodes.

master detects the loss of a slave and asks for its replacement. To ensure high availability, Hadoop offers secondary name node (SNN), to resist to failures and deal with the loss of the NM (see Fig. 2).

Hadoop includes YARN (Yet Another Resource Negotiator) a generic platform, which allows to run a program under Map Reduce. MapReduce is an innovative programming framework for big data applications. It is based on the well known functions in functional programming: map and reduce. Initially, Hadoop starts by dividing the large datasets in several splits of Key Value (KV) pairs, and each split (K, V) is associated to a map function in each node of the cluster (Fig. 3). By consequence, the different map functions cloud be executed separately and in parallel. As in HDFS, Yarn offers a master Resource Manager (RM) and several slaves Nodes Managers (NM). The RM is responsible of dividing data and calling maps functions within NM. Results of each map function are then collected and merged in a reduce function.

2.4 Cloud Computing

Cloud computing has brought the use of big data storage and analysis with Hadoop and Spark [17] at worldwide scale [18]. Cloud Computing tries to standardize the distributed storage and processing of data through Internet. One of the innovative key points in the cloud computing (CC), is the possibility for a physical machine to host several virtual machines, hence allowing the utilization at best the hardware resources. It provides on-demand, easy to use access to shared storage, services, and computing capabilities. In the CC technologies, we can distinguish between Infrastructure as a Service (IaaS), Platform as a Service (PaaS), and Software as a Service (SaaS). IaaS, is the case when CC offers data storage centers and HPC infrastructure services to the users, so that they can have access to them through Internet. It is possible to build virtual machines using Linux or MS Windows, to scale up or down the configuration

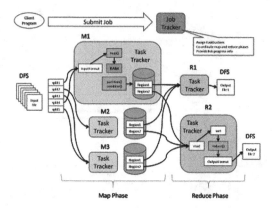

Fig. 3. The Map Reduce framework. The data are divided in pairs of Key, Values on several map functions. The final results are merged in reduce functions.

of chosen infrastructure on demand, by a payment contract. Actually, Amazon Web Services (AWS), are the top providers of IaaS in the world. Their most popular IaaS are Elastic Compute Cloud (EC2) and Simple Storage Service (S3). PaaS, is a top level of encapsulation on IaaS, thus offering to users not only to use the infrastructure, but also the already installed development platforms. Finally, SaaS represents the complete package, in which the user can have both a storage facility, a pre-configured cluster of machines, and set of programs and software on Map Reduce. All what he needs is an Internet connection and a web navigator, since the global environment is visible through Rich Internet Application (RIA). Microsoft Azure Data Analytics is a good example of the SaaS in data science and Business Intelligence.

2.5   Hadoop Most Important Components

An example of the Hadoop ecosystem is illustrated in Fig. 4. This non-exhaustive list is composed of a lot of important Apache projects. Among them we have: Hbase, Hive, Oozie, Pig, Mahout, Sqoop, ZooKeeper, Flume.

Hive and HBase Databases. Firstly, HBase (http://hbase.apache.org) is a NoSql fully distributed and column-oriented database system. It is based on Google's BigTable project [13]. HBase is column-based rather than row-based, which enables high-speed execution of operations performed over similar values across massive datasets [10]. Hive (http://hive.apache.org) is a data warehouse system that is implemented in HDFS and developed by Facebook. It supports SQL-based queries using HQL (Hive QL) to interrogate big databases. Hive queries are executed with Map Reduce.

Zoo Keeper. It is a coordination system for Hadoop applications. It ensures a shared memory for efficient distributed synchronization between nodes.

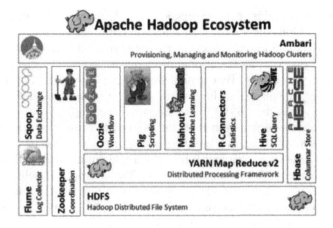

Fig. 4. The most important components of the Hadoop ecosystem.

Pig. It is a framework developed by Yahoo, which consists of a high-level data analysis with a scripting language called Pig Latin. It transforms data processing as a sequence of map reduce job. This language provides various operators using which programmers can develop their own functions for reading, writing, and processing data.

Mahout. It is a scalable machine-learning and data mining library. It includes several MapReduce enabled clustering implementations, recommendation methods, and classification techniques. It is a key library for data scientists.

Oozie. It is a work-flow scheduler and a job coordinator. It supports different Hadoop jobs, such as: Map Reduce (Java or Streaming entries), Pig scripts, Hive or Sqoop queries.

Flume. It is an Apache project used for data and logs collection, extraction, aggregation and transfer in Hadoop. It is based on distributed agent architecture.

Sqoop. Acronym to (SQL-to-Hadoop) is a tool developed by Cloudera, which is used to import and export data from and to HBase or Hive with relational systems.

3   Big Data in E-Health: A Smooth Review

As reported above, Data analytics and Big Data technologies have allowed medicine practitioners and researchers to perform high-throughput analyses, and make suitable decision support in biomedical domain. With the power of big data and data science, we are one step closer to a world where genetic diseases are more effectively managed and more frequently cured, changing patient lives forever[2]. The available datasets in health are complex, heterogeneous, and covering

---

[2] insidebigdata.com.

a wide range of domains: gene expression data [19], electronic medical record (EMR) [20,21], biomedical data [22], semantic data [23], sequence data [24], or medical images [25]. Such complex massive data are hard to manage and difficult to analyze and query using traditional mechanisms, especially when the queries themselves are complicated. By decomposing the problem with MapReduce algorithm, not only we can make the processing fast through parallelism, but also make data acces easy due to HDFS architecture. The mapped components of the query can be processed simultaneously or reduced to rapidly return results [26].

There are a lot of increasing numbers of important use case examples, where combining big data technology with health-care, bioinformatics and life sciences has become meaningfully beneficial [27–30]. These proposed solutions can be either in sequences alignment [31], sequences mapping and assembly [32,33], gene expression data analysis and SNP analysis [34], or even in neuroscience [35], medical imaging [36] and biometric domains [37].

Genomic Sequence Mapping and Analysis. One of the first applications with MapReduce in the biomedical domain is the genome processing and DNA sequencing [29]. They are attractively growing in the industrial markets, since the generated data produced by DNA sequencers, and the increasing number of individuals being sequenced. Genome Analysis Tool Kit (GATK) is an example of a genome-sequencing project under Hadoop platform [38]. Schatz et al. in [32], proposed Cloudburst, which is a hadoop tool for SNP discovery and genotyping using SoapSNP. After that, the same authors, proposed other MapReduce based tools, such as: Crossbow [39] for genome sequencing and SNP genotyping starting from short reads, Contrail[3], and Myrna [34] for differentiating gene expression levels from big RNA-seq databases. The Basic Local Alignment Search Tool (BLAST) is a well known method used for identifying similar regions between sequences. Initially a parallel version of BLAST was proposed using MPI API [40]. Vouzis et al. in [41] proposed GPU-BLAST, with GPU models. However, as we explained in the past sections these technologies suffer from the lack of scalability, thus they were not well adapted for performing the batch processing of huge amounts of sequence data. Using the Hadoop MapReduce framework, a parallel version that is called CloudBlast was presented in [31].

Gene Expression and RNA Sequence Analysis. Differential analysis and RNA-seq data analysis are well known methods for gene expression classification [42]. Langmead et al. in 2010 [34] proposed Myrna, a cloud computing software for transcriptomic expression data analysis, with large-scale RNA-seq sources. Myrna can be tested either on Amazon AWS, or on a local hadoop cluster. FX is a tool for gene expression level analysis, and RNA sequence analysis [43]. It can extract expression profiles, and SNP calling. As in the case of Myrna, FX can also be executed in the cloud or a local hadoop cluster.

Neuroscience and Biomedical Signal Analysis. Actually, biomedical industries try to focus their research and technologies to map the human brain with

---

[3] https://sourceforge.net/projects/contrail-bio/.

machines. The main objective of such developments is to understand the activity of the neurons in the brain, and to hopefully discovering fundamental inherent mechanisms. Biomedical neurosciences can help understanding brain trauma and diseases. Consequently, researchers want to build instruments for brain monitoring by recording neurons activity at different time scales. The produced data is extremely large. Authors in [35,44], presented tools for EEG data acquisition, visualization, and analysis using Hadoop ecosystem and NoSQL databases, and MapReduce framework.

Biomedical Image Analysis. MapReduce has been used in several image analysis applications in the cloud. Recently the computer science department of Virginia university proposed HIPI, an image processing library, which is regarding the developer's definition, designed to be used with the Apache Hadoop MapReduce parallel programming framework[4]. HIPI facilitates efficient and high-throughput image processing with MapReduce style parallel programs typically executed on a cluster. It provides a solution for how to store a large collection of images on the Hadoop Distributed File System (HDFS) and make them available for efficient distributed processing. In a recent work, Markonis et al. [45] proposed a MapReduce medical analysis system, for texture segmentation, indexing, and directional wavelet analysis for solid texture classification.

Personalized Treatment Recommendation. The ideal manner to customize treatment for patients, is by monitoring them continuously in order to be able to detect the effects of medication. The different doses can be adapted regarding on how the medication is working for that particular individual[5]. Authors in [46] presented a interdisciplinary cloud-based model to implement personalized medicine. The system suggests the use of cloud computing and big data architecture for connected objects data acquisition and processing. It also includes (i) the generation of cost-effective high-throughput data; (ii) hybrid education and multidisciplinary teams; (iii) data storage and processing; (iv) data integration and interpretation; and (v) individual and global economic relevance.

Patient monitoring is also as important as treatment recommendation, since vital signs are observed using healthcare facilities. The data from these various monitors can be used in real time and send alerts to nurses or care providers so they know instantly about changes in a patients condition [47].

Medical Internet of Things. Dimitrov presented a work in [6], where he explained the usefulness of the concept of Medical Internet of Things (mIoT). Regarding the author, it is a critical piece of the digital transformation of healthcare, as it allows new business models to emerge and enables changes in work processes, productivity improvements, cost containment and enhanced customer experiences. Wearable and mobile apps support fitness, health education, symptom tracking, and collaborative disease management and care coordination. All

---

[4] http://hipi.cs.virginia.edu.
[5] insidebigdata.com.

those platform analytics can raise the relevancy of data interpretations, reducing the amount of time that end users spend piecing together data outputs. Insights gained from big data analysis will drive the digital disruption of the healthcare world, business processes and real-time decision-making.

4 Conclusion and Future Trends

In this paper we have presented a global overview of the big data ecosystem, and some data analytics methods. Medicine and computational biology are among the most widespread domains that have seen a strong emergence of big data and cloud computing. Actually, hadoop is the widely used cloud computing and MapReduce framework. We have shown the advantages of HDFS file system for big data management, scalability and fault tolerance. We also have presented several big data implementations in E-Health and computational biology.

Nowadays, the use of data analytics in medicine has been hailed as a solution for saving time and money and improving patient outcomes for a healthcare organization. The role of big data in medicine is one where we can build better health profiles and better predictive models around individual patients so that we can better diagnose and treat disease[6]. However, many health systems have a hard time capturing and using data from patients that can make a real impact on patient outcomes. Part of the issue lies in EHR data, which can provide an incomplete picture of patient behavior[7]. It must be also noted that Hadoop is not a catch all technology but rather is best suited to batch processing applications, as opposed to real time ad-hoc queries. The application of this technology to suitable high impact areas, such as metagenomics, personalized medicine, systems biology and protein function and structure prediction has the potential for killer applications [30].

References

1. Inmon, W.H., Linstedt, D.: A brief history of big data. In: Inmon, W.H., Linstedt, D. (eds.) Data Architecture: A Primer for the Data Scientist, pp. 45–48. Morgan Kaufmann, Boston (2015)
2. Secchi, P., Paganoni, A.M.: Advances in Complex Data Modeling. Springer, Heidelberg (2014)
3. Fawcett, T., Provost, F.: Data Science for Business What You Need to Know about Data Mining and Data-Analytic Thinking. OReilly Media, Sebastopol (2013)
4. Zou, Q., Li, X.-B., Jiang, W.-R., Lin, Z.-Y., Li, G.-L., Chen, K.: Survey of mapreduce frame operation in bioinformatics. Brief. Bioinf. 15(4), 637–647 (2014)
5. Linstedt, D., Inmon, W.H.: Data Architecture: A Primer for the Data Scientist, Big Data, Data Warehouse and Data Vault. OReilly Media, Sebastopol (2014)
6. Dimitrov, D.V.: Medical internet of things and big data in healthcare. Healthc. Inf. Res. 22(3), 156–163 (2016)

---

[6] http://www.mckinsey.com/.

[7] http://managedhealthcareexecutive.modernmedicine.com.

7. Coulouris, G., Dollimore, J., Kindberg, T., Blair, G.: Distributed Systems: Concepts and Design, 5th edn. Addison-Wesley Publishing Company, Boston (2011)
8. Rajaraman, A., Ullman, J.D.: Mining of Massive Datasets. Cambridge University Press, New York (2011)
9. Berman, F., Fox, G., Hey, A.J.G.: Grid Computing: Making the Global Infrastructure a Reality. Wiley, New York (2003)
10. Mohammed, E.A., Far, B.H., Naugler, C.: Applications of the mapreduce programming framework to clinical big data analysis: current landscape and future trends. BioData Min. **7**(1), 22 (2014)
11. Ghemawat, S., Gobioff, H., Leung, S.T.: The Google file system. In: Proceedings of the Nineteenth ACM Symposium on Operating Systems Principles, SOSP 2003, pp. 29–43. ACM, New York (2003)
12. White, T.: Hadoop: The Definitive Guide, 1st edn. O'Reilly Media Inc., Sebastopol (2009)
13. Chang, F., Dean, J., Ghemawat, S., Hsieh, W.C., Wallach, D.A., Burrows, M., Chandra, T., Fikes, A., Gruber, R.E.: Bigtable: a distributed storage system for structured data. ACM Trans. Comput. Syst. **26**(2) (2008)
14. Lam, C.: Hadoop in Action, 1st edn. Manning Publications Co., Greenwich (2010)
15. Dean, J., Ghemawat, S.: Mapreduce: Simplified data processing on large clusters. Commun. ACM **51**(1), 107–113 (2008)
16. Dean, J., Ghemawat, S.: Mapreduce: a flexible data processing tool. Commun. ACM **53**(1), 72–77 (2010)
17. Zaharia, M., Chowdhury, M., Franklin, M.J., Shenker, S., Stoica, I.: Spark: cluster computing with working sets. In: Proceedings of the 2Nd USENIX Conference on Hot Topics in Cloud Computing, HotCloud 2010, Berkeley, CA, USA, p. 10. USENIX Association (2010)
18. Larus, J.R.: The cloud will change everything. SIGPLAN Not. **46**(3), 1–2 (2011)
19. Juan, H.F., Huang, H.C.: Bioinformatics. Humana Press, Totowa (2007). pp. 405–416
20. Hoogendoorn, M., Szolovits, P., Moons, L.M.G., Numans, M.E.: Utilizing uncoded consultation notes from electronic medical records for predictive modeling of colorectal cancer. Artif. Intell. Med. **69**, 53–61 (2016)
21. Siuly, S., Li, Y., Zhang, Y.: EEG Signal Analysis and Classification - Techniques and Applications. Health Information Science. Springer, Heidelberg (2016)
22. Kafkas, S., Kim, J.H., Pi, X., McEntyre, J.R.: Database citation in supplementary data linked to europe pubmed central full text biomedical articles. J. Biomed. Semant. **6**, 1 (2015)
23. Benabderrahmane, S., Smaïl-Tabbone, M., Poch, O., Napoli, A., Devignes, M.-D.: IntelliGO: a new vector-based semantic similarity measure including annotation origin. BMC Bioinform. **11**, 588 (2010)
24. Yu, N., Li, B., Pan, Y.: A cloud-assisted application over apache spark for investigating epigenetic markers on DNA genome sequences. In: 2016 IEEE International Conferences on Big Data and Cloud Computing (BDCloud), Social Computing and Networking (SocialCom), Sustainable Computing and Communications (SustainCom), BDCloud-SocialCom-SustainCom 2016, Atlanta, GA, USA, 8–10 October 2016, pp. 67–74 (2016)
25. Ahmed, Z., Saman, Z., Dandekar, T.: Mining biomedical images towards valuable information retrieval in biomedical and life sciences. Database **2016** (2016)
26. Fiore, S., DAnca, A., Palazzo, C., Foster, I., Williams, D.N., Aloisio, G.: Ophidia: towardbig data analytics for escience. Procedia Comput. Sci. **18**, 2376–2385 (2013)

27. Schumacher, A., Pireddu, L., Niemenmaa, M., Kallio, A., Korpelainen, E., Zanetti, G., Heljanko, K.: SeqPig: simple and scalable scripting for large sequencing data sets in Hadoop. Bioinformatics **30**(1), 119–120 (2014)
28. Pireddu, L., Leo, S., Soranzo, N., Zanetti, G.: A Hadoop-galaxy adapter for user-friendly and scalable data-intensive bioinformatics in galaxy. In: Proceedings of the 5th ACM Conference on Bioinformatics, Computational Biology, and Health Informatics, BCB 2014, Newport Beach, California, USA, 20–23 September 2014, pp. 184–191 (2014)
29. Leo, S., Santoni, F., Zanetti, G.: Biodoop: bioinformatics on hadoop. In: International Conference on Parallel Processing Workshops, ICPPW 2009, Vienna, Austria, 22–25 September 2009, pp. 415–422 (2009)
30. ODriscoll, A., Daugelaite, J., Sleator, R.D.: Big data, Hadoop and cloud computing in genomics. J. Biomed. Inf. **46**(5), 774–781 (2013)
31. Matsunaga, A.M., Tsugawa, M.O., Fortes, J.A.B.: Cloudblast: combining mapreduce and virtualization on distributed resources for bioinformatics applications. In: e-Science 2008 Fourth International Conference on e-Science, Indianapolis, IN, USA, 7–12 December 2008, pp. 222–229 (2008)
32. Schatz, M.C.: Cloudburst: highly sensitive read mapping with mapreduce. Bioinformatics **25**(11), 1363–1369 (2009)
33. Venkata, V., Prasad, S., Loshma, G.: HPC-MAQ: a parallel short-read reference assembler
34. Langmead, B., Hansen, K.D., Leek, J.T.: Cloud-scale RNA-sequencing differential expression analysis with myrna. Genome Biol. **11**(8), R83 (2010)
35. Berrada, G., Keulen, M., Habib, M.B.: Hadoop for EEG storage and processing: a feasibility study. In: Ślęzak, D., Tan, A.-H., Peters, J.F., Schwabe, L. (eds.) BIH 2014. LNCS (LNAI), vol. 8609, pp. 218–230. Springer, Heidelberg (2014). doi:10.1007/978-3-319-09891-3_21
36. Markonis, D., Schaer, R., Eggel, I., Müller, H., Depeursinge, A.: Using mapreduce for large-scale medical image analysis. In: 2012 IEEE Second International Conference on Healthcare Informatics, Imaging and Systems Biology, HISB 2012, La Jolla, CA, USA, 27–28 September 2012, p. 1 (2012)
37. Mangla, S., Raghava, N.S.: Iris recognition on hadoop: a biometrics system implementation on cloud computing. In: 2011 IEEE International Conference on Cloud Computing and Intelligence Systems, pp. 482–485, September 2011
38. McKenna, A., Hanna, M., Banks, E., Sivachenko, A., Cibulskis, K., Kernytsky, A., Garimella, K., Altshuler, D., Gabriel, S., Daly, M., DePristo, M.A.: The genome analysis toolkit: a mapreduce framework for analyzing next-generation DNA sequencing data. Genome Res. **20**(9), 1297–1303 (2010)
39. Gurtowski, J., Schatz, M.C., Langmead, B.: Genotyping in the cloud with crossbow (2002)
40. Brock, M., Goscinski, A.: Execution of compute intensive applications on hybrid clouds (case study with mpiblast). In: Sixth International Conference on Complex, Intelligent, and Software Intensive Systems, CISIS 2012, Palermo, Italy, 4–6 July 2012, pp. 995–1000 (2012)
41. Vouzis, P.D., Sahinidis, N.V.: GPU-BLAST: using graphics processors to accelerate protein sequence alignment. Bioinformatics **27**(2), 182–188 (2011)
42. Benabderrahmane, S.: Enhancing transcriptomic data mining with semantic ranking: towards a new functional spectral representation. In: Rojas, I., Guzman, F.M.O.(eds.) Proceedings of the International Work-Conference on Bioinformatics and Biomedical Engineering, IWBBIO 2013, Granada, Spain, 18–20 March 2013, pp. 721–730. Copicentro Editorial (2013)

43. Hong, D., Rhie, A., Park, S.S., Lee, J., Ju, Y.S., Kim, S., Yu, S.B., Bleazard, T., Park, H.S., Rhee, H., Chong, H., Yang, K.S., Lee, Y.S., Kim, I.H., Lee, J.S., Kim, J.I., Seo, J.S.: FX: an RNA-Seq analysis tool on the cloud. Bioinformatics **28**(5), 721–723 (2012)
44. Wang, L., Chen, D., Ranjan, R., Khan, S.U., Kolodziej, J., Wang, J.: Parallel processing of massive EEG data with mapreduce. In: 18th IEEE International Conference on Parallel and Distributed Systems, ICPADS 2012, Singapore, 17–19 December 2012, pp. 164–171 (2012)
45. Markonis, D., Schaer, R., Eggel, I., Müller, H., Depeursinge, A.: Using mapreduce for large-scale medical image analysis. CoRR, abs/1510.06937 (2015)
46. Alyass, A., Turcotte, M., Meyre, D.: From big data analysis to personalized medicine for all: challenges and opportunities. BMC Med. Genomics **8**(1), 33 (2015)
47. Naseer, A., Alkazemi, B.Y., Waraich, E.U.: A big data approach for proactive healthcare monitoring of chronic patients. In: 2016 Eighth International Conference on Ubiquitous and Future Networks (ICUFN), pp. 943–945, July 2016

On the Ability to Reconstruct Ancestral Genomes from *Mycobacterium* Genus

Christophe Guyeux[1], Bashar Al-Nuaimi[1,2(✉)], Bassam AlKindy[3], Jean-François Couchot[1], and Michel Salomon[1]

[1] FEMTO-ST Institute, UMR 6174 CNRS, DISC Computer Science Department, Univ. Bourgogne Franche-Comté (UBFC), Besançon, France
christophe.guyeux@univ-fcomte.fr, bashartalib6@gmail.com
[2] Department of Computer Science, University of Diyala, Baqubah, Iraq
[3] Department of Computer Science, Al-Mustansiriyah University, Baghdad, Iraq

Abstract. Technical signs of progress during the last decades has led to a situation in which the accumulation of genome sequence data is increasingly fast and cheap. The huge amount of molecular data available nowadays can help addressing new and essential questions in Evolution. However, reconstructing evolution of DNA sequences requires models, algorithms, statistical and computational methods of ever increasing complexity. Since most dramatic genomic changes are caused by genome rearrangements (gene duplications, gain/loss events), it becomes crucial to understand their mechanisms and reconstruct ancestors of the given genomes. This problem is known to be NP-complete even in the "simplest" case of three genomes. Heuristic algorithms are usually executed to provide approximations of the exact solution. We state that, even if the ancestral reconstruction problem is NP-hard in theory, its exact resolution is feasible in various situations, encompassing organelles and some bacteria. Such accurate reconstruction, which identifies too some highly homoplasic mutations whose ancestral status is undecidable, will be initiated in this work-in-progress, to reconstruct ancestral genomes of two *Mycobacterium* pathogenetic bacterias. By mixing automatic reconstruction of obvious situations with human interventions on signaled problematic cases, we will indicate that it should be possible to achieve a concrete, complete, and really accurate reconstruction of lineages of the *Mycobacterium tuberculosis* complex. Thus, it is possible to investigate how these genomes have evolved from their last common ancestors.

Keywords: Mycobacterium tuberculosis · Genome rearrangements · Ancestral reconstruction

1 Introduction

Mycobacterium tuberculosis is presently still one of the principal causes of death worldwide. Approximately one-third of the world population is infected by the *Mycobacterium tuberculosis complex* (MTBC), with about 9 million event cases

© Springer International Publishing AG 2017
I. Rojas and F. Ortuño (Eds.): IWBBIO 2017, Part I, LNBI 10208, pp. 642–658, 2017.
DOI: 10.1007/978-3-319-56148-6_57

annually, leading to estimated a million deaths each year. Due to their different host tropism and phenotypes, members of MTB complex display various pathogenicities ranging from particularly human (*M. tuberculosis*, *M. africanum*, and *M. canetti*) or rodent pathogens (*M. microti*) to *Mycobacteria* with a broad host spectrum (*M. bovis*) [1–3]. *Mycobacterium tuberculosis* has been in the human population around for thousands of years, as fragments of the spinal column of Egyptian mummies from 2300 BCE show definite pathological signs of tubercular decay. It has been recognized as the leading cause of mortality by 1650, while using a new staining technique, Robert Koch identified the bacterium responsible for causing consumption in 1882.

The MTB complex belongs to the slow-growing sublineage of *Mycobacteria*. Based on topographical characteristics, MTBC can be categorized into six clusters, including species such as *M. tuberculosis*, *M. africanum*, *M. bovis*, *M. microti*, and *M. canettii*. Members in MTBC share 99.95% of their genomic sequences and a rigorously clonal population structure [4]. Compared to more ancient species (*e.g.*, *M. marinum*), MTBC has shorter but more virulent chromosomes [5,6]. Considering that they all are derived from a common ancestor, it is interesting that some are human or rodent pathogens, whereas others have a wide host spectrum [7]. The genome of *M. tuberculosis* was studied using the strain *M. tuberculosis H37Rv*. It has a circular chromosome of about 4,200,000 nucleotides long, while containing about 4,000 genes [8]. The different species of the *Mycobacterium tuberculosis* complex show a 95–100% DNA relatedness based on studies of DNA homology, and the sequences of the 16 S rRNA gene are the same for all the species.

MTBC genomes have been modified during the evolution by mutation, insertion-deletion of nucleotides, by large-scale changes (inversion, duplication or deletion of large DNA strands), or by other modifications specific to repetition (insertion sequences, etc.). Being able predict both its past or its future evolution may have multiple applications: to reconstruct the past history and the ancestors of bacteria, or to better understand their mechanism of virulence and resistance acquisition. The relatively short timescale (tuberculosis disease is relatively recent, as its most recent common ancestor evolved ≈40,000 years ago [9]), the relatively reasonable sizes of considered genomes, the relative rarity of recombination events, and the recent possibility to have access to old and present bacterial DNA sequences, may lead to the possibility to model the evolution of these genomes, in order to reconstruct and to understand their ancient history and to predict their future evolution.

To do so, new algorithms of detection and of evolution regarding genomic modifications must be written. People working on this problematic mainly focus on predicting the evolution of nucleotide mutations, and by assuming specific forms for matrix mutations which seem incompatible with recent experimental measures [10]. These models for evolution must be designed differently, in order to better reflect the reality. Additionally, the serious impact of other modifications operating on the genomes (as insertions and deletions of nucleotides (indels), inter and intra chromosomic recombinations, or modifications specific to repetition), must be taken into account more deeply, while a concrete ancestral reconstruction of bacterial lineage must be finally achieved.

The objective of this work-in-progress is to prove that, given a set of close bacterial genomes, it is possible to reconstruct in practice their recent sequence evolution history, by mixing state-of-the-art tools with a pragmatic manual completion and cross-validation. We will illustrate that, in practice, it should be possible to reconstruct ancestral genomes for some lineages of the *Mycobacterium* genus, using all available complete genomes of such a lineage (for instance, 65 complete genomes of the MTB complex are currently available, and we have more than 1,000 archives of reads).

The remainder of this article is organized as follows. In Sect. 2, we start by giving reviews of computational approaches and tools for analyzing the evolution of DNA sequences. We propose in the next section a set of methodological principles that can be used for ancestral genome reconstructions, and how to apply them on *M. canettii* and *M. tuberculosis* data. Obtained results and further perspectives are discussed in Sect. 4. This research work ends with a conclusion section in which the article is summarized and intended future work is outlined.

2 Scientific Background

2.1 On Genomic Evolution

It is well-known that DNA sequences change over time due to local mutations, which are either single nucleotide polymorphism (SNP) or insertion-deletion (indel) of one nucleotide. Mutations that affect the organization of genes are called genome rearrangements, which include inversions, transpositions, and chromosome fusions and fissions. Example of such large scale modifications are illustrated in Fig. 1. During evolution, such large-scale mutations rearranging the genome have occurred, and both gene order and content have been modified accordingly, which may represent a meaningful role in speciation [11].

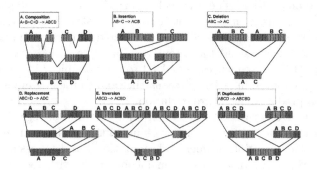

Fig. 1. Various genome rearrangement events.

One important problem in molecular evolution, which is targeted by this study, is that of reconstructing ancestral genomic sequences. In this problem, an evolutionary tree of organisms is provided, together with genomic sequences for the leaf species. The aim is to infer the genomic sequences of the ancestral nodes in the tree, that is, those of the organisms that no longer exist. Various methods have been developed to infer such ancestral sequences and they already have been used in various biological studies.

More precisely, evolution of biomolecules over time have mainly been computationally studied in two directions, namely through ancestral genome reconstruction problem and through the evolution of pan and core genomes over time. A short overview of these topics is provided below.

2.2   Ancestral Genome Reconstruction

Ancestral reconstruction may focus at sequence level or at gene order level, the former being quite resolved [12–20], at least if we do not consider indels and mutation neighborhood, while the latter is more difficult in general, due to its combinatorial complexity. More precisely, given an alignment of DNA sequences and a tree, ancestral nucleotides of extant species can be obtained by modeling the evolution of a trait through time as a stochastic process (Markov chain). Using it as the basis for statistical inference, both maximum likelihood or Bayesian inference approaches can be applied to estimate ancestral configuration.

Well known software like RAxML [21], BEAST2 [22], or PAML [23] can be used for such reconstructions. However, most of the time, like in the R package [24], indels are not considered in such ancestral state reconstruction, even if researches have recently been realized via the so-called "Poisson Indel Process" [25]. Such process is a significant improvement, if we compare it with the parsimony approach that can be found in PHAST software, or with the Thorne-Kishino-Felsenstein model of indel evolution. Large scale modifications, for its part, is most of the time regarded in a combinatorial framework by modeling genomes as permutations of genes or homologous regions. Indeed, this genome rearrangement problem [26] is usually formulated as follows: "given two genomes (permutations) and a set of allowable operations (like inversion, deletion, or transposition), what is the shortest sequence of operations that will transform one genome into the other?". However, even in the case of three genomes, such a problem is NP-hard [27], although it has received much attention in mathematics and computer science [11].

An important remark, motivating our proposal, is that the NP-hard character of this problem only appears if we consider a very large number of operations in very large sequences. On our side, we will consider quite small sequences and a relatively small number of large scale recombinations. So we face tractable problems in various real situations, on which simple and pragmatic approaches may work.

2.3   Core and Pan Genome Extraction

An early study about finding the common genes in chloroplasts has been realized by Stoebe et al. in 1998 [28]. They established the distribution of 190 identified genes and 66 hypothetical protein-coding genes (ysf) in all nine photosynthetic algal plastid genomes available (excluding non-photosynthetic Astasia tonga) from the last update of plastid genes nomenclature and distribution. The distribution reveals a set of approximately 50 core protein-coding genes retained in all taxa. In 2003, Grzebyk et al. [29] have studied the core genes among 24 chloroplast sequences extracted from public databases, 10 of them being algae plastid genomes. They broadly clustered the 50 genes from Stoebe et al. into three major functional domains: (1) genes encoded for ATP synthesis (atp genes); (2) genes encoded for photosynthetic processes (psa and psb genes); and (3) housekeeping genes that include the plastid ribosomal proteins (rpl and rps genes). The study shows that all plastid genomes were rich in housekeeping genes with the rbcL gene involved in photosynthesis. Other examples of such core and pan studies can be found in, e.g., [30–32].

Concerning bacterias, many studies have recently achieved the extraction of core and pan genomes using NCBI annotations, which are mainly based on generic annotation tools like Glimmer, GeneMarkS, or Prodigal (see for instance [33–35], the Pseudomonas aeruginosa case being resolved by us in [36]). In most of these studies, considered genomes have been annotated with various different annotation algorithms, mixing human curated and automated coding sequence prediction tools that are not specific to the genus under consideration. This large variety of manners to detect coding sequences and their functionality leads to large variability in gene boundaries (start and stop codons) and naming process, which obviously severely biases the core and pan genomes determination.

3   A Concrete Semi-automatic Ancestral Reconstruction

3.1   General Presentation

By a phylogenetic study, it is possible to reconstruct the evolutionary relationship of a set of organisms in the form of a binary tree, in which the given set of organisms are descendants placed at the leaves, while internal nodes stand for extinct ancestors connected by edges. We argue that, knowing this tree, ancestral genomes can be completely reconstructed in some easy cases, by aligning extant genomes and finding homologies between them, and then inferring various scenarii of evolutionary events during history [37]. This ancestral reconstruction can be achieved by mixing state-of-the-art algorithms and manual investigations, if the considered genomes have not evolved so much. To illustrate the feasibility of the proposal, an example of such reconstruction is provided in this section, in the case of the MTB complex.

To illustrate this claim, the complete sequences of 65 *Mycobacterium* genomes, which are available on the NCBI[1] have been downloaded. Listed according to their species, 42 genomes of *tuberculosis*, 15 *bovis*, 2 *africanum*, 5 *canettii*, and 1 *microti* have been recovered. Table 1 shows information about

---

[1] ftp://ftp.ncbi.nih.gov/genomes.

Table 1. Information about some *Mycobacterium* genomes

Organism name	Accession	Sequence length	Number of genes
Mycobacterium tuberculosis W-148	NZ_CP012090.1	4,418,548 bp	4,133
Mycobacterium tuberculosis H37Rv	NC_018143.2	4,411,709 bp	4,132
Mycobacterium africanum GM041182	NC_015758.1	4,389,314 bp	4,089
Mycobacterium africanum strain 25	CP010334.1	4,386,422 bp	4,798
Mycobacterium microti strain 12	CP010333.1	4,370,115 bp	4,321
Mycobacterium canettii CIPT 140010059	NC_015848.1	4,482,059 bp	4,137
Mycobacterium canettii CIPT 140070008	NC_019965.1	4,420,197 bp	4,103
Mycobacterium bovis strain ATCC BAA-935	NZ_CP009449.1	4,358,088 bp	4,095
Mycobacterium bovis BCG str. Tokyo 172	NZ_CP014566.1	4,371,707 bp	4,076

some of these *Mycobacterium* genomes. Among this MTBC, we particularly focused on *tuberculosis* and on *canettii*, as there are enough of them, and because the virulent *tuberculosis* species is supposed to have emerged from *canettii* forty thousand years ago. To verify such an evolutionary hypothesis, the first task of our approach, proposed to achieve an ancestral reconstruction of close genomes, is to perform a multiple sequence alignment of the sequences. This task is described in the next section.

3.2 Multiple Sequence Alignment

The first stage of this alignment stage, is to identify a common starting point in these complete circular genomes. In order to do so, we searched for a reference sequence of 200 nucleotides from *M. tuberculosis H37Rv*, and we found it or its transconjugate in each genome using a local blast. Then, a circular rotation (together with a transconjugate operation if needed) has been performed on each complete genome, so that each sequence starts with the same 200 nucleotides, if we except SNPs. Once these sequences have been operated to share the same orientation and starting location, the overall alignment of each chromosome has been performed.

Alignment of large sets of sequences is a common task during biological investigations and has a wide variety of applications incorporating homology detection [38], finding evolutionarily relevant sites, and phylogenetics. A multiple sequence alignment, as depicted in Fig. 2, may explain many aspects about a gene: which regions are constrained, which sites undergo positive selection [39], and potentially the structure of its gene product [40]. Furthermore, aligning sequences can help to detect events of mutations or recombination in couples of close genomes, which is valuable for what we intend to do. To achieve such

```
Sequence1    -TCAGGA-TGAAC----
Sequence2    ATCACGA-TGAACC---
Sequence3    ATCAGGAATGAATCC--
Sequence4    -TCACGATTGAATCGC-
Sequence5    -TCAGGAATGAATCGCM
```

Fig. 2. Representation of a multiple sequence alignment.

an alignment, we thus have considered the *AlignSeqs* function from Decipher R package [41]. Indeed, after various tests on well known alignment tools, this latter was the only one that achieved to align complete bacterial genomes with a good accuracy.

This *AlignSeqs* function takes as input two aligned sets of DNA sequences and returns a merged alignment. It can be used to achieve multiple sequence alignment in a progressive or iterative manner on sequences of the same kind. Indeed, multiple alignments are accomplished by aligning two sequences, merging with another sequence, combining with another set of sequences, and so on until all the sequences are aligned [42,43]. We thus obtained a first representation of synteny of the whole 65 *Mycobacterium* genomes, which is depicted in Fig. 3. It can be observed that these 65 genomes have a high sequence similarity with low recombination events.

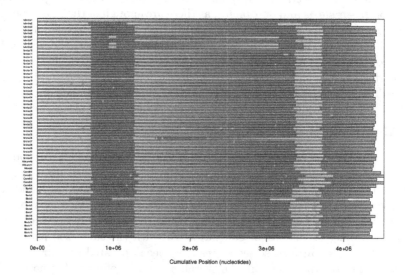

Fig. 3. A synteny representation of all available *Mycobacterium* strains

3.3 Phylogenetic Study

This first representation of synteny blocks, obtained thanks to the multiple sequence alignment of the whole *Mycobacterium* genus, has allowed us to detect the location of a few large scale inversions. We thus have been able to manually invert again these inversions, so that the multiple alignment became quite perfect, if we except small indels and SNPs. It is then possible to use all the 65 complete genomes in the next stage, namely the phylogenetic study.

Indeed, the evolutionary history of our population of genomes can be represented as a phylogenetic tree using the multiple sequence alignment combined with manual local inversions previously obtained. Various methods are well established in the literature to investigate the best phylogenetic tree for a given set of aligned sequences. Well-known techniques for phylogenetic analysis

(a)

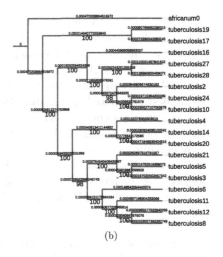

(b)

Fig. 4. Well-supported phylogenies: (a) *M. canettii* species using a *M. tuberculosis* as outgroup, (b) *M. tuberculosis* species with *M. africanum* as outgroup. Phylogenetic trees have been calculated on the entire genomes with RAxML and GTR Gamma model.

include parsimony methods, maximum likelihood, distance-based methods, and even artificial intelligence based ones [44,45]. On our side, we decided to consider the use of RAxML as a default phylogenetic tree reconstruction toolkit, a well known and reputed software based on maximum likelihood [21,46].

As we reversed the inversions, our phylogenetic investigations are based on the whole genome. This leads to well supported and trustworthy trees of strains, on which we can reliably consider to reconstruct ancestral states. As an illustrative example, we represent the phylogenetic trees of *M. canettii* species with a relevant outgroup in Fig. 4a. This very well supported tree has been obtained using RAxML with GTR Gamma model as advised by JModelTest 2.0. The *Mycobacterium tuberculosis* phylogeny, for its part, leads to bootstrap supports larger than 98%, as shown in Fig. 4b.

Having a confidential representation of the general evolution of MTBC strains due to this phylogenetic study, we are then left to reconstruct the ancestral states of the alignment at each internal node of the tree. This final ancestral reconstruction will be applied in two stages, considering first the variants of length 1 in the alignment (namely, single nucleotide polymorphism and indels of 1 nucleotide), and then larger variants that mainly consist of insertion or deletion of a subsequence at a location in the tree.

3.4 Ancestral Reconstruction: Mononucleotidic Variants

Focusing on mononucleotidic variants, we separated the treatment of single nucleotide polymorphisms (SNPs) versus insertion-deletions (indels). For the former, the situation seems quite simple, the only problem being to prevent confusion between a "true" SNP and a SNP induced by a recombination of the indel kind. For the latter, future challenges encompass to determine which indels are related to tandem repeats, which are associated with mobile elements, or which are due to repeated sequences. Let us detail each case hereafter.

Regarding SNPs, the ancestral reconstruction is achieved as follows. The marginal probability distributions for bases at ancestral nodes in the phylogenetic tree are first calculated. These distributions are obtained using the sum-product message passing algorithm [47], assuming independence of sites. The ancestral reconstruction is done by using PHAST software [48], which reconstructs indels too by parsimony, also assuming site independence. Obtained results on mononucleotidic variants are then carefully visually checked, as the number of such variants is not excessive, see Tables 2 and 3.

At the end, 2,956 SNPs and 166 indels have been found in the alignment of the clade constituted by the 5 strains of *M. canettii*, as shown in Fig. 5a. Figure 5b,

Table 2. Number of alignment columns with polymorphism, by pair of strains, on *M. canettii* genomes. Note that, when a large string is deleted at some location in the tree, all the characters of this deletion are counted here.

	canettii0	canettii1	canettii2	canettii3	canettii4	tuberculosis1
canettii0	0	3524	27256	60957	4833	3354
canettii1	3524	0	27260	61233	7971	1150
canettii2	27256	27260	0	62717	27468	27437
canettii3	60957	61233	62717	0	60987	61346
canettii4	4833	7971	27468	60987	0	7510
tuberculosis1	3354	1150	27437	61346	7510	0

Table 3. Variations in the alignment of *M. tuberculosis*

	tuberculosis4	tuberculosis19	tuberculosis17	M. canettii	tuberculosis27	tuberculosis28	tuberculosis24	tuberculosis10
tuberculosis4	0	199770	214401	219205	216387	217235	216919	217186
tuberculosis19	199770	0	212403	219039	216908	216672	216726	216953
tuberculosis17	214401	212403	0	216808	216534	217011	216786	216882
tuberculosis16	219205	219039	216808	0	216669	216916	216251	216678
tuberculosis27	216387	216908	216534	216669	0	142974	189148	199505
tuberculosis28	217235	216672	217011	216916	142974	0	189460	199412
tuberculosis24	216919	216726	216786	216251	189148	189460	0	194315
tuberculosis10	217186	216953	216882	216678	199505	199412	194315	0

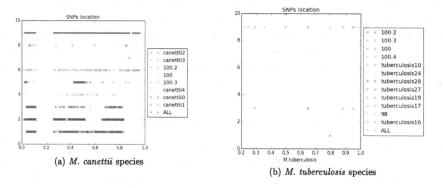

(a) *M. canettii* species

(b) *M. tuberculosis* species

Fig. 5. SNPs location of mononucleotidic variants

for its part, represents the location of the 394 SNPs and of the 25 indels that have been found in the alignment of the clade constituted by 8 genomes of *M. tuberculosis*.

3.5   Ancestral Reconstruction of Larger Variants

Mycobacterium species considered in this article are highly conserved, with really similar regions and without rearrangement. As previously evoked, we found only a few significant inversions, like the one at the last common ancestor of strains *CIPT 140010059, 140070010, 140060008, 140070017*, and *140070008*, as shown in Fig. 6a. Figure 6b, for its part, is a dotplot representing these homologous regions, as identified by the FindSynteny function in R. synteny blocks of the 42 *M. tuberculosis* are finally depicted in Fig. 7, where we have obtained 99% of DNA sequence identity. To sum up, if we except a large scale inversion, we can only report some small indels at this recombination level.

Ad hoc algorithms have then been designed to deal with mid size variants. More specifically, we have written first a string algorithm that detect small and noisy inversions, but the latter, distributed on our supercomputer facilities, was only able to detect artifacts. So either the MTBC genomes have not faced inversion events during its recent history, or this recombination case still needs further investigations. Authors tend to prefer the first possibility, as *Mycobacterium* genomes evolve in a clonal manner (which is not the case, for instance, with *Yersinia* genus, in which a large amount of mobile elements has led to a large number of reported inversions) [49]. Duplication, for its part, has not yet been investigated but, as for inversions, the analysis of synteny blocks tends to show that such events are rare, at least if we consider the large scale ones.

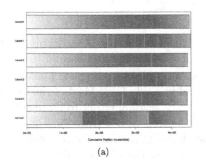

(a)

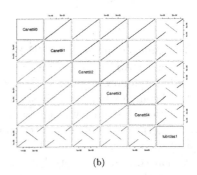

(b)

Fig. 6. (a) synteny blocks in *canettii*. Each genome is colored according to the position of the corresponding region in the first genome (gray if a region is unshared). (b) Dot plots provide an alternative representation of the synteny map of *M. canettii*. Black diagonal lines show syntenic regions sharing the same orientation, whereas red anti-diagonal ones represent blocks of synteny between opposite strands. The description of all of these species tends to show a high sequence similarity with little recombination events. (Color figure online)

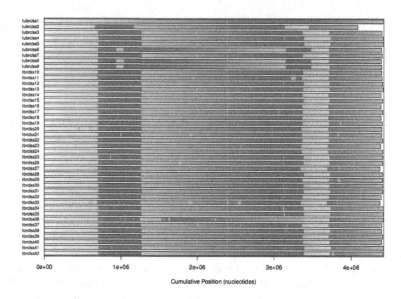

Fig. 7. A representation of *M. tuberculosis* genomes species tends to show more than 95% nucleotide similarity with little recombination events.

Table 4. Number of SNPs in the considered species (100.X refers to an ancestral node, as in the tree)

M. canettii SNPs		
Fathers	**Children**	**No. of SNPs**
100.2	*canettii2*	1041
	canettii3	12398
100	*canettii0*	1
	canettii1	9
100.3	*100*	28
	100.2	735
100.X	*100.3*	111
	canettii4	438

M. tuberculosis SNPs		
Fathers	**Children**	**No. of SNPs**
100	*tuberculosis19*	5
	tuberculosis17	14
100.2	*tuberculosis24*	1
	tuberculosis24	0
100.3	*tuberculosis27*	0
	tuberculosis28	0
98	*100.2*	1
	100.3	0
100.4	*98*	0
	tuberculosis16	1
100.X	*100*	5
	100.4	1

Both indels of midsize and SNPs are rare, for its part, has been deeply studied, using PHAST software as detection tool. From obtained results, we can conclude the following points. (1) Such events are quite rare in some lineages of the MTB complex like *tuberculosis*, as described in Table 4. (2) Most of the times, the situation is very easy to understand manually, leading either to an insertion or to a deletion at an obvious internal node of the tree, as illustrated in Fig. 9. (3) Most of the times, the inserted motif has not faced mutations during evolution: leaves that contain the motif have no mutation in it, thereby contributing to an easy to resolve situation. (4) Surprisingly, ancestral states recovered by PHAST and its parsimony approach leads to disappointing results. Similarly, obviously wrong results have been obtained with state-of-the-art competitor software. To sum up, a manual reconstruction of mid size indels is possible, due to the low number of these recombinations that are mainly very easy to resolve, while automatic tools from the literature are not currently able to do it.

All these steps are summarized in Fig. 10. In this one, gray boxes correspond to manual steps whereas all the other ones are automatically executed.

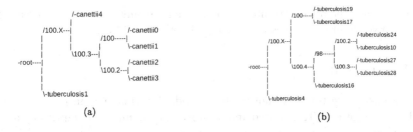

Fig. 8. Example of phylogenetic tree, show the ancestor nodes (internal node), and the relation with their children (a) *M. canettii* species, (b) *M. tuberculosis* species.

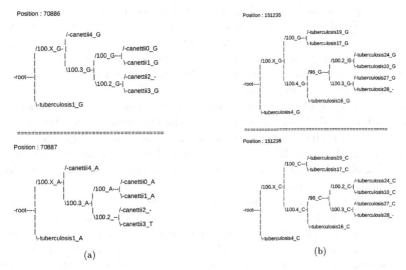

Fig. 9. The insertions and deletions of nucleotides (indels) on the internal node of the tree (a) represent the nucleotides contain the ancestor nodes and their children on *M. canettii* species, (b) *M. tuberculosis* species.

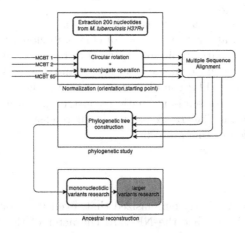

Fig. 10. Flowchart of the proposed approach.

Figure 8 shows the ancestral internal nodes as an example of phylogenetic trees between two different species comparing with their children.

4   Discussion

The obtained ancestors have not yet been studied in this work-in-progress. They will be investigated with updated and improved algorithms, encompassing mobile elements and gene content evolution analyzes.

Indeed, an important category of genome modification is the loss of functional genes, for instance because they become ineffective. In order to do so, we will consider the phylogenetic tree whose leaves will contain sets of genes, and we will compute core and pan genomes at each internal node of the tree. Having this core and pan tree, we will design an algorithm to investigate more deeply the evolution of these pan and core genomes over the tree, to see if some branches can be related to hot spots of evolution. We thus intend to determine at which rate such loss or gain occur, and which kinds of functionality are concerned. We will finally compute how much mutations fall inside a coding sequence, by studying which kind of genes has evolved on the phylogenetic tree, by wondering if the mutation rate has changed over time, and if such mutability can be related to environmental events. In other words, we will wonder which variations have been potentially significant among the numerous events that have been found when aligning these sequences.

With such a pipeline, we intend to investigate the following questions. Are some recombinations at the origin of severe tuberculosis epidemics? Are transposases responsible of such recombinations like inversions [50,51]? Are transposases in general more present in *M. tuberculosis* (affecting humans) than in *M. africanum*, *M. bovis*, or *M. bovis BCG*? Are they related to the virulence of the strain? How core and pan genomes have evolved over time in this complex? Finally, we will compare the last common ancestor of this complex to a *M. canettii*, to see if the *canettii* ancestor hypothesis can be verified by the ancestral reconstruction way.

At this point, our partial conclusion is that the reconstruction of ancestral sequences is possible, at least in the case of close and clonal bacterias. Furthermore, elements being part of this reconstruction have already be designed, at least in their first revision (for instance to detect and deal with mononucleotidic variants). However, the MTB complex seems to be a little too complicated for a first deep investigation of semi-automatic reconstruction of ancestral sequences of bacteria, and a genus like *Brucella* may be more easy to deal with in a first concrete investigation of this problem.

5   Conclusion

In this article, we have firstly emphasized that, even if various algorithms and software already exist to face the NP-hard character of the ancestral genome reconstruction problem, they do not work perfectly, in particular when SNPs or indels fall into repeated sequences. We have then argued that, when regarding the relatively low number of mutation and recombination events in such *Mycobacterium*, a pragmatic approach is possible. We have proposed to reconstruct all ancestors of all complete available genomes of *Mycobacterium tuberculosis* and of *M. canettii*. The study has started by investigating single nucleotide polymorphism level, while indels and large scale recombination are regarded in a second stage. Our conclusion is that, by mixing automatic reconstruction of obvious situations with human interventions on signaled problematic cases, it may be possible to achieve a concrete, complete, and really accurate reconstruction of some specific bacteria lineages. We can thus investigate how these genomes have evolved from their last common ancestors.

In future work, we intend to reconstruct all ancestors of all complete available genomes of specific bacteria strains, namely, and ordered by complexity: *Brucella* genus, *Yersinia pestis*, and *Pseudomonas aeruginosa*. Moreover, we intend to compare them with ancient DNA when available (like for *Y. pestis*). In parallel, original mathematical description of some recombination mechanisms will be proposed, encompassing branching process and partial differential equation approaches for modeling mobile elements. Finally, we may try to correlate the evolutionary history of microorganisms to epidemiological data: events of genomic recombination may be related to epidemic outbreaks. And such putative correlations may be learnt by deep learning algorithms, leading to a new way to predict epidemic risks.

Acknowledgments. Computations presented in this article were realised on the supercomputing facilities provided by the Mésocentre de calcul de Franche-Comté.

References

1. Smith, N.H., Gordon, S.V., de la Rua-Domenech, R., Clifton-Hadley, R.S., Hewinson, R.G.: Bottlenecks and broomsticks: the molecular evolution of mycobacterium bovis. Nat. Rev. Microbiol. **4**(9), 670–681 (2006)
2. Shamputa, I.C., SangNae, C., Lebron, J., Via, L.E.: Introduction and epidemiology of mycobacterium tuberculosis complex in humans. In: Mukundan, H., Chambers, M.A., Waters, W.R., Larsen, M.H. (eds.) Tuberculosis, Leprosy and Mycobacterial Diseases of Man and Animals: The Many Hosts of Mycobacteria, pp. 1–16. CABI (2015). http://www.cabi.org/cabebooks/ebook/20153322769
3. Brosch, R., Gordon, S.V., Marmiesse, M., Brodin, P., Buchrieser, C., Eiglmeier, K., Garnier, T., Gutierrez, C., Hewinson, G., Kremer, K., et al.: A new evolutionary scenario for the mycobacterium tuberculosis complex. Proc. Natl. Acad. Sci. **99**(6), 3684–3689 (2002)
4. Gutacker, M.M., Smoot, J.C., Migliaccio, C.A.L., Ricklefs, S.M., Hua, S., Cousins, D.V., Graviss, E.A., Shashkina, E., Kreiswirth, B.N., Musser, J.M.: Genome-wide analysis of synonymous single nucleotide polymorphisms in mycobacterium tuberculosis complex organisms: resolution of genetic relationships among closely related microbial strains. Genetics **162**(4), 1533–1543 (2002)
5. Mostowy, S., Cousins, D., Brinkman, J., Aranaz, A., Behr, M.A.: Genomic deletions suggest a phylogeny for the mycobacterium tuberculosis complex. J. Infect. Dis. **186**(1), 74–80 (2002)

6. Yamada-Noda, M., Ohkusu, K., Hata, H., Shah, M.M., Nhung, P.H., Sun, X.S., Hayashi, M., Ezaki, T.: Mycobacterium species identification-a new approach via dnaJ gene sequencing. Syst. Appl. Microbiol. **30**(6), 453–462 (2007)
7. Fabre, M., Hauck, Y., Soler, C., Koeck, J.-L., Van Ingen, J., Van Soolingen, D., Vergnaud, G., Pourcel, C.: Molecular characteristics of mycobacterium canettii the smooth mycobacterium tuberculosis bacilli. Infect. Genet. Evol. **10**(8), 1165–1173 (2010)
8. Fleischmann, R.D., Alland, D., Eisen, J.A., Carpenter, L., White, O., Peterson, J., DeBoy, R., Dodson, R., Gwinn, M., Haft, D., et al.: Whole-genome comparison of mycobacterium tuberculosis clinical and laboratory strains. J. Bacteriol. **184**(19), 5479–5490 (2002)
9. Wirth, T., Hildebrand, F., Allix-Béguec, C., Wölbeling, F., Kubica, T., Kremer, K., van Soolingen, D., Rüsch-Gerdes, S., Locht, C., Brisse, S., et al.: Origin, spread and demography of the mycobacterium tuberculosis complex. PLoS Pathog **4**(9), e1000160 (2008)
10. Lang, G.I., Murray, A.W.: Estimating the per-base-pair mutation rate in the yeast saccharomyces cerevisiae. Genetics **178**(1), 67–82 (2008)
11. Fertin, G.: Combinatorics of Genome Rearrangements. MIT Press, Cambridge (2009)
12. Ma, J., Ratan, A., Raney, B.J., Suh, B.B., Zhang, L., Miller, W., Haussler, D.: DUPCAR: reconstructing contiguous ancestral regions with duplications. J. Comput. Biol. **15**(8), 1007–1027 (2008)
13. Gagnon, Y., Blanchette, M., El-Mabrouk, N.: A flexible ancestral genome reconstruction method based on gapped adjacencies. BMC Bioinform. **13**(Suppl 19), S4 (2012)
14. Jones, B.R., Rajaraman, A., Tannier, E., Chauve, C.: ANGES: reconstructing ancestral genomes maps. Bioinformatics **28**(18), 2388–2390 (2012)
15. Ma, J., Zhang, L., Suh, B.B., Raney, B.J., Burhans, R.C., Kent, W.J., Blanchette, M., Haussler, D., Miller, W.: Reconstructing contiguous regions of an ancestral genome. Genome Res. **16**(12), 1557–1565 (2006)
16. Fei, H., Zhou, J., Zhou, L., Tang, J.: Probabilistic reconstruction of ancestral gene orders with insertions and deletions. IEEE/ACM Trans. Comput. Biol. Bioinform. **11**(4), 667–672 (2014)
17. Blanchette, M., Diallo, A.B., Green, E.D., Miller, W., Haussler, D.: Computational reconstruction of ancestral DNA sequences. In: Murphy, W.J. (ed.) Phylogenomics, pp. 171–184. Springer, Heidelberg (2008)
18. Rascol, V.L., Pontarotti, P., Levasseur, A.: Ancestral animal genomes reconstruction. Curr. Opin. Immunol. **19**(5), 542–546 (2007)
19. Larget, B., Simon, D.L., Kadane, J.B., Sweet, D.: A Bayesian analysis of metazoan mitochondrial genome arrangements. Mol. Biol. Evol. **22**(3), 486–495 (2005)
20. Hannenhalli, S., Chappey, C., Koonin, E.V., Pevzner, P.A.: Genome sequence comparison and scenarios for gene rearrangements: a test case. Genomics **30**(2), 299–311 (1995)
21. Stamatakis, A.: RAxML version 8: a tool for phylogenetic analysis and post-analysis of large phylogenies. Bioinformatics **30**(9), 1312–1313 (2014)
22. Bouckaert, R., Heled, J., Kühnert, D., Vaughan, T., Chieh-Hsi, W., Xie, D., Suchard, M.A., Rambaut, A., Drummond, A.J.: BEAST 2: a software platform for Bayesian evolutionary analysis. PLoS Comput. Biol. **10**(4), e1003537 (2014)
23. Yang, Z.: Phylogenetic analysis by maximum likelihood (PAML) (2000)
24. Paradis, E., Claude, J., Strimmer, K.: APE: analyses of phylogenetics and evolution in R language. Bioinformatics **20**(2), 289–290 (2004)

25. Bouchard-Côté, A., Jordan, M.I.: Evolutionary inference via the Poisson Indel Process. Proc. Natl. Acad. Sci. **110**(4), 1160–1166 (2013)
26. Watterson, G.A., Ewens, W.J., Hall, T.E., Morgan, A.: The chromosome inversion problem. J. Theoret. Biol. **99**(1), 1–7 (1982)
27. Even, S., Goldreich, O.: The minimum-length generator sequence problem is NP-hard. J. Algorithms **2**(3), 311–313 (1981)
28. Stoebe, B., Martin, W., Kowallik, K.V.: Distribution and nomenclature of protein-coding genes in 12 sequenced chloroplast genomes. Plant Mol. Biol. Reporter **16**(3), 243–255 (1998)
29. Grzebyk, D., Schofield, O., Vetriani, C., Falkowski, P.G.: The mesozoic radiation of eukaryotic algae: the portable plastid hypothesis1. J. Phycol. **39**(2), 259–267 (2003)
30. Sharon, I., Alperovitch, A., Rohwer, F., Haynes, M., Glaser, F., Atamna-Ismaeel, N., Pinter, R.Y., Partensky, F., Koonin, E.V., Wolf, Y.I., et al.: Photosystem I gene cassettes are present in marine virus genomes. Nature **461**(7261), 258–262 (2009)
31. De Chiara, M., Hood, D., Muzzi, A., Pickard, D.J., Perkins, T., Pizza, M., Dougan, G., Rino Rappuoli, E., Moxon, R., Soriani, M., et al.: Genome sequencing of disease and carriage isolates of nontypeable haemophilus influenzae identifies discrete population structure. Proc. Natl. Acad. Sci. **111**(14), 5439–5444 (2014)
32. Kurtz, S., Phillippy, A., Delcher, A.L., Smoot, M., Shumway, M., Antonescu, C., Salzberg, S.L.: Versatile and open software for comparing large genomes. Genome Biol. **5**(2), 1 (2004)
33. Touchon, M., Hoede, C., Tenaillon, O., Barbe, V., Baeriswyl, S., Bidet, P., Bingen, E., Bonacorsi, S., Bouchier, C., Bouvet, O., et al.: Organised genome dynamics in the escherichia coli species results in highly diverse adaptive paths. PLoS Genet **5**(1), e1000344 (2009)
34. Boissy, R., Ahmed, A., Janto, B., Earl, J., Hall, B.G., Hogg, J.S., Pusch, G.D., Hiller, L.N., Powell, E., Hayes, J., et al.: Comparative supragenomic analyses among the pathogens staphylococcus aureus, streptococcus pneumoniae, and haemophilus influenzae using a modification of the finite supragenome model. BMC Genom. **12**(1), 1 (2011)
35. Tettelin, H., Masignani, V., Cieslewicz, M.J., Donati, C., Medini, D., Ward, N.L., Angiuoli, S.V., Crabtree, J., Jones, A.L., Durkin, A.S., et al.: Genome analysis of multiple pathogenic isolates of streptococcus agalactiae: implications for the microbial pan-genome. In: Proceedings of the National Academy of Sciences of the United States of America **102**(39), pp. 13950–13955 (2005)
36. Valot, B., Guyeux, C., Rolland, J.Y., Mazouzi, K., Bertrand, X., Hocquet, D.: What it takes to be a Pseudomonas aeruginosa? The core genome of the opportunistic pathogen updated. PLoS One **10**(5), e0126468 (2015)
37. Yang, J., Li, J., Dong, L., Grünewald, S.: Analysis on the reconstruction accuracy of the fitch method for inferring ancestral states. BMC Bioinform. **12**(1), 18 (2011)
38. Wang, Y., Sadreyev, R.I., Grishin, N.V.: PROCAIN server for remote protein sequence similarity search. Bioinformatics **25**(16), 2076–2077 (2009)
39. Kemena, C., Notredame, C.: Upcoming challenges for multiple sequence alignment methods in the high-throughput era. Bioinformatics **25**(19), 2455–2465 (2009)
40. Warnow, T.: Large-scale multiple sequence alignment and phylogeny estimation. In: Chauve, C., El-Mabrouk, N., Tannier, E. (eds.) Models and Algorithms for Genome Evolution, pp. 85–146. Springer, Heidelberg (2013)
41. R Development Core Team: R: a language and environment for statistical computing. R foundation for statistical computing, Vienna, Austria (2014)

42. Gentleman, R.C., Carey, V.J., Bates, D.M., Bolstad, B., Dettling, M., Dudoit, S., Ellis, B., Gautier, L., Ge, Y., Gentry, J., et al.: Bioconductor: open software development for computational biology and bioinformatics. Genome Biol. 5(10), 1 (2004)

43. Wright, E.S.: The art of multiple sequence alignment in R (2014)

44. Alkindy, B., Guyeux, C., Couchot, J.-F., Salomon, M., Bahi, J.: Using genetic algorithm for optimizing phylogenetic tree inference in plant species. In: MCEB15, Mathematical and Computational Evolutionary Biology, Porquerolles Island, France, June 2015

45. Alkindy, B., Al-Nuaimi, B., Guyeux, C., Couchot, J.-F., Salomon, M., Alsrraj, R., Philippe, L.: Binary particle swarm optimization versus hybrid genetic algorithm for inferring well supported phylogenetic trees. In: Angelini, C., Rancoita, P.M.V., Rovetta, S. (eds.) CIBB 2015. LNCS, vol. 9874, pp. 165–179. Springer, Heidelberg (2016). doi:10.1007/978-3-319-44332-4_13

46. AlKindy, B., Guyeux, C., Couchot, J.-F., Salomon, M., Parisod, C., Bahi, J.M.: Hybrid genetic algorithm and lasso test approach for inferring well supported phylogenetic trees based on subsets of chloroplastic core genes. In: Dediu, A.-H., Hernández-Quiroz, F., Martín-Vide, C., Rosenblueth, D.A. (eds.) AlCoB 2015. LNCS, vol. 9199, pp. 83–96. Springer, Heidelberg (2015). doi:10.1007/978-3-319-21233-3_7

47. Pearl, J.: Reverend bayes on inference engines: a distributed hierarchical approach. In: AAAI, pp. 133–136 (1982)

48. Hubisz, M.J., Pollard, K.S., Siepel, A.: PHAST and RPHAST: phylogenetic analysis with space/time models. Briefings Bioinform. 12(1), 41–51 (2011). doi:10.1093/bib/bbq072

49. Behr, M.A.: Evolution of mycobacterium tuberculosis. In: Divangahi, M. (ed.) The New Paradigm of Immunity to Tuberculosis, vol. 783, pp. 81–91. Springer, Heidelberg (2013)

50. Siguier, P., Filée, J., Chandler, M.: Insertion sequences in prokaryotic genomes. Curr. Opin. Microbiol. 9(5), 526–531 (2006)

51. Bergman, C.M., Quesneville, H.: Discovering and detecting transposable elements in genome sequences. Briefings Bioinform. 8(6), 382–392 (2007)

Representativeness of a Set of Metabolic Pathways

José F. Hidalgo[1(✉)], Jose A. Egea[2], Francisco Guil[1], and José M. García[1]

[1] Grupo de Arquitectura y Computación Paralela,
Universidad de Murcia, Murcia, Spain
{jhidalgo,fguil,jmgarcia}@um.es
[2] Dpto. de Matemática Aplicada y Estadística,
Universidad Politécnica de Cartagena, Cartagena, Spain
josea.egea@upct.es, http://www.um.es/gacop

Abstract. Pathways and more precisely Elementary Flux Modes (EFM) are artefacts extracted from metabolic networks that are very useful to achieve the comprehension of a very specific metabolic function or dysfunction. Many methods to extract pathways have already been developed and all of them have to deal with common problems like the production of infeasible subnetworks and the production of the same solution repetitively. Although some strategies have been incorporated to those methods in order to mitigate the problems, they get already a high ratio of repetitions and the insistent presence of the same reactions in the solutions. We do a proposal focused on linear programming (LP) methods for pathway extraction. It aims to improve the representation of every reaction in the set of computed pathways by penalizing the most often included reactions during the extraction.

Keywords: Metabolic networks · Linear programming · EFM · Flux modes · Pathways · Systems biology

1 Motivation

The complexity of a biologic system inside a cell is comprised into a set of stoichiometric equations that explains how metabolites are produced and consumed during metabolic reactions. Several mathematical methods modelling metabolism that are able to incorporate datasets provided by different *omics* technologies are emerging. Many of these methods are encompassed within constraint-based models, in which a set of mathematical constraints are defined using a genome-scale metabolic network (GSMN) reconstruction as a starting point. Several curated GSMNs can be found in the literature [1]. However, being able to automatically characterise the biochemical reactions present in a particular metabolism through *omics* data truly constitutes a challenge [2].

GSMNs are not monolithic. They are composed by many and interrelated sub-networks where specific biological functions are done to support the life. We are interested in a specific kind of sub-network called pathways. Pathways are

© Springer International Publishing AG 2017
I. Rojas and F. Ortuño (Eds.): IWBBIO 2017, Part I, LNBI 10208, pp. 659–667, 2017.
DOI: 10.1007/978-3-319-56148-6_58

steady-state and thermodynamically feasible so the reactions produce and consume all the compounds keeping the mass balance unaltered. An elementary flux mode (EFM) [3] is a special type of metabolic pathway whose difference with regular pathways is that it cannot be decomposed into smaller pathways (i.e., a subset of an EFM is not a feasible pathway). That is called non-decomposability condition. EFMs are subsets of the metabolic network with the minimum necessary support to operate in stoichiometric steady-state balance with all reactions in the appropriate direction. The importance of EFMs has promoted the apparition of many methods to extract them from GSMNs. According to EFMs definition a straightforward way to extract them is by means of an optimization problem. Methods section is devoted to remind how EFMs extraction is translated into a linear program with an objective function to be minimized.

Linear program optimization is a very well-known tool with very efficient implementations [4,5] called *solvers*. The behaviour of all of those solvers are very similar but deterministic. The way of influencing in the obtention of different solutions for a problem expressed in terms of a linear program is to constraint the program with a different set of conditions. Stoichiometric equations determine not only the occurrence of one specific reaction but also the relationship of dependence between different reactions along the network. Flux coupling [6,7] and metabolic coupling [8] are some tryings to quantify those hidden relationships among different metabolic network nodes. They act as attractors that repetitively can drive the solver to the same set of participant reactions despite the fact that the set of constraints is strictly disjoint.

The rest of the paper is structured as follows. In Section Background, we comment briefly the use of linear programming in the EFM extraction methods. In Section Methods, we comment how the extraction of EFM is transformed into a linear program. Section Discussion presents our contribution by the modification of the optimization problem for each iteration. Section Results presents some case studies of our approach. The paper concludes by offering our conclusions and suggestions for future work.

2   Background

A plenty of works have remarked the advantages of analysing metabolic networks using EFMs [9,10]. There are some limitations in using EFMs such as how computationally demanding is to enumerate them [11] or how to initially choose a set of them with relevant biological significance from the full set of the possible EFMs. Several methods have been recently proposed to determine a subset of EFMs in GSMNs [12–15].

Computational approaches to metabolic pathways can be classified into two groups: stoichiometric approaches and path-finding approaches [16]. The first ones use the stoichiometric data to do calculations during the process. Linear Programming and Null-Space Algorithm [17] are some of the mathematical strategies applied to find pathways, mainly solving the system of linear equations proposed by the stoichiometric matrix. The second ones translate the network into a directed graph to explore it [18]. Path-finding approaches have a major drawback because they do not use stoichiometric coefficients during the exploration process.

Our proposal boost the extraction of EFMs and pathways by mitigating the insistent apparition of the same reactions in the solutions and thus promoting the participation of every reaction in the extracted solutions. It is mainly applicable to linear programming based methods. We focus on how much representative is the set of solutions we have computed for a specific number of iterations.

3   Methods

GSMN has the equivalent expression of a stoichiometric matrix S, where each metabolic reaction is represented in one column. The values inside the matrix are the stoichiometric coefficients for metabolites (rows) on each reaction. Each pathway includes or not a reaction from the full GSMN. To mathematically model a pathway every reaction is represented by one variable which numerically gives the amount of times that a reaction occurs for the wanted balance or equivalently the rate at which the substrate metabolites are converted to the product metabolites. The vector that contains the reaction rates is called flux rate.

Be v a vector of flux rates that represents a pathway and R the full set of metabolic reactions, therefore fulfilling the steady-state (Eq. 1) and the thermodynamic feasibility constraints (Eq. 2).

$$S \cdot v = 0 \tag{1}$$

$$v_r \geq 0, \quad \forall r \in R \tag{2}$$

The steady-state condition means that internal metabolites are balanced and their concentration remain constant, and the feasibility constraint means that each irreversible reaction only participates with a positive rate when it is part of the solution. Finally, v represents an EFM if it is non-decomposable, that is, v is not a positive linear combination of other flux rate vectors corresponding to pathways. This is precisely the condition that let us to consider the extraction method like an optimization problem instead of just a system of equations. Linear programming is the most used optimisation tool technique for the state of the art EFMs extraction methods.

There are many references to detailed explanations of how to convert a stoichiometric matrix into a linear program [13] given the primary and the minimality constraints. The direct translation of a stoichiometric matrix into a linear program defines a *clean linear program* (Eq. 3).

$$\text{Minimize} \quad \sum_{i=1}^{n} v_i \tag{3}$$

$$\text{subject to} \quad S \cdot v = 0$$

$$v_i \geq 0 \quad \forall r_i \in R$$

The posed linear program can be solved using LP solvers that implement the Simplex Algorithm. But, in this case, the unique minimal solution is set all the variables to zero, so we must restrict the problem to get non-trivial solutions. If we are interested in obtaining different solutions and due to the fact that LP solvers are deterministic, different conditions must be used to modify the clean linear program and additionally to constraint it. An additional constraint is a set of reactions in which representative variables will be forced to be greater than 0 (positive restrictions) so the pathway obtained must have them activated, together with others that must be exactly 0 (negative restrictions), that is, deactivated in the resulting pathway. Nothing is said about the reactions not included in the set of conditions. Let us call *seed* to this set of conditions because is the precursor of the solution. The goal of a seed is to induce non trivial and feasible solutions, and also different from the previous ones.

Each extraction method based on linear program is basically a method to produce seeds. Each constraint must be different from the previous ones but this do not guarantee that the solutions are going to be different. The seed will be a part of the full computed solution. Extraction methods use some kind of dispersion strategies to reduce the apparition of repeated solutions. As well as the repetitions, the distribution of the frequency of apparition for each reaction is also a characteristic of each extraction method. The pattern of frequencies is in some manner a consequence of the extraction method, but also of the coupling relations, the size of the GSMN and the amount of runs done for an experiment.

As a neutral (but adverse) scenery, let us consider a random seed generator as part of an invented extraction method. The random generator will produce constraints for the construction of a linear program on every iteration. We are interested in evaluating two issues: the frequency of apparition for each reaction in the full set of solutions and the similarity of the frequencies between experiments.

We propose the use of a penalization to modify the previous results trying to boost the experiment to get more relevant results or to get them sooner. The strategy that has been proposed is to modify the objective function using some set of positive *weights* $\{w_i\}$ and converting it from $\sum_{i=1}^{n} v_i$ to $\sum_{i=1}^{n} w_i \cdot v_i$. Using the same seed, different sets of weights can provide different solutions of our problem. The results of the different experiments can be compared using the Wilcoxon signed-rank test [19] and a statistic that resemble the well-known *chi-square* test.

We are interested in studying how the use of a set of weights impacts in the set of solutions we get.

4 Discussion

Our proposal is to drive the LP solver to consider less used reactions by the way of modifying the objective function using weights. It seems natural to think that choosing the most used reactions once and again favours the apparition of repeated solutions. The proposal is to penalize a reaction when it is included in one previously obtained solution. Equation 3 shows the clean linear program. The coefficients of the objective function have initially the value 1. The idea is to increase the coefficients of the objective function for some reactions, and in this way to influence to the LP solver.

Before any conclusion about the introduction of a penalization in the linear program, it is required a way to compare different pathway extraction experiments. Be e the amount of iterations we run in one optimization problem and O_i the amount of occurrences of the reaction r_i along the experiment. The frequency of apparition for each R_i is $F_i = \frac{O_i}{e}$. This variable F let us observe if there is noticeable impact applying penalization and it can be compared between experiments with different amount of iterations with or without penalization. Be F and F' the respective resulting frequencies in two different experiments. Two natural questions arise here: is there a significant differences between these two frequencies? if this is so, how can we measure the difference between them? To answer the first question, we can use the Wilcoxon signed rank test [20]. Once we are certain that there is a significant differences between using or not weights in our function, we can study the influence of these weights in the difference between F and F'. To do so, we will use the following statistic (denoted by $\overline{\chi}^2$) that comes from the well-known *chi-square* test (Eq. 4) to get the answer.

$$\overline{\chi}^2 = \sum_{i=1}^{n} \frac{(F'_i - F_i)^2}{F_i} \tag{4}$$

As in the usual χ^2 test, $\overline{\chi}^2$ provides a good measure of the differences between the values of F and F'. Observe that we cannot use it to perform a standard chi-square-test because the values F_i are not independent in general (we could have, for instance, flux-coupling between two reactions, forcing the two related frequencies to be the same). This statistic talks about all the factors that can influence the experiment, that is, the seed generator, the solver, the coupling relations, the size of the GSMN, the iterations for the experiment and also the impact of the hypothetical penalization. It can be supposed that a long enough experiment for a GSMN would extract from it the most of the possible knowledge in the produced set of solutions. In that case, it can be said that the experiment reaches some level of representativeness. The corresponding F would be such that no other longer experiments frequency distribution (F') should fail the chi-test in relation with F. Our proposal wants to reduce the needed iterations that let us consider one experiment as representative enough.

With this purpose we introduce a strategy based on penalizing the coefficients for the objective function in the linear program in those reactions with higher presence in the found solutions. The penalization can be done in different ways. The simplest one could be to consider a fixed penalty to be proportionally applied to the frequency of apparition of a reaction. Be p the penalty. Now the coefficient for v_i would be $w_i = 1 + p \cdot F_i$. Equation 5 shows the new linear program with penalties.

$$\text{Minimize} \quad \sum_{i=1}^{n} w_i \cdot v_i \tag{5}$$

$$\text{subject to} \quad S \cdot \boldsymbol{v} = \boldsymbol{0}$$

$$v_i \geq 0 \qquad \forall r_i \in R$$

$$\text{where} \quad w_i = 1 + p \cdot F_i \qquad \forall r_i \in R$$

Due to the fact that the linear program has an objective function to minimize, the higher cost of the most often included reactions will modify the solver election and it will invite it to consider to include some other less weighted reactions in the solution. As well as the repeated solutions problem, the frequency of apparition for every reaction can be affected.

If effective, our modification of the objective function should have an impact that produces difference between the frequencies of a non-penalized experiment (F_i) and the penalized one (F'_i).

5   Results

Often, a non-random seed generator has some kind of temporal affinity to a set of reactions. This is remarkable when the extraction method uses graph exploration techniques. Graph exploration methods apply in some manner the adjacency concept to put nodes into a seed. So, we think that in order to study the influence of the modification of the objective function, the best approach is to avoid any bias produced by a specific way to choose our seeds by using a uniform random generator. Good results can be interpreted as promising for other non-random ones.

We know that a short seed gives the simplex solver more freedom to build the solution, because less amount of reactions are mandatory in the output of the linear program. On the other hand, we need to have enough amount of different seeds in order to have statistically significant results. Taking into account these two requisites, in this study, we have decided to use randomly generated seeds consisting in sets of four reactions.

We start by characterizing the frequency distribution similarity between two experiments with a reasonable number of executions of our peculiar random pathway extractor without any penalization. The experiments consist of iterations of a random generation of a seed, linear program resolution and finally the accounting of the present reactions in the solution. The standard Linux random generator will be used to generate the seeds with size 4. At the end of the experiment we have the frequency of apparition for each reaction. This is the variable we compare between experiments.

The selected metabolic reconstructions for our study is *iAF1260*, the reconstruction of the *E. coli K-12 MG1655* organism [21]. iAF1260 stoichiometric matrix, once decoupled, has 3234 reactions. The second reconstruction is *core E. coli metabolic model* [22]. It is a subset of the genome-scale metabolic reconstruction iAF1260. It is described as an educational guide and includes 154 decoupled reactions.

To visualize the effect of the penalization, we apply the Wilcoxon signed-rank test. Table 1 shows the results. Each experiment consists of 50.000 iterations generating random seeds and solving the linear program. The first row shows the result of the test for two experiments without penalization and the following rows show the test for experiments with different penalizations.

Table 1. Wilcoxon signed-rank test experiments over core E.coli GSMN.

Penal. 1	Penal. 2	Wilcoxon p-value
0	2	0.0001926
0	5	5.843e-06
2	5	3.262e-07

The results of the Wilcoxon test seem to support the basis of our approach. Higher Wilcoxon p-values indicates a high similarity between the two frequencies while lower values means the results are very little related. So we see that changing the objective function by introducing weights has an impact ant that there is also a clear difference between different weights.

We may be interested in knowing the impact of the penalization over the vector of frequencies in experiments with different amount of reactions. Table 2 shows the comparison for some couple of experiments for each reconstruction with an specific amount of iterations. In this case a comparison is done using the chi-square based test. It also shows how the penalization accelerates the falling of chi-square with the same experiment length.

Table 2. Evolution of chi-square test depending on the length of the experiment and the penalization for the core E.coli reconstruction.

Iterations	Penal. 1	Penal. 2	Chi-square
10K	0	2	0.422773
10K	0	5	0.736042
10K	0	20	1.060799
10K	2	20	0.273936
50K	0	2	0.373511
50K	0	5	0.682528
50K	0	20	1.010308
100K	0	2	0.352029
100K	0	5	0.703740

It can also be observed that the distance between doing the experiment with or without any penalization is remarkable. Longer executions have to be ran to get more precise conclusions. The last comparative can be done for GSMN of the reconstruction iAF1260.

6   Conclusions and Future Work

In this paper we propose a new strategy that complements to other EFM extraction methods based on linear programming. Penalizing the apparition of each reaction in any obtained solution the solver is conditioned to avoid the inclusion of the same reaction constantly. The main effect is that the solution obtained is different from the previous ones. Another effect is that reactions that rarely are included in solutions appear more often. Statistically we measure a random method to constrain the linear program without penalization. The produced effects depend on the size of the GSMN and the amount of iterations.

Regarding future work, more accuracy information can be obtained doing longer experiments. We are considering to propose a measure of "how much amount of knowledge gives a subset of pathways already extracted". Futhermore, the shape of the distribution of solutions and reaction presence of existent extraction methods can be measured using some of the proposed ideas.

Acknowledgments. This work was supported in part by the Spanish Ministerio de Economia y Competitividad (MINECO) and European Commission FEDER under grant TIN2015-66972-C5-3-R.

References

1. Thiele, I., Palsson, B.Ø.: A protocol for generating a high-quality genome-scale metabolic reconstruction. Nat. Protoc. **5**(1), 93–121 (2010)
2. Schmidt, B.J., Ebrahim, A., Metz, T.O., Adkins, J.N., Palsson, B.Ø., Hyduke, D.R.: GIM3E: condition-specific models of cellular metabolism developed from metabolomics and expression data. Bioinformatics **29**(22), 2900–2908 (2013)
3. Schuster, S., Hilgetag, C.: On elementary flux modes in biochemical reaction systems at steady state. J. Biol. Syst. **2**(02), 165–182 (1994)
4. IBM: IBM ILOG CPLEX Optimizer (2010). http://www-01.ibm.com/software/integration/optimization/cplex-optimizer/
5. Forrest, J.: CLP-coin-or linear program solver. In: DIMACS Workshop on COIN-OR, 17–20 July (2006)
6. Burgard, A.P., Nikolaev, E.V., Schilling, C.H., Maranas, C.D.: Flux coupling analysis of genome-scale metabolic network reconstructions. Genome Res. **14**(2), 301–312 (2004)
7. Larhlimi, A., David, L., Selbig, J., Bockmayr, A.: F2C2: a fast tool for the computation of flux coupling in genome-scale metabolic networks. BMC Bioinform. **13**(1), 57 (2012)
8. Becker, S.A., Price, N.D., Palsson, B.Ø.: Metabolite coupling in genome-scale metabolic networks. BMC Bioinform. **7**(1), 1 (2006)
9. De Figueiredo, L.F., Schuster, S., Kaleta, C., Fell, D.A.: Can sugars be produced from fatty acids? A test case for pathway analysis tools. Bioinformatics **24**(22), 2615–2621 (2008)
10. Rezola, A., Pey, J., Tobalina, L., Rubio, Á., Beasley, J.E., Planes, F.J.: Advances in network-based metabolic pathway analysis and gene expression data integration. Brief. Bioinform. **16**(2), 265–279 (2015)

11. Klamt, S., Stelling, J.: Combinatorial complexity of pathway analysis in metabolic networks. Mol. Biol. Rep. **29**(1–2), 233–236 (2002)
12. De Figueiredo, L.F., Podhorski, A., Rubio, A., Kaleta, C., Beasley, J.E., Schuster, S., Planes, F.J.: Computing the shortest elementary flux modes in genome-scale metabolic networks. Bioinformatics **25**(23), 3158–3165 (2009)
13. Pey, J., Planes, F.: Direct calculation of elementary flux modes satisfying several biological constraints in genome-scale metabolic networks. Bioinformatics **30**(15), 2197–2203 (2014). (Oxford, England)
14. Rezola, A., Pey, J., de Figueiredo, L.F., Podhorski, A., Schuster, S., Rubio, A., Planes, F.J.: Selection of human tissue-specific elementary flux modes using gene expression data. Bioinformatics **29**(16), 2009–2016 (2013)
15. Gagneur, J., Klamt, S.: Computation of elementary modes: a unifying framework and the new binary approach. BMC Bioinform. **5**(1), 1 (2004)
16. Planes, F.J., Beasley, J.E.: A critical examination of stoichiometric and path-finding approaches to metabolic pathways. Brief. Bioinform. **9**(5), 422–436 (2008)
17. Jevremovic, D., Boley, D., Sosa, C.P.: Divide-and-conquer approach to the parallel computation of elementary flux modes in metabolic networks. In: 2011 IEEE International Symposium on Parallel and Distributed Processing Workshops and Ph.D. Forum (IPDPSW), pp. 502–511. IEEE (2011)
18. Hidalgo, J.F., Guil, F., Garcia, J.M.: A new approach to obtain EFMs using graph methods based on the shortest path between end nodes. In: Ortuño, F., Rojas, I. (eds.) IWBBIO 2015. LNCS, vol. 9043, pp. 641–649. Springer, Heidelberg (2015). doi:10.1007/978-3-319-16483-0_62
19. Wilcoxon, F., Katti, S., Wilcox, R.A.: Critical values and probability levels for the Wilcoxon rank sum test and the Wilcoxon signed rank test. Sel. Tables Math. Stat. **1**, 171–259 (1970)
20. Wilcoxon, F.: Individual comparisons by ranking methods. Biom. Bull. **1**(6), 80–83 (1945)
21. Feist, A.M., Henry, C.S., Reed, J.L., Krummenacker, M., Joyce, A.R., Karp, P.D., Broadbelt, L.J., Hatzimanikatis, V., Palsson, B.Ø.: A genome-scale metabolic reconstruction for escherichia coli k-12 mg1655 that accounts for 1260 orfs and thermodynamic information. Mol. Syst. Biol. **3**(1), 121 (2007)
22. Orth, J.D., Fleming, R.M., Palsson, B.: Reconstruction and use of microbial metabolic networks: the core escherichia coli metabolic model as an educational guide. EcoSal Plus 4(1) (2010)

Author Index